The Black Flies of Subtropical and Tropical Asia

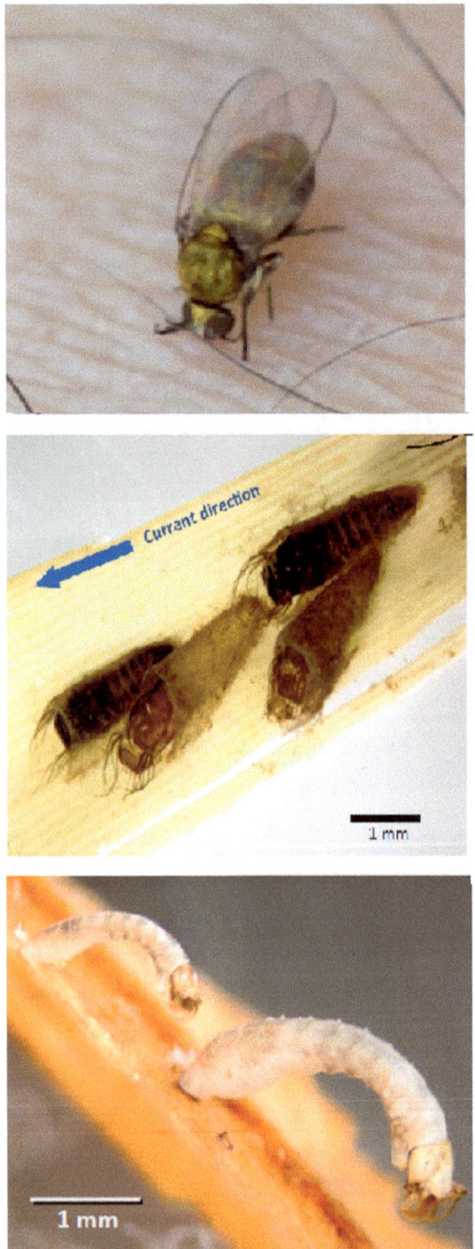

Top: Blood-feeding female of *S.* (*S.*) *nigrogilvum* (Photographed by Dr. W. Srisuka). Middle: Pupae and pupal exuviae in their cocoons of *S.* (*S.*) *oitanum* [reproduced from Takaoka (2016)]. Bottom: Larvae of *S.* (*S.*) *nodosum* [reproduced from Takaoka (2016)]

Hiroyuki Takaoka

The Black Flies of Subtropical and Tropical Asia

Taxonomy and Biology

Springer

Hiroyuki Takaoka
Oita University, Oita, Japan & Tropical Infectious Diseases
Research and Education Centre (TIDREC)
University of Malaya
Kuala Lumpur, Malaysia

ISBN 978-981-99-5218-2 ISBN 978-981-99-5219-9 (eBook)
https://doi.org/10.1007/978-981-99-5219-9

This Springer imprint is published by the registered company Springer Nature Singapore Pte Ltd.
The registered company address is: 152 Beach Road, #21-01/04 Gateway East, Singapore 189721, Singapore

Foreword

From its dry northwestern reaches to its southern tropical rainforests, the Oriental Region and its natural riches have long fascinated the world. The land of orangutans, rhinoceroses, tarsiers, peacocks, birdwing butterflies, and other wonderous creatures has drawn naturalists, such as Alfred Russel Wallace, in search of its wild treasurers. But, the smaller, more secretive denizens of the region have been slow to show themselves to the world. Among them are the black flies. Comprising a family of more than 2,400 miniature species globally, black flies develop in nearly every stretch of flowing water in the Oriental Region.

The first black fly known from the Oriental Region was described by Eduard Becher (1856–1886), an Austrian entomologist who died at the young age of 30. Although Becher might be said to have initiated the study of Oriental black flies with his 1885 description of the female of *Simulium indicum* from Assam, India, he is best remembered as the author of the familiar dipteran taxa Cyclorrhapha, Aschiza, and Schizophora. From Becher's inaugural description of the black fly, published without fanfarc in the *Journal of the Asiatic Society of Bengal*, the number of species of black flics known from the Oriental Region slowly increased to slightly more than 100 by1980—roughly at the rate of one new species per year. From the 1990s to the present, the number of new species discovered and formally described from the Oriental Region surged at a rate of about 15 species per year.

One of the last great zoogeographic regions to be explored for its simuliids, the Oriental Region today has about 700 nominal species—more than any other region. In 2008, 17% of the world's black flies were known from the Oriental Region. Today, that figure is nearly 29%, and there is good reason to expect that the number of species will continue to increase. Comprised of 21 countries in whole or in part, the Oriental Region still holds vast areas of unexplored habitat for black flies. Thailand, for instance, has 150 nominal species and Vietnam has 76, but only 23 species are shared. Thus, when areas such as Cambodia, currently with no recorded species, and Laos, with only two recorded species, are surveyed, the number of species known in the Oriental Region should increase substantially. And as cryptic species continue to be revealed by integrated taxonomic methods, the actual number of known species in the Oriental Region will more nearly approach reality.

The rapid increase in formally named species in the Oriental Region was not the work of an army of taxonomists, but rather the result of the dedicated efforts led by one man: Hiroyuki Takaoka. Professor Takaoka—Hiro to his colleagues—has worked tirelessly for the past 50 years to collect, describe, and classify the black flies of the Oriental Region, and to inspire a new generation of workers to learn the skills and carry on the research while expanding the scope to include molecular systematics, ecology, and parasitology. His work shows the value of the morphotaxonomic approach. Each formally described species stands as a testable hypothesis that can be further investigated through evaluations of chromosomes, DNA, and life histories. Each nominal species in the Oriental Region is accessible through the identification keys and illustrations provided in this book. Professor Takaoka's illustrations are reminiscent of those of the world's great insect morphologist, Robert Snodgrass. These excellent line-drawings, free of extraneous details, show the diagnostic differences among species. Professor Takaoka's comprehensive taxonomic synthesis of all known black flies in the Oriental Region provides the imperative for accurate species identification that will facilitate future investigations of these remarkable organisms.

Department of Plant and Environmental Sciences, Clemson University, Clemson, SC, USA

Peter H. Adler

Preface

In this book, a newly uncovered world of black flies, tiny biting insects, of subtropical and tropical Asia (zoogeographically called as Oriental Region, in which 21 countries are involved in whole or in part) is introduced. It is characterized by the high diversity in lineages and morphological features, in particular, female terminalia, male genitalia, pupal gills, and cocoons, all of which are finely shown by 1,691 line-drawings.

In Chapter 1 (Introduction), black flies are briefly described on their biology and effects on human and animal health and welfare, and studies on black flies in this region are historically summarized. In Chapter 2, all 675 named and 44 unnamed species of black flies recorded from the Oriental Region are classified into 11 subgenera of the genus *Simulium* s. l. and further in species-groups and subgroups in certain large subgenera, such as *Gomphostilbia*, *Nevermannia,* and *Simulium* s. str. Revised diagnoses are provided for all subgenera, species-groups and subgroups. Keys to identify each subgenus and species-group are constructed. In Chapter 3, synoptic accounts for all named and unnamed species are given. A brief history, distribution, bionomics (aquatic habitats, biting habits, and natural infections with parasites), morphological characteristics, and taxonomic problems are noted for each species. In Chapter 4, an overview, a species list and identification keys for females, males, pupae, and mature larvae are provided for each of 15 countries or areas [India, Nepal, Bhutan, Myanmar, Sri Lanka, Thailand, Vietnam, Peninsular Malaysia, Borneo (Sabah, Sarawak, and Kalimantan), Sunda Archipelago (Sumatra, Java, Bali, Lombok, Sumba, Sumbawa, Flores, and Timor), East Indonesia (Sulawesi, Maluku Islands, and Irian Jaya), South China, Nansei Islands (Japan), Taiwan, and Philippines]. In Epilog, my motive and aim to write this book are mentioned as a message to young researchers and students, emphasizing how much work on the Oriental black flies remains to be investigated, and the importance of accurate species identification as a basis of future studies of black flies.

Yufu City, Oita, Japan — Hiroyuki Takaoka
21 June 2023

Acknowledgments

I am most grateful to Dr. P.H. Adler, Professor Emeritus, Clemson University, USA for his kindness in reading the draft of this book, providing valuable suggestions, and particularly in writing "Foreword" for this book. Thanks are due to Dr. M. Fukuda, Oita University, Japan, who kindly scanned all the line-drawings used for this book. My gratitude goes to the following colleagues and friends, who supported me in supplying valuable references and/or simuliid specimens: Dr. M. Yang, Professor, Guizhou Medical University, China; Dr. Y. Otsuka, Kagoshima University, Japan; Dr. V.L. Low and Dr. Z. Ya'cob, Tropical Infectious Diseases Research and Education Centre, Universiti Malaya, Malaysia; Dr. P. Pramual, Professor, Mahasarakham University, Dr. A. Saeung, Chiang Mai University, and Dr. W. Srisuka, Queen Sirikit Botanic Garden, Thailand; and Dr. S. Anbalagan, Bharathidasan University, India.

I wish to thank Ms. F. Yamaguchi, Springer, for her great effort in enabling the publication of this book, and Ms. M. Harding, Ms. P. Arulmurugan, and Ms. L. Srinivasan, Production Team, Springer, for their excellent work during the publication process of this book.

While reviewing the works on the Oriental black flies, I remember many eminent professors and researchers who provided encouragements, supports, and collaborations to my works during the past 52 years. On this special occasion, I like to appreciate all of them: the late Dr. H. Takahasi, Tokyo, Dr. I. Tada, Professor Emeritus, Kyushu University; Dr. I. Miyagi, Professor Emeritus, Ryukyu University; Dr. J. Yukawa, Professor Emeritus, Kyushu University; Dr. H. Suzuki, Nagasaki University; Dr. K. Saito, Yokohama City University; Dr. M. Baba, Dr. M. Fukuda, and Ms. C. Aoki, Oita University, Japan; the late Dr. R.W. Crosskey, British Museum (Natural History), UK; the late Dr. D.M. Davies, Professor Emeritus, McMaster University; the late Dr. D.A. Craig, Professor Emeritus, Alberta University, Canada; Dr. P.H. Adler, Professor Emeritus, Clemson University; Dr. M.S. Mulla, Professor Emeritus, University of California, USA; Dr. C. Yule, Professor, University of the Sunshine Coast, Australia; Dr. A. Saeung, the late Dr. W. Choochote, and Dr. P. Somboon, Chiang Mai University; Dr. W. Srisuka,

Queen Sirikit Botanic Garden, Dr. C. Kuvangkadilok, Mahidol University; Dr. P. Pramual, Professor, Mahasarakham University; Dr. S. Thongsahuan, Prince Songkla University, Thailand; Dr. U.K. Hadi and the late Dr. S.H. Sigit, Professors, Bogor Agricultural University; Dr. I.W. Suana, University of Mataram, Indonesia; Dr. L. De Las Llagas, Professor, University of the Philippines; Mr. V.F. Tenedero, Manila, Philippines; Dr. X.D. Pham, Ministry of Science and Technology, Vietnam; Dr. H.S. Yong, Dr. M. Sofian-Azirun, Dr. R. Hashim, all Professor Emeritus, Dr. S. Uni and Dr. R. Ramli, Professors, Dr. C.D. Chen and Dr. K.W. Lau, Institute of Biological Sciences, Universiti Malaya; Dr. A.B. Sazaly, Professor and Director, Dr. V.L. Low and Dr. Z. Ya'cob, Tropical Infectious Diseases Research and Education Centre, Universiti Malaya; and Dr. C. Leh, Sarawak Museum, Malaysia; Dr. W. Henry and Dr. S. Thapa, Darjeeling Government College, India; Ms. S. Shrestha, Tribhuvan University, Nepal; and the late Dr. C.L. Chung, Center For Disease Control, and Dr. Y.T. Huang, Taiwan.

Finally, I thank my wife, Asako, for her endless support to my works on black flies.

Contents

Chapter 1
Introduction

Black flies are small insects (adult body 1.5–6.0 mm long) belonging to the family Simuliidae in the order Diptera. Females of most species of black flies take blood meals from mammals and birds to mature their eggs. Due to this biting habit, black flies are among the medically and veterinarily most important insects. Females of certain species affect human and animal health and welfare, not only causing dermatoses but also transmitting filarial and protozoan parasites such as *Onchocerca volvulus* and *O. dewittei*, the causative agents of human and zoonotic onchocerciasis, respectively (Crosskey, 1990; Takaoka, 2015a; Takaoka et al., 2012a), and *Leucocytozoon* and *Trypanosoma* species, which are blood parasites of domestic birds (Crosskey, 1990).

The life of black flies consists of four developmental stages: egg, larva, pupa, and adult. Larvae and pupae of black flies live in flowing streams and rivers. The duration of development from egg to adult emergence varies from 10 days to several months depending on the species and water temperatures. The distributions of black flies are worldwide, from the tropics to the Arctic (except Antarctica) and from lowlands to high elevations, where running water is available. Filter-feeding larvae of black flies play an important role in food webs in the terrestrial ecosystems as processors of organic materials as well as prey of fish and other macro-organisms.

Studies on black flies have been conducted mainly in relation to their effects on human and cattle health and welfare, and many works have been done in Africa (Ethiopian Region), Europe (including the former USSR) (Palearctic Region), North America (Nearctic Region), Central and South Americas (Neotropical Region), and Australia (Australasian Region), whereas relatively few works had been done until recently in subtropical and tropical Asia (Oriental Region) (Takaoka, 2015b).

The ranges of six zoogeographical regions follow those of Wallace (1876). The Oriental Region (Fig. 1) consists of continental and insular areas, the latter further divided into continental and oceanic islands. It is delimited in the north from the Palearctic Region by the Himalayan mountains and the Yangtze River, in the east by the Weber line, and in the west by the dry zone of Kashmir. The Weber line (Audley-Charles, 1981; Halt et al., 2013) is suitable for black flies rather than the Wallace

H. Takaoka, *The Black Flies of Subtropical and Tropical Asia*,
https://doi.org/10.1007/978-981-99-5219-9_1

line, another famous boundary between the Oriental and Australasian Regions, as suggested by Takaoka (1996a). Countries included in the Oriental Region are India (except Kashmir and Himachal Pradesh), Nepal, Bhutan, Bangladesh, Myanmar, Sri Lanka, Maldives, Thailand, Laos, Vietnam, Cambodia, Malaysia, Singapore, Brunei, Indonesia (except Halmahera, Ambon, Seram, and Irian Jaya), Timor (East), South China (Yunnan, Guizhou, Hainan, Guangxi, Jiangxi, Zhejiang, Hunan, Fujian, Guangdong, and Hong Kong), Japan (only parts of the Nansei Islands: Tokara Islands, Amami Islands, and Ryukyu Islands), Taiwan, Philippines, and Palau. The Oriental Region is climatologically subtropical and tropical and is rich in diversity of natural environments reflected by the long geological and climatological history of continental and insular areas.

The history of studies of black flies in the Oriental Region began in 1884 when *Simulium (Simulium) indicum* was described from India by Becher (1885). The number of nominal species of black flies in the Oriental Region was 78 among about 1100 species around the world, when Crosskey (1973) first listed the species in this region. It was the fewest among the six zoogeographical regions. The number of black fly species somewhat increased to 118 by the early 1980s (Datta, 1983), but drastically increased to 246 by the middle of the 1990s (Takaoka and Davies, 1995a), 335 by the early 2000s (Takaoka, 2003), 496 by the end of 2014 (Adler and Crosskey, 2015), and 675 by May 2023 (Adler, 2022; Aupalee et al., 2022; Pramual et al., 2022b; Srisuka et al., 2022a, b, 2023a, b; Takaoka et al., 2022a). The number of named species in the Oriental Region currently accounts for 28% of the total number of the world black fly species (2398) (Adler, 2022). Notably, it exceeds 504 estimated by Currie and Adler (2008).

Studies on black flies in the Oriental Region were reviewed (Takaoka, 2015b), and a historical revision of supra-specific taxa (genera, subgenera, and species-groups) was outlined; faunal characteristics such as poverty of supra-generic taxa, species richness, and high diversity of phylogenetic lineages in terms of the number of subgenera (without species-groups) and species-groups combined were highlighted; changes in the number of filaments and extent of inflation of the pupal gills from plesiomorphic to apomorphic states were suggested in relation to speciation histories in certain lineages; a wealth of diversity in morphological characteristics including several novel ones was illustrated; several different geographical distribution patterns of phylogenetic lineages (i.e., continental land type, continental land and continental island type, and insular type) were categorized; three black fly species, *S. (Gomphostilbia) asakoae* Takaoka and Davies, *S. (S.) nigrogilvum* Summers, and *S. (S.) nodosum* Puri, were noted for their natural infections with zoonotic filariae in Thailand; and current taxonomic problems were discussed.

The fauna of black flies in the Oriental Region was first compared with those of the five other zoogeographical regions by Takaoka (2017b). It revealed the highest speciation index (15.4), though the number of phylogenetic lineages was moderate, 34/165 total world lineages. The index of faunal affinities at the lineage level was the highest (31.9%) for the Palearctic Region, followed by 6.7% for the Australasian Region, 6.0% for the Nearctic Region, 3.7% for the Ethiopian Region, and 1.4% for the Neotropical Region. In addition, southern or northern dispersals of certain lineages during and after the ice ages were inferred.

In this book, a revised classification of black flies in the Oriental Region, synoptic accounts of 675 named and 44 unnamed species reported by May 2023, and a species list and keys to identify adult females, males, pupae, and mature larvae in each of 15 countries or areas, are provided.

Morphological features and their terms (Figs. 2 and 3) followed Takaoka (2003) and partially Adler et al. (2004). All figures showing the morphological characteristics of subgenera, species-groups, and subgroups were arranged by using 1691 pieces of illustrations, all (but one) of which were originally drawn by myself.

Chapter 2
Classification of Black Flies in the Oriental Region

1 Overview

Crosskey (1973) first classified 78 species of black flies in the Oriental Region into three subgenera of the genus *Simulium* Latreille s. l.: eight species in *Eusimulium* Roubaud s. l., 14 species in *Gomphostilbia* Enderlein, 53 species in *Simulium* Latreille s. str., and three species in *Simulium* s. l. species unplaced to subgenus. Lewis (1973) proposed a new subgenus *Himalayum* for *S.* (*S.*) *indicum* Becher. However, it was sunk as a synonym of the subgenus *Simulium* s. str. by Otsuka et al. (2003). The number of the subgenera in the Oriental Region increased to five since the subgenus *Eusimulium* s. l. was split into *Eusimulium* s. str., *Montisimulium* Rubtsov and *Nevermannia* Enderlein, until Takaoka (1983) created a new subgenus *Wallacellum* and added the subgenus *Morops* Enderlein for some of black flies in the Philippines. Takaoka (2003) reviewed the definition of the two subgenera *Gomphostilbia* and *Morops* when working on black flies in Sulawesi, Maluku, and Irian Jaya, and subsequently all species previously placed in the subgenus *Morops* in the Oriental Region were transferred to the subgenus *Gomphostilbia*. Davies and Györkös (1988a) and Chen and Zhang (1998) recorded the subgenus *Byssodon* Enderlein from Sri Lanka, and the subgenus *Wilhelmia* Enderlein from South China, respectively. Takaoka and Davies (1995a) recorded the genus *Sulcicnephia* Rubtsov for a single species from Peninsular Malaysia. However, Takaoka and Adler (1997) reevaluated the species placed in the genus *Sulcicnephia* and transferred it to a newly created subgenus *Daviesellum* within the genus *Simulium* s. l. Takaoka and Choochote (2005b) created a new subgenus *Asiosimulium* based on specimens collected from Thailand. Chen (2016) recorded the subgenus *Boophthora* Enderlein from Guizhou, China. Thus, all species of black flies in the Oriental Region are placed in 11 subgenera in the genus *Simulium* s. l. (Table 1).

Under the three larger subgenera, *Simulium* s. str., *Gomphostilbia* and *Nevermannia*, 21, 12, and three species-groups were used, respectively, and species

H. Takaoka, *The Black Flies of Subtropical and Tropical Asia*,
https://doi.org/10.1007/978-981-99-5219-9_2

Table 1 Classification of world fauna of Simuliidae[a]

Family	Subfamily	Tribe	Genus	Subgenus[b]
Simuliidae	Parasimuliinae		*Parasimulium*	*Astoneomyia*, *Parasimulium* s. str.
	Simuliinae	Prosimuliini	*Gymnopais*	
			Helodon	*Distosimulium*, *Helodon*, *Parahelodon*
			Levitinia	
			Prosimulium	
			Twinnia	
			Urosimulium	
		Simuliini	*Araucnephia*	
			Araucnephioides	
			Austrocnephia	
			Austrosimulium	*Austrosimulium* s. str., *Novaustrosimulium*
			Bunyipellum	
			Cnephia	
			Cnesia	
			Cnesiamima	
			Crozetia	
			Ectemnia	
			Ectemnoides	
			Gigantodax	
			Greniera	
			Lutzsimulium	
			Metacnephia	
			Nothogreniera	
			Paracnephia	*Paracnephia* s. str., *Procnephia*
			Paraustrosimulium	
			Pedrowygomyia	
			Protaustrosimulium	
			Simulium	*Afrosimulium*, *Anasolen*, ***Asiosimulium***, *Aspathia*, ***Boophthora***, *Boreosimulium*, ***Byssodon***, *Chirostilbia*, *Crosskeyellum*, ***Daviesellum***, *Edwardsellum*, *Eremulium*, ***Eusimulium***, *Freemanellum*, ***Gomphostilbia***, *Hebridosimulium*, *Hellichiella*, *Inselliellum*, *Lewisellum*, *Meilloniellum*, *Metomphalus*, ***Montisimulium***, *Morops*, ***Nevermannia***, *Notolepria*, *Phoretomyia*, *Pomeroyellum*, *Psaroniocompsa*, *Psilopelmia*, *Psilozia*, *Pternaspatha*, *Rubzovia*, *Shoenbaueria*, ***Simulium* s. str.**, *Trichodagmia*, ***Wallacellum***, ***Willhelmia***, *Xenosimulium*
			Stegopterna	
			Sulcicnephia	
			Tlalocomyia	

[a]Reconstructed from Adler (2022).
[b]All species found in the Oriental Region are assigned to these 11 subgenera of the genus *Simulium* shown in bold faced type.

in certain species-groups in the subgenera *Gomphostilbia* and *Simulium* s. str. were further placed in subgroups (Takaoka, 2012, 2017a, b).

A classification scheme follows the arrangement of Adler (2022) except some cytoforms and molecular forms, which are not included.

2 Keys to Identify 11 Subgenera of the Genus *Simulium* s. l. in the Oriental Region*

The keys provided by Takaoka et al. (2017c) are somewhat modified as follows.

Females

1. Postnotum haired .... *Eusimulium* (except one species)
   Postnotum bare 2
2. Katepisternum haired 3
   Katepisternum bare 4
3. Hind tibia with long, narrow ridge on anterior inner surface (Fig. 92G). Calcipala elongate, reaching apex of the second tarsomere (Fig. 92I, J) .... *Wallacellum*
   Hind tibia without long, narrow ridge on anterior inner surface. Calcipala medium-long, not reaching apex of the second tarsomere (Fig. 32C) .... *Gomphostilbia*
4. Pleural membrane haired 5
   Pleural membrane bare 6
5. Claw without tooth (Fig. 96C) .... *Wilhelmia*
   Claw with small subbasal tooth .... *Simulium* s. str. (*S. ornatum* species-group)
6. Paraproct with cluster of dark spines anteriorly (Fig. 8G) ....*Daviesellum*
   Paraproct without cluster of dark spines anteriorly 7
7. Cibarium with numerous spinous processes medially (Fig. 4D) .....*Asiosimulium*
   Cibarium without spinous processes medially 8
8. Claw with large basal tooth 9
   Claw without tooth or with small or medium-sized tooth 12
9. Basal section of radial vein bare .... *Byssodon*
   Basal section of radial vein haired 10
10. Paraproct almost entirely setose or narrowly bare on ventral surface (Figs. 43G and 47I), and as long as cercus (Fig. 47J). Spermatheca globular or ovoidal (Figs. 43I and 47K) .... *Nevermannia* (*S. feuerborni* species-group and *S. vernum* species-group)
    Paraproct widely bare on ventral surface (Figs. 10E and 45D), and much shorter than cercus (Figs. 10F and 45E). Spermatheca ellipsoidal (Figs. 10G and 45F,G) 11

11. Genital fork with small projection directed anteriorly from each lateral arm (Fig. 10D) .... *Eusimulium* (one species)
    Genital fork without projection directed anteriorly from each lateral arm (Fig. 45C) .... *Nevermannia* (*S. ruficorne* species-group)
12. Basal section of radial vein bare ....
    .... *Simulium* s. str. (except several species) & *Boophthora*
    Basal section of radial vein haired 13
13. Claw without tooth .... *Simulium* s. str. (several species of *S. griseifrons* species-group, *S. multistriatum* species-group and *S. striatum* species-group)
    Claw with small or medium-sized basal tooth 14
14. Claw with medium-sized tooth (Fig. 42C) ....
    .... *Montisimulium* (except two species)
    Claw with small tooth (Fig. 42D) 15
15. Cibarium with dorsomedian process (Fig. 62B) ....
    .... *Simulium* s. str. (four species of *S. indicum* species-group)
    Cibarium without dorsomedian process 16
16. Mandible with inner teeth but without outer teeth ....
    .... *Montisimulium* (two species)
    Mandible with inner and outer teeth ....
    .... *Nevermannia* (one species of *S. vernum* species-group)

Males

1. Style longer than coxite (Fig. 49I) 2
   Style shorter than coxite (Fig. 17O) 4
2. Coxite much longer than wide (Fig. 8J) .... *Daviesellum*
   Coxite as long as, or slightly shorter or longer than, wide (Fig. 49I) 3
3. Style with many spines subapically (Fig. 7I) .... *Boophthora*
   Style with or without single spine apically or subapically (Fig. 49J) ....
   .... *Simulium* s. str.
4. Katepisternum haired 5
   Katepisternum bare 6
5. Hind tibia with long, narrow ridge on anterior inner surface (similar to Fig. 92G). Calcipala elongate, reaching apex of the second tarsomere (Fig. 93A, B) .... *Wallacellum*
   Hind tibia without long, narrow ridge on anterior inner surface. Calcipala medium-long, not reaching apex of the second tarsomere (Fig. 32I) .... *Gomphostilbia*
6. Pleural membrane haired .... *Wilhelmia*
   Pleural membrane bare 7
7. Basal section of radial vein bare .... *Byssodon*
   Basal section of radial vein haired 8
8. Paramere without hooks (Fig. 5J) .... *Asiosimulium*
   Paramere with hook(s) 9
9. Hind basitarsus slender (Fig. 10H). Ventral plate with median keel (Figs. 10I and 45I) 10

Hind basitarsus enlarged (Fig. 43J); ventral plate without median keel (e.g., Fig. 43K) 11

10. Ventral plate narrow, with divergent long arms when viewed ventrally (Fig. 10I) .... *Eusimulium*
    Ventral plate broad, with parallel-sided short arms when viewed ventrally (Fig. 45I) .... *Nevermannia* (*S. ruficorne* species-group)
11. Paramere with one long hook (Fig. 47Q). Median sclerite forked apically (Fig. 47P) .... *Nevermannia* (*S. vernum* species-group)
    Paramere with three or more hooks (Figs. 42O & 43P). Median sclerite club-shaped, not forked apically (Fig. 42N) .... *Montisimulium* and *Nevermannia* (*S. feuerborni* species-group)

Pupae

1. Gill filaments with numerous minute black spots (Fig. 42T) .... *Montisimulium*
   Gill filaments without minute black spots 2
2. Abdominal segments with spine-combs dorsally (at least on segment 8) (Fig. 3C) 3
   Abdominal segments 5–9 without spine-combs dorsally (Fig. 95E) 20
3. Last abdominal segment with grapnel-like hooklets on each side .... *Gomphostilbia* (most species)
   Last abdominal segment without grapnel-like hooklets on each side 4
4. Last abdominal segment without terminal hooks .... *Simulium* s. str. (most species) and *Boophthora*
   Last abdominal segment with terminal hooks 5
5. Terminal hooks wide, plate-like, with outer margins crenulated (similar to Fig. 27F) .... *Gomphostilbia* (two species of *S. ceylonicum* species-group and three species of *S. epistum* species-group)
   Terminal hooks simple, cone-like, with outer margins smooth (Fig. 44H) 6
6. Gill of inflated, dumbbell-like structure, without filaments (Fig. 48G) .... *Nevermannia* (*S. taulingense* of *S. vernum* species-group)
   Gill filamentous, or of one inflated structure with six slender filaments (Fig. 88K, L), or of one large and two small globular structures and two thumb-like short filaments (88I,J) 7
7. Gill with three filaments .... *Gomphostilbia* (*S. trirugosum* of *S. varicorne* species-group)
   Gill with four, five, six eight, nine, or 10 filaments, or of one inflated structure with six slender filaments (Fig. 88K, L) or of one large and two small globular structures and two thumb-like short filaments (Fig. 88I, J) 8
8. Gill with four filaments 9
   Gill with five, six, eight, nine, or 10 filaments, or of one inflated structure with six slender filaments (Fig. 88K, L) or of one large and two small globular structures and two thumb-like short filaments (Fig. 88I, J) 11

9. Gill with short to medium-long common basal stalk (Fig. 48E) ....*Nevermannia* (most species of *S. vernum* species-group and *S. borneoense* of *S. feuerborni* species-group)
   Gill with extremely short common basal stalk (Figs. 10P and 46D–G) 10
10. Cocoon with anterodorsal projection (Fig. 46H) ....*Nevermannia* (three species of *S. ruficorne* species-group)
    Cocoon without anterodorsal projection (Fig. 10Q) ....*Eusimulium*
11. Gill with five slender filaments (Fig. 44F) ....*Nevermannia* (*S. phami* of *S. feuerborni* species-group)
    Gill with six, eight, nine or 10 filaments, or of one inflated structure with six slender filaments (Fig. 88K, L) or of one large and two small globular structures and two thumb-like short filaments (Fig. 88I, J) 12
12. Gill with six filaments, of which four are directed forward and two are directed upward (Fig. 88A–H), or of one inflated structure with six slender filaments (Fig. 88K, L), or of one large, two small globular structures and two thumb-like short filaments (Fig. 88I, J) ....*Simulium* s. str. (six species of *S. variegatum* species-group)
    Gill with six, eight, nine, or 10 filaments all directed forward 13
13. Gill with six filaments 14
    Gill with eight, nine, or 10 filaments 18
14. Gill filaments much shorter than pupal body (e.g., Fig. 60D–K) 15
    Gill filaments as long as or longer than pupal body 17
15. Cocoon wall-pocket-shaped, without anterodorsal projection (Fig. 60M), or shoe-shaped, with or without anterolateral open space(s) on each side (Fig. 60N–Q), or boot-shaped, roughly woven, entirely covering pupal body and gill filaments (Fig. 60S) ....*Simulium* s. str. (most species of *S. griseifrons* species-group, several species of *S. tuberosum* species-group and many species of *S. variegatum* species-group)
    Cocoon wall-pocket-shaped, with short to long anterodorsal projection (Fig. 87E, G) 16
16. Gill filaments of same thickness (Fig. 46C) ....*Nevermannia* (*S. aureohirtum* of *S. ruficorne* species-group)
    Gill filaments slender except basal portions somewhat inflated (Fig. 87E, G) ....*Simulium* s. str. (two species of *S. variegatum* species-group)
17. Cocoon long (6.0–7.0 mm), thinly woven, covering pupal body and gill filaments (Fig. 89F). Frontal trichomes bifid or trifid .... *Simulium* s. str. (*S. molliculum* of *S. variegatum* species-group)
    Cocoon medium-long (less than 5.0 mm), thickly woven, covering pupal body. Frontal trichomes unbranched ....*Nevermannia* (most species of *S. feuerborni* species-group)

18. Gill filaments arranged as (3+3)+2 from dorsal to ventral ....*Gomphostilbia* (two species of *S. batoense* species-group
    Gill filaments arranged as 3+2+3 or 2+3+3 or 2+2+2+2 from dorsal to ventral ... 19
19. Gill filaments arranged as 3+2+3 or 2+3+3 from dorsal to ventral (Fig. 46A and B) ....*Nevermannia* (two species of *S. ruficorne* species-group)
    Gill filaments arranged as 2+2+2+2 from dorsal to ventral (e.g., Fig. 49Q) .... *Simulium* s. str. (several species of *S. argentipes* species-group, *S. takense* of *S. multistriatum* species-group, *S. ornatum* of *S. ornatum* species-group and a few species of *S. striatum* species-group)
20. Gill with 19–56 filaments (Fig. 6A, B) ....*Asiosimulium*
    Gill with three, four, six, eight, nine, or 10 filaments 21
21. Gill with three filaments .... *Byssodon*
    Gill with four, six, eight, nine, or 10 filaments 22
22. Gill with four filaments 23
    Gill with six, eight, nine, or 10 filaments 24
23. Grapnel-like hooklets present ....*Gomphostilbia* (*S. palauense* of *S. palauense* species-group)
    Grapnel-like hooklets absent ....*Wallacellum*
24. Gill with eight filaments 25
    Gill with six, nine, or 10 filaments filaments 26
25. Eight gill filaments arranged in four pairs and tapered apically (Fig. 91J) .... *Simulium* s. str. (*S. tamorense*)
    Eight gill filaments arranged otherwise, and not tapered apically (Fig. 97A, B) .... *Wilhelmia*
26. Gill with nine or 10 filaments (Fig. 57M–P) ....*Simulium* s. str. (all four species of *S. eximium* species-group)
    Gill with six filaments 27
27. Abdominal segment 2 dorsally with five small spinous setae on each side ....*Simulium* s. str. (two species of *S. melanopus* species-group and three species of *S. indicum* species-group)
    Abdominal segment 2 dorsally with five stout hooks on each side (Fig. 9C) ....*Daviesellum*

Mature larvae

1. Hypostoma wide, with prominent median tooth accompanied by six small teeth on each side (Fig. 9L, N) ....*Daviesellum*
   Hypostoma with median tooth accompanied by one corner tooth and three intermediate teeth on each side (Fig. 32U) 2
2. Mandibular serration composed of two large or medium-sized teeth (Fig. 10R) 3
   Mandibular serration composed of one large tooth and one small tooth (Fig. 44M) 5
3. Pharate pupal gill with four, six, or eight short somewhat inflated filaments ....*Nevermannia* (most species of *S. ruficorne* species-group)

Pharate pupal gill with four long slender filaments 4

4. Postgenal cleft absent ....*Nevermannia* (one species of *S. ruficorne* species-group)
   Postgenal cleft small to medium-long, as long as or shorter than postgenal bridge ....*Eusimulium*
5. Main tooth of mandibular serration at acute angle to mandible apically (Fig. 32T) ....*Gomphostilbia* (except two species of *S. baisasae* species-group)
   Main tooth of mandibular serration at right or obtuse angle to mandible apically (Fig. 42V) 6
6. Postgenal cleft absent or vestigial (Fig. 42X, Y) .... *Montisimulium*
   Postgenal cleft small to large (Figs. 40K and 44O) 7
7. Pectination of uppermost primary ray of labral fan (Fig. 95V) different from those of other primary rays (Fig. 95W) .... *Wallacellum*
   Pectination of uppermost primary ray of labral fan similar to those of other primary rays 8
8. Ventral papillae absent 9
   Ventral papillae small to large (Fig. 89Z) 10
9. Hypostoma with intermediate teeth nearly as long as each corner tooth (Fig. 97G) .... *Wilhelmia*
   Hypostoma with intermediate teeth shorter than each corner tooth .... *Simulium* s. str. (most species)
10. Hypostoma with lateral margins smooth (similar to Fig. 32U) ....*Gomphostilbia* (two species of *S. baisasae* species-group)
    Hypostoma with lateral margins serrated (Fig. 44N) 11
11. Postgenal cleft more than twice as long as postgenal bridge 12
    Postgenal cleft less than 1.5 times as long as postgenal bridge 15
12. Pharate pupal gill with three filaments .... *Byssodon*
    Pharate pupal gill with four, five, six, or 19–56 filaments 13
13. Pharate pupal gill with 19–56 filaments ....*Asiosimulium*
    Pharate pupal gill with four, five, or six slender filaments 14
14. Ventral papillae of moderate size. Rectal organ simple or compound with few secondary lobules per lobe .... *Boophthora*
    Ventral papillae of small size. Rectal organ compound, with several secondary lobules per lobe ....*Simulium* s. str. (several species of *S. tuberosum* species-group)
15. Ventral papillae of moderate or large size ....*Nevermannia* (species of *S. feurerboni* species-group & *S. vernum* species-groups)
    Ventral papillae small (Fig. 89Z), or invisible when viewed laterally .... *Simulium* s. str. (11 species of the *S. variegatum* species-group)

* The characteristics of the subgenus *Byssodon* used above are those of *S. languidum*, the only species of the subgenus in the Oriental Region and are not necessarily shared with other species of the subgenus in other regions.

3 Systematic Treatment of 11 Subgenera of the Genus *Simulium* s. l. in the Oriental Region

Subgenus Asiosimulium *Takaoka and Choochote*

Asiosimulium Takaoka and Choochote, 2005 (as subgenus of *Simulium* s. l.). Type species. *oblongum* (original designation).

Diagnosis. Female and male: Antenna composed of scape, pedicel, and nine flagellomeres. Both pleural membrane and katepisternum bare. Basal portion of radial vein fully haired. Female: Frons shiny. Cibarium with numerous spinous processes medially (Fig. 4D). Tarsal claw with large basal tooth and with wide space between it and basal tooth (Fig. 4F). Abdominal segments 2 and 5–8 (or 6–8) shiny. Ovipositor valve triangular (Fig. 4E). Genital fork with or without projection directed anteriorly on each arm (Fig. 4G–J). Paraproct shorter than wide (Fig. 4K–P). Cercus short to long (Fig. 4K–P). Spermatheca globular, ovoid, or pear-shaped (Fig. 4Q–T). Male: Hind basitarsus slightly to greatly enlarged (Fig. 5A, B). Abdominal segments 2 and 5–7 (or 8) each with pair of shiny dorsolateral patches (except one species lacking such shiny patches on segments 5–7, and one species with shiny patches also on segments 3 and 4). Coxite subquadrate, much longer than wide (Fig. 5C, M). Style shorter than coxite (Fig. 5C, M). Ventral plate in form of transverse haired plate and produced ventromedially (appearing to be median keel in some species), with posterior margin with or without incision (Fig. 5C, M). Paramere lacking hooks (Fig. 5J). Median sclerite plate-like, widened apically (Fig. 5I, L, P). Aedeagal membrane with fine setae (Fig. 5K) except one species with spinous setae (Fig. 5Q). Pupa: Head with two frontal trichomes and one facial trichome on each side (Fig. 6C). Gill of arborescent type, with 19–56 slender filaments (Fig. 6A, B). Dorsal surface of abdominal segments 5–9 without spine-combs. Ventral surface of abdominal segments 5–7 each with pair of unbranched hooks on each side. Abdominal segment 9 with pair of cone-shaped terminal hooks (Fig. 6D) and a few grapnel-like hooklets (Fig. 6E). Cocoon wall-pocket-shaped (Fig. 6F). Larva: Hypostoma with distinct median and corner teeth much longer than three small intermediate teeth on each side, and with lateral margins serrated apically (Fig. 6H). Mandibular serration composed of one large tooth and one small tooth, of which large tooth at right angle to mandible (Fig. 6G). Supernumerary serration absent (present in one species). Postgenal cleft deep, with its apex reaching or close to posterior margin of hypostoma and with or without pigmented sheath of subesophageal ganglion (Fig. 6I). Ventral papillae small.

Remarks. The diagnosis given above is somewhat modified from the original diagnosis of the subgenus *Asiosimulium*, defined by Takaoka and Choochote (2005b). It is one of the two small black fly subgenera endemic in the Oriental Region and is represented by only seven species.

The geographic distribution pattern of the subgenus *Asiosimulium* is continental, though widely disjunctive between central-east Myanmar and western Nepal.

The phylogenetic relationships of this subgenus are not well resolved, though a few molecular and morphological studies have been conducted. In multigene analyses using three mitochondrial genes (COI, ND4, and 16S rRNA), and two segments of the nuclear 28S ribosomal RNA (the D1 to D2 and the D4 expansion regions), Phayuhasena et al. (2010) showed that among four other subgenera in Thailand, the subgenus *Asiosimulium* was positioned the nearest to the subgenus *Simulium* s. str., whereas Otsuka et al. (2007) using analysis of the sequences of the mitochondrial 16S rRNA gene showed that the subgenus *Asiosimulium* had no sister relationship with any of eight other subgenera, though it was positioned next to the most basal subgenus *Wallacellum*, forming a clade with seven other subgenera, i.e., *Daviesellum*, *Gomphostilbia*, *Hellichiella* Rubtsov, *Montisimulium*, *Morops* Enderlein, *Nevermannia* Enderlein, and *Simulium* s. str. In a cladistic analysis of morphological characteristics, Takaoka (2012) showed that among 10 subgenera examined, the subgenus *Asiosimulium* has a closer relationship with a large clade composed of five subgenera, i.e., *Daviesellum*, *Gomphostilbia*, *Inseliellum* Rubtsov, *Morops*, and *Simulium* s. str., than with four other subgenera, i.e., *Hebridosimulium* Grenier and Rageau, *Montisimulium*, *Nevermannia*, and *Wallacellum*.

The biting habit and medical or veterinary importance of the subgenus *Asiosimulium* remain to be investigated, though females of *S. wanchaii* were caught while flying around a human in Doi Pui National Park, Chiang Mai, Thailand (Takaoka and Choochote, 2006a).

Species Included. Simulium furvum Takaoka and Srisuka, 2013, *S. oblongum* Takaoka and Choochote, 2005, *S. phurueaense* Tangkawanit, Wongpakam and Pramual, 2018, *S. saeungae* Takaoka and Srisuka, 2018, and *S. wanchaii* Takaoka and Choochote, 2006, all described from Thailand; *S. shanene* Takaoka, Srisuka and Saeung, 2017 from Myanmar; and *S. suchitrae* Takaoka, 2010 from Nepal.

Subgenus **Boophthora** *Enderlein*

Boophthora Enderlein, 1921 (as genus). Type species: *erythrocephlum*, as *argyreata* (original designation).

Pseudosimulium Baranov, 1926 (preocc.) (as subgenus of *Simulium* s. l.). Type species: *erythrocephalum*, as *argyreata* [misidentified] (original designation).

Diagnosis. Female and male: Antenna composed of scape, pedicel, and nine flagellomeres. Scutum with silvery spot on each shoulder. Both pleural membrane and katepisternum bare. Basal section of radial vein bare. Female: Frons shiny. Cibarium unarmed, with or without tiny processes (Fig. 7A). Tarsal claw simple. Ovipositor valve triangular (Fig. 7B). Genital fork with projection directed anteriorly on each arm (Fig. 7C). Paraproct shorter than wide (Fig. 7D, E). Cercus short (Fig. 7E). Spermatheca oval (Fig. 7F). Male: Hind basitarsus slender and nearly parallel-sided (Fig. 7G). Abdominal segment 2 with pair of shiny dorsolateral patches. Coxite subquadrate, shorter than wide (Fig. 7H). Style 1.2 times as long as coxite, flat ventrodorsally, and with 12–15 spines subapically (Fig. 7I). Ventral plate with body subquadrate, with haired ventral surface, elongate projection directed posteriorly from its base on each side, and arms short, directed forward (Fig. 7J). Median sclerite strap-like and not bifurcate (Fig. 7K). Paramere with many small short hooks (Fig. 7L). Pupa: Head with two frontal trichomes and one facial trichome on each side. Gill as long as or slightly longer than pupal body, with four to six filaments (Fig. 7N–P). Abdominal segments 7–9 with spine-combs dorsally. Ventral surface of abdominal segments 5–7 each with pair of bifd hooks on each side. Abdominal segment 9 without terminal hooks. Cocoon wall-pocket-shaped. Larva: Mandible with mandibular serration composed of one large tooth and one small tooth, of which large tooth at right angle to mandible; supernumerary serration absent (Fig. 7Q). Hypostoma with distinct median and corner teeth much longer than three small intermediate teeth on each side, lateral margins serrated apically, and three or four hypostomal bristles (divergent posteriorly from lateral margin on each side) (Fig. 7R). Postgenal cleft deep, 2.0–2.5 times as long as postgenal bridge (Fig. 7S). Rectal organ simple or with one short secondary lobule per lobe (last abdominal segment with ventral papillae widely separated).

Remarks. The above diagnosis is based on characteristics of *S. guiyangense* Chen, Liu and Yang (Chen, 2016), the only species of this subgenus in the Oriental Region, except a few characteristics in parentheses, which are cited from Crosskey (1969). All line-drawings of the characteristics in Fig. 7 are based on slide-mounted specimens of *S. yonagoense* Okamoto and *S. makunbei* Ono, both from Japan.

Species Included. Simulium guiyangense Chen, Liu and Yang, 2016 from China (Guizhou).

Subgenus **Byssodon** *Enderlein*

Byssodon Enderlein, 1925 (as genus). Type species: *meridionale,* as *forbesi* (original designation).

Psilocnetha Enderlein, 1935 (as genus). Type species: *griseicolle*, as *scapulata* (original designation).

Titanopteryx Enderliein, 1935 (as genus). Type species: *maculatum* (original designation).

Echinosimulium Baranov, 1938 (as genus). Type species: *maculatum, as echinatum* (original description).

Gibbinsiellum Rubtsov, 1962 (as genus). Type species: *griseicolle* (original designation).

Diagnosis. Female and male: Antenna composed of scape, pedicel and nine flagellomeres. Both pleural membrane and katepisternum bare. Basal portion of radial vein bare. Female: Cibarium unarmed. Scutum with dark lyre-shaped pattern. Tarsal claw with large basal tooth. Abdominal segments 2 and 5–9 shiny. Ovipositor valve triangular. Genital fork with two projections, each directed anteriorly and posteromedially, on each arm. Paraproct shorter than wide. Cercus short. Spermatheca oval, (without internal hairs): Male: Hind basitarsus slender and nearly parallel-sided. Coxite subquadrate, much longer than wide. (Style as long as or shorter than coxite). Ventral plate in form of transverse haired plate. Paramere with three or more hooks (median sclerite strap-like or large dilated and bifurcate). Pupa: Head with three frontal trichomes and one facial trichome on each side. Gill shorter than pupal body, with three filaments. Dorsal surface of abdominal segments 5–9 without spine-combs. Ventral surface of abdominal segments 5–7 each with pair of bifid to quadrifid hooks on each side. Abdominal segment 9 with pair of cone-shaped terminal hooks. Cocoon shoe-shaped. Larva: Hypostoma with distinct median and corner teeth longer than three intermediate teeth on each side, lateral margins serrated apically, and four hypostomal bristles (lying nearly parallel-sided to lateral margin on each side). Mandible with mandibular serration composed of one large tooth and one small tooth. Postgenal cleft deep, reaching posterior margin of hypostoma. Abdominal segments 1–6 each without conical protuberances dorsally. Ventral papillae small.

Remarks. The above diagnosis of this subgenus is based on the description of *S. languidum* by Davies and Györkös (1988a), though some characteristics in parentheses, of which information is not available from the description of *S. languidum,* are cited from the diagnosis given by Crosskey (1969). The angle formed by the large mandibular tooth against the inner margin of the mandible is unknown.

Species Included. Simulium languidum Davies and Györkös, 1988 from Sri Lanka.

Subgenus **Daviesellum** *Takaoka and Adler*

Daviesellum Takaoka and Adler, 1997 (as subgenus of *Simulium* s. l.). Type species: *pahangense* (original designation).

Diagnosis. Female and male: Antenna composed of scape, pedicel and nine flagellomeres. Pleural membrane and katepisternum bare. Basal section of radius bare. Female: Frons shiny, with several hairs along each lateral margin. Cibarium with numerous microtubercles along posterodorsal margin (Fig. 8A). Scutum white pruinose but not patterned. Tarsal claw with small subbasal tooth (Fig. 8B). Abdomen with dorsolateral pair of white pruinose patches on segment 2. Tergite of segments 6–8 shiny. Ovipositor valve produced posteriorly and tongue-like (Fig. 8C). Genital fork with well-developed arms, each with or without projection directed anteriorly (Fig. 8C, D). Paraproct of unique form, widely concave and bare anteriorly and with a cluster of numerous dark stout spines anteromedially

(Fig. 8E–G). Spermatheca globular or ovoid, with internal setae (Fig. 8H). Male: Scutum white pruinose, with three faint longitudinal vittae. Abdomen with dorsolateral pair of white pruinose patches on segments 2 and 4–8. Coxite much longer than wide, gradually narrowed posteriorly (Fig. 8J, M). Style elongate, 1.5–1.7 times length of coxite, slender and gradually tapered toward apex, with apical spine (Fig. 8J, M). Ventral plate in form of transverse haired plate, with narrow longitudinal median process (Fig. 8J–O). Paramere narrow, with numerous hooks (Fig. 8P). Median sclerite wide, plate-like (Fig. 8Q). Pupa: Head with two frontal trichomes and one facial trichome on each side (Fig. 9A). Gill with six short filaments of subequal length and thickness (Fig. 9B). Dorsum of abdominal segment 2 with four stout spinous hooks on each side, equivalent in size to those on dorsum of abdominal segments 3 and 4 (Fig. 9C, D). Abdominal segments 5–9 without spine-combs dorsally. Abdominal segments 5–7 each with pair of unbranched stout hooks ventrally on each side (Fig. 9E). Terminal hooks absent. Cocoon boot-shaped (Fig. 9F). Larva: Cephalic apotome widest just before posterior margin (Fig. 9H). Hypostoma extremely wide, with 13 apical teeth composed of one prominent median tooth accompanied at its base by two small teeth on each side and four small individual teeth on each side (Fig. 9L, N). Postgenal cleft deep (Fig. 9M, O). Mandible with well-developed comb-teeth, mandibular serration composed of one tooth, and supernumerary serration absent (Fig. 9K). Labral fan with uppermost primary ray (Fig. 9I) having different pectination from that of other primary rays (Fig. 9J). Proleg with sclerotized rod on each side. Rectal organ of three lobes, each with numerous secondary lobules. Anal sclerite of usual X-shape, posterior arms much longer than anterior ones (Fig. 9P). Posterior circlet with over 400 rows of up to over 40 hooklets per row (Fig. 9Q). Last abdominal segment remarkably bulged laterally or somewhat ventrolaterally, forming large papilla and small one, which are visible when viewed ventrally but not when viewed laterally (Fig. 9Q).

Remarks. The above diagnosis of this subgenus is cited from Takaoka and Adler (1997) with a slight modification.

This subgenus was established for two species from Peninsular Malaysia and Thailand by Takaoka and Adler (1997). It is endemic to the Oriental Region.

It is distinguished from the 10 other subgenera of the genus *Simulium* s. l. by the following characteristics: in the female by the paraproct with a cluster of dark spines (Fig. 8G), in the male by the elongate coxite, which is longer than wide, and is shorter than the style (Fig. 8J, M); in the pupa by the dorsum of abdominal segment 2 with four stout hooks on each side (Fig. 9C); and in the larva by the hypostoma extremely wide, with 13 apical teeth (Fig. 9L, N), mandibular serration composed of a single tooth (Fig. 9K) and posterior circlet with over 400 rows of up to over 40 hooklets per row.

Species Included. Simulium courtneyi Takaoka and Adler, 1997 from Thailand; and *S. pahangense* Takaoka and Davies, 1995 from Malaysia (Peninsular) and Thailand.

Subgenus Eusimulium *Roubaud*

Eusimulium Roubaud, 1906 (as subgenus of *Simulium*). Type species: *aureum* (monotypy).

Diagnosis. Female and male: Scutum not patterned. Pleural membrane and katepisternum bare (except katepisternum with several hairs in female of *S. satsumense* from Japan). Basal portion of radial vein haired. Female: Frons not shiny, densely covered with fine hairs. Mandible with teeth on inner margin and lacking teeth on outer margin. Postnotum with hairs (except *S. weiningense* from Guizhou, China, which has no hairs on postnotum). Claw with long tooth (Fig. 10B). Abdomen dull except segments 8 and 9 slightly shiny dorsally when illuminated at certain angles. Abdominal segment 7 with large sternal plate medially. Ovipositor valve triangular or tongue-like (Fig. 10C). Genital fork with arms each having wide lateral plate with small projection directed anteriorly (Fig. 10D). Paraproct shorter than wide, with ventral surface widely bare (Fig. 10E). Cercus semicircular, longer than paraproct (Fig. 10F). Spermatheca ellipsoidal, with pigmentation reaching junction with spermathecal duct or extended onto duct (Fig. 10G). Male: Postnotum bare. Hind basitarsus parallel-sided or somewhat enlarged and spindle-shaped (Fig. 10H). Coxite enlarged, longer than wide (Fig. 10I). Style much shorter than coxite, inwardly twisted (Fig. 10I). Ventral plate Y-shaped, with haired median keel (Fig. 10I–K). Median sclerite simple, club-shaped (Fig. 10L). Paramere with one hook (Fig. 10M). Aedeagal membrane covered with spinous setae and dorsal plate present (Fig. 10N). Cercus large, with relatively long hairs (Fig. 10O). Pupa: Frons with two pairs of trichomes. Gill with four slender filaments (Fig. 10P). Abdomen with dorsal spine-combs. Ventral surface of abdominal segments 5–7 each with pair of bifid inner and bifid (on segment 5) or unbranched (on segments 6 and 7) outer hooks on each side. Abdominal segment 9 with pair of cone-shaped terminal hooks but without grapnel-like hooklets on each side. Cocoon wall-pocket-shaped (Fig. 10Q). Larva: Mandible with mandibular serration composed of two medium-sized teeth (Fig. 10R). Supernumerary serration absent. Hypostoma with median and corner teeth prominent, and lateral margins serrated apically. Postgenal cleft short to medium-long. Accessory sclerite absent. Ventral papillae large.

Remarks. The subgenus *Eusimulium* was once treated as the *S. aureum* species-group in the genus *Eusimulium* by Rubtsov (1956) or in the subgenus *Eusimulium* by Crosskey (1969). The above diagnosis of this subgenus follows those of Crosskey (1969) and Adler et al. (2004) with slight modifications.

This subgenus is morphologically homogeneous and is distinguished from other subgenera by the unique shape of the female terminalia (in particular short paraproct) and male genitalia (Fig. 10C–G, I–O) and two pairs of chromosomes (rather than three pairs of chromosomes) (Adler et al., 2004).

It is composed of 40 named species, of which 36 are distributed in the Holarctic Region, three in the Oriental Region, and one in the Neotropical Region (Adler, 2022).

Species Included. *Simulium jomsomense* Takaoka and Shrestha, 2010 from Nepal; *S. taipei* (Shiraki, 1935) from Taiwan; and *S. weiningense* Chen and Zhang, 1997 and *S.* sp. D Zhang, Wen and Chen, 1999, both from China (Guizhou).

Subgenus Gomphostilbia *Enderlein*

Gomphostilbia Enderlein, 1921 (as genus). Type species: *ceylonicum* (original designation)

Diagnosis. Female and male: Antenna consisting of scape, pedicel, and nine flagellomeres, or rarely seven or eight flagellomeres (Fig. 39A, C). Pleural membrane bare, or rarely with few to ca. 80 hairs. Katepisternum haired. Basal section of radius haired. Vein R_1 with spinules as well as hairs. Scutum with or without patterns. Female: Frons dull, moderately or densely covered with hairs, or slightly to moderately shiny and sparsely to moderately covered with hairs. Proboscis short (labrum shorter than clypeus) except *S. tahanense*, which has long proboscis (labrum much longer than clypeus). Mandible serrated apically on both sides, or serrated on inner margin and partially or not serrated on outer margin. Cibarium with or without produced posterior margin, without any processes (Fig. 26D, E). Third palpal segment shorter than fourth segment, with sensory vesicle of various sizes. Claw with large basal tooth, or rarely simple, or with small tooth or medium-sized tooth. Subcosta with hairs or very rarely bare. Abdominal segments 6–8 or rarely segments 5–8 shiny; seventh sternite undeveloped or rarely developed (Fig. 41C), as in *S. guniki*. Genital fork with or without short small projection directed anteriorly, and with or without triangular projection directed posteromedially from each arm. Spermatheca ellipsoidal or rarely globose, without dark pigmented neck or rarely with dark pigmented neck (Fig. 12B), as in *S. ambigens*, and without internal setae or rarely with internal setae, as in three species of the *S. baisasae* species-group. Paraproct normal, somewhat longer than cercus. Male: Hind basitarsus slender and parallel-sided, or slightly to greatly expanded, and spindle-shaped or wedge-shaped. Abdomen with pair of dorsal or dorsolateral shiny patches on segments 2 and 5–7, or rarely, shiny patches on segments 2–8 or on segments 2–9. Coxite much longer than style (Fig. 12E). Style somewhat broad basally, then gradually tapered toward apex, or broad from base to near apex and with single apical spine (or rarely lacking apical spine). Ventral plate simple and rather flat, in form of transverse plate with subparallel basal arms, with or without ventrally produced process, without any median keel, except *S. ambigens*, which has laterally compressed median keel (Fig. 12E, H), and with hairs both on ventral and posterior surfaces. Paramere with few to several parameral hooks of different sizes (or rarely without hooks, as in *S. hiroshii* in South Pacific). Pupa: Frons usually covered with tubercles or rarely bare, and with three pairs of trichomes. Antennal sheath without any spine-like projection but with or without tubercles. Gill with eight simple short to long thread-like filaments, usually arranged as 3+3+2 or 2+2+2+2 filaments, or rarely gill with three, four, six, 13, or 16 filaments, and also rarely gill formed of inflated structure with or without slender filaments, as in species of the *S. banauense* (Fig. 18D, E, H) and *S. gombakense* species-groups (Fig. 37A–I). Abdomen with spine-combs (except *S. palauense* lacking spine-combs) dorsally. Ventral surface of abdominal segments 5–7 each with a pair of branched inner and branched (on segment 5) or unbranched (on segments 6 and 7) hooks on each side. Abdominal segment 9 with grapnel-like hooklets on each side (except several species lacking grapnel-like hooklets, as three species in the *S. miblosi* subgroup); terminal hooks plate-like extended laterally and usually with their outer margins crenulated, or small and cone-like, or rarely small, flat, and bent anteriorly, appearing to be mushroom-like when viewed laterally. Cocoon usually simple, wall-pocket-shaped, covering most of pupal body, or rarely short covering only posterior half of pupal body, and rarely with short to long anterodorsal projection, and without neck or rarely with low anteroventral neck. Larva: Hypostoma with usual nine apical teeth, three to six bristles in row, subparallel to each lateral margin, which is usually smooth (Fig. 32U) except *S. palauense* and few other species, which have serrated lateral margins (Fig. 38P). Head-spots positive or sometimes negative. Antenna longer than stem of labral fan, without any pale annulation. Postgenal cleft variable in length, usually subequal to, or longer than, postgenal bridge, sometimes long reaching posterior margin of hypostoma, or rarely shorter than postgenal bridge (Fig. 41N), as in *S. guniki*, or very rarely absent (Fig. 38Q), as in *S. palauense*. Mandible with major tooth of mandibular serration at an acute angle to mandible on apical side (except *S. baisasae*), and without supernumerary serration (except *S. palauense*). Abdomen with or without dorsolateral protuberances, and with or without minute spinous setae dorsally on posterior segments. Ventral papillae well developed. Anal sclerite X-shaped, usually widely sclerotized basally; accessory sclerite absent or rarely present (Fig. 38R), as in *S. palauense*.

Remarks. The diagnosis given by Takaoka (2003) is slightly modified as above.

The subgenus *Gomphostilbia* was first defined by Crosskey (1967), and modified by Datta (1973), Takaoka (1983), Davies and Györkös (1987a), Takaoka and Davies (1996), and Takaoka (2003, 2012). This subgenus was divided into 15 species-groups, of which 11 species-groups including the *S. palauense* species-group were distributed in the Oriental Region and the remaining four species-groups in the Australasian Region (Takaoka, 2012). A new species-group, *S. alienigenum* species-group, is here proposed to accommodate two species from Palawan and Borneo. Therefore, a total of 12 species-groups are recognized in the Oriental Region.

The subgenus *Gomphostilbia* is the second largest in the Oriental Region, being represented by 268 named and 15 unnamed species.

The pupal gills are remarkable in the subgenus *Gomphostilbia* showing a wide range of variations in their shape, and in the number and arrangement of filaments (Takaoka, 2012). The different number of pupal gill filaments (four, six, eight, and 10) was most clearly expressed among species in the *S. ceylonicum* species-group (Takaoka and Davies, 1996; Takaoka, 2003), and even the intraspecific variation in the number of pupal gill filaments was recognized in *S. visayaense* Takaoka, 1983 (eight, nine and 10) (Fig. 16I, J) and in *S. barioense* Takaoka, 2008 (13 and 16) (Fig. 35D–F) (Takaoka, 1983, 2008c). The process of the reduction of pupal gill filaments from six to five, as well as the direction of speciation, was suggested by comparing pupal gills of six species of the *S. sherwoodi* species-group in the Australasian Region (Takaoka, 1995). The complicated configurations of pupal gills consisting of an inflated structure and slender filaments were also encountered in 10 species in the *S. gombakense* species-group (Fig. 37A–I), and several species of the *banauense* species-group (Fig. 18C–H). These broad variations are probably ascribed to a great number of species and their successful adaptations to various types of ecological environments, as shown by their wide latitudinal and longitudinal distributions. The presumed long evolutionary history of this subgenus may be another explanation.

Key to Identify 16 Species-groups of the Subgenus *Gomphostilbia* *

Females

1. Antenna with seven or eight flagellomeres (Fig. 39A–C) ....*S. varicorne* species-group
   Antenna with nine flagellomeres 2
2. Scutum orange ....*S. hiroshii* species-group**
   Scutum brownish to black 3
3. Claw simple, without tooth .... *S. baisasae* species-group
   Claw with small to large tooth 4
4. Spermatheca with darkened neck (Fig. 12B) ....*S. ambigens* species-group
   Spermatheca without darkened neck 5
5. Spermatheca with internal hairs 6
   Spermatheca without internal hairs 8
6. Pleural membrane haired. Claw with small tooth ....*S. sherwoodi* species-group**
   Pleural membrane bare. Claw with large tooth 7
7. Each arm of genital fork with stout ridge directed anteriorly ....*S. heldsbachense* species-group**
   Each arm of genital fork without stout ridge directed anteriorly .... *S. hemicyclium* species-group**
8. Pleural membrane or its vicinity haired 9
   Pleural membrane bare 10
9. Claw with small tooth (Fig. 11C) .... *S. alienigenum* species-group (*S. alienigenum*)
   Claw with medium-sized or large tooth (Fig. 17E) .... *S. banauense* species-group

10. Hind tibia whitish on basal half or more 11
 Hind tibia entirely darkened or darkened except basal two-fifths or less yellow 14
11. Cibarium with simple dorsal margin (Fig. 26E) 12
 Cibarium with dorsal margin bent forwardly in form of short plate (Fig. 26D) 13
12. Antenna yellow except five apical flagellomeres darkened (Fig. 11A) ....*S. alienigenum* species-group (*S. kalimantanense*)
 Antenna pale brown except scape, pedicel and basal half of first flagellomere grayish yellow ....*S. ceylonicum* species-group (*S. atratoides*) and *S. epistum* species-group (*S. miblosi* subgroup)
13. Base of radius with tuft of dark hairs ....*S. ceylonicum* species-group
 Base of radius with tuft of yellow hairs .... *S. asakoae* species-group, *S. gombakense* species-group and *S. epistum* species-group
14. Mandible lacking outer teeth (Fig. 38B) ....*S. palauense* species-group
 Mandible with outer teeth 15
15. Fore coxa dark ....*S. darjeelingense* species-group
 Fore coxa whitish-yellow or yellow .... *S. batoense* species-group

*The key provided by Takaoka (2012) is somewhat modified.
**These four species-groups are distributed only in the Australasian Region.

Simulium (*Gomphostilbia*) *alienigenum* Species-group

Diagnosis. Female and male: Antenna with nine flagellomeres. Pleural membrane haired or bare. Upper mesopleuron region haired or bare. Female: Cibarium unarmed (Fig. 11B). Claw with small or long basal tooth (Fig. 11C, D). Male: Hind basitarsus slender, nearly parallel-sided (Fig. 11J). Ventral plate transverse when viewed ventrally, much produced ventrally (Fig. 11K, M). Pupa: Gill with four long filaments (Fig. 11Q). Abdominal segment 9 with pair of cone-shaped terminal hooks (Fig. 11R). Cocoon wall-pocket-shaped. Larva: Postgenal cleft medium-long, bullet-shaped (Fig. 11S).

Remarks. This species-group is newly created to accommodate two species, which were previously placed in the *S. banauense* species-group. It is distinguished from the *S. banauense* species-group in the female and male by having no or a few hairs on the pleural membrane, and in the pupa by having four gill filaments (Fig. 11Q).

Species Included. *Simulium alienigenum* Takaoka, 1983 from Philippines (Palawan); and *S. kalimantanense* Takaoka and Sofian-Azirun, 2016 from Indonesia (Kalimantan) and Malaysia (Sabah).

Simulium (*Gomphostilbia*) *ambigens* Species-group

Diagnosis. Female and male: Antenna with nine flagellomeres. Pleural membrane bare. Female: Mandible with inner teeth but lacking outer teeth. Cibarium simple, with posterodorsal margin somewhat extended dorsally and nearly straight and with narrow horizontal slit medially near posterodorsal margin (Fig. 12A). Claw with small subbasal tooth. Spermatheca with dark pigmented neck (Fig. 12B) and inconspicuous internal hairs. Male: Hind basitarsus slender, nearly parallel-sided. Ventral plate with haired median keel ventrally (Fig. 12E, G, H). Pupa: Gill with six slender filaments (Fig. 12J). Terminal hooks simple, cone-shaped.

Remarks. This species-group was established by Takaoka (1983).

The female cibarium with its posterodorsal margin somewhat extended dorsally and nearly straight (Fig. 12A) and the male ventral plate with a median keel (Fig. 12E, G, H) are unique among species of the subgenus *Gomphostilbia*.

Species Included. *Simulium ambigens* Delfinado, 1969 from Philippines (Mindanao).

Simulium (*Gomphostilbia*) *asakoae* Species-group

Diagnosis. Female and male: Antenna with nine flagellomeres. Pleural membrane bare. Base of the radial vein with tuft of yellowish hairs (except *S. inthanonense*, *S. rutengense*, and *S. yunnanense*, which have tuft of dark hairs). Fore coxae yellowish. Female: Cibarium with posterodorsal margin bent forward and mediolongitudinal ridge with bifid apex (Fig. 13F). Hind tibia usually whitish-yellow on basal half to three-fourths, and dark on rest. Claw with large basal tooth (Fig. 13G). Each arm of genital fork with or without short projection directed posteromedially (Fig. 13J). Male: Hind basitarsus moderately to greatly enlarged, spindle-shaped, or wedge-shaped (Fig. 13N–P). Ventral plate transverse, moderately produced ventrally, and with lateral margins somewhat emarginated subbasally or medially when viewed ventrally (Fig. 13Q) (except in few species, of which ventral plate has lateral margins nearly parallel-sided when viewed ventrally), and ventral margin rounded or nearly straight or slightly concave when viewed posteriorly (Fig. 13S–V). Pupa: Gill with eight slender long filaments (Fig. 14A–F) except *S. hoiseni* and *S. udomi*, which have six filaments, and *S. brevilabrum*, which has four filaments. Terminal hooks slightly to moderately widened, with outer margins crenulated (Fig. 14G–J). Larva: Postgenal cleft medium-long (Fig. 14O–Q). Body often with various color patterns (Fig. 15A–T).

Remarks. The *S. asakoae* species-group was derived from the former *S. ceylonicum* species-group (Takaoka, 2012). This is the largest species-group in the subgenus *Gomphostilbia*, being represented by 68 species.

It is characterized by a combination of the yellow tuft hair of the base of the radial vein, yellow fore coxa, enlarged male hind basitarsus, and ventral plate with lateral margins somewhat emarginated subbasally or medially or rarely parallel-sided when viewed ventrally (Fig. 13Q) (Takaoka, 2012).

Species Included. Simulium asakoae Takaoka and Davies, 1995 from Malaysia (Peninsular), Thailand, Taiwan, China (Hong Kong) and Vietnam; *S. banluangense* Takaoka, Srisuka and Fukuda, 2020, *S. chaowaense* Takaoka, Srisuka and Saeung, 2020, *S. chiangdaoense* Takaoka and Srisuka, 2009, *S. huaimorense* Takaoka, Srisuka and Saeung, 2020, *S. junkumae* Takaoka, Srisuka and Saeung, 2020, *S. kiewfinense* Takaoka, Srisuka and Fukuda, 2020, *S. kiewlomense* Takaoka, Srisuka and Saeung, 2021, *S. klonglanense* Takaoka, Srisuka and Saeung, 2020, *S. loeiense* Takaoka, Srisuka and Fukuda, 2020, *S. maehongsonense* Takaoka, Srisuka and Saeung, 2020, *S. maelanoiense* Takaoka, Srisuka and Saeung, 2020, *S. maewongense* Takaoka, Srisuka and Saeung, 2020, *S. mokroense* Takaoka, Srisuka and Saeung, 2022, *S. muangpanense* Takaoka, Srisuka and Fukuda, 2020, *S. namdanense* Takaoka, Srisuka and Saeung, 2020, *S. nanoiense* Takaoka, Srisuka and Saeung, 2020, *S. nanthaburiense* Takaoka, Srisuka and Fukuda, 2020, *S. phapeungense* Takaoka, Srisuka and Fukuda, 2020, *S. pitasawatae* Takaoka, Srisuka and Saeung, 2020, *S. puaense* Takaoka, Srisuka and Saeung, 2020, *S. sutheppuiense* Takaoka, Srisuka and Saeung, 2020, *S. rampae* Takaoka, Srisuka and Saeung, 2019, *S. songense* Takaoka, Srisuka and Fukuda, 2020, *S. teerachanense* Takaoka, Srisuka and Fukuda, 2020, *S. thungchangense* Takaoka, Srisuka and Saeung, 2020 and *S. udomi* Takaoka and Choochote, 2006, all from Thailand; *S. brinchangense* Takaoka, Sofian-Azirun and Hashim, 2014, *S. hoiseni* Takaoka, 2008, *S. izuae* Takaoka, Sofian-Azirun and Hashim, 2013, *S. sofiani* Takaoka and Hashim, 2011, *S. tanahrataense* Takaoka, Sofian-Azirun and Ya'cob, 2014, and *S. roslihashimi* Takaoka and Sofian-Azirun, 2011, all from Peninsular Malaysia; *S. chaudinhense* Takaoka and Sofian-Azirun, 2017, *S. confertum* Takaoka and Sofian-Azirun, 2015, *S. fuscidorsum* Takaoka and Ya'cob, 2015, *S. hmongense* Takaoka and Ya'cob, 2017, *S. hongthaii* Takaoka, Sofian-Azirun and Ya'cob, 2014, *S. phulocense* Takaoka and Chen, 2015, *S. sanchayense* Takaoka and Lau, 2017, *S. quychauense* Takaoka and Chen, 2017, *S. tamdaoense* Takaoka, Sofian-Azirun and Ya'cob, 2014, *S. thituyenae* Takaoka and Pham, 2015 and *S. vinhphucense* Takaoka and Low, 2017, all from Vietnam; *S. gyorkosae* Takaoka and Davies, 1996 from Indonesia (Java, Sumatra, Sulawesi); *S. brevilabrum* Takaoka, Hadi and Sigit, 2006, *S. rutengense* Takaoka, Hadi and Sigit, 2006 and *S. sunapii* Takaoka, Sofian-Azirun and Wayan, 2016. all from Indonesia (Flores); *S. inthanonense* Takaoka and Suzuki, 1984 from Thailand, China (Yunnan) and Vietnam (?); *S. epauletum* Guo, Zhang and An, 2011, *S. jianfengense* An and Long, 1994 from China (Hainan); *S. fanjingshanense* Chen, Zhang and Wen, 2000, *S. mengi* Chen, Zhang and Wen, 2000 and *S. miaolingense* Wen and Chen, 2000, all from China (Guizhou); *S. yunnanense* Chen and Zhang, 2004 from China (Yunnan); *S. kaohsiungense* Takaoka, Huang and Fukuda, 2021, *S. taitungense* Huang and Takaoka, 2011, *S. shoufengense*

Takaoka, Low and Ya'cob, 2021 and *S. tuenense* Takaoka, 1979, all from Taiwan; *S. lurauense* Takaoka, Sofian-Azirun and Hashim, 2011 from Malaysia (Peninsular) and Myanmar; *S. monglaense* Takaoka, Srisuka and Saeung, 2017 and *S. namense* Takaoka, 1989, both from Myanmar; *S. myanmarense* Takaoka, Srisuka and Saeung, 2017 from Myanmar and Thailand; *S. lamjungense* Takaoka and Shrestha, 2020, *S. nareshi* Takaoka and Shrestha, 2020 and *S. nepalense* Lewis, 1964, all from Nepal; and *S. okinawense* Takaoka, 1976 and *S. tokarense* Takaoka, 1973, both from Japan (Nansei Islands).

Simulium (*Gomphostilbia*) *baisasae* Species-group

Diagnosis. Female and male: Antenna with nine flagellomeres. Pleural membrane bare. Fore coxae brown. Female: Mandible with inner teeth and with or without outer teeth. Cibarium simple, unarmed, with posterodorsal margin somewhat convex and strongly sclerotized (Fig. 16A). Claw simple. Abdominal segments 5–8 shiny dorsally. Genital fork with small projection directed anteriorly from each lateral arm (Fig. 16B). Spermatheca with reticulate surface patterns (Fig. 16B) and internal hairs. Male: Hind basitarsus slender, nearly parallel-sided. Ventral plate much produced ventrally (Fig. 16E, F). Pupa: Gill with six, eight, nine, or ten filaments (Fig. 16H–K). Terminal hooks cone-shaped. Larva: Postgenal cleft medium-long (Fig. 16M, O).

Remarks. This species-group was defined by Takaoka (1983).

It is characterized in the female by the simple claw and spermatheca with internal hairs.

Species Included. *Simulium baisasae* Delfinado, 1962 [1960], *S. visayaense* Takaoka, 1983 and *S. bicolense* Takaoka, 1983, all from Philippines.

Simulium (*Gomphostilbia*) *banauense* Species-group

Diagnosis. Female and male: Antenna with nine flagellomeres. Pleural membrane haired with few to about 30 hairs. Female: Mandible with teeth on inner margin and lacking teeth on outer margin except *S. lilianae*, which has teeth on outer margin (Fig. 17C). Cibarium simple, unarmed (Fig. 17D). Claw with large basal tooth (Fig. 17E). Each arm of genital fork with short projection directed posteromedially (Fig. 17F). Male: Hind basitarsus slender, parallel-sided or somewhat tapered apically (Fig. 17M, N). Ventral plate transverse, moderately to greatly produced ventrally, nearly parallel-sided or narrowed posteriorly when viewed ventrally (Fig. 17O). Pupa: Gill with six slender or somewhat inflated filaments, or of inflated structure with six slender filaments (Fig. 18A–H). Terminal hooks wide, plate-like, with outer margins crenulated.

Remarks. This species-group was created under the subgenus *Morops* (Takaoka, 1983) and later transferred to the subgenus *Gomphostilbia* (Takaoka, 2003).

It is characterized in the female and male by the haired pleural membrane.

Species Included. *Simulium atuganense* Takaoka, 2009 and *S. katangladense* Takaoka, 2009, both from Philippines (Mindanao); *S. banauense* Takaoka, 1983, *S. bansonae* Takaoka, 1983, *S. cagayanense* Takaoka, 2005, *S. ifugaoense* Takaoka, 1983, *S. lilianae* Takaoka, 2000, *S. liliwense* Takaoka, 1983 and *S. quezonense* Takaoka, 2009, all from Philippines (Luzon); *S. manbucalense* Takaoka, 1983 and *S. riverai* Takaoka, 1983, both from Philippines (Negros); *S. salazarae* Takaoka, 1983 from Philippines (Bohol, Luzon, Mindanao and Samar); and *S. disneyi* Takaoka and Roberts, 1988 from Indonesia (Sulawesi).

Simulium (*Gomphostilbia*) *batoense* Species-group

Diagnosis. Female and male: Antenna with nine flagellomeres. Pleural membrane bare. Hind tibia entirely darkened or darkened on apical three-fifths or more (except *S. dentistylum*). Base of radial vein with tuft of dark brown hairs (except *S. baglungense*). Female: Frons sparsely or moderately covered with hairs (Fig. 21A, B). Mandible with inner and outer teeth. Cibarium with forwardly bent plate from dorsal margin and mediolongitudinal ridge having darkened split narrow apex (Fig. 21C–E) or U-shaped apex (Fig. 23A). Claw with large basal tooth, or rarely with small or medium-sized subbasal tooth (Fig. 21G–I). Each arm of genital fork with or without short projection directed posteromedially (Fig. 21K, L). Male: Scutum with color patterns of various shapes (Fig. 21P–S). Hind basitarsus slender, parallel-sided (Fig. 21T), or rarely spindle-shaped, as in *S. pattoni*. Ventral plate transverse, not produced ventrally, or slightly to moderately produced, and nearly parallel-sided or somewhat narrowed posteriorly when viewed ventrally (Fig. 21U), as in *S. brevitruncatum*, or rarely, widened posteriorly (Fig. 21A'), as in *S. davaoense* and *S. samarense*, or emarginated medially, as in *S. butwalense*. Pupa: Antennal sheath with or without tubercles (Fig. 23P). Gill with eight (rarely six or 10) filaments, or very rarely of inflated structure with eight slender filaments (Fig. 22D). Terminal hook small and cone-like, or small and flat, mushroom-like when viewed laterally (Fig. 25T–V), or slightly to somewhat widened, triangular, with outer margin smooth or undulate, and less than 2.5 times as long as inner margin (Fig. 22G, H). Larva: Postgenal cleft long (Fig. 22M, N).

Remarks. This species-group was heterogeneous and thus divided into seven subgroups, of which six are distributed in the Oriental Region, and one is restricted to Honshu and Kyushu, Japan (*S. omutaense* subgroup, characterized by the pupal gill with eight filaments, of which one of two filaments of the ventral pair is much longer than the seven other filaments) (Takaoka et al., 2021e).

Simulium (*Gomphostilbia*) *binuanense* Subgroup

Diagnosis. Female: Frons sparsely to moderately covered with hairs (Fig. 19A). Cibarium with darkened narrow apex of mediolongitudinal ridge (Fig. 19B, C). Claw with large tooth (Fig. 19E). Genital fork without projection directed posteromedially on each arm (Fig. 19G). Male: Ventral plate flat (Fig. 19M, P). Pupa: Gill with eight slender long filaments, of which two ventral paired filaments slightly different in length and thickness from each other, and longer than pupal body (Fig. 19R). Terminal hooks triangular or somewhat broadened laterally (Fig. 19S–U). Grapnel-like hooklets present except *S. mangasepi*, which has setae in place of grapnel-like hooklets.

Species Included. *Simulium babuyanense* Takaoka and Tenedero, 2007, *S. binuanense* Takaoka and Tenedero, 2008 and *S. mangasepi* Takaoka, 1983, all from Philippines (Palawan); *S. jerantutense* Takaoka, Sofian-Azirun and Ya'cob, 2014 and *S. tekamense* Takaoka, Sofian-Azirun and Ya'cob, 2014, both from Malaysia (Peninsular); and *S. pegalanense* Smart and Clifford, 1969, from Malaysia (Peninsular and Sabah).

Simulium (*Gomphostilbia*) *decuplum* Subgroup

Diagnosis. Female: Sensory vesicle long (Fig. 20A). Cibarium with U-shaped apex of mediolongitudinal ridge (Fig. 20B). Genital fork without projection directed posteromedially on each arm (Fig. 20F). Male: Style with narrowest at apical one-third when viewed ventrolaterally (Fig. 20L). Ventral plate flat (Fig. 20N). Pupa: Gill with 10 slender filaments, which are shorter than pupal body (Fig. 20Q). Cocoon widely extended ventrolaterally (Fig. 20S). Larva: Postgenal cleft long, with its apex close to posterior margin of hypostoma (Fig. 20T).

Species Included. *Simulium bannaense* Chen and Zhang, 2003 from China (Yunnan); *S. jiulianshanense* Kang, Zhang and Chen, 2007 from China (Jiangxi); *S. decuplum* Takaoka and Davies, 1995 from Malaysia (Peninsular), Myanmar and India (Assam) (?); *S. lamdongense* Takaoka and Sofian-Azirun, 2015 from Vietnam; *S. chiangraiense* Takaoka, Srisuka and Saeung, 2018, *S. huaikaeoense* Takaoka, Srisuka and Low, 2018, *S. khaokhorense* Takaoka, Srisuka and Low, *S. maeklangense* Takaoka, Srisuka and Saeung, 2018 and *S. pamiangense* Takaoka, Srisuka and Saeung, 2018, all from Thailand; and *S.* Pabho sp. Lewis, 1974 from India (Assam).

Simulium (*Gomphostilbia*) *duolongum* Subgroup

Diagnosis. Female: Frons narrow to wide (Fig. 21A, B). Cibarium with dark narrow apex of mediolongitudinal ridge (Fig. 21C–E). Tarsal claw with small or medium-sized or large tooth (Fig. 21G–I). Genital fork with or without projection directed posteromedially on each arm (Fig. 21K, L). Male: Scutum with color patterns of different shapes (Fig. 21P–S). Pupa: Gill with eight slender filaments, of which two ventral paired filaments subequal in length to each other and more than 1.5 times as long as six other filaments (Fig. 22A–C) except *S. tumidum*, which has the gill composed of two inflated tubular structures each with a slender filament apically, and six short slender filaments basally (Fig. 22D).

Species Included. Simulium abadii Takaoka, 2003 *S. hadiae* Takaoka, 2003, *S. kolakaense* Takaoka, 2003, *S. manadoense* Takaoka, 2003, *S. marosense* Takaoka, Sofian-Azirun and Suana, 2019, *S. singgihi* Takaoka, 2003, *S. syafruddini* Takaoka, 2003, *S. torautense* Takaoka and Roberts, 1988 and *S. tumidum* Takaoka, 2003, all from Indonesia (Sulawesi); *S. batoense* Edwards, 1934 from Indonesia (Java); *S. friederichsi* Edwards, 1934 from Indonesia (Java and Sumatra); *S. lemborense* Takaoka and Sofian-Azirun, 2017 from Indonesia (Flores and Sumba); *S. bhutanense* Takaoka and Somboon, 2008 from Bhutan and Nepal; *S. dachaisense* Takaoka and Lau, 2015 and *S. yvonneae* Takaoka and Low, 2019, both from Vietnam; *S. davaoense* Takaoka, 1983 from Philippines (Mindanao); *S. samarense* Takaoka and Tenedero, 2008 from Philippines (Samar); *S. duolongum* Takaoka and Davies, 1995 from Malaysia (Peninsular), Indonesia (Sumatra), Thailand, and Vietnam (?); *S. johorense* Takaoka, Sofian-Azirun and Ya'cob, 2014 from Malaysia (Peninsular); *S. siamense* Takaoka and Suzuki, 1984 (complex) from Thailand, Nepal (?), and Vietnam (?); *S. syuhaiense* Huang and Takaoka, 2008 from Taiwan; *S. tahanense* Takaoka and Davies, 1995 from Malaysia (Peninsular and Sarawak) and Indonesia (Kalimantan); *S. yaeyamaense* Takaoka, 1991 from Japan (Ryukyu Islands); and *S. laosense* Takaoka, Srisuka and Saeung, 2017 from Laos.

Simulium (Gomphostilbia) parahiyangum Subgroup

Diagnosis. Female: Cibarium with dark U-shaped apex of mediolongitudinal ridge (Fig. 23A). Male: Style narrow from middle to apex (Fig. 23K). Ventral plate flat (Fig. 23M). Pupa: Antennal sheath covered with tubercles (Fig. 23P). Gill with eight short slender filaments of subequal length (less than 1 mm) (Fig. 23Q, R). Larva: Postgenal cleft long, reaching posterior margin of hypostoma (Fig. 23V). Abdomen with or without paired protuberances dorsally (Fig. 23T, U). Abdomen markedly narrowed between segments 4 and 5 (Fig. 23T).

Species Included. Simulium dentistylum Takaoka and Davies, 1995 from Malaysia (Peninsular) and Thailand; *S. sazalyi* Takaoka, Ya'cob and Low, 2018 from Malaysia (Peninsular); *S. longlanhense* Takaoka and Ya'cob, 2015 from Vietnam; *S. ngaoense* Takaoka, Srisuka and Saeung, 2018 from Thailand and Vietnam; and *S. parahiyangum* Takaoka and Sigit, 1991 from Indonesia (Java and Sumatra), Malaysia (Sabah and Sarawak), and India (Assam) (?).

Simulium (*Gomphostilbia*) *sextuplum* Subgroup

Diagnosis. Female: Frons nearly bare except several hairs along each lateral margin (Fig. 24A). Cibarium with mediolongitudinal ridge having narrow dark apex deeply incised (Fig. 24B). Male: Paramere with base covered with fine setae (Fig. 24P, Q). Pupa: Gill with six slender filaments of same length, and longer than pupal body (Fig. 24R). Cocoon wall-pocket-shaped, with short anterodorsal projection (Fig. 24T, U).

Species Included. Simulium paukatense Takaoka, 2008 from Malaysia (Sarawak); *S. sextuplum* Takaoka and Davies, 1995 from Malaysia (Peninsular); and *S. fukudae* Takaoka, Srisuka and Saeung, 2017 from Thailand.

Simulium (*Gomphostilbia*) *terengganuense* Subgroup

Diagnosis. Female and male: Hind tibia darkened except basal two-fifths whitish and with light brown subbasal marking (Fig. 25C, E). Female: Frons narrow. Sensory vesicle short (Fig. 25A). Cibarium with mediolongitudinal ridge having dark narrow apex (Fig. 25B). Male: Ventral plate wide, transverse, with posterior margin produced posteromedially (Fig. 25M). Pupa: Gill with eight long slender filaments of almost same length, nearly as long as or slightly longer or shorter than pupal body (Fig. 25S). Terminal hooks small, flat and mushroom-like when viewed laterally (Fig. 25V). Larva: Abdomen with distinct pattern of reddish-brown markings dorsally and dorsolaterally (Fig. 25W, X).

Species Included. Simulium terengganuense Takaoka, Sofian-Azirun and Ya'cob, 2012 from Malaysia (Peninsular and Sarawak).

Simulium (*Gomphostilbia*) *batoense* Species-group Species Unplaced to Subgroup

Species Included. Simulium agasthyamalaiense Anbalagan, 2019, *S. dinakarani* Anbalagan, 2020, *S. krishnani* Anbalagan, 2020, *S. kumbakkaraiense* Anbalagan, 2019, and *S. panagudiense* Anbalagan, 2015, all from India (Tamil Nadu); *S. kottoorense* Anbalagan, 2015, *S. peteri* Anbalagan, 2014, and *S. takaokai* Anbalagan, 2014, all three from India (Kerala); *S. brevitruncatum* Takaoka, 2003, *S. miyagii* Takaoka, 2003, and *S. sulawesiense* Takaoka and Roberts, 1988, all from Indonesia (Sulawesi); *S. baglungense* Takaoka and Shrestha, 2010 and *S. butwalense* Takaoka and Shrestha, 2010, *S. dadeldhuraense* Takaoka and Shrestha, 2020 and *S. purandharaense* Takaoka and Shrestha, 2020, all from Nepal; *S. brivetruncum* Chen and Chen, 2000 and *S. hainanense* Long, An, Hao, Yan and Yuan, 1994, both from China (Hainan); *S. chuzargangense* Takaoka and Somboon, 2008 from Bhutan; *S. dola* Davies and Györkös, 1987 and *S. krombeini* Davies and Györkös, 1987, both from Sri Lanka; and *S. pattoni* Senior-White, 1922 from India (Tamil Nadu), Sri Lanka and China (Hainan) (?); and *S. bakalalanense* Ya'cob, Takaoka and Sofian-Azirun, 2015 from Malaysia (Sarawak).

Simulium (*Gomphostilbia*) *ceylonicum* Species-group

Diagnosis. Female and male: Antenna with nine flagellomeres. Pleural membrane bare. Fore coxa yellow. Base of the radius with tuft of dark hairs (except *S. atratoides*, *S. longitruncum*, and *S. unii,* all with tuft of yellow hairs). Female: Cibarium with short plate forwardly bent from posterodorsal margin, with mediolongitudinal ridge having bifid apex (Fig. 26D), or rarely unarmed (Fig. 26E), as in *S. atratoides*. Genital fork with projection directed posteromedially on each arm (Fig. 26H). Male: Scutum with color pattern of different shapes (Fig. 26L–O). Hind basitarsus enlarged, wedge-shaped (Fig. 26P), or spindle-shaped (Fig. 26Q). Ventral plate transverse, somewhat to moderately produced ventrally, with lateral margins slightly or moderately narrowed posteriorly (Fig. 26R) when viewed ventrally. Pupa: Gill with four, six, eight, and 10 slender long filaments (Fig. 27A–E). Terminal hooks simple triangular or widened, with outer margins crenulated (Fig. 27F–I). Larva: Postgenal cleft usually long (Fig. 27L) but rarely medium-long (Fig. 27K) or extremely long reaching posterior margin of hypostoma (Fig. 27M). Thoracic proleg bare but exceptionally haired (Fig. 27N), as in *S. capillatum*. Abdomen normal in shape but exceptionally markedly constricted between segments 4 and 5 (Fig. 27J), as in *S. capillatum*.

Remarks. This species-group was first established by Takaoka (1983), and finally redefined by Takaoka (2012). It is characterized by a combination of the tuft of dark hairs at the base of the radius and the ventral plate with lateral margins narrowed posteriorly when viewed ventrally (Fig. 26R).

It is further divided into four subgroups by the number of pupal gill filaments.

Simulium (*Gomphostilbia*) *atratoides* Subgroup

Diagnosis. Female and male: Base of radius with tuft of yellow hairs. Female: Mandible without outer teeth (Fig. 26C). Cibarium simple, unarmed (Fig. 26E). Pupa: Gill with 10 slender long filaments (Fig. 27E). Larva: Postgenal cleft medium-long (Fig. 27K).

Species Included. *Simulium atratoides* Takaoka and Davies, 1996 from Indonesia (Java).

Simulium (*Gomphostilbia*) *sheilae* Subgroup

Diagnosis. Female and male: Base of radius with tuft of dark hairs, or rarely with tuft of yellow hairs as in *S. longitruncum*. Female: Sensory vesicle usually long (Fig. 26B). Pupa: Gill with eight slender long filaments (Fig. 27A).

Species Included. *Simulium napuense* Takaoka, 2003 from Indonesia (Sulawesi); *S. apoense* Takaoka, 1983 from Philippines (Mindanao); *S. bataksense* Takaoka and

Tenedero, 2007 from Philippines (Palawan); *S. mindoroense* Takaoka and Tenedero, 2007 from Philippines (Mindoro); *S. capillatum* Takaoka, 2009 from Malaysia (Sabah and Sarawak); *S. leparense* Takaoka, Sofian-Azirun and Ya'cob, 2014 and *S. longitruncum* Takaoka and Davies, 1995, both from Malaysia (Peninsular); *S. ceylonicum* (Enderlein), 1921 and *S. ela* Davies and Györkös, 1987, both from Sri Lanka; *S. curtatum* Jitklang et al., 2008, *S. pangsidaense* Takaoka, Srisuka and Saeung, 2021 and *S. wijiti* Takaoka, Srisuka and Saeung, 2023, all from Thailand; *S. hunanense* Zhang and Chen, 2004 from China (Hunan); *S. nigrofemoralum* Chen and Zhang, 2001 from China (Hainan); *S. wuyishanense* Chen and Zhang, 2009 from China (Fujian); *S. sheilae* Takaoka and Davies, 1995 from Indonesia (Sumatra), Malaysia (Peninsular and Sarawak), and Thailand; *S. trangense* Jitklang et al., 2008 from Thailand and Malaysia (Peninsular); and *S. unii* Takaoka and Pham, 2017 from Vietnam.

***Simulium* (*Gomphostilbia*) *rosemaryae* Subgroup**

Diagnosis. Female: Sensory vesicle short (Fig. 26A). Pupa: Gill with six slender long filaments (Fig. 27B, C). Grapnel-like hooklets present except *S. kamimurai*, which lacks grapnel-like hooklets.

Species Included. Simulium kamimurai Takaoka, 2003 from Indonesia (Sulawesi and Seram); *S. rangatense* Takaoka, Sofian-Azirun and Wayan, 2017 from Indonesia (Flores); and *S. rosemaryae* Takaoka and Roberts, 1988 from Indonesia (Sulawesi).

***Simulium* (*Gomphostilbia*) *mogii* Subgroup**

Diagnosis. Pupa: Gill with four slender long filaments (Fig. 27D).

Species Included. Simulium gimpuense Takaoka, 2003 and *S. mogii* Takaoka, 2003, both from Indonesia (Sulawesi).

***Simulium* (*Gomphostilbia*) *darjeelingense* Species-group**

Diagnosis. Female and male: Antenna with nine flagellomeres. Pleural membrane bare. Legs (including the fore coxae) mostly dark. Hair tuft of base of radial vein brownish or blackish. Female: Sensory vesicle long (Fig. 28A). Cibarium with mediolongitudinal ridge with bifid apex (Fig. 28B). Claw with large tooth (Fig. 28C). Genital fork without projection directed posteromedially on each arm (Fig. 28E). Male: Hind basitarsus much enlarged and spindle-shaped (Fig. 28I). Style narrowed from base to apical one-third, then nearly parallel-sided (Fig. 28K). Ventral plate transverse, not produced ventrally and slightly narrowed posteriorly when viewed ventrally (Fig. 28J). Pupa: Face with additional trichome on each side in two species (Fig. 29A). Gill with eight slender short filaments, much shorter than pupal body, having no annular ridges on their surface (Fig. 29C–E). Terminal hooks simple, triangular, or somewhat widened, with smooth outer margins (Fig. 29F, G). Cocoon wall-pocket-shaped, widely extended ventrolaterally (Fig. 29H, I). Larva: Postgenal cleft long (Fig. 29J).

Remarks. This species-group was derived from the former *S. ceylonicum* species-group (Takaoka, 2012).

Species Included. *Simulium chayamaritae* Takaoka and Srisuka, 2000 from Thailand; *S. curvastylum* Chen and Zhang, 2001 from China (Hainan); *S. guizhouense* Chen, Zhang and Yang, 2003 from China (Guizhou); *S. darjeelingense* Datta, 1973 from India (Assam, Arnachal Pradesh, and West Bengal); *S. eshimai* Takaoka and Adler, 2017 from Vietnam; and *S. shivapuriense* Takaoka and Shrestha, 2020 and *S. shresthae* Takaoka and Dangi, 2020, both from Nepal.

Simulium (Gomphostilbia) epistum species-group

Diagnosis. Female and male: Antenna with nine flagellomeres. Pleural membrane bare. Hair tuft on base of radial vein yellowish (except *S. miblosi* having tuft of yellow hairs mixed with dark brown ones). Hind tibia whitish or yellowish on basal two-fifths or more, with subbasal dark marking (Fig. 33B, C) or rarely without such dark marking (Fig. 32H). Female: Cibarium with forwardly bent plate and mediolongitudinal ridge with bifid apex (Fig. 30B) [except species of the *S. miblosi* subgroup having unarmed cibarium (Fig. 32B)]. Claw with large basal tooth (Fig. 30C). Male: Hind basitarsus slender, parallel-sided (Fig. 32I). Ventral plate transverse, moderately produced ventrally, and nearly parallel-sided or slightly narrowed posteriorly when viewed ventrally (Fig. 32J) [except *S. auratum*, of which ventral plate is widened posteriorly (Fig. 30Q), and *S. repentinum,* of which ventral plate is angulated basally, then narrowed posteriorly (Fig. 30N)]. Pupa: Gill with eight slender filaments (Figs. 31A–E and 32Q), or with one inflated tubular filament and seven slender filaments (Fig. 31F), or with 13 or 16 short filaments (Fig. 35B–D). Terminal hook widened and plate-like, with outer margin crenulated and more than 2.8 times as long as inner margin (Fig. 31H) [except three species of the *S. montiblense* subgroup, one species of the *S. zonatum* subgroup, of which terminal hooks are simple, cone-like, or somewhat widened laterally, with smooth outer margins (Fig. 31G)]. Cocoon wall-pocket-shaped, rarely with narrow anteroventral collar, as in *S. miblosi* and *S. kelabitense*, or much shortened in length (Fig. 35H, I), as in the *S. montiblense* subgroup. Larva: Postgenal cleft long (Fig. 31J) but rarely short (Fig. 31I), as in *S. adleri*, *S. zonatum* and *S. merapiense*.

Remarks. This species-group is further divided into four subgroups by the number and arrangement of pupal gill filaments, shape of the terminal hooks, and presence or absence of grapnel-like hooklets.

Simulium (*Gomphostilbia*) *zonatum* Subgroup

Diagnosis. Female and male: Hind tibia with dark subbasal marking. Female: Mandible with both inner and outer teeth (Fig. 30A). Male: Ventral plate narrowed or widened posteriorly when viewed ventrally (Fig. 30I, N, Q), and thin to thick when viewed posteriorly (Fig. 30L, P, S). Style usually slender (Fig. 30J, O) and occasionally thick (Fig. 30R) when viewed ventrolaterally. Pupa: One of two ventral paired filaments thicker and longer than seven other filaments (Fig. 31A–F). Terminal hooks widened with crenulated outer margins (Fig. 31H) [except *S. zonatum* which has simple triangular terminal hooks (Fig. 31G)]. Grapnel-like hooklets present. Larva: Postgenal cleft usually long (Fig. 31J), but rarely short (Fig. 31I) as in *S. adleri* and *S. zonatum*.

Species Included. Simulium adleri Jitklang and Kuvangkadilok, 2008 [2007] from Thailand and Malaysia (Peninsular); *S. auratum* Takaoka, 2009, *S. kelabitense* Takaoka, 2008 and *S. sarawakense* Takaoka, all from Malaysia (Sarawak); *S. rayohense* Smart and Clifford from Malaysia (Sabah); *S. cheongi* Takaoka and Davies, 1995 from Malaysia (Peninsular), Indonesia (Kalimantan), and Thailand; *S. datfaense* Takaoka and Otsuka, 2009 from Thailand; *S. epistum* Delfinado, 1971 from Philippines (Palawan); *S. lagunaense* Takaoka, 1983 and *S. luzonicum* Takaoka, 1983, both from Philippines (Luzon); *S. mindanaoense* Takaoka, 1983 from Philippines (Mindanao); *S. repentinum* Takaoka from Philippines (Samar); *S. floresense* Takaoka, Hadi and Sigit, 2006 from Indonesia (Flores); *S. zonatum* Edwards, 1934 and *S. padangense* Takaoka and Sigit, 1997, both from Indonesia (Sumatra); and *S. whartoni* Takaoka and Davies, 1995 from Malaysia (Peninsular) and Indonesia (Sumatra).

Simulium (*Gomphostilbia*) *miblosi* Subgroup

Diagnosis. Female and male: Hind tibia without subbasal dark marking (Fig. 32H). Hind basitarsus with relatively long calcipala extended beyond middle of second tarsomere (Fig. 32C, I). Female: Mandible with inner teeth but without outer teeth (Fig. 32A). Cibarium simple, without forwardly bent plate and mediolongitudinal ridge (Fig. 32B). Genital fork without projection directed posteromedially on each arm (Fig. 32E). Pupa: Gill with eight slender filaments of almost same length, arranged as 3+3+2 from dorsal to ventral (Fig. 32Q). Abdominal segment 9 with two setae (Fig. 32S) in place of grapnel-like hooklets on each lateral surface. Cocoon wall-pocket-shaped, or rarely shoe-shaped with low anteroventral neck. Larva: Postgenal cleft long but not reaching posterior margin of hypostoma (Fig. 32V).

Species Included. Simulium lehi Takaoka 2001, from Malaysia (Sarawak); *S. miblosi* Takaoka, 1983 from Philippines (Mindanao); and *S. rarum* Takaoka and Tenedero, 2008 from Philippines (Palawan).

Simulium (*Gomphostilbia*) *atratum* Subgroup

Diagnosis. Female and male: Hind tibia with dark subbasal marking (Fig. 33B, C). Female: Mandible with both inner and outer teeth. Cibarium with mediolongitudinal ridge having bifid apex (Fig. 33A). Genital fork with projection directed posteromedially on each arm (Fig. 33F). Male: Style slender or widened (Fig. 33K) when viewed ventrolaterally. Ventral plate much produced ventrally (Fig. 33M). Pupa: Two filaments of ventral pair of gill subequal in length to each other, and as long as or much longer than six other filaments (Fig. 33Q–S). Terminal hooks widened, plate-like (Fig. 33T). Grapnel-like hooklets present. Larva: Postgenal cleft long, its apex close to posterior margin of hypostoma (Fig. 33U).

Species Included. Simulium angulistylum Takaoka and Davies, 1995 from Malaysia (Peninsular) and Thailand; *S. atratum* De Meijere, 1913 (senior synonym of *S. sundaicum* Edwards, 1934) from Indonesia (Java, Sumatra, Bali, Lombok, Sumbawa, Flores, and Timor); *S. isanense* Takaoka, Srisuka and Saeung, 2017, *S. otsukai* Takaoka and Choochote, 2009, and *S. prayooki* Pramual, Jomkumsing, Wongpakam and Takaoka, 2022, all from Thailand.

Simulium (*Gomphostilbia*) *montiblense* Subgroup

Diagnosis. Female: Mandible with or without outer teeth (Fig. 34A). Cibarium with mediolongitudinal ridge having darkened apex (Fig. 34B). Genital fork with long or medium-long projection directed posteromedially on each arm (Fig. 34E). Male: Ventral plate thin or thick when viewed posteriorly (Fig. 34L, Q). Pupa: Dorsomedial and ventrolateral groups of thoracic trichomes arising close together (Fig. 35A). Gill with 8 or 13 or 16 short slender filaments (Fig. 35B–D). Terminal hooks simple, triangular (Fig. 35G). Grapnel-like hooklets present. Cocoon short, covering only posterior half of pupal body (Fig. 35H, I). Larva: Postgenal cleft long (Fig. 35J).

Species Included. Simulium barioense Takaoka, 2008 and *S. fulgidum* Takaoka, 2009, both from Malaysia (Sarawak); *S. montiblense* Takaoka, 1983 from Philippines (Palawan), China (Guizhou) (?), and Vietnam (?); and *S. paserense* Takaoka and Sofian-Azirun, 2017 from Indonesia (Kalimantan).

Simulium (*Gomphostilbia*) *epistum* Species-group Species Unplaced to Subgroup

Species Included. Simulium flavocinctum Edwards, 1934 from Indonesia (Sumatra); and *S. merapiense* Takaoka and Sofian-Azirun, 2016 from Indonesia (Java and Lombok).

Simulium (*Gomphostilbia*) *gombakense* Species-group

Diagnosis. Female and male: Antenna with nine flagellomeres. Female: Sensory vesicle medium-long to long (Fig. 36A, B). Cibarium with V or U-shaped dark apex of mediolongitudinal ridge (Fig. 36C, D). Claw with medium-sized or long tooth (Fig. 36E, F). Genital fork with distinct projection directed posteromedially on each arm (Fig. 36H, I). Male: Hind basitarsus slender, parallel-sided, or slightly to somewhat enlarged (Fig. 36M–O), which is, though, much narrower than hind tibia and femur. Ventral plate transverse, widened posteriorly from base, and moderately produced ventrally (Fig. 36P, T) except *S. maleewongae*, *S. nuwakotense*, *S. sachini*, and *S. thuathienense*, of which ventral plate is only slightly produced ventrally and its ventral margin is depressed medially when viewed posteriorly (Fig. 36Y). Pupa: Gill composed of inflated structure with eight short slender filaments (Fig. 37A–I), or with 10 short slender filaments (as in *S. antlerum*). Terminal hooks simple, cone-shaped, or wide, plate-like, with outer margins crenulated. Larva: Postgenal cleft long but not reaching posterior margin of hypostoma.

Remarks. This species-group was created by Takaoka (2012).

It is characterized by the pupal gill composed of an inflated structure with eight or 10 short slender filaments (Fig. 37A–I).

Species Included. *Simulium antlerum* Chen, 2001 from China (Hainan); *S. dudgeoni* Takaoka and Davies, 1995 from China (Hong Kong); *S. aziruni* Takaoka, Hashim and Chen, 2012 and *S. langkawiense* Takaoka, Sofian-Azirun and Ya'cob, 2013, both from Malaysia (Peninsular); *S. gombakense* Takaoka and Davies, 1995 from Malaysia (Peninsular), Myanmar, and Thailand; *S. maleewongae* Takaoka, Srisuka and Saeung, 2017, *S. paiense* Takaoka, Srisuka and Saeung, 2018 (complex), and *S. prayongi* Takaoka and Choochote, 2005, all from Thailand; *S. nuwakotense* Takaoka and Shrestha, 2010 from Nepal; *S. williei* Takaoka and Thapa, 2010 from India (West Bengal); *S. sachini* Takaoka and Henry, 2010 from India (West Bengal) and Nepal; and *S. thuathienense* Takaoka and Sofian-Azirun, 2015 from Vietnam.

Simulium (*Gomphostilbia*) *palauense* Species-group

Diagnosis. Similar to the *S. batoense* species-group except following characteristics. Female: Mandible with or without outer teeth (Fig. 38B). Each arm of genital fork with distinct projection directed posteromedially (Fig. 38E). Male: Ventral plate greatly produced ventrally and much narrowed posteriorly (Fig. 38I, J). Pupa: Gill with four or six slender filaments (Fig. 38M). Dorsum of abdomen with or without spine-combs. Terminal hook small, flat, with blunt apex (Fig. 38N). Larva: Mandible with supernumerary serration (Fig. 38O). Hypostoma with lateral margins smooth or serrated apically (Fig. 38P). Postgenal cleft absent (Fig. 38Q) or long, as in *S. curvum*. Accessary sclerite present (Fig. 38R) or absent, as in *S. curvum*.

Remarks. This species-group was created by Takaoka (2012) for two species, one from Palau and another (*S. curvum* Takaoka) from Halmahera.

Species Included. *Simulium palauense* Stone, 1964 from Palau.

Simulium (*Gomphostilbia*) *varicorne* Species-group

Diagnosis. Female and male: Antenna with seven or eight flagellomeres (Fig. 39A–C). Pleural membrane haired or bare. Female: Sensory vesicle short to long (Fig. 39D, E). Cibarium with pair of projections produced from dorsal margin (Fig. 39F). Subcosta haired or bare. Claw with large basal tooth (Fig. 39G). Each arm of genital fork strongly angulated anterodorsally, with distinct projection directed posteromedially and slightly to moderately developed projection directed anterodorsally (Fig. 39I, J). Spermatheca globular or ellipsoidal (Fig. 39M). Male: Hind basitarsus slender, parallel-sided. Ventral plate flat or moderately produced ventrally (Fig. 39N–V). Pupa: Gill with three somewhat inflated filaments, or six or eight slender filaments, or composed of U-shaped inflated structure having 12 slender filaments (Fig. 40A–E). Terminal hooks simple cone-like, or wide, plate-like (Fig. 40F, G). Grapnel-like hooklets present or absent. Larva: Hypostoma with lateral margins smooth or rarely serrated (Fig. 40I). Postgenal cleft medium-long, or long, reaching posterior margin of hypostoma (Fig. 40J, K). Abdomen with or without protuberances on dorsal surface (Fig. 40L–N).

Remarks. This species-group was first defined by Takaoka and Davies (1996), and later modified by Takaoka (2012). The *S. trirugosum* species-group was merged with the *S. varicorne* species-group and treated as a subgroup by Takaoka (2012).

Simulium (*Gomphostilbia*) *chumpornense* Subgroup

Diagnosis. Female and male: Antenna with eight flagellomeres (Fig. 39B). Pleural membrane bare. Female: Subcosta bare. Spermatheca ellipsoidal. Male: Ventral plate flat, not produced ventrally (Fig. 39Q–S). Pupa: Gill with eight slender long filaments, of which six dorsal filaments arranged as 3+3, or 2+(2+2) or 1+1+1+1+2 from dorsal to ventral (Fig. 40C), or gill with six filaments arranged as (2+2)+2 from dorsal to ventral (Fig. 40B), or gill composed of U-shaped inflated structure having 12 slender filaments (Fig. 40D). Terminal hooks small, cone-like, and with rounded apex (similar to Fig. 40F). Larva: Abdomen with protuberances dorsally (Fig. 40L). Hypostoma with lateral margins smooth or serrated apically (Fig. 40I). Postgenal cleft long, reaching posterior margin of hypostoma (Fig. 40J).

Species Included. *Simulium chumpornense* Takaoka and Kuvangkadilok, 2000, *S. piroonae* Takaoka and Srisuka, 2014, *S. khelangense* Takaoka, Srisuka and Saeung, 2022, and *S. kuvangkadilokae* Pramual and Tangkawanit, 2008, all from Thailand; *S. sumbaense* Takaoka and Suana, 2018 from Indonesia (Sumba); *S. tomae* Takaoka, 2003 from Indonesia (Sulawesi); and *S. varicorne* Edwards, 1929 from Indonesia (Java and Sumatra) and Malaysia (Peninsular).

Simulium (*Gomphostilbia*) *burtoni* Subgroup

Diagnosis. Female and male: Antenna with eight flagellomeres (Fig. 39A). Pleural membrane bare. Female: Subcosta haired. Spermatheca ellipsoidal. Male: Ventral plate moderately produced ventrally (Fig. 39T–V). Pupa: Gill with eight slender long filaments arranged as (3+3)+2 from dorsal to ventral (Fig. 40A); stalk of dorsal triplet subequal in thickness to that of middle triplet. Terminal hooks small, cone-like (similar to Fig. 40F), or somewhat widened and plate-like, with undulate or crenulated outer margins (Fig. 40G). Larva: Postgenal cleft medium-long to long, not reaching posterior margin of hypostoma (Fig. 40K).

Species Included. *Simulium breviflagellum* Takaoka and Sofian-Azirun, 2015 from Vietnam; *S. burtoni* Takaoka and Davies, 1995 from Malaysia (Peninsular) and Thailand; and *S. huangi* Takaoka, 2017 from Taiwan.

Simulium (*Gomphostilbia*) *novemarticulatum* Subgroup

Diagnosis. Female and male: Antenna with seven flagellomeres (Fig. 39C). Pleural membrane bare. Female: Subcosta haired. Spermatheca ellipsoidal. Male: Ventral plate moderately produced ventrally (Fig. 39N). Pupa: Gill with eight slender long filaments arranged as 3+3+2 from dorsal to ventral (Fig. 40E). Terminal hooks small and cone-like, with round apex (Fig. 40F). Larva: Postgenal cleft long, reaching posterior margin of hypostoma (similar to Fig. 40J). Abdomen with protuberances dorsally (Fig. 40M, N). Abdominal segment 8 with dark setae of various shapes dorsally (Fig. 40O, Q).

Species Included. *Simulium charlesi* Takaoka, 2008 from Malaysia (Peninsular and Sarawak); and *S. novemarticulatum* Takaoka and Davies, 1995 from Malaysia (Peninsular) and Thailand.

Simulium (*Gomphostilbia*) *trirugosum* Subgroup

Diagnosis. Female and male: Antenna with eight flagellomeres. Pleural membrane haired. Female: Subcosta haired. Each arm of genital fork strongly angulated anterodorsally and with distinct projection directed posteromedially. Spermatheca globular. Male: Ventral plate flat, not produced ventrally. Pupa: Gill with three somewhat inflated filaments. Grapnel-like hooklets absent. Terminal hooks small and simple. Larva: Postgenal cleft long, reaching posterior margin of hypostoma.

Species Included. Simulium trirugosum Davies and Györkös, 1988 from Sri Lanka.

Simulium (*Gomphostilbia*) Species Unplaced to Species-group

Species Included. Simulium azhari Takaoka, Sofian-Azirun and Ya'cob, 2014 from Malaysia (Peninsular); *S. guniki* Takaoka, 2001 from Malaysia (Sabah); *S. hiroyukii*, Ya'cob and Sofian-Azirun, 2015 from Malaysia (Sarawak); *S. bucolicum* Datta, 1975, *S. fidum* Datta, 1975, and *S. litoreum* Datta, 1975, all from India (Assam); *S. tenuistylum* Datta, 1994 from India (West Bengal and Sikkim); *S. cauveryense* Anbalagan, 2015 from India (Tamil Nadu); *S. metatarsale* Brunetti, 1911 from India (West Bengal) and China (?); *S. dhangi* Takaoka and Sofian-Azirun, 2018 from Indonesia (Timor and Sumba); *S. mamasaense* Takaoka, 2003 and *S. serratum* Takaoka, 2003, both from Indonesia (Sulawesi); *S. doisaketense* Jitklang et al., 2008 from Thailand; *S. guangxiense* Sun, 2009, *S. longshengense* Chen, Zhang and Zhang, 2007, *S. penis* Sun, 2009, *S. pingxiangense* An, Hao and Mai, 1990, and *S. yuanbaoshanense* Chen, Zhang and Zhang, 2007, all from China (Guangxi Zhuang); *S. heteropara* Sun, 2009, *S. huguangense* Sun, 2009, *S. jinbianense* Zhang and Chen, 2004, and *S. xiangxiense* Sun, 2009, all from China (Hunan); *S. tumum* Chen and Zhang, 2001 and *S.* sp. [*S. longitruncum* Zhang and Chen, 2003 (junior primary homonym of *S. longitruncum* Takaoka and Davies, 1995, but no available substitute name)], both from China (Hainan); *S.* sp. 'A' Zhang, Wen and Chen, 1999 from China (Guizhou); *S.* ? *sundaicum* Edwards, Datta, 1991 and *S.* sp. nr. *litoreum* Datta, 1975, both from India; and *S.* nr. *asakoae* 2 Jitklang et al., 2008, *S.* nr. *asakoae* 3 Jitklang et al., 2008, *S.* nr. *asakoae* 4 Jitklang et al., 2008, *S.* nr. *sheilae* 3 Jitklang et al., 2008, *S.* sp. A Phasuk et al., 2005, *S.* sp. B Phasuk et al., 2005, and *S.* sp. C Phasuk et al., 2005, all from Thailand; *S.* Ranga sp. Lewis, 1974 and *S.* Lakhimpur sp. Lewis, 1974, both from India (Assam); and *S.* sp. [*S. sumatraense* An, Nie and Chen, 2006 (junior primary homonym of *S. sumatraense* Takaoka and Sigit, 1997, but no available substitute name) from Indonesia (Sumatra) (?).

Subgenus Montisimulium *Rubtsov*

Montisimulium Rubtsov, 1974 (as genus). Type species: *schevyakovi* (original designation).

Diagnosis. Female and male: Basal portion of radial vein haired. Both pleural membrane and katepisternum bare. Female: Sensory vesicle medium-long to long

(Fig. 42A, B). Mandible with teeth on inner margin but lacking teeth on outer margin. Cibarium unarmed. Claw with short to medium-long basal tooth (Fig. 42C, D). Abdominal segments 2 and 6–8 shiny dorsally when illuminated at certain angles. Abdominal segment 7 with large sternal plate medially. Ovipositor valve tongue-like (Fig. 42E). Genital fork with projection directed anteriorly on each arm (Fig. 42F). Spermatheca ovoid or globular (Fig. 42I). Male: Hind basitarsus enlarged, spindle-shaped (Fig. 42J). Style shorter than coxite, inwardly twisted (Fig. 42K). Ventral plate shorter than wide, lamellate (Fig. 42K). Paramere with few to several hooks (Fig. 42O). Pupa: Frons with two pairs of short trichomes. Gill filamentous or inflated (Fig. 42R, S), with black minute pigment spots inside the outer layer of cuticle (Fig. 42T). Abdomen with dorsal spine-combs and cone-like terminal hooks. Ventral surface of abdominal segments 5–7 each with pair of bifid inner and bifid (on segment 5) or unbranched (on segments 6 and 7) outer hooks on each side. Abdominal segment 9 with or without grapnel-like hooklets on each side. Larva: Antennal segment 2 with or without unpigmented secondary annulations (Fig. 42W). Hypostoma with median and corner teeth prominent, and lateral margins serrated apically (Fig. 42U). Postgenal cleft small, rudimentary, or absent (Fig. 42X, Y). Ventral papillae present.

Remarks. The subgenus *Montisimulium* was originally defined as a genus by Rubtsov (1974). It was also treated as the *S. montium* species-group, one of the five species-groups in the subgenus *Eusimulium* s. l. (Datta, 1973, 1974b).

It is characterized by the pupal gill with numerous minute dark pigment spots in the outer layer of the cuticle (Fig. 42T), a characteristic not found in other subgenera, and larval head capsule with or without a small or vestigial postgenal cleft (Fig. 42X, Y).

This subgenus is composed of 78 formally named and 11 unnamed species, of which 67 species are recorded from the Palearctic Region, 18 species are from the Oriental Region, and four species are from both regions. In the Oriental Region, its distribution is restricted to high elevations (1500–3700 m) in northern areas of the region, in particular, the southern mountain foothills of the Himalaya.

The females and males of most species of this subgenus are similar morphologically to one another, but the pupae are well distinguished by the number and arrangement of gill filaments.

Species Included. *Simulium angkaense* Takaoka and Choochote, 2005, *S. laoleense* Takaoka and Choochote, 2005, *S. merga* Takaoka and Choochote, 2005, *S. nanense* Takaoka and Srisuka, 2010, *S. phahompokense* Takaoka and Srisuka, 2010, and *S. surachaii* Takaoka and Choochote, 2005, all from Thailand; *S. chowi* Takaoka, 1979 from Taiwan; *S. nemorivagum* Datta, 1973 from India (West Bengal and Arnachal Pradesh), Bhutan, China (Tibet), Nepal, and Pakistan; *S. dattai* Takaoka and Somboon, 2008 from Bhutan, India (West Bengal), and Nepal; *S. dasguptai* Datta, 1974 from India (West Bengal) and China (Sichuan); *S. ghoomense* Datta, 1975 from India (West Bengal) and China (Tibet); *S. jumlaense* Takaoka and Adler, 2020 from Nepal; *S. kambaitense* Takaoka, 1989 from Myanmar; *S. kirgisorum* (Rubtsov, 1956) from China (Yunnan, Hubei, Beijing, and Tibet) and Kyrgyzstan; *S. separatum* Chen, Xie and Zhang, 2012 from China (Guizhou); *S. weisiense* Deng, Xue and Chen, 2005 from China (Yunnan); *S. nigrofilum* Takaoka and Sofian-Azirun, 2017 from Vietnam; *S. taktsangense* Takaoka and Somboon, 2008 from Bhutan and India (West Bengal) (?); *S.* sp. [*Titanopteryx tanae* Xue, 1993 (secondary homonym of *S.* (*S.*) *tanae* Xue, 1992 but no available substitute name)] from China (Yunnan); *S.* sp. C Datta, Dey and Paul, 1975 from India (West Bengal); *S.* sp. B Zhang, Wen and Chen, 1999 from China (Guizhou); and *S.* sp. G Kuvangkadilok, Boonkemtong and Baimai, 1998 from Thailand.

Subgenus Nevermannia *Enderlein*

Nevermannia Enderlein, 1921 (as genus). Type species: *ruficorne*, as *annulipes* (original designation).
Cnetha Enderlein, 1921 (as genus). Type species: *vernum* (designation by ICZN Opinion 1416, Gen. List).
Stilboplax Enderlein, 1921 (as genus). Type species: *speculiventre* (original designation).
Pseudonevermannia Baranov, 1926 (as subgenus of *Nevermannia*). Type species: *vernum* (designation by ICZN Opinion 1416, Index).
Cryptectemnia Enderlein, 1936 (as genus). Type species: *orsovae*, as *laticalx* (original designation).
Chelocnetha Enderlein, 1936 (as genus). Type species: *ornatipes*, as *biroi* (original designation).
Dexomyia Crosskey, 1969 (as subgenus of *Simulium* s. l.). Type species: *atlanticum* (original designation).

Diagnosis. Female and male: Pleural membrane and katepisternum bare (though katepisternum is haired in three species in the *S. vernum* species-group in Japan). Basal portion of the radial vein haired. Female: Mandible with teeth on its inner and outer margins (most species in the *S. feuerborni* species-group lacking teeth on outer margin). Claw with large tooth (Fig. 43C) or rarely small tooth (Fig. 47E). Abdominal segments 2 and 6–8 shiny or not shiny dorsally when illuminated at certain angles. Abdominal segment 7 with large sternal plate medially. Ovipositor valve triangular (Fig. 47F) or tongue-like (Fig. 43D). Genital fork with or without projection directed anteriorly on each arm (Figs. 43E, F and 45C). Spermatheca ovoid or globular or ellipsoidal (Figs. 43I and 45F, G). Male: Hind basitarsus parallel-sided (Fig. 45H) or enlarged, spindle-shaped (Fig. 47L). Style shorter than coxite, inwardly twisted (Fig. 43K). Ventral plate shorter than wide, lamellate, with or without median keel (Figs. 43K and 45I, K). Paramere with one to several hooks (Figs. 43P and 45L). Pupa: Frons with two or three pairs of trichomes (Fig. 48A). Gill filamentous, with four, five, six, or eight filaments (Figs. 44A–G and 46A–G), or rarely of inflated structure (Fig. 48G). Abdomen with dorsal spine-combs. Ventral surface of abdominal segments 5–7 each with pair of bifid inner and bifid (on segment 5) or unbranched (on segments 6 and 7) outer hooks on each side.

Abdominal segment 9 with pair of distinct cone-like terminal hooks (Fig. 44H) but without grapnel-like hooklets. Larva: Mandible with mandibular serration composed of two medium-sized teeth (Fig. 46I, J), or one medium-sized tooth and one small tooth (Fig. 44M). Supernumerary serration present (Fig. 48M) or absent. Hypostoma with median and corner teeth prominent, lateral margins serrated apically (Fig. 44N). Postgenal cleft short to medium-long (Figs. 44O and 46M), or absent. Accessory sclerites present (Fig. 48P) or absent. Ventral papillae present, medium-sized or large.

Remarks. The subgenus *Nevermannia* was originally defined as a genus based on *S. annulipes* Becker, 1908 (=*S. ruficorne* Macquart, 1834) by Enderlein (1921).

It is composed of 247 species, the third largest in the genus *Simulium* s. l. and is widely distributed in the Australasian, Ethiopian, Nearctic, Oriental, and Palearctic Regions. It is heterogeneous, consisting of the *S. atlanticum*, *S. loutetense*, *S. feuerborni*, *S. ruficorne*, and *S. vernum* species-groups. All these species-groups were defined by Crosskey (1969) except the *S. feuerborni* species-group defined by Datta (1973).

In the Oriental Region, the *S. feuerborni*, *S. ruficorne*, and *S. vernum* species-groups are known.

Keys to Identify Three Species-groups of the Subgenus *Nevermannia*

Females

1. Paraproct much shorter than cercus (Fig. 45E) ....*S. ruficorne* species-group
   Paraproct longer than cercus (Fig. 43H) 2
2. Genital fork with distinct projection directed anteriorly on each arm (Fig. 43E, F) .... *S. feuerborni* species-group
   Genital fork with no or small projection directed anteriorly on each arm (Fig. 47G, H) .... *S. vernum* species-group

Males

1. Hind basitarsus slender (Fig. 45H). Ventral plate with median keel (Fig. 45K) .... *S. ruficorne* species-group
   Hind basitarsus enlarged (Fig. 47L). Ventral plate without median keel 2
2. Paramere with three or more hooks (Fig. 43P) .... *S. feuerborni* species-group
   Paramere with one hook (Fig. 47Q) .... *S. vernum* species-group

Pupae

1. Dorsum of abdominal segments 7 and 8 (or only 8) with spine-combs ....*S. ruficorne* species-group
   Dorsum of abdominal segments 6–9 (or 5–8) with spine combs 2
2. Gill with six filaments (or rarely four or five filaments) (Fig. 44A–G) .... *S. feuerborni* species-group
   Gill with four filaments (Fig. 48E) [or inflated, dumbbell-like structure (Fig. 48G)] .... *S. vernum* species-group

Mature larvae

1. Mandibular serration composed of two medium-sized teeth (Fig. 46I, J) ....*S. ruficorne* species-group
   Mandibular serration composed of one medium-sized tooth and one small tooth (Fig. 48M) 2
2. Supernumerary serration present (Fig. 48M) (or rarely absent) .... *S. vernum* species-group
   Supernumerary serration absent (Fig. 44M) ....*S. feuerborni* species-group

Simulium (Nevermannia) feuerborni Species-group

Diagnosis. Female: Third palpal segment usually enlarged, with long sensory vesicle (Fig. 43A). Mandible with teeth on inner margin and usually lacking teeth on outer margin (though, rarely with teeth on outer margin). Cibarium with or without tiny processes near the ventral margin (Fig. 43B). Claw with large basal tooth (Fig. 43C). Spermatheca globular or ovoid, with unpigmented neck (Fig. 43I). Male: Hind basitarsus enlarged (Fig. 43J). Style broad, with truncated apex (Fig. 43L). Ventral plate large, broad, without median keel (Fig. 43K). Paramere with 3–11 hooks (Fig. 43P). Median sclerite long, rod-like (Fig. 43O). Pupa: Frons with two pairs of trichomes. Gill filamentous, with six filaments (rarely four or five filaments) (Fig. 44A–G). Cocoon wall-pocket-shaped, with or without anterodorsal projection (Fig. 44I, J). Larva: Mandibular serration composed of one medium-sized tooth and one small tooth (Fig. 44M). Supernumerary serration absent (except in one species). Postgenal cleft small, usually quadrate (Fig. 44O).

Remarks. This group includes 36 formally described and five unnamed species, of which 33 species (28 named and five unnamed) are recorded from the Oriental Region, six named species are from the Palearctic Region and two named species are from both regions.

Species Included. Simulium bachmaense Takaoka, Sofian-Azirun and Ya'cob, 2014, *S. langbiangense* Takaoka, Sofian-Azirun and Ya'cob, 2014, *S. phami*

Takaoka, Sofian-Azirun and Ya'cob, 2014 and *S. pumatense* Takaoka, Low and Pham, 2019, all from Vietnam; *S. chiangklangense* Takaoka and Choochote, 2005, *S. fangense* Takaoka and Choochote, 2006, *S. fruticosum* Takaoka and Choochote, 2005, *S. vessabutrae* Takaoka and Srisuka, 2010, and *S. wichaii* Takaoka, 2010, all from Thailand; *S. bryopodium* Delfinado, 1971 and *S. perlucidulum* Takaoka, 1983, both from Philippines; *S. ledangense* Ya'cob, Takaoka and Sofian-Azirun, 2014 and *S. pairoti* Ya'cob, Takaoka and Sofian-Azirun, 2017, both from Malaysia (Peninsular); *S. fuscinervis* Edwards, 1933 and *S. borneoense* Takaoka, 2001, both from Malaysia (Sabah); *S. leigongshanense* Chen and Zhang, 1997 from China (Guizhou); *S. qiongzhouense* Chen, Zhang and Yang, 2003 from China (Hainan); *S. wangxianense* Chen, Zhang and Bi, 2004 from China (Hunan); *S. mie* Ogata and Sasa, 1954 from Japan (Honshu, Kyushu, Shikoku and Nansei Islands), Korea (South), and China (Fujian, Guizhou, Sichuan, Yunnan and Zhejiang); *S. maeaiense* Takaoka and Srisuka, 2011 from Thailand and Vietnam; *S. feuerborni* Edwards 1934 (complex) from Indonesia (Java, Sumatra and Bali) and Thailand; *S. minahasaense* Takaoka, 2003 and *S. chairuddini* Takaoka, 2003, both from Indonesia (Sulawesi); *S. chitoense* Takaoka, 1979 from Taiwan and China (Hainan); *S. morisonoi* Takaoka, 1973 from Japan (Nansei Islands); *S. mongarense* Takaoka and Somboon, 2008 from Bhutan; *S. nivritiae* Takaoka and Shrestha, 2020 from Nepal; *S. praelargum* Datta, 1973 (complex) from India (West Bengal), Nepal, and China (Tibet); *S. rufithorax* Brunetti, 1911 from India (Maharashtra, Karnataka, and West Bengal); *S. subratai* Takaoka, Thapa and Henry, 2011 from India (West Bengal); *S.* sp. [*Titanopteryx kirgisorum* Xue, 1991 (secondary homonym of *S.* (*M.*) *kirgisorum* Rubtsov, but no available substitute name)] from China (Yunnan); *S.* sp. A Takaoka, 1989 from Myanmar; and *S.* sp. "A" Datta, 1973, *S.* sp. D Datta, Dey and Paul, 1975 and *S.* sp. 2 (nr. *fruticosum*) Henry, Dey, Varma, Thapa and Procunier, 2010 (probably the same species as *S.* sp. D), all from India (West Bengal).

Simulium (*Nevermannia*) *ruficorne* Species-group

Diagnosis. Female and male: Hind tibiae with dark subbasal marking (Fig. 45B). Female: Cibarium unarmed (Fig. 45A). Claw with medium-long or long basal tooth (rarely simple or with small tooth in certain species of the Australasian and Ethiopian Regions). Abdomen not shiny dorsally. Ovipositor valve rounded posteriorly (Fig. 45C). Genital fork with arms each having lateral plate (Fig. 45C). Spermatheca ellipsoidal, with sclerotized neck or rarely unsclerotized neck (Fig. 45F, G). Paraproct much shorter than cercus, widely bare on ventral surface, with a cluster of setae along ventromedial margin (Fig. 45D, E). Male: Hind basitarsus slender, nearly parallel-sided (Fig. 45H). Style broad, bluntly truncate (Fig. 45I). Ventral plate large, broad, with haired median keel (Fig. 45I). Paramere with one hook (Fig. 45L). Median sclerite long, rod-like (rarely bifurcated apically). Pupa: Frons with two (rarely three) pairs of trichomes. Gill filamentous, with four, six or eight filaments (Fig. 46A–G). Cocoon simple or with anterodorsal projection (Fig. 46H).

Larva: Mandibular serration composed of two medium-sized teeth (Fig. 46I, J). Supernumerary serration absent. Hypostoma with lateral margins markedly serrated (Fig. 46K, L). Postgenal cleft nearly as long as or shorter than postgenal bridge (Fig. 46M) (or small, rounded or subquadrate, forming minute notch on posteroventral margin of head capsule, or absent in certain species of the Ethiopian Region).

Remarks. The *S. ruficorne* species-group is widely distributed in the Ethiopian, Palearctic, Oriental, and Australasian Regions. It includes 66 formally described and one unnamed species, of which five species are known from the Oriental Region, and two are from the Oriental and Palearctic Regions (Adler, 2022).

This species-group is distinctive among the subgenus *Nevermannia* in having unique female terminalia and male genitalia including the paraproct much shorter than the cercus (Fig. 45E) and ventral plate with a haired median keel (Fig. 45K).

DNA sequence-based studies using the mitochondrial 16S rRNA gene also showed that the *S. ruficorne* species-group is distant from both the *S. feuerborni* and *S. vernum* species-groups, while the latter two species-groups are close to each other (Otsuka et al., 2001).

Takaoka (2017b) speculated that species of this species-group expanded their distribution from the Oriental Region westward to the Ethiopian Region through Palearctic Region, and eastward to the Australasian Region based on the reduction of the pupal gill filaments from eight to four.

Species Included. Simulium angustitarse (Lundström, 1911) from the Palearctic Region and China (Yunnan); *S. aureohirtum* Brunetti, 1911 from the Oriental Region, parts of the Palearctic Region and parts of the Australasian Region; *S. falcoe* (Shiraki, 1935) and *S. geniculare* (Shiraki, 1935), both from Taiwan; *S. glatthaari* Takaoka and Davies, 1995 from Indonesia (Sumatra); *S. wayani* Takaoka and Chen, 2018 from Indonesia (Timor); and *S. karavalliense* Anbalagan et al., 2020 from India (Tamil Nadu).

Simulium (*Nevermannia*) *vernum* Species-group

Diagnosis. Female: Sensory vesicle usually long (Fig. 47A). Cibarium unarmed (Fig. 47C). Claw with large basal tooth (Fig. 47D), or rarely with small tooth (Fig. 47E). Abdomen shiny or not dorsally. Genital fork without projection directed anteriorly on each arm (Fig. 47G, H). Paraproct longer than cercus (Fig. 47J). Spermatheca globular, with unpigmented neck (Fig. 47K). Male: Hind basitarsus enlarged (Fig. 47L). Style boot-shaped, bluntly truncate (Fig. 47N). Ventral plate large, broad, without median keel (Fig. 47M, P). Median sclerite long, bifurcated apically (Fig. 47P). Paramere with one hook (Fig. 47Q). Pupa: Frons with two or three pairs of trichomes (Fig. 48A). Frons and thorax unicolorous, or rarely partially black (Fig. 48A, B). Thorax smooth, or rarely with ridges (Fig. 48C, D). Gill with four filaments (Fig. 48E, F) (though with six or eight filaments in the Nearctic

Region), or of inflated dumbbell structure (Fig. 48G, H). Cocoon simple wall-pocket-shaped (Fig. 48H), or with single or paired anterodorsal projections (Fig. 48I, K), or round shoe-shaped, with anteroventral collar (Fig. 48J). Larva: Mandibular serration composed of one medium-sized tooth and one minute tooth and usually with supernumerary serration (Fig. 48M). Postgenal cleft as long as or shorter than postgenal bridge, rounded, or subquadrate (Fig. 48O). Accessory sclerites present (Fig. 48P) or absent.

Remarks. The *S. vernum* species-group is widely distributed in the Palearctic, Oriental, and Nearctic Regions. It includes 138 formally described and eight unnamed species, of which 25 formally named and one unnamed species are recorded from the Oriental Region and one named species is recorded from both the Oriental and Palearctic Regions.

It is characterized by the unique male genitalia including the style boot-shaped (Fig. 47N), ventral plate broad, without a haired median keel (Fig. 47M, P), and median sclerite bifurcated apically (Fig. 47P).

Species Included. Simulium aberrans Delfinado, 1989 from Philippines (Luzon); *S. bumthangense* Takaoka and Somboon, 2008 and *S. thrimshinglaense* Takaoka and Somboon, 2008, both from Bhutan; *S. burmense* Takaoka, 1989 from Myanmar; *S. cangshanense* (Xue, 1993) and *S. cheni* (Xue, 1993), both from China (Yunnan); *S. chenzhouense* Chen, Zhang and Bi, 2004 and *S. zhangjiajiense* Chen, Zhang and Bi, 2004, both from China (Hunan); *S. qiaolaoense* Chen, 2001 and *S. qingshuiense* Chen, 2001, both from China (Guizhou); *S. caudisclerum* Takaoka and Davies, 1995 and *S. kurtaki* Takaoka and Davies, 1995, both from Malaysia (Peninsular); *S. chomthongense* Takaoka, Srisuka and Choochote, 2012 and *S. khunklangense* Takaoka and Srisuka, 2013, both from Thailand; *S. gracile* Datta, 1973 and *S. purii* Datta, 1973, both from India (West Bengal), China (Tibet) and Nepal; *S. laichauense* Takaoka and Sofian-Azirun, 2017 and *S. tayense* Takaoka and Ya'cob, 2017, both from Vietnam; S. *gandakiense* Adler, Takaoka and Ruiz-Arrondo, 2020, *S. lekhaniense* Takaoka and Shrestha, 2010, *S. letense* Takaoka and Shrestha, 2010, and *S. magnacentrum* Adler, Takaoka and Ruiz-Arrondo, 2020, all from Nepal; S. *silvestre* Rubtsov, 1956 from China (Sichuan and Yunnan), Finland, Norway, Sweden, Russia, Ukraine, Mongolia, Siberia, Canada, and USA; *S. taulingense* Takaoka, 1979 and *S. yushangense* Takaoka, 1979, both from Taiwan; *S. tosariense* Edwards, 1934 and *S. tjibodense* Edwards, 1934, both from Indonesia (Java); and *S.* sp. A Datta, 1975 from India (West Bengal).

Simulium (_Nevermannia_) Species Unplaced to Species-group

Species Included. Simulium sp. 'C' Zhang, Wen and Chen, 1999 from China (Guizhou).

Subgenus Simulium *Latreille s. str.*

Simulium Latreille, 1802: 426 (as genus). Type species: *colombaschense* (Scopoli, 1780) (original designation)
Odagmia Enderlein, 1921 (as genus). Type species: *ornatum* (original designation)
Friesia Enderlein, 1922 (as genus). Type species: *bezzii*, as *tristrigata* (original designation)
Discosphyria Enderlein, 1922 (as genus). Type species: *trifasciatum*, as *odagmiina* (original designation)
Gynonychodon Enderlein, 1925 (as genus). Type species: *nobile* (original designation)
Pseudodagmia Baranov, 1926 (as subgenus of *Odagmia*). Type species: *variegatum* (original designation)
Danubiosimulium Baranov, 1935 (as subgenus of *Simulium* s. l.). Type species: *colombaschense,* as *colombaschensis* (monotypy)
Cleitosimulium Séguy and Dorier, 1936 (as subgenus of *Simulium* s. l.). Type species: *argenteostriatum*, as *rupicolum* (original designation)
Gnus Rubtsov, 1940 (as subgenus of *Simulium* s. l.). Type species: *decimatum* (original designation)
Tetisimulium Rubtsov, 1963 (as genus). Type species: *bezzii* (original designation)
Parabyssodon Rubtsov, 1964 (as genus). Type species: *transiens* (original designation)
Phosterodoros Stone and Snoddy, 1969 (as subgenus of *Simulium* s. l.). Type species: *jenningsi* (original designation)
Himalayum Lewis, 1973 (as subgenus of *Simulium* s. l.). Type species: *indicum* (original designation)
Paragnus Rubtsov and Yankovsky, 1982 (as genus). Type species: *bukovskii* (original designation)
Archesimulium Rubtsov and Yankovsky, 1982 (as subgenus of *Simulium* s. str.). Type species: *tuberosum* (original designation)
Argentisimulium Rubtsov and Yankovsky, 1982 (as subgenus of *Simulium* s. str.). Type species: *noelleri* (original designation)
Striatosimulium Rubtsov and Yankovsky, 1982 (as subgenus of *Simulium* s. str.). Type species: *japonicum* (original designation)

Diagnosis. Female and male: Antenna with scape, pedicel, and nine flagellomeres. Both pleural membrane and katepisternum bare. Female: Frons shiny, almost bare, though with several hairs along each lateral margin (rarely dull, densely covered with fine hairs as in the *S. indicum* species-group). Mandible with teeth on inner and outer margins (rarely with distinct teeth on inner teeth but with vestigial teeth on outer margin). Claw simple or with short to medium-long tooth. Basal portion of radial vein bare (rarely haired as in species in the *S. indicum*, *S. griseifrons*, *S. multistriatum*, and *S. striatum* species-groups). Abdominal segments 2 and 6–8 (or 5–8) shiny when illuminated at certain angles. Abdominal segment 7 with pair of large sternal plates. Ovipositor valves membranous or well sclerotized, of various shapes. Genital fork with distinct projection directed anteriorly on each arm. Paraproct of various shapes. Cercus semicircular or much wider than long when viewed laterally, with or without ventromedial plate. Spermatheca ovoid or globular, with or without internal setae. Male: Hind basitarsus slender, parallel-sided or enlarged and spindle-shaped. Basal portion of radial vein bare (except species in the *S. indicum* species-group). Coxite nearly quadrate. Style longer than coxite, with or without basal protuberance having spines. Ventral plate shorter than wide, lamellate, or longer than wide, bare, or haired, with or without medial process produced ventrally, having teeth or not on posterior surface. Paramere with few to several hooks. Aedeagal membrane setose (though bare in one species). Pupa: Frons with two pairs of short trichomes (except one species with three pairs of trichomes). Gill with 2–38 slender or inflated filaments. Abdomen with dorsal spine-combs (except four species of the *S. eximium* species-group, two species of the *S. melanopus* species-group, three species of the *S. indicum* species-group and *S. tamorense* lacking spine-combs), with or without cone-like terminal hooks and lacking grapnel-like hooklets on each side. Ventral surface of abdominal segments 5–7 each with pair of bifid or unbranched inner and bifid or unbranched outer hooks on each side (though lacking outer hook on each side in three species of the *S. dumogaense* species-group). Cocoon simple wall-pocket-shaped, shoe-shaped, boot-shaped, or corbicular-shaped. Larva: Hypostoma with nine anterior teeth, of which median and corner teeth prominent, with lateral margins serrated apically, and with bristles lying divergent posteriorly from lateral margin on each side. Postgenal cleft medium-long to long or reaching posterior margin of hypostoma. Abdomen with or without dorsal protuberances. Anal sclerite X-shaped, with anterior arms shorter than posterior ones. Accessory sclerite absent (rarely present as in the *S. indicum* species-group). Ventral papillae absent or present (though small).

Remarks. The subgenus *Simulium* s. str. is the largest of the 38 subgenera of the genus *Simulium* s. l. and is widely distributed in the Palearctic, Oriental, Nearctic, and Australasian Regions. This subgenus was originally defined by Crosskey (1969) and redefined by Takaoka (2003) and Adler et al. (2004).

It was first divided into 19 species-groups by Crosskey (1987) based on many of those established by Rubtsov (1956), and later among those species-groups, the *S. multistriatum* species-group was split further into four species-groups, i.e., *S. eximium*, *S. griseifrons*, and *S. multistriatum* in a strict sense and *S. striatum* species-groups, by Takaoka and Davies (1996). Four more species-groups, *S. hunteri*, *S. metallicum*, *S. slossonae*, and *S. argentipes* species-groups, were added by Crosskey and Howard (1997), Currie (1997), and Takaoka and Saito (2007). The *S. hunteri* and *S. metallicum* species-groups from the Nearctic and the Neotropical Regions were soon thought to be different from the subgenus *Simulium* s. str. and all the 25 species in both species-groups were treated under the subgenus *Aspathia* Enderlein by Adler et al. (2004). As a result, until 2010, 24 species-groups were recognized within the subgenus *Simulium* s. str. (Adler and Crosskey, 2010). The *S. christophersi* species-group was added by Takaoka et al. (2014h). Takaoka (2017a) revised species of eight species-groups (seven named and one unnamed) of the subgenus *Simulium* s. str. chiefly based on the adult female and male genitalia, and divided four species-groups into more than two lineages: the *S. christophersi* species-group into two: *S. christophersi* redefined and *S. nebulicola* species-groups; the *S. griseifrons* species-group into six: *S. griseifrons*, *S. grossifilum*, *S. chungi*, *S. crocinum*, *S. indicum*, and *S. yadongense* species-groups; the *S. malyschevi* species-group into two: *S. malyschevi* and *S. pavlovskii* species-groups; the *S. melanopus* species-group into three: *S. melanopus*, *S. dumogaense* and *S. iridescens* species-groups. In addition, the *S. crassifilum* species-group was merged with the *S. variegatum* species-group. In all, 33 species-groups (excluding species unplaced to species-group) are now recognized. The wide diversity shown by the great number of the species-groups within the single subgenus suggests that this subgenus includes heterogeneous elements as well as homogeneous ones that originated differently from one another.

In the Oriental Region, the subgenus *Simulium* s. str. is one of the two large components of *Simulium* s. l. consisting of 310 species (294 named and 16 unnamed), of which 290 species are further classified into 21 species-groups and 20 species are left ungrouped.

Keys to Identify 21 Species-groups of the Subgenus *Simulium* s. str. in the Oriental Region

Females

1. Fronto-ocular area weakly undeveloped (Fig. 74B, C) 2
   Fronto-ocular area developed 3
2. Cibarium with many sharply pointed processes (Fig. 74D) ....
   .... *S. nobile* species-group
   Cibarium with several tiny processes (Fig. 76A) ....
   .... *S. striatum* species-group (one species)
3. Claw simple, without any tooth 4
   Claw with small or medium-sized tooth 13
4. Cibarium with median projection on dorsal margin (Fig. 58A) 5
   Cibarium without median projection on dorsal margin 6

5. Median projection of cibarium furnished with many spines (Fig. 58A–F) .... *S. griseifrons* species-group
Median projection of cibarium without spines (Fig. 70A) .... *S. multistriatum* species-group
6. Ovipositor valve with ventrally-produced lobe along inner margin (Fig. 76D, E) .... *S. striatum* species-group
Ovipositor valve without ventrally-produced lobe along inner margin 7
7. Ovipositor valve with long projection directed from inner surface (Fig. 61B) .... *S. grossifilum* species-group
Ovipositor valve without long projection directed from inner surface 8
8. Ovipositor valve tapered apically (Fig. 51B) 9
Ovipositor valve not tapered apically 10
9. Paraproct in lateral view with outer surface largely bare except narrow portion along posterior margin haired (Fig. 57D) .... *S. eximium* species-group
Paraproct in lateral view with outer surface mostly haired (Fig. 51E) .... *S. chungi* species-group
10. Paraproct with ventromedial plate (Fig. 67H) 11
Paraproct without ventromedial plate 12
11. Ventromedial plate of paraproct with stout hairs (Fig. 54I) .... *S. dumogaense* species-group (three species)
Ventromedial plate of paraproct with fine hairs (Fig. 67H) .... *S. melanopus* species-group (two species)
12. Ovipositor valve bare narrowly along posterior margin and with inner margin not darkly pigmented (Fig. 52G) .... *S. crocinum* species-group
Ovipositor valve haired along posterior margin and with inner margin darkly pigmented (Fig. 78B) .... *S. tuberosum* species-group
13. Frons fully haired .... *S. indicum* species-group
Frons bare except several hairs along each lateral margin and lower margin 14
14. Paraproct with ventromedial plate (Fig. 54I) 15
Paraproct without ventromedial plate 16
15. Ventromedial plate of paraproct with stout hairs (Fig. 54I–R) .... *S. dumogaense* species-group
Ventromedial plate of paraproct with fine hairs (Fig. 67P) .... *S. melanopus* species-group
16. Ovipositor valve obliquely truncated posteriorly (Fig. 65D) .... *S. malyschevi* species-group
Ovipositor valve not truncated posteriorly 17
17. Ovipositor valve with inner margin nearly straight 18
Ovipositor valve with inner margin widely concave (Fig. 49C) 19
18. Paraproct in ventral view elongate, much longer than wide (Fig. 64E) .... *S. iridescens* species-group

Paraproct in ventral view nearly quadrate or rounded or triangular, as long as wide or little shorter than wide (Fig. 50E) .... *S. christophersi* species-group, *S. nebulicola* species-group and *S. pavlovskii* species-group

19. Ovipositor valve without medium-long or long stout hairs (Fig. 49C) .... *S. argentipes* species-group
    Ovipositor valve with medium-long or long stout hairs (Fig. 85C) 20
20. Pleural membrane haired ....*S. ornatum* species-group and *S. variegatum* species-group (one species)
    Pleural membrane bare .... *S. variegatum* species-group and *S. yadongense* species-group

Males

1. Pleural membrane haired ....*S. ornatum* species-group and *S. variegatum* species-group (one species)
   Pleural membrane bare 2
2. Scutum broadly silvery-pruinose with transverse inverted-V-shaped non-pruinose marking (Fig. 74S, T) ....*S. nobile* species-group
   Scutum otherwise 3
3. Style without basal projection (Fig. 62K) 4
   Style with distinct basal projection (Fig. 59H) 7
4. Hind basitarsus slender (Fig. 62I) 5
   Hind basitarsus somewhat to greatly enlarged 6
5. Ventral plate without toothed posterolateral margins (Fig. 62J) .... *S. indicum* species-group
   Ventral plate with toothed or crenulate posterolateral margins (Fig. 65L) .... *S. malyschevi* species-group, *S. melanopus* species-group (two species) and *S. pavlovskii* species-group
6. Body of ventral plate with ventral margin appearing bilobed when viewed laterally (Fig. 64J) .... *S. iridescens* species-group
   Body of ventral plate with ventral margin appearing unilobed when viewed laterally (Fig. 49L) ....*S. argentipes* species-group, *S. melanopus* species-group, *S. nebulicola* species-group and *S. variegatum* species-group
7. Basal protuberance of style short, rounded apically (Figs. 50K, 59H and 76Q) 8
   Basal protuberance of style long, pointed apically (Figs. 52L and 71C) 13
8. Ventral plate without toothed posterolateral margins (Fig. 59E) 9
   Ventral plate with toothed or crenulate posterolateral margins (Fig. 50M) 10
9. Ventral plate in ventral view saddle-shaped (Fig. 76P) .... *S. striatum* species-group
   Ventral plate in ventral view otherwise (Fig. 59B) .... *S. griseifrons* species-group
10. Ventral plate with ventrally produced process medially .... *S. yadongense* species-group
    Ventral plate otherwise 11

11. Ventral plate with its body narrow when viewed ventrally (Fig. 50I) .... *S. christophersi* species-group
Ventral plate quadrate or subquadrate when viewed ventrally (Fig. 80L) 12
12. Basal protuberance of style with cone-like spines (Fig. 80M) .... *S. tuberosum* species-group
Basal protuberance of style without cone-like spines (Fig. 55X) .... *S. dumogaense* species-group
13. Body of ventral plate not produced ventrally (Fig. 52N, O) .... *S. crocinum* species-group
Body of ventral plate moderately produced ventrally 14
14. Body of ventral plate as long as or longer than wide when viewed ventrally (Fig. 71I, L) .... *S. multistriatum* species-group
Body of ventral plate wider than long when viewed ventrally (Fig. 61H) ....*S. chungi* species-group and *S. grossifilum* species-group

Simulium (Simulium) argentipes Species-group

Diagnosis. Female: Cibarium without dorsomedian process (Fig. 49A). Claw with small tooth (Fig. 49B). Basal portion of radial vein bare. Ovipositor valve rounded apically and with inner margin concave (Fig. 49C). Male: Style without basal protuberance (Fig. 49J). Ventral plate Y-shaped when viewed ventrally (Fig. 49I), with posterolateral margins serrated (Fig. 49L, M). Pupa: Gill with eight filaments (Fig. 49Q), or 8–10 filaments (*S. minangkabaum*), or 38 filaments (Fig. 49P) (*S. arboreum*). Terminal hooks small, triangular, or absent. Cocoon corbicular-shaped (Fig. 49R, S). Larva: Abdomen with paired dorsal protuberances (Fig. 49W), which are, though, absent in one species. Postgenal cleft large and rounded (Fig. 49V).

Remarks. This species-group was defined by Takaoka and Saito (2007). It includes nine species, of which eight species are distributed in the Oriental Region, whereas one species, *S. bidentatum*, is distributed in both the Oriental and Palearctic Regions.

Species Included. *Simulium argentipes* Edwards, 1928 from Malaysia (Peninsular); *S. beludense* Takaoka, 1996 from Malaysia (Sabah and Sarawak); *S. arboreum* Takaoka, 2006 and *S. canlaonense* Delfinado, 1969, both from Philippines; *S. bidentatum* (Shiraki, 1935) from Japan (Nansei Islands, Honshu, Shikoku, and Kyushu) and China (Anhui, Fujian, Guizhou, Liaoning, Qinghai, Shanxi, Sichuan, and Yunnan); *S. minangkabaum* Takaoka and Sigit, 1997 from Indonesia (Sumatra); *S. haiduanense* Takaoka, Low and Huang, 2018 from Taiwan; *S. sansahoense* Takaoka and Chen, 2017 from Vietnam; and *S. tanae* Xue, 1992 from China (Yunnan, Jiangxi, and Zhejiang).

Simulium (*Simulium*) *christophersi* Species-group

Diagnosis. Female: Cibarium without dorsomedian projection (Fig. 50A). Scutum white pruinose, with three faint longitudinal non-pruinose vittae (one narrow median vitta and two somewhat wider submedian vittae) except *S. fuscopilosum*, which has no longitudinal vittae on scutum. Claw with small subbasal tooth (Fig. 50B). Ovipositor valve triangular with its inner margin nearly straight, covered with many short to medium-long hairs (Fig. 50C). Male: Style with short subbasal protuberance (Fig. 50K). Ventral plate Y-shaped, with toothed posterolateral margins (Fig. 50I, M). Pupa: Head and thoracic integument almost bare. Gill with six filaments (Fig. 50O). Terminal hooks absent. Cocoon wall-pocket-shaped, with anterolateral open spaces on each side (Fig. 50P). Larva: Postgenal cleft triangular (Fig. 50S).

Remarks. Seven named and one unnamed species are included, of which four species are known from the Oriental Region, three species are from the Palearctic Region, and one species is from both regions.

Species Included. *Simulium atipornae* Takaoka, Srisuka and Choochote, 2014 from Thailand and Vietnam; *S. howletti* Puri, 1932 from India (Bihar) and China (Tibet); and *S. fuscopilosum* Edwards, 1928, *S. rasuli* Takaoka, Ya'cob and Sofian-Azirun, 2018, and *S.* sp. C Takaoka and Davies, 1995, all from Malaysia (Peninsular).

Simulium (*Simulium*) *chungi* Species-group

Diagnosis. Female: Cibarium without dorsomedian projection (Fig. 51A). Scutum with five longitudinal vittae. Claw simple. Ovipositor valve with narrow tail-like tip posteromedially (Fig. 51B). Male: Hind basitarsus somewhat enlarged (Fig. 51G). Style with long basal protuberance (Fig. 51I). Ventral plate rounded posteriorly, lacking teeth on posterior surface (Fig. 51H). Pupa: Gill with six filaments (Fig. 51L), Terminal hooks small, triangular. Cocoon wall-pocket-shaped, with anterolateral open spaces on each side (Fig. 51M). Larva: Postgenal cleft large, mitre-shaped (Fig. 51Q).

Remarks. This species-group is represented by two species from Taiwan, Vietnam, and Hong Kong.

Species Included. *Simulium chungi* Takaoka and Huang, 2006 from Taiwan and Vietnam; and *S. taipokauense* Takaoka, Davies and Dudgeon, 1995 from China (Hong Kong).

Simulium (*Simulium*) *crocinum* Species-group

Diagnosis. Female and male: Antenna showing well-developed pit of sensilla on each flagellomere (Fig. 52A–E). Female: Cibarium without dorsomedian projection (Fig. 52F). Scutum without pattern or faintly with five longitudinal vittae. Claw simple. Basal portion of radial vein bare. Ovipositor valve broad, triangular and thick, having bare and transparent portions along inner and posterior margins (Fig. 52G). Male: Hind basitarsus somewhat enlarged (Fig. 52K). Style with long basal protuberance (Fig. 52L). Ventral plate with body nearly quadrate, flat, and without teeth (Fig. 52M–O). Pupa. Gill with six thread-like filaments (Fig. 53A–H). Terminal hooks usually absent, but present in one species. Cocoon wall-pocket-shaped, with or without anterolateral open space(s) on each side (Fig. 53I–L). Larva: Postgenal cleft medium-long, triangular (Fig. 53O).

Remarks. This species-group was one of the six species-groups split from the former *S. griseifrons* species-group by Takaoka (2017a). It includes 23 species, of which 19 are known from the Oriental Region, three are from the Palearctic Region, and one is from both regions.

Species Included. Simulium crocinum Takaoka and Choochote, 2004 and *S. mediocoloratum* Takaoka and Choochote, 2004, both from Thailand; *S. rudnicki* Takaoka and Davies, 1995 and *S. yongi* Takaoka and Davies, 1997 from Malaysia (Peninsular) and Thailand; *S. ephippioidum* Chen and Wen, 1999 from China (Guizhou and Hubei); *S. fuzhouense* Zhang and Wang, 1991 from China (Fujian and Guizhou); *S. ledongense* Yang and Chen, 2001, *S. parawaterfallum* Zhang, Yang and Chen 2003, and *S. waterfallum* Zhang, Yang and Chen, 2003, all from China (Hainan); *S. ramulosum* Chen, 2000 from China (Guizhou); *S. shangchuanense* An and Hao, 1998 from China (Guangdong and Shangchuan Island); *S. tenuatum* Chen, 2000 and *S. wulingense* Zhang and Chen, 2000, both from China (Guizhou); *S. laocaiense* Takaoka and Lau, 2017, *S. turgidum* Takaoka and Pham, 2017, *S. vietnamense* Takaoka, Sofian-Azirun and Chen, 2014, and *S. yukawai* Takaoka and Pham, 2019, all from Vietnam; and *S. serenum* Huang and Takaoka, 2009, *S. taichungense* Takaoka and Huang, 2018, and *S. ufengense* Takaoka, 1979, all from Taiwan.

Simulium (*Simulium*) *dumogaense* Species-group

Diagnosis. Female: Cibarium without dorsomedian projection (Fig. 54A). Scutum not patterned (though patterned in three species). Basal portion of radius bare (though partially haired in two species). Fore tarsus with moderate or thick hair crest (Fig. 54B). Hind basitarsus unicolored or bicolored (Fig. 54C, D). Claw with small to medium-sized subbasal tooth (Fig. 54F, G), or rarely simple (Fig. 54E). Ovipositor valve triangular or tapered apically (Fig. 54H, K, N, Q). Genital fork with or without projection directed anteriorly on each arm (Fig. 54H, K, N, Q). Paraproct with distinct ventromedial plate bearing stout hairs (Fig. 54I, L, O, R). Male: Scutum with color pattern of various shapes (Fig. 55A–H). Hind basitarsus slender or enlarged, and unicolored or bicolored (Fig. 55I–K). Style with basal protuberance (Fig. 55L, R, X, B'). Ventral plate with teeth on posterior surface (Fig. 55N, T, Z, D'), and body with ventral margin appearing unilobed when viewed laterally (Fig. 55M, S, Y, C'). Pupa: Frons with two trichomes on each side (except one species with three trichomes on each side). Gill with six thread-like filaments, though with eight filaments in one species (Fig. 56D–I). Ventral surface of abdomen segments 5–7 each with pair of unbranched hooks on each side (though segments 6 and 7 each with only unbranched inner hook and lacking outer hook on each side in three species; and segment 5 with pair of bifid hooks and segments 6 and 7 each with bifid inner and unbranched outer hooks on each side in one species). Terminal hooks absent. Cocoon shoe-shaped or boot-shaped (Fig. 56L, M). Larva: Postgenal cleft medium-long, triangular or large, mitre-shaped (Fig. 56P–R).

Remarks. This species-group was separated from the former *S. melanopus* species-group (Takaoka, 2017a). It contains ten species, of which nine species are known from Sulawesi (Oriental Region) and one species (*S. seramense* Takaoka, 2003) is from Seram, Indonesia (Australasian Region).

Species Included. *Simulium contractum* Takaoka, 2003, *S. dumogaense* Takaoka and Roberts, 1988, *S. fuscitarse* Takaoka, 2003, *S. landonoense* Takaoka, 2003, *S. mediodentatum* Takaoka, 2003, *S. nudifrons* Takaoka, 2003, *S. tomohonense* Takaoka, 2003, *S. tumpaense* Takaoka and Roberts, 1988, and *S. watanabei* Takaoka, 2003, all from Indonesia (Sulawesi).

Simulium (*Simulium*) *eximium* Species-group

Diagnosis. Female: Cibarium without dorsomedian projection (Fig. 57A). Scutum with longitudinal vittae. Claw simple, without tooth. Basal portion of radial vein fully haired. Ovipositor valve triangular, tapered posteromedially (Fig. 57C). Paraproct with ventrally raised ridge (Fig. 57C, D). Male: Basal portion of radial vein bare. Style with long basal protuberance (Fig. 57F). Ventral plate pear-shaped (Fig. 57E). Pupa: Thorax with rugose swelling anteromedially on each side (Fig. 57L). Gill with nine or 10 filaments (Fig. 57M–P). Abdominal segments 5–9 lacking spine-combs dorsally. Terminal hooks small, triangular. Cocoon boot-shaped (Fig. 57Q–S). Larva: Postgenal cleft medium-long to long, bullet-shaped (Fig. 57U).

Remarks. This was one of the four species-groups that originated from the former *S. multistriatum* species-group (Takaoka and Davies, 1996). This species-group is small, including four species, all from the Sunda Archipelago, Indonesia.

Species Included. *Simulium eximium* De Meijere, 1913 from Indonesia (Java, Bali, Flores, Lombok and Sumatra); *S. ranauense* Takaoka, Hadi and Sigit, 2006 from Indonesia (Sumatra); and *S. thienemanni* Edwards, 1934 and *S. upikae* Takaoka and Davies, 1996, both from Indonesia (Java).

Simulium (*Simulium*) *griseifrons* Species-group

Diagnosis. Female and male: Basal portion of radial vein bare or rarely haired. Female: Cibarium with distinct dorsomedian projection (Fig. 58A–F). Scutum with five longitudinal vittae. Claw simple. Ovipositor valve triangular, with inner margin nearly straight or slightly sinuous and close together with opposing margin (Fig. 58G, H). Genital fork with distinct projection directed forward on each arm (Fig. 58G, H). Male: Hind basitarsus enlarged (Fig. 59A). Style with short basal protuberance having spines (Fig. 59C). Ventral plate with body longer than wide, and without teeth (Fig. 59B, G). Pupa: Frons and thorax usually with multibranched trichomes (Fig. 60A, B) and often with relatively large tubercles with secondary projections (Fig. 60C). Gill with six filaments (Fig. 60D–K). Terminal hooks cone-shaped (Fig. 60L) or rarely absent. Cocoon of various shapes (Fig. 60M–T). Larva: Postgenal cleft medium-long, triangular (Fig. 60W).

Remarks. The *S. griseifrons* species-group, originally defined by Takaoka and Davies (1996), was characterized by a combination of the male ventral plate without teeth, and pupal gill with six filaments. It was split into six groups mainly based on the shape of the ovipositor valves and the presence or absence of the basal protuberance of the styles (Takaoka, 2017a). This species-group is one of these six groups, now used in a restricted sense, as redefined by Takaoka (2017a).

Species Included. Simulium choochotei Takaoka, 2002, *S. phukaense* Takaoka and Choochote, 2003, *S. pukaengense* Takaoka and Choochote, 2005, *S. suchariti* Takaoka and Choochote, 2004, *S. thongsahuani* Takaoka and Choochote, 2010, and *S. visuti* Takaoka and Choochote, 2006, all from Thailand; *S. digrammicum* Edwards, 1928 from Malaysia (Peninsular) and Thailand; *S. gravelyi* Puri, 1933 from India (Tamil Nadu) and China (Tibet); *S. griseifrons* Brunetti, 1911 from India (Uttar Pradesh, Himachal Pradesh, and Kashmir) and China (Tibet); *S. japonicum* Matsumura, 1931 from Japan (Honshu, Kyushu, Shikoku, Hokkaido, and Nansei Islands), China (Fujian, Jilin, Liaoning, and Tibet), Korea (South), and Siberia; *S. maenoi* Takaoka and Choochote, 2002 from Thailand and Vietnam; *S. spoonatum* An and Yan, 1998 from China (Guangdong and Shangchuang Island); and *S. tenuitarsus* Puri, 1933 from India (West Bengal).

Simulium (*Simulium*) *grossifilum* Species-group

Diagnosis. Female: Cibarium without dorsomedian projection (Fig. 61A). Scutum with five longitudinal vittae. Claw simple. Basal portion of radius fully haired. Ovipositor valve flap-like, with elongate dorsal projection (Fig. 61B). Male: Hind basitarsus somewhat enlarged (Fig. 61G). Style with long basal protuberance (Fig. 61J). Ventral plate much wider than long, and with no teeth (Fig. 61H). Pupa: Gill with six (or seven) somewhat inflated or thread-like filaments (Fig. 61M, N). Terminal hooks small, triangular, or absent. Cocoon roughly woven in particular near anterior margin (Fig. 61P). Larva: Postgenal cleft medium-long, bullet-shaped (Fig. 61T).

Remarks. This is one of the six species-groups that derived from the former *S. griseifrons* species-group (Takaoka, 2017a). It is small, including only five species, of which four species are known from the Oriental Region.

Species Included. Simulium grossifilum Takaoka and Davies, 1995 from Malaysia (Peninsular), Thailand, and Vietnam; *S. hengshanense* Cai, Chu, An and Wang, 2017 from China (Hunan and Hubei); *S. lushanense* Chen, Kang and Zhang, 2007 from China (Jiangxi); and *S. phayaoense* Takaoka and Choochote, 2005 from Thailand.

Simulium (*Simulium*) *indicum* Species-group

Diagnosis. Female and male: Basal portion of radial vein haired. Fore tarsus with thick hair crest (Fig. 62H). Female: Frons densely covered with yellow short hairs. Mandible with or without outer teeth (Fig. 62A). Cibarium with distinct dorsomedian projection (Fig. 62B). Scutum not patterned. Claw with small subbasal tooth (Fig. 62C). Ovipositor valve tongue-like or triangular, with inner margin widely concave or sinuous (Fig. 62D, E). Paraproct much longer than cercus, with outer surface widely bare (Fig. 62G). Male: Hind basitarsus slender (Fig. 62I). Style without basal protuberance (Fig. 62K). Ventral plate longer than wide and with no teeth (Fig. 62J, L, M). Pupa: Gill with six filaments (Fig. 63A). Abdominal segments 5–9 lacking spine-combs dorsally. Terminal hooks absent. Cocoon shoe-shaped (Fig. 63C). Larva: Mandible with supernumerary serration (Fig. 63F) except one species lacking supernumerary serration. Hypostoma with median tooth small, much shorter than each corner tooth (Fig. 63G). Postgenal cleft long, with its apex close to posterior margin of hypostoma (Fig. 63E). Abdomen covered with unpigmented or dark fan-like setae dorsally (Fig. 63H, I). Accessory sclerite present (Fig. 63J).

Remarks. This is a small species-group consisting of four species (Takaoka, 2017a). However, it is most distinct among six species-groups that originated from the former *S. griseifrons* species-group in having the female frons densely covered with yellow short hairs and larval abdomen with an accessory sclerite (Fig. 63J), both of which are not seen in other species of the subgenus *Simulium* s. str.

Species Included. Simulium indicum Becher, 1885 [1884] from India (Assam and West Bengal), Afghanistan, Bangladesh, Bhutan, Myanmar, China (Tibet), Nepal, and Pakistan (Kashmir); *S. nigrogilvum* Summers, 1911 from Thailand, Myanmar, and Vietnam; and *S. umphangense* Takaoka, Srisuka and Saeung, 2017 and *S. vanellum* Huang, Phasuk, Champaisaeng and Adler 2010, both from Thailand.

Simulium (*Simulium*) *iridescens* Species-group

Diagnosis. Female: Cibarium without dorsomedian projection (Fig. 64A). Scutum not patterned. Claw with small subbasal tooth (Fig. 64D). Ovipositor valve normal, triangular (Fig. 64E). Paraproct elongate without distinct ventromedial plate when viewed ventrally (Fig. 64E). Male: Hind basitarsus moderately enlarged (Fig. 64C). Style without basal protuberance (Fig. 64G). Ventral plate with teeth on posterior surface, and body having ventral margin appearing bilobed when viewed laterally (Fig. 64I, J). Pupa: Gill with six slender or inflated filaments (Fig. 64M, N). Terminal hooks absent. Cocoon wall-pocket-shaped (Fig. 64O). Larva: Postgenal cleft long, triangular (Fig. 64R).

Remarks. This species-group was separated from the former *S. melanopus* species-group (Takaoka, 2017a). It contains two species from Indonesia.

Species Included. *Simulium iridescens* De Meijere, 1913 from Indonesia (Java, Bali, Lombok, and Sumatra); and *S. javaense* Takaoka and Hadi, 1991 from Indonesia (Java and Bali).

Simulium (*Simulium*) *malyschevi* Species-group

Diagnosis. Female: Cibarium with dorsomedian projection (Fig. 65A). Scutum not patterned. Claw with small subbasal tooth (Fig. 65B). Basal portion of radial vein bare. Ventral surface of abdominal segment 7 with or without branched hairs (Fig. 65C). Ovipositor valve obliquely slanted posteriorly (Fig. 65D). Genital fork with projection directed anteriorly on each arm and with or without projection directed posteromedially on each arm (Fig. 65E, F). Male: Hind basitarsus slender (Fig. 65K). Style without basal protuberance (Fig. 65N). Ventral plate Y-shaped, with narrow body bearing toothed posterior surface (Fig. 65L, O, P). Pupa: Gill with two or six filaments (Fig. 66A–C) (in all eight Oriental species). Terminal hooks absent. Cocoon wall-pocket-shaped, with or without large anterolateral open space on each side (Fig. 66D–F). Larva: Postgenal cleft long, with its apex close to or reaching posterior margin of hypostoma (Fig. 66J, K). Abdomen with or without dorsal or dorsolateral protuberances (Fig. 66L, M).

Remarks. The *S. malyschevi* species-group was originally defined as one of the four species-groups in the genus *Gnus* Rubtsov by Rubtsov (1956), and later applied to all the species in the genus *Gnus*. This genus was sunk as a synonym of the subgenus *Simulium* s. str. (Crosskey, 1987). Recently, it was divided into two lineages based on the shape of the female ovipositor valve, which is obliquely slanted posteriorly (Fig. 65D) in one group, and is triangular in the other (Takaoka, 2017a). The first group, the *S. malyschevi* species-group in a restricted sense, accommodates 24 species including eight species in the Oriental Region, whereas the second group, the *S. pavlovskii* species-group, includes 17 species of the former *S. malyschevi* species-group, of which one species is from Guangdong, China (Takaoka, 2017a).

Species Included. Simulium baimaii Kuvangkadilok and Takaoka, 1999, *S. lomkaoense* Takaoka and Srisuka, 2014, and *S. siripoomense* Takaoka and Saito, 1996, all from Thailand; *S. damingense* Chen, Zhang and Zhang, 2007 from China (Guangxi Zhuang); *S. jacuticum* Rubtsov, 1944 from Siberia, China (Fujian, Heilongjiang, Inner Mongolia, and Shanxi); *S. hirtipannus* Puri, 1932 from India (Assam) and China (Fujian, Guizhou, and Tibet); *S. obliquum* Takaoka and Low, 2017 from Vietnam; and *S. phidimense* Takaoka and Shrestha, 2020 from Nepal.

Simulium (*Simulium*) *melanopus* Species-group

Diagnosis. Female: Antennal flagellomeres 1–9 with or without pits of sensilla (Fig. 67A). Cibarium without dorsomedian projection (Fig. 67B). Scutum not patterned. Claw with small subbasal tooth (Fig. 67E), or rarely simple, or with medium-long tooth (Fig. 67C, D). Basal portion of radial vein bare or rarely haired. Ovipositor valve variable in shape from normal, triangular, to elongate (Fig. 67F, K, M, O). Paraproct with distinct ventromedial plate bearing fine setae (Fig. 67H, P). Male: Scutum with color pattern of different shapes (Fig. 68A–C). Style without basal protuberance (Fig. 68G, N, T). Ventral plate with teeth on posterior surface (Fig. 68I, V), rarely without teeth (Fig. 68P), and body having ventral margin appearing unilobed when viewed laterally. Paramere rarely with fine setae on base (Fig. 68Q). Aedeagal membrane with fine setae (Fig. 68L) except one specie without setae (Fig. 68Y). Pupa: Gill with six filaments (Fig. 69A–H), or rarely with eight filaments (Fig. 69I). Abdomen with spine-combs dorsally at least on segment 8 except two species lacking spine-combs. Terminal hooks absent. Cocoon usually shoe-shaped (Fig. 69L–S), or rarely wall-pocket-shaped (Fig. 69J, K, T). Larva: Postgenal cleft large, rounded or bullet-shaped (Fig. 69W, X).

Remarks. This species-group was created for many species of the subgenus *Simulium* s. str. from the Philippines and Sabah, of which the female paraproct has its ventromedial plate with fine setae (Fig. 67H, P) (Takaoka, 1983). It is represented by 36 species, of which most species are distributed in the Philippines and Borneo, and one species, *S. bishopi* Takaoka and Davies, is distributed in Malaysia (Peninsular), Indonesia (Sumatra), and southern Thailand.

Species Included. Simulium abatanense Takaoka, 1983, *S. acostai* Takaoka, 1983, *S. aquilum* Takaoka, 2006, *S. atrum* Delfinado, 1969, *S. banahauense* Takaoka, 2007, *S. concavum* Takaoka, 2006, *S. conspicuum* Takaoka, 2006, *S. discrepans* Delfinado, 1969, *S. forcipatum* Delfinado, 1969, *S. gatchaliani* Takaoka, 1983, *S. halconense* Takaoka and Tenedero, 2007, *S. iwahigense* Takaoka, 1983, *S. malaibaense* Takaoka and Tenedero, 2007, *S. mayoyaoense* Takaoka, 2006, *S. melanopus* Edwards, 1929, *S. neodiscrepans* Takaoka, 2006, *S. palawanense* Delfinado, 1971, *S. pandum* Takaoka, 2006, *S. retusum* Delfinado, 1971, *S. salebrosum* Takaoka, 1983, *S. simulacrum* Delfinado, 1969, *S. subatrum* Takaoka, 1983, *S. sumigorense* Takaoka, 2006, *S. suplidoi* Takaoka, 1983, *S. taalense* Takaoka, 1983, and *S. tomentosum* Delfinado, 1969, all from Philippines;

S. bishopi Takaoka and Davies, 1995 from Malaysia (Peninsular), Thailand, and Indonesia (Sumatra); *S. cheedhangi* Takaoka, Ya'cob and Sofian-Azirun, 2015 and *S. laterale* Edwards, 1933, both from Malaysia (Sabah and Sarawak); *S. crassimanum* Edwards, 1933, *S. lardizabalae* Takaoka and Sofian-Azirun, 2015, *S. maklarini* Takaoka, 2007, *S. nigripilosum* Edwards, 1933, and *S. timpohonense* Takaoka and Sofian-Azirun, 2015, all from Malaysia (Sabah); *S. mirum* Takaoka, Sofian-Azirun and Low, 2016 from Malaysia (Sabah and Sarawak) and Indonesia (Kalimantan); and *S. murudense* Takaoka, Ya'cob and Sofian-Azirun, 2015 from Malaysia (Sarawak).

Simulium (*Simulium*) *multistriatum* Species-group

Diagnosis. Female: Cibarium with dorsomedian projection (Fig. 70A). Scutum with longitudinal vittae (Fig. 70B–D). Claw simple, without tooth. Basal portion of radial vein bare or rarely haired. Ovipositor valve with inner margin usually transparent (Fig. 70E). Male: Hind basitarsus enlarged (Fig. 71A). Style with long basal protuberance (Fig. 71C). Ventral plate with toothed posterolateral margins (Fig. 71E, K, P), or rarely without teeth on posterolateral margins (Fig. 71N). Pupa: Gill with eight filaments (Fig. 72A–H). Gill rarely with enlarged basal fenestra (Fig. 72H). Terminal hooks absent. Cocoon wall-pocket-shaped, with or without anterolateral open space(s) on each side, or shoe-shaped, with or without open space(s) anterolaterally on each side (Fig. 72I–M). Larva: Abdomen with or without paired protuberances dorsally (Fig. 72N). Postgenal cleft large, rounded (Fig. 72R).

Remarks. This species-group was first defined by Rubtsov (1956). Later, it was divided into four species-groups, as noted above. The *S. multistriatum* species-group is now used in a restricted sense (Takaoka and Davies, 1996).

Species Included. *Simulium barraudi* Puri, 1932 from India (Kashmir and Himachal Pradesh), Myanmar, China (Tibet), Bhutan, Nepal, and Pakistan; *S. bullatum* Takaoka and Choochote, 2005, *S. chainarongi* Kuvangkadilok and Takaoka, 1999, *S. chaliowae* Takaoka and Boonkemtong, 1999, *S. lampangense* Takaoka and Choochote, 2005, *S. takense* Takaoka and Choochote, 2005, *S. triglobus* Takaoka and Kuvangkadilok, 1999, *S. ubonae* Thaijarern et al., 2019, and *S. undecimum* Takaoka, Srisuka and Saeung, 2019, all from Thailand; *S. chanyae* Takaoka and Choochote, 2007 from Thailand and Myanmar; *S. daoense* Takaoka and Adler, 2017 from Thailand and Vietnam; *S. demolaense* Takaoka and Somboon, 2008, *S. deothangense* Takaoka and Somboon, 2008, and *S. thimphuense* Takaoka, Namgay and Somboon, 2021, all from Bhutan; *S. dentatum* Puri, 1932 from India (West Bengal, Assam, Meghalaya, and Sikkim), Bhutan, and China (Tibet); *S. fenestratum* Edwards, 1934 (complex) from Indonesia (Sumatra) and Thailand; *S. gokarnaense* Takaoka and Shrestha, 2020, *S. hillycum* Maskey, 1989, *S. maskeyae* Takaoka and Shrestha, 2020, and *S. ratekholaense* Takaoka and Shrestha, 2020, all from Nepal; *S. hirtinervis* Edwards, 1928 from Malaysia (Peninsular) and Vietnam; *S. kisapense* Takaoka, Sofian-Azirun and Ya'cob, 2012 from

Malaysia (Peninsular); *S. lacduongense* Takaoka and Ya'cob, 2015 from Vietnam; *S. laui* Takaoka and Sofian-Azirun, 2015 (complex) from Vietnam and Myanmar; *S. malayense* Takaoka and Davies, 1995 from Malaysia (Peninsular), Thailand, and Vietnam; *S. novolineatum* Puri, 1933 from India (West Bengal, Arunachal Pradesh, Karnataka, and Uttar Pradesh) and Myanmar; *S. rotifilis* Chen and Zhang, 1998 and *S. uncum* Zhang and Chen, 2001, both from China (Guizhou); *S. sakishimaense* Takaoka, 1977 from Japan (Nansei Islands), Taiwan, and China (Fujian, Guizhou, Hong Kong, Hubei, Jiangxi, Yunnan, and Zhejiang); and *S.* sp. nr. *novolineatum*, Datta, 1975 from India (West Bengal).

Simulium (*Simulium*) *nebulicola* Species-group

Diagnosis. Female: Cibarium without dorsomedian projection (Fig. 73A). Scutum not patterned. Claw with small subbasal tooth (Fig. 73E). Basal portion of radial vein bare. Ovipositor valve triangular (Fig. 73F). Male: Style without basal protuberance. Ventral plate Y-shaped, with teeth on posterior surface (Fig. 73H, I, K). Pupa: Gill with six slender filaments (Fig. 73N–P). Terminal hooks small, cone-like, or absent. Cocoon wall-pocket-shaped. Larva: Postgenal cleft triangular (Fig. 73R–T).

Remarks. This species-group was separated from the former *S. christophersi* species-group (Takaoka, 2017a). All three species included are morphologically similar to one another and endemic to the Sunda Archipelago in Indonesia (Takaoka and Davies, 1996, Takaoka et al., 2017a).

Species Included. *Simulium celsum* Takaoka and Davies, 1996 from Indonesia (Java); *S. nebulicola* Edwards, 1934 from Indonesia (Java, Flores and Sumatra); and *S. sumatraense* Takaoka and Sigit, 1997 from Indonesia (Sumatra).

Simulium (*Simulium*) *nobile* Species-group

Diagnosis. Female and male: Antenna usually yellow except few apical flagellomeres darkened (Fig. 74A). Hind tarsus narrow with tarsomeres 2–5 usually yellow (Fig. 74E, F). Female: Fronto-ocular area undeveloped or weakly developed (Fig. 74B, C). Cibarium without dorsomedian projection but furnished with sharp teeth (Fig. 74D). Scutum not patterned. Claw with or without small tooth (Fig. 74G, H). Ovipositor valves triangular or extended posteriorly (Fig. 74I, N). Paraproct produced ventrally, with anteroventral surface well sclerotized, having stout hairs at its apex (Fig. 74K, L, P, Q). Male: Scutum white pruinose with inverted V-shaped non-pruinose dark marking (Fig. 74S, T). Hind basitarsus slender (Fig. 74F). Style without basal protuberance (Fig. 74W). Ventral plate with its body longer than width, having vertical rows of several teeth on posterior surface, and median process produced ventrally (Fig. 74U, X, Y, B'–E'). Pupa: Gill with three or six short inflated filaments or six, eight, 10, or 12 slender filaments (Fig. 75A–F). Terminal hooks absent. Cocoon usually shoe-shaped or rarely simple, wall-pocket-shaped

(Fig. 75G–J). Larva: Postgenal cleft long, reaching posterior margin of hypostoma (Fig. 75N). Abdomen with or without paired dorsal protuberances (Fig. 75K, O).

Remarks. This was one of the 19 species-groups of the subgenus *Simulium* s. str. used by Crosskey (1987). The *S. nobile* species-group was first used to accommodate *S. nodosum* (at that time, as *S. shirakii* Takahasi and Kono) from Taiwan by Takaoka (1979).

Most of species in this species-group are insular in geographical distribution, while a few species are continental. Species of this species-group are inferred to have spread westward from the Philippines to India based on the hypothesis that the decreased number or inflation of the pupal gill filaments is apomorphic, and the increased number or slenderness of the pupal gill filaments is plesiomorphic (Takaoka, 2017b).

Species Included. Simulium asishi Datta, 1988 from India (West Bengal); *S. baltazarae* Delfinado, 1962, *S. benquetense* Takaoka, 1983, *S. cotabatoense* Takaoka, 1983, *S. delfinadoae* Takaoka, 1983, *S. latistylum* Takaoka, 1983, and *S. leytense* Takaoka, 1983, all from Philippines; *S. kiuliense* Smart and Clifford, 1969 from Malaysia (Sabah); *S. nobile* De Meijere, 1907 from Indonesia (Java, Bali, and Sumatra); *S. nodosum* Puri, 1933 from India (Karnataka, Assam, and West Bengal), Myanmar, China (Fujian, Hainan, Hong Kong, and Yunnan), Taiwan, Thailand, and Vietnam; *S. timorense* Takaoka, Hadi and Sigit, 2006 from Indonesia (West Timor, Flores, Lombok, Sumba, and Sumbawa); and *S. vanluni* Ya'cob, Takaoka and Sofian-Azirun, 2017 from Malaysia (Peninsular) and Thailand.

Simulium (*Simulium*) *ornatum* Species-group

Diagnosis. Female and male: Pleural membrane haired. Female: Cibarium with dorsomedian projection (similar to Fig. 84A) and tiny processes. Scutum with white spot on each shoulder. Claw with small tooth. Ovipositor valve with inner margin concave, rounded apically. Male: Hind basitarsus moderately enlarged. Ventral plate Y-shaped, with teeth on each posterolateral margin (similar to Fig. 84I, L). Style without basal protuberance. Pupa: Gill with eight slender filaments. Terminal hook small, triangular or cone-shaped. Cocoon wall-pocket-shaped, without anterodorsal projection. Larva: Postgenal cleft short to medium-long. Rectal organ simple, without secondary lobules. Ventral papillae small.

Species Included. Simulium ornatum Meigen, 1818 (complex) from China (Guizhou, Jilin, Liaoning, and Yunnan) and many countries in Europe and Middle East.

Simulium (*Simulium*) *pavlovskii* Species-group

Diagnosis. Female: Cibarium without dorsomedian projection. Scutum not patterned. Claw with small subbasal tooth. Basal portion of radial vein bare. Ovipositor valve rounded posteromedially. Ventral surface of abdominal segment 7 without branched hairs. Male: Hind basitarsus parallel-sided, narrower than hind tibia and femur. Style without basal protuberance. Ventral plate Y-shaped, with toothed posterior surface. Pupa: Gill with six (or 10) filaments. Cocoon wall-pocket-shaped, with large anterolateral window on each side (or shoe-shaped, with many open spaces anteriorly).

Remarks. This is one of the two species-groups that originated from the *S. malyschevi* species-group (Takaoka, 2017a). It includes 17 species of the former *S. malyschevi* species-group, of which one species is known from the Oriental Region (Takaoka, 2017a), whereas the remaining 16 species are recorded from the Palearctic and Nearctic Regions.

The cibarium in the female of *S. jieyangense* appears to have a dorsomedian projection according to the illustration (An et al., 1994).

Species Included. Simulium jieyangense An, Yan, Yang and Hao, 1994 from China (Guangdong).

Simulium (*Simulium*) *striatum* Species-group

Diagnosis. Female: Cibarium without dorsomedian projection (Fig. 76A). Scutum with longitudinal vittae (Fig. 76B, C). Claw simple, without tooth. Sternite 8 with posterior margin concave medially in form of inverted letter U (Fig. 76D). Ovipositor valve with ventrally produced lobe along inner margin (Fig. 76D, E). Paraproct much extended ventrally (Fig. 76H). Male: Hind basitarsus slightly to greatly enlarged (Fig. 76K–O). Ventral plate saddle-shaped, without teeth on posterior surface (Fig. 76P, S, T). Style with short basal protuberance (Fig. 76Q, R). Pupa: Gill with eight, nine or 10 filaments (Fig. 77A–E). Terminal hooks small, triangular, or absent. Cocoon shoe-shaped with or without small open spaces anteriorly (Fig. 77F–H). Larva: Postgenal cleft large and rounded (Fig. 77K). Abdomen with or without paired dorsal protuberances (Fig. 77L).

Remarks. This is one of the three species-groups that originated from the former *S. multistriatum* species-group (Takaoka and Davies, 1996).

Species Included. Simulium argyrocinctum De Meijere, 1913 from Indonesia (Java and Sumatra); *S. baliense* Takaoka and Sofian-Azirun, 2017 from Indonesia (Bali and Lombok); *S. bagmaticum* Maskey, 1987 [1985], *S. beniense* Takaoka and Shrestha, 2020, *S. chainpurense* Takaoka and Shrestha, 2020, *S. dadagaunense* Takaoka and Shrestha, 2020, and *S. salyanense* Takaoka and Shrestha, 2020, all from Nepal; *S. chiangmaiense* Takaoka and Suzuki, 1984 from Thailand, Myanmar, China (Yunnan), and Vietnam; *S. consimile* Puri, 1932 from India (Punjab and Himachal Pradesh), Pakistan, and Sri Lanka; *S. grisescens* Brunetti, 1911 from India (West Bengal, Arunachal Pradesh, Assam, Himachal Pradesh, Karnataka, Kerala, Meghalaya, Manipur, Nagaland, Punjab, and Tamil Nadu), Myanmar, Nepal, Pakistan, and Taiwan (?); *S. jeffreyi* Takaoka and Davies, 1995 from

Malaysia (Peninsular) and Vietnam; *S. kapuri* Datta, 1975 from India (Assam and West Bengal); *S. lineothorax* Puri, 1932 from India (Assam and Tamil Nadu), *S. pothigaiense* Anbalagan, 2018 from India (Tamil Nadu); *S. latistriatum* Senior-White, 1922 from Bhutan, India (Karnataka, Kerala, Maharashtra, and Tamil Nadu), and Nepal; *S. maeklongkeense* Takaoka, Srisuka and Saeung, 2020, *S. phraense* Takaoka, Srisuka and Saeung, 2018, *S. poolpholi* Takaoka, Srisuka and Saeung, 2022, *S. tadtonense* Takaoka, Srisuka and Saeung, 2020, *S. thilorsuense* Takaoka, Srisuka and Saeung, 2022, and *S. wangkwaiense* Takaoka, Srisuka and Saeung, 2020, all from Thailand; *S. menglaense* Chen, 2003 from China (Yunnan and Hainan); S. *bahense* Chen, 2003 and *S. nujiangense* Xue, 1993, both from China (Yunnan); *S. wuzhishanense* Chen, 2003 from China (Hainan); *S. nakhonense* Takaoka and Suzuki, 1984 from Thailand, Myanmar, China (Hainan), and Vietnam; *S. pallidum* Puri, 1932 from India (Punjab and Himachal Pradesh), Nepal, and Pakistan; *S. palmatum* Puri, 1932 from India (Maharashtra, Karnataka, and Kerala); *S. perakense* Takaoka, Ya'cob and Sofian-Azirun, 2018 from Malaysia (Peninsular); *S. pingtungense* Huang and Takaoka, 2008 and *S. yanpingense* Takaoka and Huang, 2022, both from Taiwan; *S. quinquestriatum* (Shiraki, 1935) from Taiwan, China (Fujian, Guangdong, Guizhou, Hainan, Hong Kong, Hubei, Jiangxi, Liaoning, Sichuan, Yunnan, and Tibet) (?), Korea (?), and Vietnam (?); *S. striatum* Brunetti, 1912 and *S. subpalmatum* Davies and Györkös, 1992, both from Sri Lanka; *S. tavanense* Takaoka, Sofian-Azirun, 2017 and *S. taythienense* Takaoka, Sofian-Azirun and Ya'cob, 2014, both from Vietnam; *S. thailandicum* Takaoka and Davies, 1984 from Thailand, China (Yunnan), and Vietnam; *S. valparaiense* Anbalagan, 2018 from India (Tamil Nadu); *S. yanaense* Anbalagan, 2019 from India (Karnataka); *S. xuandai* Takaoka, Sofian-Azirun and Ya'cob, 2014 from Vietnam and Myanmar; *S.* sp. B Takaoka and Davies, 1993 from Malaysia (Peninsular); *S.* sp. nr. *grisescens* Datta, 1975 from India (West Bengal); and *S.* Diju sp. Lewis, 1974 and *S.* Doigroon sp. Lewis, 1974, both from India (Assam).

Simulium (*Simulium*) *tuberosum* Species-group

Diagnosis. Female: Cibarium without dorsomedian projection (Fig. 78A). Scutum not patterned. Claw simple, without tooth. Ventral surface of abdominal segment 7 with or without paired stout hairs. Ovipositor valve triangular, with nearly straight inner margin (Fig. 78B). Male: Hind basitarsus enlarged (Fig. 78G, H). Ventral plate with body nearly quadrate when viewed ventrally (Fig. 78I), with teeth on each posterolateral margin. Style with basal protuberance covered with spines (Fig. 78J). Pupa: Frons bare, or sparsely to densely covered with tubercles (Fig. 83A–C). Gill with six filaments (Fig. 79A–G). Terminal hooks small, conical, or absent. Cocoon wall-pocket-shaped. Larva: Postgenal cleft medium-long to long, triangular (Fig. 79M).

Remarks. This species-group was defined by Rubtsov (1956). It is represented by 72 species, of which seven species are recorded from the Nearctic Region, 13 species are from the Palearctic Region, three species are from both regions, 44 species are from the Oriental Region, and five species are from both the Oriental and Palearctic Regions.

Takaoka (2017b) divided this species-group into three subgroups: *S. brevipar*, *S. nitidithorax*, and *S. rufibasis* subgroups.

Simulium (*Simulium*) *brevipar* Subgroup

Diagnosis. Female: Abdominal segment 7 without pair of tuft hairs. Pupa: Thorax with pit-like organ near base of gill (Fig. 79H). Terminal hooks absent.

Remarks. All nine species are known from the Oriental Region.

Species Included. *Simulium brevipar* Takaoka and Davies, 1995 (complex) from Malaysia (Peninsular), Indonesia (Sumatra), Thailand, and Vietnam; *S. huense* Takaoka and Ya'cob, 2017, *S. lowi* Takaoka and Adler, 2017, and *S. sapaense* Takaoka and Low, 2017, all from Vietnam; *S. sigiti* Takaoka and Hadi, 1991 from Indonesia (Java); *S. tatopaniense* Takaoka and Shrestha, 2020 from Nepal; *S. tianchi* Chen, Zhan and Yang, 2003 from China (Hainan); *S. tiomanense* Takaoka, Sofian-Azirun and Belabut, 2012 from Malaysia (Peninsular); and *S. yuphae* Takaoka and Choochote, 2005 (complex) from Thailand.

Simulium (*Simulium*) *nitidithorax* Subgroup

Diagnosis. Female: Abdominal segment 7 without pair of tuft hairs. Pupa: Thorax lacking pit-like organ near base of gill. Terminal hooks usually present, or rarely absent.

Remarks. This subgroup is represented by 44 species, of which 20 species are known from the Oriental Region, 11 species are from the Palearctic Region, three species are from both regions, seven species are from the Nearctic Region, and three species are from both the Palearctic and Nearctic Regions.

Species Included. *Simulium alberti* Takaoka, 2008, *S. keningauense* Takaoka, 2008, and *S. sabahense* Smart and Clifford, 1969, all from Malaysia (Sabah and Sarawak); *S. aeneifacies* Edwards, 1933 and *S. masilauense* Takaoka, 2008, both from Malaysia (Sabah); *S. lunduense* Takaoka, 2008 from Malaysia (Sarawak); *S. congi* Takaoka and Sofian-Azirun, 2015, *S. suoivangense* Takaoka and Pham, 2017, and *S. xuandei* Takaoka and Pham, 2015, all from Vietnam; *S. jasmoni* Takaoka, Sofian-Azirun and Belabut, 2012 from Malaysia (Peninsular); *S. jianshiense* Takaoka, Otsuka and Adler, 2021 and *S. puliense* Takaoka, 1979, both from Taiwan; *S. mainitense* Takaoka and Tenedero, 2019, *S. quasifrenum* Delfinado, 1971, and *S. tukurongense* Takaoka and Tenedero, 2019, all from Philippines; *S. nigrifacies* Datta, 1974 from India (West Bengal), Myanmar, and China (Tibet);

S. nitidithorax Puri, 1932 from India (Arunachal Pradesh and West Bengal) and China (Fujian, Hainan, and Tibet); *S. rangjungense* Takaoka and Somboon, 2008 from Bhutan; *S. ryukyuense* Ogata, 1971 from Japan (Nansei Islands); *S. sanyaense* Cai, Liu and An, 2016 from China (Hainan); *S. suzukii* Rubtsov, 1963 from Japan, Korea (South), Siberia, and China (Guizhou, Hong Kong, Jilin, Jiangxi, Sichuan, and Yunnan); *S. tani* Takaoka and Davies, 1995 (complex) from Myanmar, Malaysia (Peninsular), Thailand, and Vietnam; and *S. prominentum* Chen and Zhang, 2002 from China (Guizhou).

Simulium (*Simulium*) *rufibasis* Subgroup

Diagnosis. Female: Abdominal segment 7 with pair of tuft hairs ventrally (Fig. 82B–E). Pupa: Thorax lacking pit-like organ near base of gill. Terminal hooks usually absent (though rarely present).

Remarks. This subgroup includes 19 species, of which 15 species are known from the Oriental Region, two species from the Palearctic Region and two species are known from both the Oriental and Palearctic Regions.

Species Included. Simulium arisanum Shiraki, 1935, *S. hehuanense* Takaoka, 2021 and *S. xiulinense* Takaoka, 2021, all from Taiwan; *S. fuscicoxae* Takaoka and Ya'cob, 2017, *S. giayense* Takaoka and Low, 2017, and *S. rosliramlii* Takaoka and Chen, 2017, all from Vietnam; *S. doipuiense* Takaoka and Choochote, 2005 (complex) from Thailand and Vietnam; *S. manooni* Takaoka and Choochote, 2005, *S. weji* Takaoka, 2001, *S. tenebrosum* Takaoka, Srisuka and Saeung, 2018 [2019] (complex) and *S. setsukoae* Takaoka and Choochote, 2004, all from Thailand; *S. ramosum* Puri, 1932 from India (Himachal Pradesh and West Bengal) and Pakistan; *S. rufibasis* Brunetti, 1911 from India (West Bengal, Arunachal Pradesh, Himachal Pradesh, Meghalaya, Manipur, and Sikkim), Myanmar, China (Yunnan) and Pakistan; *S. pradyai* Takaoka, 2008 from Bhutan and Nepal; and *S. dailekhense* Takaoka and Shrestha, 2020, *S. langtangense* Takaoka and Shrestha, 2020 and *S. nakanoi* Takaoka and Shrestha, 2020, all from Nepal.

Simulium (*Simulium*) *variegatum* Species-group

Diagnosis. Female: Cibarium with dorsomedian projection (Figs. 84A and 85A) or rarely unarmed (Fig. 86B). Scutum not patterned. Claw with small tooth (Fig. 84B). Ovipositor valve with inner margin concave and usually pointed (Fig. 84C) or rarely rounded (Fig. 86D) apically. Male: Hind basitarsus slightly to moderately enlarged (Figs. 85H and 86J, K). Ventral plate Y-shaped, with teeth on each posterolateral margin (Fig. 84I, L). Style without basal protuberance (Fig. 84J). Pupa: Gill with six slender or somewhat inflated filaments (Figs. 84O, P, 87A–H and 88A–H), or rarely globose with or without six slender filaments (Fig. 88I–L). Terminal hook small, triangular or cone-shaped (Fig. 89E), or rarely absent. Cocoon wall-pocket-shaped, with or without anterodorsal projection (Fig. 89F–T). Larva: Postgenal cleft short to medium-long (Figs. 84U and 89W–Y). Ventral papillae small (Fig. 89Z) or absent.

Remarks. This species-group is one of the 19 species-groups defined by Crosskey (1987) based on many of those established by Rubtsov (1956).

Species Included. *Simulium barnesi* Takaoka and Suzuki, 1984 from Thailand and India (Assam); *S. bulla* Davies and Györkös, 1987, *S. cruszi* Davies and Györkös, 1992, *S. nubis* Davies and Györkös, 1987, and *S. paranubis* Davies and Györkös, 1992, all from Sri Lanka; *S. caohaiense* Chen and Zhang, 1997 from China (Guizhou); *S. chamlongi* Takaoka and Suzuki, 1984 from Thailand, Myanmar, China (Yunnan and Tibet), and Vietnam; *S. elongatum* Takaoka, 2003, *S. flavum* Takaoka, 2003, *S. globosum* Takaoka, 2003, *S. laxum* Takaoka, 2003, *S. linduense* Takaoka, 2003, *S. lorense* Takaoka, 2003, *S. luridum* Takaoka, 2003, *S. malinoense* Takaoka, 2003, *S. molliculum* Takaoka, 2003, *S. palopoense* Takaoka, 2003, and *S. quinquefiliatum* Takaoka, 2003, all from Indonesia (Sulawesi); *S. gurneyae* Senior-White, 1922 from India (Tamil Nadu, Karnataka, and Maharashtra); *S. nilgiricum* Puri, 1932 from India (Tamil Nadu, Karnataka, and Maharashtra) and Sri Lanka; *S. hackeri* Edwards, 1928 from Malaysia (Peninsular); *S. himalayense* Puri, 1932 from India (West Bengal, Arunachal Pradesh, Bihar, Himachal Pradesh, and Sikkim), China (Tibet), Nepal, and Pakistan; *S. karenkoense* (Shiraki, 1935) from Taiwan; *S. kiewmaepanense* Takaoka, Srisuka and Saeung, 2017, *S. luculentum* Takaoka, Srisuka and Saeung, 2019 and *S. srisukai* Takaoka and Saeung, 2017, all from Thailand; *S. oitanum* (Shiraki, 1935) from Japan (Kyushu, Hokkaido, Honshu, Shikoku, and Nansei Islands), China (Guizhou, Jilin, Liaoning, and Yunnan), Korea (South), Nepal (?) and Siberia; *S. phuluense* Takaoka and Sofian-Azirun, 2017 from Vietnam; *S. qianense* Chen and Chen, 2001 from China (Guizhou and Hubei); *S. taiwanicum* Takaoka, 1979 from Taiwan and China (Sichuan, Guizhou, Jiangsu, and Jiangxi); and *S. tulshii* Takaoka and Shrestha, 2010, *S.* sp. A Maskey, 1998 and *S.* sp. B Maskey, 1998, all from Nepal.

Simulium (*Simulium*) *yadongense* Species-group

Diagnosis. Female: Scutum without longitudinal vittae. Claw with small tooth. Ovipositor valve with inner margin concave. Male: Style with distinct triangular basal protuberance bearing no spines. Pupa: Gill with six slender filaments. Terminal hooks present or absent.

Remarks. This species-group is small, including only three species, of which *S. longchatum* Chen, Zhang and Huang from Sichuan, and *S. zunyiense* Chen, Yiu and Zhang from Guizhou, were transferred from the *S. variegatum* species-group (Takaoka, 2017a). This species-group is similar to the *S. variegatum* species-group by sharing the similar shape of the ovipositor valve and claw with a small subbasal tooth, but it differs from the latter by the style with a basal protuberance and the shape of the ventral plate, which bears a ventrally produced process from the middle of its body (Chen et al., 2005, 2012).

Species Included. S. zunyiense Chen, Xiu and Zhang, 2012 from China (Guizhou).

Simulium (_Simulium_) s. str. Species Unplaced to Species-group

Species Included. Simulium ambiguum Shiraki, 1935 and *S. katoi* Shiraki, 1935, both from Taiwan; *S. biforaminiferum* Datta. 1974, *S. palniense*, Puri, 1933, and *S. singtamense* Datta and Pal, 1975, all from India; *S. multifurcatum* Zhang and Wang, 1991, *S. neoacontum* Chen, Xiu and Zhang, 2012, and *S. shennongjiaense* Yang, Luo and Chen, 2005, all from China; *S. cremnosi* Davies and Györkös, 1987 from Sri Lanka; *S. parvulum* Takaoka, 2003 from Indonesia (Sulawesi); *S. tamorense* Takaoka and Shrestha, 2010 and *S. titarense* Takaoka and Shrestha, 2010, both from Nepal; *S.* sp. E Zhang, Wen and Chen, 1999, *S.* sp. F Zhang, Wen and Chen, 1999, *S.* sp. G Zhang, Wen and Chen, 1999, and *S.* sp. H Zhang, Wen and Chen, 1999, all from China (Guizhou); *S.* Kamti sp. Lewis, 1974 from India (Uttar Pradesh); *S.* Rengma sp. Lewis, 1974, *S.* Assam sp. Lewis, 1974, and *S.* Subansiri sp. Lewis, 1974, all from India (Assam).

Subgenus Wallacellum *Takaoka*

Wallacellum Takaoka, 1983: 20. Type species: *carinatum* (original designation).

Diagnosis. Female and male: Antenna composed of scape, pedicel and nine flagellomeres. Pleural membrane with scale-like hairs and simple hairs or yellow scale-like hairs only. Katepisternum haired. Basal section of radius haired. Hind tibia with low process narrowly produced along shaft on basal half or basal one-third of anterior internal surface, which is densely covered with small, spear-shaped, spinous scales elaborately arranged on its outer and anterior surface (Fig. 92G, H). Hind basitarsus usually lacking row of strong spines along anterior margin (Fig. 92I, J). Calcipala greatly enlarged, extended to apical end of second or third tarsomere, and nearly as wide as apical portion of hind basitarsus (Fig. 92I, J). Abdomen densely covered with yellow scale-like hairs as well as brown scale-like and simple hairs, and not shiny dorsally. Female: Frons and clypeus not shiny, densely covered with yellow scale-like hairs as well as brown hairs. Palpus with third segment enlarged, longer than fourth segment (Fig. 92A). Sensory vesicle short to long (Fig. 92B, C). Mandible lacking outer teeth (Fig. 92D). Cibarium unarmed (Fig. 92E, F). Scutum dull, with three dark longitudinal vittae. Claw with large basal tooth (Fig. 92K). Sternite 7 weakly developed, in form of broad plate medially. Ovipositor valves membranous, rounded posteriorly, with their posterior margins folded inward or not (Fig. 92L, O). Genital fork with one projection (or rarely two) directed anterodorsally on each arm, and with posterior portion of each arm extended posteriorly along outer margin of each paraproct in some species (Fig. 92R–U). Paraproct normal, or strongly produced anteroventrally (Fig. 92M, N, P, Q). Cercus much shorter than wide (Fig. 92N, Q). Spermatheca strongly sclerotized, globular or ovoid, with polygonal pattern on surface, and lacking internal setae (Fig. 92V, W). Male: Coxite much longer than wide (Fig. 93C, I, M, Q). Style shorter than coxite, wedge-shaped when viewed ventrolaterally and with simple apical spine (Fig. 93D, J, N, R). Ventral plate transverse with setae (Fig. 93C, I, M, Q). Median sclerite broad, plate-like, somewhat broadened toward tip (Fig. 93G). Paramere narrow, with four large hooks (Fig. 93H). Aedeagal membrane with fine setae or rarely spinous setae (Fig. 93H, U). Pupa: Frons with two trichomes or rarely three on each side (Fig. 95A, B). Antennal sheaths smooth or with conspicuous spines (Fig. 95C, D). Gill with four slender filaments or inflated tubular filaments (Fig. 94A–H). Abdominal segments 6–9 devoid of spine-combs dorsally (Fig. 95E). Abdominal segments 6 and 7 each with bifid or trifid inner hook but lacking outer hook ventrally on each side (Fig. 95F). Terminal hooks stout cone-shaped (Fig. 95G). Grapnel-like hooklets absent. Cocoon wall-pocket-shaped or shoe-shaped (Fig. 95H–O). Larva: Labral fan with uppermost primary ray having pectination different from other primary rays (Fig. 95V, W). Hypostoma with nine apical teeth, of which median tooth and each corner tooth prominent, with serrated lateral margins (Fig. 95U). Mandibular serration composed of one large tooth and one small one (Fig. 95T). Angle of major tooth against mandible at obtuse angle apically (Fig. 95T). Postgenal cleft of medium-size, usually with small apical extension, and with or without pigmented sheath of subesophageal ganglion (Fig. 95X, Y). Ventral papillae present (Fig. 95Z). Accessory sclerites present (Fig. 95Z) or absent. Rectal organ simple or compound.

Remarks. This subgenus was proposed to accommodate eight species from the Philippines and one species from Yonakuni Island and Lan Yu Island by Takaoka (1983). *Wallacellum* is distinctive in having some morphological characteristics including the presence of the low process on the hind tibia (Fig. 92G) and enlarged calcipala (Fig. 92I, J), as well as both pleural membrane and katepisternum haired in the female and male, as noted by Takaoka (1983). This subgenus is now represented by 17 species, of which 14 species are distributed in the Philippines, one species, *S. yonakuniense* Takaoka, is in Yonakuni Island, Japan and Lan Yu Island, Taiwan, one species, *S. celebesense* Takaoka, is in Sulawesi, Indonesia, and one species, *S. alfurense* Takaoka, is in Seram and Biak, Indonesia. In addition, one unnamed species, *S.* (*W.*) sp., was reported from Java and the Lesser Sunda Archipelago (Takaoka et al., 2006a; Almet et al., 2016; Hadi, personal communication). The geographical distribution of this subgenus is entirely insular, ranging northward to Yonakuni Island, Japan, southeastward to Biak Island, and southwestward to Java, Indonesia, which is almost the same geographic area as so-called "Wallacea" defined by Dickerson (1928).

The phylogenetic relationship of this subgenus was analyzed by Otsuka et al. (2007). Otsuka and Takaoka (2015) morphologically and molecularly analyzed the species of the subgenus *Wallacellum*, and inferred that the Philippines is a speciation center and divided the species into three groups: the first having more plesiomorphic characteristics, the second having more apomorphic ones, and the third having intermediate characteristics.

Biting habits and other biological aspects of this subgenus have not been investigated.

Species Included. Simulium celebesense Takaoka, 2003 from Indonesia (Sulawesi); *S. amplum* Takaoka, 2009, *S. cabrerai* Takaoka, 1983, *S. claveriaense* Takaoka, 2006, *S. makilingense* Takaoka, 1983, *S. molawingense* Takaoka, 2006, and *S. spinosibranchium* Takaoka, 1983, all from Philippines (Luzon); *S. marilogense* Takaoka, 2009, *S. ogonukii* Takaoka, 1983, and *S. resimum* Takaoka, 1983, all from Philippines (Mindanao); *S. recurvum* Takaoka, 1983 and *S. carinatum* Delfinado, 1969, both from Philippines (Luzon and Negros); *S. suyoense* Takaoka, 2006 from Philippines (Luzon and Mindoro); *S. tenederoi* Takaoka, 2009 from Philippines (Samar); *S. tuyense* Takaoka, 1983 from Philippines (Luzon, Samar and Palawan); *S. yonakuniense* Takaoka, 1972 from Japan (Yonakuni Island) and Taiwan (Lan Yu Island); and *S.* (*W.*) sp. Takaoka, Hadi and Sigit, 2006 from Indonesia (Java, Flores and Timor).

Subgenus Wilhelmia *Enderlein*

Wilhelmia Enderlein, 1921: 199. Type species: *lineatum* (original designation).

Diagnosis. Diagnosis of this subgenus defined by Crosskey (1969) is somewhat modified as follows. Female and male: Basal section of radius haired. Pleural membrane and part of upper mesopleuron region haired. Katepisternum bare. Fore tarsus slender. Female: Frons densely covered with hairs. Cibarium unarmed (Fig. 96B). Scutum gray-pruinose, with three dark longitudinal vittae. Tarsal claws elongate, without basal tooth (Fig. 96C). Abdomen uniformly silvery-scaled. Seventh sternite undeveloped. Ovipositor valve forming slender curled acuminate process (Fig. 96D). Spermatheca without internal setae (Fig. 96H). Male: Scutum grayish black to velvet black with or without pair of large silvery gray triangular areas anterolaterally. Hind basitarsus slender, nearly parallel-sided (Fig. 96I). Coxite very large (Fig. 96J). Style small, usually folded down against body of coxite, with one apical spine (Fig. 96K). Ventral plate subtriangular, with elongate basal arms widely divergent, and body flat, plate-like, and not toothed (Fig. 96J, L, M). Median sclerite toothed, usually bifurcate or clubbed apically (Fig. 96N). Paramere long and slender, with numerous hooks (Fig. 96O). Aedeagal membrane setose (Fig. 96P). Pupa: Gill formed of enlarged thin-walled unornamented tubular basal arms, oriented dorso-ventrally, between which usually six tubular filaments arise and extend forward (Fig. 97A, B). Abdominal segments 5–9 without spine-combs dorsally and abdominal segments 5–7 each with pair of unbranched hooks ventrally on each side. Terminal hooks present (Fig. 97C). Cocoon shoe-shaped, with well-formed neck (Fig. 97D, E). Larva: Head spots positive. Hypostoma with usual nine apical teeth, all short and blunt, with three to six hypostomal bristles divergent posteriorly from lateral margin on each side (Fig. 97G). Postgenal cleft much longer than postgenal bridge (Fig. 97H). Mandible with three comb-teeth decreasing in length, and with serration composed of two teeth (Fig. 97F). Antenna short, nearly as long as or longer than basal stalk of labral fan. Ventral papillae absent. Accessory sclerite absent. Rectal organ simple, without secondary lobules. Posterior circlet with 90–120 rows of 17–30 hooklets per row.

Remarks. The majority of this species-group is Palearctic, although one species is recorded from the Oriental Region and one species is from both regions. There were two species-groups: *S. equinum* species-group and *S. lineatum* species-group (at that time, as *S. salopiense* species-group) (Rubtsov, 1956). However, the latter was merged with the *S. equinum* species-group by Adler and Crosskey (2008). At present, 30 species are placed in the *S. equinum* species-group, whereas one species, *S. buettikeri*, from Middle East is placed in the *S. buettikeri* species-group (Adler, 2022).

Species Included. Simulium takahasii (Rubtsov, 1962) from Japan and China (Beijing, Guizhou, Qinghai, Shanxi, Xinjiang, and Yunnan); and *S. xingyiense* Chen and Zhang, 1998 from China (Guizhou).

Chapter 3
Synoptic Accounts of Species of Black Flies in the Oriental Region

In this chapter, synoptic accounts of all 719 species (675 named and 44 unnamed) of black flies recorded in the Oriental Region by May 2023 are given under each subgenus, species-group, and subgroup. Two species, *S. seramense* Takaoka in the *S. dumogaense* species-group of the subgenus *Simulium* s. str. and *S. alfurense* Takaoka in the subgenus *Wallacellum*, are added for reference, though both species are distributed in the Australasian Region.

A synoptic account of each black fly species includes the reference with the original description, subsequent related references, if any, geographic distribution, bionomics such as aquatic habitats and biting habits, and remarks, in which morphological characteristics of the species are noted, and taxonomic problems, if any, are discussed.

1 Synoptic Accounts of All 719 Species of Black Flies in the Oriental Region

Subgenus Asiosimulium *Takaoka and Choochote*

Simulium (Asiosimulium) furvum Takaoka and Srisuka, 2013
(Figs. 4E, I, M, N, S, 5B, M–Q and 6G, H, I)

Simulium (*Asiosimulium*) *furvum* Takaoka and Srisuka [in Takaoka et al., 2013c]: 494–499 (female, male, pupa, and larva).

Distribution. Thailand.

Bionomics. Aquatic habitat: A small stream (flow slow, bottom rocky, 27.6 °C, exposed to the sun, elevation 468 m) in Maewa National Park, Lampang Province (Takaoka et al., 2013c).

Remarks. *Simulium furvum* was described from Lampang Province, Thailand (Takaoka et al., 2013c).

H. Takaoka, *The Black Flies of Subtropical and Tropical Asia*,
https://doi.org/10.1007/978-981-99-5219-9_3

It is characterized in the female by the genital fork with a stout projection directed anteriorly on each arm (Fig. 4I), cercus 1.3 times as long as its basal width when viewed laterally (Fig. 4N) and spermatheca pear-shaped (Fig. 4S); in the male by the hind basitarsus enlarged and wedge-shaped (Fig. 5B), ventral plate with a laterally compressed median keel directed ventrally (Fig. 5O) and a deep notch posteromedially (Fig. 5M) and aedeagal membrane with stout spines (Fig. 5Q); in the pupa by the head and thoracic integument with tubercles and gill with 22 filaments; and in the larva by the gray-pigmented sheath of the subesophageal ganglion (Fig. 6I).

Simulium (Asiosimulium) oblongum Takaoka and Choochote, 2005

(Figs. 4D, F, G, O, P, R, 5A, C–K and 6C, E)

Simulium (Asiosimulium) oblongum Takaoka and Choochote, 2005b: 33–41 (female, male, pupa, and larva).

Distribution. Thailand.

Bionomics. Aquatic habitat: A small stream (width 0.8 m, depth 3 cm, bottom rocky, covered with reddish muds, 29 °C, exposed to the sun, elevation 270 m) in Muang District, Muk Da Han Province, Thailand (Takaoka and Choochote, 2005b).

Remarks. This species was described from Thailand (Takaoka and Choochote, 2005b)

It is characterized in the female by the cerci elongated, about twice as long as its basal width when viewed laterally (Fig. 4O, P) and genital fork with an incision at the base of arms (Fig. 4G); in the male by the hind basitarsus slightly enlarged and spindle-shaped, 5.4 times as long as its greatest width (Fig. 5A) and ventral plate broad and rounded posteriorly, with its posterior margin having a small medial notch (Fig. 5C); in the pupa by the gill with 31–33 filaments; and in the larva by the labral fan with 43–45 primary rays, and postgenal cleft long but not reaching the posterior margin of the hypostoma.

Simulium (Asiosimulium) phurueaense Tangkawanit, Wongpakam and Pramual, 2018

Simulium (Asiosimulium) phurueaense Tangkawanit, Wongpakam, and Pramual 2018: 111–121 (female, male, pupa, and larva).

Distribution. Thailand.

Bionomics. Aquatic habitats: Streams at an elevation of 1100 m or above (Takgkawanit et al., 2018).

Remarks. This species was described from Thailand (Tangkawanit et al., 2018).

It is similar to *S. oblongum* in many characteristics but is distinguished by the larval head capsule with a pigmented sheath of the subesophageal ganglion on the ventral surface.

According to Tangkawanit et al. (2018), this species differs genetically from *S. oblongum*: a minimum genetic distance between both species was 3.51% when COI sequences were compared.

***Simulium (Asiosimulium) saeungae* Takaoka and Srisuka, 2018**

(Fig. 4H, Q)

Simulium (Asiosimulium) saeungae Takaoka and Srisuka [in Takaoka et al., 2018k]: 569–575 (female, male, pupa, and larva).

Distribution. Thailand.

Bionomics. Aquatic habitat: A small stream (flow slow, bottom rocky, 19.6 °C, partially shaded, elevation 1316 m) in Khunsathan National Park, Nan Province (Takaoka et al., 2018k).

Remarks. *Simulium saeungae* was described from northern Thailand (Takaoka et al., 2018k).

It is characterized in the female by the antenna dark yellow to light brown except the scape, pedicel, and the base of the first flagellomere yellow and a few apical flagellomeres darkened, genital fork with a distinct projection directed anteriorly on each arm (Fig. 4H), cercus nearly as long as its basal width and spermatheca globular, widely unpigmented around the base of its duct (Fig. 4Q); in the male by the antenna yellow except the base of the first flagellomere whitish yellow and the apical flagellomere light brown, hind basitarsus enlarged, and ventral plate with its body broad, with its posterior margin shallowly concave when viewed ventrally; in the pupa by the head and thoracic integument bare, gill of an arborescent type, with 42–56 filaments; and in the larva by the small number of primary rays of the labral fan (30–33).

***Simulium (Asiosimulium) shanense* Takaoka, Srusuka and Saeung, 2017**

Simulium (Asiosimulium) shanense Takaoka, Srisuka and Saeung [in Takaoka et al., 2017h]: 137–140 (male).

Distribution. Myanmar.

Bionomics. Unknown.

Remarks. *Simulium shanense* was described based on a male caught by a sweeping net in central Myanmar (Takaoka et al., 2017h). Its female, pupa and larva are unknown.

The male of this species is characterized by abdominal segments 2–8 each having a pair of shiny dorsolateral patches, a characteristic not found in the males of the other species of the subgenus *Asiosimulium*.

Simulium (Asiosimulium) suchitrae Takaoka, 2010
(Figs. 4B, C, J–L, T and 6B, D, F)

Simulium (*Asiosimulium*) *suchitrae* Takaoka [in Takaoka and Shrestha, 2010]: 4–8 (female, pupa, and larva).

Distribution. Nepal.

Bionomics. Aquatic habitat: A small stream (bottom rocky, flow slow, shaded, 15 °C, elevation 1826 m) near Titar, Mustang (Takaoka and Shrestha, 2010).

Remarks. This species was described from western Nepal (Takaoka and Shrestha, 2010). Its male is unknown.

It is characterized in the female by the genital fork lacking a projection directed anteriorly from each arm (Fig. 4J), and spermatheca pear-like (Fig. 4T); in the pupa by the head integument bare and gill with an enlarged transparent basal fenestra basally and 28 filaments (Fig. 6B); and in the larva by the labral fan with 30 primary rays.

Simulium (Asiosimulium) wanchaii Takaoka and Choochote, 2006
(Figs. 4A, 5L and 6A)

Simulium (*Asiosimulium*) *wanchaii* Takaoka and Choochote, 2006a: 45–48 (female); Srisuka et al., 2015a: 504–509 (male, pupa, and larva).

Distribution. Thailand.

Bionomics. Aquatic habitat: A small stream (bottom rocky, flow slow, 27.6 °C, exposed to the sun, elevation 468 m) in Maewa National Park, Lampang Province (Srisuka et al., 2015a).

Remarks. *Simulium wanchaii* was described from females collected by a hand net while attracted to a human in Doi Pui National Park, northern Thailand (Takaoka and Choochote, 2006a). The male, pupa, and mature larva of this species were later described by Srisuka et al. (2015a).

It is characterized in the female by the cercus 1.2 times as long as its basal width; in the male by the hind basitarsus enlarged and spindle-shaped, slightly narrower than the hind tibia and abdominal segments 5–7 lacking a pair of shiny dorsolateral patches; and in the pupa by the gill with 19 filaments (Fig. 6A).

Subgenus Boophthora *Enderlein*

Simulium (Boophthora) guiyangense Chen, Liu and Yang, 2016

Simulium (*Boophthora*) *guiyangense* Chen, Liu and Yang [in Chen, 2016]: 148–152 (female, male, pupa, and larva).

Distribution. China (Guizhou).

Bionomics. Aquatic habitats: Small streams (width 0.3–1.0 m, depth 0.1–0.4 m, flow slow, 24 °C, shaded, elevation 1251 m) (Chen, 2016).

Remarks. *Simulium guiyangense* was described from Guizhou Province, China by Chen (2016). This is the only species of the subgenus *Boophthora* recorded in the Oriental Region.

It is characterized in the female by the sensory vesicle 0.48–0.52 times as long as the third palpal segment and cibarium smooth, without any processes; in the male by the number of upper-eye (large) facets in 17 vertical columns and 17 horizontal rows and style with 12–15 subapical spines; in the pupa by the gill with four to six filaments, abdominal segments 7–9 each with spine-combs dorsally and terminal hooks absent; and in the larva by the postgenal cleft 2.0–2.5 times as long as the postgenal bridge.

In the original description (Chen, 2016), the presence or absence of the ventral papillae on the last larval abdominal segment was not mentioned, though other species of this subgenus have ventral papillae widely separated from each other. In addition, the basal portion of the radius was noted to have a tuft of brown hairs, but it is confirmed that the basal portion of the radius is bare by the observation of the type specimen by Prof. M. Yang, Guizhou Medical University (Personal communication).

Subgenus Byssodon *Enderlein*

***Simulium* (*Byssodon*) *languidum* Davies and Györkös, 1988**
Simulium (*Byssodon*) *languidum* Davies and Györkös, 1988a: 2150–2150 (female, male, pupa, and larva).

Distribution. Sri Lanka.

Bionomics. Aquatic habitats: Two slow-flowing streams (width 0.9–6.0 m, 25.0–26.5 °C) (Davies and Györkös, 1988a).

Remarks. *Simulium languidum* was described from Sri Lanka by Davies and Györkös (1988a). This is the only species in the subgenus *Byssodon* in the Oriental Region.

It is characterized in the female by the scutum dark brown, having three blackish brown longitudinal vittae, basal section of the radius and subcosta bare, claw with a large basal tooth 0.4 times as long as the claw, ovipositor valve triangular, genital fork with two projections, one directed forward and the other directed posteromedially, on each arm and paraproct not produced ventrally beyond the tip of the cercus; in the male by the number of upper-eye (large) facets in 13 vertical columns and 18 horizontal rows, fore and hind basitarsi 7.0 and 5.6 times as long as their greatest width, respectively, and ventral plate with its body broad and not so produced ventrally; in the pupa by the gill with three short filaments, abdomen lacking spine-combs dorsally and cocoon shoe-shaped; and in the larva by the postgenal cleft reaching the posterior margin of the hypostoma and rectal organ simple, without secondary lobules.

In the original description (Davies and Györkös, 1988a), a pair of hooks were noted to be present on sterna 6–8 on each side of the pupal abdomen. "Sterna 6–8" may be mistaken, being corrected as "sterna 5–7."

Subgenus **Daviesellum** *Takaoka and Adler*

Simulium (Daviesellum) courtneyi Takaoka and Adler, 1997

(Figs. 8D, H, M–Q and 9G, N–P)

Simulium (*Daviesellum*) *courtneyi* Takaoka and Adler, 1997: 24–26 (female, male, pupa, and larva).

Distribution. Thailand.

Bionomics. Aquatic habitats: Two shaded streams (width 5–8 m, bottom rocky, 16–20 °C, elevation 730 m) flowing over a series of cascades in Monthathan Waterfall, Chiang Mai Province (Takaoka and Adler, 1997).

Remarks. This species was described from Thailand by Takaoka and Adler (1997).

It is characterized in the female by the genital fork with a projection directed anteriorly on each arm (Fig. 8D), paraproct slightly longer than wide and spermatheca globular (Fig. 8H); in the male by the number of upper-eye (large) facets in 16 vertical columns and 17 horizontal rows and ventral plate with its body rectangular, much longer than wide when viewed ventrally (Fig. 8M); in the pupa by the frons and face densely covered with tubercles and gill with six short filaments subequal in length and thickness to one another and their maximum diameter slightly thinner than that of the interspiracular trunk; and in the larva by the body 9.0–9.5 mm long and dark grayish brown, hypostoma with nine bristles on each side, postgenal cleft not reaching the posterior margin of the hypostoma (Fig. 9O), rectal organ compound, with eight or nine finger-like secondary lobules per lobe and posterior circlet with ca. 420 rows of up to 44 hooklets per row.

Simulium (Daviesellum) pahangense Takaoka and Davies, 1995

(Figs. 8A–C, E–G, I–L and 9A–F, H–M, Q)

Simulium (*Simulium*) *pahangense* Takaoka and Davies, 1995a: 150–151 (female).

Sulcicnephia unidens Takaoka and Davies, 1995a: 10–13 (pupa and larva). Synonymized by Takaoka and Adler, 1997.

Simulium (*Daviesellum*) *pahangense*: Takaoka and Adler, 1997: 19–24 (female, male, pupa, and larva).

Distribution. Malaysia (Peninsular) and Thailand.

Bionomics. Aquatic habitats: A stream (width 3–4 m, 21.0 °C, exposed to the sun, elevation ca. 1000 m) at Fraser's Hill (type locality of this species) in Peninsular Malaysia; a rock surface of a cascading stream (width 2–5 m, 26.0 °C, elevation ca. 230 m) in a lowland in Selangor, Peninsular Malaysia (Takaoka and Adler,

1997); well-shaded waterfalls and torrential areas of streams (18–20 °C) flowing over rock, in Thailand.

The guts of six larvae of this species collected from Phangna and Kanchanaburi Provinces in Thailand were infected with the trichomycete fungus, *Harpella melusinae* Legeri and Duboscq (Takaoka and Adler, 1997).

Remarks. This species was described from a single female captured by a sweeping net near a stream in Peninsular Malaysia by Takaoka and Davies (1995a). The pupa and larva, which were thought as *S.* (*S.*) *pahangense* at that time, were in fact those of *S.* (*S.*) *yongi* Takaoka and Davies (Takaoka and Davies, 1997), and the pupa and larva described as *Sulcicnephia unidens* by Takaoka and Davies (1995a) were found to be the same species as *S.* (*D.*) *pahangense* (Takaoka and Adler, 1997).

It is similar to *S. courtneyi* but is distinguished by the following characteristics: in the female by the genital fork without a projection directed anteriorly on each arm, paraproct nearly as long as wide and spermatheca ovoid; in the male by the number of upper-eye (large) facets in 18 vertical columns and 19 horizontal rows and ventral plate with its body wide, rounded posteriorly, and much shorter than wide when viewed ventrally (Fig. 8J); in the pupa by the frons almost bare (Fig. 9A) and gill with six short filaments subequal in length and thickness to one another and their maximum diameter as thick as or slightly thicker than that of the interspiracular trunk (Fig. 9B); and in the larva by the body ca. 6.5 mm long and reddish brown, hypostoma with 11 or 12 bristles on each side (Fig. 9L), postgenal cleft reaching the posterior margin of the hypostoma (Fig. 9M), rectal organ compound, with 18–25 finger-like secondary lobules per lobe and posterior circlet with ca. 460 rows of up to 53 hooklets per row (Fig. 9Q).

Subgenus Eusimulium *Roubaud*

Simulium (*Eusimulium*) *jomsomense* Takaoka and shrestha, 2010

(Fig. 10A–Q)

Simulium (*Eusimulium*) *jomsomense* Takaoka and Shrestha, 2010: 8–15 (female, male, and pupa).

Distribution. Nepal.

Bionomics. Aquatic habitat: A stream (flow moderate, 10 °C, exposed to the sun, elevation 2745 m), Mustang (Takaoka and Shrestha, 2010).

Remarks. This species was described from Nepal by Takaoka and Shrestha (2010). Its larva is unknown.

It is characterized in the female by the sensory vesicle 0.50–0.55 times as long as the third palpal segment (Fig. 10A), postnotum with fine hairs and genital fork without a projection directed forward on each arm (Fig. 10D); and in the male by the number of upper-eye (large) facets in 18 vertical columns and 20 horizontal rows and hind basitarsus entirely medium brown, somewhat enlarged, spindle-shaped, 4.8 times as long as its greatest width and 0.80 and 0.83 times as wide as the hind tibia and femur, respectively (Fig. 10H).

Simulium (_Eusimulium_) _taipei_ (Shiraki, 1935)
Eusimulium taipei Shiraki, 1935: 15–18 (female).
Simulium (*Eusimulium*) *taipei*: Takaoka, 1979: 384.

Distribution. Taiwan.

Bionomics. A single female was captured while biting Dr. Shiraki in his laboratory of the Taipei University (Shiraki, 1935). Aquatic habitats are unknown.

Remarks. This species was described from a single adult female captured in Taipei, Taiwan by Shiraki (1935). Its male, pupa and larva are unknown.

It is characterized in the female by the back of the head covered with yellow hairs intermixed with a few dark hairs, scutum reddish brown, with three dark longitudinal vittae, postnotum with hairs (a diagnostic characteristic of the subgenus *Eusimulium*), and hind femur and tibia yellow except their apical one-fourth and one-third darkened, respectively.

Simulium (_Eusimulium_) _weiningense_ Chen and Zhang, 1997
Simulium (*Eusimulium*) *weiningense* Chen and Zhang, 1997b: 303–306 (female, male, pupa, and larva).

Distribution. China (Guizhou).

Bionomics. Aquatic habitat: A stream (20 °C) at an elevation of 2250 m (Chen and Zhang 1997b).

Remarks. This species was described from Guizhou, China by Chen and Zhang (1997b).

It is characterized in the female by the postnotum lacking fine hairs (departing from the diagnosis of the subgenus *Eusimulium*) and genital fork without a projection directed forward on each arm; and in the larva by the labral fan with 40–42 primary rays and posterior circlet with ca. 80 rows of 10–12 hooklets per row.

The accessory sclerites were noted to be present but these are in fact rectal scales according to Fig. 16 in Chen and Zhang (1997b).

Simulium (_Eusimulium_) sp. D Zhang, Wen and Chen, 1999
Simulium (*Eusimulium*) sp. D Zhang, Wen and Chen, 1999: 233.

Distribution. China (Guizhou).

Bionomics. Unknown.

Remarks. This unnamed species was reported from Guizhou, China by Zhang et al. (1999). However, its morphological characteristics were not noted. This species was not treated in the monographs on Chinese black flies by Chen and An (2003) and Chen (2016).

Subgenus Gomphostilbia *Enderlein*

Simulium (*Gomphostilbia*) *alienigenum* Species-group

Simulium (*Gomphostilbia*) *alienigenum* Takaoka, 1983

(Fig. 11C)

Simulium (*Morops*) *alienigenum* Takaoka, 1983: 55–58 (female, male, pupa, and larva).
Simulium (*Gomphostilbia*) *alienigenum* Takaoka 2003: 546.

Distribution. Philippines (Palawan).

Bionomics. Aquatic habitats: Small to medium-sized streams in lowlands (Takaoka, 1983).

Remarks. This species was described under the subgenus *Morops* from Palawan, the Philippines (Takaoka, 1983). It was later transferred to the subgenus *Gomphostilbia* (Takaoka, 2003).

It is characterized in the female and male by having the antenna yellow except flagellomeres 1 and 5–9 brownish black (similar to Fig. 11A), ca. 10 hairs on the upper mesopleuron region and no or a few hairs also on the pleural membrane; in the female by the sensory vesicle 0.25 times as long as the third palpal segment and claw with a small basal tooth (Fig. 11C); in the pupa by the gill with four long slender filaments arising from a short common basal stalk; and in the larva by the head spots on the cephalic apotome, of which at least posterolateral spots and a posterior spot of mediolongitudinal spots are somewhat darkened.

Simulium (*Gomphostilbia*) *kalimantanense* Takaoka and Sofian-Azirun, 2016

(Fig. 11A, B, D–S)

Simulium (*Gomphostilbia*) *kalimantanense* Takaoka and Sofian-Azirun [in Takaoka et al., 2016b]: 798–804 (female, male, pupa, and larva).

Distribution. Indonesia (Kalimantan) and Malaysia (Sabah).

Bionomics. Aquatic habitats: Small streams (width 1–2 m, depth 5–11 cm, 24–25 °C, partially shaded, elevation 62–230 m) slow flowing in natural forests in Longikis, Kalimantan (Takaoka et al., 2016b).

Remarks. *Simulium kalimantanense* was described from Kalimantan, Indonesia (Takaoka et al., 2016b). This species was also recorded from Sabah, Malaysia (Ya'cob et al., 2020).

It has close similarities to *S. alienigenum* from Palawan, the Philippines (Takaoka, 1983) including the distinct color pattern of the antenna of the female and male (Fig. 11A) but is distinguished by the following characteristics: in the female and male by lacking hairs on the pleural membrane or its vicinity; in the female by the much narrower frons (frons:head ratio 1.0:7.4 in this species versus 1.0:5.7 in *S. alienigenum*), sensory vesicle medium-long (0.38 times as long as third palpal segment) and claw with a long basal tooth 0.4 times as long as the claw (Fig. 11D); in the pupa by the gill with four long filaments arising from the elongate common basal stalk (Fig. 11Q); and in the larva by the indistinct or negative head spots on the cephalic apotome.

Simulium (Gomphostilbia) ambigens Species-group

Simulium (Gomphostilbia) ambigens Delfinado, 1969

(Fig. 12A–J)

Simulium (*Eusimulium*) *ambigens* Delfinado, 1969: 206 (male).
Simulium (*Gomphostilbia*) *ambigens*: Crosskey, 1973: 424; Takaoka, 1983: 83–85 (female, male, pupa, and larva).

Distribution. Philippines (Mindanao).

Bionomics. Aquatic habitats: Slow-flowing streams (2–4 m wide) in lowland areas in Davao, Mindanao (Takaoka, 1983).

Remarks. This species was described from a single male collected from Mindanao, the Philippines (Delfinado, 1969). Its female, pupa and larva were described by Takaoka (1983).

This species is distinctive among species of the subgenus *Gomphostilbia* in having a dark-pigmented neck of the spermatheca (Fig. 12B) in the female and a median keel on the ventral surface of the ventral plate (Fig. 12H) in the male.

Simulium (Gomphostilbia) asakoae Species-group

Simulium (Gomphostilbia) asakoae Takaoka and Davies, 1995

(Figs. 13B, E, F, I, U, 14B, K, P and 15D)

Simulium (*Gomphostilbia*) *asakoae* Takaoka and Davies, 1995a: 55–60 (female, male, pupa, and larva); Takaoka et al., 2017c: 15–23 (female, male, pupa, and larva).
Simulium (*Gomphostilbia*) *metatarsale* (*nec* Brunetti): Takaoka, 1979: 384–386 (female, male, pupa, and larva).
Simulium (*Gomphostilbia*) sp. C Takaoka and Suzuki, 1984: 21–22 (male and pupa).

Distribution. China (Hong Kong), Malaysia (Peninsular), Taiwan, Thailand, and Vietnam.

Bionomics. Aquatic habitats: Small streams at high elevations (1500–1570 m) and a stream at a medium elevation (536 m) in Cameron's Highlands, Peninsular Malaysia (Takaoka and Davies, 1995a).

This species is a human biter (Choochote et al., 2005) and a vector of unknown filaria in Thailand (Fukuda et al., 2003; Ishii et al., 2008). However, a reconfirmation of the vector species is needed because there are more than 25 member species of the *S. asakoae* species-group, of which females are morphologically similar to one another (Takaoka et al., 2020h).

Remarks. This species was described from Peninsular Malaysia (Takaoka and Davies, 1995a). It was recorded from Thailand (Takaoka and Saito, 1996), Hong Kong (Takaoka et al., 1995), and Vietnam (Pham, 1998). The species previously regarded as *S. metatarsale* Brunetti in Taiwan (Takaoka, 1979) was morphologically and molecularly identified as *S. asakoae* (Takaoka et al., 2021b). The redescription of this species was given based on specimens from Vietnam (Takaoka et al., 2017c).

It is characterized in the female by the medium-long sensory vesicle (Fig. 13B), mandible with several teeth on the outer margin (Fig. 13E) and hind tibia yellowish on its basal two-thirds; in the male by the bright vermilion upper-eye (large) facets, subcosta with hairs, hind basitarsus as wide as or slightly wider than the hind tibia and ventral plate with its ventral margin nearly straight when viewed posteriorly (Fig. 13U); in the pupa by abdominal segments 1–3 (rarely 1–4) each darkened on the dorsal surface; and in the larva by the medium-sized postgenal cleft (Fig. 14P) and unicolored abdominal segments without dark transverse bands, though reddish brown pigments present on the dorsal surface of abdominal segments 5 and 6 (Fig. 15D).

***Simulium* (*Gomphostilbia*) *banluangense* Takaoka, Srisuka and Fukuda, 2020**

Simulium (*Gomphostilbia*) *banluangense* Takaoka, Srisuka and Fukuda [in Takaoka et al., 2020h]: 110–112 (male and pupa).

Distribution. Thailand.

Bionomics. Aquatic habitat: A stream (width 1.0 m, 25 °C, exposed to the sun, elevation 804 m) at Ban Luang, Chiang Mai Province (Takaoka et al., 2020h).

Remarks. This species was described from Thailand (Takaoka et al., 2020h). Its female and larva are unknown.

It is characterized in the male by the number of upper-eye (large) facets in 14 vertical columns and 14 or 15 horizontal rows on each side and hind basitarsus spindle-shaped; and in the pupa by the common basal stalk of the gill short, the stalk of the ventral pair of gill filaments thicker than the interspiracular trunk and terminal hooks triangular.

Simulium doisaketense described based on the pupal and larval specimens from Thailand (Jitklang et al., 2008) has the short common basal stalk of the gill, like this species, but the stalk of the ventral pair of filaments is medium-long, 1.3 times as long as the common basal stalk and thinner than the interspiracular trunk, based on my observation of the type specimen (Unpublished data).

***Simulium* (*Gomphostilbia*) *brevilabrum* Takaoka, Hadi and Sigit, 2006**

Simulium (*Gomphostilbia*) *brevilabrum* Takaoka, Hadi and Sigit, 2006a: 8–13 (female, male, and pupa).

Distribution. Indonesia (Flores).

Bionomics. Aquatic habitat: A slow-flowing small stream in Flores (Takaoka et al., 2006a).

Remarks. *Simulium brevilabrum* was described from Flores, Indonesia (Takaoka et al., 2006a). Its larva is unknown. This species, which was once placed in the *S. ceylonicum* species-group by Takaoka (2012), is here transferred to the *S. asakoae* species-group by having the base of the female and male radius with a tuft of yellow hairs (though their basal portions are darkened), male hind tibia yellowish white on its basal two-fifths or little more, and ventral plate with its body nearly parallel-sided when viewed ventrally.

It is characterized in the female by the narrow frons (frons:head ratio 1:6), medium-long sensory vesicle (0.32 times as long as the third palpal segment), and short labrum (0.54 times as long as the clypeus); in the male by the fore basitarsus 7.0 times as long as its greatest width and hind basitarsus yellow on its basal two-fifths and a little more; and in the pupa by the gill with four slender filaments.

Simulium (Gomphostilbia) brinchangense Takaoka, Sofian-Azirun and Hashim, 2014

Simulium (*Gomphostilbia*) *brinchangense* Takaoka, Sofian-Azirun and Hashim [in Takaoka et al., 2014f]: 55–61 (female, male, pupa, and larva).

Distribution. Malaysia (Peninsular).

Bionomics. Aquatic habitat: A small stream at an elevation of 1800 m in Cameron's Highlands (Takaoka et al., 2014f).

Remarks. This species was described from Peninsular Malaysia (Takaoka et al., 2014f).

It is similar to *S. asakoae* but is distinguished from the latter species in the male by the brown upper-eye (large) facets; in the pupa by the dorsum of abdominal segments 1 and 2 not darkened; and in the larva by the head capsule not darkened around the postgenal cleft.

Simulium (Gomphostilbia) chaowaense Takaoka, Srisuka and Saeung, 2020

(Fig. 15G)

Simulium (*Gomphostilbia*) *chaowaense* Takaoka, Srisuka and Saeung [in Takaoka et al., 2020h]: 103–106 (male, pupa, and larva).

Distribution. Thailand.

Bionomics. Aquatic habitat: a small stream (width 40 cm, depth 3.5 cm, bottom sandy, flow moderate, pH 7.3, 21.9 °C, partially shaded, elevation 582 m), Chao Wa Waterfall, Song District, Phrae Province (Takaoka et al., 2020h).

Remarks. This species was described from Thailand (Takaoka et al., 2020h). Its female is unknown.

It is characterized in the male by the number of upper-eye (large) facets in 14 or 15 vertical columns and 14 or 15 horizontal rows and hind basitarsus 0.9 times as wide as the hind femur; and in the larva by the postgenal cleft medium-long, 1.2–1.3 times as long as the postgenal bridge and all abdominal segments grayish (Fig. 15G).

This species is similar to *S. tanahrataense* described from males and their associated pupal exuviae from Peninsular Malaysia (Takaoka et al., 2014f) in many characteristics including the number of male upper-eye (large) facets. However, it is barely distinguished from the latter species in the male by the first flagellomere of the antenna 1.8 times as long as the second (2.1 times in *S. tanahrataense*), hind basitarsus 0.9 times as wide as the hind femur (1.0 time in *S. tanahrataense*) and ventral plate with its ventral margin rounded when viewed posteriorly (straight or slightly concave in *S. tanahrataense*). The pupa of this species is almost indistinguishable from that of *S. tanahrataense*, although there is a slight difference in the length of filaments of the dorsal and middle triplets including their stalks and the common basal stalk (2.5–2.6 mm in this species versus 1.6–2.2 mm in *S. tanahrataense*).

***Simulium* (*Gomphostilbia*) *chaudinhense* Takaoka and Sofian-Azirun, 2017**
(Fig. 13C)

Simulium (*Gomphostilbia*) *chaudinhense* Takaoka and Sofian-Azirun [in Takaoka et al., 2017c]: 23–26 (female, male, pupa, and larva).

Distribution. Vietnam.

Bionomics. Aquatic habitat: A small stream (width 1 m, depth 5–10 cm, bottom sandy, 18.0 °C, shaded, elevation of 80 m) in Nghe An Province, northern Vietnam (Takaoka et al., 2017c).

Remarks. This species was described from Vietnam (Takaoka et al., 2017c).

It is similar to *S. asakoae* but is distinguished in the female by the short sensory vesicle (Fig. 13C) and the longer claw tooth; in the male by the smaller number of upper-eye (large) facets in 10 (rarely 9) vertical columns and 12 horizontal rows; and in the larva by abdominal segments 1 and 2 each encircled with a grayish band.

***Simulium* (*Gomphostilbia*) *chiangdaoense* Takaoka and Srisuka, 2009**
(Figs. 14H, N and 15R)

Simulium (*Gomphostilbia*) *chiangdaoense* Takaoka and Srisuka, 2009: 269–276 (female, male, and pupa).

Distribution. Thailand.

Bionomics. Aquatic habitats: A small stream (width 0.2–1.0 m, shaded, elevation 1560 m) in a forest, Doi Chiang Dao, and a stream (width 1.2 m, shaded, elevation 1667 m), Doi Inthanon National Parks, Chiang Mai Province (Takaoka and Srisuka, 2009).

Remarks. This species was described from Thailand (Takaoka and Srisuka, 2009). Its larva is unknown.

It is characterized in the female by the mandible with its outer margin serrated and hind tibia whitish on its basal half or little more; in the male by the subcosta haired and ventral plate with its ventral margin rounded when viewed posteriorly; and in the pupa by the cocoon with an elongate anterodorsal projection much extended forward beyond the anteroventral tips of the cocoon (Fig. 14N). Srisuka et al. (2022a) reported that this species is genetically diverse, with five lineages, and is morphologically variable in the number of male upper-eye (large) facets, which is in 13 to 17 vertical columns and 14 to 17 horizontal rows on each side.

Simulium (Gomphostilbia) confertum Takaoka and Sofian-Azirun, 2015
Simulium (*Gomphostilbia*) *confertum* Takaoka and Sofian-Azirun [in Takaoka et al., 2015c]: 7–13 (female, male, and pupa).

Distribution. Vietnam.

Bionomics. Aquatic habitat: A small stream (width 1–3 m, exposed to the sun, elevation 1492 m), Lam Dong Province (Takaoka et al., 2015c).

Remarks. This species was described from Vietnam (Takaoka et al., 2015c). The larva of this species is unknown.

It is characterized in the female by the mandible with a few teeth on its outer margin and hind tibia whitish on its basal half; in the male by the number of upper-eye (large) facets in 14 vertical columns and 15 horizontal rows; and in the pupa by the stalk of the dorsal triplet of the gill filaments at an angle of 45–60° against the stalk of the ventral pair when viewed laterally.

Simulium (Gomphostilbia) epauletum Guo, Zhang and An, 2011
Simulium (*Gomphostilbia*) *epauletum* Guo, Zhang and An, 2011: 87–92 (female, male, pupa, and larva).

Distribution. China (Hainan).

Bionomics. Aquatic habitat: A small shaded stream at an elevation of 830 m (Guo et al., 2011).

Remarks. *Simulium epauletum* was described from Hainan, China by Guo et al. (2011). Here this species is placed in the *S. asakoae* species group since it has the brownish yellow fore coxae in the female and male, male hind basitarsus enlarged and ventral plate with its lateral margins slightly concave medially when viewed ventrally, and larval postgenal cleft medium-long (twice as long as the postgenal bridge), though the color of a tuft of hairs at the base of the radius was not mentioned in the original description (Guo et al., 2011).

It is characterized in the female by the mandible with 21 inner and 35 outer teeth (probably, the mandible was wrongly observed, since it is unusual that the outer teeth are greater in number than the inner teeth) and sensory vesicle medium-long, 0.3 times as long as the third palpal segment; in the male by the number of upper-eye (large) facets in 14 vertical columns and 12 horizontal rows and scutum with three dark longitudinal vittae; and in the pupa by the gill with eight filaments arranged as 3+3+2 from dorsal to ventral, of which two filaments of the ventral pair are much longer than the six other filaments (3.4 mm versus 2.2–2.6 mm), terminal hooks irregularly triangular when viewed caudally and cocoon simple, wall-pocket-shaped (Guo et al., 2011).

Simulium (Gomphostilbia) fanjingshanense Chen, Zhang and Wen, 2000
Simulium (*Gomphostilbia*) *fanjingshanense* Chen, Zhang and Wen, 2000: 21–24 (female, male, pupa, and larva).

Distribution. China (Guizhou).

Bionomics. Aquatic habitat: A small stream at an elevation of 600 m (Chen et al., 2000).

Remarks. *Simulium fanjingshanense* was described from Guizhou by Chen et al. (2000). This species was once placed in the *S. ceylonicum* species-group by Takaoka (2012). However, here it is transferred to the *S. asakoae* species group since it has the base of the radius with a tuft of brownish yellow hairs and ventral plate with its lateral margins slightly concave or nearly straight when viewed ventrally and larval postgenal cleft nearly as long as the postgenal bridge.

It is characterized in the female by the sensory vesicle short, 0.25 times as long as the third palpal segment and mandible with ca. 14 outer teeth; in the male by the number of upper-eye (large) facets in 15 vertical columns and 12 horizontal rows and hind basitarsus enlarged, spindle-shaped, and ca. 4.8 times as long as its greatest width; and in the pupa by the gill with eight slender filaments arranged as [(2+1) +(1+2)]+2 from dorsal to ventral and cocoon wall-pocket-shaped, with a short anterodorsal projection.

***Simulium* (*Gomphostilbia*) *fuscidorsum* Takaoka and Ya'cob, 2015**
(Fig. 14C)

Simulium (*Gomphostilbia*) *fuscidorsum* Takaoka and Ya'cob [in Takaoka et al., 2015c]: 13–17 (female, pupa, and larva).

Distribution. Vietnam.

Bionomics. Aquatic habitat: A small stream (width 0.4 m, partially shaded, elevation 1722 m) slowly flowing in a swampy area, Lam Dong Province (Takaoka et al., 2015c).

Remarks. This species was described from Vietnam (Takaoka et al., 2015c). The male of this species remains unknown.

It is characterized in the female by the sensory vesicle long, 0.57 times as long as the third palpal segment; and in the pupa by the gill with an elongate common basal stalk and elongate stalk of the ventral pair of the gill (Fig. 14C), abdominal segments 1–5 darkened on the dorsal surface and cocoon with a medium-long anterodorsal projection; and in the larva by the postgenal cleft medium-long, 1.2 times as long as the postgenal bridge.

***Simulium* (*Gomphostilbia*) *gyorkosae* Takaoka and Davies, 1996**

Simulium (*Eusimulium*) *metatarsale* var. Edwards, 1934: 119–121 (female, male, and pupa).

Simulium (*Gomphostilbia*) *gyorkosae* Takaoka and Davies, 1996: 28–32 (female, male, pupa, and larva); Takaoka et al., 2000: 161; Takaoka, 2003: 178–179 (male and pupa).

Distribution. Indonesia [Java, Sumatra and Sulawesi (?)].

Bionomics. Aquatic habitats: Small streams (0.2–0.6 m wide) in forests and in open fields, at an elevation of 500–1400 m, in Puncak and Cibodas, West Java (Takaoka and Davies, 1996).

Remarks. This species was first recorded from Java, Indonesia, as an unnamed variety of *S.* (*Eusimulium*) *metatarsale* Brunetti, 1911 from India (Edwards, 1934). It was later described as a new species, *S. gyorkosae* by Takaoka and Davies (1996). This species was recorded from Sumatra (Takaoka et al., 2000) and Sulawesi

(Takaoka, 2003). The record of this species from Sulawesi was based on a male reared from a pupa (Takaoka, 2003). However, it should be reevaluated since there is a possibility that the male collected from Sulawesi is different from that of *S. gyorkosae* because this male has a tuft of dark hairs at the base of the radial vein in place of a tuft of yellow hairs in the male of *S. gyorkosae* from Java (Takaoka and Davies, 1996).

It is characterized in the female by the mandible with a few outer teeth at some distance from the tip and hind tibia yellow on its basal two-thirds; and in the pupa by the anterior margin of the cocoon resembling the letter W when viewed dorsally.

***Simulium* (*Gomphostilbia*) *hmongense* Takaoka and Ya'cob, 2017**

(Fig. 14G, M)

Simulium (*Gomphostilbia*) *hmongense* Takaoka and Ya'cob [in Takaoka et al., 2017c]: 20–29 (male, pupa, and larva).

Distribution. Vietnam.

Bionomics. Aquatic habitats: A small stream (width 0.2 m, depth 2 cm, flow slow, bottom rocky, 7.0 °C, partially shaded, elevation 1983 m) and a small stream (width 0.2 m, depth 5 cm, flow slow, bottom sandy and muddy, 8.0 °C, exposed to the sun, elevation 1925 m) in Lai Chau, Sapa, Lao Cai Province (Takaoka et al., 2017c).

Remarks. This species was described from Vietnam (Takaoka et al., 2017c). Its female is unknown.

It is characterized in the male by the greater number of upper-eye (large) facets in 17 vertical and 17 horizontal rows; in the pupa by the thoracic integument sparsely covered with tubercles, middle triplet of the gill with a long primary stalk, terminal hooks small and narrow (Fig. 14G), and cocoon with a medium-long anterodorsal projection (Fig. 14M); and in the larva by the postgenal cleft short, 0.7 times as long as the postgenal bridge.

***Simulium* (*Gomphostilbia*) *hoiseni* Takaoka, 2008**

(Fig. 14F)

Simulium (*Gomphostilbia*) *hoiseni* Takaoka, 2008a: 9–14 (male and pupa).

Distribution. Malaysia (Peninsular).

Bionomics. This is a rare species being found once only from the type stream (elevation 1570 m) in Cameron's Highlands (Takaoka, 2008a).

Remarks. This species was described from a male reared from a pupa collected from Cameron's Highlands (Takaoka, 2008a). Its female and larva remain unknown.

It is characterized in the male by abdominal segments 2–8 each with a pair of shiny dorsolateral or lateral patches; and in the pupa by the gill with six filaments (Fig. 14F).

***Simulium* (*Gomphostilbia*) *hongthaii* Takaoka, Sofian-Azirun and Ya'cob, 2014**

Simulium (*Gomphostilbia*) *hongthaii* Takaoka, Sofian-Azirun and Ya'cob [in Takaoka et al., 2014d]: 348–351 (male and pupa); Takaoka et al., 2017c: 30–33 (female and larva).

Distribution. Vietnam.

Bionomics. Aquatic habitats: A small stream (width 0.5 m, shaded, elevation ca. 1000 m) in Vinh Phuc Province, and a stream (width 1.5 m, depth 10 cm, 13.0 °C, partially shaded, elevation 1105 m) in Lao Cai Province (Takaoka et al., 2014d, 2017c).

Remarks. This species was described from Vietnam by Takaoka et al. (2014d). Its female and larva were subsequently described by Takaoka et al. (2017c).

It is characterized in the male by the greater number of upper-eye (large) facets in 16 vertical columns and 16 horizontal rows and ventral plate with its body trapezoid when viewed posteriorly.

***Simulium* (*Gomphostilbia*) *huaimorense* Takaoka, Srisuka and Saeung, 2020**
(Figs. 13O and 15I)

Simulium (*Gomphostilbia*) *huaimorense* Takaoka, Srisuka and Saeung [in Takaoka et al., 2020h]: 119–123 (female, male, pupa, and larva).

Distribution. Thailand.

Bionomics. Aquatic habitats: A stream (width 1.2 m, depth 10 cm, bottom sandy, flow moderate, pH 6.6, 17.1 °C, exposed to the sun, elevation 1154 m) at a coffee plantation, Huai Mor, Chiang Mai Province, and a small stream (width 30 cm, depth 2 cm, bed sandy, moderate flow, pH 7.3, 19.8 °C, partially shaded, elevation 1440 m) at a coffee plantation, Pha Lung Village, Chiang Rai Province (Takaoka et al., 2020h).

Remarks. This species was described from Thailand (Takaoka et al., 2020h).

It is similar to *S. myanmarense* from Myanmar (Takaoka et al., 2017g) but is distinguished in the male by the number of upper-eye (large) facets in 16 vertical columns and 17 horizontal rows (14 or 15 vertical columns and 15 or 16 horizontal rows in *S. myanmarense*) and relative length of the hind basitarsus against its greatest width (Fig. 13O), which is 3.6–4.0 in this species versus 4.3–4.4 in *S. myanmarense*; in the pupa by the extremely short primary stalks of the dorsal and middle triplets of gill filaments, which are 0.1–0.3 and 0.6–0.7 times as long as their common stalk (the dorsal and middle primary stalks are nearly as long as their common stalk in *S. myanmarense*); and in the larva by abdominal segments 1–4 gray (Fig. 15I) (abdominal segments 1 and 2 gray in *S. myanmarense*).

This species is distinguished from *S. doisaketense* from Doi Saket, Chiang Mai Province (Jitklang et al., 2008), by the primary stalks of the dorsal and middle triplets of the pupal gill filaments much shorter than their common stalk (much longer in *S. doisaketense*).

***Simulium* (*Gomphostilbia*) *inthanonense* Takaoka and Suzuki, 1984**
(Figs. 14A, O and 15S, T)

Simulium (*Gomphostilbia*) *inthanonense* Takaoka and Suzuki, 1984: 18–21 (female, male, pupa, and larva); Takaoka et al., 2020h: 135–140.

Distribution. China (Yunnan), Thailand, and Vietnam (?).

Bionomics. Aquatic habitat: A stream at an elevation of ca. 1700 m in Doi Inthanon National Parks (Takaoka and Suzuki, 1984).

Remarks. This species was described from Thailand by Takaoka and Suzuki (1984). The revised description was given by Takaoka et al. (2020h)

It was recorded from Vietnam (Pham, 1998, 1999). There remains a possibility that the species regarded as *S. inthanonense* in Vietnam was either *S. phulocense* or *S. unii* or other related species, because its previous identification was based on only two pupae and three larvae (Pham, 1998, 1999). These three and a few other related species have a similar cocoon with an anterodorsal projection and a similar short larval postgenal cleft.

Takaoka et al. (2020h) transferred this species from the *S. ceylonicum* species-group to the *S. asakoae* species-group, based on the observation on the male ventral plate emarginated on both sides when viewed ventrally, and analysis of the COI gene sequences. This species is exceptional in the *S. asakoae* species-group in having the dark tuft hairs on the base of the radial vein of the female and the male, departing from the diagnosis of the species-group (Takaoka, 2012).

It is characterized in the female by the sensory vesicle long, 0.58–0.61 times as long as the third palpal segment; in the male by the greater number of upper-eye (large) facets in 17 vertical columns and 17 (rarely 18) horizontal rows and sensory vesicle long, 0.40–0.47 times as long as the third palpal segment; in the pupa by the cocoon with a short anterodorsal projection; and in the larva by the postgenal cleft nearly as long as or little shorter than the postgenal bridge (Fig. 14O) (Takaoka et al., 2020h).

***Simulium* (*Gomphostilbia*) *izuae* Takaoka, Sofian-Azirun and Hashim, 2013**
Simulium (*Gomphostilbia*) *izuae* Takaoka, Sofian-Azirun and Hashim, 2013a: 1179–1189 (female, male, pupa, and larva).

Distribution. Malaysia (Peninsular).

Bionomics. Aquatic habitat: An irrigation channel (width 0.3–0.5 m, 17.0 °C, elevation 1250 m) moderately flowing in a cultivated field in Cameron's Highlands (Takaoka et al., 2013a).

Remarks. This species was described from Peninsular Malaysia (Takaoka et al., 2013a).

It is similar to *S. asakoae* but is distinguished in the female by the mandible without outer teeth; and in the pupa by the ventral paired filaments with a short stalk shorter than the common basal stalk and the dorsum of abdominal segments 1 and 2 not darkened.

***Simulium* (*Gomphostilbia*) *jianfengense* Long, An and Hao, 1994**
Simulium (*Gomphostilbia*) *jianfengense* Long, An and Hao [in Long et al., 1994]: 407–408 (female, male, pupa, and larva).

Distribution. China (Hainan).

Bionomics. Aquatic habitats: Streams at an elevation of 500–950 m (Long et al., 1994).

Remarks. This species was described from Hainan, China by Long et al. (1994).

It is similar to *S. asakoae* but is distinguished by the number of male upper-eye (large) facets in 15 horizontal rows and ventral plate produced ventromedially, appearing to be triangular when viewed caudally.

In the original description (Long et al., 1994), the mandible was noted to have 31 outer teeth and no inner teeth. Probably it is wrong, and it reads that the mandible has 31 inner teeth and no outer tooth.

***Simulium* (*Gomphostilbia*) *junkumae* Takaoka, Srisuka and Saeung, 2020** (Fig. 15K)

Simulium (*Gomphostilbia*) *junkumae* Takaoka, Srisuka and Saeung [in Takaoka et al., 2020h]: 113–116 (female, male, pupa, and larva).

Distribution. Thailand.

Bionomics. Aquatic habitat: A small stream (width 40 cm, depth 3 cm, bottom sandy, flow moderate, pH 6.2, 19 °C, partially shaded, elevation 1395 m) at Doi Pui Temple, Chiang Mai Province (Takaoka et al., 2020h).

Remarks. This species was described from Thailand (Takaoka et al., 2020h).

It is characterized in the female by the relatively narrower frons against the head width (1.0:4.7–5.5) and mandible without outer teeth; in the male by the greater number of upper-eye (large) facets in 15–17 vertical columns and 15–17 horizontal rows on each side; and in the larva by the small number of the primary rays of the labral fan (24 or 25), postgenal cleft as long as or little shorter than the postgenal bridge and abdominal segments 1–3 entirely gray (Fig. 15K).

The larva of *S.* nr. *asakoae* 2 reported from Thailand by Jitklang et al. (2008) has a medium-sized postgenal cleft and abdominal segments 1–3 each with a greenish transverse band, both characteristics resembling those of *S. junkumae*, but it differs by lacking distinct color markings on abdominal segments 5–8.

***Simulium* (*Gomphostilbia*) *kaohsiungense* Takaoka, Huang and Fukuda, 2021**

Simulium (*Gomphostilbia*) *kaohsiungense* Takaoka, Huang and Fukuda [in Takaoka et al., 2021a]: 2–6 (female, male, and pupa).

Distribution. Taiwan.

Bionomics. Aquatic habitat: A small stream (width 3 m, 19.0 °C, shaded, elevation 805 m), Namasia District, Kaohsiung City (Takaoka et al., 2021a).

Remarks. This species was described from Taiwan (Takaoka et al., 2021a). Its larva is unknown.

It is characterized in the female by the elongate sensory vesicle; in the male by the number of male upper-eye (large) facets in 15 vertical columns and 15 or 16 horizontal rows and hind basitarsus much widened, 1.0–1.2 times the width of the hind femur; and in the pupa by the terminal hooks widened. These characteristics combined distinguish this species from most species of the *S. asakoae* species-group.

Simulium (Gomphostilbia) kiewfinense Takaoka, Srisuka and Fukuda, 2020
Simulium (*Gomphostilbia*) *kiewfinense* Takaoka, Srisuka and Fukuda [in Takaoka et al., 2020h]: 117–119 (male and pupa).

Distribution. Thailand.

Bionomics. Aquatic habitat: A stream (width 1.1 m, depth 12 cm, bottom sandy, flow moderate, pH 6.9, 18.3 °C, partially shaded, elevation 1446 m) at Kiew Fin, Lampang Province (Takaoka et al., 2020h).

Remarks. This species was described from Thailand (Takaoka et al., 2020h). Its female and larva are unknown.

It is characterized in the male by the greater number of upper-eye (large) facets in 16 or 17 vertical columns and 16 or 17 horizontal rows on each side.

This species is similar to *S. junkumae* from Thailand (Takaoka et al., 2020h) and *S. hongthaii* from Vietnam (Takaoka et al., 2014d) in having the greater number of male upper-eye (large) facets. However, it is distinguished from both species by the relative length of the fore basitarsus against its greatest width (6.6–7.1 in this species versus 7.8–8.1 in *S. junkumae* and 7.5–8.4 in *S. hongthaii*).

Simulium (Gomphostilbia) kiewlomense Takaoka, Srisuka and Saeung, 2021
Simulium (*Gomphostilbia*) *kiewlomense* Takaoka, Srisuka and Saeung [in Srisuka et al., 2021]: 2–7 (female, male, pupa, and larva).

Distribution. Thailand.

Bionomics. Aquatic habitat: A stream (width 30 cm, depth 5 cm, flow slow, 14.0 °C, shaded, elevation 1966 m) at Kiew Lom, Chiang Mai Province (Srisuka et al., 2021).

Remarks. This species was described from Thailand (Srisuka et al., 2021). It was placed in the *S. asakoae* species-group, though the ventral plate is narrowed posteriorly when viewed ventrally, departing from the diagnosis of the *S. asakoae* species-group. It is because this species was genetically positioned in subgroup III in the *S. asakoae* species-group, and its other morphological characteristics fit to the diagnosis of the *S. asakoae* species-group (Srisuka et al., 2021).

It is characterized in the female by the elongate sensory vesicle; in the male by the hind basitarsus wider than the hind femur; in the pupa by the small triangular terminal hooks; and in the larva by the medium-long postgenal cleft and abdominal segments 1–3 light greenish.

This species has a close similarity to *S. nareshi* from Nepal (Takaoka et al., 2020f) but is distinguished in the female by the relative length of the height of the frons to its narrowest width (2.3–2.4 in this species versus 2.8 in *S. nareshi*) and relative length of the labrum to the clypeus (0.7 in this species versus 0.6 in *S. nareshi*); and in the male by the relative length of the fore basitarsus to its greatest width (9.2–9.4 in this species versus 10.2 in *S. nareshi*).

Simulium (_Gomphostilbia_) _klonglanense_ Takaoka, Srisuka and Saeung, 2020
(Fig. 15E)

Simulium (*Gomphostilbia*) *klonglanense* Takaoka, Srisuka and Saeung [in Takaoka et al., 2020h]: 126–130 (female, male, pupa, and larva).

Distribution. Thailand.

Bionomics. Aquatic habitat: A stream of Klong Nam Lai (width 1.4 m, depth 13 cm, bottom sandy, flow moderate, pH 6.2, 25.8 °C, exposed to the sun, elevation 196 m), Klong Lan District, Kham Phaeng Phet Province (Takaoka et al., 2020h).

Remarks. This species was described from Thailand (Takaoka et al., 2020h).

It is characterized in the female by the elongate sensory vesicle and mandible with four distinct outer teeth; in the male by the number of upper-eye (large) facets in 13 or 14 vertical columns and 14 or 15 horizontal rows on each side and hind basitarsus moderately widened, 0.9 times as wide as the hind tibia and femur; in the pupa by the small terminal hooks; and in the larva by the postgenal cleft long, 3.7–4.0 times as long as the postgenal bridge and abdominal segments 1–4 gray (Fig. 15E).

Simulium (_Gomphostilbia_) _lamjungense_ Takaoka and Shrestha, 2020

Simulium (*Gomphostilbia*) *lamjungense* Takaoka and Shrestha [in Takaoka et al., 2020f]: 177–183 (female, male, and pupa).

Distribution. Nepal.

Bionomics. Aquatic habitat: A stream in Lamjung (Takaoka et al., 2020f).

Remarks. This species was described from Nepal by Takaoka et al. (2020f). Its larva is unknown.

It is characterized in the female and male by the hind tibia with a faint dark subbasal marking; in the female by the sensory vesicle small, 0.18 times as long as the third palpal segment; in the male by the hind basitarsus enlarged, 3.7 times as long as its greatest width, and 1.0 and 1.3 times as wide as the hind tibia and femur, respectively; and in the pupa by the terminal hooks wide, plate-like, with their outer margins 2.7–3.0 times as long as their inner margins and crenulated.

Simulium (_Gomphostilbia_) _loeiense_ Takaoka, Srisuka and Fukuda, 2020
(Fig. 15H)

Simulium (*Gomphostilbia*) *loeiense* Takaoka, Srisuka and Fukuda [in Takaoka et al., 2020h]: 77–81 (female, male, pupa, and larva).

Distribution. Thailand.

Bionomics. Aquatic habitat: A small stream (width 40 cm, depth 2 cm, bed sandy, flow slow, pH 5.8, 19.5 °C, partially shaded, elevation 1525 m) at Khok Nok Kra Ba, Loei Province (Takaoka et al., 2020h).

Remarks. This species was described from Thailand (Takaoka et al., 2020h).

It is characterized in the female by the mandible with several outer teeth; in the male by the small number of upper-eye (large) facets in 11 vertical columns and 13 horizontal rows, subcosta with many hairs and hind basitarsus much widened, 1.2–1.3 times as wide as the hind femur; in the pupa by the dorsal surface of abdominal segments 1 and 2 with several tubercles on each side; and in the larva by the postgenal cleft 1.2–1.3 times as long as the postgenal bridge and abdominal segments 1–5, 7, and 8 gray (Fig. 15H).

Simulium (_Gomphostilbia_) _lurauense_ Takaoka, Sofian-Azirun and Hashim, 2011

(Fig. 13N)

Simulium (*Gomphostilbia*) *lurauense* Takaoka, Sofian-Azirun and Hashim [in Takaoka et al., 2011b]: 63–71 (female, male, pupa, and larva).

Distribution. Malaysia (Peninsular).

Bionomics. Aquatic habitats: Small streams at low to medium elevations (380–580 m) in Cameron's Highlands and other areas (Takaoka et al., 2011b).

Remarks. This species was described from females, males, pupae and larvae (Takaoka et al., 2011b).

It is characterized in the female by the elongate sensory vesicle; in the male by the number of upper-eye (large) facets in 14 or 15 vertical columns and 14 or 15 horizontal rows and hind basitarsus almost entirely darkened, spindle-shaped and narrower than the hind tibia (Fig. 13N); and in the pupa by the small terminal hooks.

Simulium (_Gomphostilbia_) _maehongsonense_ Takaoka, Srisuka and Saeung, 2020

(Figs. 13D and 15P)

Simulium (*Gomphostilbia*) *maehongsonense* Takaoka, Srisuka and Saeung [in Takaoka et al., 2020h]: 99–102 (female, male, pupa, and larva).

Distribution. Thailand.

Bionomics. Aquatic habitat: A stream (width 120 cm, depth 2 cm, bottom sandy, flow fast, pH 7.6, 21.2 °C, partially shade, elevation 1034 m), Pha Peung, Mae Hong Son Province (Takaoka et al., 2020h).

Remarks. This species was described from Thailand (Takaoka et al., 2020h).

It is characterized in the female by the short sensory vesicle and mandible lacking distinct outer teeth; in the male by the number of upper-eye (large) facets in 13 vertical columns and 14 horizontal rows; in the pupa by the relatively short gill filaments (longest 2.1–2.3 mm long) and relatively wide terminal hooks; and in the larva by the postgenal cleft medium-long, 1.4–2.2 times as long as the postgenal bridge and abdominal segment 1 light gray (Fig. 15P).

Simulium (_Gomphostilbia_) _maelanoiense_ Takaoka, Srisuka and Saeung, 2020

Simulium (*Gomphostilbia*) *maelanoiense* Takaoka, Srisuka and Saeung [in Takaoka et al., 2020h]: 81–85 (female, male, and pupa).

Distribution. Thailand.

Bionomics. Aquatic habitats: A small stream (width 25 cm, depth 2.5 cm, bottom sandy, flow moderate, pH 6.8, 21.9 °C, partially shaded, elevation 974 m) at Tee La Chan Waterfall, Mae Hong Son Province, and a stream (width 1.1 m, depth 12 cm, bottom sandy, flow moderate, pH 6.9, 18.3 °C, partially shaded, elevation 1446 m) at Kiew Fin, Lampang Province (Takaoka et al., 2020h).

Remarks. This species was described from Thailand (Takaoka et al., 2020h). Its larva is unknown.

It is characterized in the female by the mandible with one tooth on the outer margin; and in the male by the small number of upper-eye (large) facets in 11 vertical columns and 13 horizontal rows and hind basitarsus 1.0–1.1 times as wide as the hind femur.

This species is similar in the male and pupa to *S. sutheppuiense* from Thailand (Takaoka et al., 2020h) but is barely distinguished in the female by the number of the outer teeth of the mandible (one tooth in this species versus three teeth in *S. sutheppuiense*) and relative length of the labrum against the clypeus (0.7 in this species versus 0.6 in *S. sutheppuiense*).

***Simulium* (*Gomphostilbia*) *maewongense* Takaoka, Srisuka and Saeung, 2020** (Fig. 15M)

Simulium (*Gomphostilbia*) *maewongense* Takaoka, Srisuka and Saeung [in Takaoka et al., 2020h]: 74–77 (female, male, pupa, and larva).

Distribution. Thailand.

Bionomics. Aquatic habitats: A small stream (width 20 cm, depth 2 cm bottom sandy, flow slow, 19.5 °C, partially shaded, elevation 1322 m), Chon Yen, Mae Wong National Park, and a stream of Klong Nam Lai (width 1.4 m, depth 13 cm, bottom sandy, flow moderate, pH 6.23, 25.8 °C, exposed to the sun, elevation 196 m), Klong Lan District, Kham Phaeng Phet Province (Takaoka et al., 2020h).

Remarks. This species was described from Thailand (Takaoka et al., 2020h).

It is characterized in the female by the mandible lacking outer teeth; in the male by the number of upper-eye (large) facets in 11 vertical columns and 13 or 14 horizontal rows; in the pupa by the thorax bare except the anterior two-fifths or half and a small area of the dorsal surface near the posterior margin moderately covered with tubercles and cocoon with a short anterodorsal projection or bulge; and in the larva by abdominal segments 1 and 2 grayish (Fig. 15M) and the postgenal cleft 0.9–1.2 times as long as the postgenal bridge.

***Simulium* (*Gomphostilbia*) *mengi* Chen, Zhang and Wen, 2000**

Simulium (*Gomphostilbia*) *mengi* Chen, Zhang and Wen, 2000: 24–28 (female, male, pupa, and larva).

Distribution. China (Guizhou).

Bionomics. Aquatic habitat: A fast-flowing stream (elevation 600 m) in Fanjing mountain, Guizhou (Chen et al., 2000).

Remarks. This species was described from Guizhou, China by Chen et al. (2000).

It is characterized in the female by the mandible lacking outer teeth, sensory vesicle 0.25 times as long as the third palpal segment and claw with a long basal tooth 0.6 times as long as the claw; in the male by the small number of upper-eye (large) facets in eight vertical columns and nine horizontal rows and hind basitarsus enlarged, 2.9 times as long as its greatest width; in the pupa by the gill with eight slender filaments arranged as [(1+2)+(1+2)]+2 from dorsal to ventral; and in the larva by the postgenal cleft ca. 3.0 times as long as the postgenal bridge and abdominal segments 6–8 each with unbranched dark spinous setae dorsally.

Grapnel-like hooklets were noted to be present ventrolaterally on sternite 8 of the pupa. It may be wrong because such grapnel-like hooklets are present on each lateral surface of pupal abdominal segment 9 in most species of the subgenus *Gomphostilbia*.

***Simulium* (*Gomphostilbia*) *miaolingense* Wen and Chen, 2000**
Simulium (*Gomphostilbia*) *miaolingense* Wen and Chen, 2000: 112–115 (female, male, pupa, and larva).

Distribution. China (Guizhou).

Bionomics. Aquatic habitat: A rapid of a mountain stream at an elevation of 1200 m (Wen and Chen, 2000).

Remarks. This species was described from Guizhou, China by Wen and Chen (2000). It was placed in the *S. asakoae* species-group by Takaoka (2012) based on the ventral plate with its body emarginated subbasally on each side when viewed ventrally, although information on the color of the tuft hairs on the base of the radius was not available.

It is characterized in the female by the mandible with eight inner and 10 outer teeth and genital fork with a small projection directed anteriorly on each arm; in the male by the ventral plate with its body flat, not produced ventrally but produced outward basally on each side, all these characteristics being unique in the *S. asakoae* species-group.

***Simulium* (*Gomphostilbia*) *mokroense* Takaoka, Srisuka and Saeung, 2022**
Simulium (*Gomphostilbia*) *mokroense* Takaoka, Srisuka and Saeung [in Srisuka et al., 2022a]: 2–8 (female, male, pupa, and larva).

Distribution. Thailand.

Bionomics. Aquatic habitat: A stream (width 2 m, depth 20 cm, flow moderate, 19.3 °C, partially shaded, elevation 1262 m), Umphang wildlife sanctuary, Mo Kro, Tak Province (Srisuka et al., 2022a).

Remarks. This species was described from Thailand (Srisuka et al., 2022a).

It is almost indistinguishable from *S. chiangdaoense* in many characteristics except the hair tuft at the base of the radius of the female and male, which is composed of yellow hairs mixed with dark ones.

***Simulium* (*Gomphostilbia*) *monglaense* Takaoka, Srisuka and Saeung, 2017**
Simulium (*Gomphostilbia*) *monglaense* Takaoka, Srisuka and Saeung [in Takaoka et al., 2017g]: 45–48 (female, male, pupa, and larva).

Distribution. Myanmar.

Bionomics. Aquatic habitat: A small forest stream (0.5 m wide) at an elevation of 767 m, in Shan State (Takaoka et al., 2017g).

Remarks. This species was described from Myanmar (Takaoka et al., 2017g). Jomkumsing et al. (2019) and Aupalee et al. (2020b) suggested the presence of *S. monglaense* in Thailand, based on the close similarity of the COI gene sequences among females of "*S. asakoae* complex" caught using a human attractant. The record of this species from Thailand should be morphologically confirmed.

It is characterized in the female by the mandible lacking outer teeth; in the male by the number of upper-eye (large) facets in 15 vertical columns and 15 horizontal rows; in the pupa by the gill with eight filaments arranged as 3+3+2 from dorsal to ventral; and in the larva by the postgenal cleft medium-long, 1.7–1.9 times as long as the postgenal bridge and abdominal segments 1 and 4 each encircled with a gray broad band.

***Simulium* (*Gomphostilbia*) *muangpanense* Takaoka, Srisuka and Fukuda, 2020**
(Fig. 15O)

Simulium (*Gomphostilbia*) *muangpanense* Takaoka, Srisuka and Fukuda [in Takaoka et al., 2020h]: 95–99 (female, male, pupa, and larva).

Distribution. Thailand.

Bionomics. Aquatic habitat: A stream (width 30 cm, depth 5 cm, bottom sandy, flow moderate, pH 6.3, 22.6 °C, exposed to the sun, elevation 1097 m), Pa Miang Village, Lampang Province (Takaoka et al., 2020h).

Remarks. This species was described from Thailand (Takaoka et al., 2020h).

It is characterized in the female by the mandible with two or three outer teeth; in the male by the number of upper-eye (large) facets in 13 vertical columns and 14 horizontal rows and hind basitarsus 0.9 times as wide as the hind femur; and in the larva by abdominal segments 1, 4, 5, 7, and 8 gray at least on the dorsal and dorsolateral surface (Fig. 15O).

***Simulium* (*Gomphostilbia*) *myanmarense* Takaoka, Srisuka and Saeung, 2017**
(Fig. 15L)

Simulium (*Gomphostilbia*) *myanmarense* Takaoka, Srisuka and Saeung [in Takaoka et al., 2017g]: 40–45 (female, male, pupa, and larva); Takaoka et al., 2020h: 135.

Distribution. Myanmar and Thailand.

Bionomics. Aquatic habitats: A river (7 m wide) at an elevation of 614 m, in Shan State, Myanmar (Takaoka et al., 2017g); and a small stream (width 55 cm, depth 3 cm, bottom sandy, flow moderate, pH 7.2, 18.4 °C, partially shaded, elevation 901 m), Pha Dang Village, Pha Yao Province, and a small stream (width 30 cm, depth 5 cm, bottom sandy, flow moderate, pH 6.3, 23.6 °C, exposed to the sun, elevation 1097 m), Pa Miang Village, Lampang Province, Thailand (Takaoka et al., 2020h).

Remarks. This species was described from Myanmar (Takaoka et al., 2017g). It was also recorded from Thailand (Takaoka et al., 2020h). Jomkumsing et al. (2019) suggested the presence of *S. myanmarense* in Thailand, based on the close similarity of the COI gene sequences among females of "*S. asakoae* complex" caught using a human attractant. It was confirmed morphologically and molecularly (Takaoka et al., 2020h).

It is characterized in the female by the mandible lacking outer teeth; in the male by the number of upper-eye (large) facets in 15 (or rarely 14) vertical columns and 15 horizontal rows; in the pupa by the cocoon with a medium-long anterodorsal projection; and in the larva by the medium-long postgenal cleft and abdominal segments 1 and 2 each encircled with a gray band (Fig. 15L).

Simulium (Gomphostilbia) namdanense Takaoka, Srisuka and Saeung, 2020
(Figs. 13A and 15F)

Simulium (*Gomphostilbia*) *namdanense* Takaoka, Srisuka and Saeung [in Takaoka et al., 2020h]: 130–135 (female, male, pupa, and larva).

Distribution. Thailand.

Bionomics. Aquatic habitats: A medium-sized stream of Nam Khwang (width 4 m, depth 30 cm, bottom sandy, flow fast, pH 7.2, 19.5 °C, partially shaded, elevation 1192 m), Nam Dan Village, Nan Province, and a medium-sized stream (width 3 m, depth 32 cm, bottom sandy and rocky, flow moderate, pH 6.5, 23.2 °C, partially shaded, elevation 503 m) at Huai Kaew Waterfall, Chiang Mai Province (Takaoka et al., 2020h).

Remarks. This species was described from Thailand (Takaoka et al., 2020h).

It is characterized in the female by the elongate sensory vesicle (Fig. 13A); in the male by the higher number of upper-eye (large) facets in 15 or 16 vertical columns and 16 or 17 horizontal rows and hind basitarsus spindle-shaped, narrower than the hind tibia and femur; in the pupa by the small terminal hooks; and in the larva by the postgenal cleft 2.5 times as long as the postgenal bridge and abdominal segments 1–4 each encircled with a light grayish band (Fig. 15F).

This species is distinguished from *S. klonglanense* from Thailand by the female wing length (2.0 mm in this species versus 1.6 mm in *S. klonglanense*) and the number of male upper-eye (large) facets in 15 or 16 vertical columns and 16 or 17 horizontal rows (13 or 14 vertical columns and 14 or 15 horizontal rows in *S. klonglanense*).

Simulium (Gomphostilbia) namense Takaoka, 1989

Simulium (*Gomphostilbia*) *namense* Takaoka, 1989b: 249–252 (female and male).

Distribution. Myanmar.

Bionomics. Unknown.

Remarks. This species was described from males and females caught by Malaise traps in Kambaiti, Myanmar (Takaoka, 1989b). Its pupa and larva are unknown.

It is characterized in the female by the mandible with three or four outer teeth and hind tibia whitish on its basal two-thirds; and in the male by the number of upper-eye (large) facets in 14 vertical columns and 15 horizontal rows, fore basitarsus slender, 9.0 times as long as its greatest width and ventral plate with its body slightly narrowed posteriorly when viewed ventrally.

Simulium (Gomphostilbia) nanoiense Takaoka, Srisuka and Saeung, 2020
(Fig. 15J)

Simulium (*Gomphostilbia*) *nanoiense* Takaoka, Srisuka and Saeung [in Takaoka et al., 2020h]: 91–95 (female, male, pupa, and larva).

Distribution. Thailand.

Bionomics. Aquatic habitat: A stream (width 1 m, depth 3 cm, bottom sandy, flow moderate, pH 6.2, 19.6 °C, partially shaded, elevation 1349 m), Khun Sathan Village, Na Noi District, Nan Province, Thailand (Takaoka et al., 2020h).

Remarks. This species was described from Thailand (Takaoka et al., 2020h).

It is characterized in the female by the sensory vesicle short, 0.22–0.24 times as long as the third palpal segment and mandible with two outer teeth; in the male by the number of upper-eye (large) facets in 12 or 13 vertical columns and 14 (rarely 13) horizontal rows; and in the larva by the postgenal cleft as long as or little longer than the postgenal bridge and abdominal segments 1–3, 7, and 8 grayish (Fig. 15J).

***Simulium* (*Gomphostilbia*) *nanthaburiense* Takaoka, Srisuka and Fukuda, 2020**
Simulium (*Gomphostilbia*) *nanthaburiense* Takaoka, Srisuka and Fukuda [in Takaoka et al., 2020h]: 88–91 (female, male, and pupa).

Distribution. Thailand.

Bionomics. Aquatic habitats: A stream (width 80 cm, depth 2.5 cm, flow moderate, pH 7.2, 20.1 °C, partially shaded, elevation 1157 m), Nam Dan Village, Nan Province, and a stream (width 1.2 m, depth 10 cm, bottom sandy, flow moderate, pH 6.63, 17.1 °C, partially shaded, elevation 1154 m), Huai Mor Nuea Village, Chiang Mai Province (Takaoka et al., 2020h).

Remarks. This species was described from Thailand (Takaoka et al., 2020h). Its larva is unknown.

It is characterized in the female by the mandible with two distinct outer teeth; and in the male by the small number of upper-eye (large) facets in 12 vertical columns and 13 or 14 horizontal rows

This species is similar to *S. brinchangense* from Peninsular Malaysia (Takaoka et al., 2014f) but is distinguished by the relative length of the stalk of the ventral pair of gill filaments against the common basal stalk, which is 0.6–0.9 in this species but 1.1–1.2 in *S. brinchangense,* and relative length of the outer margin against the inner margin of the pupal terminal hooks, which is 1.9–2.1 in this species but 3.0–3.6 in *S. brinchangense.*

***Simulium* (*Gomphostilbia*) *nareshi* Takaoka and Shrestha, 2020**
Simulium (*Gomphostilbia*) *nareshi* Takaoka and Shrestha [in Takaoka et al., 2020f]: 183–186 (female, male, and pupa).

Distribution. Nepal.

Bionomics. Aquatic habitat: A small stream (elevation 1600 m) in Shivapuri National Park (Takaoka et al., 2020f).

Remarks. This species was described from Nepal by Takaoka et al. (2020f). Its larva is unknown. This species is exceptional among species of the *S. asakoae* species-group by having the ventral plate narrowed basally, then nearly parallel-sided, when viewed ventrally.

It is characterized in the female and male by the hind tibia with a dark subbasal marking; in the female by the sensory vesicle long, 0.5 times as long as the third palpal segment; in the male by the number of upper-eye (large) facets in 15 vertical columns and 16 horizontal rows and hind basitarsus enlarged, 3.5 times as long as its greatest width, and 1.1 and 1.4 times as wide as the hind tibia and femur, respectively; and in the pupa by the terminal hooks triangular, with their outer margins 1.4 times as long as their inner margins and crenulated.

***Simulium* (*Gomphostilbia*) *nepalense* Lewis, 1964**
Simulium (*Gomphostilbia*) *nepalense* Lewis, 1964: 291–293 (female and male).

Distribution. Nepal.

Bionomics. Adult females and males of this species were captured at an elevation of 540–1.950 m, at Sangu, Nepal (Lewis, 1964).

Remarks. This species was described from Nepal by Lewis (1964). Its pupa and larva are unknown.

According to the original description and illustrations (Lewis, 1964), it is characterized in the female and male by the scutum with brassy short hairs; in the female by the antenna gray except the scape, pedicel and first flagellomere orange, sensory vesicle short, 0.33 times as long as the third palpal segment, cibarium with a mediolongitudinal ridge with a bifid tip, hind tibia yellow except its apical one-third darkened and claw with a large basal tooth 0.5 times as long as the claw; and in the male by the hind tibia and basitarsus yellow except the apical three-fifths and seven-tenths darkened, respectively, hind basitarsus enlarged (though its relative length to the greatest width was not mentioned) and ventral plate with its lateral margins emarginated medially when viewed ventrally and its ventral margin somewhat concave medially when viewed caudally.

***Simulium* (*Gomphostilbia*) *okinawense* Takaoka, 1976**
Simulium (*Gomphostilbia*) *okinawense* Takaoka, 1976b: 385–388 (female, male, pupa and larva); Takaoka et al., 2023a: 2–8 (female, male, pupa, and larva).

Distribution. Japan (Amamioshima Island, Kakeroma Island, Tokunoshima Island, Okinawa Island, Ishigaki Island, Iriomote Island, Yoro Island, and Uke Island, in Nansei Islands).

Bionomics. Aquatic habitats: Small streams near and in forests (Takaoka et al., 2023a).

Remarks. This species was described from Okinawa, the Nansei Islands, Japan (Takaoka, 1976b). This species was placed in the *S. ceylonicum* species-group (Takaoka, 2012) but was later transferred to the *S. asakoae* species-group based on the result of morphological reexamination and genetic analysis (Takaoka et al., 2023a).

It is characterized in the female by the sensory vesicle 0.27–0.30 times as long as the third palpal segment and the base of the radial vein with a tuft of yellow hairs; in the male by the number of upper-eye (large) facets in 15 vertical columns and 16 horizontal rows; in the pupa by the terminal hooks triangular and plate-like, of which their outer margins are 1.3–1.7 times as long as their inner margins and crenulated; and in the larva by the body entirely gray with a reddish brown transverse band each on the dorsum of abdominal segments 5 and 6, and postgenal cleft medium-long, 1.5–2.0 times as long as the postgenal bridge.

Simulium (_Gomphostilbia_) _phapeungense_ Takaoka, Srisuka and Fukuda, 2020
(Fig. 15N)

Simulium (*Gomphostilbia*) *phapeungense* Takaoka, Srisuka and Fukuda [in Takaoka et al., 2020h]: 85–88 (male, pupa, and larva).

Distribution. Thailand.

Bionomics. Aquatic habitats: A stream (width 120 cm, depth 2 cm, bottom sandy, flow fast, pH 7.6, 21.2 °C, partially shade, elevation 1034 m), Pha Peung, Muang District, Mae Hong Son Province, Thailand, and a stream (width 80 cm, depth 2.5 cm, flow moderate, pH 7.2, 20.1 °C, partially shaded, elevation 1157 m), Nam Dan Village, Pua District, Nan Province, Thailand (Takaoka et al., 2020h).

Remarks. This species was described from Thailand (Takaoka et al., 2020h). Its female is unknown.

It is similar to *S. loeiense* from Thailand but is readily distinguished by the dorsum of pupal abdominal segments 1 and 2 without tubercles and larval abdominal segments 1, 3, 4 and 5 light grayish (Fig. 15N).

Simulium (_Gomphostilbia_) _phulocense_ Takaoka and Chen, 2015
(Figs. 13T and 14L)

Simulium (*Gomphostilbia*) *phulocense* Takaoka and Chen [in Takaoka et al., 2015c]: 18–24 (female, male, and pupa).

Distribution. Vietnam.

Bionomics. Aquatic habitat: A small stream (width 0.5–1.0 m, shaded, elevation 1186 m) in a forest, Bach Ma National Park, Thua Thien Hue Province (Takaoka et al., 2015b).

Remarks. This species was described from Vietnam (Takaoka et al., 2015c). Its larva is unknown.

It is characterized in the female by the hind tibia yellow on its basal half and medium to dark brown on its apical half; in the male by the ventral plate with its ventral margin rounded when viewed caudally (Fig. 13T); and in the pupa by the anterior margin of the cocoon resembling the letter W when viewed posterodorsally (Fig. 14L).

Simulium (_Gomphostilbia_) _pitasawatae_ Takaoka, Srisuka and Saeung, 2020
(Figs. 14J and 15B)

Simulium (*Gomphostilbia*) *pitasawatae* Takaoka, Srisuka and Saeung [in Takaoka et al., 2020h]: 106–110 (female, male, pupa, and larva).

Distribution. Thailand.

Bionomics. Aquatic habitat: A stream (width 1.2 m, depth 10 cm, bottom sandy, flow moderate, pH 6.6, 17.1 °C, partially shaded, elevation 1154 m), Huai Mor Nuea Village, Chiang Mai Province (Takaoka et al., 2020h).

Remarks. This species was described from Thailand (Takaoka et al., 2020h).

It is characterized in the female by the small sensory vesicle and mandible with three to five distinct outer teeth; in the male by the number of upper-eye (large) facets in 14 vertical columns and 15 horizontal rows; in the pupa by the dorsal triplet of the gill filaments without their stalk or with an extremely short stalk and wide terminal hooks (Fig. 14J); and in the larva by the postgenal cleft as long as or little longer than the postgenal bridge and abdominal segments 1–4 light ocherous (Fig. 15B).

***Simulium* (*Gomphostilbia*) *puaense* Takaoka, Srisuka and Saeung, 2020**
Simulium (*Gomphostilbia*) *puaense* Takaoka, Srisuka and Saeung [in Takaoka et al., 2020h]: 64–67 (female, male, and pupa).

Distribution. Thailand.

Bionomics. Aquatic habitats: A small waterfall (width 80 cm, depth 3 cm, flow fast, pH 7.2, 20.1 °C, exposed to the sun, elevation 1157 m), Nam Dan Village, Nan Province, and a stream (width 30 cm, depth 5 cm, bottom sandy, flow moderate, pH 6.3, 22.6 °C, exposed to the sun, elevation 1097 m), Pa Miang Village, Lampang Province (Takaoka et al., 2020h).

Remarks. This species was described from Thailand (Takaoka et al., 2020h).

It is characterized in the female by the sensory vesicle short, 0.24 times as long as the third palpal segment and mandible with two or three outer teeth; in the male by the small number of upper-eye (large) facets in nine or 10 vertical columns and 12 horizontal rows and hind basitarsus 3.2–3.6 times as long as its greatest width and 0.9–1.1 and 1.1–1.2 times as wide as the hind tibia and femur, respectively.

***Simulium* (*Gomphostilbia*) *quychauense* Takaoka and Chen, 2017**
Simulium (*Gomphostilbia*) *quychauense* Takaoka and Chen [in Takaoka et al., 2017c]: 34–37 (male and pupa).

Distribution. Vietnam.

Bionomics. Aquatic habitat: A small streams (width 1.5–3.0 m, bottom with pebbles and sands, flow moderate, 18 °C, elevation 73 m), Nghe An Province (Takaoka et al., 2017c).

Remarks. This species was described from Vietnam (Takaoka et al., 2017c). The female and larva of this species are unknown.

It is characterized in the male by the number of upper-eye (large) facets in 14 vertical and 14 horizontal rows; and in the pupa by the small terminal hooks.

***Simulium* (*Gomphostilbia*) *rampae* Takaoka, Srisuka and Saeung, 2019**
Simulium (*Gomphostilbia*) *rampae* Takaoka, Srisuka and Saeung [in Srisuka et al., 2019]: 2–6 (female, pupa, and larva).

Distribution. Thailand.

Bionomics. Aquatic habitat: A small stream (width 0.6 m, depth 0.1 m, bottom sandy, 14.5 °C, pH 6.6, partially shaded, elevation 1685 m) in Doi Inthanon National Parks, Chiang Mai Province (Srisuka et al., 2019).

Remarks. This species was described from Thailand by Srisuka et al. (2019). Its female is unknown.

It is characterized in the male by the large number of upper-eye (large) facets in 17 vertical columns and 18 horizontal rows and hind basitarsus enlarged, 3.8 times as long as its greatest width, and 0.9 and 1.1 times as wide as the hind tibia and femur, respectively; in the pupa by the gill with seven or eight filaments arising from a long common basal stalk, last abdominal segment with a pair of triangular terminal hooks and cocoon with a short anterodorsal projection; and in the larva by the postgenal cleft medium-long, 0.9–1.2 times as long as the postgenal bridge.

***Simulium* (*Gomphostilbia*) *roslihashimi* Takaoka and Sofian-Azirun, 2011**
Simulium (*Gomphostilbia*) *roslihashimi* Takaoka and Sofian-Azirun [in Takaoka et al., 2011b]: 55–63 (female, male, pupa, and larva).

Distribution. Malaysia (Peninsular).

Bionomics. Aquatic habitats: Small streams at low elevations in Cameron's Highlands and in other areas (Takaoka et al., 2011b).

Remarks. This species was described from Peninsular Malaysia (Takaoka et al., 2011b).

The male of this species is characterized by the antenna almost entirely yellow. The pupa of this species is similar to that of *S. lurauense* except terminal hooks, which are widened and have crenulated outer margins.

***Simulium* (*Gomphostilbia*) *rutengense* Takaoka, Hadi and Sigit, 2006**
Simulium (*Gomphostilbia*) *rutengense* Takaoka, Hadi and Sigit, 2006a: 13–17 (male, pupa, and larva).

Distribution. Indonesia (Flores).

Bionomics. Aquatic habitat: A small shaded stream in a forest (Takaoka et al., 2006a).

Remarks. This species was described from Flores, Indonesia (Takaoka et al., 2006a). Its female is unknown. This species was placed in the *S. ceylonicum* species-group by Takaoka (2012). However, it is here transferred to the *S. asakoae* species-group by the male ventral plate with its lateral margins slightly emarginated when viewed ventrally. This species is exceptional in this species-group by having the base of the male radius with a tuft of dark hairs (Takaoka et al., 2006a).

It is characterized in the male by the subcosta haired except near its apex and the base of the radius with a tuft of dark hairs; in the pupa by the dorsal surface of abdominal segments 1, 2, and 9 tuberculate; and in the larva by the postgenal cleft 2.7 times as long as the postgenal bridge, and rectal organ with two or three thumb-like secondary lobules per lobe.

***Simulium* (*Gomphostilbia*) *sanchayense* Takaoka and Lau, 2017**
Simulium (*Gomphostilbia*) *brinchangense* (*nec* Takaoka, Sofian-Azirun and Hashim) [in Takaoka et al., 2014d]: 364–365.
Simulium (*Gomphostilbia*) *sanchayense* Takaoka and Lau [in Takaoka et al., 2017c]: 38–42 (male and pupa).

Distribution. Vietnam.

Bionomics. Aquatic habitat: A small stream (width 0.5 m, bottom rocky, 17 °C, shaded, elevation of 975 m) in Tam Dao National Park, Vinh Phuc Province (Takaoka et al., 2017c).

Remarks. This species was described from Vietnam (Takaoka et al., 2017c). Its female and larva are unknown. This species was once recorded from Vietnam under the name *S. brinchangense* from Peninsular Malaysia (Takaoka et al., 2014d).

This species is morphologically indistinguishable from *S. brinchangense* except the ventral margin of the ventral plate nearly straight when viewed posteriorly. The DNA analysis using CO1 gene sequences showed the distinctness of this species from *S. brinchangense* (Takaoka et al., 2017c).

***Simulium* (*Gomphostilbia*) *shoufengense* Takaoka, Low and Ya'cob, 2021**

Simulium (*Gomphostilbia*) *shoufengense* Takaoka, Low and Ya'cob [in Takaoka et al., 2021a]: 6–8 (female, male, and pupa).

Distribution. Taiwan.

Bionomics. Aquatic habitat: A stream (width 1 m, 23.0 °C, partially shaded, elevation 4 m), Shounong Road, Shoufeng Township (Takaoka et al., 2021a).

Remarks. *Simulium shoufengense* was described from Taiwan (Takaoka et al., 2021a). Its larva is unknown.

This species is similar to *S. asakoae* in having a similar number of male upper-eye (large) facets and enlarged male hind basitarsus much wider than the hind femur but is distinguished from the latter species in the pupa by the primary and secondary stalks of the dorsal triplet of the gill combined, which is as long as or somewhat longer than the common basal stalk (always shorter in *S. asakoae*), dorsum of abdominal segments 1 and 2 yellow (light to medium brown in *S. asaokae*), and cocoon without an anterodorsal bulge (usually with an anterodorsal bulge in *S. asaokae*).

***Simulium* (*Gomphostilbia*) *sofiani* Takaoka and Hashim, 2011**

(Fig. 13V)

Simulium (*Gomphostilbia*) *sofiani* Takaoka and Hashim [in Takaoka et al., 2011a]: 390–398 (female, male, pupa, and larva).

Distribution. Malaysia (Peninsular).

Bionomics. Aquatic habitat: A small stream at high elevation (1470 m) in Cameron's Highlands (Takaoka et al., 2011a).

Remarks. This species was described from Peninsular Malaysia (Takaoka et al., 2011a).

It is characterized in the female by the elongate sensory vesicle; in the male by the number of upper-eye (large) facets (15 or 16 vertical columns and 15 or 16 horizontal rows), hind basitarsus spindle-shaped and entirely darkened and ventral plate with its ventral margin slightly concave when viewed posteriorly (Fig. 13V); and in the pupa by the long common basal stalk, which is longer than the interspiracular trunk, and the small terminal hooks.

***Simulium* (*Gomphostilbia*) *songense* Takaoka, Srisuka and Fukuda, 2020**

Simulium (*Gomphostilbia*) *songense* Takaoka, Srisuka and Fukuda [in Takaoka et al., 2020h]: 124–126 (female and pupa).

Distribution. Thailand.

Bionomics. Aquatic habitats: A small stream (width 80 cm, depth 2.5 cm, flow fast, pH 7.2, 20.1 °C, partially shaded, elevation 1157 m), Nam Dan Village, Nan Province, and a small stream (width 40 cm, depth 3.5 cm, bottom sandy, flow moderate, pH 7.36, 21.9 °C, partially shaded, elevation 582 m) at Chao Wa Waterfall, Phrae Province (Takaoka et al., 2020h).

Remarks. This species was described from Thailand (Takaoka et al., 2020h). Its male and larva are unknown.

It is characterized in the female by the mandible without outer teeth; and in the pupa by the short stalk of the ventral pair of gill filaments, which is half the length of the interspiracular trunk.

Lacking teeth on the outer margin of the mandible, the female of this species seems to be most similar to *S. myanmarense* and *S. monglaense* originally described from Myanmar (Takaoka et al., 2017g) but is distinguished from the latter two species by the relative length of the fore basitarsus against its greatest width (5.5 in this species versus 6.0 in *S. myanmarense* and 6.7 in *S. monglaense*).

***Simulium* (*Gomphostilbia*) *sunapii* Takaoka, Sofian-Azirun and Wayan, 2017**

Simulium (*Gomphostilbia*) *sunapii* Takaoka, Sofian-Azirun and Wayan [in Takaoka et al., 2017a]: 576–582 (female, male, pupa, and larva).

Distribution. Indonesia (Flores).

Bionomics. Aquatic habitats: A small stream (width 0.2–1.0 m, exposed to the sun, elevation 577 m) in Cuncalolos, and a small stream (partially shaded, elevation 838 m) in Kempo (Takaoka et al., 2017a).

Remarks. This species was described from Flores, the Lesser Sunda Archipelago, Indonesia (Takaoka et al., 2017a).

It is characterized in the female by the mandible with its outer margin serrated and hind tibia whitish on its basal half or little more; in the male by the small number of upper-eye (large) facets in nine vertical columns and 12 horizontal rows and subcosta with a few hairs; and in the larva by the postgenal cleft long, 4.7 times as long as the postgenal bridge.

***Simulium* (*Gomphostilbia*) *sutheppuiense* Takaoka, Srisuka and Saeung, 2020** (Figs. 14E and 15C)

Simulium (*Gomphostilbia*) *sutheppuiense* Takaoka, Srisuka and Saeung [in Takaoka et al., 2020h]: 67–70 (female, male, pupa, and larva).

Distribution. Thailand.

Bionomics. Aquatic habitat: A small stream (width 60 cm, depth 15 cm, bottom sandy, flow moderate, pH 6.9, 20 °C, partially shaded, elevation 1395 m) at Doi Pui Temple, Chiang Mai Province (Takaoka et al., 2020h).

Remarks. This species was described from Thailand (Takaoka et al., 2020h).

It is characterized in the female by the mandible with three teeth on its outer margin; in the male by the small number of upper-eye (large) facets in 11 vertical columns and 13 horizontal rows; in the pupa by the dorsal triplet of the gill filaments with an extremely short stalk (Fig. 14E); and in the larva by the postgenal cleft 1.2–2.4 times as long as the postgenal bridge and abdominal segments 1–4 light ocherous (Fig. 15C).

This species is similar to *S. asakoae* but is distinguished in the female by the shorter sensory vesicle relative to the third palpal segment; in the male by the dark-brown upper-eye (large) facets; in the pupa by the dorsum of abdominal segments 1–3 light yellow; and in the larva by abdominal segments 1–4 light ocherous (Fig. 15C).

Simulium (_Gomphostilbia_) _taitungense_ Huang and Takaoka, 2011
Simulium (*Gomphostilbia*) *taitungense* Huang and Takaoka [in Huang et al., 2011: 578–585 (female, male, pupa, and larva).

Distribution. Taiwan.

Bionomics. Aquatic habitats: A small stream (width 0.3 m, exposed to the sun, elevation 320 m), and a small stream (width 0.3 m, flow slow, exposed to the sun, elevation 781 m), Taitung County, Taiwan (Huang et al., 2011).

Remarks. This species was described from Taiwan (Huang et al., 2011).

It is characterized in the female by the mandible with its outer margin serrated and hind tibia whitish yellow on a little less than its basal half; in the male by the large number of upper-eye (large) facets in 15 vertical columns and 15 or 16 horizontal rows; and in the larva by the head capsule with a darkened X-shaped marking medially on the ventral surface.

Simulium (_Gomphostilbia_) _tamdaoense_ Takaoka, Sofian-Azirun and Ya'cob, 2014
Simulium (*Gomphostilbia*) *tamdaoense* Takaoka, Sofian-Azirun and Ya'cob [in Takaoka et al., 2014d]: 351–353 (male and pupa).

Distribution. Vietnam.

Bionomics. Aquatic habitat: A roadside ditch (0.5–1.0 m wide, exposed to the sun, elevation 173 m) in Vinh Phuc Province (Takaoka et al., 2014d).

Remarks. This species was described from Vietnam (Takaoka et al., 2014d). Its female and larva are unknown.

It is characterized in the male by the number of upper-eye (large) facets in 14 vertical columns and 14 horizontal rows and paramere with fine hairs basally.

Simulium (_Gomphostilbia_) _tanahrataense_ Takaoka, Sofian-Azirun and Ya'cob, 2014
Simulium (*Gomphostilbia*) *tanahrataense* Takaoka, Sofian-Azirun and Ya'cob [in Takaoka et al., 2014f]: 61–63 (male and pupa).

Distribution. Malaysia (Peninsular).

Bionomics. Aquatic habitat: A small stream at high elevation (1470 m) in Cameron's Highlands (Takaoka et al., 2014f).

Remarks. This species was described from Peninsular Malaysia (Takaoka et al., 2014f). Its female and larva are unknown.

The male of *S. tanahrataense* is similar to those of *S. sofiani* and *S. lurauense* in the number of upper-eye (large) facets, the absence of longitudinal vittae on the dark scutum, and the shape of the ventral plate but is distinguished from the latter two species by the hind basitarsus which is bicolored and much widened.

Simulium (_Gomphostilbia_) _teerachanense_ Takaoka, Srisuka and Fukuda, 2020
(Figs. 14Q and 15A)

Simulium (*Gomphostilbia*) *teerachanense* Takaoka, Srisuka and Fukuda [in Takaoka et al., 2020h]: 71–74 (male, pupa, and larva).

Distribution. Thailand.

Bionomics. Aquatic habitats: A small stream (width 25 cm, depth 2.5 cm, bottom sandy, flow moderate, pH 6.8, 21.9 °C, partially shaded, elevation 974 m) at Tee Ra Chan Waterfall, and a stream (elevation 817 m), Huai Hee Village, Mae Hong Song Province (Takaoka et al., 2020h).

Remarks. This species was described from Thailand (Takaoka et al., 2020h). Its female is unknown.

It is characterized in the male by the small number of upper-eye (large) facets in 11 vertical columns and 13 horizontal rows and antenna almost entirely yellow; and in the larva by the postgenal cleft 2.8–3.3 times as long as the postgenal bridge (Fig. 13Q) and abdominal segments 1–4 dull ocherous (Fig. 15A).

This species is similar to *S. roslihashimi* from Peninsular Malaysia (Takaoka et al., 2011b) in many characteristics including the small number of the male upper-eye (large) facets and male antenna almost entirely yellow. However, it is distinguished by the male fore basitarsus 7.3–7.5 times as long as its greatest width (6.6–6.8 times in *S. roslihashimi*) and ventral plate with its ventral margin nearly straight when viewed posteriorly (rounded ventrally in *S. roslihashimi*).

Simulium (_Gomphostilbia_) _thituyenae_ Takaoka and Pham, 2015

Simulium (*Gomphostilbia*) *thituyenae* Takaoka and Pham [in Takaoka et al., 2015c]: 24–27 (female and pupa).

Distribution. Vietnam.

Bionomics. Aquatic habitat: A stream (width 4–6 m, exposed to the sun, elevation 1452 m) flowing through an open land, Lam Dong Province (Takaoka et al., 2015c).

Remarks. This species was described from Vietnam (Takaoka et al., 2015c). Its male and larva are unknown.

It is characterized in the female by the sensory vesicle long, 0.63–0.64 times as long as the third palpal segment; and in the pupa by the small narrow terminal hooks.

Simulium (_Gomphostilbia_) _thungchangense_ Takaoka, Srisuka and Saeung, 2020
(Figs. 13G, H, J–M, P–S, W, X and 15Q)

Simulium (*Gomphostilbia*) *thungchangense* Takaoka, Srisuka and Saeung [in Takaoka et al., 2020h]: 55–64 (female, male, pupa, and larva).

Distribution. Thailand.

Bionomics. Aquatic habitats: A stream (width 60 cm, depth 10 cm, bottom sandy, flow moderate, pH 6.9, 17.9 °C, exposed to the sun, elevation 1376 m) at Siribhume Waterfall, Chiang Mai Province, and a stream (18 °C, elevation 1470 m), Phu Kha, Nan Province (Takaoka et al., 2020h).

Remarks. This species was described from Thailand (Takaoka et al., 2020h).

It is characterized in the female by the mandible with several teeth on its outer margin and hind tibia yellowish on a little more than its basal half; in the male by the small number of upper-eye (large) facets in nine vertical columns and 12 horizontal rows, subcostal with many hairs and hind basitarsus much widened, 1.5–1.7 times as wide as the hind femur (Fig. 13P); in the pupa by the dorsal surface of abdominal segments 1 and 2 sparsely covered with minute tubercles; and in the larva by the postgenal cleft as long as or little longer than the postgenal bridge and abdominal segments 1–4 light greenish or greenish gray (Fig. 15Q).

Simulium (Gomphostilbia) tokarense Takaoka, 1973

(Fig. 14D)

Simulium (*Gomphostilbia*) *tokarense* Takaoka, 1973: 201–203 (female, male, pupa and larva); Takaoka et al., 2023a: 8–12 (female, male, pupa, and larva).

Distribution. Japan (Nakanoshima Island in Nansei Islands).

Bionomics. Aquatic habitats: Small mountainous streams (width 1–3 m) (Takaoka et al., 2023a).

Remarks. This species was described from Nakanoshima Island, the Nansei Islands, Japan (Takaoka, 1973). This species was placed in the *S. ceylonicum* species-group (Takaoka, 2012). However, it was transferred to the *S. asakoae* species-group based on the result of morphological reexamination and genetic analysis (Takaoka et al., 2023a).

It is characterized in the female by the small sensory vesicle and the base of the radius with a tuft of yellow hairs mixed with few dark hairs; in the pupa by the gill with eight filaments arranged as (1+2)+(1+2)+2, arising close together from a short common basal stalk (Fig. 14D); and in the larva by the postgenal cleft 1.2 times as long as the postgenal bridge and rectal organ simple, without secondary lobules.

Simulium (Gomphostilbia) tuenense Takaoka, 1979

Simulium (*Gomphostilbia*) *tuenense* Takaoka, 1979: 387–388 (pupa and larva); Huang et al., 2011: 585–587 (male).

Distribution. Taiwan.

Bionomics. Aquatic habitats: A small stream (0.3–1.0 m wide) flowing down from the Sun Moon Lake, and a small irrigation canal (0.6 m wide, exposed to the sun) in Yilan County (Takaoka, 1979).

Remarks. This species was described from Taiwan (Takaoka, 1979). It male was described by Huang et al. (2011). Its female is unknown.

It is characterized in the male by the large number of upper-eye (large) facets in 16 vertical columns and 16 horizontal rows; and in the larva by abdominal segments 1 and 2 each with a dark gray band.

Simulium (Gomphostilbia) udomi Takaoka and Choochote, 2006

Simulium (*Gomphostilbia*) *udomi* Takaoka and Choochote, 2006e: 229–233 (male and pupa); Saeung et al., 2017: 15–19 (female and larva).

Distribution. Thailand.

Bionomics. Aquatic habitat: A small stream (width ca. 1.0 m, depth ca. 0.5 m, flow moderate, 17.0 °C, shaded, elevation 1350 m) in Nan Province (Takaoka and Choochote, 2006e).

Remarks. *Simulium udomi* was described from Thailand by Takaoka and Choochote (2006e). Its female and larva were described by Saeung et al. (2017). This species was placed in the *S. ceylonicum* species-group (Takaoka and Choochote, 2006e) but later transferred to the *S. asakoae* species-group (Saeung et al., 2017).

It is similar to *S. chiangdaoense* from Thailand in many characteristics including the elongate anterodorsal projection of the cocoon (similar to Fig. 14N). However, it differs from *S. chiangdaoense* and other related species by having the pupal gill with six filaments.

***Simulium* (*Gomphostilbia*) *vinhphucense* Takaoka and Low, 2017**

Simulium (*Gomphostilbia*) *vinhphucense* Takaoka and Low [in Takaoka et al., 2017c]: 49–52 (male and pupa).

Distribution. Vietnam.

Bionomics. Aquatic habitat: a small stream (width 0.5 m, bottom rocky, 17 °C, shaded, elevation 975 m) in Vinh Phuc Province (Takaoka et al., 2017c).

Remarks. This species was described from Vietnam by Takaoka et al. (2017c). Its female and larva are unknown.

It is characterized in the male by the relatively small number of upper-eye (large) facets in nine vertical columns and 11 or 12 horizontal rows and hind basitarsus greatly enlarged, 3.3 times as long as its greatest width, and 1.1 and 1.5 times as wide as the hind tibia and femur, respectively; and in the pupa by the terminal hooks wide, plate-like, with their outer margins three times as long as their inner margins.

***Simulium* (*Gomphostilbia*) *yunnanense* Chen and Zhan, 2004**

Simulium (*Gomphostilbia*) *yunnanense* Chen and Zhan, 2004: 87–89 (female, male, pupa, and larva).

Distribution. China (Yunnan).

Bionomics. Aquatic habitat: A small stream (shaded, elevation 700 m) in Mengla County, Yunnan Province (Chen and Zhang 2004).

Remarks. This species was described from Yunnan, China by Chen and Zhang (2004).

It is characterized in the female and male by the scutum with three dark longitudinal vittae and the base of the radius with a tuft of dark hairs; in the female by the sensory vesicle 0.3 times as long as the third palpal segment and mandible lacking outer teeth; in the male by the number of upper-eye (large) facets in 10 vertical columns and 14 horizontal rows and fore and hind basitarsi ca. 7.0 and 4.3 times as long as their greatest widths, respectively; in the pupa by the cocoon with a short to medium-long anterodorsal projection; and in the larva by the postgenal cleft long, 3.5–4.0 times as long as the postgenal bridge.

Simulium (Gomphostilbia) baisasae Species-group

Simulium (Gomphostilbia) baisasae Delfinado, 1962 [1960]

(Fig. 16A–C, H)

Simulium baisasae Delfinado, 1962 [1960]: 53–57 (female, male, pupa, and larva).
Simulium (Gomphostilbia) baisasae Delfinado, 1971: 137; Takaoka, 1983: 77–80 (female, male, pupa, and larva).

Distribution. Philippines (Luzon and Mindoro).

Bionomics. Aquatic habitats: Small to medium-sized streams in lowlands, Luzon (Takaoka, 1983).

Remarks. This species was described from Luzon, the Philippines by Delfinado (1962). Its female, male, pupa, and larva were redescribed by Takaoka (1983). The distribution of this species is recognized from Luzon Island and Mindoro Island. The previous records of this species from other islands based on only adult females (Delfinado, 1971) were thought to be incorrect judging from the findings of two other related species (Takaoka, 1983).

It is characterized by the different color of a tuft hair at the base of the radius between the female and the male (yellow in the female but dark in the male), pupal gill with six slender filaments (Fig. 16H) and cocoon with an anterodorsal margin strongly woven and somewhat bulged anteromedially.

The number of male upper-eye (large) facets was noted as ca. 14 horizontal rows (Takaoka, 1983) but is 17–18 vertical columns and 18 or 19 horizontal rows in 10 specimens collected from Luzon and Mindoro (Unpublished data).

Simulium (Gomphostilbia) bicolense Takaoka, 1983

(Fig. 16D–G, K, N, O)

Simulium (Gomphostilbia) bicolense Takaoka, 1983: 82–82 (female, male, pupa, and larva).

Distribution. Philippines (Luzon and Samar).

Bionomics. Aquatic habitats: Small streams and shallow rivers in lowlands in the southern area of Luzon, and Samar (Takaoka, 1983).

Remarks. This species was described from Luzon, the Philippines (Takaoka, 1983). It was later recorded from Samar (Takaoka and Tenedero, 2008c).

It is characterized by the pupal gill with 10 slender filaments arranged as 2+2+3+3 from dorsal to ventral (Fig. 16K). The female, male, and larva of this species are similar to those of *S. baisasae*.

Simulium (Gomphostilbia) visayaense Takaoka, 1983

(Fig. 16I, J, L, M)

Simulium (Gomphostilbia) visayaense Takaoka, 1983: 80–83 (female, male, pupa, and larva).

Distribution. Philippines (Cebu, Bohol, Leyte, Mindanao, Negros, and Panay).

Bionomics. Aquatic habitats: Small streams, irrigation canals and shallow rivers in lowlands (Takaoka, 1983).

Remarks. This species was described from the Visayan Islands and Mindanao, the Philippines (Takaoka, 1983).

It is characterized by the pupal gill with eight, nine, and 10 slender filaments (Fig. 16I, J). An intraspecific variation in the number of pupal gill filaments is rare in the subgenus *Gomphostilbia*.

The female, male, and larva of this species are similar to those of *S. baisasae*. The number of male upper-eye (large) facets was not mentioned in the original description (Takaoka, 1983). It is 14–17 vertical columns and 17 or 18 horizontal rows in four specimens collected from Canlaon, Negros (Unpublished data). The base of the radius has a tuft of yellow hairs mixed with dark hairs in the female but has a tuft of dark hairs in the male.

Simulium (Gomphostilbia) banauense Species-group

Simulium (*Gomphostilbia*) *atuganense* Takaoka, 2009

(Fig. 17A, O–R)

Simulium (*Gomphostilbia*) *atuganense* Takaoka, 2009b: 65–72 (female, male, pupa, and larva).

Distribution. Philippines (Mindanao).

Bionomics. Aquatic habitats: Small to medium-sized streams (width 2–5 m, elevation 1300–1400 m) in Mt. Katanglad (Takaoka, 2009b).

Remarks. This species was described from Mindanao, the Philippines (Takaoka, 2009b).

It is characterized in the female by abdominal segment 5 paler than segments 3 and 4 dorsolaterally; and in the pupa by the frons lacking tubercles and gill with six slender filaments arranged as (2+2)+2 from dorsal to ventral, of which two filaments of the ventral pair are much longer than those of the middle and dorsal pairs and the stalk of the dorsal pair is divergent basally from the stalk of the ventral pair at an angle of ca. 150 degrees.

Simulium (*Gomphostilbia*) *banauense* Takaoka, 1983

(Fig. 18H)

Simulium (*Morops*) *banauense* Takaoka, 1983: 43–46 (female, male, pupa, and larva).

Simulium (*Gomphostilbia*) *banauense* Takaoka, 2003: 52.

Distribution. Philippines (Luzon).

Bionomics. Aquatic habitat: A small stream at an elevation of ca. 2000 m (Takaoka, 1983).

Remarks. This species was described from Banaue, Luzon Island, the Philippines (Takaoka, 1983).

It is characterized by the pupal gill with six short inflated filaments each with a rounded apex (Fig. 18H).

***Simulium* (*Gomphostilbia*) *bansonae* Takaoka, 1983**

(Fig. 18A)

Simulium (*Morops*) *bansonae* Takaoka, 1983: 50–51 (female, male, and pupa).
Simulium (*Gomphostilbia*) *bansonae* Takaoka, 2003: 52; Takaoka, 2009b: 76.

Distribution. Philippines (Luzon).

Bionomics. Aquatic habitat: A small stream at an elevation of ca. 2000 m (Takaoka, 1983).

Remarks. This species was described from Luzon, the Philippines (Takaoka, 1983). Its larva is unknown. Some of the female and pupal characteristics were revised by Takaoka (2009b).

It is characterized in the female by the abdominal segments 2 and 5–8 shiny dorsally; and in the pupa by the bare frons and gill with six slender filaments arranged as 1+1+2+2 from dorsal to ventral, all of which are subequal in length to one another (Fig. 18A).

***Simulium* (*Gomphostilbia*) *cagayanense* Takaoka, 2005**

(Figs. 17D, E and 18C)

Simulium (*Gomphostilbia*) *cagayanense* Takaoka, 2005: 211–218 (female, male, pupa, and larva).

Distribution. Philippines (Luzon).

Bionomics. Aquatic habitat: A small cascading stream (width 10–50 cm, bottom rocky, elevation ca. 40 m) (Takaoka, 2005).

Remarks. This species was collected from Cagayan, a northern area of Luzon Island, the Philippines (Takaoka, 2005).

It is characterized in the female by abdominal segments 2 and 5–9 shiny dorsally; and in the pupa by the gill composed of three inflated tubular filaments arranged as 2+1 from dorsal to ventral, each with two slender filaments apically (Fig. 18C).

***Simulium* (*Gomphostilbia*) *disneyi* Takaoka and Roberts, 1988**

Simulium (*Morops*) *disneyi* Takaoka and Roberts, 1988: 196–198 (female)
Simulium (*Gomphostilbia*) *disneyi*: Takaoka 2003: 61–63 (female).

Distribution. Indonesia (Sulawesi).

Bionomics. Unknown.

Remarks. This species was described from a single female collected by a light trap in Dumoga-Bone, North Sulawesi (Takaoka and Roberts, 1988). The revised description of the female was provided by Takaoka (2003). The male, pupa, and larva remain unknown.

The female of this species is characterized by the narrow frons (frons:head ratio 1.0:6.4), mandible with two teeth on its outer margin and calcipala elongated (1.7 times as long as wide).

***Simulium* (*Gomphostilbia*) *ifugaoense* Takaoka, 1983**

(Fig. 18E)

Simulium (*Morops*) *ifugaoense* Takaoka, 1983: 46–47 (female, male, and pupa).
Simulium (*Gomphostilbia*) *ifugaoense*: Takaoka, 2003: 52.

Distribution. Philippines (Luzon).

Bionomics. Aquatic habitat: A small stream at an elevation of ca. 2000 m (Takaoka, 1983).

Remarks. This species was described from Luzon, the Philippines (Takaoka, 1983). Its larva is unknown.

It is characterized by the pupal gill composed of two inflated tubular filaments, of which the dorsal one bears four slender filaments apically and the ventral one has two slender filaments apically (Fig. 18E).

***Simulium* (*Gomphostilbia*) *katangladense* Takaoka, 2009**

(Fig. 17K, L, N)

Simulium (*Gomphostilbia*) *katangladense* Takaoka, 2009b: 73–76 (male and pupa).

Distribution. Philippines (Mindanao).

Bionomics. Aquatic habitat: A medium-sized stream (3–5 m wide) at an elevation of 1300 m in Mt. Katanglad, Mindanao (Takaoka, 2009b).

Remarks. This species was described from Mindanao, the Philippines (Takaoka, 2009b). The female and larva remain unknown.

It is characterized in the male by the sensory vesicle with an enlarged opening; and in the pupa by the gill with six slender filaments arranged as (2+2)+2 from dorsal to ventral, of which two filaments of the ventral pair are much longer than those of the middle and dorsal pairs.

***Simulium* (*Gomphostilbia*) *lilianae* Takaoka, 2000**

(Figs. 17B, C, F–H and 18G)

Simulium (*Morops*) *lilianae* Takaoka, 2000b: 361–364 (female and pupa).
Simulium (*Gomphostilbia*) *lilianae*: Takaoka, 2003: 52.

Distribution. Philippines (Luzon).

Bionomics. Aquatic habitat: A fast-flowing stream (5 m wide) at an elevation of ca. 1100 m (Takaoka, 2000b).

Remarks. This species was described from Banaue, Luzon, the Philippines (Takaoka, 2000b). Its male and larva are unknown.

It is characterized in the female by the mandible serrated on both sides; and in the pupa by the anterior two-thirds of the thorax bare and the gill with six short inflated antler-like filaments, arranged as (2+2)+2 from dorsal to ventral (Fig. 18G).

***Simulium* (*Gomphostilbia*) *liliwense* Takaoka, 1983**

(Figs. 17I, J, M and 18B)

Simulium (*Morops*) *liliwense* Takaoka, 1983: 51–53 (female, male, pupa, and larva).
Simulium (*Gomphostilbia*) *liliwense*: Takaoka, 2003: 52; Takaoka, 2009b: 76–77.

Distribution. Philippines (Luzon).

Bionomics. Aquatic habitats: A small stream (width 0.5–1.6 m) and a creek (width 4–6 m) at a mountain foot (Takaoka, 1983).

Remarks. This species was described from Laguna, Luzon, the Philippines (Takaoka, 1983). Some characteristics of the female, male, and pupa of this species were revised by Takaoka (2009b).

It is characterized by the pupal gill with six slender filaments arranged as (2+2)+2 from dorsal to ventral, all of which are subequal in length to one another.

***Simulium (Gomphostilbia) manbucalense* Takaoka, 1983**
(Fig. 18F)

Simulium (*Morops*) *manbucalense* Takaoka, 1983: 47–48 (male, pupa, and larva).
Simulium (*Gomphostilbia*) *manbucalense*: Takaoka, 2003: 52.

Distribution. Philippines (Negros).

Bionomics. Aquatic habitat: A shaded creek (4–6 m wide) in a forest (Takaoka, 1983).

Remarks. This species was described from Manbucal, Negros, the Philippines (Takaoka, 1983). Its female is unknown.

The pupal gill of this species is essentially similar to that of *S. ifugaoense* except the inflated tubular filaments lacking any patterns on their cuticular surface (Fig. 18F).

***Simulium (Gomphostilbia) quezonense* Takaoka, 2009**

Simulium (*Gomphostilbia*) *quezonense* Takaoka, 2009b: 73–72 (female, male, pupa, and larva).

Distribution. Philippines (Luzon).

Bionomics. Aquatic habitat: A river (5–10 m wide) at an elevation of ca. 600 m (Takaoka, 2009b).

Remarks. This species was described from Quezon, Luzon (Takaoka, 2009b). Its larva remains unknown.

It is characterized in the female by abdominal segment 5 paler than segments 3 and 4 dorsolaterally; in the male by the hind basitarsus spindle-shaped; and in the pupa by the gill with six slender filaments arranged as (2+2)+2 from dorsal to ventral, of which two filaments of the ventral pair are much longer than those of the middle and dorsal pairs.

***Simulium (Gomphostilbia) riverai* Takaoka, 1983**
(Fig. 18D)

Simulium (*Morops*) *riverai* Takaoka, 1983: 48–50 (female, male, pupa, and larva).
Simulium (*Gomphostilbia*) *riverai*: Takaoka, 2003: 52.

Distribution. Philippines (Negros).

Bionomics. Aquatic habitat: A shaded creek (width 4–6 m) in a forest (Takaoka, 1983).

Remarks. This species was described from Manbucal, Negros (Takaoka, 1983).

It is characterized by the pupal gill composed of four inflated filaments arranged as 2+2 from dorsal to ventral, of which each filament of dorsal pair has two slender filaments apically, and each filament of the ventral pair has one slender filament apically (Fig. 18D).

***Simulium* (*Gomphostilbia*) *salazarae* Takaoka, 1983**

Simulium (*Morops*) *salazarae* Takaoka, 1983: 53–55 (female, male, pupa, and larva).

Simulium (*Gomphostilbia*) *salazarae*: Takaoka, 2003: 52; Takaoka, 2009b: 77.

Distribution. Philippines (Bohol, Luzon, and Mindanao).

Bionomics. Aquatic habitats: Small forest streams (width 0.3–0.5 m) in Mindanao, a mountain creek (width 4–5 m) in Luzon and a headwater of the spring (only ca. 5 m in flowing length) in Bohol (Takaoka, 1983).

Remarks. This species was described from Bohol, Luzon and Mindanao, the Philippines (Takaoka, 1983). It was later recorded from Samar (Takaoka and Tenedero, 2008c). Some characteristics of the female, male, and pupa of this species were revised by Takaoka (2009b).

It is characterized by the pupal gill with six slender filaments arranged as (2+2)+2 from dorsal to ventral (although stalks of the dorsal and middle pairs lie almost side by side), all of which are subequal in length to one another.

***Simulium (Gomphostilbia) batoense* Species-group**

***Simulium (Gomphostilbia) binuanense* Subgroup**

***Simulium* (*Gomphostilbia*) *babuyanense* Takaoka and Tenedero, 2007**
(Fig. 19K–Q, U)

Simulium (*Gomphostilbia*) *babuyanense* Takaoka and Tenedero, 2007c: 208–212 (male and pupa).

Distribution. Philippines (Palawan Island).

Bionomics. Aquatic habitat: A small stream (width 3–5 m, 29 °C, exposed to the sun, elevation 5 m) near the Anilawan bridge of the road to El Nido from Puerto Princesa (Takaoka and Tenedero, 2007c).

Remarks. This species was described from a single male reared from a pupa collected from Palawan Island, the Philippines (Takaoka and Tenedero, 2007c). Its female and larva are unknown.

It is characterized in the male by the number of upper-eye (large) facets in 14 vertical columns and 14 horizontal rows on each side and scutal color pattern consisting of a non-pruinose median vitta and two non-pruinose submedian spots contrasting to the whitish-gray pruinose area of the scutum, when illuminated anterodorsally and viewed dorsally; and in the pupa by the gill with eight long slender filaments, of which the inner filament of the ventral pair is the longest (5.3 mm), twice the length of the pupal body, and terminal hooks widened, with their outer margins weakly serrate or undulate.

***Simulium* (*Gomphostilbia*) *binuanense* Takaoka and Tenedero, 2008**
(Fig. 19A, B, D–J, R, T)

Simulium (*Gomphostilbia*) *binuanense* Takaoka and Tenedero, 2008a: 1–7 (female and pupa); Takaoka *et al.*, 2009b: 213–216 (male).

Distribution. Philippines (Palawan).

Bionomics. Aquatic habitat: A small stream (width 2 m, 24–26 °C, exposed to the sun, elevation 40 m) near the Binuan bridge at the diverging point to Montible from the main road from Puerto Princesa to Brooks Point, Palawan (Takaoka and Tenedero, 2008a).

Remarks. This species was described based on a single female reared from a pupa collected from Palawan Island, the Philippines (Takaoka and Tenedero, 2008a). Its male was described by Takaoka et al. (2009b).

It is characterized in the female by the frons almost bare except several hairs along each lateral margin (Fig. 19A), sensory vesicle long, 0.56 times as long as the third palpal segment, and subcosta without hairs; in the male by the number of upper-eye (large) facets in 12 or 13 vertical columns and 12 or 13 horizontal rows; and in the pupa by the elongate stalk of the ventral pair and elongated secondary stalks of the middle and dorsal triplets of the gill (Fig. 19R).

***Simulium* (*Gomphostilbia*) *jerantutense* Takaoka, Sofian-Azirun and Ya'cob, 2014**
(Fig. 19S)

Simulium (*Gomphostilbia*) *jerantutense* Takaoka, Sofian-Azirun and Ya'cob, 2014b: 477–479 (female and pupa).

Distribution. Malaysia (Peninsular).

Bionomics. Aquatic habitat: A small stream (width 0.5–1.0 m, 27 °C, shaded, elevation 280 m) slowly flowing in an oil palm plantation in Pahang (Takaoka et al., 2014b).

Remarks. This species was described from two females reared from pupae from Peninsular Malaysia (Takaoka et al., 2014b). Its male and larva are unknown.

It is characterized in the female by the claw with a basal tooth 0.46 times the length of the claw; and in the pupa by the gill with eight filaments arranged as [2+1+(2+1)]+2 from dorsal to ventral.

***Simulium* (*Gomphostilbia*) *mangasepi* Takaoka, 1983**

Simulium (*Gomphostilbia*) *mangasepi* Takaoka, 1983: 73–74 (female, pupa, and head capsule of larva); Takaoka and Tenedero, 2007c: 218–221 (male).

Distribution. Philippines (Palawan Island).

Bionomics. Aquatic habitat: A small stream (width 2 m, 24–26 °C, exposed to the sun, elevation 40 m), near the Binuan bridge at the diverging point to Montible from the main road from Puerto Princesa to Brooks Point (Takaoka, 1983).

Remarks. This species was originally described from a single female reared from a pupa, and a head capsule of the larva attached to this pupa, collected from Palawan, the Philippines (Takaoka, 1983). Later, its male was fully described and its pupa was revised for its certain characteristics (Takaoka and Tenedero, 2007c). Its larva remains unknown.

It is characterized in the female by the genital fork with its stem bearing a somewhat inflated apex; in the male by the number of upper-eye (large) facets in 11 or 12 vertical columns and 13 or 14 horizontal rows; and in the pupa by the gill with eight long slender filaments arranged as (3+3)+2, of which two triplets lie horizontally and the ventral pair has a long stalk three to five times as long as the common basal stalk, and abdominal segment 9 with a pair of cone-like terminal hooks dorsally and a few slender unbranched and branched setae (in place of grapnel-like hooklets) on each lateral side.

***Simulium* (*Gomphostilbia*) *pegalanense* Smart and Clifford, 1969**

Simulium (*Gomphostilbia*) *pegalanense* Smart and Clifford, 1969: 39 (male, pupa, and larva); Takaoka and Tenedero, 2008a: 7 (brief description of female).
Simulium (*Gomphostilbia*) sp. A: Takaoka and Davies, 1995a: 79–81 (pupa and larva).

Distribution. Malaysia (Sabah and Peninsular).

Bionomics. Aquatic habitats: Small streams in lowlands, Sabah (Smart and Clifford, 1969).

Remarks. This species was described from Sabah, Malaysia (Smart and Clifford, 1969). Some morphological characteristics of the female was briefly mentioned by Takaoka and Tenedero (2008a). *Simulium* sp. A described from the pupa and larva from Peninsular Malaysia was identified as *S. pegalanense* (Takaoka et al., 2018m).

It is similar to *S. binuanense* in many characteristics including the arrangement of the pupal gill filaments but is barely distinguished in the female by the frons with 10 hairs just above the antennal base (no hairs in *S. binuanense*); in the male by the number of upper-eye (large) facets in 15 or 16 vertical columns (12 or 13 vertical columns in *S. binuanense*); and in the pupa by the shape of the pupal terminal hooks, of which the outer margin is nearly as long as the inner margin (the outer margin is about twice as long as the inner margin in *S. binuanense*) (Takaoka and Tenedero, 2008a; Takaoka et al., 2009b).

***Simulium* (*Gomphostilbia*) *tekamense* Takaoka, Sofian-Azirun and Ya'cob, 2014** (Fig. 19C)

Simulium (*Gomphostilbia*) *tekamense* Takaoka, Sofian-Azirun and Ya'cob, 2014b: 473–476 (female and pupa).

Distribution. Malaysia (Peninsular).

Bionomics. Aquatic habitat: A small stream (width 0.5–1.0 m, 27 °C, shaded, elevation 280 m) slowly flowing an oil palm plantation in Pahang (Takaoka et al., 2014b).

Remarks. This species was described from two females reared from pupae from Peninsular Malaysia (Takaoka et al., 2014b). Its male and larva are unknown.

It is similar to *S. binuanense* by having the subcosta bare or with two hairs but it differs from the latter species in the female by the frons moderately covered with hairs and the mediolongitudinal ridge of the cibarium having a deep notch apically (Fig. 19C); and in the pupa by the gill filaments arranged as [3+(1+2)]+2 from dorsal to ventral, of which the ventral pair has a medium-long stalk.

Simulium (Gomphostilbia) decuplum Subgroup

Simulium (*Gomphostilbia*) *bannaense* Chen and Zhang, 2003

Simulium (*Gomphostilbia*) *bannaense* Chen and Zhang, 2003: 542–544 (female, male, pupa, and larva).

Distribution. China (Yunnan).

Bionomics. Aquatic habitat: A small stream (shaded, elevation 700 m) in Yunnan (Chen and Zhang, 2003).

Remarks. This species was described from Yunnan Province, China by Chen and Zhang (2003).

It is characterized in the female by the sensory vesicle long, 0.6 times as long as the third palpal segment and claw with a small subbasal tooth; in the male by the small number of upper-eye (large) facets in 12 vertical columns and 13 horizontal rows; in the pupa by the gill with 10 short filaments arranged as 3+5+2 from dorsal to ventral and the dorsum of abdominal segment 1 sparsely covered with tubercles; and in the larva by the postgenal cleft long, with its apex reaching the posterior margin of the hypostoma.

Simulium (*Gomphostilbia*) *chiangraiense* Takaoka, Srisuka and Saeung, 2018 (Fig. 20A–I)

Simulium (*Gomphostilbia*) *chiangraiense* Takaoka, Srisuka and Saeung [in Takaoka et al., 2018f]: 272–277 (female, male, pupa, and larva).

Distribution. Thailand.

Bionomics. Aquatic habitat: A small stream (width 1 m, depth 2.5 cm, bottom sandy, flow moderate, 20.8 °C, pH 6.5, exposed to the sun, elevation 1321 m), Pang Khon Village, Muang District, Chiang Rai Province, northern Thailand (Takaoka et al., 2018f).

Remarks. This species was described from Thailand by Takaoka et al. (2018f).

It is similar to *S. decuplum* but is distinguished in the male by the number of upper-eye (large) facets in 16 vertical columns and 16 horizontal rows (in 14 or 15 vertical columns and 14 or 15 horizontal rows in *S. decuplum*); in the pupa by the inner filament of the ventral pair of the gill 1.5–1.6 mm long (ca. 0.7 mm in *S. decuplum*) and the dorsal surface of abdominal segments 1 and 2 darkened (not darkened in *S. decuplum*); and in the larva by the thorax densely covered with dark multibranched setae (moderately in *S. decuplum*) (Takaoka and Davies 1995a; Takaoka, unpublished data).

***Simulium* (*Gomphostilbia*) *decuplum* Takaoka and Davies, 1995**
Simulium (*Gomphostilbia*) *decuplum* Takaoka and Davies, 1995a: 46–50 (female, pupa, and larva); Takaoka et al., 2018m (male).

Distribution. Malaysia (Peninsular) and India (Assam) (?).

Bionomics. Aquatic habitats: Small streams (width 0.2–1.5 m, 20.5–24.0 °C) moderate to fast flowing in lowlands (Takaoka and Davies, 1995a).

Remarks. This species was described from Peninsular Malaysia (Takaoka and Davies, 1995a). It was also recorded from Thailand (Takaoka and Saito, 1996) and India (Borah et al., 2012). The male of this species was described by Takaoka et al. (2018m). The species formerly regarded as *S. decuplum* in Thailand was genetically shown to include more than two lineages (Pramual et al., 2011), and finally divided into five species based on morphological and genetic analyses (Takaoka et al., 2018f), of which *S. huaikaeoense* was based on specimens collected from Huai Kaeo, Chiang Mai Province, the area where *S. decuplum* was previously recorded from Thailand by Takaoka and Saito (1996). Thus, the previous record of *S. decuplum* might be interpreted as that of *S. huaikaeoense*. The species identified as *S. decuplum* in Assam, India should be reevaluated in the future.

It is characterized in the female by the sensory vesicle long, 0.54 times as long as the third palpal segment, and claw with a relatively long basal tooth 0.61 times as long as the claw; in the male by the number of upper-eye (large) facets in 14 or 15 vertical columns and 14 or 15 horizontal rows; in the pupa by the gill with 10 short filaments arranged as 4+4+2 from dorsal to ventral and abdominal segments 1 and 2 not darkened and lacking tubercles on their dorsal surface; and in the larva by the postgenal cleft long but not reaching the posterior margin of the hypostoma.

***Simulium* (*Gomphostilbia*) *huaikaeoense* Takaoka, Srisuka and Low, 2018**
Simulium (*Gomphostilbia*) *huaikaeoense* Takaoka, Srisuka and Low [in Takaoka et al., 2018f]: 282–283 (female, male, pupa, and larva).

Distribution. Thailand.

Bionomics. Aquatic habitats: A stream (width 3.5 m, depth 35 cm, bottom sandy, flow moderate, 23.2 °C, partially shaded, elevation 503 m), Huai Kaeo, Doi Suthep-Pui National Park, Chiang Mai Province, and a stream (width 4 m, depth 30 cm, bottom sandy, flow fast, 19.5 °C, elevation 1192 m), Namdan Village, Doi Phuka National Park, Nan Province (Takaoka et al., 2018f).

Remarks. This species was described from Thailand by Takaoka et al. (2018f).

It is distinguished from *S. chiangraiense* and *S. pamiangense* by the smaller number of the male upper-eye (large) facets in 13 or 14 vertical columns and 14 or 15 horizontal rows, shorter pupal gill filaments (0.7–1.0 mm long), pale dorsal surface of pupal abdominal segments 1 and 2, and larval thorax and abdomen densely covered with black, multibranched setae dorsally and dorsolaterally.

This species is distinguished from *S. maeklangense* in the female by the lighter color of the mid and hind femora; and in the male by the smaller number of upper-eye (large) facets in 13 or 14 vertical columns and 14 or 15 horizontal rows; and from *S. khaokhoense* in the female by the fore basitarsus 5.3 times as long as its greatest width; in the male by the fore basitarsus 7.4 times as long as its greatest width; and in the larva by the posterior circlet with 86 rows of hooklets.

***Simulium* (*Gomphostilbia*) *jiulianshanense* Kang, Zhang and Chen, 2007**

Simulium (*Gomphostilbia*) *jiulianshanense* Kang, Zhang and Chen, 2007: 185–186 (pupa and larva).

Distribution. China (Jiangxi).

Bionomics. Aquatic habitat: A trickle at an elevation of 750 m, in Jiulianshan, Jiangxi (Kang et al., 2007).

Remarks. This species was described from Jiangxi, China by Kang et al. (2007). Its adult female and male are unknown.

It is characterized in the pupa by the gill with 10 short filaments arranged as 3+5+2 from dorsal to ventral; and in the larva by the postgenal cleft reaching the posterior border of the hypostoma, by which it is distinguished from *S. decuplum* from Peninsular Malaysia (Takaoka and Davies, 1995a).

***Simulium* (*Gomphostilbia*) *khaokhorense* Takaoka, Srisuka and Low, 2018**

Simulium (*Gomphostilbia*) *khaokhorense* Takaoka, Srisuka and Low [in Takaoka et al., 2018f]: 280–282 (female, male, pupa, and larva).

Distribution. Thailand.

Bionomics. Aquatic habitat: A small stream (width 70 cm, depth 4 cm, bottom sandy, flow moderate, 24.2 °C, partially shaded, elevation 540 m), Ton Nampasak Research Station, Khao Kho District, Phetchabun Province (Takaoka et al., 2018f).

Remarks. This species was described from Thailand by Takaoka et al. (2018f).

It is similar to *S. maeklangense* but is distinguished in the female by the lighter color of the mid and hind femora and in the male by the smaller number of upper-eye (large) facets in 14 vertical columns and 14 horizontal rows.

This species is distinguished from *S. chiangraiense* and *S. pamiangense* by the shorter wing length (1.4–1.6 mm), smaller number of male upper-eye (large) facets in 14 vertical columns and 14 horizontal rows, shorter pupal gill filaments (0.7–1.0 mm long), pale dorsal surface of pupal abdominal segments 1 and 2, and larval thorax and abdomen densely covered with black, multibranched setae dorsally and dorsolaterally.

***Simulium* (*Gomphostilbia*) *lamdongense* Takaoka and Sofian-Azirun, 2015**
(Fig. 20J–U)

Simulium (*Gomphostilbia*) *lamdongense* Takaoka and Sofian-Azirun (in Takaoka et al., 2015c): 35–41 (female, male, pupa, and larva).

Distribution. Vietnam.

Bionomics. Aquatic habitats: Two small streams (width 1.0–2.0 m, depth 10–20 cm, flow moderate, 15 °C, shaded, elevation 1539 m, and width 0.5–1.0 m, depth 3–5 cm, 15 °C, exposed to the sun, elevation 1746 m) (Takaoka et al., 2015c).

Remarks. This species was described from Lam Dong Province in southern Vietnam (Takaoka et al., 2015c).

It is characterized in the female by the sensory vesicle medium-long, 0.43–0.45 times as long as the third palpal segment; in the male by the greater number of upper-eye (large) facets in 16 or 17 vertical columns and 17 or 18 horizontal rows; in the pupa by the thoracic integument sparsely covered with tubercles; and in the larva by the postgenal cleft long, with its apex close to the posterior margin of the hypostoma (Fig. 20T).

***Simulium* (*Gomphostilbia*) *maeklangense* Takaoka, Srisuka and Saeung, 2018**
Simulium (*Gomphostilbia*) *maeklangense* Takaoka, Srisuka and Saeung [in Takaoka et al., 2018f]: 278–280 (female, male, pupa, and larva).

Distribution. Thailand.

Bionomics. Aquatic habitat: A small stream (width 45 cm, depth 24 cm, bottom sandy, 20.2 °C, partially shaded, elevation 1314 m), moderately flowing, Mae Klang Watershed Management Unit, Doi Inthanon National Parks, Chiang Mai Province (Takaoka et al., 2018f).

Remarks. This species was described from Thailand by Takaoka et al. (2018f). It is distinguished from *S. chiangraiense* and *S. pamiangense* by the shorter pupal gill filaments (0.7–1.0 mm long), pale dorsal surface of pupal abdominal segments 1 and 2, and larval thoracic and abdominal segments densely covered with dark multibranched setae dorsally and dorsolaterally.

***Simulium* (*Gomphostilbia*) *pamiangense* Takaoka, Srisuka and Saeung, 2018**
Simulium (*Gomphostilbia*) *pamiangense* Takaoka, Srisuka and Saeung [in Takaoka et al., 2018f]: 277–278 (female, male, pupa, and larva).

Distribution. Thailand.

Bionomics. Aquatic habitat: A small stream (width 30 cm, depth 5 cm, bottom sandy, flow moderate, 22.6 °C, pH 6.3, exposed to the sun, elevation 1097 m), Pa Miang Village, Chae Hom District, Lampang Province (Takaoka et al., 2018f).

Remarks. This species was described from Thailand by Takaoka et al. (2018f). It is similar to *S. chiangraiense* but is distinguished from the latter species in the male by the length of the fore basitarsus relative to its greatest width (6.6 in this species versus 7.6–7.8 in *S. chiangraiense*); and in the larva by the number of the primary rays of the labral fan (32–36 in this species versus 24–27 in *S. chiangraiense*).

***Simulium* (*Gomphostilbia*) Pabho sp. Lewis, 1974**
Simulium Pabho sp. Lewis, 1974: 38 (female and pupa).

Distribution. India (Assam).

Bionomics. Unknown.

Remarks. This unnamed species was reported from Assam, India, by Lewis (1974). Its male and larva are unknown. Morphological characteristics of its female and pupa are not available except that the female spermatheca is elongated, the pupal gill is composed of 10 filaments arranged 4+4+2 from dorsal to ventral. It is provisionally placed in the *S. decuplum* subgroup of the *S. batoense* species-group of the subgenus *Gomphostilbia* judging from the number and arrangement of the gill filaments and the shape of the spermatheca.

Simulium (Gomphostilbia) duolongum Subgroup

Simulium (Gomphostilbia) abadii Takaoka, 2003

Simulium (*Gomphostilbia*) *abadii* Takaoka, 2003: 76–81 (female, male, and pupa).

Distribution. Indonesia (Sulawesi).

Bionomics. Aquatic habitat: A stream (width ca. 3.0 m, 28.0 °C, partially shaded, elevation 40 m), Kendari, Southeast Sulawesi (Takaoka, 2003).

Remarks. This species was described from Sulawesi, Indonesia (Takaoka, 2003). Its larva is unknown.

It is characterized in the female by the sensory vesicle medium-long, 0.38 times as long as the third palpal segment; in the male by the number of upper-eye (large) facets in 13 or 14 vertical columns and 14 horizontal rows; and in the pupa by the gill with eight slender filaments arranged as (2+1)+3+2 or [(2+1)+3]+2 from dorsal to ventral, of which two filaments of the ventral pair are twice as long as the six other filaments and have a short stalk about half the length of the interspiracular trunk and abdominal segments 1 and 2 light yellow and bare dorsally.

Simulium (Gomphostilbia) batoense Edwards, 1934

(Fig. 22P)

Simulium (*Eusimulium*) *batoense* Edwards, 1934: 124–125 (female, male, and pupa).

Simulium (*Gomphostilbia*) *batoense*: Crosskey, 1973: 424; Takaoka and Davies, 1996: 22–26 (female, male, pupa, and larva).

Distribution. Indonesia (Java).

Bionomics. Aquatic habitat: A small stream (width ca. 1 m) flowing in a tea plantation at the foot of Mt. Tangkubanperahu, Bandung (Takaoka and Davies, 1996).

Remarks. This species was described from Java, Indonesia by Edwards (1934), and redescribed by Takaoka and Davies (1996).

It is characterized in the female by the claw with a small basal tooth (similar to Fig. 21I); in the male by the number of upper-eye (large) facets in 17 vertical columns and 17 horizontal rows, scutum with yellow short hairs and ventral plate with its body slightly narrowed posteriorly when viewed ventrally; in the pupa by the gill with eight slender filaments arranged as 2+[1+(1+2)]+2 dorsal to ventral, of which two filaments of the ventral pair are 1.9–2.2 times as long as the six other filaments, and abdominal segments 1 and 2 unpigmented and bare dorsally; in the larva by the postgenal cleft arrow-head-shaped and long, 6.5–10.0 times as long as the postgenal bridge.

Simulium (Gomphostilbia) bhutanense Takaoka and Somboon, 2008

Simulium (*Gomphostilbia*) *bhutanense* Takaoka and Somboon, 2008: 213–216 (female); Shrestha and Takaoka, 2009: 253–254; Takaoka et al., 2020f: 195–198 (male and pupa).

Distribution. Bhutan and Nepal.

Bionomics. Aquatic habitats: An irrigation channel (width 0.8 m, 29.6 °C, shaded, elevation 218 m) in Chuzargang, Gelephu in Bhutan (Takaoka and Somboon, 2008) and a stream at an elevation of 660 m, Pachaskilo, Pachkhal, Nepal (Shrestha and Takaoka, 2009).

Remarks. This species was described from females reared from pupae collected in Bhutan (Takaoka and Somboon, 2008) and was also recorded from Nepal (Shrestha and Takaoka, 2009). Its male and pupa were described from specimens collected from Nepal by Takaoka et al. (2020f). Its larva is unknown.

It is characterized in the female by the claw with a medium-sized tooth 0.32 times the length of the claw and abdominal segments 2 and 5–9 shiny dorsally; in the male by the number of upper-eye (large) facets in 14 or 15 vertical columns and 16 or 17 horizontal rows and hind basitarsus parallel-sided and somewhat enlarged, 4.7 times as long as its greatest width and 0.7 times as wide as the hind tibia and femur; and in the pupa by the gill with eight filaments arranged as (1+2)+(1+2)+2 or 3+(1+2)+2 from dorsal to ventral, of which two filaments of the ventral pair are much longer and thicker than the six other filaments and have a short stalk, about half the length of the interspiracular trunk.

***Simulium* (*Gomphostilbia*) *dachaisense* Takaoka and Lau, 2015**

(Figs. 21E, F, I–K, M–O and 22G, I)

Simulium (*Gomphostilbia*) *dachaisense* Takaoka and Lau [in Takaoka et al., 2015c]: 28–34 (female, male, pupa, and larva).

Distribution. Vietnam.

Bionomics. Aquatic habitat: A small shallow stream (width 3–6 m, depth 10–20 cm, bottom sandy, 20 °C, exposed to the sun, elevation 1440 m) slowly flowing in an open land (Takaoka et al., 2015c).

Remarks. This species was described from Lam Dong Province in southern Vietnam (Takaoka et al., 2015c).

It is characterized in the female by the claw with a small basal tooth (Fig. 21I); in the male by the number of upper-eye (large) facets in 18 vertical columns and 18 horizontal rows and scutum with yellow short hairs; in the pupa by the gill with eight slender filaments arranged as (2+1)+(1+2)+2 from dorsal to ventral, of which two filaments of the ventral pair are 2.0–3.0 times as long as the six other filaments, abdominal segments 1 and 2 grayish dorsally and abdominal segment 1 weakly tuberculate dorsally; and in the larva by the postgenal cleft long, 6.3–6.9 times as long as the postgenal bridge and abdominal segments 5–9 densely covered with unbranched or bifid dark spinous setae dorsally.

***Simulium* (*Gomphostilbia*) *davaoense* Takaoka, 1983**

Simulium (*Gomphostilbia*) *davaoense* Takaoka, 1983: 76–77 (male and pupa).

Distribution. Philippines (Mindanao).

Bionomics. Aquatic habitat: A stream (width 5–6 m) moderately running through a rice field and a coconut plantation in lowland, Davao, Mindanao (Takaoka, 1983).

Remarks. This species was described from Davao, Mindanao, the Philippines (Takaoka, 1983). Its female and larva are unknown.

It is characterized in the male by the number of upper-eye (large) facets in 12 vertical columns and 12 horizontal rows, style narrow, nearly parallel-sided when viewed ventrolaterally and ventral plate with its body transverse, widened posteriorly when viewed ventrally; and in the pupa by the gill composed of eight filaments arranged as 3+3+2 from dorsal to ventral, of which two filaments of the ventral pair are about twice as long as the six other filaments.

***Simulium* (*Gomphostilbia*) *duolongum* Takaoka and Davies, 1995**

Simulium (*Gomphostilbia*) *duolongum* Takaoka and Davies, 1995a: 19–24 (female, male, pupa, and larva).

Distribution. Malaysia (Peninsular), Indonesia (Sumatra), and Vietnam (?).

Bionomics. Aquatic habitats: Small streams (25 °C) at low elevation, Janda Baik Reserve near Bentong, Pahang (Takaoka and Davies, 1995a).

Remarks. This species was originally described from Peninsular Malaysia (Takaoka and Davies 1995a). It was later recorded from Sumatra, Indonesia (Takaoka et al., 2000) and Vietnam (Pham, 1999). The identity of the species regarded as *S. duolongum* based on two pupae and two larvae in Vietnam should be confirmed based on the adult males and females.

It is characterized in the female by the sensory vesicle medium-long, 0.3 times as long as the third palpal segment; in the male by the number of upper-eye (large) facets in 15 vertical columns and 15 horizontal rows; in the pupa by the gill composed of eight filaments arranged as 3+3+2 from dorsal to ventral, of which two filaments of the ventral pair are 2.5 times as long as the six other filaments of dorsal and middle triplets and abdominal segments 1 and 2 almost transparent dorsally; and in the larva by abdominal segments 5–8 densely covered with dark setae of various sizes, of which most are flattened, with fine comb-like transparent branches apically.

***Simulium* (*Gomphostilbia*) *friederichsi* Edwards, 1934**

Simulium (*Eusimulium*) *friederichsi* Edwards, 1934: 118–119 (male)
Simulium (*Gomphostilbia*) *friederichsi*: Takaoka et al., 2000: 160–161.

Distribution. Indonesia (Java and Sumatra).

Bionomics. Unknown.

Remarks. This species was described from Java, Indonesia by Edwards (1934). Its female, pupa, and larva are unknown. The male specimens reared from pupae collected from Sumatra were tentatively identified as *S. friederichsi* by Takaoka et al. (2000) since the scutal patterns were variable: some were identical to that of *S. friederichsi* and some others were similar to those of *S. siamense* from Thailand (Takaoka and Suzuki, 1984) and *S. yaeyamaense* from the Nansei Islands, Japan (Takaoka, 1991).

The male of this species is characterized by the scutum with three dark non-pruinose vittae (one median, long and two submedian, oval), the hind tibia black except its base pale and the ventral plate with its body transverse.

***Simulium* (*Gomphostilbia*) *hadiae* Takaoka, 2003**
(Fig. 21R)

Simulium (*Gomphostilbia*) *hadiae* Takaoka, 2003: 102–107 (female, male, and pupa).

Distribution. Indonesia (Sulawesi).

Bionomics. Aquatic habitat: A stream (width 2.0–3.0 m, 26.0 °C, shaded, elevation 10 m) in a sparsely wooded area, Tangkoko, North Sulawesi (Takaoka, 2003).

Remarks. This species was described from Sulawesi, Indonesia (Takaoka, 2003). Its larva is unknown.

It is characterized in the female by the wide frons (frons:head ratio 1.0:4.1); in the male by the number of upper-eye (large) facets in 12 or 13 vertical columns and 13 horizontal rows, scutal pattern consisting of the dark medial vita and two dark submedian spots on the shiny pruinose background (Fig. 21R) and covered with dark short hairs intermixed with white short hairs on the shoulders, along the lateral margins and on the prescutellar area and ventral plate slightly narrowed posteriorly when viewed ventrally; and in the pupa by the gill with eight slender filaments arranged as 3+(1+2)+2 from dorsal to ventral, of which two filaments of the ventral pair are 1.5–2.0 times as long as the six other filaments, and abdominal segments 1 and 2 light yellow dorsally.

***Simulium* (*Gomphostilbia*) *johorense* Takaoka, Sofian-Azirun and Ya'cob, 2014**
(Figs. 21D, P, Q, S–Z and 22N)

Simulium (*Gomphostilbia*) *johorense* Takaoka, Sofian-Azirun and Ya'cob [in Takaoka et al., 2014a]: 17–24 (female, male, pupa, and larva).

Distribution. Malaysia (Peninsular).

Bionomics. Aquatic habitats: A small lowland stream (width 0.2–0.5 m, depth 10 cm, flow slow, bottom sandy, 28.5 °C, exposed to the sun, elevation 79 m) running along the edge of an oil plantation at the foothill of Mt. Ledang, Johor (type locality) and a stream (width 0.5–3.0 m, depth 14–20 cm, flow moderate, bottom sandy, 27.8 °C, exposed to the sun, elevation 67 m) running along an oil plantation in Negeri Sembilan (Takaoka et al., 2014a).

Remarks. This species was described from Peninsular Malaysia (Takaoka et al., 2014a).

It is characterized in the female by the subcosta, which is bare or with one to five hairs; in the male by the number of upper-eye (large) facets in 13 vertical columns and 14 horizontal rows and scutum with various color patterns (Fig. 21P, Q, S): in the pupa by the gill with eight slender filaments arranged as 3+3+2 from dorsal to ventral, of which the longest inner filament of the ventral pair is 1.5 (or little more) times the length of the six other filaments, and abdominal segments 1 and 2 gray dorsally; and in the larva by the postgenal cleft long, with its apex reaching the posterior margin of the hypostoma (Fig. 22N).

***Simulium* (*Gomphostilbia*) *kolakaense* Takaoka, 2003**
Simulium (*Gomphostilbia*) *kolakaense* Takaoka, 2003: 88–95 (female, male, pupa, and larva).

Distribution. Indonesia (Sulawesi).

Bionomics. Aquatic habitat: A roadside stream (width 0.5–1.0 m, 27.0 °C, partially shaded, elevation 100 m), Kolaka, Southeast Sulawesi (Takaoka, 2003).

Remarks. This species was described from Sulawesi, Indonesia (Takaoka, 2003).

It is characterized in the male by the number of upper-eye (large) facets in 13–15 vertical columns and 14 horizontal rows, scutum with yellow short hairs intermixed with dark short hairs and ventral plate with its body narrowed posteriorly when viewed ventrally; in the pupa by the gill with eight slender filaments arranged as 3+3+2 from dorsal to ventral, of which two filaments of the ventral pair are 2.4–3.0 times as long as the six other filaments and abdominal segments 1 and 2 pale and bare dorsally; in the larva by the postgenal cleft arrow-head-shaped, 2.4–3.0 times as long as the postgenal bridge.

***Simulium* (*Gomphostilbia*) *laosense* Takaoka, Srisuka and Saeung, 2017**
Simulium (*Gomphostilbia*) *laosense* Takaoka, Srisuka and Saeung, 2017n: 1544–1551 (female, male, pupa, and larva).

Distribution. Laos.

Bionomics. Aquatic habitat: The Huai Yang stream (width 2.5 m, depth 15 cm, bottom sandy, flow moderate, 29.4 °C, pH 6.3, exposed to the sun, elevation 96 m), Uloompon District, Champasak Province (Takaoka et al., 2017n).

Remarks. This species was described from Laos as a species in the *S. batoense* species-group by Takaoka et al. (2017n).

It is characterized in the female by the fore basitarsus 5.6–5.7 times as long as its greatest width; in the male by the number of upper-eye (large) facets in 15 vertical columns and 15 horizontal rows and scutum covered with brassy short hairs; in the pupa by the arrangement and relative length of the eight gill filaments, stalks of the two triplets and one pair arising independently from the short common basal stalk, the inner filament of the ventral pair slightly longer than the counter filament and 1.7–1.8 times as long as three filaments of the middle triplet and abdominal segments 1 and 2 weakly tuberculate partially on their dorsal surface; and in the larva by the postgenal cleft long, with its apex close to the posterior margin of the hypostoma.

***Simulium* (*Gomphostilbia*) *lemborense* Takaoka and Sifian-Azirun, 2017** (Figs. 21L and 22C, O)
Simulium (*Gomphostilbia*) *lemborense* Takaoka and Sofian-Azirun [in Takaoka et al., 2017b]: 150–156 (female, male, pupa, and larva).

Distribution. Indonesia (Flores and Sumba).

Bionomics. Aquatic habitat: A slow-flowing stream (width 20–30 m, depth 30–40 cm, bottom rocky, 23.0 °C, exposed to the sun, elevation 53 m), Wae Loagge, Lembor, Flores Island (Takaoka et al., 2017b).

Remarks. This species was described from Flores, the Lesser Sunda Archipelago, Indonesia (Takaoka et al., 2017b). It was recorded from Sumba (Takaoka et al., 2018d).

It is characterized in the female by the narrow frons (frons:head ratio 1.0:6.2–6.3); in the male by the number of upper-eye (large) facets in 13 vertical columns and 13 horizontal rows, scutum with yellow short hairs, style wide basally, about half as wide as its length when viewed ventrolaterally and ventral plate with its body narrowed posteriorly when viewed ventrally; in the pupa by the gill with eight filaments arranged as (2+1)+(1+2)+2 from dorsal to ventral, of which two filaments of the ventral pair are three to four times as long as the six other filaments (Fig. 22C) and abdominal segments 1 and 2 entirely dark gray and bare dorsally; and in the larva by the postgenal cleft bullet-shaped, 2.6–2.9 times as long as the postgenal bridge.

***Simulium* (*Gomphostilbia*) *manadoense* Takaoka, 2003**
(Fig. 21L and Fig. 22C, O)

Simulium (*Gomphostilbia*) *manadoense* Takaoka, 2003: 71–76 (female, male, and pupa).
Simulium (*Gomphostilbia*) sp. B Takaoka and Roberts, 1988: 211 (pupa).

Distribution. Indonesia (Sulawesi).

Bionomics. Aquatic habitat: A stream (width ca. 3.0 m, 26.0 °C, shaded, elevation 70 m), near Amuran, Southeast Sulawesi (Takaoka, 2003).

Remarks. This species was described from Sulawesi, Indonesia (Takaoka, 2003). Its larva is unknown.

It is characterized in the male by the scutum with yellowish white short hairs interspersed with dark short hairs, the number of upper-eye (large) facets in 12 vertical columns and 13 horizontal rows and ventral plate with its body slightly widened from the base to the middle when viewed from ventrally; and in the pupa by the gill with eight slender filaments, of which two filaments of the ventral pair are 1.8 times as long as the six other filaments, abdominal segments 1 and 2 light yellow and bare dorsally and terminal hooks widened laterally, with their outer margins undulate.

***Simulium* (*Gomphostilbia*) *marosense* Takaoka, Sofian-Azirun and Suana, 2019**

Simulium (*Gomphostilbia*) *marosense* Takaoka, Sofian-Azirun and Suana [Takaoka et al., 2019d]: 392–399 (female, male, and pupa).

Distribution. Indonesia (Sulawesi).

Bionomics. Aquatic habitat: A slow-flowing stream (width 20–30 m, depth 30–40 cm, bottom rocky, 23.0 °C, partially shaded, elevation 53 m) in Maros, South Sulawesi (Takaoka et al., 2019d).

Remarks. This species was described from Sulawesi, Indonesia (Takaoka et al., 2019d).

It is characterized in the female by the narrow frons (frons:head ratio 1.0:6.3) and sensory vesicle medium-long, 0.34–0.38 times as long as the third palpal segment; in the male by the number of upper-eye (large) facets in 12 or 13 vertical columns and 13 or 14 horizontal rows, scutum with yellow short hairs and ventral plate with its ventral margin gently rounded ventrally when viewed posteriorly; in the pupa by the gill with eight filaments arranged as 3+3+2 from dorsal to ventral, of which the stalks of the dorsal and middle triplets are short and two filaments of the ventral pair are 1.9–3.0 times as long as the six other filaments, and abdominal segments 1 and 2 entirely dark gray to light brown and bare or sparsely covered with tubercles dorsally; and in the larva by the postgenal cleft long, 2.9–3.1 times as long as the postgenal bridge.

***Simulium* (*Gomphostilbia*) *samarense* Takaoka and Tenedero, 2008**
(Fig. 21A'–C')

Simulium (*Gomphostilbia*) *samarense* Takaoka and Tenedero, 2008c: 330–335 (male and pupa).

Distribution. Philippines (Samar).

Bionomics. Aquatic habitat: A small stream (width 1.5–2.5 m, 24 °C, shaded, elevation 50–75 m) moderately flowing in a coconut plantation in western Samar (Takaoka and Tenedero, 2008c).

Remarks. This species was described from a single male reared from a pupa collected from Samar Island, the Philippines (Takaoka and Tenedero, 2008c). Its female and larva are unknown.

This species is characterized in the male by the number of upper-eye (large) facets in 12 vertical columns and 12 or 13 horizontal rows, sensory vesicle with an enlarged opening, scutum with yellow short hairs and ventral plate slightly widened from the base to the middle when viewed ventrally, and moderately produced ventromedially, triangular and relatively thin when viewed posteriorly (Fig. 21C'); and in the pupa by the gill with eight slender filaments arranged as [(1+2)+(1+2)]+2 from dorsal to ventral, of which two filaments of the ventral pair are 1.4–1.7 times as long as the six other filaments, and abdominal segments 1 and 2 light brown and with tubercles dorsally.

***Simulium* (*Gomphostilbia*) *siamense* Takaoka and Suzuki, 1984 (complex)**

Simulium (*Gomphostilbia*) *siamense* Takaoka and Suzuki, 1984: 14–18 (female, male, pupa, and larva).

Distribution. Thailand, Nepal (?), and Vietnam (?).

Bionomics. Aquatic habitats: Small streams in Mae Hongson Province (Takaoka and Suzuki, 1984).

Remarks. This species was described from Thailand (Takaoka and Suzuki, 1984). It was later recorded from Nepal and Vietnam by Crosskey and Howard (1997). No detailed information of the record was given, leaving the localities in Nepal and Vietnam unknown. Kuvangkadilok et al. (2008) reported that *S. siamense* in Thailand was represented by five cytoforms. Pramual and Wongpakam (2011) genetically analyzed this species by using mitochondrial DNA sequences.

It is characterized in the male by the scutal color pattern consisting of a non-pruinose narrow median vitta and two non-pruinose elliptical submedian spots contrasting to the pruinose other areas; in the pupa by the gill with eight slender filaments, of which two filaments of the ventral pair are 1.5–2.0 times as long as the six other filaments; and in the larva by the postgenal cleft long, with its apex close to the posterior margin of the hypostoma.

***Simulium* (*Gomphostilbia*) *singgihi* Takaoka, 2003**

Simulium (*Gomphostilbia*) *singgihi* Takaoka, 2003: 95–102 (female, male, pupa, and larva).

Distribution. Indonesia (Sulawesi).

Bionomics. Aquatic habitats: Three streams (width 0.5–2.0 m, 22.0–28.0 °C, partially shaded, elevation 440–970 m), South Sulawesi (Takaoka, 2003).

Remarks. This species was described from Sulawesi, Indonesia (Takaoka, 2003). It is characterized by the pupal gill with eight slender filaments, of which two filaments of the ventral pair is 3.7–4.3 times as long as the six other filaments.

***Simulium* (*Gomphostilbia*) *syafruddini* Takaoka, 2003**

(Fig. 22B, J)

Simulium (*Gomphostilbia*) *syafruddini* Takaoka, 2003: 81–88 (female, male, pupa, and larva).

Distribution. Indonesia (Sulawesi).

Bionomics. Aquatic habitat: A forest stream (width 0.8 m, 22.0 °C, shaded, elevation 600 m) and another stream (width ca. 3.0 m, 28.2 °C, partially shaded, elevation 460 m), both in Gimpu, Central Sulawesi (Takaoka, 2003).

Remarks. This species was described from Sulawesi, Indonesia (Takaoka, 2003). It is characterized in the female and male by the scutum with copper-colored hairs; in the male by the number of upper-eye (large) facets in 16 vertical columns and 16 horizontal rows; and in the pupa by the gill with eight slender filaments arranged as (3+3)+2 from dorsal to ventral, of which two filaments of the ventral pair is about three times as long as the six other filaments (Fig. 22B); and in the larva by the postgenal cleft bullet-shaped and long, 3.3–4.6 times as long as the postgenal bridge.

***Simulium* (*Gomphostilbia*) *syuhaiense* Huang and Takaoka, 2008**

Simulium (*Gomphostilbia*) *syuhaiense* Huang and Takaoka, 2008a: 171–178 (female, male, pupa, and larva).

Distribution. Taiwan.

Bionomics. Aquatic habitats: Two slow-flowing streams (width 2 m and 3–4 m, shaded or partially shaded, elevation 210 m) in Pingtung County, Taiwan (Huang and Takaoka, 2008a).

Remarks. This species was described from Taiwan (Huang and Takaoka, 2008a). It is characterized in the female by the sensory vesicle long, 0.6 times as long as the third palpal segment; in the male by the number of upper-eye (large) facets in 17 or 18 vertical columns and 17 horizontal rows; and in the pupa by the dorsal surface of abdominal segments 1 and 2 light brown.

Simulium (*Gomphostilbia*) *tahanense* Takaoka and Davies, 1995

(Figs. 21A, H and 22E, F, L)

Simulium (*Gomphostilbia*) *tahanense* Takaoka and Davies, 1995a: 24–28 (female, pupa, and larva); Takaoka et al., 2018m: 16–18 (male).

Distribution. Malaysia (Peninsular and Sarawak) and Indonesia (Kalimantan).

Bionomics. Aquatic habitats: A small stream (width 2–3 m, 25 °C, shaded, elevation 67 m) in Tahan, Pahang (Takaoka and Davies, 1995a), and a small lowland stream (width 2–5 m), Negri Sembilan, Peninsular Malaysia; and a stream (width 2 m, 25 °C, elevation 118 m) in Kalimantan (Takaoka et al., 2016b).

Remarks. This species was described from Peninsular Malaysia (Takaoka and Davies, 1995a). It was later recorded from Sarawak (Takaoka, 2012) and Kalimantan, Indonesia (Takaoka et al., 2016b). The male of this species was described by Takaoka et al. (2018m).

It is characterized in the female and male by the elongate proboscis (the length ratio of the labrum against the clypeus is 1.2 in the female and 0.9 in the male); in the female by the frons nearly parallel-sided and almost bare except several hairs along both lateral sides and near the lower part (Fig. 21A) and claw with a medium-sized tooth 0.43 times as long as the claw (Fig. 21H); in the male by the number of upper-eye (large) facets in 11 vertical columns and 11 horizontal rows and ventral plate slightly widened posteriorly when viewed ventrally; in the pupa by the head integument with elongate tubercles (Fig. 22E) and gill with eight filaments arranged as [2+1+(1+2)]+2 from dorsal to ventral, of which two filaments of the ventral pair are twice as long as the six other filaments; and in the larva by thoracic segment 1 and abdominal segments 1 and 5–8 each with a grayish black marking dorsally (Fig. 22L).

Simulium (*Gomphostilbia*) *torautense* Takaoka and Roberts, 1988

(Figs. 21B, C, G and 22K)

Simulium (*Gomphostilbia*) *torautense* Takaoka and Roberts, 1988: 203–208 (female, male, pupa, and larva); Takaoka, 2003: 64–71 (female, male, pupa, and larva).

Distribution. Indonesia (Sulawesi).

Bionomics. Aquatic habitats: A river (width ca. 8 m, 25.0 °C, shaded, elevation ca. 500 m) and small streams (width 0.2–3.0 m, 25.0–26.5 °C, partially shaded, elevation 20–650 m) in northern Sulawesi, and a small stream (width 1.0–2.0 m, 20.0 °C, exposed to the sun, elevation 1180 m) in Central Sulawesi (Takaoka, 2003).

Remarks. This species was described from Sulawesi, Indonesia (Takaoka and Roberts, 1988). It was redescribed by Takaoka (2003).

It is characterized in the male by the number of upper-eye (large) facets in 15 or 16 vertical columns and 16 or 17 horizontal rows and ventral plate nearly parallel-sided when viewed ventrally; in the pupa by the gill with eight slender filaments arranged as 3+(3+2) from dorsal to ventral, of which two triplets of filaments arise side by side nearly horizontally, abdominal segments 1 and 2 light yellowish brown and bare dorsally and cocoon with a short anterodorsal projection (Fig. 22K).

***Simulium* (*Gomphostilbia*) *tumidum* Takaoka, 2003**

(Fig. 22D, H)

Simulium (*Gomphostilbia*) *tumidum* Takaoka, 2003: 120–126 (female, male, pupa, and larva).

Distribution. Indonesia (Sulawesi).

Bionomics. Aquatic habitat: A lowland stream (width 3–4 m, flow slow, 24.0 °C, exposed to the sun, elevation 100 m), Kendari, Southeast Sulawesi (Takaoka, 2003).

Remarks. This species was described from Sulawesi, Indonesia (Takaoka, 2003).

It is characterized in the female by the spermatheca long, ca. 1.9 times as long as its width; in the male by the number of upper-eye (large) facets in 11 vertical columns and 12 horizontal rows; in the pupa by the gill composed of a short, inflated, Y-shaped structure and eight slender short filaments, of which six filaments arise close together at the base and each of two others arises at the apex of each branch of the inflated structure (Fig. 22D). This species is exceptional among the species of the *S. batoense* species-group in having the pupal gill with an inflated structure.

***Simulium* (*Gomphostilbia*) *yaeyamaense* Takaoka, 1991**

Simulium (*Eusimulium*) *batoense* (*nec* Edwards): Ogata, 1956: 59–61 (female, male, and pupa).

Simulium (*Gomphostilbia*) *batoense* (*nec* Edwards): Takaoka, 1976b: 389 (larva).

Simulium (*Gomphostilbia*) *yaeyamaense* Takaoka, 1991: 266–272 (female, male, pupa, and larva).

Distribution. Japan (Ishigaki Island and Iriomote Island in Nansei Islands).

Bionomics. Aquatic habitats: Small streams in lowlands in Ishigaki Island and Iriomote Island (Takaoka, 1991).

Remarks. This species was first recorded as *S.* (*E.*) *batoense* from Ishigaki Island, the Nansei Islands, Japan by Ogata (1956). It was described as a new species, *S. yaeyamaense*, by Takaoka (1991) after examining the type specimens of *S. batoense* from Java, preserved in Natural History Museum, London, UK. Hadi et al. (1995) observed salivary gland chromosomes and showed that sex-chromosomes of this species are heterozygous in females.

It is characterized in the female by the medium-long sensory vesicle 0.36 times as long as the third palpal segment; in the male by the non-pruinose scutal pattern composed of a median vitta and two elliptical submedian spots; and in the pupa by the gill with eight slender filaments, of which two filaments of the ventral pair are nearly twice as long as the six other filaments of the middle and dorsal triplets. The large basal tooth of the claw separates this species from *S. batoense*, of which the female claw has a small subbasal tooth (Takaoka and Davies, 1996).

***Simulium* (*Gomphostilbia*) *yvonneae* Takaoka and Low, 2019**

(Fig. 22A, M, Q)

Simulium (*Gomphostilbia*) *yvonneae* Takaoka and Low [in Takaoka et al., 2019c]: 433–440 (female, male, pupa, and larva).

Distribution. Vietnam.

Bionomics. Aquatic habitats: A river (width 1.4–3.0 m, depth 10–15 cm, bottom sands, pebbles and stones, 18.0 °C, exposed to the sun, elevation 73 m), moderately flowing in a natural forest, and a stream (width 2–5 m, depth 10–20 cm, bottom rocks and pebbles, 17.0 °C, exposed to the sun, elevation 93 m), fast flowing in an agricultural area, Chau Dinh, Quy Chau, Nghe An Province, northern Vietnam (Takaoka et al., 2019c).

Remarks. This species was described from Vietnam (Takaoka et al., 2019c).

It is characterized by the relatively larger number of male upper-eye (large) facets in 16 vertical columns and 16 horizontal rows, pupal gill with eight filaments arranged as 3+(1+2)+2 from dorsal to ventral, of which two filaments of the ventral pair are 1.8 times as long as the longest filament of the middle and dorsal triplets (Fig. 22A), and pupal abdominal segments 1 and 2 darkened dorsally.

Simulium (Gomphostilbia) parahiyangum Subgroup

Simulium (*Gomphostilbia*) *dentistylum* Takaoka and Davies, 1995

Simulium (*Gomphostilbia*) *dentistylum* Takaoka and Davies, 1995a: 51–55 (male, pupa, and larva); Phasuk et al., 2005: 54 (larva).

Distribution. Malaysia (Peninsular) and Thailand.

Bionomics. Aquatic habitat: A small stream (21 °C) flowing in an open grassy and swampy area in Fraser's Hill (Takaoka and Davies, 1995a).

Remarks. This species was described from Peninsular Malaysia (Takaoka and Davies, 1995a). Its female is unknown. The larvae of this species were recorded from Khao Yai National Park, Thailand by Phasuk et al. (2005). This species was once recorded from Thailand based on a pupa (Kuvangkadilok and Takaoka, 2000). However, this was later found to be the pupa of *S. ngaoense* described from Thailand (Takaoka et al., 2018a). Similarly, one male reared from a pupa and three larvae from Sabah were wrongly identified as *S. dentistylum* (Takaoka, 2001a). These larvae were finally identified to be *S. capillatum* (Takaoka, 2009c). The male and its associated pupa are here re-identified as those of *S. parahiyangum*.

It is characterized in the male by the number of upper-eye (large) facets in 14 vertical columns and 14 horizontal rows, mid and hind tibiae yellow except the apical one-third brown and abdominal segments 2 and 3 yellow; in the pupa by the antennal sheath with tubercles (similar to Fig. 23P), thoracic integument with cone-shaped tubercles and gill with eight short filaments arranged as 2+1+3+2 from dorsal to ventral; and in the larva by the postgenal cleft long, reaching the posterior margin of the hypostoma and abdomen deeply constricted between segments 4 and 5 when viewed dorsally (similar to Fig. 23T) and without dorsolateral protuberances. The larva of this species is distinguished from that of *S. capillatum* by lacking hairs on the thoracic proleg.

Simulium (Gomphostilbia) longlanhense Takaoka and Ya'cob, 2015
(Fig. 23A–Q, S, V, W, Y)

Simulium (*Gomphostilbia*) *longlanhense* Takaoka and Ya'cob [in Takaoka et al., 2015c]: 41–48 (female, male, pupa, and larva).

Distribution. Vietnam.

Bionomics. Aquatic habitat: A stream (width 4–6 m, 18 °C, exposed to the sun, elevation 1452 m) moderately flowing in an open land, Lam Dong Province (Takaoka et al., 2015c).

Remarks. This species was described from Lam Dong Province in southern Vietnam (Takaoka et al., 2015c). It is similar to *S. dentistylum* but is differentiated by the number of male upper-eye (large) facets in 15 vertical columns and 16 horizontal rows and male abdominal segments 2, 6, and 7 each with a pair of shiny dorsolateral patches.

Simulium (Gomphostilbia) ngaoense Takaoka, Srisuka and Saeung, 2018
(Fig. 23X, Z)

Simulium (*Gomphostilbia*) *parahiyangum* (*nec* Takaoka and Sigit, 1992): Takaoka and Saito, 1996: 165; Takaoka et al., 2017c: 53.

Simulium (Gomphostilbia) ngaoense Takaoka, Srisuka and Saeung [in Takaoka et al., 2018a]: 1455–1462 (female, male, pupa, and larva).

Distribution. Thailand and Vietnam.

Bionomics. Aquatic habitat: A stream (width 1.0 m, depth 5–10 cm, bottom sandy, flow slow, 18.0 °C, shaded, elevation 80 m), Nghe An Province (Takaoka et al., 2018a).

Remarks. This species was described from Thailand and Vietnam by Takaoka et al. (2018a).

It is characterized in the male by the number of upper-eye (large) facets in 13 or 14 vertical columns and 14 horizontal rows; and in the pupa by the frons moderately covered with tubercles.

Simulium (Gomphostilbia) parahiyangum Takaoka and Sigit, 1992
(Fig. 23R, T, U)

Simulium (*Gomphostilbia*) *parahiyangum* Takaoka and Sigit, 1992: 135–142 (female, male, pupa, and larva).

Simulium (*Gomphostilbia*) sp. nr. *varicorne* Datta, 1975a: 35 (larva). Regarded as *S.* (*G.*) *parahiyangum* by Borah *et al.*, 2012.

Distribution. Indonesia (Java and Sumatra), Malaysia (Sabah and Sarawak), and India (Assam) (?).

Bionomics. Aquatic habitats: A small stream (width 1 m, shaded, elevation ca. 1000 m) flowing in a tea plantation near National Park of Mt. Tangkubanperahu, Java (Takaoka and Sigit, 1992), and small streams at low to medium-high elevations in Sabah (Takaoka, 2001a).

Remarks. This species was described from Java (Takaoka and Sigit, 1992). It was recorded from Peninsular Malaysia (Takaoka and Davies, 1995a), Sabah (Takaoka, 2001a), Sarawak (Takaoka, 2008c), Thailand (Takaoka and Saito, 1996), India (Arunachal Pradesh) (Borah et al., 2012), and Vietnam (Takaoka et al., 2017c). However, populations from Thailand and Vietnam were described as *S. ngaoense*, and those from Peninsular Malaysia were described as *S. sazalyi* (Takaoka et al., 2018a). The identity of the larva recorded from Assam as *S.* sp. nr. *varicorne* (Datta, 1975a) and later thought as *S. parahiyangum* (Borah et al., 2012) should be confirmed based on the detailed examination of the adults and pupal specimens.

It is characterized in the male by the style without an apical spine; in the pupa by the gill with eight short slender filaments (Fig. 23R); and in the larva by the body with dorsolateral protuberances (Fig. 23T, U), postgenal cleft with its apex reaching the posterior margin of the hypostoma (similar to Fig. 23V) and enlarged dark spinous setae dorsally on a few posterior abdominal segments (similar to Fig. 23X).

***Simulium (Gomphostilbia) sazalyi* Takaoka, Ya'cob and Low, 2018**

Simulium (*Gomphostilbia*) *parahiyangum*: (*nec* Takaoka and Sigit, 1992): Takaoka and Davies, 1995a: 50–51.

Simulium (*Gomphostilbia*) *sazalyi* Takaoka, Ya'cob and Low [in Takaoka et al., 2018a]: 1462–1463 (female, male, pupa, and larva).

Distribution. Malaysia (Peninsular).

Bionomics. Aquatic habitats: Small to medium-sized streams at low to medium elevations in many localities in Peninsular Malaysia (Takaoka et al., 2018a).

Remarks. This species was first recorded from Peninsular Malaysia as *S. parahiyangum* originally described from Java (Takaoka and Sigit, 1992), based on the female, pupa, and larvae collected from Gombak, Fraser's Hill and elsewhere (Takaoka and Davies, 1995a). Takaoka et al. (2018a) described this species as a new species, *S. sazalyi*, based on the smaller number of male upper-eye (large) facets in 12–14 vertical columns and 13–15 horizontal rows (17 vertical columns and 15 horizontal rows in *S. parahiyangum*) and pupal frons sparsely covered with tubercles (densely covered with tubercles in *S. parahiyangum*).

This species is distinguished from *S. dentistylum* by the style lacking an apical spine in the male, antennal sheath with nine ridges with tubercles in the pupa (ten ridges with tubercles in *S. dentistylum*), and abdominal segments 1–5 each with paired dorsal protuberances (such protuberances are absent in *S. dentistylum*).

Simulium (Gomphostilbia) sextuplum Subgroup

Simulium (*Gomphostilbia*) *fukudae* Takaoka, Srisuka and Saeung, 2017

(Fig. 24B, O, P)

Simulium (*Gomphostilbia*) *fukudae* Takaoka, Srisuka and Saeung, 2017m: 1552–1559 (female, male, pupa, and larva).

Distribution. Thailand.

Bionomics. Aquatic habitat: A moderately flowing stream (width 3 m, 21.7 °C, pH 4.71, partially shaded, elevation 337 m), Along Korn Waterfall, Surin Province (Takaoka et al., 2017m).

Remarks. This species was described from Thailand (Takaoka et al., 2017m).

It is characterized in the female by the frons almost bare (similar to Fig. 24A); in the male by the number of upper-eye (large) facets in 14 or 15 vertical columns and 15 horizontal rows, ventral plate with its body widened posteriorly when viewed ventrally and paramere covered with minute setae (Fig. 24P); in the pupa by the stalk of the ventral pair of gill filaments 0.9–1.0 times as long as the interspiracular trunk; and in the larva by the labral fan with 40–45 primary rays and posterior circlet with 95 rows of hooklets.

Simulium (*Gomphostilbia*) *paukatense* Takaoka, 2008

(Fig. 24A, C–N, Q–W)

Simulium (*Gomphostilbia*) *paukatense* Takaoka, 2008c: 196–203 (female, male, pupa, and larva).

Distribution. Malaysia (Sarawak).

Bionomics. Aquatic habitat: A slow flowing stream (width 0.3–0.4 m, 20 °C, partially shaded, elevation of 1010 m) in a forest, Bario (Takaoka, 2008c).

Remarks. This species was described from Sarawak, Malaysia (Takaoka, 2008c).

It is similar to *S. fukudae* in sharing the female frons almost bare (Fig. 24A) and male paramere covered with minute setae (Fig. 24Q) but is distinguished in the male by the number of upper-eye (large) facets in 17 or 18 vertical columns and 16 or 17 horizontal rows, ventral plate with its body nearly parallel-sided when viewed ventrally (Fig. 24K); in the pupa by the stalk of the ventral pair of gill filaments 1.6 times as long as the interspiracular trunk (Fig. 24R); and in the larva by the posterior circlet with 68–74 rows of hooklets.

Simulium (*Gomphostilbia*) *sextuplum* Takaoka and Davies, 1995

Simulium (*Gomphostilbia*) *sextuplum* Takaoka and Davies, 1995a: 28–32 (female, pupa, and larva).

Distribution. Malaysia (Peninsular).

Bionomics. Aquatic habitats: Tributaries of the Jemampong River, Bentong, Pahang (Takaoka and Davies, 1995a).

Remarks. This species was described from Peninsular Malaysia (Takaoka and Davies, 1995a). Its male is unknown.

It is characterized in the female by the frons almost bare except several hairs along both lateral margins (similar to Fig. 24A); and in the pupa by the gill with six long slender filaments arranged as (2+2)+2 from dorsal to ventral.

Simulium (Gomphostilbia) terengganuense Subgroup

Simulium (*Gomphostilbia*) *terengganuense* Takaoka, Sofian-Azirun and Ya'cob, 2012

(Fig. 25A–Y)

Simulium (*Gomphostilbia*) *terengganuense* Takaoka, Sofian-Azirun and Ya'cob [in Takaoka et al., 2012b]: 803–810 (female, male, pupa, and larva).

Distribution. Malaysia (Peninsular and Sarawak).

Bionomics. Aquatic habitats: A stream (width 0.5 m, flow slow, bottom muddy, 24.0 °C, shaded, elevation 310 m) flowing in an open area recently logged near a natural forest, Terengganu Province, Peninsular Malaysia (Takaoka et al., 2012b).

Remarks. This species was described from Peninsular Malaysia (Takaoka et al., 2012b). It was also recorded from Sarawak (Ya'cob et al., 2015b).

It is characterized in the female by the narrow frons with its narrowest width medially and small sensory vesicle (Fig. 25A); in the male by the ventral plate with its posterior margin convex when viewed ventrally (Fig. 25M); in the pupa by the gill with eight slender filaments of almost the same length (Fig. 25S) and terminal hooks of the mushroom-shape when viewed laterally (Fig. 25V); and in the larva by the remarkable color pattern on the thorax and abdomen (Fig. 25W, X) (Takaoka et al., 2012b).

Simulium (Gomphostilbia) batoense Species-group Species Unplaced to Subgroup

Simulium (Gomphostilbia) agasthyamalaiense Vijayan, Anbalagan, Rekha, Dinakaran and Krishnan, 2019

Simulium (*Gomphostilbia*) *agasthyamalaiense* Vijayan, Anbalagan, Rekha, Dinakaran and Krishnan, 2019: 569–573 (female, male, pupa, and larva).

Distribution. India (Tamil Nadu).

Bionomics. Aquatic habitat: A moderately-flowing stream (width 8–10 m, depth 15–20 cm, 18.1 °C, elevation 1250 m), Tamil Nadu (Vijayan et al., 2019).

Remarks. This species was described from India (Vijayan et al., 2019).

It is characterized in the female by the relatively narrower frons against the width of the head (frons:head ratio 1.0:4.3); in the male by the number of upper-eye (large) facets in 20 vertical columns and 18 or 19 horizontal rows and hind basitarsus spindle-shaped; and in the pupa by the gill with eight short slender filaments arranged as [(1+2)+(1+2)]+2 from dorsal to ventral [though (2+1)+(1+2)+2 in

Fig. 3B in Vijayan et al. (2019)] and abdominal segments 1 and 2 covered with tubercles dorsally.

The length of filaments of the pupal gill was noted to be 1.8–1.9 mm for the dorsal triplet, 2.1–2.2 mm for the middle triplet and 2.5–2.7 mm for the ventral pair; thus, this species appears to be placed in the *S. duolongum* subgroup, but the dorsal and middle triplets are only slightly shorter than the ventral pair in Fig. 3A in Vijayan et al. (2019), making it difficult to place this species in any subgroup. The postgenal cleft was also noted to be 2.1 times as long as the postgenal bridge, but it is 1.5 times as long as the postgenal bridge in Fig. 3E in Vijayan et al. (2019).

***Simulium* (*Gomphostilbia*) *baglungense* Takaoka and Shrestha, 2010**

Simulium (*Gomphostilbia*) *baglungense* Takaoka and Shrestha, 2010: 15–18 (female and pupa).

Distribution. Nepal.

Bionomics. Aquatic habitat: A small stream (flow slow, 19.5 °C, exposed to the sun, elevation 1469 m), Padhera, Lekhani, Baglung (Takaoka and Shrestha, 2010).

Remarks. This species was described from a single female reared from a pupa collected in Baglung, Nepal (Takaoka and Shrestha, 2010). Its male and larva are unknown. This species is left unplaced to subgroup because the length of the pupal gill filaments is unknown.

It is characterized by the pupal frons and thorax bare (except the posterior one-third) and the pupal gill with a relatively large transparent bulb-like basal fenestra.

***Simulium* (*Gomphostilbia*) *bakalalanense* Ya'cob, Takaoka, and Sofian-Azirun, 2015**

Simulium (*Gomphostilbia*) *bakalalanense* Ya'cob, Takaoka, and Sofian-Azirun, 2015b: 783–788 (male, pupa, and larva).

Distribution. Malaysia (Sarawak).

Bionomics. Aquatic habitat: A small stream (width 0.5–0.6 m, depth 10 cm, partially shaded, 14.4 °C, elevation 2115 m) at Mt. Murud (Ya'cob et al., 2015).

Remarks. This species was described from Sarawak, Malaysia, by Ya'cob et al. (2015). Its female is unknown.

It is characterized in the male by the small number of upper-eye (large) facets in 11 vertical and 13 horizontal rows, ventral plate subquadrate, 0.4 times as long as wide when viewed ventrally, and moderately produced ventromedially when viewed posteriorly; in the pupa by the gill with eight slender long filaments of the same length, arranged as [2+1+(1+2)]+2 from dorsal to ventral, swollen transparent basal fenestra, abdominal segment 9 without grapnel-like hooklets, terminal hooks triangular, with outer margins partially crenulated, and cocoon shoe-shaped; and in the larva by the postgenal cleft narrow, medium-long, 2.7 times as long as the postgenal bridge.

***Simulium* (*Gomphostilbia*) *brevitruncatum* Takaoka, 2003**

Simulium (*Gomphostilbia*) *brevitruncatum* Takaoka, 2003: 107–114 (female, male, pupa, and larva).

Distribution. Indonesia (Sulawesi).

Bionomics. Aquatic habitat: A small stream (width ca. 2 m, 22.0 °C, exposed to the sun, elevation 970 m) in South Sulawesi (Takaoka, 2003).

Remarks. This species was described from Sulawesi, Indonesia (Takaoka, 2003).

It is characterized by the pupal gill with eight slender short filaments of the same length (0.5–0.7 times as long as the pupal body), all arising close together from a short common basal stalk.

***Simulium* (*Gomphostilbia*) *brivetruncum* Chen and Chen, 2000**

Simulium (*Gomphostilbia*) *brivetruncum* Chen and Chen, 2000: 129–132 (female, male, pupa, and larva).

Distribution. China (Hainan).

Bionomics. Aquatic habitat: A small stream (shaded, elevation 650–820 m) in Jiangfengling County, Hainan (Chen and Chen, 2000).

Remarks. This species was described from Hainan, China by Chen and Chen (2000).

It is characterized in the female by the sensory vesicle long, 0.7 times as long as the third palpal segment, mandible lacking outer teeth, scutum with whitish yellow fine hairs, claw with a relatively long tooth 0.6 times as long as the claw; in the male by the number of upper-eye (large) facets in 12 vertical columns and 13 horizontal rows; in the pupa by the gill with eight slender filaments arranged as [(2+1)+(1+2)] +2 from dorsal to ventral and slightly shorter than the pupal body; and in the larva by the antenna about twice as long as the stalk of the labral fan and postgenal cleft medium-long, 1.8 times as long as the postgenal bridge.

***Simulium* (*Gomphostilbia*) *butwalense* Takaoka and Shrestha, 2010**

Simulium (*Gomphostilbia*) *butwalense* Takaoka and Shrestha, 2010: 18–23 (male, pupa, and larva).

Distribution. Nepal.

Bionomics. Aquatic habitat: A small, rapid-flowing stream (15.0 °C, exposed to the sun, elevation 241 m), Shidhdhababa, near the gate to Palpa, Butwal (Takaoka and Shrestha, 2010).

Remarks. This species was described from Nepal (Takaoka and Shrestha, 2010). Its female is unknown. This species is left unplaced to subgroup because the length of the pupal gill filaments is unknown.

It is characterized in the male by the ventral plate, which is not markedly produced ventrally and is angulate on each posterolateral corner; in the pupa by the gill with eight filaments arranged as [(2+1)+(2+1)]+2 or {2+[1+(1+2)]}+2 from dorsal to ventral, of which the inner filament of the ventral pair is apparently thicker than the seven other filaments, and terminal hooks simple, triangular; and in the larva by the postgenal cleft long, reaching the posterior margin of the hypostoma.

***Simulium* (*Gomphostilbia*) *chuzargangense* Takaoka and Somboon, 2008**

Simulium (*Gomphostilbia*) *chuzargangense* Takaoka and Somboon, 2008: 216–218 (female).

Distribution. Bhutan.

Bionomics. Aquatic habitat: An irrigation channel (width 0.8 m, 29.6 °C, shaded, elevation 218 m) in Chuzargang, Gelephu in Bhutan (Takaoka and Somboon, 2008).

Remarks. This species was described from a single female reared from a pupa collected from Bhutan (Takaoka and Somboon, 2008). This species is left unplaced to subgroup because its pupa is unknown.

The female of this species is characterized by the subcosta bare or with one hair.

Simulium (Gomphostilbia) dadeldhuraense **Takaoka and Shrestha, 2020**
Simulium (Gomphostilbia) dadeldhuraense Takaoka and Shrestha [In Takaoka et al., 2020f]: 191–195 (female and pupa).

Distribution. Nepal.

Bionomics. Aquatic habitat: A stream (20.5 °C, elevation 1732 m), Dadeldhura, Nepal (Takaoka et al., 2020f).

Remarks. This species was described from a single female reared from a pupa collected in Nepal by Takaoka et al. (2020f). Its male and larva are unknown.

It is similar to *S. baglungense* described from Nepal (Takaoka and Shrestha, 2010) in many characteristics including the female leg color, female terminalia, pupal thorax almost bare, and pupal gill with a swollen round basal fenestra. However, it is barely distinguished from the latter species in the female by the relative length of the sensory vesicle against the third palpal segment (0.33–0.35 in this species versus 0.43–0.52 in *S. baglungense*) and relative length of the fore basitarsus against its greatest width (7.4 in this species versus 6.8 in *S. baglungense*), and in the pupa by the arrangement of the gill filaments (2+1+(2+1)+2 in this species versus (2+1)+(1+2)+2 in *S. baglungense*), the relative width of the round basal fenestra against the width of the interspiracular trunk (3 in this species versus 2 in *S. baglungense*).

Simulium (Gomphostilbia) dinakarani **Anbalagan, Vijayan, Balachandran, Thiyonila and Surya, 2020**
Simulium (Gomphostilbia) dinakarani Anbalagan, Vijayan, Balachandran, Thiyonila and Surya, 2020b: 58–64 (female, male, pupa, and larva).

Distribution. India (Tamil Nadu).

Bionomics. Aquatic habitat: A fast-flowing stream (width 5–12 m, depth 5–20 cm, 19.7 °C, elevation 1186 m), Tamil Nadu (Anbalagan et al., 2020b).

Remarks. This species was described from India (Anbalagan et al., 2020b).

It is characterized in the female by the relatively wide frons against the width of the head (frons:head ratio 1.0:2.7) and relatively shorter claw tooth (0.4 times as long as the claw); in the male by the number of upper-eye (large) facets in 13 vertical columns and 16 horizontal rows; and in the pupa by the gill with eight short slender filaments arranged as [(1+2)+(1+2)]+2 from dorsal to ventral [though [(2+1)+(2+1)] +2 in Fig. 3A in Anbalagan et al. (2020b)] and abdominal segments 1 and 2 covered with tubercles dorsally.

The length of filaments of the pupal gill was noted as the dorsal triplet 1.9–2.1 mm, the middle triplet 2.1–2.3 mm and the ventral pair 3.2–3.4 mm; thus, this species appears to be placed in the *S. duolongum* subgroup, but the dorsal and middle triplets are only slightly shorter than the ventral pair in Fig. 3A in Anbalagan et al. (2020b), making it difficult to place this species in any subgroup. In addition, there are differences in the descriptions and relevant figures in Anbalagan et al. (2020b). For example, the coxite was noted to be 1.1 times as long as its greatest width, but it is 1.3 times as long as its greatest width in Fig. 2D, and the postgenal cleft was also noted to be 4.1 times as long as the postgenal bridge, but it is ca. 10 times as long as the postgenal bridge in Fig. 3E in Anbalagan et al. (2020b).

***Simulium* (*Gomphostilbia*) *dola* Davies and Györkös, 1987**

Simulium (*Gomphostilbia*) *dola* Davies and Györkös, 1987a: 1500–1501 (female, male, pupa, and larva).

Distribution. Sri Lanka.

Bionomics. Aquatic habitat: The Gilimale Creek (width 8–12 m, elevation ca. 150 m), 10 km northeast of Ratnapupura (Davies and Györkös, 1987a).

Remarks. This species was described from Sri Lanka (Davies and Györkös, 1987a).

It is characterized in the female by abdominal segments 2 and 5–9 shiny dorsally; in the male by the reduced number of upper-eye (large) facets in 11 vertical columns and 11 horizontal rows and subcosta with a few hairs ventrally; and in the pupa by the gill with eight filaments, of which six filaments of the dorsal and middle triplets are somewhat inflated basally, then abruptly tapered. The length of the pupal gill is not clear, though it was noted as "longest filament (1.25 mm) more than 1.5 times the length of pupa" in the original description (Davies and Györkös, 1987a).

***Simulium* (*Gomphostilbia*) *hainanense* Long, An, Hao, Yan and Yuan, 1994**

Simulium (*Gomphostilbia*) *hainanense* Long, An, Hao, Yan and Yuan, 1994: 405–406 (female, male, pupa, and larva).

Distribution. China (Hainan).

Bionomics. Aquatic habitats: Two streams at an elevation of 220 m and 850 m.

Remarks. This species was described from Hainan, China, by Long et al. (1994).

It is characterized in the female by the sensory vesicle half as long as the third palpal segment, scutum covered with yellowish white short hairs, mid femur and tibia yellow on their basal three-fourths and hind tibia black except its base yellow; in the male by the scutum with yellow short hairs and ventral plate flat, with lateral margins parallel-sided when viewed ventrally; in the pupa by the gill with eight filaments arranged as 3+3+2 from dorsal to ventral, arising from a short common basal stalk; and in the larva by the cephalic apotome with positive head spots, labral fan with 28 primary rays, postgenal cleft medium-long, 2.5 times as long as the postgenal bridge and posterior circlet with 75 rows of 3–11 hooklets per row (Long et al., 1994).

This species is placed in the *S. batoense* species-group by having the female and male hind tibiae darkened except the base and ventral plate flat but it cannot be further placed in any subgroup because no information is available on the color of tuft hairs at the base of the female and male radius, size of the tooth of the claw and size of trichomes on the head and thorax, relative length of the eight gill filaments and presence or absence of pupal terminal hooks and grapnel-like hooklets.

***Simulium* (*Gomphostilbia*) *kottoorense* Anbalagan, Prasanna, Kannan, Dinakaran and Krishnan, 2015**

Simulium (*Gomphostilbia*) *kottoorense* Anbalagan, Prasanna, Kannan, Dinakaran and Krishnan, 2015b: 99–103 (female, male, pupa, and larva).

Distribution. India (Kerala).

Bionomics. Aquatic habitat: The Anchu Nazhika River (elevation 275 m), Kerala (Anbalagan et al., 2015b).

Remarks. This species was described from Kerala, India (Anbalagan et al., 2015b).

It is characterized in the female by the relatively wide frons against the width of the head (frons:head ratio 1.0:2.4) and claw with a relatively long tooth 0.66 times as long as the claw [though 0.3 times the length of the claw in Fig. 4D in Anbalagan et al. (2015b)]; in the male by the number of upper-eye (large) facets in 20 vertical columns and 12 horizontal rows, abdominal segments 2–7 each with a pair of shiny dorsolateral patches and ventral plate concave ventromedially when viewed caudally; and in the pupa by the gill with eight short slender filaments arranged as [(1+2) +(1+2)]+2 from dorsal to ventral [though (2+1)+(1+2)+2 in Fig. 6A in Anbalagan et al. (2015b)], arising from a short common basal stalk, abdominal segments 1 and 2 covered with tubercles dorsally and terminal hooks triangle, of which their outer margins 2.3–2.5 times as long as their inner margins and crenulated [though their outer margins as long as their inner margins and not crenulate in Fig. 6B in Anbalagan et al. (2015b)].

The length of filaments of the pupal gill was noted as "the dorsal triplet 1.3–1.4 mm, the middle triplet 1.8–1.9 mm and the ventral pair 2.0–2.1 mm," making it difficult to place this species in subgroup. The postgenal cleft was also noted to be 1.7 times as long as the postgenal bridge, but it is 4.3 times as long as the postgenal bridge in Fig. 6D in Anbalagan et al. (2015b).

***Simulium (Gomphostilbia) krishnani* Anbalagan, 2020**

Simulium (*Gomphostilbia*) *krishnani* Anbalagan [in Anbalagan et al., 2020b]: 64–70 (female, male, pupa, and larva).

Distribution. India (Tamil Nadu).

Bionomics. Aquatic habitat: A stream (width 3–5 m, depth 5–30 cm, flow fast, 23.0 °C, elevation 527 m), Tamil Nadu (Anbalagan et al., 2020b).

Remarks. This species was described from India (Anbalagan et al., 2020b).

It is characterized in the female by the relatively wide frons against the width of the head (frons:head ratio 1.0:2.6) and relatively large claw tooth (0.68 times as long as the claw) [though 0.46 in Fig. 4D in Anbalagan et al. (2020b)]; in the male by the number of upper-eye (large) facets in 13 vertical columns and 14 horizontal rows; and in the pupa by the gill with eight short slender filaments arranged as [(1+2) +(1+2)]+2 from dorsal to ventral [though (1+2)+(1+2)+2 in Fig. 6A in Anbalagan et al. (2020b)] (the dorsal triplet 1.4–1.5 mm, the middle triplet 1.3–1.5 mm and the ventral pair 1.6–1.7 mm) and abdominal segments 1 and 2 covered with tubercles dorsally. The ventral plate is V-shaped, with a small body only half the length of its arms, which is quite different from those of other species in the *S. batoense* species-group, making it difficult to place this species in any subgroup. The postgenal cleft was noted to be 3.04 times as long as the postgenal bridge, but it is 12.5 times as long as the postgenal bridge in Fig. 6E in Anbalagan et al. (2020b).

***Simulium* (*Gomphostilbia*) *krombeini* Davies and Györkös, 1987**
Simulium (*Gomphostilbia*) *krombeini* Davies and Györkös, 1987a: 1491–1495 (female, male, pupa, and larva).

Distribution. Sri Lanka.

Bionomics. Aquatic habitat: A jungle tributary (width 0.75–1.40 m, elevation ca. 150 m) of Gilimale Creek, 10 km northeast of Ratnapupura (Davies and Györkös, 1987a).

Remarks. This species was described from Sri Lanka (Davies and Györkös, 1987a).

It is characterized in the female by the sensory vesicle long, half as long as the third palpal segment, and claw with a large tooth two-thirds as long as the claw; in the male by the scutum with three faint longitudinal vittae and the number of upper-eye (large) facets in 12 or 13 vertical columns and 12 horizontal rows; in the pupa by the short gill filaments arranged as 3+3+2 from dorsal to ventral (0.7 times as long as the body); and in the larva by the rectal organ of three simple lobes (without secondary lobules).

***Simulium* (*Gomphostilbia*) *kumbakkaraiense* Anbalagan, Vijayan, Dinakaran and Krishnan 2019**
Simulium (*Gomphostilbia*) *kumbakkaraiense* Anbalagan, Vijayan, Dinakaran and Krishnan, 2019b: 480–486 (female, male, pupa, and larva).

Distribution. India (Tamil Nadu).

Bionomics. Aquatic habitat: A fast-flowing stream (width 5–8 m, depth 5–20 cm, 23.2 °C, exposed to the sun, elevation 402 m), Tamil Nadu (Anbalagan et al., 2019b).

Remarks. This species was described from India (Anbalagan et al., 2019b).

It is characterized in the female by the mandible with 16 inner and 13 outer teeth; in the male by the number of upper-eye (large) facets in 13 vertical columns and 13 or 14 horizontal rows; and in the pupa by the gill with eight short slender filaments (1.4–1.7 mm long, about half as long as the pupal body) and abdominal segments 1 and 2 covered with tubercles dorsally.

The pupal gill was described as having eight short filaments arranged as [(1+2) +(1+2)]+2 from dorsal to ventral, but its arrangement is not clearly shown in Fig. 4C in Anbalagan et al. (2019b), thus making it difficult to place it in any subgroup. In addition, there are differences in the descriptions and relevant figures in Anbalagan et al. (2019b). For example, the coxite was noted to be 1.1 times as long as its greatest width, but it is 1.6 times as long as its greatest width in Fig. 3C, and the postgenal cleft was also noted to be 1.7 times as long as the postgenal bridge, but it is about seven times as long as the postgenal bridge in Fig. 4E in Anbalagan et al. (2019b).

***Simulium* (*Gomphostilbia*) *miyagii* Takaoka, 2003**
Simulium (*Gomphostilbia*) *miyagii* Takaoka, 2003: 114–119 (male, pupa, and larva).

Distribution. Indonesia (Sulawesi).

Bionomics. Aquatic habitats: A stream (width ca. 2 m, 22.0 °C, exposed to the sun, elevation 970 m) and a slow-flowing stream (width 0.5–1.0 m, 28.0 °C, exposed to the sun, elevation 850 m) (Takaoka, 2003).

Remarks. This species was described from South Sulawesi, Indonesia (Takaoka, 2003). Its female is unknown.

It is characterized by the pupal gill with eight slender filaments (2.0–2.5 mm long, nearly as long as or slightly shorter than the pupal body), arranged as 3+5 from dorsal to ventral, arising from a long common basal stalk, and cocoon with a short anterodorsal bulge or projection.

***Simulium (Gomphostilbia) panagudiense* Anbalagan, Prasanna, Kannan, Dinakaran and Krishnan, 2015**

Simulium (*Gomphostilbia*) *panagudiense* Anbalagan, Prasanna, Kannan, Dinakaran and Krishnan, 2015b: 96–99 (female, male, pupa, and larva).

Distribution. India (Tamil Nadu).

Bionomics. Aquatic habitat: The Kuthirapanchan Waterfalls (elevation 110 m), Tamil Nadu (Anbalagan et al., 2015b).

Remarks. This species was described from Tamil Nadu, India (Anbalagan et al., 2015b).

It is characterized in the female by the relatively short labrum against the clypeus (0.4) and scutum without longitudinal vittae; in the male by the number of upper-eye (large) facets in 12 vertical columns and 10 horizontal rows and abdominal segments 2–7 each with a pair of shiny dorsolateral patches; in the pupa by the gill with eight slender filaments of the same length and thickness, arranged as [(1+2)+(1+2)]+2 from dorsal to ventral [though (1+2)+(1+2)+2 in Fig. 3A in Anbalagan et al. (2015b)], arising from a short basal common stalk, spreading at an angle of 40–50 degrees between the stalks of the dorsal triplet and the ventral pair when viewed laterally [though 90 degrees in Fig. 3A in Anbalagan et al. (2015b)] and abdominal segments 1 and 2 covered with tubercles dorsally. The length of filaments of the pupal gill was noted as "0.8–2.3 mm for the dorsal triplet, 2.4–2.5 mm for the middle triplet, and 2.6–2.7 mm for the ventral paired filaments, much longer than the pupal body (1.9–2.1 mm long)."

The postgenal cleft was noted to be 2.1 times as long as the postgenal bridge, but it is 8.6 times as long as the postgenal bridge in Fig. 3D in Anbalagan et al. (2015b).

***Simulium (Gomphostilbia) pattoni* Senior-White, 1922**

Simulium pattoni Senior-White, 1922: 126–131 (female).

Simulium (*Gomphostilbia*) *pattoni*: Davies and Györkös, 1987a: 1486–1489 (female, male, pupa, and larva).

Distribution. India (Tamil Nadu), Sri Lanka, and China (Hainan) (?).

Bionomics. This is the most abundant and widespread species in Sri Lanka (Davies and Györkös, 1987a). The holotype female was collected from a bull's belly at dusk (Senior-White, 1922).

Remarks. This species was collected from Tamil Nadu, southern India by Senior-White (1922) and was later recorded from Sri Lank by Davies and Györkös (1987a). This species was noted to be distributed from Hainan, China, based on an unidentifiable reference by Chen (2016). This ambiguous record should be confirmed in the future. The description of its male, pupa and larva, together with the redescription of the female was provided by Davies and Györkös (1987a).

It is characterized in the female by the sensory vesicle long, half as long as the third palpal segment and claw with a small subbasal tooth; in the male by the number of upper-eye (large) facets in 17 vertical columns and 17 horizontal rows, scutum with three longitudinal vittae, hind basitarsus narrow, spindle-shaped and subcosta bare or with a few hairs ventrally; and in the pupa by the gill with eight slender filaments arranged as 1+2+(1+2)+2 from dorsal to ventral and slightly longer than the pupal body.

Simulium (Gomphostilbia) peteri **Anbalagan, 2014**

Simulium (*Gomphostilbia*) *peteri* Anbalagan [in Anbalagan et al., 2014]: 452–458 (female, male, pupa, and larva).

Distribution. India (Kerala).

Bionomics. Aquatic habitat: The Anchu Nazhika River (elevation 275 m), Kerala (Anbalagan et al., 2014).

Remarks. This species was described from Kerala, India (Anbalagan et al., 2014).

It is characterized in the female by the claw with a relatively long tooth 0.57 times as long as the claw [though 0.50 times the length of the claw in Fig. 1D in Anbalagan et al. (2014)]; in the male by the number of upper-eye (large) facets in 12 vertical columns and 12 horizontal rows, abdominal segments 2–7 each with a pair of shiny dorsolateral patches and ventral plate concave ventromedially when viewed caudally; and in the pupa by the gill with eight slender filaments (as long as or slightly shorter than the pupal body) arranged as [(1+2)+(1+2)]+2 or (3+3)+2 from dorsal to ventral [though (1+2)+(2+1)+2 in Fig. 3A in Anbalagan et al. (2014)], arising basally from the short common basal stalk [though no common basal stalk in Fig. 3A in Anbalagan et al. (2014)], abdominal segments 1 and 2 covered with tubercles dorsally and abdominal segment 9 with a pair of terminal hooks, of which the outer margin is 3.0–3.5 times as long as the inner margin and crenulated [though the outer margin is as long as the inner margin and not crenulated in Fig. 3B in Anbalagan et al. (2014)]. The postgenal cleft was noted to be 2.6 times as long as the postgenal bridge, but it is 3.0 times as long as the postgenal bridge in Fig. 3D in Anbalagan et al. (2014).

In the original description (Anbalagan et al., 2014), the length of filaments of the pupal gill was noted as "the dorsal and middle triplets 2.0–2.3 mm, and the ventral pair 2.7–3.0 mm," though all filaments appear to be subequal in length in Fig. 3A in Anbalagan et al. (2014), and the thickness of the two filaments of the ventral pair was noted to be subequal, though the lower filament is 1.5 times as thick as the upper filament when compared basally in Fig. 3A in Anbalagan et al. (2014).

Simulium **(*****Gomphostilbia*****)** ***purandharaense*** **Takaoka and Shrestha, 2020**

Simulium (*Gomphostilbia*) *purandharaense* Takaoka and Shrestha [In Takaoka et al., 2020c]: 201–205 (female and pupa).

Distribution. Nepal.

Bionomics. Aquatic habitat: A stream (25 °C, elevation 514 m), Purandhara, Dang, Nepal (Takaoka et al., 2020c).

Remarks. This species was described from a single female reared from a pupa collected in Nepal by Takaoka et al. (2020c). Its male and larva are unknown.

It is similar to *S. chuzargangense* described from Bhutan (Takaoka and Somboon, 2008) in many characteristics including the female wing with the subcosta bare or with one hair, but is distinguished from the latter in the female by the relative length of the sensory vesicle against the third palpal segment (0.43–0.46 in this species versus 0.38 in *S. chuzargangense*), relative width of the head against the greatest width of the frons (4.8 in this species versus 5.2 in *S. chuzargangense*), and relative length of the spermatheca against its greatest width (1.42 in this species versus 1.56 in *S. chuzargangense*).

***Simulium* (*Gomphostilbia*) *sulawesiense* Takaoka and Roberts, 1988**

Simulium (*Gomphostilbia*) *sulawesiense* Takaoka and Roberts, 1988: 198–203 (female); Takaoka, 2003: 222–224 (female).

Distribution. Indonesia (Sulawesi).

Bionomics. Unknown.

Remarks. This species was described from a single female caught by a light trap in Sulawesi, Indonesia (Takaoka and Roberts, 1988), and its revised description was given by Takaoka (2003). Its male, pupa, and larva are unknown.

It is characterized by the relatively wide frons (frons:head ratio 1.0:3.6), small sensory vesicle (0.2 times as long as the third palpal segment), mandible with 16 inner teeth and lacking outer teeth, unarmed cibarium and hind tibia white on the basal one-third, then gradually darkened to the apex.

***Simulium* (*Gomphostilbia*) *takaokai* Anbalagan, 2014**

Simulium (*Gomphostilbia*) *takaokai* Anbalagan [in Anbalagan et al., 2014]: 458–463 (female, male, pupa, and larva).

Distribution. India (Kerala).

Bionomics. Aquatic habitat: A stream in Kulathupuzha (elevation 127 m), Kerala (Anbalagan et al., 2014).

Remarks. This species was described from Kerala, India (Anbalagan et al., 2014).

It is characterized in the female by the relatively few number of the inner teeth of the mandible (16) and claw with a relatively short tooth 0.4 times as long as the claw; in the male by the number of upper-eye (large) facets in 12 vertical columns and 10 horizontal rows and ventral plate concave ventromedially when viewed caudally; and in the pupa by the gill with eight slender filaments as long as or slightly longer than the pupal body (1.9–2.1 mm long) arranged as [(1+2)+(1+2)]+2 from dorsal to ventral, arising from a short common basal stalk, two filaments of the ventral pair slightly shorter than the six other filaments, abdominal segments 1–7 pigmented and with small to large tubercles dorsally, and abdominal segment 9 with a pair of terminal hooks, of which the outer margin is 3.0–3.1 times as long as the inner margin and crenulated [though the outer margin is subequal in length to the inner margin and not crenulated in Fig. 6B in Anbalagan et al. (2014)]. The postgenal cleft was noted to be 1.77 times as long as the postgenal bridge, but it is 2.4 times as long as the postgenal bridge in Fig. 6E in Anbalagan et al. (2014).

Simulium (*Gomphostilbia*) *ceylonicum* Species-group

Simulium (Gomphostilbia) atratoides Subgroup

Simulium (*Gomphostilbia*) *atratoides* Takaoka and Davies, 1996

(Figs. 26B, C, E and 27E, K)

Simulium (*Gomphostilbia*) *atratoides* Takaoka and Davies, 1996: 33–35 (female, male, pupa, and larva).

Distribution. Indonesia (Java).

Bionomics. Aquatic habitats: Small streams (width 0.2–0.4 m, elevation 1100–1400 m) in Gunung Mas, Puncak, West Java (Takaoka and Davies, 1996).

Remarks. This species was described from Java, Indonesia (Takaoka and Davies, 1996). The larva of this species was once wrongly associated with *S. atratum* by Edwards (1934). This species was placed in the *S. ceylonicum* species-group because the ventral plate of this species has its body narrowed posteriorly when viewed ventrally (Takaoka, 2012). However, this species is exceptional among species in the *S. ceylonicum* species-group by having several other characteristics, such as the base of the female and male radial vein with a tuft of yellow hairs, female mandible lacking outer teeth (Fig. 26C), female cibarium simple, without a mediolongitudinal ridge (Fig. 26E), male hind tibia yellowish white on the basal half or little less and larval postgenal cleft medium-long (1.3 times as long as the postgenal bridge) (Fig. 27K), all departing from the definition of the *S. ceylonicum* species-group (Takaoka, 2012).

Apart from these characteristics, it is characterized in the female by the narrow frons (frons:head ratio 1.0:6.9) and sensory vesicle long (0.66 times as long as the third palpal segment) (Fig. 26B); in the male by the number of upper-eye (large) facets in 11 vertical columns and 14 horizontal rows; and in the pupa by the gill with 10 slender filaments arranged as (3+3)+(2+2) from dorsal to ventral, arising from a long common basal stalk (Fig. 27E).

Simulium (Gomphostilbia) sheilae Subgroup

Simulium (*Gomphostilbia*) *apoense* Takaoka, 1983

Simulium (*Gomphostilbia*) *apoense* Takaoka, 1983: 74–76 (female, pupa, and larva); Takaoka and Tenedero, 2022a: 334–337 (male).

Distribution. Philippines (Mindanao).

Bionomics. Aquatic habitats: A mountain stream (width 2–3 m, exposed to the sun) in Mt. Apo, Davao City (Takaoka, 1983), and a small stream (width 0.2–0.4 m,

20.0 °C, exposed to the sun, elevation 1180 m), Marilog, Davao del Sur (Takaoka and Tenedero, 2022a).

Remarks. This species was described from Mindanao, the Philippines (Takaoka, 1983). Its male was described by Takaoka and Tenedero (2022a).

It is characterized in the female by the sensory vesicle long (0.67 times as long as the third palpal segment); in the male by the number of upper eye (large) facets in 11 vertical columns and 13 horizontal rows, sensory vesicle short (0.16–0.19 times as long as the third palpal segment) and hind basitarsus enlarged, 1.0 and 1.2–1.3 times as wide as the hind tibia and femur, respectively; and in the pupa by the gill with eight filaments arranged as 3+3+2 from dorsal to ventral, arising from a short common basal stalk.

***Simulium* (*Gomphostilbia*) *bataksense* Takaoka and Tenedero, 2007**

Simulium (*Gomphostilbia*) *bataksense* Takaoka and Tenedero, 2007c: 213–217 (female and pupa).

Distribution. Philippines (Palawan).

Bionomics. Aquatic habitat: A small stream (width 2–3 m, 25 °C, shaded, elevation of 47 m) near the road to El Nido from Puerto Princesa (Takaoka and Tenedero, 2007c).

Remarks. This species was described from a single female reared from a pupa in Palawan, the Philippines (Takaoka and Tenedero, 2007c). Its male and larva are unknown.

The female of this species is characterized by the sensory vesicle medium-long (0.43 times as long as the third palpal segment).

***Simulium* (*Gomphostilbia*) *capillatum* Takaoka, 2009**

(Figs. 26Q and 27J, N, P)

Simulium (*Gomphostilbia*) *capillatum* Takaoka, 2009c: 97–100 (larva); Takaoka et al., 2014g: 522–525 (male and pupa).

Distribution. Malaysia (Sarawak and Sabah).

Bionomics. Aquatic habitats: A stream (width 0.2 m, flow slow, 23 °C, partially shaded, elevation 550 m) running in a forest, Hot spring Park, Sabah (Takaoka, 2009c): a small stream (width 0.3–0.6 m, flow slow, 20 °C, shaded, elevation ca. 1000 m) running in a forest, and a stream (width 0.1–0.3 m, flow slow, 21 °C, partially shaded, elevation 980 m) running along the border of a pasture, Bakalalan, Sarawak (Takaoka et al., 2014g).

Remarks. This species was described from three larvae collected from Sarawak and Sabah, Malaysia (Takaoka, 2009c). Its male and pupa were later described by Takaoka et al. (2014g). Its female is unknown. These three larvae of this species from Sabah were wrongly identified as *S. dentistylum* (Takaoka, 2001a).

The male of this species is characterized by the large number of upper-eye (large) facets in 14 or 15 vertical columns and 16 or 17 horizontal rows, hind tibia darkened except its base and abdominal segments 2–8 each with a pair of shiny dorsolateral or lateral patches.

The larva of this species is distinctive in having the thoracic proleg fully haired with unbranched and branched colorless setae (Fig. 27N), a deep constriction between abdominal segments 4 and 5 (Fig. 27J) and thorax and abdomen densely covered with variously shaped setae (Fig. 27P).

***Simulium* (*Gomphostilbia*) *ceylonicum* (Enderlein, 1921)**
Gomphostilbia ceylonica Enderlein, 1921: 77–78 (female).
Simulium (*Gomphostilbia*) *ceylonicum*: Davies and Györkös, 1987a: 1489–1491 (female, male, pupa, and larva).

Distribution. Sri Lanka.

Bionomics. Aquatic habitats: Streams in various elevations from 150 to 2000 m (Davies and Györkös, 1987a).

Remarks. This species was described from Sri Lank (Enderlein, 1921). The description of the male, pupa, and larva of this species, together with the redescription of the female, was given by Davies and Györkös (1987a).

It is characterized in the female by the sensory vesicle medium-long (0.36 times the length of the third palpal segment), the hind tibia yellowish on its basal two-thirds and abdominal segments 2 and 5–9 shiny dorsally; in the male by the number of upper-eye (large) facets in 13 vertical columns and 14 horizontal rows, hind basitarsus enlarged, spindle-shaped, 3.4 times as long as its greatest width, narrower than the hind tibia and femur, and pale yellow on the basal three-fifths; and in the larva by the rectal organ simple (without secondary lobules).

Davies and Györkös (1987a) noted that the pupa has a pair of hooklets on each side of the ventral surface of abdominal segments 6–8 in this species and *S. ela*. This may be wrong since all other species of the subgenus *Gomphostilbia* have a similar pair of hooklets on each side of the ventral surface of pupal abdominal segments 5–7.

***Simulium* (*Gomphostilbia*) *curtatum* Jitklang, Kuvangkadilok, Baimai, Takaoka and Adler, 2008**
Simulium (*Gomphostilbia*) *curtatum* Jitklang, Kuvangkadilok, Baimai, Takaoka and Adler, 2008: 14–15 (female, male, pupa, and larva).

Distribution. Thailand.

Bionomics. Aquatic habitats: Streams (width 0.2–2.5 m, depth 5–12 cm, 14.4–20.0 °C, elevation 950–1615 m) (Jitklang et al., 2008).

Remarks. This species was described from Thailand by Jitklang et al. (2008).

It is characterized in the female by the sensory vesicle long (0.6 times as long as the third palpal segment); in the male by the number of upper-eye (large) facets in 13 vertical columns and 16 or 17 horizontal rows; in the pupa by the gill with eight slender filaments arranged as [3+(1+2)]+2 or [(1+2)+(1+2)]+2 from dorsal to ventral, arising from a medium-long common basal stalk, and terminal hooks with serrated outer margins; and in the larva by the head spots faintly positive, postgenal cleft medium-long, 1.1–1.4 times as long as the postgenal bridge, and abdominal segments 1–3 each with a greenish transverse band and abdominal segment 4 with a reddish brown transverse band dorsally (Jitklang et al., 2008).

Chromosomally, it is characterized by seven fixed inversions (IS-4, IIS-5, IIL-7, IIL-8, IIL-9, IIIS-2, and multiple inversions in IIIL) and two floating inversions (IL-2 and IIIL-9) (Jitklang et al., 2008).

Simulium (Gomphostilbia) ela Davies and Györkös, 1987

Simulium (*Gomphostilbia*) *ela* Davies and Györkös, 1987a: 1495–1500 (female, male, pupa, and larva).

Distribution. Sri Lanka.

Bionomics. Aquatic habitat: A small stream (width 60–75 cm, 150 m) at the top of the watershed of Gin River (Davies and Györkös, 1987a).

Remarks. This species was described from Sri Lanka (Davies and Györkös, 1987a).

It is characterized in the female by the sensory vesicle long (0.7 times as long as the third palpal segment), scutum covered with brassy short hairs and hind tibia dark brown except its basal one-third pale; in the male by the number of upper-eye (large) facets in 11 vertical columns and 13 horizontal rows, scutum with three longitudinal vittae, hind tibia whitish yellow on its basal one-third, hind basitarsus enlarged and wedge-shaped, 2.7 times as long as its greatest width, equal in width to the hind tibia and pale medium brown on its basal one-third to two-fifths; and in the larva by the rectal organ simple (without secondary lobules).

Simulium (Gomphostilbia) hunanense Zhang and Chen, 2004

Simulium (*Gomphostilbia*) *hunanense* Zhang and Chen, 2004: 374–376 (female, male, pupa, and larva).

Distribution. China (Hunan).

Bionomics. Aquatic habitats: Small streams at an elevation of 480 m (Zhang and Chen, 2004).

Remarks. This species was described from Hunan, China by Zhang and Chen (2004).

It is characterized in the female by the sensory vesicle medium-long (0.32 times as long as the third palpal segment); in the male by the number of upper-eye (large) facets in 12 vertical columns and 13 horizontal rows, scutum with three fine longitudinal vittae, hind basitarsus yellow on its basal two-thirds, and enlarged, 4.0 times as long as its greatest width; in the pupa by the gill with eight slender filaments arranged as (3+3)+2 from dorsal to ventral, of which two filaments of the ventral pair are 1.4 times as long as the six other filaments, and last abdominal segment with horn-like terminal hooks; and in the larva by the postgenal cleft medium-long, 2.0–2.5 times as long as the postgenal bridge.

Simulium (Gomphostilbia) leparense Takaoka, Sofian-Azirun and Ya'cob, 2014

Simulium (*Gomphostilbia*) *leparense* Takaoka, Sofian-Azirun and Ya'cob [in Takaoka et al., 2014g]: 518–522 (female, male, and pupa).

Distribution. Malaysia (Peninsular).

Bionomics. Aquatic habitat: A small stream (width 0.2–0.7 m, 24 °C, shaded, elevation 400 m) slowly flowing along an old logging road in a natural forest, Jerantut, Pahang (Takaoka et al., 2014g).

Remarks. This species was described from Peninsular Malaysia (Takaoka et al., 2014g). Its larva is unknown.

It is characterized in the female by the sensory vesicle long (0.55 times as long as the third palpal segment) and hind tibia darkened except its basal one-fourth of the anterior surface and basal two-fifths of the posterior surface whitish; in the male by the small number of upper-eye (large) facets in seven or eight vertical columns and 10 or 11 horizontal rows, hind tibia darkened except its basal one-fourth whitish and abdominal segments 2–8 each with a pair of shiny dorsolateral or lateral patches; and in the pupa by lacking grapnel-like hooklets on abdominal segment 9.

***Simulium* (*Gomphostilbia*) *longitruncum* Takaoka and Davies, 1995**

Simulium (*Gomphostilbia*) *longitruncum* Takaoka and Davies, 1995a: 65–70 (female, male, pupa, and larva).

Distribution. Malaysia (Peninsular).

Bionomics. Aquatic habitat: A small stream (width 1.0–1.6 m, 20.5 °C) in West Frazer's Hill, Pahang (Takaoka and Davies, 1995a).

Remarks. This species was described from Peninsular Malaysia (Takaoka and Davies, 1995a). It is characterized in the female and male by the base of the radius with a tuft of yellow hairs; in the female by the sensory vesicle short, 0.22 times the length of the third palpal segment; in the male by the number of upper-eye (large) facets in 13 vertical columns and 14 horizontal rows; in the pupa by the gill with eight filaments arranged as 3+[1+(3+1)] or (2+1)+[(2+1)+2] or 3+(3+1+1), arising from a long common basal stalk; and in the larva by the postgenal cleft medium-long, 1.4–1.9 times as long as the postgenal bridge.

***Simulium* (*Gomphostilbia*) *mindoroense* Takaoka and Tenedero, 2007**
(Figs. 26P, R–W and 27A, F)

Simulium (*Gomphostilbia*) *mindoroense* Takaoka and Tenedero, 2007a: 30–37 (female, male, pupa, and larva).

Distribution. Philippines (Mindoro).

Bionomics. Aquatic habitat: A small stream (width 0.2–0.3 m, flow slow, 25.0 °C, shaded, elevation 213 m) just above Tamalaw Falls, Mindoro Oriental (Takaoka and Tenedero, 2007a).

Remarks. This species was described from Mindoro, the Philippines (Takaoka and Tenedero, 2007a).

It is characterized in the female by the claw with a relatively short tooth 0.44 times the length of the claw; in the male by the number of upper-eye (large) facets in 12 vertical columns and 13 horizontal rows and scutum grayish white pruinose on each shoulder, narrowly along each lateral margin and on the prescutellar area; and in the pupa by the terminal hooks wide, plate-like, with their outer margins crenulated (Fig. 27F), and cocoon without any anterodorsal bulge or projection.

Simulium (Gomphostilbia) napuense Takaoka, 2003

Simulium (*Gomphostilbia*) *napuense* Takaoka, 2003: 179–183 (male and pupa).

Distribution. Indonesia (Sulawesi).

Bionomics. Aquatic habitat: A forest stream (width ca. 3 m, 17.0 °C, partially shaded, elevation 1520 m) in Napu (Takaoka, 2003).

Remarks. This species was described from Central Sulawesi, Indonesia (Takaoka, 2003). Its female and larva are unknown.

This species is characterized in the male by the number of upper-eye (large) facets in 11 vertical columns and 14 or 15 horizontal rows and scutum with three faint longitudinal vittae; and in the pupa by the gill with eight slender filaments arranged as [(2+1)+2+1]+2 from dorsal to ventral and terminal hooks wide, plate-like, and with their outer margins crenulated.

Simulium (Gomphostilbia) nigrofemoralum Chen and Zhang, 2001

Simulium (*Gomphostilbia*) *nigrofemoralum* Chen and Zhang, 2001: 364–366 (female, male, pupa, and larva).

Distribution. China (Hainan).

Bionomics. Aquatic habitat: A trickle (elevation 800 m) in Jiangfenging County (Chen and Zhang, 2001).

Remarks. This species was described from Hainan, China by Chen and Zhang (2001).

It is characterized in the female by the mandible lacking outer teeth and sensory vesicle medium-long, 0.4 times as long as the third palpal segment; in the male by the number of upper-eye (large) facets in 13 vertical columns and 12 horizontal rows, hind basitarsus black except the basal one-third pale yellow [though the basal two-thirds pale yellow in Fig. 18 in Chen and Zhang (2001)] and enlarged, about four times as long as its greatest width (though nearly parallel-sided) and ventral plate flat when viewed caudally; in the pupa by the gill with eight slender filaments arranged as [(1+2)+(1+2)]+2 from dorsal to ventral, angle between the stalk of the ventral pair and the common stalk of middle and dorsal triplets ca. 30 degrees and cocoon wall-pocket-shaped, with an anterodorsal projection; and in the larva by the postgenal cleft long, 3.5–4.0 times as long as the postgenal bridge, and abdominal segments 6–8, each with branched dark spinous setae dorsally.

Sternite 8 of the pupa was noted to have grapnel-like hooklets ventrolaterally. It might be wrong because such grapnel-like hooklets in most species of the subgenus *Gomphostilbia* are always located on each lateral surface of abdominal segment 9.

Simulium (Gomphostilbia) pangsidaense Takaoka, Srisuka and Saeung, 2021 (Figs. 26F–K and 27G)

Simulium (*Gomphostilbia*) *pangsidaense* Takaoka, Srisuka and Saeung [in Takaoka et al., 2021e]: 1–6 (female, male, pupa, and larva).

Distribution. Thailand.

Bionomics. Aquatic habitat: A stream (width 2 m, depth 30 cm, bottom rocky, 27 °C, partially shaded, elevation 240 m) in Pang Sida National Park, Sa Kaew Province (Takaoka et al., 2021e).

Remarks. This species was described from Thailand (Takaoka et al., 2021e).

It is distinguished from three Thai members of the *S. ceylonicum* species-group by the following characteristics: from *S. curtatum* and *S. sheilae* by the wide pupal terminal hooks (triangular terminal hooks in the latter two species), and from *S. sheilae* and *S. trangense* by the larger number of male upper-eye (large) facets in 13 vertical columns and 14 or 15 horizontal rows (10 or 11 vertical columns and 12 or 13 horizontal rows in the latter two species) (Jitklang et al., 2008; Takaoka and Davies, 1995a; Takaoka et al., 2018m).

This species is distinguished from all other members of the *S. ceylonicum* species-group by a combination of the elongate female sensory vesicle, male upper-eye (large) facets in 13 vertical columns and 14 or 15 horizontal rows, pupal gill with eight filaments and cocoon without an anterodorsal projection.

***Simulium* (*Gomphostilbia*) *sheilae* Takaoka and Davies, 1995**

(Fig. 27H, I, M)

Simulium (*Gomphostilbia*) *sheilae* Takaoka and Davies, 1995a: 60–65 (female, male, pupa, and larva); Kuvangkadilok and Takaoka, 2000: 172–173; Takaoka et al., 2000: 161; Takaoka, 2001a; 229; Takaoka, 2001b: 250.

Distribution. Malaysia (Peninsular, Sabah and Sarawak), Indonesia (Sumatra), and Thailand.

Bionomics. Aquatic habitats: Small streams (width 0.6–1.3 m, 26 °C) in lowlands in Peninsular Malaysia (Takaoka and Davies, 1995a).

Remarks. This species was collected from Peninsular Malaysia (Takaoka and Davies, 1995a). It was later recorded from Sabah and Sarawak (Takaoka, 2001a, b), Sumatra, Indonesia (Takaoka et al., 2000), and Thailand (Kuvangkadilok and Takaoka, 2000).

The populations of *S. sheilae* in Thailand and Sumatra, Indonesia, were genetically different from those in Peninsular Malaysia (Hew et al., 2023a).

It is characterized in the female by the sensory vesicle long, 0.68 times the length of the third palpal segment; in the male by the number of upper-eye (large) facets in 10 vertical columns and 13 horizontal rows and scutum almost entirely white pruinose; in the pupa by the gill with eight filaments arranged as (1+2)+(1+2)+2 from dorsal to ventral, arising from the short common basal stalk, and terminal hooks triangular and plate-like (Fig. 27H, I); and in the larva by the postgenal cleft long, reaching the posterior margin of the hypostoma (Fig. 27M).

***Simulium* (*Gomphostilbia*) *trangense* Jitklang, Kuvangkadilok, Baimai, Takaoka and Adler, 2008**

(Fig. 27L)

Simulium (*Gomphostilbia*) *trangense* Jitklang, Kuvangkadilok, Baimai, Takaoka and Adler, 2008: 19–20 (female, male, pupa, and larva); Takaoka et al. 2018m: 21–27 (female, male, pupa and larva).

Simulium (*Gomphostilbia*) sp. nr. *sheilae* Phasuk et al., 2005: 52–57 (male and larva).

Simulium (*Gomphostilbia*) *sheilae* (*nec* Takaoka and Davies): Phasuk et al., 2005: 52–57 (larva).

Distribution. Malaysia (Peninsular) and Thailand.

Bionomics. Aquatic habitats: Small to medium-sized streams at low to medium elevations in the mainland of Peninsular Malaysia, Langkawi Island, Pangkor Island, and Tioman Island (Takaoka et al., 2018m).

Remarks. This species was briefly described from females, males, pupae, and larvae collected from southern parts of Thailand (Jitklang et al., 2008). It was later redescribed by Takaoka et al. (2018m).

It is similar to *S. sheilae* but is distinguished by the following characteristics (characteristics of *S. sheilae* in parentheses): the common basal stalk of the pupal gill medium-long (short), terminal hooks wide, plate-like, with their outer margins crenulated (triangular, with their outer margins smooth or slightly undulate), and the larval postgenal cleft not reaching the posterior margin of the hypostoma (Fig. 27L) (reaching the posterior margin of the hypostoma), and thoracic segment 1 encircled by a reddish-brown band (lacking a reddish-brown marking).

***Simulium* (*Gomphostilbia*) *unii* Takaoka and Pham, 2017**

Simulium (*Gomphostilbia*) *unii* Takaoka and Pham [in Takaoka et al., 2017c]: 42–49 (female, male, pupa, and larva).

Distribution. Vietnam.

Bionomics. Aquatic habitat: A small stream (width 1.0 m, depth 5–10 cm, bottom sandy, flow slow, 18.0 °C, shaded, elevation 8 0 m) in Nghe An Province (Takaoka et al., 2017c).

Remarks. This species was described from Vietnam and placed in the *S. asakoae* species-group by Takaoka et al. (2017c). It was because many characteristics including the base of the female and male radial vein with a tuft of yellow hairs, female mandible with two or three outer teeth at some distance from the apex, male hind tibia whitish yellow on a little less than the basal half, and larval postgenal cleft rounded and medium-long (1.4 times as long as the postgenal bridge), were in good agreement with the diagnosis of the *S. asakoae* species-group defined by Takaoka (2012). However, given the more weight to the shape of the male ventral plate for the assignment to species-group, this species is here transferred to the *S. ceylonicum* species-group because it has the ventral plate with its body slightly narrowed posteriorly when viewed ventrally. Future genetic studies are needed to confirm this assignment to the species-group since this species is exceptional among species in the *S. ceylonicum* species-group in having such an unexpected combination of morphological characteristics, as shown in *S. atratoides*.

It is characterized in the female by the sensory vesicle short (0.2 times as long as the third palpal segment); in the male by the number of upper-eye (large) facets in 13 or 14 vertical columns and 14 horizontal rows, hind basitarsus moderately enlarged, spindle-shaped, 4.3–4.6 times as long as its greatest width, and 0.8–0.9 and 0.8 times as wide as the hind tibia and femur, respectively; in the pupa by the gill with eight filaments arranged as [(2+1)+(2+1)]+2 or [3+(2+1)]+2 or [(2+1)+3]+2 or

(3+3)+2 from dorsal to ventral and terminal hooks triangular, with their outer margins 1.8 times as long as their inner margins and crenulated, and cocoon with an anterodorsal projection (similar to Fig. 14L); and in the larva by abdominal segments 1 and 2 each with a grayish transverse band.

***Simulium* (*Gomphostilbia*) *wijiti* Takaoka, Srisuka and Saeung, 2023**

Simulium (*Gomphostilbia*) *wijiti* Takaoka, Srisuka and Saeung [in Srisuka et al., 2023b]: 2–8 (female, male, pupa, and larva).

Distribution. Thailand.

Bionomics. Aquatic habitat: A stream (width 30 cm, depth 1 cm, bottom muddy, 22 °C, exposed to the sun, elevation 979 m) in Nam Pla Mong Village, Pai District, Mae Hong Son Province (Srisuka et al., 2023b).

Remarks. This species was described based on specimens from Mae Hong Son Province, Thailand (Srisuka et al., 2023b).

It is distinguished from four Thai members of the *S. ceylonicum* species-group: *S. curtatum* Jitklang et al., *S. pangsidaense* Takaoka, Srisuka and Saeung, *S. sheilae* Takaoka and Davies, and *S. trangense* Jitklang et al., in the female by the short to medium-long sensory vesicle; in the male by the large number of upper-eye (large) facets in 15 vertical columns and 15 or 16 horizontal rows; in the pupa by the dorsum of all abdominal segments darkened; and in the larva by the antenna as long as or slightly shorter than the stem of the labral fan (longer than the stem of the labral fan in four other species). Phylogenetic analysis based on the COI gene sequences revealed that this species is genetically closely related to *S. leparense* of the *S. ceylonicum* species-group, but is clearly separated from the latter species and the three Thai-related species: *S. curtatum*, *S. sheilae* and *S. trangense*, with interspecific genetic distances ranging from 9.65% to 12.67%.

***Simulium* (*Gomphostilbia*) *wuyishanense* Chen and Zhang, 2009**

Simulium (*Gomphostilbia*) *wuyishanense* Chen and Zhang [in Zhang et al., 2009]: 107–110 (female, male, pupa, and larva).

Distribution. China (Fujian).

Bionomics. Aquatic habitat: A stream (flow fast, exposed to the sun, elevation 1258 m) in Wuyi Mountain (Zhang et al., 2009).

Remarks. This species was described from Fujian, China by Zhang et al. (2009).

It is characterized in the female by the frons and clypeus covered with several yellowish white hairs, sensory vesicle short, 0.28 times as long as the third palpal segment; in the male by the number of upper-eye (large) facets in 12 vertical columns and 12 horizontal rows, clypeus covered with some pale hairs and fore and hind basitarsi ca. 8.0 and ca. 4.0 times as long as their greatest width, respectively; in the pupa by the gill with eight slender filaments arranged as [(1+2)+(1+2)] +2 from dorsal to ventral, of which two filaments of the ventral pair are twice as long as and as thick as the six other filaments, last abdominal segment with conical terminal hooks and cocoon with an anterodorsal projection; and in the larva by the labral fan with ca. 28 primary rays and postgenal cleft long, 2.5–3.0 times as long as the postgenal bridge.

Simulium (Gomphostilbia) rosemaryae Subgroup

Simulium (Gomphostilbia) kamimurai Takaoka, 2003

(Fig. 26L, M)

Simulium (*Gomphostilbia*) *kamimurai* Takaoka, 2003: 191–199 (female, male, pupa, and larva).

Distribution. Indonesia (Seram and Sulawesi).

Bionomics. Aquatic habitats: A trickle (width 2–10 cm, depth 0.2 cm, shaded, 25.5 °C, elevation 3 m) slow-flowing from the seepage to the seashore, central Seram, and small slow-flowing streams (width 0.2–3.0 m, 20–26 °C, elevation 3–1450 m) in Seram and Sulawesi (Takaoka, 2003).

Remarks. This species was described based on specimens from Seram and Sulawesi (Takaoka, 2003).

It is characterized by the elongate female sensory vesicle and pupal abdominal segment 9 without grapnel-like hooklets on each lateral surface.

Simulium (Gomphostilbia) rangatense Takaoka, Sofian-Azirun and Wayan, 2017

(Fig. 27C)

Simulium (*Gomphostilbia*) *rangatense* Takaoka, Sofian-Azirun and Wayan [in Takaoka et al., 2017a]: 582–586 (male and pupa).

Distribution. Indonesia (Flores).

Bionomics. Aquatic habitat: A moderate to fast-flowing stream (width 1.0–1.5 m, depth 10–20 cm, bottom of pebbles and sands, 21.0 °C, exposed to the sun, elevation 785 m), Rangat, Kempo, Flores, Indonesia (Takaoka et al., 2017a).

Remarks. This species was described from Flores, the Lesser Sunda Archipelago (Takaoka et al., 2017a). Its female and larva are unknown.

The pupa of this species is characterized by the gill with six filaments, of which two filaments of the middle pair have a long stalk (Fig. 27C) and abdominal segments 1, 2 and 6–9 covered with minute spines or tubercles dorsally.

This species is distinguished from *S. kamimurai* and *S. rosemaryae* in the male by the hind basitarsus, which is whitish yellow on the basal half or slightly less (yellow or dark yellow on the basal one-third or two-fifths in the latter two species); and in the pupa by the gill with an elongate stalk of the middle pair (short in the latter two species), and abdominal segments 1, 2, 8, and 9 each with tubercles dorsally (without tubercles in the latter two species).

Simulium (Gomphostilbia) rosemaryae Takaoka and Roberts, 1988

(Figs. 26A, D, N, O and 27B, O)

Simulium (*Gomphostilbia*) *rosemaryae* Takaoka and Roberts, 1988: 208–211 (female, male, pupa, and larva); Takaoka, 2003: 199–206 (female, male, pupa, and larva).

Distribution. Indonesia (Sulawesi).

Bionomics. Aquatic habitats: Streams with slow to moderate flow at various elevations from 100 to 1480 m in Sulawesi (Takaoka, 2003).

Remarks. This species was described from Sulawesi by Takaoka and Roberts (1988) and redescribed by Takaoka (2003).

It is similar to *S. kamimurai* from Sulawesi and Seram (Takaoka, 2003) but is easily distinguished by the medium-long female sensory vesicle (Fig. 26A), relative length of the labrum against the clypeus (0.87–1.05 in this species versus 0.67–0.69 in *S. kamimurai*), and pupal abdominal segment 9 with grapnel-like hooklets on each lateral side.

There are variations in the size of female tergite 5, and color pattern of the male scutum (Fig. 26N, O) (Takaoka, 2003).

Simulium (Gomphostilbia) mogii Subgroup

Simulium (*Gomphostilbia*) *gimpuense* Takaoka, 2003

(Fig. 27D)

Simulium (*Gomphostilbia*) *gimpuense* Takaoka, 2003: 212–217 (female, male, and pupa).

Distribution. Indonesia (Sulawesi).

Bionomics. Aquatic habitat: A small stream (width 0.8 m, 22 °C, shaded, elevation 600 m) slow-flowing in natural forest, in Gimpu (Takaoka, 2003).

Remarks. This species was described from Central Sulawesi, Indonesia (Takaoka, 2003). Its larva is unknown.

It is characterized in the female and male by the scutum light brown; in the male by the number of upper-eye (large) facets in nine or 10 vertical columns and 13 horizontal rows; and in the pupa by the common basal stalk of the gill shorter than the stalk of the ventral pair (Fig. 27D).

Simulium (*Gomphostilbia*) *mogii* Takaoka, 2003

Simulium (*Gomphostilbia*) *mogii* Takaoka, 2003: 206–212 (female, male, pupa, and larva).

Distribution. Indonesia (Sulawesi).

Bionomics. Aquatic habitats: Small streams (width 0.4–1.0 m, 18–21 °C, shaded, elevation 900–1480 m), in South and Central Sulawesi (Takaoka, 2003).

Remarks. This species was described based on specimens from South and Central Sulawesi (Takaoka, 2003).

It is similar to *S. gimpuense* from Sulawesi (Takaoka, 2003) but is barely distinguished by the relative length of the female fore basitarsus against its greatest width (4.9 in this species versus 5.7 in *S. gimpuense*), the number of the vertical columns of male upper-eye (large) facets (11 or 12 in this species versus nine or 10 in *S. gimpuense*), the pupal frons densely covered with tubercles (moderately in *S. gimpuense*), and the common basal stalk of the pupal gill medium-long, longer than the stalk of the ventral pair of filaments (short in *S. gimpuense*).

Simulium (Gomphostilbia) darjeelingense Species-group

Simulium (Gomphostilbia) chayamaritae Takaoka and Srisuka, 2010

(Fig. 29B, D, G)

Simulium (*Gomphostilbia*) *chayamaritae* Takaoka and Srisuka, 2010b: 265–270 (male, pupa, and larva); Takaoka et al. (2018l): 976–979 (female).

Distribution. Thailand.

Bionomics. Aquatic habitats: A small stream (width 1.0 m, depth 10 cm, 15 °C, shaded, elevation 1966 m) and a small stream (width 0.8 m, depth 30 cm, flow moderate, 15 °C, exposed to the sun, elevation 1460 m) in Phahompok National Park, Chiang Mai Province (Takaoka and Srisuka, 2010b).

Remarks. This species was described from Thailand (Takaoka and Srisuka, 2010b). Its female was described by Takaoka et al. (2018l).

It is characterized in the female and male by the base of the radius with a tuft of dark hairs; in the female by the sensory vesicle elongate (0.54–0.59 times as long as the third palpal segment), scutum without dark longitudinal vittae and fore coxa yellow; in the male by the large number of upper-eye (large) facets in 16 vertical columns and 17 horizontal rows, fore coxa darkened and ventral plate flat; and in the pupa by the frons and thorax with sharply-pointed tubercles (Fig. 29B) (in place of round tubercles) and the face with an additional pair of trichomes (similar to Fig. 29A).

Simulium (Gomphostilbia) curvastylum Chen and Zhang, 2001

Simulium (*Gomphostilbia*) *curvastylum* Chen and Zhang, 2001: 361–363 (female, male, pupa, and larva).

Distribution. China (Hainan).

Bionomics. Aquatic habitat: A small stream (shaded, elevation 750–850 m) (Chen and Zhang, 2001).

Remarks. This species was described from Hainan, China by Chen and Zhang (2001). It was placed in the *S. ceylonicum* species-group by Takaoka (2012) but is here transferred to the *S. darjeelingense* species-group. The yellow male fore coxa departs the diagnosis of this species-group but other characteristics such as the flat ventral plate, twisted style, and short gill filaments fit to the diagnosis of the *S. darjeelingense* species-group.

It is similar to all six other species of the *S. darjeelingense* species-group in many characteristics including the elongate female sensory vesicle, flat ventral plate, twisted style and long postgenal cleft with its apex close to the posterior border of the hypostoma. However, it is distinguished by having the lower number of male upper-eye (large) facets in 11 vertical columns and 11 horizontal rows and yellow male fore coxa.

Simulium (Gomphostilbia) darjeelingense Datta, 1973

Simulium (*Gomphostilbia*) *darjeelingense* Datta, 1973: 380–382: (female, male, and pupa); Shrestha and Takaoka, 2009: 254.

Simulium (*Gomphostilbia*) *unum* Datta, 1975a: 38 (pupa). Synonymized by Borah et al., 2012: 93.

Distribution. India.

Bionomics. Aquatic habitats: Small streams flowing down the Lebong Cart Road (elevations (2030–2050 m), in Darjeeling (Datta, 1973).

Remarks. This species was described from Darjeeling, India (Datta, 1973). It was later recorded from Nepal (Shrestha and Takaoka, 2009). However, the species regarded as this species was re-identified as a new species, *S. shivapuriense* (Takaoka et al., 2020f). The larva of *S. darjeelingense* is unknown.

It is characterized by the number of male upper-eye (large) facets in 14 vertical columns and 14 horizontal rows and pupal gill with eight slender short filaments arranged as 2+(2+2)+2 from dorsal to ventral, all arising basally from a short common basal stalk. The terminal hooks were noted to be absent in the original description (Datta, 1973). Probably they were overlooked.

Simulium (_Gomphostilbia_) _eshimai_ Takaoka and Adler, 2017
(Figs. 28A–N and 29A, C, I, J)

Simulium (*Gomphostilbia*) *eshimai* Takaoka and Adler [in Takaoka et al., 2017c]: 53–60: (female, male, pupa, and larva).

Distribution. Vietnam.

Bionomics. Aquatic habitats: A small cascading stream (width 2.5 m, depth 1 cm, bottom rocky, 11.0 °C, exposed to the sun, elevation 1853 m) in a forest, and a river (width 6 m, depth 0.6 m, bottom rocky, flow fast, 11.0 °C, exposed to the sun, elevation 1728 m) in a forest (Takaoka et al., 2017c).

Remarks. This species was described from Sapa, northern Vietnam (Takaoka et al., 2017c).

It is characterized in the female by the sensory vesicle long (half as long as the third palpal segment); in the male by the large number of upper-eye (large) facets in 17 vertical columns and 17 horizontal rows; and in the pupa by the presence of an additional pair of facial trichomes (Fig. 29A), like *S. chayamaritae*.

Simulium (_Gomphostilbia_) _guizhouense_ Chen, Zhang and Yang, 2003

Simulium (*Gomphostilbia*) *guizhouense* Chen, Zhang and Yang, 2003a: 48–50 (male and pupa).

Distribution. China (Guizhou and Hubei).

Bionomics. A single pupa was collected from a stream (flow fast, elevation 1300 m) in Leigong Mountain (Chen et al., 2003a).

Remarks. This species was described from Guizhou, China by Chen et al. (2003a). Its female and larva are unknown.

It is characterized in the male by the legs including the fore coxa darkened, and hind basitarsus enlarged, about four times as long as its greatest width, and ventral plate flat and narrowed posteriorly when viewed ventrally; and in the pupa by the gill with eight short filaments arranged as 2+4+2 from dorsal to ventral, of which the outer filament of the ventral pair is much thicker than the seven other filaments.

The male genitalia and the arrangement of the pupal gill filaments are similar to those of *S. darjeelingense*. However, it is difficult to compare this species with *S. darjeelingense* because the information on the number of male upper-eye (large) facets of this species is lacking.

***Simulium* (*Gomphostilbia*) *shivapuriense* Takaoka and Shrestha, 2020**

Simulium (*Gomphostilbia*) *shivapuriense* Takaoka and Shrestha [in Takaoka et al., 2020f]: 190–191: (male and pupa).
Simulium (*Gomphostilbia*) *darjeelingense* (*nec* Datta): Shrestha and Takaoka, 2009: 254 (male and pupa).

Distribution. Nepal.

Bionomics. Aquatic habitat: A small stream at an elevation of 1600 m in Shivapuri National Park (Shrestha and Takaoka, 2009).

Remarks. This species was described from Nepal (Takaoka et al., 2020f). Its female and larva are unknown. This is the same species once misidentified as *S. darjeelingense* by Shrestha and Takaoka (2009).

It is similar to *S. darjeelingense* but is distinguished in the pupa by the gill with eight slender short filaments arranged as [(2+1)+(1+2)]+2 from dorsal to ventral and abdominal segment 9 with a pair of triangular, somewhat wide terminal hooks.

***Simulium* (*Gomphostilbia*) *shresthae* Takaoka and Dangi, 2020**

(Fig. 29E, F, H)

Simulium (*Gomphostilbia*) *shresthae* Takaoka and Dangi [in Takaoka et al., 2020f]: 186–189: (male and pupa).

Distribution. Nepal.

Bionomics. Aquatic habitat: A small stream (20.5 °C, elevation 1732 m) in Dadeldhura, Nepal (Takaoka et al., 2020f).

Remarks. This species was described from Nepal (Takaoka et al., 2020f). Its female and larva are unknown.

It is characterized in the male by the number of upper-eye (large) facets in 12 vertical columns and 13 horizontal rows; and in the pupa by the gill with eight slender short filaments arranged as [3+(1+2)]+2 from dorsal to ventral (Fig. 29E) and abdominal segment 9 with a pair of small, cone-like terminal hooks.

***Simulium* (*Gomphostilbia*) *epistum* Species-group**

***Simulium* (*Gomphostilbia*) *zonatum* Subgroup**

***Simulium* (*Gomphostilbia*) *adleri* Jitklang and Kuvangkadilok, 2008 [2007]**

(Fig. 31G)

Simulium (*Gomphostilbia*) *adleri* Jitklang and Kuvangkadilok, 2008 [2007]: 556–560 (male, pupa, and larva); Takaoka et al., 2018m: 27–31 (female and pupa).

Distribution. Malaysia (Peninsular) and Thailand.

Bionomics. Aquatic habitat: A small stream (width 0.9 m, flow moderate, 23.3 °C, partially shaded, elevation 423 m), Jalan Sungai Koyan, Raub, Pahang, Malaysia (Takaoka et al., 2018m).

Remarks. This species was originally described from males, pupae and larvae collected from Thailand (Jitklang and Kuvangkadilok, 2008). It was recorded from Peninsular Malaysia by Ya'cob et al. (2016). The female of this species was described by Takaoka et al. (2018m).

It is characterized in the female by the elongate sensory vesicle; in the pupa by the frons and thorax without tubercles, stalks of two triplets and that of the ventral pair of gill filaments not lying in the same plane and abdominal segment 9 with a pair of conical terminal hooks; and in the larva by the short postgenal cleft, which is shorter than the postgenal bridge, and dark gray transverse bands on the abdomen except segment 6 unpigmented.

***Simulium* (*Gomphostilbia*) *auratum* Takaoka, 2009**

(Figs. 30Q–S and 31B)

Simulium (*Gomphostilbia*) *auratum* Takaoka, 2009c: 100–104 (male and pupa).

Distribution. Malaysia (Sarawak).

Bionomics. Aquatic habitats: A stream (width 0.2–0.3 m, flow slow, 20 °C, shaded, elevation 1000 m) flowing in a natural forest, and a stream (width 0.2–0.5 m, flow slow, 21 °C, exposed to the sun, elevation 980 m) flowing in a grassy hill, Bakalalan (Takaoka, 2009c).

Remarks. This species was described from Sarawak, Malaysia (Takaoka, 2009c). Its female and larva are unknown.

The male of this species is characterized by the number of upper-eye (large) facets in 15 or 16 vertical columns and 15 horizontal rows, style broad and twisted inward (Fig. 30R), abdominal segments 2–6 mostly yellow and ventral plate widened posteriorly when viewed ventrally (Fig. 30Q).

***Simulium* (*Gomphostilbia*) *cheongi* Takaoka and Davies, 1995**

Simulium (*Gomphostilbia*) *cheongi* Takaoka and Davies, 1995a: 37–42 (female, male, pupa, and larva).

Distribution. Malaysia (Peninsular), Indonesia (Kalimantan and Sumatra), and Thailand.

Bionomics. Aquatic habitats: Small, forest streams (24 °C) in lowlands (Takaoka and Davies, 1995a).

Remarks. This species was described from Peninsular Malaysia (Takaoka and Davies, 1995a). It was later recorded from Thailand (Takaoka et al., 2009a), Sumatra (Takaoka et al., 2000), and Kalimantan, Indonesia (Takaoka et al., 2017e).

It is characterized in the female by the narrow frons (frons:head ratio 1.0:6.3); in the male by the antenna entirely yellow; and in the pupa by the gill with eight filaments, of which the outer filament of the ventral pair is a little less than twice as long as the seven other filaments.

Simulium (_Gomphostilbia_) _datfaense_ Takaoka and Otsuka, 2009
(Fig. 31H)

Simulium (*Gomphostilbia*) *datfaense* Takaoka and Otsuka [in Takaoka et al. 2009a]: 259–263 (female and pupa).

Distribution. Thailand.

Bionomics. Aquatic habitat: A cascading stream (width 1 m, 23.5 °C, shaded, elevation 250 m) in Tai Rom Yen National Park, Surat Thani Province (Takaoka et al., 2009a).

Remarks. This species was described from a single female reared from a pupa collected from Surat Thani Province, southern Thailand (Takaoka et al., 2009a). Its male and larva are unknown.

The pupa of this species is characterized by the gill with eight filaments arranged as 3+3+2 from inside to outside and seven of eight filaments tapered basally.

Simulium (_Gomphostilbia_) _epistum_ Delfinado, 1971

Simulium (*Gomphostilbia*) *epistum* Delfinado, 1971: 136–137 (male); Takaoka, 1983: 63–66 (female, male, pupa, and larva).

Distribution. Philippines (Balabac and Palawan).

Bionomics. Aquatic habitats: Small streams slowly flowing in lowlands near mountains (Takaoka, 1983).

Remarks. This species was described from Balabac Island, south of Palawan Island, the Philippines (Delfinado, 1971). The female, pupa, and larva of this species were described based on specimens collected from Palawan by Takaoka (1983).

It is characterized in the male by the style broad and twisted inward; in the pupa by the gill with eight filaments, of which the outer filament of the ventral pair is twice as long as the seven other filaments.

Simulium (_Gomphostilbia_) _floresense_ Takaoka, Hadi and Sigit, 2006
(Fig. 31D)

Simulium (*Gomphostilbia*) *floresense* Takaoka, Hadi and Sigit, 2006a: 2–8 (female, male, pupa, and larva).

Distribution. Indonesia (Flores).

Bionomics. Aquatic habitat: A small stream (width 0.3 m, flow slow, 18 °C, shaded, elevation 1060 m) flowing in a forest (Takaoka et al., 2006a).

Remarks. This species was described from Flores, the Lesser Sunda Archipelago (Takaoka et al., 2006a).

It is characterized in the female by the narrow frons (frons:head ratio 1:6.7), sensory vesicle medium-long, 0.4–0.5 times as long as the third palpal segment; and in the male by the number of upper-eye (large) facets in 13 vertical columns and 14 horizontal rows and antenna yellow except four or five apical flagellomeres light brown.

***Simulium* (*Gomphostilbia*) *kelabitense* Takaoka, 2008**

(Figs. 30A, C–M and 31A, J)

Simulium (*Gomphostilbia*) *kelabitense* Takaoka, 2008c: 181–189 (female, male, pupa, and larva).

Distribution. Malaysia (Sarawak).

Bionomics. Aquatic habitats: Small streams (width 0.3–1.0 m, flow slow to moderate, 20 °C, shaded, elevation 1000 m) flowing in a forest, Bario (Takaoka, 2008c).

Remarks. This species was described from Sarawak, Malaysia (Takaoka, 2008c).

It is characterized in the male by the antenna almost yellow with the apex darkened and the number of upper-eye (large) facets in 12 vertical columns and 13 horizontal rows; and in the pupa by the gill with eight slender filaments, of which the outer filament of the ventral pair is 1.4 times as long as the seven other filaments.

***Simulium* (*Gomphostilbia*) *lagunaense* Takaoka, 1983**

Simulium (*Gomphostilbia*) *lagunaense* Takaoka, 1983: 67–68 (female, male, pupa, and larva).

Distribution. Philippines (Luzon).

Bionomics. Aquatic habitat: A cascading Molawin Creek (width 6 m) in Los Baños, Laguna (Takaoka, 1983).

Remarks. This species was described from Luzon, the Philippines (Takaoka, 1983).

It is similar to *S. epistum* but is distinguished from the latter species by the male ventral plate, of which the posterior half is narrowed and parallel-sided when viewed ventrally; and the pupal gill with eight filaments, of which the outer filament of the ventral pair is 1.5 times as long as and 1.7 times as thick as the seven other filaments.

***Simulium* (*Gomphostilbia*) *luzonicum* Takaoka, 1983**

Simulium (*Gomphostilbia*) *luzonicum* Takaoka, 1983: 66–67 (female, male, pupa, and larva).

Distribution. Philippines (Luzon).

Bionomics. Aquatic habitat: A small stream slowly flowing in a forest (Takaoka, 1983).

Remarks. This species was described from Luzon, the Philippines (Takaoka, 1983).

It is similar to *S. epistum* but is distinguished from the latter species by the male style broad basally, then tapered apically and the pupal gill with eight filaments, of which the outer filament of the ventral pair is twice as long as and 1.4 times as thick as the seven other filaments.

***Simulium* (*Gomphostilbia*) *mindanaoense* Takaoka, 1983**

Simulium (*Gomphostilbia*) *mindanaoense* Takaoka, 1983: 69–70 (female, male, pupa, and larva).

Distribution. Philippines (Luzon).

Bionomics. Aquatic habitat: A mountain stream (width 2–3 m) in Mt. Apo, Mindanao (Takaoka, 1983).

Remarks. This species was described from Mindanao, the Philippines (Takaoka, 1983).

It is similar to *S. luzonicum* but is distinguished from the latter species by the male ventral plate slightly narrowed posteriorly when viewed ventrally and the pupal gill with eight filaments, of which the outer filament of the ventral pair is 1.5 times as long as the inner counter filament and is twice as long as and 1.7 times as thick as the six filaments of the dorsal and middle triplets.

***Simulium* (*Gomphostilbia*) *padangense* Takaoka and Sigit, 1997**

(Fig. 31F)

Simulium (*Gomphostilbia*) *padangense* Takaoka and Sigit, 1997: 69–72 (female, male, pupa, and larva).

Distribution. Indonesia (Sumatra).

Bionomics. Aquatic habitat: A small stream (width 0.3–0.5 m, bottom rocky, flow moderate, 19 °C, shaded, elevation 1060 m) flowing in a forest, West Sumatra (Takaoka and Sigit, 1997).

Remarks. This species was described from Sumatra, Indonesia (Takaoka and Sigit, 1997).

It is characterized by the pupal gill composed of an inflated trunk and seven slender filaments arising from the base of the inflated trunk (Fig. 31F).

***Simulium* (*Gomphostilbia*) *rayohense* Smart and Clifford, 1969**

Simulium (*Gomphostilbia*) *rayohense* Smart and Clifford, 1969: 35–39 (female, male, pupa, and larva).

Distribution. Malaysia (Sabah).

Bionomics. Aquatic habitats: Rapidly flowing streams at low elevations (Smart and Clifford, 1969).

Remarks. This species was described from Sabah, Malaysia by Smart and Clifford (1969).

It is characterized in the female and male by the hind tibia yellowish on the basal three-fourth and with a dark subbasal marking (according to the Figs. 16H and 17G); in the female by the base of the radius with a tuft of yellow hairs mixed with dark hairs; in the male by the base of the radius with a tuft of yellow hairs only and ventral plate lamellate, much wider than long (according to Fig. 13E); in the pupa by the gill with eight slender filaments arranged as (3+3)+2 from dorsal to ventral, of which the outer filament of the ventral pair is much thicker and longer than the seven other filaments; and in the larva by the postgenal cleft long, with its apex close to the posterior margin of the hypostoma, labral fan with ca. 38 primary rays, abdomen covered with dark multi-branched setae on posterior segments, rectal organ with about eight secondary lobules per lobe and posterior circlet with 76–78 rows of 8–13 hooklets per row.

***Simulium* (*Gomphostilbia*) *repentinum* Takaoka and Tenedero, 2008**
(Figs. 30N–P and 31C)

Simulium (*Gomphostilbia*) *repentinum* Takaoka and Tenedero, 2008c: 325–330 (male and pupa).

Distribution. Philippines (Samar).

Bionomics. Aquatic habitats: A small stream (1–2 m wide, bottom rocky, flow moderate, 24 °C, shaded, elevation 30–50 m) flowing in a coconut plantation, northern Samar, and a stream (width 0.2–0.5 m, bottom of rocks and stones, 25 °C, exposed to the sun, elevation 110–130 m) flowing in a coconut plantation, western Samar (Takaoka and Tenedero, 2008c).

Remarks. This species was described from Samar, the Philippines (Takaoka and Tenedero, 2008c). Its female and larva are unknown.

The pupa of this species is characterized by the gill with eight filaments arranged as two paired filaments and six individual filaments, all arising from a long common basal stalk (Fig. 31C).

***Simulium* (*Gomphostilbia*) *sarawakense* Takaoka, 2001**
(Fig. 30B)

Simulium (*Gomphostilbia*) *sarawakense* Takaoka, 2001b; 247–250 (female and pupa); Takaoka et al., 2017d: 442–445 (male and larva).

Distribution. Malaysia (Sarawak) and Indonesia (Kalimantan).

Bionomics. Aquatic habitat: A small stream (width 0.5–3.0 m, bottom rocky, 25 °C, shaded) in a natural forest, Pueh, Sarawak (Takaoka, 2001b).

Remarks. This species was described from a female and its associated pupal exuviae collected in Sarawak, Malaysia (Takaoka 2001b). Its male and larva were described based on specimens from Kalimantan, Indonesia by Takaoka et al. (2017d).

This species is similar to *S. rayohense* but is distinguished in the larva by lacking setae on the thorax and abdominal segments 1–4.

***Simulium* (*Gomphostilbia*) *whartoni* Takaoka and Davies, 1995**

Simulium (*Gomphostilbia*) *whartoni* Takaoka and Davies, 1995a: 32–37 (female, male, pupa, and larva).

Simulium (*Gomphostilbia*) *zonatum* (*nec* Edwards): Crosskey, 1973: 425.

Distribution. Malaysia (Peninsular) and Indonesia (Sumatra).

Bionomics. Aquatic habitats: Streams (width 0.4–2.5 m, 20–24 °C) in lowlands (Takaoka and Davies, 1995a).

Remarks. This species was described from Peninsular Malaysia (Takaoka and Davies, 1995a). It was later recorded from Sumatra, Indonesia (Takaoka et al., 2000).

It is characterized in the male by the antenna yellow except a few apical flagellomeres somewhat darkened; and in the pupa by the gill with eight filaments, of which the outer filament of the ventral pair is three times as long as the seven other filaments.

Simulium (*Gomphostilbia*) *zonatum* Edwards, 1934

(Fig. 31E, I)

Simulium (*Gomphostilbia*) *zonatum* Edwards, 1934: 126 (female, male, pupa, and larva).

Distribution. Indonesia (Sumatra).

Bionomics. Aquatic habitat: A stream in a forest, east of Surabaya Village (Edwards, 1934).

Remarks. This species was briefly described from South Sumatra, Indonesia by Edwards (1934).

It is characterized in the male by the number of upper-eye (large) facets in ca. 10 horizontal rows and antenna yellow except three or four apical flagellomeres darkened; in the pupa by the gill with eight filaments arranged as [3+(1+2)]+2 from dorsal to ventral, of which the outer filament of the ventral pair is a little more than twice as long as and three times as thick as the seven other filaments, which are subequal in length and thickness; and in the larva by the body yellowish except thoracic segment 3 and abdominal segments 1–3 brownish, and abdominal segments 4, 5, 7, and 8 purplish dorsally, postgenal cleft small, a little less than the length of the postgenal bridge and the second antennal article much shorter than the first and third articles (Edwards, 1934).

Examination of the slide-mounted specimens deposited in Natural History Museum, London, shows the following characteristics: in the female, sensory vesicle enlarged, half as long as the third palpal segment, mandible with 22 inner and eight outer teeth and claw with a large basal tooth 0.62 times as long as the claw; in the pupa, gill with a short common basal stalk, ventral pair with a short stalk, and two triplets arising from an extremely short stalk close together (Fig. 31E), grapnel-like hooklets present, and terminal hooks simple, cone-like; and in the larva, body length ca. 4.0 mm, hypostoma with four or five bristles per side, postgenal cleft 0.8 times as long as the postgenal bridge (Fig. 31I), anal sclerite X-shaped, with its anterior arms slightly shorter than its posterior ones, and posterior circlet with 74 rows of up to 11 hooklets per row (Unpublished data).

This species is unique in the *S. zonatum* subgroup by having an enlarged female sensory vesicle, an elongate female basal claw tooth, a pair of simple pupal terminal hooks and a short postgenal cleft (Fig. 31I).

Simulium (Gomphostilbia) miblosi Subgroup

Simulium (*Gomphostilbia*) *lehi* Takaoka, 2001

(Fig. 32A–G, T–V)

Simulium (*Gomphostilbia*) *lehi* Takaoka, 2001b: 243–247 (female, male, pupa, and larva).

Distribution. Malaysia (Sarawak) and Indonesia (Kalimantan).

Bionomics. Aquatic habitats: A small stream (width 0.5–3.0 m, bottom rocky, 21 °C, shaded, elevation 690 m) in a natural forest of Mt. Pueh, Sematan, Sarawak, and a small stream (width 0.2 m, flow slow, 23–25 °C) at Hot spring Park in Poring, Sabah (Takaoka, 2001b).

Remarks. This species was described from Sarawak, Malaysia (Takaoka, 2001b). It was recorded from Kalimantan, Indonesia (Takaoka et al., 2017e).

It is characterized in the female and male by the elongated calcipala extended beyond the middle of the second tarsomere (similar to Fig. 32I); in the female by the mandible with 12–14 teeth only on its inner margin (Fig. 32A) (the smallest number of the inner teeth in the subgenus *Gomphostilbia*) and cibarium unarmed (Fig. 32B); in the male by the ventral plate much produced ventrally and dorsally; in the pupa by the gill with eight filaments arranged as (2+1)+(1+2)+2 from dorsal to ventral and abdominal segment 9 with two unbranched setae (in place of grapnel-like hooklets) on each lateral surface; and in the larva by the entirely grayish green body.

Simulium (_Gomphostilbia_) _miblosi_ Takaoka, 1983

(Fig. 32H–P, S)

Simulium (*Gomphostilbia*) *miblosi* Takaoka, 1983: 70–71 (female and pupa); Takaoka and Tenedero, 2022a: 332–334 (male).

Distribution. Philippines (Mindanao).

Bionomics. Aquatic habitat: A mountain stream (width 2–3 m) in Mt. Apo, Davao City, Mindanao (Takaoka, 1983).

Remarks. This species was described from two females reared from pupae collected from Mindanao, the Philippines (Takaoka, 1983). The male of this species was described by Takaoka and Tenedero (2022a). Its larva is unknown.

It is characterized in the female by the mandible lacking teeth on its outer margin, cibarium unarmed, hind tibia white on the basal three-fifths and brown on the rest, base of the radius with a tuft of yellow hairs mixed with several brown hairs and sternite 8 with ca. 100 hairs on each side; in the male by the ventral plate much produced ventrally (Fig. 32M); in the pupa by the frons and most of the thorax with tubercles of irregular forms, most of which have tiny secondary projections, frons with two pairs of trichomes, pair of all three dorsomedial trichomes on the thorax short and stout, abdominal segment 9 with two unbranched setae (Fig. 32S) (in place of grapnel-like hooklets) on each lateral surface and cocoon shoe-shaped, with a low anteroventral collar.

Simulium (_Gomphostilbia_) _rarum_ Takaoka and Tenedero, 2008

(Fig. 32Q, R)

Simulium (*Gomphostilbia*) *rarum* Takaoka and Tenedero, 2008b: 319–324 (male and pupa).

Distribution. Philippines (Palawan).

Bionomics. Aquatic habitats: A stream (width 0.5–2.0 m, flow slow, 21 °C, shaded, elevation 690 m) and a stream (width 2–3 m, bottom of rocks and stones, flow moderate, 22.3 °C, shaded, elevation 250 m) in a natural forest of Mt. Taltalaba, Napsan, Palawan (Takaoka and Tenedero, 2008b).

Remarks. This species was described from Palawan, the Philippines (Takaoka and Tenedero, 2008b). Its female and larva are unknown.

It is characterized in the male by the relatively long calcipala extended beyond the middle of the second tarsomere and ventral plate much produced ventrally and dorsally; and in the pupa by the frons sparsely covered with tubercles, gill with eight filaments arranged as (2+1)+(1+2)+2 from dorsal to ventral (Fig. 32Q), abdominal segment 9 with two setae (in place of grapnel-like hooklets) on each lateral surface and cocoon simple, wall-pocket-shaped.

Simulium (Gomphostilbia) atratum Subgroup

Simulium (_Gomphostilbia_) _angulistylum_ Takaoka and Davies, 1995 (complex)
Simulium (*Gomphostilbia*) *angulistylum* Takaoka and Davies, 1995a: 42–46 (female, male, pupa, and larva).

Distribution. Malaysia (Peninsular) and Thailand.

Bionomics. Aquatic habitats: Small streams (width 0.3–2.5 m, 23–24 °C) in lowlands, Peninsular Malaysia (Takaoka and Davies, 1995a).

Remarks. This species was described from Peninsular Malaysia (Takaoka and Davies, 1995a). In Thailand, this species was found to be composed of three cytoforms: A, B, and C (Pramual and Kuvangkadilok, 2012).

It is characterized in the male by the style broad and twisted inward and ventral plate transverse and widest subbasally when viewed ventrally; in the pupa by the gill with eight slender filaments arising from a short common basal stalk; and in the larva by the postgenal cleft long, with its apex close to the posterior margin of the hypostoma.

Simulium (_Gomphostilbia_) _atratum_ De Meijere, 1913
(Fig. 33B, C, R)
Simulium atratum De Meijere, 1913: 331–332 (female and male).
Simulium (*Gomphostilbia*) *atratum*: Crosskey, 1973: 424; Takaoka and Davies, 1996: 18.
Simulium (*Gomphostilbia*) *sundaicum* Edwards, 1934: Takaoka and Davies, 1996: 19–22 (female, male, pupa, and larva). Synonymized by Takaoka (2012).

Distribution. Indonesia (Bali, Flores, Java, Lombok, Sumatra, Sumbawa, and Timor) and Malaysia (Peninsular) (?).

Bionomics. Aquatic habitats: Small streams (width 0.2–3.0 m, 20–24 °C, shaded or partially shaded, elevation 135–1093 m) in the Lesser Sunda Archipelago (Takaoka et al., 2017e).

Remarks. This species was described from two females and one male collected from Java, Indonesia (De Meijere, 1913). The larva regarded as *S. atratum* by Edwards (1934) was finally identified to be that of *S. atratoides* Takaoka and Davies (Takaoka and Davies, 1996). *Simulium sundaicum* described from Java and Sumatra by Edwards (1934) was synonymized with *S. atratum* by Takaoka (2012) according to the suggestion by Takaoka and Davies (1996). Hew et al. (2023b) genetically showed that *S. atratum* is represented by two lineages (A and B), of which lineage A is distributed in Lombok, Bali, Java, and Sumatra, whereas lineage B is distributed in Lombok, Sumbawa, Flores, and Timor.

This species was once reported from Peninsular Malaysia (Crosskey, 1973) but no further information on this species is available. There is a possibility that the species regarded as *S. atratum* (at that time as *S. sundaicum*) is *S. cheongi* judging from the similarity in many characteristics including the color of legs and the arrangement of the pupal gill filaments. Its distribution record from Peninsular Malaysia should be reconfirmed.

It is characterized by the mid and hind tibiae in the female and male each with a dark subbasal band (Fig. 33B, C) and the number of male upper-eye (large) facets in 14 vertical columns and 16 horizontal rows.

***Simulium* (*Gomphostilbia*) *isanense* Takaoka, Srisuka and Saeung, 2017**
(Fig. 33A, D–Q, T, U)

Simulium (*Gomphostilbia*) *isanense* Takaoka, Srisuka and Saeung [in Takaoka et al., 2017f]: 374–379 (female, male, pupa, and larva).

Distribution. Thailand.

Bionomics. Aquatic habitat: A stream (width 3 m, depth 5 cm, bottom sandy and rocky, flow moderate, 21.7 °C, pH 4.71, partially shaded, elevation 337 m), Along Korn Waterfall, Surin Province (Takaoka et al., 2017f).

Remarks. This species was described from Thailand (Takaoka et al., 2017f).

It is similar to *S. angulistylum* from Peninsular Malaysia (Takaoka and Davies, 1995a), but is distinguished from the latter species by the following characteristics (those of *S. angulistylum* in parentheses): in the female by the length of the fore basitarsus relative to its greatest width, which is 7.3–7.4 (6.5); in the male by the length of the fore basitarsus relative to its greatest width, which is 7.7–7.8 (5.5); and in the pupa by abdominal segments 1–5 darkened on the dorsal surface (not darkened).

***Simulium* (*Gomphostilbia*) *otsukai* Takaoka and Choochote, 2009**
(Fig. 33S)

Simulium (*Gomphostilbia*) *otsukai* Takaoka and Choochote [in Takaoka et al. 2009a]: 263–267 (male and pupa).

Distribution. Thailand.

Bionomics. Aquatic habitat: A small stream (width 2 m, flow moderate, 28 °C, shaded, elevation 280 m) in Tai Rom Yen National Park, Surat Thani Province (Takaoka et al., 2009a).

Remarks. This species was described from a single male reared from a pupa collected from Surat Thani Province, southern Thailand (Takaoka et al., 2009a). Its female and larva are unknown.

It is characterized in the male by the broad inwardly twisted style; and in the pupa by the gill with eight filaments arranged as 1+1+2+[1+(1+2)] from dorsal to ventral (Fig. 33S).

***Simulium* (*Gomphostilbia*) *prayooki* Pramual, Jomkumsing, Wongpakam and Takaoka, 2022**

Simulium (*Gomphostilbia*) *prayooki* Pramual, Jomkumsing, Wongpakam and Takaoka, 2022b: 281–290 (female, male, pupa, and larva).

Distribution. Thailand.

Bionomics. Aquatic habitat: A stream (width 2 m, depth 4 cm, flow moderate (0.94 m/s), bottom rocky, shaded, elevation 1141 m) (Pramual et al., 2022b).

Remarks. This species was described from Thailand by Pramual et al. (2022b). This is the same as cytoform C of the *S. angulistylum* complex, which is chromosomally distinguished from cytoforms A and B (Pramual and Kuvangkadilok, 2012).

It is morphologically similar to *S. angulistylum* from Peninsular Malaysia and *S. isanense* from Thailand but is distinguished from the latter two species in the female by the relatively slender hind basitarsus 7.5–8.0 times as long as its greatest width (6.4–6.5 times in the latter two species), claw with a relatively long basal tooth 0.63–0.66 times as long as the claw (0.50–0.51 in the latter two species), and genital fork lacking a projection directed posteromedially from each arm (present in the latter two species); and in the larva by the greater number (16–18) of secondary lobules per lobe of the rectal organ (8 or 9 in *S. angulistylum* and 9–14 in *S. isanense*) (Takaoka and Davies, 1995a; Takaoka et al., 2017f).

Simulium (Gomphostilbia) montiblense Subgroup

***Simulium* (*Gomphostilbia*) *barioense* Takaoka, 2008**

(Figs. 34A, B, O–R and 35D–G, J, K)

Simulium (*Gomphostilbia*) *barioense* Takaoka, 2008c: 189–196 (female, male, pupa, and larva).

Distribution. Malaysia (Sarawak and Sabah).

Bionomics. Aquatic habitats: Slow to moderate flowing streams (width 0.3–6.0 m, 18.2–20.0 °C, shaded, elevation 1000–1080 m), and a trickle (width 0.1–0.2 m, 21 °C, exposed to the sun, elevation ca. 1040 m) in a forest, Bario, Sarawak (Takaoka, 2008c).

Remarks. This species was described from Sarawak, Malaysia (Takaoka, 2008c).

It is characterized by the pupal gill with 13 or 16 slender filaments (Fig. 35D, F) and a shortened cocoon (similar to Fig. 35H, I).

***Simulium* (*Gomphostilbia*) *fulgidum* Takaoka, 2009**

(Figs. 34C–N and 35A, B, H, I)

Simulium (*Gomphostilbia*) *fulgidum* Takaoka, 2009c: 104–111 (female, male, pupa, and larva).

Distribution. Malaysia (Sarawak).

Bionomics. Aquatic habitat: A stream (width 1.0–3.0 m, flow moderate, 24 °C, shaded, elevation 50–60 m) flowing in a forest of Mt. Pueh, Sematan, Sarawak (Takaoka, 2009c).

Remarks. This species was described from Sarawak, Malaysia (Takaoka, 2009c).

It is characterized in the male by the antenna yellow except two apical flagellomeres, abdominal segments 1 and 2 yellow, abdominal segments 2–9 each with a pair of shiny dorsolateral or lateral areas; and in the pupa by the thorax with two groups (dosomedial and ventrolateral) of three trichomes arising close together (Fig. 35A) on each side, gill with eight slender filaments arising together from a short common basal stalk (Fig. 35B) and cocoon shortened (Fig. 35H, I).

***Simulium* (*Gomphostilbia*) *montiblense* Takaoka, 1983**

(Fig. 35C)

Simulium (*Gomphostilbia*) *montiblense* Takaoka, 1983: 71–73 (female, male, pupa, and larva).

Distribution. Philippines (Palawan), China (Fujian) (?), and Vietnam (?).

Bionomics. Aquatic habitat: A cascading mountain stream (width 3–5 m) with many large stones in Montible, Palawan (Takaoka, 1983).

Remarks. This species was described from Palawan, the Philippines (Takaoka, 1983). The record of this species from Vietnam by Crosskey and Howard (1997) was doubted due to lack of information on specimens and collection data by Takaoka et al. (2017c).

It was also recorded from Fujian, China by Zhang et al. (2009). However, it was not listed in the monograph of Chinese black flies (Chen, 2016). It is possible that *S. guizhouense* described from Guizhou Province (Chen et al., 2003a) was misidentified as *S. montiblense* because the pupal gills of both species are almost indistinguishable from each other.

It is characterized in the male by the antenna yellow except four or five apical flagellomeres and abdominal segments 1 and 2 yellow; and in the pupa by the gill with eight slender short filaments, arranged as 3+3+2 from dorsal to ventral, all arising from a short common basal stalk (Fig. 35C).

***Simulium* (*Gomphostilbia*) *paserense* Takaoka and Sofian-Azirun, 2017**

Simulium (*Gomphostilbia*) *paserense* Takaoka and Sofian-Azirun [in Takaoka et al., 2017d]: 438–441 (pharate pupa and larva).

Distribution. Indonesia (Kalimantan).

Bionomics. Aquatic habitat: A small stream (width 1.5 m, depth 7–10 cm, flow moderate to fast, 22.0 °C, partially shaded, elevation 435 m) in Kalimantan (Takaoka et al., 2017d).

Remarks. This species was described from two pharate pupae and three larvae in Kalimantan, Indonesia (Takaoka et al., 2017d). Its adults and cocoon are unknown.

It is characterized in the pupa by the thorax on each side with two groups (dosomedial and ventrolateral) groups of three trichomes arising close together, and gill with 13 short slender filaments, all arising together from an extremely short common basal stalk. This species seems to be related to *S. barioense* and *S. fulgidum* from Sarawak (Takaoka, 2008c, 2009c) in sharing a similar arrangement of the short pupal gill filaments and of the thoracic trichomes.

Simulium (Gomphostilbia) epistum Species-group Species unplaced to subgroup

Simulium (Gomphostilbia) flavocinctum Edwards, 1934

Simulium (Gomphostilbia) flavocinctum Edwards, 1934: 125–126 (female, male, pupa, and larva).

Distribution. Indonesia (Sumatra).

Bionomics. Unknown.

Remarks. This species was briefly described from Sumatra, Indonesia (Edwards, 1934). *Simulium flavocinctum* is placed in the *epistum* species-group based on the pale male hind tibia (Edwards, 1934). However, its placement to subgroup is difficult for lack of information on the grapnel-like hooklets. In addition, the association of the male (designated as holotype) of this species with female, pupa and larva was said to be quite uncertain in the original description (Edwards, 1934). The pupa, thought to be *S. flavocinctum*, has the gill with eight filaments of the same length, and one of two ventral paired filaments is distinctly thicker than the other filament. Thus, the arrangement of the pupal gill filaments agrees with that of either the *S. miblosi* subgroup or the *S. atratum* subgroup.

It is characterized in the female by the antenna yellow except the apical flagellomere darkened, and hind tibia with a dark subbasal marking; and in the male by the antenna entirely yellow, and hind tibia without a subbasal dark marking; in the pupa by the gill with eight slender filaments arranged vertically as (3+3)+2 from dorsal to ventral, all subequal in length to one another, and the outer filament of the ventral pair thicker than the inner filament; in the larva by the body with dark bands on abdominal segments 1–4, and dark greenish dorsally on abdominal segments 5–8, labral fan with ca. 25 primary rays, rectal organ with two to four secondary lobules per lobe, and posterior circlets with ca. 70 rows of up to 12 hooklets per row (Edwards, 1934).

The female and male of this species are similar to those of *S. cheongi* in the *S. zonatum* subgroup by having the yellowish antenna. However, the pupa of this species differs from that of *S. cheongi* and those of other species in the *S. zonatum* subgroup by having the gill with eight slender filaments all subequal in length to one another, and two triplets of filaments lying vertically and sharing a distinct stalk.

Simulium (Gomphostilbia) merapiense Takaoka and Sofian-Azirun, 2016

Simulium (Gomphostilbia) merapiense Takaoka and Sofian-Azirun [in Takaoka et al., 2016c]: 76–82 (female, male, pupa, and larva); Takaoka et al., 2017e: 173 (larva).

Distribution. Indonesia (Java and Lombok).

Bionomics. Aquatic habitat: A small stream (width 0.2 m, flow slow, 20 °C, shaded, elevation 920 m) flowing for a short distance in shrubs, then going underground, in Merapi National Park, Java (Takaoka et al., 2016c).

Remarks. This species was described from Java, Indonesia (Takaoka et al., 2016c). It was also recorded from Lombok, the Lesser Sunda Archipelago, Indonesia (Takaoka et al., 2017e).

It is characterized in the female by the elongate sensory vesicle; in the pupa by the head and thoracic integument entirely bare, gill with eight short slender filaments all arising from a short common basal stalk, which has a somewhat enlarged basal fenestra; and in the larva by the postgenal cleft short, 0.27 times as long as the postgenal bridge.

Simulium (*Gomphostilbia*) *gombakense* Species-group

Simulium (*Gomphostilbia*) *antlerum* Chen, 2001

Simulium (*Morops*) *antlerum* Chen, 2001a: 565–568 (male and pupa).
Simulium (*Gomphostilbia*) *antlerum*: Takaoka, 2003: 546; Takaoka, 2012: 15.

Distribution. China (Hainan).

Bionomics. Aquatic habitat: A fast-flowing small stream (elevation 650 m).

Remarks. This species was described from a single pharate male and a pupa collected from Hainan, China (Chen, 2001a). Its female and larva are unknown. This species was first assigned in the subgenus *Morops* by Chen (2001a), later treated under the *S. banauense* species-group of the subgenus *Gomphostilbia* by Takaoka (2003), and recently transferred to the *S. gombakense* species-group of the same subgenus by Takaoka (2012).

It is characterized in the male by the pleural membrane haired; and in the pupa by the gill composed of an inflated structure having 10 slender short filaments, of which eight and two filaments arise from the basal portion and the dorsal branch of the inflated structure, respectively.

Simulium (*Gomphostilbia*) *aziruni* Takaoka and Chen, 2012
(Fig. 37B)

Simulium (*Gomphostilbia*) *aziruni* Takaoka and Chen [in Takaoka et al., 2012c]: 810–812 (pupa and larva); Ya'cob et al., 2019: 67–70 (female).

Distribution. Malaysia (Peninsular).

Bionomics. Aquatic habitats: A small stream (width 0.5–1.0 m, flow slow, 27 °C, shaded, elevation 280 m) running in a palm oil plantation and a small stream (width 20–30 cm, depth 10 cm, 23.5 °C, exposed to the sun, elevation 450 m) running in a grassland (Takaoka et al., 2012c).

Remarks. This species was described from a pupal exuviae and a larva collected in Peninsular Malaysia (Takaoka et al., 2012c). The adult female was described by Ya'cob et al. (2019). Its male remains unknown.

The female is characterized by the hind tibia yellow on the basal half and with a dark subbasal marking, and claw with a medium-sized basal tooth 0.4 times as long as the claw. The pupa is characterized by the gill consisting of an inflated structure tapered toward the apex and eight slender filaments (Fig. 37B).

Simulium (Gomphostilbia) dudgeoni Takaoka and Davies, 1995

Simulium (*Gomphostilbia*) *dudgeoni* Takaoka and Davies [in Takaoka et al., 1995]: 189–192 (female and male); An et al. (2007): 249–251 (pupa and larva).

Distribution. China (Hong Kong).

Bionomics. Unknown.

Remarks. This species was described from a female and five males, caught by a light trap in Hong Kong (Takaoka et al., 1995). The pupa and larva of this species were described by An et al. (2007). However, the pupal gill is composed of eight filaments arranged as 3+3+2 from dorsal to ventral, differing from those of all other members of the *S. gombakense* species-group. It is needed to confirm whether the pupa and larva were correctly associated with the adults.

This species is characterized in the female by the medium-long sensory vesicle (0.4 times as long as the third maxillary palpal segment); and in the male by the hind tibia darkened except the basal one-fifth to one-fourth yellow.

Simulium (Gomphostilbia) gombakense Takaoka and Davies, 1995
(Figs. 36M, P–R and 37E)

Simulium (*Morops*) *gombakense* Takaoka and Davies, 1995a: 82–84 (pharate pupa and larva).
Simulium (*Gomphostilbia*) *gombakense*: Takaoka, 2000a: 111–114 (male and pupa); Takaoka et al., 2018g: 165–166 (female).

Distribution. Malaysia (Peninsular), Myanmar, and Thailand.

Bionomics. Aquatic habitats: Small streams in low and medium elevations (388–1750 m) in Cameron's Highlands (Ya'cob et al., 2016).

Remarks. This species was described from a mature larva under the subgenus *Morops* from Peninsular Malaysia (Takaoka and Davies, 1995a). This species was transferred to the subgenus *Gomphostilbia* when its male and pupa were described by Takaoka (2000a). The female of this species was later described by Takaoka et al. (2010c). The female described as *S. gombakense* based on females reared from pupae collected from Mae Klang Waterfall, Chiang Mai Province, Thailand (Takaoka et al., 2010c) was corrected as that of *S. paiense* Takaoka, Srisuka and Saeung (complex) (Takaoka et al., 2018g).

It is characterized by the pupal gill consisting of an inflated structure directed forward, from which a dorsal branch is arising medially, and an apical branch is produced forwardly, with six slender filaments arranged as 1+2+3 arising from the basal portion, and three finger-like projections and one slender filament on each branch (Fig. 37E). The apical branch is constricted basally. The number of male upper-eye (large) facets is 13–15 vertical columns and 14–16 horizontal rows.

Simulium (Gomphostilbia) langkawiense Takaoka, Sofian-Azirun and Ya'cob, 2013
(Figs. 36S and 37F)

Simulium (*Gomphostilbia*) *langkawiense* Takaoka, Sofian-Azirun and Ya'cob, 2013b: 702–708 (female, male, pupa, and larva).

Distribution. Malaysia (Langkawi Island).

Bionomics. Aquatic habitats: A drainage and small streams (width 0.3–1.0 m, 24–25 °C, partially shaded or exposed to the sun, elevation of 33–176 m) in Langkawi Island (Takaoka et al., 2013b).

Remarks. This species was described from Langkawi Island, Peninsular Malaysia (Takaoka et al., 2013b).

It is similar to *S. gombakense* but is distinguished in the female by the genital fork with lateral plates moderately angulated laterally; and in the male by the number of upper-eye (large) facets in 11 or 12 vertical columns and 12 or 13 horizontal rows.

***Simulium* (*Gomphostilbia*) *maleewongae* Takaoka, Srisuka and Saeung, 2017**
(Figs. 36A, D, H and 37A)

Simulium (*Gomphostilbia*) *maleewongae* Takaoka, Srisuka and Saeung, 2017i: 91–97 (female, male, pupa, and larva).

Distribution. Thailand.

Bionomics. Aquatic habitat: A small stream (bottom of pebbles, flow slow, 21.9 °C, partially shaded, elevation 605 m) in Mae Hong Son Province (Takaoka et al., 2017i).

Remarks. This species was described from Thailand (Takaoka et al., 2017i).

It is characterized in the female by the medium-long sensory vesicle (Fig. 36A); in the male by the ventral plate with its ventral margin depressed medially when viewed caudally (similar to Fig. 36Y); in the pupa by the gill consisting of an inflated structure divided into two, a long branch directed forward and a short branch directed upward, and eight slender filaments, of which six filaments are arising from a common basal stalk and two filaments each arising apically and basally from the long branch (Fig. 37A), cone-like terminal hooks and cocoon with a short anterodorsal projection; and in the larva by the postgenal cleft 0.7–1.1 times as long as the postgenal bridge.

***Simulium* (*Gomphostilbia*) *nuwakotense* Takaoka and Shrestha, 2010**
(Figs. 36W, X, Y and 37H)

Simulium (*Gomphostilbia*) *nuwakotense* Takaoka and Shrestha, 2010: 23–27 (male and pupa).

Distribution. Nepal.

Bionomics. Aquatic habitat: A small stream (flow rapid, 23.0 °C, exposed to the sun, elevation 1811 m) flowing along the road, near Ranipauwa, Nuwakot (Takaoka and Shrestha, 2010).

Remarks. This species was described from a male reared from a pupa collected from Nepal (Takaoka and Shrestha, 2010). Its female and larva remain unknown.

This species is characterized by the male ventral plate with its ventral margin depressed medially when viewed posteriorly (Fig. 36Y) and pupal gill consisting of an inflated structure with a dorsomedial branch and an apical branch, each having two constrictions basally and a small nipple-like projection and two finger-like projections (Fig. 37H).

***Simulium* (*Gomphostilbia*) *paiense* Takaoka, Srisuka and Saeung, 2018 (complex)**
(Fig. 36B, C, G, I–L)

Simulium (*Gomphostilbia*) *paiense* Takaoka, Srisuka and Saeung [in Takaoka et al., 2018g]: 158–164 (female, male, pupa, and larva).

Distribution. Thailand.

Bionomics. Aquatic habitats: Small streams at an elevation of 503–1537 m in Mae Hong Son Province (Takaoka et al., 2018g).

Remarks. This species was described from Thailand (Takaoka et al., 2018g). It is a species complex consisting of three genetic forms (genoforms A, B, and C), and morphologically almost indistinguishable from *S. gombakense* although genetically well separated from the latter species (Takaoka et al., 2018g).

***Simulium* (*Gomphostilbia*) *prayongi* Takaoka and Choochote, 2005**
(Fig. 37G)

Simulium (*Gomphostilbia*) *prayongi* Takaoka and Choochote, 2005e: 123–127 (male and pupa).

Distribution. Thailand.

Bionomics. Aquatic habitat: A stream (width 1 m, depth ca. 5 cm, flow moderate, 24 °C, shaded, elevation 420 m) in Nang Khruan Waterfall, Tak Province (Takaoka and Choochote, 2005e).

Remarks. This species was described from Thailand (Takaoka and Choochote, 2005e). Its female and larva are unknown.

It is characterized by the pupal gill composed of an inflated V-shaped structure and eight slender filaments (Fig. 37G).

***Simulium* (*Gomphostilbia*) *sachini* Takaoka and Henry, 2010**
(Figs. 36E, O and 37I)

Simulium (*Gomphostilbia*) *sachini* Takaoka and Henry [in Takaoka et al., 2010e]: 108–110 (pupa); Takaoka and Shrestha, 2010: 27–29 (male and larva); Takaoka et al., 2022b: 79–82 (female and male).
Simulium (*Gomphostilbia*) *gombakense* (*nec* Takaoka and Davies, 1995a): Shrestha and Takaoka 2009: 253–258 (pupa, misidentification).

Distribution. India (West Bengal) and Nepal.

Bionomics. Aquatic habitats: A small shallow stream (20 °C, elevation 1576 m) in Tatopani, and a small stream (18.5 °C, elevation 1627 m), in Kodari, in Nepal: and a small stream at an elevation of 2354 m in Darjeeling, India (Takaoka et al., 2010e).

Remarks. This species was described from a pupa collected from Darjeeling, India (Takaoka et al., 2010e). Its male and larva were described based on samples from Nepal (Takaoka and Shrestha, 2010), and its female and male were described based on samples from India (Takaoka et al., 2022b).

The female of this species is characterized by the medium-long basal tooth on the claw (Fig. 36E). The male of this species is similar to that of *S. nuwakotense* in having a similar ventral plate, but the pupa is distinguished by the dorsomedial branch of the inflated structure of the gill with only one constriction basally (Fig. 37I).

***Simulium* (*Gomphostilbia*) *thuathienense* Takaoka and Sofian-Azirun, 2015**
(Fig. 37D)

Simulium (*Gomphostilbia*) *thuathienense* Takaoka and Sofian-Azirun [in Takaoka et al., 2015c]: 48–52 (female, pupa, and larva); Takaoka et al., 2018g: 164–165 (male).

Distribution. Thailand and Vietnam.

Bionomics. Aquatic habitats: A small stream (width 0.2–0.3 m, depth 2–3 cm, flow slow, 17.0 °C, shaded, elevation 1313 m) running from a forest near the top of Bach Ma National Park, Thua Thien Hue Province, central Vietnam, and two streams at an elevation of 1972 and 2157 m in Doi Phahompok National Park, Chiang Mai Province, Thailand (Takaoka et al., 2015c, 2018g).

Remarks. This species was described from Vietnam (Takaoka et al., 2015c). The male was described based on specimens from Thailand (Takaoka et al., 2018g).

It is characterized in the male by the ventral plate with its ventral margin deeply depressed medially when viewed posteriorly (similar to Fig. 36Y); and in the pupa by the terminal hooks cone-like (not plate-like) and gill composed of an inflated structure with two branches, each having a constriction basally (Fig. 37D); and in the larva by the postgenal cleft medium-long, nearly as long as the postgenal bridge.

***Simulium* (*Gomphostilbia*) *williei* Takaoka and Thapa, 2010**
(Figs. 36F, N, T–V and 37C)

Simulium (*Gomphostilbia*) *williei* Takaoka and Thapa [in Takaoka et al., 2010e]: 106–108 (pupa and larva); Takaoka et al., 2022b: 83–85 (female and male).

Distribution. India (West Bengal).

Bionomics. Aquatic habitat: A small stream at an elevation of 2354 m, Dali, Darjeeling (Takaoka et al., 2010e).

Remarks. This species was described from a pupa and a larva collected from Darjeeling, India (Takaoka et al., 2010e). Its adult female and male were described by Takaoka et al. (2022b).

This species is characterized in the male by the ventral plate produced posteroventrally (Fig. 36V); in the pupa by the gill with an inflated structure having eight slender filaments (Fig. 37C), and wide, plate-like terminal hooks; and in the larva by the elongate postgenal cleft, with its apex close to the posterior margin of the hypostoma.

Simulium (Gomphostilbia) palauense Species-group

Simulium (*Gomphostilbia*) *palauense* Stone, 1964

(Fig. 38A–R)

Simulium (*Eusimulium*) *palauense* Stone, 1964: 633 (female and male).
Simulium (*Inseliellum*) *palauense*: Crosskey, 1989: 224.
Simulium (*Gomphostilbia*) *palauense*: Takaoka and Davies, 1995a: 161; Takaoka and Craig, 1999: 195–201 (female, male, pupa, and larva).

Distribution. Palau.

Bionomics. Aquatic habitat: A river (width 8 m, depth 1–7 cm, flow rapid, 27.0 °C, exposed to the sun, elevation 10 m) (Takaoka and Craig, 1999).

Remarks. This species was described on the basis of the female and male specimens from Palau by Stone (1964). Its pupa and larva were described by Takaoka and Craig (1999).

This species is characterized in the female by the sensory vesicle short (0.25 times as long as the third palpal segment) (Fig. 38A), mandible lacking outer teeth (Fig. 38B), cibarium with a distinct dark X-shaped ridge medially (Fig. 38C), and genital fork with two projections from each arm, one directed forwardly and another directed posteromedially (Fig. 38E); and in the male by the ventral plate much produced ventrally (Fig. 38J).

It is exceptional among the subgenus *Gomphostilbia* in the pupa by the abdomen without spine-combs dorsally; and in the larva by the hypostoma with a serrated lateral margins (Fig. 38P), mandible with a supernumerary serration (Fig. 38O), postgenal cleft absent (Fig. 38Q), and accessory sclerite present (Fig. 38R).

Simulium (Gomphostilbia) varicorne Species-group

Simulium (Gomphostilbia) chumpornense Subgroup

Simulium (*Gomphostilbia*) *chumpornense* Takaoka and Kuvangkadilok, 2000

(Fig. 39F)

Simulium (*Gomphostilbia*) *chumpornense* Takaoka and Kuvangkadilok, [in Kuvangkadilok and Takaoka, 2000]: 167–172 (female, male, pupa, and larva).

Distribution. Thailand.

Bionomics. Aquatic habitat: A wide stream (width 5–6 m, 26 °C, pH 9.3, elevation 40 m), Ka Po Waterfall, Chumporn Province (Kuvangkadilok and Takaoka, 2000). Females of *S. chumpornense* were captured using human attractants at low and medium elevations in Doi Inthanon National Parks, Chiang Mai Province (Choochote et al., 2005). This species is a natural vector of protozoan parasites of the genera *Leucocytozoon* Berestneff and *Trypanosoma* Gruby in Thailand (Thaijarern et al., 2019a; Pramual et al., 2020).

Remarks. This species was described from Thailand by Kuvangkadilok and Takaoka (2000).

It is characterized in the female by the spermatheca with the base of its duct narrowly sclerotized, in the male by the number of upper-eye (large) facets in 15 vertical columns and 15 horizontal rows; in the pupa by the dorsal group of six filaments composed of four individual and two paired filaments; and in the larva by the postgenal cleft long but not reaching the posterior margin of the hypostoma.

***Simulium* (*Gomphostilbia*) *khelangense* Takaoka, Srisuka and Saeung, 2022**

Simulium (*Gomphostilbia*) *khelangense* Takaoka, Srisuka and Saeung [in Srisuka et al., 2022b]: 5–8 (female).

Distribution. Thailand.

Bionomics. Aquatic habitats are unknown.

Remarks. This species was described on the basis of females collected by a sweeping net in Lampang, Phitsanulok, and Chiang Mai Provinces (Srisuka et al., 2022b). Its male, pupa, and larva are unknown.

The female of this species is similar to that of *S. kuvangkadilokae* from Thailand in the same subgroup but is barely distinguished from the latter species by the head width relative to the greatest width of the frons and length of the labrum relative to the clypeus. A genetic analysis using the COI gene sequences similarly shows that *S. khelangense* is most closely related to *S. kuvangkadilokae*, with a genetic distance of 1.23–2.81%.

***Simulium* (*Gomphostilbia*) *kuvangkadilokae* Pramual and Tangkawanit, 2008**
(Figs. 39B, J, Q–S and 40D, L)

Simulium (*Gomphostilbia*) *kuvangkadilokae* Pramual and Tangkawanit, 2008: 297–303 (male, pupa, and larva); Takaoka and Srisuka, 2010a: 39–47 (female, male, pupa, and larva).

Distribution. Thailand.

Bionomics. Aquatic habitats: A small stream (width 20 cm, depth 5 cm, flow slow, elevation 145 m) at Tad Yai Waterfall, Amnat Charoen Province, and a small stream (width 20 cm, depth 30 cm, bottom rocky, flow moderate, exposed to the sun, elevation 123 m), Huai Yang, Ubon Ratchathani Province (Takaoka and Srisuka, 2010a).

Remarks. This species was described from the male, pupa and larva in Thailand (Pramual and Tangkawanit, 2008), and its male and pupa were redescribed when its female was first described by Takaoka and Srisuka (2010a).

It is characterized in the male by the number of upper-eye (large) facets in 11 or 12 vertical columns and 12 or 13 horizontal rows; in the pupa by the gill composed of an inflated U-shaped structure with 12 slender filaments (Fig. 40D); and in the larva by the postgenal cleft long, reaching the posterior margin of the hypostoma and thoracic segments 1–3 and abdominal segments 1–5 each with two to four pairs of protuberances (Fig. 40L).

Simulium (_Gomphostilbia_) _piroonae_ Takaoka and Srisuka, 2014
(Fig. 40B, I)

Simulium (*Gomphostilbia*) *piroonae* Takaoka and Srisuka [in Takaoka et al., 2014i]: 314–322 (female, male, pupa, and larva).

Distribution. Thailand.

Bionomics. Aquatic habitat: A stream (width 1.7 m, depth 10 cm, bottom sandy, flow slow, 31.1 °C, exposed to the sun, elevation 451 m) in Mae Hong Son Province (Takaoka et al., 2014i).

Remarks. This species was described from Thailand (Takaoka et al., 2014i).

It is characterized in the female by the mandible with teeth on both margins, and bare subcosta; in the male by the number of upper-eye (large) facets in 11 or 12 vertical columns and 12 horizontal rows; in the pupa by the gill composed of six slender filaments arranged as (2+2)+2 or [1+(1+2)]+2 from dorsal to ventral (Fig. 40B); and in the larva by the hypostoma with its lateral margins serrated apically (Fig. 40I), postgenal cleft reaching the posterior margin of the hypostoma, and body with two pairs of conical protuberances on the thoracic segments 1 and 2, and three pairs of conical protuberances on the thoracic segment 3 and abdominal segments 1–5.

Simulium (_Gomphostilbia_) _sumbaense_ Takaoka and Suana, 2018

Simulium (*Gomphostilbia*) *sumbaense* Takaoka and Suana [in Takaoka et al., 2018d]: 957–963 (female, male, and pupa).

Distribution. Indonesia (Sumba).

Bionomics. Aquatic habitat: A stream (width 0.5–0.6 m, bottom of pebbles and sands, 30 °C, exposed to the sun, elevation 3 m below the sea-level) in Waingapu, Sumba (Takaoka et al., 2018d).

Remarks. This species was described from Sumba, the Lesser Sunda Archipelago, Indonesia (Takaoka et al., 2018d). Its larva is unknown.

It is characterized in the female by the wide frons (frons:head ratio 1.0:3.9); in the male by the number of upper-eye (large) facets in 13 vertical columns and 13 horizontal rows; and in the pupa by the dorsal group of the gill composed of two individual and four paired filaments.

Simulium (_Gomphostilbia_) _tomae_ Takaoka, 2003

Simulium (*Gomphostilbia*) *tomae* Takaoka, 2003: 217–222 (female and pupa).

Distribution. Indonesia (Sulawesi).

Bionomics. Aquatic habitat: A stream (width 3 m, flow slow, 26 °C, shaded, elevation 170 m) near Amurang, North Sulawesi (Takaoka, 2003).

Remarks. This species was described from a single female reared from a pupa collected from Sulawesi, Indonesia (Takaoka, 2003). Its male and larva are unknown.

It is characterized in the female by the narrow frons (frons:head ratio 1.0:6.7); and in the pupa by the dorsal group of the gill composed of three pairs of six filaments.

***Simulium* (*Gomphostilbia*) *varicorne* Edwards, 1929**
(Figs. 39D and 40C, J, P)

Simulium (*Gomphostilbia*) *varicorne* Edwards, 1929: 159–160 (male); Takaoka and Davies, 1995a: 70–72 (male and pupa); Takaoka et al., 2018m: 33–40 (female, male, pupa, and lava).

Distribution. Indonesia (Sumatra and Java) and Malaysia (Peninsular).

Bionomics. Aquatic habitat: A stream (width 2–4 m, 26.5 °C, exposed to the sun, elevation 100 m) in Negri Sembilan, Peninsular Malaysia (Takaoka et al., 2018m).

Remarks. This species was described from a single male collected from Sumatra (Edwards, 1929). The pupa was described and the male was redescribed based on specimens from Peninsular Malaysia (Takaoka and Davies, 1995a). The redescription of all stages (except the egg) was made by Takaoka et al. (2018m).

This species is characterized in the female by the small sensory vesicle, 0.25–0.26 times as long as the third palpal segment (Fig. 39D), and spermatheca with sclerotized neck, and in the larva by the hypostoma with serrated lateral margins, and body with protuberances on the dorsal and dorsolateral surfaces of abdominal segments 1–5.

Nepal was one of the three distribution countries for *S. varicorne* in the world list of black flies (Adler and Crosskey, 2012). This was probably based on the thesis of Prof. Mala Maskey, in which *S. varicorne* was recorded based on larval specimens collected in Nepal. However, the larva of "*S. varicorne*" illustrated in her thesis, that was personally shown to me by Prof. Mala Maskey in 2011, reveals that by lacking protuberances on the dorsal surface of the abdomen, it differs not only from *S. varicorne* recently collected from Peninsular Malaysia and Sumatra, Indonesia (Takaoka et al., 2018m) but also from *S. parahiyangum*, once misidentified as *S. varicorne* by Edwards (1934). In fact, it is similar to the larva of *S. shivapuriense* (in the *S. darjeelingense* species-group) described from Nepal (Takaoka et al., 2020f). Nepal should be deleted as a distribution country of *S. varicorne* at least until its distribution is confirmed.

Simulium (Gomphostilbia) burtoni Subgroup

***Simulium* (*Gomphostilbia*) *breviflagellum* Takaoka and Sofian-Azirun, 2015**

Simulium (*Gomphostilbia*) *breviflagellum* Takaoka and Sofian-Azirun [in Takaoka et al., 2015c]: 53–57 (male and pupa).

Distribution. Vietnam.

Bionomics. Aquatic habitat: A stream (width 4–6 m, flow slow, 18 °C, exposed to the sun, elevation 1452 m) in Lamdong Province (Takaoka et al., 2015c).

Remarks. This species was described from Vietnam (Takaoka et al., 2015c). Its female and larva are unknown.

It is characterized in the male by the number of upper-eye large facets in 13 or 14 vertical columns and 14 horizontal rows and elongate sensory vesicle 0.44 times as long as the third palpal segment; and in the pupa by the wide plate-like terminal hooks with crenulated margins and cocoon with a short anterodorsal projection.

***Simulium* (*Gomphostilbia*) *burtoni* Takaoka and Davies, 1995**

Simulium (*Gomphostilbia*) *burtoni* Takaoka and Davies, 1995a: 72–77 (female, male, pupa, and larva).

Distribution. Malaysia (Peninsular) and Thailand.

Bionomics. Aquatic habitat: A drainage canal (width 0.6 m, 25 °C) from Champeroh River (width 3.6–5.5 m), Janda Baik Reserve (Takaoka and Davies, 1995a).

Remarks. This species was described from Peninsular Malaysia by Takaoka and Davies (1995a) and later was recorded from Thailand by Takaoka and Choochote (2004a).

It is characterized by the female abdomen shiny on segments 5–8 (wrongly noted as "shiny on segments 6–8" in the original description), male upper-eye (large) facets in 12 or 13 vertical columns and 13 horizontal rows, pupal gill with eight filaments, of which the dorsal and middle triplets each with an elongate secondary stalk, and broad, plate-like pupal terminal hooks with crenulated outer margins. In populations of Thailand, the number of male upper-eye (large) facets is variable: 10–14 vertical columns and 12–15 horizontal rows (Unpublished data).

***Simulium* (*Gomphostilbia*) *huangi* Takaoka, 2017**

(Fig. 40A, G, K)

Simulium (*Gomphostilbia*) *huangi* Takaoka [in Takaoka and Huang, 2017]: 439–446 (female, male, pupa, and larva).

Distribution. Taiwan.

Bionomics. Aquatic habitat: The Keelung River (width 2–3 m, 14 °C, elevation 300–310 m) in New Taipei City (Takaoka and Huang, 2017).

Remarks. This species is characterized in the female by the antenna with darkened flagellomeres (except the basal one-third of flagellomere 1 yellow) (Fig. 39A), abdominal segments 5–8 shiny dorsally; in the male by the number of upper-eye (large) facets in 12 or 13 vertical columns and 13 or 14 horizontal rows; in the pupa by the gill with eight filaments arranged as [(1+2)+(1+2)]+2 from dorsal to ventral (Fig. 40A), broad, plate-like pupal terminal hooks with crenulated outer margins (Fig. 40G), and cocoon wall-pocket-shaped, rounded, with its anterodorsal margin bulged medially; and in the larva by the postgenal cleft 1.3 times as long as the postgenal bridge (Fig. 40K).

***Simulium (Gomphostilbia) novemarticulatum* Subgroup**

***Simulium* (*Gomphostilbia*) *novemarticulatum* Takaoka and Davies, 1995**

(Fig. 40F, H, M–O)

Simulium (*Gomphostilbia*) *novemarticulatum* Takaoka and Davies, 1995a: 77–79 (female); Phasuk et al., 2005: 58 (female, male, pupa, and larva); Takaoka et al., 2010a: 59–67 (male, pupa, and larva).

Distribution. Malaysia (Peninsular) and Thailand.

Bionomics. Aquatic habitats: A river (width 5–8 m, depth 30–50 cm, flow moderate, 26.0 °C, shaded, elevation 220 m), in Gua Musang, Kelantan Province, Peninsular Malaysia, and a river (width ca. 20 m, depth 10–30 cm, flow slow, 26.5 °C, exposed to the sun, elevation 60 m) in Khao Sok National Park, Surat Thani Province, southern Thailand (Takaoka et al., 2010a).

Remarks. This species was described from a female in Peninsular Malaysia (Takaoka and Davies, 1995a). Its male, pupa, and larva were described based on Malaysian and Thai specimens by Takaoka et al. (2010a). Phasuk et al. (2005) first recorded this species from southern Thailand.

It is characterized in the male by the slender style; and in the larva by the bifid thick dark setae on abdominal segments 5–8 (Fig. 40O).

***Simulium* (*Gomphostilbia*) *charlesi* Takaoka, 2008**

(Figs. 39C, G, H, K, L, N–P and 40E, Q)

Simulium (*Gomphostilbia*) *charlesi* Takaoka, 2008c: 203–210 (female, male, pupa, and larva).

Distribution. Malaysia (Peninsular and Sarawak).

Bionomics. Aquatic habitat: A stream (width 8–10 m, flow moderate, 21.6 °C, exposed to the sun, elevation ca.1000 m), Bario, Sarawak (Takaoka, 2008c).

Remarks. This species was described from Sarawak, Malaysia (Takaoka, 2008c). It was recorded from Peninsular Malaysia (Takaoka, 2012).

It is characterized in the male by the style broad and twisted inward (Fig. 39N); and in the larva by the unbranched thick dark setae on abdominal segments 5–8 (Fig. 40Q).

***Simulium (Gomphostilbia) trirugosum* Subgroup**

***Simulium* (*Gomphostilbia*) *trirugosum* Davies and Györkös, 1988**

Simulium (*Gomphostilbia*) *trirugosum* Davies and Györkös, 1988b: 605–609 (female, male, pupa, and larva).

Distribution. Sri Lanka.

Bionomics. Aquatic habitats: Small streams (width 0.3–6.0 m) in eastern Sri Lanka (Davies and Györkös, 1988b). One male was caught by a black-light trap (Davies and Györkös, 1988b).

Remarks. This species was originally described as *S.* (*Morops*) *trirugosum* from Sri Lanka (Davies and Györkös, 1988b). It was transferred to the subgenus *Gomphostilbia* by Takaoka (2003) and was placed in the *S. varicorne* species-group by Takaoka (2012).

It is characterized by the female and male pleural membrane haired, pupal gill with three filaments, which are somewhat inflated on their basal half, and pupal abdomen lacking grapnel-like hooklets.

Simulium (Gomphostilbia) Species Unplaced to Species-group

Simulium (Gomphostilbia) azhari Takaoka, Sofian-Azirun and Ya'cob, 2014

(Fig. 41A, D–H)

Simulium (*Gomphostilbia*) *azhari* Takaoka, Sofian-Azirun and Ya'cob [in Takaoka et al., 2014a]: 10–17 (female, male, pupa, and larva).

Distribution. Malaysia (Peninsular).

Bionomics. Aquatic habitats: A stream (width 1 m, depth 5 cm, bottom muddy, flow slow, 23.0 °C, shaded, elevation 106 m) and a small stream (width ca. 60 cm, depth 2 cm, bottom sandy, flow slow, 24.0 °C, partially shaded, elevation 118 m) flowing at the edge of a natural forest in a swampy area, Kedah Province (Takaoka et al., 2014a).

Remarks. This species was described from Peninsular Malaysia, and placed in the *S. parahiyangum* subgroup of the *S. batoense* species-group (Takaoka et al., 2014a). However, this species is here transferred to *Gomphostilbia* species unplaced to species-group, because this species is distinguished from species of the *S. batoense* species-group by the female cibarium with mediolongitudinal ridge with a broad, bifid apex (Fig. 41A), male ventral plate much produced ventrally (0.61 as high as the width) (Fig. 41D, F, G), and male style broad when viewed ventrolaterally (Fig. 41E), these characteristics being seen in species of the *S. epistum* species-group. However, dark legs of this species depart from the definition of the *S. epistum* species-group.

It is characterized in the female by the narrow female frons (frons:head ratio 1.0: 7.1); in the male by the broad style (Fig. 41E) and ventral plate much produced ventrally; and in the pupa by the gill with eight slender filaments arranged as 3+3+2 from dorsal to ventral and 0.7–0.8 times as long as the pupal body (Fig. 41H), and terminal hooks simple, cone-shaped.

Simulium (Gomphostilbia) bucolicum Datta, 1975

Simulium (*Gomphostilbia*) *bucolicum* Datta, 1975a: 35–37 (female, pupa, and larva).

Distribution. India (Assam).

Bionomics. Aquatic habitat: A shallow stream in the Sonai Rupai forest (exposed to the sun, elevation ca. 70 m) (Datta, 1975a).

Remarks. This species was described from Assam, India (Datta, 1975a). This species remains in *Gomphostilbia* species unplaced to species group for lack of the male. It may be placed in the *S. epistum* species-group because its female has yellow tuft hairs at the base of the radial vein, although its fore coxa and hind tibia, which were noted to be brownish (Datta, 1975a), depart from the definition of the *S. epistum* species-group.

The pupal gill with eight slender filaments is arranged as (3+3)+2 from dorsal to ventral, of which the dorsal and middle triplets are arising from a relatively thick stalk, about twice as thick as the stalk of the ventral paired filaments, and the larval postgenal cleft is three times the length of the postgenal bridge, according to Fig. 21 and Fig. 44 in Datta (1975).

Information on many features including the frontal ratio, frons:head ratio, mandible, maxillary lacinia, cibarium, relative length of the fore and hind basitarsi against their greatest widths, relative lengths of the pupal gill filaments, terminal hooks (no illustration, though these were noted as short and blunt), number of secondary lobules of the larval rectal organ, and number of rows and hooklets of the larval posterior circlet is lacking, making its identification difficult.

***Simulium* (*Gomphostilbia*) *cauveryense* Anbalagan, 2015**

Simulium (*Gomphostilbia*) *cauveryense* Anbalagan [in Anbalagan et al., 2015a]: 554–561 (female, male, pupa, and larva).

Distribution. India (Tamil Nadu).

Bionomics. Aquatic habitat: The Cauvery River (width 15 m, depth 30 cm, 16 °C (yearly average), Tamil Nadu (Anbalagan et al., 2015a).

Remarks. This species was described from Karnataka, South India (Anbalagan et al., 2015a).

It is characterized in the female by the mandible with 14–16 inner teeth but lacking outer teeth; in the male by the number of upper-eye (large) facets in 13 vertical columns and 15 horizontal rows, and hind basitarsus somewhat enlarged, spindle-shaped; and in the pupa by the gill with 10 short filaments arranged as 2+3+3+2 from dorsal to ventral, of which six upper filaments are somewhat inflated on their basal half, being two or three times as thick as the four lower filaments [according to Fig. 3A in Anbalagan et al. (2015a)], and dorsum of abdominal segments 1 and 2 covered with tubercles.

The gaps between the descriptions and relevant figures in Anbalagan et al. (2015a) are found in the following features: the female hind tibia was noted to be light to dark brown except the basal half white to yellowish white but it is entirely darkened in Fig. 1C; the ventral plate in the ventral view was noted to be widened posteriorly but it is narrowed posteriorly in Fig. 2D and G; and the postgenal cleft was noted to be 2.71 times as long as the postgenal bridge but it is 10 times as long as the postgenal bridge according to Fig. 3E in Anbalagan et al. (2015a).

This species was placed in the *S. batoense* species-group by Anbalagan et al. (2015a). However, the male hind basitarsus, which is somewhat enlarged (4.2 times as long as its greatest width) and spindle-shaped, and the shoe-shaped cocoon depart from the definition of the *S. batoense* species-group (Takaoka, 2012). The somewhat enlarged male hind basitarsus is reported in the *S. gombakense* species-group, and the shoe-shaped cocoon very rarely occurs only in the *S. epistum* species-group in the subgenus *Gomphostilbia*. Strikingly, the pupal gill with 10 short filaments arranged as 2+3+3+2 from dorsal to ventral is similar to those of species of the *S. striatum* species-group in the subgenus *Simulium* s. str.

Simulium (*Gomphostilbia*) *dhangi* Takaoka and Sofian-Azirun, 2018 (Fig. 41I, M)

Simulium (*Gomphostilbia*) *dhangi* Takaoka and Sofian-Azirun [in Takaoka et al., 2018d]: 954–957 (pupa and larva).

Distribution. Indonesia (Timor and Sumba).

Bionomics. Aquatic habitats: A shallow tributary flowing from the main channel of a stream (width 1–3 m, depth 2–5 cm, bottom rocky with pebbles, 23.5 °C, shaded, elevation 679 m) in Timor, and a stream (width 2–5 m, depth 5–10 cm, bottom rocky with pebbles, 23.3 °C, partially shaded, elevation 505 m) in Sumba (Takaoka et al., 2018d).

Remarks. This species was described from Indonesia by Takaoka et al. (2018d). Its adult female and male are unknown. It is difficult to place this species in any species-group for lack of the adult stage. The configuration of the pupal gill (Fig. 41I) suggests that it may be placed in the *S. sherwoodi* species-group, though the presence of grapnel-like hooklets departs from the definition of the *S. sherwoodi* species-group defined by Takaoka (1995).

It is characterized in the pupa by the gill composed of two inflated tubular trunks arising from a short common basal stalk and two pairs of four short slender filaments arising basally from the dorsal surface of the dorsal inflated tubular trunk (Fig. 41I), and terminal hooks wide and plate-like, with their outer margins crenulated (Fig. 41M); and in the larva by the short second and long third antennal articles, 0.5–0.7 and 1.4–1.5 times as long as the first article, respectively, postgenal cleft medium-long, 1.4 times as long as the postgenal bridge, and rectal organ compound, with one or two short, thumb-like secondary lobules per lobe.

Simulium (*Gomphostilbia*) *doisaketense* Jitklang, Kuvangkadilok, Baimai, Takaoka and Adler, 2008

Simulium (*Gomphostilbia*) *doisaketense* Jitklang, Kuvangkadilok, Baimai, Takaoka and Adler, 2008: 20–21 (pupa and larva).

Distribution. Thailand.

Bionomics. Aquatic habitats: Streams (width 1.5–3.7 m, 15.0–22.3 °C, elevation 300–850 m) (Jitklang et al., 2008).

Remarks. This species was described from Thailand by Jitklang et al. (2008). Its adult female and male are unknown. It is difficult to place this species in any species-group for lack of the adult stage, though the medium-long larval postgenal cleft suggests that it may be placed in the *S. asakoae* species-group.

It is characterized in the pupa by the gill with eight filaments arranged as [(2+1)+3]+2 from dorsal to ventral, and arising from a short common basal stalk, terminal hooks wide, with their outer margins crenulated, and cocoon with an anterodorsal margin somewhat produced anteriorly; and in the larva by the distinct head spots, postgenal cleft 1.8–2.0 times as long as the postgenal bridge, and abdominal segments 1–4 each with a greenish transverse band (Jitklang et al., 2008).

It is chromosomally characterized by three fixed inversions (IIS-8, IIL-11, and IIIL-complex 6) and five floating inversions (IIS-7, IIL-7, IIIL-6, IIIL-7, and IIIL-8) (Jitklang et al., 2008).

Simulium (Gomphostilbia) fidum Datta, 1975
Simulium (*Gomphostilbia*) *fidum* Datta, 1975a: 38–39 (male, pupa, and larva).

Distribution. India (Assam).

Bionomics. Aquatic habitat: A shallow stream in the Sonai Rupai forest (exposed to the sun, elevation ca. 70 m) (Datta, 1975a).

Remarks. This species was described from Assam, India (Datta, 1975a). This species remains in *Gomphostilbia* species unplaced to species-group for lack of the female. It may be placed in the *S. epistum* species-group since the outer filament of the ventral pair is apparently thicker than the inner filament, and the hind tibia is yellowish with a dark subbasal marking (Datta, 1975a). However, the brownish fore coxa and the dark hair tuft at the base of the radius depart from the definition of the *S. epistum* species-group.

It is characterized in the male by the number of upper-eye (large) facets in 13 vertical columns and 14 horizontal rows, scutum with golden short hairs, fore coxa brownish, and hind tibia yellowish with a dark subbasal band and a dark apical cap; in the pupa by the gill with eight filaments arranged as [(1+2)+(1+2)]+2 from dorsal to ventral, of which the dorsal and middle triplets share an extremely short stalk, and the lower filament of the ventral pair is thickest, and in the larva by the postgenal cleft long, almost reaching the posterior margin of the hypostoma.

Simulium (Gomphostilbia) guangxiense Sun, 2009
Simulium (*Gomphostilbia*) *guangxiense* Sun, 2009: 547 (female, male, pupa, and larva).

Distribution. China (Guangxi Zhuang).

Bionomics. Unknown.

Remarks. This species was described from Guangxi Zhuang, China by Sun (2009). It is difficult to place this species in any species-group because of the lack of information for key features such as the color of the tuft hairs at the base of the radius, shape of the male hind basitarsus, and relative length and thickness of the gill filaments. In the original description (Sun, 2009), this species was noted to be similar to *S. miblosi* Takaoka (in the *S. epistum* species-group) and *S. tenuistylum* Datta (in the *S. asakoae* or *S. ceylonicum* species-group). However, what characteristics they share were not mentioned.

It is characterized in the female by the mandible lacking outer teeth, sensory vesicle short, 0.3 times as long as the third palpal segment, fore basitarsus 6.1 times as long as its greatest width, and hind tibia yellow on the basal two-thirds; in the male by the fore basitarsus 7.0 times as long as its greatest width, hind tibia and basitarsus yellow on the basal half, and ventral plate flat, not produced ventrally, and with lateral margins slightly emarginated medially when viewed ventrally; in the pupa by the gill with eight filaments arranged as 3+(1+2)+2 from dorsal to ventral, terminal hooks small, and cocoon wall-pocket-shaped; and in the larva by the postgenal cleft long, 4.0 times as long as the postgenal bridge, rectal organ compound, with six to nine finger-like secondary lobules per lobe, and posterior circlet with 74 rows of seven hooklets per row (Sun, 2009).

The frontal ratio of the female was shown as 2.3:1.0:1.5 in the original description (Sun, 2009). Probably it was wrong, since the height of the frons is always longer than the greatest width of the frons in the subgenus *Gomphostilbia*. It should read as 1.5:1.0:2.3.

***Simulium* (*Gomphostilbia*) *guniki* Takaoka, 2001**

(Fig. 41B, C, J–L, N)

Simulium (*Gomphostilbia*) *guniki* Takaoka, 2001a: 221–225 (female, male, pupa, and larva).

Distribution. Malaysia (Sabah).

Bionomics. Aquatic habitat: A clean small stream (width 0.5–1.0 m, bottom rocky, flow moderate, 13 °C, shaded, elevation 1970 m) in a natural forest, just above Carson Falls in Kinabalu National Park (Takaoka, 2001a).

Remarks. This species was described from Sabah, Malaysia (Takaoka, 2001a). This species was once assigned to the *S. darjeelingense* species-group by sharing the darkened fore coxae of the female and male, and the enlarged male hind basitarsus (Takaoka, 2012). However, it is distinguished from species of the *S. darjeelingense* species-group by the cibarium with the mediolongitudinal ridge narrowed apically without a bifid apex (Fig. 41B), female abdominal segment 7 with a triangular sclerotized sternal plate medially (Fig. 41C), ventral plate produced ventrally, and postgenal cleft short. This species is here transferred to *Gomphostilbia* species unplaced to species-group.

It is characterized in the female by the narrow frons (frons:head ratio 1.0:6.1), abdominal segment 7 with a triangular sclerotized sternal plate medially (Fig. 41C) (very rare in the subgenus *Gomphostilbia*); in the male by the number of upper-eye (large) facets in 14 vertical columns and 16 or 17 horizontal rows; in the pupa by the frons and thorax with relatively large tubercles having tiny secondary projections (Fig. 41K), and gill with eight filaments all arising close together from a short common basal stalk (Fig. 41J); and in the larva by the postgenal cleft short (0.4 times the length of the postgenal bridge) (Fig. 41N).

***Simulium* (*Gomphostilbia*) *heteropara* Sun, 2009**

Simulium (*Gomphostilbia*) *heteropara* Sun, 2009: 547–548 (female, male, pupa, and larva).

Distribution. China (Hunan).

Bionomics. Unknown.

Remarks. This species was described from Hunan, China by Sun (2009). It is difficult to place this species in any species-group for lack of information for key features such as the color of the tuft hairs at the base of the radius, shape of the male hind basitarsus, and relative length and thickness of the gill filaments. In the original description (Sun, 2009), this species was noted to be similar to *S. torautense* Takaoka and Roberts (in the *S. batoense* species-group) and *S. visayaense* Takaoka (in the *S. baisasae* species-group). However, what characteristics they share were not mentioned.

It is characterized in the female by the mandible lacking outer teeth, sensory vesicle short, 0.3 times as long as the third palpal segment, fore basitarsus 6.5 times as long as its greatest width, and hind tibia yellow on the basal two-thirds; in the male by the fore basitarsus 6.5 times as long as its greatest width, hind tibia and basitarsus yellow on the basal half, and ventral plate flat, somewhat produced ventrally, and narrowed posteriorly when viewed ventrally; in the pupa by the gill with eight filaments arranged as [3+(1+2)]+2 from dorsal to ventral, terminal hooks small, and cocoon wall-pocket-shaped; and in the larva by the postgenal cleft medium-long, 1.5 times as long as the postgenal bridge, rectal organ compound, with six to nine finger-like secondary lobules per lobe, and posterior circlet with 73 rows of 12 hooklets per row (Sun, 2009).

The frontal ratio of the female was shown as 2.3:1.0:1.6 in the original description (Sun, 2009). Probably it was wrong, since the height of the frons is always longer than the greatest width of the frons in the subgenus *Gomphostilbia*. It should read as 1.6:1.0:2.3.

***Simulium* (*Gomphostilbia*) *hiroyukii* Ya'cob and Sofian-Azirun, 2015**

Simulium (*Gomphostilbia*) *hiroyukii* Ya'cob and Sofian-Azirun [in Ya'cob et al., 2015a]: 425–432 (female, male, pupa, and larva).

Distribution. Malaysia (Sarawak).

Bionomics. Aquatic habitat: A clean small stream (width 0.5–0.6 m, depth 10 cm, partially shaded, 14.4 °C, elevation ca. 2115 m) at Church Camp, Mt. Murud, Sarawak (Ya'cob et al., 2015a).

Remarks. This species was described from Sarawak, Malaysia (Ya'cob et al., 2015a). This species was once assigned to the *S. darjeelingense* species-group by sharing the darkened fore coxae of the female and male and the enlarged hind basitarsus, but it is distinguished from species of the *S. darjeelingense* species-group by the cibarium with the mediolongitudinal ridge narrowed apically without a bifid apex, ventral plate produced ventrally, and pupal gill filaments somewhat longer than the pupal body. This species is here transferred to *Gomphostilbia* species unplaced to species-group.

It is characterized in the female by the sensory vesicle 0.36–0.46 times as long as the third palpal segment; in the male by the number of upper-eye (large) facets in 13 or 14 vertical columns and 16 or 17 horizontal rows; and in the pupa by the gill with eight filaments arranged as [(2+1)+(2+1)]+2 from dorsal to ventral, and with the secondary stalk of the middle triplet medium to extremely long (1.4–2.2 mm), and terminal hooks wide and plate-like, with their outer margins crenulated.

***Simulium* (*Gomphostilbia*) *huguangense* Sun, 2009**

Simulium (*Gomphostilbia*) *huguangense* Sun, 2009: 546 (female, male, pupa, and larva).

Distribution. China (Hunan).

Bionomics. Aquatic habitats: Small streams (flow 0.6 m/s,14–18 °C, pH 5.0–5.5) (Sun, 2009).

Remarks. This species was described from Hunan, China by Sun (2009). It is difficult to place this species in any species-group for lack of information for key features such as color of the tuft hairs at the base of the radius, shape of the male hind basitarsus, and relative length and thickness of the gill filaments. In the original description (Sun, 2009), this species was noted to be similar to *S. jianfengense* (in the *S. asakoae* species-group), *S. mengi* (in the *S. asakoae* species-group), and *S. siamense* (in the *S. batoense* species-group). However, what characteristics they share were not mentioned.

It is characterized in the female by the mandible lacking outer teeth, sensory vesicle short, 0.27 times as long as the third palpal segment, fore basitarsus 6.7 times as long as its greatest width, and hind tibia yellow on the basal four-fifths; in the male by the fore basitarsus 8.4 times as long as its greatest width, hind basitarsus yellow on the basal three-fourths, and ventral plate flat, moderately produced ventrally, and nearly parallel-sided when viewed ventrally; in the pupa by the gill with eight filaments arranged as [(2+1)+3]+2 from dorsal to ventral, terminal hooks small, and cocoon wall-pocket-shaped; and in the larva by the postgenal cleft long, 4.0 times as long as the postgenal bridge, rectal organ compound, with six finger-like secondary lobules per lobe, and posterior circlet with 50 rows of 13 hooklets per row (Sun, 2009).

The frontal ratio of the female was shown as 2.7:1.0:1.7 in the original description (Sun, 2009). Probably it was wrong, since the height of the frons is always longer than the greatest width of the frons in the subgenus *Gomphostilbia*. It should read as 1.7:1.0:2.7.

***Simulium* (*Gomphostilbia*) *jinbianense* Zhang and Chen, 2004**

Simulium (*Gomphostilbia*) *jinbianense* Zhang and Chen, 2004: 784–785 (female, pupa, and larva).

Distribution. China (Hunan).

Bionomics. Aquatic habitats: Small streams (shaded, elevation 480 m) (Zhang and Chen, 2004).

Remarks. This species was described from Hunan, China by Zhang and Chen (2004). Its male is unknown.

It is difficult to place this species in any species-group because its male is unknown.

The female hind tibia was noted to be yellowish on the basal three-fourths, medium to dark brown on the apical one-fourth and with the brownish subbasal ring, fitting to the diagnosis of the *S. epistum* species-group, but the tuft of dark hairs at the base of the radius departs from the diagnosis of the same species-group. According to Fig. 7 in Zhang and Chen (2004), the posterior surface of the hind tibia is entirely darkened except the base, suggesting that this species is placed in the *S. batoense* species-group.

It is characterized in the female by the sensory vesicle medium-long, 0.32 times as long as the third palpal segment, scutum with three dark longitudinal vittae, covered with yellow short hairs, and genital fork with a projection directed anteriorly on each arm; in the pupa by the gill with eight long filaments arranged as [(1+2)+(1+2)]+2 from dorsal to ventral, and cocoon wall-pocket shaped, with a short anterodorsal projection; and in the larva by the postgenal cleft medium-long, 1.0–1.3 times as long as the postgenal bridge.

The pupal terminal hooks were described to be definite but their shape was not illustrated.

***Simulium* (*Gomphostilbia*) *litoreum* Datta, 1975**
Simulium (*Gomphostilbia*) *litoreum* Datta, 1975a: 37 (pupa and larva).

Distribution. India (Assam).

Bionomics. Aquatic habitat: A shallow stream in the Sonai Rupai forest (exposed to the sun, elevation ca. 70 m) (Datta, 1975a).

Remarks. This species was described from Assam, India (Datta, 1975a). This species remains unplaced in any species-group for lack of the female and male. It may be placed in the *S. binuanense* subgroup of the *S. batoense* species-group since the upper filament of the ventral pair is apparently thicker and longer than its counter filament.

***Simulium* (*Gomphostilbia*) *longshengense* Chen, Zhang and Zhang, 2007**
Simulium (*Gomphostilbia*) *longshengense* Chen, Zhang and Zhang, 2007c: 784–785 (male, pupa, and larva).

Distribution. China (Guangxi Zhuang).

Bionomics. Aquatic habitats: A small stream (shaded, elevation 800 m) in Sanmen, and a stream at an elevation of 1462 m, in Shiwandashan (Chen et al., 2007c).

Remarks. This species was described from Guangxi Zhuang, China by Chen et al. (2007c). Its female is unknown.

It is difficult to place this species in any species-group. The tuft of dark hairs at the base of the radius, flat ventral plate and slender basitarsus in the male, and gill with eight slender filaments, of which two filaments of the ventral pair are 2.5–3.0 times as long as the six other filaments of the dorsal and middle triplets, agree with the diagnosis of the *S. batoense* species-group. However, the yellowish legs in the male depart from the diagnosis of the *S. batoense* species-group.

It is characterized in the male by the number of upper-eye (large) facets in 11 vertical columns and 13 horizontal rows, scutum with yellow short hairs, yellowish legs, dark tuft hairs at the base of the radius, and flat ventral plate; in the pupa by the gill with eight long filaments arranged as (2+1)+[(1+2)+2] from dorsal to ventral, and cocoon wall-pocket-shaped; and in the larva by the postgenal cleft medium-long, ca. 1.5 times as long as the postgenal bridge. The shape of the pupal terminal hook was not described.

***Simulium* (*Gomphostilbia*) *mamasaense* Takaoka, 2003**
Simulium (*Gomphostilbia*) *mamasaense* Takaoka, 2003: 227–229 (pupa).

Distribution. Indonesia (Sulawesi).

Bionomics. Aquatic habitat: A small stream (width 0.2–0.3 m, flow slow, 21 °C, shaded, elevation 1050 m) in a coffee plantation, near Mamasa, South Sulawesi (Takaoka, 2003).

Remarks. This species was described from Sulawesi, Indonesia by Takaoka (2003). Its female, male, and larva are unknown. It is difficult to place it in any species-group for lack of the adult stage.

It is characterized in the pupa by the gill with eight slender filaments arranged as [(1+2)+(1+2)]+2 from dorsal to ventral, of which the middle triplet has the long secondary stalk 2.1 times as long as the interspiracular trunk, and terminal hooks moderately widened, with their outer margins about three times as long as their inner margins, and abdominal segment 9 lacking grapnel-like hooklets.

***Simulium (Gomphostilbia) metatarsale* Brunetti, 1911**
Simulium metatarsale Brunetti, 1911: 284–285 (male).
Simulium (*Gomphostilbia*) *metatarsale*: Crosskey, 1967; Datta, 1973: 382; Datta, 1997: 143.

Distribution. India (West Bengal) and China (?).

Bionomics. Aquatic habitat: A trickle at an elevation of ca. 500 m, Jalpaiguri District, West Bengal (Datta, 1997).

Remarks. This species was described from a single male collected from Kurseong, Darjeeling, norther India by Brunetti (1911). Its female, pupa, and larva are unknown.

It was reported from Java, Indonesia by Edwards (1934), Taiwan by Takaoka (1979), and China by Chen and An (2003). However, the species previously thought to be *S. metatarsale* in Indonesia and Taiwan were reidentified as *S. gyorkosae* and *S. asakoae*, respectively (Takaoka and Davies, 1996; Takaoka et al., 2021a).

The species regarded as *S. metatarsale* in various provinces in China (Chen and An, 2003) should be reevaluated for its species status.

According to the original description by Brunetti (1911), it is characterized in the male by the hind tibia and basitarsus yellowish on the basal half, and hind basitarsus enlarged and as wide as the hind tibia.

It is strange that Datta (1973) compared the female, male, pupa, and larva of this species with those of *S. tenuistylum* when the latter species was described as a new species, although he mentioned in the same paper that no material of *S. metatarsale* was collected from the type locality (Kurseong). The place where his material of *S. metatarsale* came from was not stated.

Datta (1997) identified one male and one female reared from pupae collected together with one larva from Rajabhatkhawa, Jalpaiguri District, in 1986, as *S. metatarsale*, and differentiated it from *S. tenuistylum* in the keys (characteristics of *S. tenuistylum* in parentheses): "in the female by hind femur entirely with dark pollinosity on outer surface (hind femur gradually with dark pollinosity on outer surface); in the male by basimere (=coxite) less than twice as long as wide, but longer than distimere (=style) (basimere twice as long as wide, nearly as long as distimere); in the pupa by the upper triplet of gill branching almost at a point (upper triplet branching close to base into a single dorsal filament and two ventral filaments on a long stalk); in the larva by abdominal cuticle with dark bands on segments 1–4 dorsally (abdominal cuticle with a dark dorsal band with two pairs of dark lateral spots posterior to it)". In *S. tenuistylum*, the coxite is ca. 1.5 times as long as the style, and the upper triplets are branching into an individual filament and two paired filaments almost at the same point according to Plate XIII-12 and Plate XIII-16 in

Datta (1973). The other differences appear to be undecisive to separate the two species.

Since the leg color and the size of the male hind tibia and basitarsus are in good agreement between these two species, and *S. tenuistylum* and *S. metatarsale* are the only species in the *S. asakoae* or *S. ceylonicum* species-group in West Bengal, there is a strong possibility that *S. tenuistylum* is synonymous with *S. metatarsale*.

Simulium* (*Gomphostilbia*) *penis **Sun, 2009**
Simulium (*Gomphostilbia*) *penis* Sun, 2009: 546 (pupa and larva).

Distribution. China (Guangxi Zhuang).
Bionomics. Unknown.
Remarks. This species was described from Guangxi Zhuang, China by Sun (2009). Its adult female and male are unknown. It is difficult to place this species in any species-group for lack of the adult stage and information on the relative length and thickness of the pupal gill filaments. In the original description (Sun, 2009), this species was noted to be similar to *S. jianfengense* and *S. laoshanstum* Ren, An and Kan (both in the *S. asakoae* species-group). However, it was not mentioned on what characteristics they share.

It is characterized in the pupa by the gill with eight filaments arranged as [3+(1+2)]+2 from dorsal to ventral, of which the dorsal triplet has an extremely short stalk, terminal hooks small, and cocoon wall-pocket-shaped; and in the larva by the postgenal cleft twice as long as the postgenal bridge, rectal organ compound, with 8–11 finger-like secondary lobules per lobe, and posterior circlet with ca. 83 rows of ca. 10 hooklets per row. The pupal gill filaments of this species are arranged as [(2+1)+3]+2 from dorsal to ventral, according to Fig. 12–76 by Chen (2016). Probably, the original figure was wrongly redrawn.

Simulium* (*Gomphostilbia*) *pingxiangense **An, Hao and Mai, 1990**
Simulium (*Gomphostilbia*) *pingxiangense* An, Hao and Mai, 1990: 100–102 (female and male).

Distribution. China (Guangxi Zhuang).
Bionomics. Unknown.
Remarks. This species was described from China by An et al. (1990). Its pupa and larva are unknown. It may be placed in either the *S. asakoae* species-group or the *S. ceylonicum* species-group by having the male hind basitarsus enlarged, much wider than the hind tibia and femur. Information on the color of the tuft hairs on the base of the radius is not available, and it is also difficult to judge whether the lateral margins of the ventral plate are depressed medially or parallel-sided or narrowed posteriorly when viewed ventrally, though illustrated in An et al. (1990).

It is characterized in the female by the sensory vesicle 0.4 times as long as the third palpal segment, mandible without outer teeth, hind tibia yellow on the basal two-thirds; and in the male by the number of upper-eye (large) facets in 13 horizontal rows, hind tibia and basitarsus yellow on the basal one-third and half, respectively, and hind basitarsus enlarged, 1.25 times as wide as the hind tibia and femur [according to Fig. 1d in An et al. (1990)].

Simulium (_Gomphostilbia_) _serratum_ Takaoka, 2003

Simulium (*Gomphostilbia*) *serratum* Takaoka, 2003: 224–227 (pupa and larva).

Distribution. Indonesia (Sulawesi).

Bionomics. Aquatic habitats: A small stream (width 0.2–0.3 m, 21 °C, shaded, elevation 1050 m) in a coffee plantation, Mamasa, and a stream (width 0.2–0.3 m, 21 °C, shaded, elevation 870 m) flowing along the road from Palopo to Makale just reaching the mountain pass (Takaoka, 2003).

Remarks. This species was described from Sulawesi, Indonesia by Takaoka (2003). Its female and male are not known. It is difficult to place it in any species-group for lack of the adult stage.

It is characterized in the pupa by the gill with eight slender filaments arranged as (1+2)+[(1+1)+(1+2)] from dorsal to ventral, arising from a long common basal stalk 1.1 times as long as the interspiracular trunk, and terminal hooks much widened, with their outer margins 6.1 times as long as their inner margins; and in the larva by the postgenal cleft bullet-shaped, ca. 2.7 times as long as the postgenal bridge.

Simulium (_Gomphostilbia_) _tenuistylum_ Datta, 1973

Simulium (*Gomphostilbia*) *tenuistylum* Datta, 1973: 377–380 (female, male, pupa, and larva).

Distribution. India (West Bengal and Sikkim).

Bionomics. Aquatic habitats: Small streams at high elevations (1150–2025 m), in Darjeeling (Datta, 1973). Females of this species feed humans, horse and cow (Datta, 1983).

Remarks. This species was described from Darjeeling, North India (Datta, 1973). This species is here transferred from the *S. ceylonicum* species-group to *Gomphostilbia* species unplaced to species-group because the shape of the ventral plate when viewed ventrally is unknown, though it is certain that this species belongs to either the *S. asakoae* species-group or the *S. ceylonicum* species-group.

It is characterized in the female by the sensory vesicle medium-long, 0.37 times as long as the third palpal segment, and hind tibia yellowish on more than the basal two-thirds; in the male by the number of upper-eye (large) facets in 12 vertical columns and 14 horizontal rows, hind tibia and basitarsus yellowish on the basal half, hind basitarsus enlarged, wider than the hind tibia, and ventral plate only slightly produced ventrally when viewed posteriorly; in the pupa by the gill with eight long filaments arranged as (2+1)+(1+2)+2 from dorsal to ventral; and in the larva by the postgenal cleft 3.0 times as long as the postgenal bridge.

As noted under *S. metatarsale*, this species is suspected to be synonymous with *S. metatarsale*.

Simulium (_Gomphostilbia_) _tumum_ Chen and Zhang, 2001

Simulium (*Gomphostilbia*) *tumum* Chen and Zhang, 2001: 366–368 (male, pupa, and larva).

Distribution. China (Hainan).

Bionomics. Aquatic habitat: A trickle (elevation 630 m) in Hainan Island, China (Chen and Zhang, 2001).

Remarks. *Simulium tumum* was described from Hainan, China by Chen and Zhang (2001). Its female is unknown. It was provisionally placed in the *epistum* species-group (Takaoka, 2012). This placement was based on two adult characteristics, i.e., the yellow hair tuft and the male hind tibia yellow on the basal two-thirds. However, its other characteristics, such as the flat ventral plate, and cocoon with an anterodorsal projection, depart from the *S. epistum* species-group but appear to agree with the *batoense* species-group.

It is characterized in the male by the number of upper-eye (large) facets in 12 vertical columns and 10 horizontal rows; in the pupa by the frons and thorax sparsely covered with tubercles, and gill with eight slender filaments arranged as [(1+2)+(1+2)]+2 from dorsal to ventral; and in the larva by the postgenal cleft about eight times as long as the postgenal bridge and posterior abdominal segments sparsely covered with dark unbranched spinous setae dorsally.

***Simulium* (*Gomphostilbia*) *xiangxiense* Sun, 2009**

Simulium (*Gomphostilbia*) *xiangxiense* Sun, 2009: 548–549 (female, male, pupa, and larva).

Distribution. China (Hunan).

Bionomics. Unknown.

Remarks. This species was described from Hunan, China by Sun (2009). It is difficult to place this species in any species-group because of the lack of descriptions of key features such as color of legs and tuft hairs at the base of the radius, shape of the male hind basitarsus, and relative length and thickness of the gill filaments. In the original description (Sun, 2009), this species was noted to be similar to *S. tenuistylum*. However, it was not mentioned on what characteristics both species share.

It is characterized in the female by the mandible with 23 inner and eight outer teeth, sensory vesicle 0.27 times as long as the third palpal segment and fore basitarsus 5.4 times as long as its greatest width; in the male by the fore basitarsus 9.0 times as long as its greatest width and ventral plate flat, with its lateral margins narrowed medially, then nearly parallel-sided posteriorly when viewed ventrally; in the pupa by the gill with eight filaments [though six filaments in Fig. 57 in Sun (2009)], terminal hooks small and cocoon wall-pocket-shaped; and in the larva by the postgenal cleft twice as long as the postgenal bridge, rectal organ compound, with nine or 10 finger-like secondary lobules per lobe and posterior circlet with ca. 87 rows of about eight hooklets per row.

***Simulium* (*Gomphostilbia*) *yuanbaoshanense* Chen, Zhang and Zhang, 2007**

Simulium (*Gomphostilbia*) *yuanbaoshanense* Chen, Zhang and Zhang, 2007c: 782–784 (female, male, pupa, and larva).

Distribution. China (Guangxi).

Bionomics. Aquatic habitats: Small streams at an elevation of 2086 m, in Yanbao Mountain (Chen et al., 2007c).

Remarks. This species was described from Guangxi, China by Chen et al. (2007c).

As in *S. longshengense*, it is difficult to place this species in any species-group. The tuft of dark hairs at the base of the radius in the female and male, flat ventral plate and slender basitarsus in the male, and gill with eight slender filaments, of which two filaments of the ventral pair are twice as long as the six other filaments of the dorsal and middle triplets in the pupa fit the diagnosis of the *S. batoense* species-group. However, the yellowish legs in the female and male depart from the diagnosis of the *S. batoense* species-group.

It is characterized in the female and male by scutum with yellow short hairs, yellowish legs, and dark tuft hairs at the base of the radius; in the female by the short sensory vesicle 0.25 times as long as the third palpal segment; in the male by the number of upper-eye (large) facets in 10 vertical columns and 10 horizontal rows and flat ventral plate; in the pupa by the gill with eight long filaments arranged as [(2+1)+(1+2)]+2 from dorsal to ventral and cocoon wall-pocket-shaped, with an anterodorsal projection; and in the larva by the postgenal cleft 1.0–1.5 times as long as the postgenal bridge. The shape of pupal terminal hooks was not described.

Simulium (*Gomphostilbia*) sp. nr. *litoreum* Datta, 1975

Simulium (*Gomphostilbia*) sp. nr. *litoreum* Datta, 1975a: 37 (larva).

Distribution. India (Assam).

Bionomics. Aquatic habitat: A shallow stream in the Sonai Rupai forest (exposed to the sun, elevation ca. 70 m) (Datta, 1975a).

Remarks. This unnamed species was described from a single larva collected from Assam, India (Datta, 1975a). Its female, male, and pupa are unknown. This species remains unplaced in any species-group for lack of the female and male.

Datta (1975a) noted that it may be the larva of *S.* (*G.*) *litoreum* described based on specimens collected from the same locality.

It is characterized in the larva by the cephalic apotome with positive head spots, mandible with three bristles on each side, labral fan with ca. 30 primary rays, postgenal cleft long, with its apex close to the posterior margin of the hypostoma and pharate pupal gill with eight thread-like filaments arranged as 3+3+2 from dorsal to ventral.

Simulium (*Gomphostilbia*) Lakhimpur sp. Lewis, 1974

Simulium (*Gomphostilbia*) Lakhimpur sp. Lewis, 1974: 38 (pupa).

Distribution. India (Assam).

Bionomics. Unknown.

Remarks. This unnamed species was reported from Assam, India by Lewis (1974). Its female, male, and larva are unknown.

Morphological characteristics of its pupa are not available except that the gill is composed of eight filaments arranged 6+2 from dorsal to ventral. The arrangement of the gill filaments of this species is similar to those of species in the

S. darjeelingense species-group of the subgenus *Gomphostilbia*. However, it is difficult to place this species in any species-group due to the lack of detailed information for its morphology.

Simulium (*Gomphostilbia*) Ranga sp. Lewis, 1974

Simulium (*Gomphostilbia*) Ranga sp. Lewis, 1974: 38 (pupa).

Distribution. India (Assam).

Bionomics. Unknown.

Remarks. This unnamed species was reported from Assam, India by Lewis (1974). Its female, male, and larva are unknown.

Morphological characteristics of its pupa are not available except that the gill is composed of eight filaments arranged 3+3+2 from dorsal to ventral. It is provisionally placed in the subgenus *Gomphostilbia* judging from the number and arrangement of the gill filaments.

Simulium (*Gomphostilbia*) sp. Zhang and Chen, 2003

Simulium (*Gomphostilbia*) *longitruncum* (*nec* Takaoka and Davies) Zhang and Chen, 2003: 354–355 (male, pupa, and larva).

Distribution. China (Hainan).

Bionomics. Aquatic habitat: A stream at an elevation of 600 m (Zhang and Chen 2003).

Remarks. This species was described from Hainan, China by Zhang and Chen (2003). Its female and larva are unknown. This is a junior synonym of *S.* (*G.*) *longitruncum* Takaoka and Davies, 1995 but no substitute name is available (Adler, 2022).

It is difficult to place this species in any species-group for lack of the adult male, though this species is placed in the *S. ceylonicum* species-group (Adler, 2022).

It is characterized in the female by the sensory vesicle long, 0.6 times as long as the third palpal segment; in the pupa by the gill with eight long filaments arranged as 2+1+2+1+2 from dorsal to ventral, of which both the middle and ventral pairs have an extremely long stalk, and cocoon wall-pocket-shaped.

The mandible was noted to have 14 inner and 17 outer teeth in the original description (Zhang and Chen, 2003). Probably, it may be wrong because the number of inner teeth is always greater than that of outer teeth.

Simulium (*Gomphostilbia*) sp. An, Nie and Chen, 2006

Simulium sumatraense An, Nie and Chen [in Nie et al., 2006]: 158–160 (female).

Distribution. Indonesia (Sumatra) (?).

Bionomics. Unknown.

Remarks. *Simulium* (*G.*) *sumatraense* was described based on a female caught while trapped inside a lampshade of the ship that arrived at Qinhuangdao City, China, from Palembang and Dumai, Sumatra, Indonesia by Nie et al. (2006). It was a junior primary homonym of *S.* (*S.*) *sumatraense* Takaoka and Sigit, 1997, but its substitute name was not available (Adler, 2022).

The female of this species is characterized by the scutum without longitudinal vittae, katepisternum haired, basal portion of the radius haired, claw simple, genital fork without a projection directed anteriorly from each lateral arm, spermatheca with a pale reticulate pattern on its surface and without internal hairs. This species agrees with the diagnosis of the *S. baisasae* species-group in having the simple claw and spermatheca with a pale reticulate surface pattern, but disagrees in lacking longitudinal vittae on the scutum, a projection from each lateral arm of the genital fork and internal hairs of the spermatheca.

There is no information on the departure date and sea route of the ship from Sumatra in the original description (Nie et al., 2006). In addition, no species of the *S. baisasae* species-group has been recorded from Sumatra (Takaoka et al., 2000). There is a possibility that this female black fly was attracted to the ship on its way to China after departing from the initial port of Sumatra. Thus, its distribution record in Sumatra is dubious.

Simulium (*Gomphostilbia*) ? *sundaicum* Datta, 1991

Simulium (*Gomphostilbia*) ? *sundaicum* Edwards: Datta, 1991b: 12 (pharate male, pupa, and larva); Datta, 1992a: 48–49.

Distribution. India (West Bengal).

Bionomics. Aquatic habitats: Trickles and irrigation channels at an elevation of 50–750 m (Datta 1991b).

Remarks. This species was reported from West Bengal, India by Datta (1991b). Its female is unknown.

According to Datta (1991b, 1992a), it is characterized in the male by the hind basitarsus slender and ventral plate wide; in the pupa by the gill with eight filaments arranged as 3+3+2 from dorsal to ventral, of which two filaments of the ventral pair are twice as long as the six other filaments and cocoon simple, without an anterodorsal projection; and in the larva by the postgenal cleft very long.

The relative length of eight gill filaments in the pupa of this species suggests that it may be placed in the *S. duolongum* subgroup in the *S. batoense* species-group or the *S. atratum* subgroup in the *S. epistum* species-group. However, information of the color of tuft hairs at the base of the radius and of legs, and the genitalia in the male is needed to decide which species-group this species belongs to.

Simulium (_Gomphostilbia_) sp. 'A' Zhang, Wen and Chen, 1999
Simulium (_Gomphostilbia_) sp. 'A' Zhang, Wen and Chen, 1999: 234

Distribution. China (Guizhou).
Bionomics. Unknown.
Remarks. This unnamed species was reported from Guizhou, China by Zhang et al. (1999). However, its morphological characteristics are not noted. This species was not treated in the monographs on Chinese black flies by Chen and An (2003) and Chen (2016).

Simulium (_Gomphostilbia_) nr. _asakoae_ 2 Jitklang, Kuvangkadilok, Baimai, Takaoka and Adler, 2008
Simulium (_Gomphostilbia_) nr. _asakoae_ 2 Jitklang, Kuvangkadilok, Baimai, Takaoka and Adler, 2008: 14 (pupa and larva).
Simulium (_Gomphostilbia_) _asakoae_ (_nec_ Takaoka and Davies): Phasuk et al. (2005): 57 (pupa, and larva).

Distribution. Thailand.
Bionomics. Aquatic habitat: Streams (width 3.7–5.0 m, elevation 300–700 m) in Kanchanaburi and Nakhon Patchasima Provinces (Jitklang et al., 2008).
Remarks. This unnamed species was reported from Thailand with its chromosomal characteristics by Jitklang et al. (2008). Its female and male are unknown.

It is characterized in the larva by the postgenal cleft medium-long, somewhat longer than the postgenal bridge, with its apex rounded, abdomen yellow except segments 1–3 greenish and segment 5 without a distinct pigmented band and abdominal setae unbranched.

Simulium (_Gomphostilbia_) nr. _asakoae_ 3 Jitklang, Kuvangkadilok, Baimai, Takaoka and Adler, 2008
Simulium (_Gomphostilbia_) nr. _asakoae_ 3 Jitklang, Kuvangkadilok, Baimai, Takaoka and Adler, 2008: 15 (pupa and larva).

Distribution. Thailand.
Bionomics. Aquatic habitat: A stream (width 1.5 m, elevation 600 m) in Phayao Province (Jitklang et al., 2008).
Remarks. This unnamed species was reported from Thailand with its chromosomal characteristics by Jitklang et al. (2008). Its female and male are unknown.

It is characterized in the pupa by the terminal hooks wide, with their outer margins crenulated, and an acute angle (60 degrees) formed by the stalks of the ventral pair and dorsal triplet and in the larva by the postgenal cleft medium-long, twice as long as the postgenal bridge, and with its apex pointed, abdominal segments 1 and 4 each with a greenish band, and segments 5–8 each with a reddish-brown transverse band.

Simulium (*Gomphostilbia*) nr. *asakoae* 4 Jitklang, Kuvangkadilok, Baimai, Takaoka and Adler, 2008

Simulium (*Gomphostilbia*) nr. *asakoae* 4 Jitklang, Kuvangkadilok, Baimai, Takaoka and Adler, 2008: 17 (pupa and larva).

Distribution. Thailand.

Bionomics. Aquatic habitat: a stream (width 0.3 m, shaded, bottom with pebbles and gravel, elevation 330 m) in Sakon Nakhon Province (Jitklang et al., 2008).

Remarks. This unnamed species was reported from Thailand with its chromosomal characteristics by Jitklang et al. (2008). Its female and male are unknown.

It is characterized in the pupa by the terminal hooks with their outer margins smooth, and ca. 90 degrees formed by the stalks of the ventral pair and the dorsal triplet; and in the larva by the postgenal cleft long, ca. 5.4 times as long as the postgenal bridge, with its apex pointed and abdomen lacking colored markings dorsally.

Simulium (*Gomphostilbia*) nr. *sheilae* 3 Jitklang, Kuvangkadilok, Baimai, Takaoka and Adler, 2008

Simulium (*Gomphostilbia*) nr. *sheilae* 3 Jitklang, Kuvangkadilok, Baimai, Takaoka and Adler, 2008: 21 (pupa and larva).

Distribution. Thailand.

Bionomics. Aquatic habitat: a stream (width 5.5–6.0 m, bottom with pebbles and gravel, elevation 300 m) in Sakon Nakhon Province (Jitklang et al., 2008).

Remarks. This unnamed species was reported from Thailand with its chromosomal characteristics by Jitklang et al. (2008). Its female and male are unknown.

It is characterized in the pupa by the stalk of the dorsal triplet of the gill as thick as that of the middle triplet, ca. 90 degrees formed by the stalks of the ventral pair and the dorsal triplet and terminal hooks with their outer margins smooth; and in the larva by the postgenal cleft medium-long, 1.5–2.3 times as long as the postgenal bridge, with its apex rounded, and abdominal segments 1–4 each with a greenish transverse band.

Simulium (*Gomphostilbia*) sp. A Phasuk, Chanpaisaeng, Adler and Courtney, 2005

Simulium (*Gomphostilbia*) sp. A Phasuk, Chanpaisaeng, Adler and Courtney, 2005: 55–57 (pupa, and larva).

Distribution. Thailand.

Bionomics. Aquatic habitats: Temporary roadside creeks to permanent waterfalls in Khao Yai National Park (Phasuk et al., 2005).

Remarks. This unnamed species was reported from Thailand with its chromosomal characteristics by Phasuk et al. (2005). It is difficult to place this species in any species-group for lack of information for most morphological characteristics of the adult female and male.

It is characterized in the adults by the antennal with nine flagellomeres; in the pupa by the gill with eight filaments arranged as [3+(1+2)]+2 from dorsal to ventral; in the pupa by the terminal hooks with their outer margins smooth; and in the larva by the postgenal cleft long, 4.7–7.0 times as long as the postgenal bridge, with its apex pointed, and abdomen covered with thick, dark setae each with three to six branches dorsally.

Simulium (*Gomphostilbia*) sp. B Phasuk, Chanpaisaeng, Adler and Courtney, 2005

Simulium (*Gomphostilbia*) sp. B Phasuk, Chanpaisaeng, Adler and Courtney, 2005: 58 (pupa, and larva).

Distribution. Thailand.

Bionomics. Aquatic habitats: Temporary roadside creeks to permanent waterfalls in Khao Yai National Park (Phasuk et al., 2005).

Remarks. This unnamed species was reported from Thailand with its chromosomal characteristics by Phasuk et al. (2005). Its female and male are unknown. It is difficult to place this species in any species-group for lack of the adult stage.

It is characterized in the larva by the postgenal cleft long, almost reaching the posterior margin of the hypostoma and abdomen covered with dark setae each unbranched or split into two or three branches dorsally.

Simulium (*Gomphostilbia*) sp. C Phasuk, Chanpaisaeng, Adler and Courtney, 2005

Simulium (*Gomphostilbia*) sp. C Phasuk, Chanpaisaeng, Adler and Courtney, 2005: 58–59 (pupa, and larva).

Distribution. Thailand.

Bionomics. Aquatic habitats: Temporary roadside creeks to permanent waterfalls in Khao Yai National Park (Phasuk et al., 2005).

Remarks. This unnamed species was reported from Thailand with its chromosomal characteristics by Phasuk et al. (2005). Its female and male are unknown. It is difficult to place this species in any species-group for lack of the adult stage.

It is characterized in the larva by the postgenal cleft medium-long, 2.0–2.8 times as long as the postgenal bridge and abdomen nearly lacking setae dorsally.

Subgenus Montisimulium *Rubtsov*

Simulium (*Montisimulium*) *angkaense* Takaoka and Choochote, 2005 (Fig. 42Q, S, W)

Simulium (*Montisimulium*) *angkaense* Takaoka and Choochote, 2005j: 209–212 (pupa and larva); Takaoka and Choochote, 2009: 113–117 (female and male).

Distribution. Thailand.

Bionomics. Aquatic habitats: The pupae of this species were found in small depressions formed on the surface of rocks in a small forest stream of Siribhume Waterfall, whereas the larvae were collected from fallen leaves and the surface of rocks at Siribhume Waterfall and Ang Ka (Takaoka and Choochote, 2005j).

Remarks. This species was described from pupae and larvae collected from Thailand (Takaoka and Choochote, 2005j) and its female and male were described by Takaoka and Choochote (2009).

The pupa is characterized by the gill with 12 slender filaments arising from a medium-long common basal stalk, arranged as 2+[2+(1+2)+2+(1+2)] or [2+(1+2)+2+[2+(1+2)] or [1+1+(1+2)]+2+2+(1+2) or 2+2+[(1+2)+2]+(1+2) from inner to outer (Fig. 42Q, S), abdominal segment 9 without grapnel-like hooklets on each lateral side, and cocoon without an anterodorsal projection (Fig. 42Q).

Simulium (*Montisimulium*) *chowi* Takaoka, 1979

Simulium (*Eusimulium*) *chowi* Takaoka, 1979: 374–377 (female, pupa, and larva).

Distribution. Taiwan.

Bionomics. Aquatic habitat: A cascading stream (width 1 m, shaded, elevation ca. 2600 m) (Takaoka, 1979).

Remarks. This species was described from Ta U Lin, Taiwan (Takaoka, 1979). Its male is unknown.

It is characterized by the pupal gill with 12 filaments arranged as 2+[(2+1)+(2+1)]+(2+2) from inner to outer, cocoon with a short anterodorsal projection, larval antennal article 2 with three or four secondary annulations, and larval postgenal cleft small, vestigial.

Simulium (*Montisimulium*) *dasguptai* Datta, 1974

Simulium (*Eusimulium*) *dasguptai* Datta, 1974b: 459–462 (pupa and larva).
Simulium (*Nevermannia*) *dasguptai* Datta: Takaoka, 2003; Adler and Crosskey, 2009.
Simulium (*Montisimulium*) *dasguptai* Datta: Takaoka *et al.*, 2009c: 217–220 (male).

Distribution. India (West Bengal) and China (Sichuan).

Bionomics. Aquatic habitats: Trickles at an elevation of 1660–2480 m, Darjeeling (Datta, 1974b).

Remarks. This species was first described from pupae and larvae collected from India under the subgenus *Eusimulium* by Datta (1974b). After treated under the subgenus *Nevermannia* by Takaoka (2003) and by Adler and Crosskey (2009), it was transferred to the subgenus *Montisimulium* when its male was described by Takaoka et al. (2009c). The female of this species remains unknown.

It is characterized by the pupal gill composed of a long, inflated, horn-like structure, with 14 slender filaments.

***Simulium* (*Montisimulium*) *dattai* Takaoka and Somboon, 2008**
(Fig. 42T, V, Y)

Simulium (*Eusimulium*) sp. B Datta et al., 1975: 95–96 (pupa).
Simulium (*Montisimulium*) *dattai* Takaoka and Somboon, 2008: 218–220 (larva).

Distribution. Bhutan, India (West Bengal) and Nepal.

Bionomics. Aquatic habitat: A small stream at an elevation of 2292 m in Mongar, Bhutan (Takaoka and Somboon, 2008).

Remarks. The pupa of this species was described as *S.* (*E.*) sp. B by Datta et al. (1975). This species was named as *S. dattai* when its larva was described based on larvae collected from Bhutan (Takaoka and Somboon, 2008). The female and male of this species are unknown.

This species is characterized by the gill with 14 slender filaments arising from a medium-long common basal stalk, and arranged as 2+(2+2)+[(2+2)+2]+2 from dorsal to ventral, and larval postgenal cleft, which is small, quadrate or M-shaped, and ca. 0.3 times as long as the postgenal bridge (Fig. 42Y).

***Simulium* (*Montisimulium*) *ghoomense* Datta, 1975**

Simulium (*Montisimulium*) *ghoomense* Datta, 1975b: 67–74 (female, male, pupa, and larva).

Distribution. India (West Bengal) and China (Tibet).

Bionomics. Aquatic habitats: Trickles at an elevation of 2250–2480 m, Darjeeling (Datta, 1975b).

Remarks. This species was described from India (Datta, 1975b), and later was reported from Tibet (Chen and An, 2003).

It is characterized by the male paramere with seven hooks; pupal gill with 12 slender filaments arising from a short common basal stalk, and arranged as 2+[(2+1)+(2+1)]+(2+2) from dorsal to ventral, cocoon with an anterodorsal projection, larval antennal article 2 with four secondary annulations, and postgenal cleft absent.

The giant chromosomes of larval salivary glands of this species were studied by Henry et al. (2011).

***Simulium* (*Montisimulium*) *jumlaense* Takaoka and Adler, 2020**

Simulium (*Montisimulium*) *jumlaense* Takaoka and Adler [in Takaoka et al., 2020f]: 198–200 (female).

Distribution. Nepal.

Bionomics. A single female of this species was captured by a Malaise trap at an elevation of 2675 m, in Jumla, Nepal (Takaoka et al., 2020f). Aquatic habitats are unknown.

Remarks. This species was described from a single female in Nepal (Takaoka et al., 2020f). The male, pupa, and larva are unknown.

The female of this species is characterized by the claw with an extremely small tooth (similar to Fig. 42D), which is shared by two species, *S. phahompokense* Takaoka and Srisuka from Thailand and *S. decimfiliatum* (Rubtsov) from Uzbekistan. It is distinguished from the two latter species by the relative length of the sensory vesicle against the third palpal segment (0.42 in this species versus 0.32–0.34 in *S. phahompokense* and 0.25 in *S. decimfiliatum*).

***Simulium* (*Montisimulium*) *kambaitense* Takaoka, 1989**

Simulium (*Montisimulium*) *kambaitense* Takaoka, 1989b: 245–248 (female and male).

Distribution. Myanmar.

Bionomics. Aquatic habitats are unknown.

Remarks. This species was described from females and males captured by a Malaise trap in Myanmar (Takaoka 1989b). Its pupa and larvae remain unknown.

This species is characterized in the female by the sensory vesicle long, over half the length of the third palpal segment; and in the male by the number of upper-eye (large) facets in 19 vertical columns and 19 horizontal rows, ventral plate with its posterior margin markedly depressed medially, and paramere with four hooks.

***Simulium* (*Montisimulium*) *kirgisorum* (Rubtsov, 1956)**

Eusimulium kirgisorum Rubtsov, 1956: 829–830 (male, pupa, and larva); Xue, 1987: 110.

Simulium (*Montisimulium*) *kirgisorum*: Crosskey, 1988: 456; Chen and An, 2003: 157–158.

Distribution. Kyrgyzstan and China (Beijing, Hubei, Tibet, and Yunnan).

Bionomics. Aquatic habitats: Small springs (flow fast, bottom stony, 9–10 °C) near their source at an elevation of 1200–1400 m (Rubtsov, 1956).

Remarks. This species was described from Kyrgyzstan by Rubtsov (1956). Its female is unknown.

It is characterized in the male by the paramere with three long and four small hooks; in the pupa by the gill with 12 filaments arranged as 2+(3+3)+(2+2) from dorsal to ventral, arising from a medium-long common basal stalk, and cocoon wall-pocket-shaped, covering only pupal abdomen; and in the larva by the postgenal cleft small, nearly quadrate in shape.

It was recorded from Yunnan, China by Xue (1987). *Simulium* (*M.*) *kirgisorum* differs from *S.* (*M.*) sp., which Xue (1993c) described under the name of *Titanopterix tanae* (homonym of *S.* (*S.*) *tanae* Xue, 1991), by having the different arrangement of the pupal gill filaments and cocoon without an anterodorsal projection.

Simulium (_Montisimulium_) _laoleense_ Takaoka and Choochote, 2005
(Fig. 42K–O)

Simulium (*Montisimulium*) *laoleense* Takaoka and Choochote, 2005j: 212–214 (larva); Takaoka et al., 2010d: 365–371 (female, male, and pupa).

Distribution. Thailand.

Bionomics. Aquatic habitat: A small seasonal stream (width 0.5 m, flow slow, 18.5 °C, shaded, elevation ca. 1500 m) flowing in a forest, Siribhume Waterfall (Takaoka and Choochote, 2005j).

Remarks. This species was described from larvae from Thailand (Takaoka and Choochote, 2005j). Its adults and pupa were described by Takaoka et al. (2010d).

It is characterized by the pupal gill with 14 slender filaments arising from a long common basal stalk and arranged as four groups: dorsal (2+2), inner (2), outer (2+2 or 1+1+1+1), and ventral (2+2).

Simulium (_Montisimulium_) _merga_ Takaoka and Choochote, 2005
(Fig. 42A, C, J, P, R, U, X)

Simulium (*Montisimulium*) *merga* Takaoka and Choochote, 2005a: 21–28 (female, male, pupa, and larva).

Distribution. Thailand.

Bionomics. Aquatic habitat: A small forest stream (6.5–14.0 °C, shaded, elevation 2560 m), Ang Ka, Doi Inthanon National Parks, Chiang Mai (Takaoka and Choochote, 2005a).

Remarks. This species was described from Thailand (Takaoka and Choochote, 2005a).

The pupa of this species is characterized by the gill consisting of two inflated structures, each with eight and six slender filaments near its apex (Fig. 42R), abdominal segment 9 with grapnel-like hooklets, and cocoon with a short anterodorsal projection (Fig. 42P).

Simulium (_Montisimulium_) _nanense_ Takaoka and Srisuka, 2010

Simulium (*Montisimulium*) *nanense* Takaoka and Srisuka [in Takaoka et al., 2010d]: 371–374 (female, male, pupa, and larva).

Distribution. Thailand.

Bionomics. Aquatic habitat: A small stream (width 25 cm, depth 5 cm, flow slow, 20.0 °C, partially shaded, elevation 1554 m), Doi Phu Ka National Park (Takaoka et al., 2010d).

Remarks. This species was described from females, males, pupae and larvae from Thailand (Takaoka et al., 2010d).

This species is similar to *S. laoleense* in sharing many characteristics including the number and arrangement of the pupal gill filaments, though the outer group of gill filaments arises more basally than the dorsal and inner groups.

***Simulium* (*Montisimulium*) *nemorivagum* Datta, 1973**

Simulium (*Eusimulium*) *nemorivagum* Datta, 1973: 373–375 (male, pupa and larva); Datta, 1974b: 457–459 (female).

Simulium (*Montisimulium*) *nemorivagum*: Datta, 1992: 47–48; An, 1996: 471.

Distribution. India (West Bengal and Arunachal Pradesh), China (Tibet), Pakistan, Bhutan, and Nepal.

Bionomics. Aquatic habitats: Trickles at an elevation of 1900–2250 m, Darjeeling, India (Datta, 1973) and a small forest stream at an elevation of 3720 m in Mongar, Bhutan (Takaoka and Somboon, 2008).

Remarks. This species was described from males, pupae and larvae from Darjeeling, India (Datta, 1973). Its female was described by Datta (1974b). It was also reported from Bhutan (Takaoka and Somboon, 2008) and Nepal (Shrestha and Takaoka, 2009).

It is characterized in the female by the sensory vesicle long, 0.65 times as long as the third palpal segment and claw with a medium-long tooth 0.29 times as long as the claw; in the male by the number of upper-eye (large) facets in 12 vertical columns and 14 horizontal rows, and paramere with seven hooks; in the pupa by the gill with 12 slender filaments arising from a medium-long common basal stalk, and arranged as 2+[(2+1)+(2+1)]+(2+2) from inner to outer, and cocoon without an anterodorsal projection; and in the larva by the antennal article 2 with three faint secondary annulations (similar to Fig. 42W), rectal organ with eight or nine finger-like secondary lobules per lobe and postgenal cleft small and vestigial (similar to Fig. 42X).

***Simulium* (*Montisimulium*) *nigrofilum* Takaoka and Sofian-Azirun, 2017**

Simulium (*Montisimulium*) *nigrofilum* Takaoka and Sofian-Azirun [in Takaoka et al., 2017c]: 61–65 (male, pupa, and larva).

Distribution. Vietnam.

Bionomics. Aquatic habitat: A small stream (width 0.5 m, bottom rocky, 11.0 °C, exposed to the sun, elevation 1750 m), moderately flowing in a secondary forest, Sapa (Takaoka et al., 2017c).

Remarks. This species was described from Vietnam (Takaoka et al., 2017c). Its female is unknown.

It is characterized by the pupal gill with 12 filaments arising from a medium-long common basal stalk, and arranged as those of *S. nemorivagum*, abdominal segment 9 with grapnel-like hooklets, cocoon without an anterodorsal projection, and larval antennal article 2 with three secondary annulations (similar to Fig. 42W).

***Simulium* (*Montisimulium*) *phahompokense* Takaoka and Srisuka, 2010**

(Fig. 42B, D, E–I)

Simulium (*Montisimulium*) *phahompokense* Takaoka and Srisuka, 2010d: 261–264 (female).

Distribution. Thailand.

Bionomics. Females of this species were captured by a Malaise trap at an elevation of 2174 m, in Doi Pha Hom Pok National Parks, and at an elevation of

2500 m, in a forest near the summit, Doi Inthanon National Parks, Chiang Mai (Takaoka and Srisuka, 2010d). Aquatic habitats are unknown.

Remarks. This species was described from females in Thailand (Takaoka and Srisuka, 2010d). The male, pupa, and larva are unknown.

The female of this species is characterized by the claw with a small tooth 0.06 times as long as the claw (Fig. 42D), a rare characteristic among species of the subgenus *Montisimulium*.

***Simulium* (*Montisimulium*) *separatum* Chen, Xiu and Zhang, 2012**
Simulium (*Montisimulium*) *separatum* Chen, Xiu and Zhang, 2012: 382–388 (female, pupa, and larva).

Distribution. China (Guizhou).

Bionomics. Aquatic habitat: A small, fast-flowing stream at an elevation of 1483 m, in Dadong, Guizhou (Chen et al., 2012).

Remarks. This species was described from Guizhou Province, China (Chen et al., 2012). Its male is unknown.

It is characterized in the female by the sensory vesicle long, 0.6 times as long as the third palpal segment and claw with a medium-sized tooth 0.33 times as long as the claw; in the pupa by the gill with 12 slender filaments arranged as 2+(3+3)+(2+2) from dorsal to ventral, arising from a short common basal stalk, and cocoon wall-pocket-shaped; and in the larva by the second antennal article with two pale annulations, postgenal cleft extremely small, labral fan with ca. 40 primary rays, rectal organ with 14–16 secondary lobules per lobe and posterior circlet with ca. 94 rows of up to 15 hooklets per row.

***Simulium* (*Montisimulium*) *surachaii* Takaoka and Choochote, 2005**
Simulium (*Montisimulium*) *surachaii* Takaoka and Choochote, 2005a: 29 (female).

Distribution. Thailand.

Bionomics. No information on aquatic habitats.

Remarks. This species was described from a single female caught by a hand net (Takaoka and Choochote, 2005a). The male, pupa, and larva are unknown.

The female of this species is characterized by the shorter claw tooth 0.28 times as long as the claw.

***Simulium* (*Montisimulium*) *taktsangense* Takaoka and Somboon, 2008**
Simulium (*Montisimulium*) *taktsangense* Takaoka and Somboon, 2008: 220–222 (larva).

Distribution. Bhutan and India (West Bengal) (?).

Bionomics. Aquatic habitat: A small forest stream at an elevation of 3103 m in Paro, Bhutan (Takaoka and Somboon, 2008).

Remarks. This species was described from larvae collected from Bhutan (Takaoka and Somboon, 2008). Its female, male, and pupa remain unknown. Henry et al. (2010) recorded this species from Darjeeling, India. However, no information on specimens and their morphological characteristics was given. It may be the same species recorded as *S*. sp. C from West Bengal by Datta et al.

(1975). The pupa of S. sp. C has 14 slender filaments arising from a long common basal stalk, like *S. taktsangense*, but has a different arrangement of gill filaments.

Simulium taktsangense is characterized by the pharate pupal gill with 14 slender filaments arranged as 2+(2+2)+(2+2)+[(1+2)+1] from dorsal to ventral, arising from a long common basal stalk and larval postgenal cleft small and vestigial.

Simulium (_Montisimulium_) _weisiense_ Deng, Xue and Chen, 2005

Simulium (*Montisimulium*) *weisiense* Deng, Xue and Chen, 2005: 191–192 (female, male, pupa, and larva).

Distribution. China (Yunnan).

Bionomics. Aquatic habitat: A small stream at an elevation of 2300 m (Deng et al., 2005).

Remarks. This species was described from Yunnan Province, China by Deng et al. (2005).

It is characterized in the male by the number of upper-eye (large) facets in 18 horizontal rows, and paramere with five hooks; in the pupa by the gill with 14 filaments arising from a short common basal stalk, and cocoon with a short anterodorsal projection; and in the larva by the postgenal cleft small, and posterior circlet with 50 rows of 12–14 hooklets per row.

Simulium (_Montisimulium_) sp. Xue, 1993

Titanopteryx tanae Xue, 1993c: 466–468 (female, male, pupa, and larva).

Simulium (*Montisimulium*) *tanae* Xue: An, 1996: 471 [junior homonym of *S.* (*S.*) *tanae* Xue, 1992, but no substitute name is available (Adler, 2022)].

Distribution. China (Yunnan).

Bionomics. Aquatic habitat: A stream at high elevation of 2300 m (Xue, 1993c).

Remarks. This species was described from Yunnan under the name of *Titanopterix tanae* by Xue (1993c). It was a homonym of *S.* (*S.*) *tanae* described by Xue (1992).

It is characterized in the female by the mandible with 27 inner teeth and lacking outer teeth and sensory vesicle long, over half as long as the third palpal segment; in the male by the paramere with four hooks; in the pupa by the gill with 12 filaments arranged as (5+2)+2+3 from dorsal to ventral, arising from a short common basal stalk and cocoon with an anterodorsal projection; and in the larva by the postgenal cleft extremely small.

Simulium (_Montisimulium_) sp. B Zhang, Wen and Chen, 1999

Simulium (*Montisimulium*) sp. B Zhang, Wen and Chen, 1999: 232.

Distribution. China (Guizhou).

Bionomics. Unknown.

Remarks. This unnamed species was reported from Guizhou, China by Zhang et al. (1999). However, its morphological characteristics were not noted. This species was not treated in the monographs on Chinese black flies by Chen and An (2003) and Chen (2016).

***Simulium* (*Montisimulium*) sp. C Datta, Dey and Paul, 1975**

Simulium (*Montisimulium*) sp. C Datta, Dey and Paul, 1975: 96–97 (pupa).

Distribution. India (West Bengal).

Bionomics. Aquatic habitat: A small trickle (elevation 2025 m), Darjeeling (Datta et al., 1975).

Remarks. This species was recorded from Darjeeling, India by Datta et al. (1975). Its female, male, and larva are unknown.

It is characterized by the pupal gill with 14 slender filaments arising from a long common basal stalk, and cocoon wall-pocket-shaped, with thick anterior margin. The arrangement of the pupal gill filaments is [(2+1)+(2+1)]+[2+(2+2)]+2 from dorsal to ventral.

***Simulium* (*Montisimulium*) sp. G Kuvangkadilok, Boonkemtong and Baimai, 1998**

Simulium (*Montisimulium*) sp. G Kuvangkadilok, Boonkemtong and Baimai, 1998: 224 (larva).

Distribution. Thailand.

Bionomics. Aquatic habitat: A small high mountain stream in Doi Inthanon National Parks (Kuvangkadilok et al., 1998).

Remarks. This unnamed species was reported from Doi Inthanon National Parks, Chiang Mai by Kuvangkadilok et al. (1998). Its female, male, and pupa are unknown. The morphological characteristics of this species are unknown. It is therefore difficult to confirm whether this species is a distinct species or one of six related species described from Thailand.

Subgenus Nevermannia *Enderlein*

Simulium (*Nevermannia*) *feuerborni* Species-group

***Simulium* (*Nevermannia*) *bachmaense* Takaoka, Sofian-Azirun and Ya'cob, 2014**

Simulium (*Nevermannia*) *bachmaense* Takaoka, Sofian-Azirun and Ya'cob [in Takaoka et al., 2014e]: 566–570 (female, male, pupa, and larva).

Distribution. Vietnam.

Bionomics. Aquatic habitat: A small stream (width 0.2–0.3 m, depth 2–3 cm, 17.0 °C, shaded, elevation 1414 m) in Bach Ma National Park, Vietnam (Takaoka et al., 2014e).

Remarks. This species was described from Vietnam (Takaoka et al., 2014e).

It is characterized by the female genital fork with a triangular lobe-like projection pointed posteromedially on each arm. The lobe-like projection is rounded posteromedially in most species of this species-group. The pupal gill has six slender filaments arranged as 4+2 from dorsal to ventral, arising from a short common basal stalk.

***Simulium* (*Nevermannia*) *borneoense* Takaoka, 2001**

(Fig. 44G)

Simulium (*Nevermannia*) *borneoense* Takaoka, 2001a: 225–228 (male, pupa, and larva).

Distribution. Malaysia (Sabah).

Bionomics. Aquatic habitat: A small stream (width 0.5–1.0 m, bottom rocky, flow moderate, 13 °C, shaded, elevation 1970 m) just above Carson Falls in Kinabalu National Park (Takaoka, 2001a).

Remarks. This species was described from Sabah, Malaysia (Takaoka, 2001a). The female of this species remains unknown.

It is remarkable that this species has the pupal gill with four slender filaments (Fig. 44G). The pupal gill has six slender filaments in most species of the *feuerborni* species-group.

***Simulium* (*Nevermannia*) *bryopodium* Delfinado, 1971**

Simulium (*Nevermannia*) *bryopodium* Delfinado, 1971: 135–136 (male).

Distribution. Philippines (Palawan).

Bionomics. Unknown.

Remarks. This species was described based on a single male collected from Palawan, the Philippines by Delfinado (1971). The female, pupa, and larva are unknown.

Detailed morphological characteristics are not available.

***Simulium* (*Nevermannia*) *chairuddini* Takaoka, 2003**

Simulium (*Nevermannia*) *chairuddini* Takaoka, 2003: 28–35 (female, male, pupa, and larva).

Distribution. Indonesia (Sulawesi).

Bionomics. Aquatic habitats: Small streams at an elevation of 1100–1500 m, in Sulawesi, Indonesia. Natural infections with *Coelomycidium* sp. and microsporidans were reported in a few larvae of this species (Takaoka, 2003).

Remarks. This species was described from Sulawesi, Indonesia by Takaoka (2003).

The pupa of this species is characterized by the gill with six filaments arranged as 4+2 from dorsal to ventral, two filaments of the ventral pair with a short stalk, and common basal stalk with an enlarged transparent oblong basal fenestra.

***Simulium* (*Nevermannia*) *chiangklangense* Takaoka and Choochote, 2005**
(Fig. 44A)

Simulium (*Nevermannia*) *chiangklangense* Takaoka and Choochote, 2005g: 139–141 (female and pupa).

Distribution. Thailand.

Bionomics. Aquatic habitat: A small stream (width 30–50 cm, depth 2–3 cm, flow slow, 15.0 °C, partially shaded, elevation 790 m) running in a forest, in Chiang Klang, Nan Province (Takaoka and Choochote, 2005g).

Remarks. This species was described from a single female reared from a pupa collected from Thailand by Takaoka and Choochote (2005g). Its male and larva remain unknown.

The pupa of this species is characterized by the gill with six filaments arranged as 2+(1+1)+2 from dorsal to ventral, arising from a short common basal stalk, and two filaments of the ventral pair with a short stalk (Fig. 44A).

***Simulium* (*Nevermannia*) *chitoense* Takaoka, 1979**

Simulium (*Eusimulium*) *chitoense* Takaoka, 1979: 377–382 (female, male, pupa, and larva).

Distribution. Taiwan and China (Hainan).

Bionomics. Aquatic habitat: A small stream (width 0.2–1.0 m, shaded, elevation ca. 1000 m) in a mountainous area, in Shi To, Taiwan (Takaoka, 1979).

Remarks. This species was described from Taiwan by Takaoka (1979).

It is characterized in the female and male by the blackish brown scutum; in the female by the sensory vesicle 0.6 times as long as the third palpal segment; in the male by the number of upper-eye (large) facets in ca. 16 horizontal rows, and paramere with seven hooks; in the pupa by the gill with six filaments arranged as 4+2 from dorsal to ventral, arising from a short common basal stalk, two filaments of the ventral pair with a medium-long stalk, and cocoon wall-pocket-shaped, without an anterodorsal projection; and in the larva by the postgenal cleft half as long as the postgenal bridge.

***Simulium* (*Nevermannia*) *fangense* Takaoka and Choochote, 2006**
(Fig. 44M, N)

Simulium (*Nevermannia*) *fangense* Takaoka and Choochote, 2006c: 83–90 (female, male, pupa, and larva).

Distribution. Thailand.

Bionomics. Aquatic habitat: A small stream (width 1.5–2.0 m, 19.0 °C, exposed to the sun, elevation 1400 m) in a sparsely forested area, Doi Ang Khang, Chiang Mai Province (Takaoka and Choochote, 2006c).

Remarks. This species was described from Thailand by Takaoka and Choochote (2006c).

It is characterized in the pupa by the gill with six filaments, of which one or two filaments of the middle group are directed outward independently from a short common basal stalk, and two filaments of the ventral pair have a short to medium-long stalk, and cocoon with a short anterodorsal projection; and in the larva by the rectal organ compound, with 17–21 finger-like secondary lobules per lobe.

***Simulium* (*Nevermannia*) *feuerborni* Edwards, 1934 (complex)**

Simulium feuerborni Edwards, 1934: 129–132 (male, pupa, and pupa).
Simulium (*Eusimulium*) *feuerborni*: Crosskey, 1973: 424; Datta, 1973: 226.
Simulium (*Nevermannia*) *feuerborni*: Crosskey, 1987: 458: Takaoka and Davies, 1996: 10–13 (female, male, pupa, and larva); Kuvangkadilok et al., 1998: 218 (larva).
Simulium (*Eusimulium*) sp. B Takaoka and Suzuki, 1984: 13–14 (pupa and larva).

Distribution. Indonesia (Java and Sumatra) and Thailand.

Bionomics. Aquatic habitats: Small streams at an elevation of 900–1200 m in West Java and irrigation ditches or brooks (width 0.2–1.0 m, 17–20 °C, flow slow to moderate, shaded or exposed to the sun, elevation 950–1450 m) in Sumatra (Takaoka and Davies, 1996; Takaoka et al., 2000). Natural infection with *Coelomycidium* sp. was reported from a larva of this species from Sumatra (Takaoka et al., 2000).

Remarks. This species was described from East Java and Bali, Indonesia by Edwards (1934). The female of this species was described by Takaoka and Davies (1996). This species was reported from Sumatra, Indonesia (Takaoka et al., 2000), Thailand (Kuvangkadilok et al., 1998), and Peninsular Malaysia (Takaoka and Davies, 1995a). It was divided into four cytoforms by chromosome analyses (Pramual and Wongpakam 2013; Pramual et al., 2015): cytoforms A and B from Thailand, cytoform C from Peninsular Malaysia, and cytoform D from West Java, Indonesia. Cytoform C was described as a new species, *S.* (*N.*) *pairoti* Ya'cob, Takaoka and Sofian-Azirun, by Ya'cob et al. (2017b).

This species is characterized by the male paramere with three or four hooks and pupal gill with six slender filaments arranged as 4+2 from dorsal to ventral, and cocoon with an anterodorsal projection.

***Simulium* (*Nevermannia*) *fruticosum* Takaoka and Choochote, 2005 (Fig. 44C, D)**

Simulium (*Nevermannia*) *fruticosum* Takaoka and Choochote, 2005g: 133–139 (female, male, pupa, and larva).

Distribution. Thailand.

Bionomics. Aquatic habitat: A small forest stream (width 50 cm, depth 2 cm, 25.0 °C, shaded, elevation 1420 m) in Doi Pui, Chiang Mai Province (Takaoka and Choochote 2005g).

Remarks. This species was described from Thailand by Takaoka and Choochote (2005g)

The pupa of this species is characterized by the gill with six filaments arranged as 4+2 from dorsal to ventral, arising from a medium-long common basal stalk, two filaments of the ventral pair with a long stalk (Fig. 44C, D), and cocoon wall-pocket-shaped, without an anterodorsal projection.

***Simulium* (*Nevermannia*) *fuscinervis* Edwards, 1933**

Simulium fuscinervis Edwards, 1933: 256–257 (male); Smart and Clifford, 1969: 46–47 (male).
Simulium (*Eusimulium*) *fuscinervis*: Datta, 1983: 226.
Simulium (*Nevermannia*) *fuscinervis*: Takaoka, 2001a: 228.

Distribution. Malaysia (Sabah).

Bionomics. No information.

Remarks. This species was described from a single male collected from Sabah, Malaysia by Edwards (1933). The female, pupa, and larva of this species remain unknown.

The male of this species is characterized by the reduced number of upper-eye (large) facets in 12 vertical columns and 12–14 horizontal rows, and paramere with 10 hooks.

***Simulium* (*Nevermannia*) *langbiangense* Takaoka, Sofian-Azirun and Ya'cob, 2014**

(Fig. 43A, E)

Simulium (*Nevermannia*) *langbiangense* Takaoka, Sofian-Azirun and Ya'cob [in Takaoka et al., 2014e]: 556–563 (female, male, pupa, and larva).

Distribution. Vietnam.

Bionomics. Aquatic habitat: A small stream (width 0.5–1.5 m, depth 10–20 cm, flow moderate, 15.0 °C, partially shaded, elevation 1540 m) flowing in a coffee plantation, Lam Dong Province (Takaoka et al., 2014e).

Remarks. This species was described from Vietnam (Takaoka et al., 2014e).

It is characterized in the female by the sensory vesicle with an enlarged opening (Fig. 43A); in the pupa by the gill with six slender filaments arranged in three pairs (dorsal outer, dorsal inner, and ventral), of which the dorsal inner pair is sessile, the dorsal outer pair has a short stalk, and the ventral pair has a short to medium-long stalk, all pairs arising from a short common basal stalk, and cocoon wall-pocket-shaped, without an anterodorsal projection; and in the larva by the postgenal cleft 0.6–0.7 times as long as the postgenal bridge.

***Simulium* (*Nevermannia*) *ledangense* Ya'cob, Takaoka and Sofian-Azirun, 2014**

Simulium (*Nevermannia*) *ledangense* Ya'cob, Takaoka and Sofian-Azirun, 2014: 228–235 (female, male, pupa, and larva).

Distribution. Malaysia (Peninsular).

Bionomics. Aquatic habitat: A small stream (width 0.5–0.6 m, depth 12 cm, flow slow, 18.9 °C, shaded, elevation 1000 m) at Mt. Ledang, Johor Province (Ya'cob et al., 2014).

Remarks. This species was described from Peninsular Malaysia by Ya'cob et al. (2014).

It is characterized by the pupal gill with six slender filaments, of which two filaments of the ventral pair have an elongate stalk, about half the entire length of filaments.

***Simulium* (*Nevermannia*) *leigongshanense* Chen and Zhang, 1997**

Simulium (*Nevermannia*) *leigongshanense* Chen and Zhang, 1997b: 301–306 (female, male, pupa, and larva).

Distribution. China (Guizhou).

Bionomics. Aquatic habitat: A small streams at an elevation of 700 m (Chen and Zhang, 1997b).

Remarks. This species was described from Guizhou, China by Chen and Zhang (1997b).

It is characterized in the male by the number of upper-eye (large) facets in 16 horizontal rows, and paramere with six or seven hooks; in the pupa by the gill with six slender filaments arranged as 4+2 from dorsal to ventral, and cocoon with a long anterodorsal projection; and in the larva by the postgenal cleft short, 0.29 times as long as the postgenal bridge.

***Simulium* (*Nevermannia*) *maeaiense* Takaoka and Srisuka, 2011**

(Figs. 43B, C, D, F–Q and 44E, I–L, O)

Simulium (*Nevermannia*) *maeaiense* Takaoka and Srisuka, 2011: 58–67 (female, male, pupa, and larva).

Distribution. Thailand and Vietnam.

Bionomics. Aquatic habitat: A small stream (width 20 cm, depth 10 cm, flow slow, 14.0 °C, partially shaded, elevation 1556 m) in Doi Pha Hom Pok National Park, Chiang Mai Province, Thailand (Takaoka and Srisuka, 2011).

Remarks. This species was described from Thailand by Takaoka and Srisuka (2011). It was recorded from Vietnam by Takaoka et al. (2017c).

It is characterized in the female by the genital fork with a strongly sclerotized narrow horizontal bar at the base of the anteriorly directed projection on each arm (Fig. 43F); in the male by the large number of upper-eye (large) facets in 19 vertical columns and 20 horizontal rows; in the pupa by the gill with six slender filaments arranged horizontally as 2+1+2+1 from inside to outside, arising from a medium-long common basal stalk (Fig. 44E), and cocoon with an anterodorsal projection; and in the larva by abdominal segment 3 with six reddish-brown spots on each side (Fig. 44K, L).

***Simulium* (*Nevermannia*) *mie* Ogata and Sasa, 1954**

Simulium (*Eusimulium*) *mie* Ogata and Sasa, 1954: 327–328 (female); Ogata et al., 1956: 72; Takaoka, 1976a: 176–178 (female, male, pupa, and larva).

Eusimulium mie: Rubtsov, 1959–1964: 309–310; Orii et al., 1969: 1–13 (larva).

Simulium J-7 Bentinck: 1955: 6 (male and pupa).

Distribution. Japan (Honshu, Shikoku, Kyushu and Nansei Islands), Korea (South), and China (Fujian, Guizhou, Yunnan, Sichuan, and Zhejiang).

Bionomics. Aquatic habitats: Small slow-flowing streams (Takaoka, 1976a).

Remarks. This species was described based on females from Honshu, Japan (Ogata and Sasa, 1954) and revised descriptions of its female, male, pupa, and larva were given by Takaoka (1976a). In the Oriental Region, it was recorded from Amamioshima Island and Okinawa Island, in the Nansei Islands (Takaoka, 1976a) and Guizhou, Fujian, and Yunnan Provinces in China (Chen and An, 2003).

This species is characterized in the male by the number of upper-eye (large) facets in 14 horizontal rows; in the pupa by the gill with six slender filaments arranged as 2+2+2 from dorsal to ventral, arising from a medium-long common basal stalk, and cocoon wall-pocket-shaped, without an anterodorsal projection; and in the larva by the rectal organ with 8–13 finger-like secondary lobules per lobe.

***Simulium* (*Nevermannia*) *minahasaense* Takaoka, 2003**

Simulium (*Nevermannia*) sp. A Takaoka and Roberts, 1988: 195 (larva).
Simulium (*Nevermannia*) *minahasaense* Takaoka, 2003: 35–37 (female, male, pupa, and larva).

Distribution. Indonesia (Sulawesi).

Bionomics. Aquatic habitats: Streams at an elevation of 820–1050 m, Sulawesi (Takaoka, 2003).

Remarks. This species was described from Sulawesi, Indonesia by Takaoka (2003).

The pupa of this species is characterized by the gill with six filaments arranged as 2+2+2 from dorsal to ventral, of which the ventral and middle pairs of filaments have a medium-long or long stalk, and common basal stalk with an enlarged transparent oblong basal fenestra.

***Simulium* (*Nevermannia*) *mongarense* Takaoka and Somboon, 2008**

Simulium (*Nevermannia*) *mongarense* Takaoka and Somboon, 2008: 223–229 (male, pupa, and larva).

Distribution. Bhutan.

Bionomics. Aquatic habitat: A small stream at an elevation of 2344 m in Bhutan (Takaoka and Somboon, 2008).

Remarks. This species was described from Bhutan by Takaoka and Somboon (2008). Its female is unknown.

It is characterized in the male by the large number of upper-eye (large) facets in 20 horizontal rows; in the pupa by the gill with six slender filaments arranged as 4+2 from dorsal to ventral, arising from a short common basal stalk, and cocoon with an anterodorsal projection; and in the larva by the rectal organ compound, with 17–19 finger-like secondary lobules per lobe.

***Simulium* (*Nevermannia*) *morisonoi* Takaoka, 1973**

Simulium (*Eusimulium*) *morisonoi* Takaoka, 1973: 203–207 (female, male, pupa, and larva).

Distribution. Japan (Nakanoshima Island in Nansei Islands).

Bionomics. Aquatic habitat: A small slow-flowing stream (Takaoka, 1973).

Remarks. This species was described from Nakanoshima Island in the Nansei Islands, Japan (Takaoka, 1973).

It resembles *S. mie* but is barely distinguished from the latter species in the female by the genital fork with a small projection directed anteriorly on each arm (a distinct projection in *S. mie*); and in the larva by the rectal organ with three to five thumb-like secondary lobules per lobe (8–13 finger-like secondary lobules per lobe in *S. mie*), and labral fan with 22–26 primary rays (30–34 primary rays in *S. mie*).

***Simulium* (*Nevermannia*) *nivritiae* Takaoka and Shrestha, 2020**
Simulium (*Nevermannia*) *nivritiae* Takaoka and Shrestha, 2020c: 205–211 (female, male, and pupa).

Distribution. Nepal.

Bionomics. Aquatic habitats: A stream (elevation 2137 m) and another stream (elevation 1541 m) in Silgadi (Takaoka et al., 2020c).

Remarks. This species was described from Nepal by Takaoka et al. (2020c). Its larva is unknown.

It is characterized in the female by the small number (11) of tiny processes on the cibarium; in the male by the relatively high number of upper-eye (large) facets in 18 vertical columns and 20 horizontal rows; and in the pupa by the cocoon simple, wall-pocket-shaped.

***Simulium* (*Nevermannia*) *pairoti* Ya'cob, Takaoka and Sofian-Azirun, 2017**
Simulium (*Nevermannia*) *pairoti* Ya'cob, Takaoka and Sofian-Azirun [in Ya'cob et al., 2017b]: 134–139 (female, male, pupa, and larva).

Distribution. Malaysia (Peninsular).

Bionomics. Aquatic habitats: Small streams at an elevation of 1500–1813 m only in Cameron's Highlands (Ya'cob et al., 2017b).

Remarks. This species was first reported from Peninsular Malaysia as *S. feuerborni* originally described from Java (Takaoka and Davies, 1995a). Pramual et al. (2015) showed chromosomally that *S. feuerborni* is a species complex consisting of four distinct lineages: cytoforms A and B from Thailand, cytoform C from Peninsular Malaysia, and cytoform D from East Java. After specimens from Cameron's Highlands were morphologically and genetically examined, cytoform C was described as a new species, *S. pairoti* (Ya'cob et al., 2017b).

This species is characterized in the male by the number of upper-eye (large) facets in 15 or 16 vertical columns and 16 or 17 horizontal rows, and paramere with five hooks; and in the pupa by the gill composed of six slender filaments, of which four dorsal filaments usually lie horizontally, and two ventral paired filaments have a stalk variable in length from short to long, and cocoon with or without a short anterodorsal projection.

***Simulium* (*Nevermannia*) *perlucidulum* Takaoka, 1983**
Simulium (*Nevermannia*) *perlucidulum* Takaoka, 1983: 17–20 (female, male, pupa, and larva).

Distribution. Philippines.

Bionomics. Aquatic habitat: A stream (2–3 m wide) at an elevation of ca. 1000 m, Mt. Apo, Mindanao (Takaoka, 1983).

Remarks. This species was described from Mindanao, the Philippines by Takaoka (1983).

The pupa of this species is characterized by the gill with six filaments arising at the same level from a short common basal stalk having an enlarged transparent round basal fenestra.

***Simulium* (*Nevermannia*) *phami* Takaoka, Sofian-Azirun and Ya'cob, 2014**
(Fig. 44F)

Simulium (*Nevermannia*) *phami* Takaoka, Sofian-Azirun and Ya'cob [in Takaoka et al., 2014e]: 563–566 (female, male, pupa, and larva).

Distribution. Vietnam.

Bionomics. Aquatic habitat: A small stream (width 1–2 m, depth 10–20 cm, flow moderate, 15 °C, partially shaded, elevation 1539 m) in Lam Dong Province, Vietnam (Takaoka et al., 2014e).

Remarks. This species was described from Vietnam (Takaoka et al., 2014e).

The pupa of this species is distinctive among species of this species-group in having the gill with five filaments (Fig. 44F) rather than six filaments. The cocoon has a short anterodorsal projection.

***Simulium* (*Nevermannia*) *praelargum* Datta, 1973 (complex)**

Simulium (*Eusimulium*) *praelargum* Datta, 1973: 365–368 (female, male, pupa, and larva).

Simulium (*Nevermannia*) *praelargum*: Datta, 1992a: 45; Takaoka et al., 2020c: 211–212: (female, male, and pupa).

Distribution. India (West Bengal), Nepal and China (Tibet).

Bionomics. According to Datta (1992a), this species breeds in small trickles, feeds on human and certain other mammals, and is bivoltine, overwintering in the larval stage.

Remarks. This species was described from Darjeeling, India (Datta, 1973) and placed in the *S. feuerborni* species-group, newly proposed by Datta (1973). This species was chromosomally studied by Henry et al. (2010) and was found to be a species complex consisting of three cytoforms (Thapa et al., 2014, 2017). The female, male, and pupa of this species were redescribed based on specimens from Nepal by Takaoka et al. (2020c).

It is characterized in the male by the number of upper-eye (large) facets in 16 vertical columns and ca. 17 horizontal rows (16–18 vertical columns and 18 or 19 horizontal rows in males from Nepal), scutum orange-red, and paramere with six or seven hooks; in the pupa by the gill with six slender filaments arranged as 2+1+1+2 from dorsal to ventral, and cocoon with a long anterodorsal projection; and in the larva by the postgenal cleft 0.32 times as long as the postgenal bridge.

***Simulium* (*Nevermannia*) *pumatense* Takaoka, Low and Pham, 2019**
Simulium (*Nevermannia*) *pumatense* Takaoka, Low and Pham [in Takaoka et al., 2019b]: 321–328 (female, male, pupa, and larva).

Distribution. Vietnam.
Bionomics. Aquatic habitat: A small roadside ditch (width 15–50 cm, depth 5 cm, bottom muddy, 15.0 °C, partially shaded, elevation 223 m), Pu Mat National Park, Nghe An Province, central Vietnam (Takaoka et al., 2019b).
Remarks. This species was described from Vietnam (Takaoka et al., 2019b).
It is characterized in the female by the relatively smaller numbers of inner teeth of the mandible (15–18), and of tiny conical processes (16) on the cibarium; in the male by the number of upper-eye (large) facets in 15 vertical columns and 16 horizontal rows; and in the pupa by the cocoon with an anterodorsal projection. A molecular analysis of the DNA barcoding COI gene shows that this species is most closely related to cytoform A of the *S. feuerborni* complex from Thailand (Takaoka et al., 2019b).

***Simulium* (*Nevermannia*) *qiongzhouense* Chen, Zhang and Yang, 2003**
Simulium (*Nevermannia*) *qiongzhouense* Chen, Zhang and Yang, 2003b: 745–747 (female, male, pupa, and larva).

Distribution. China (Hainan).
Bionomics. Aquatic habitat: A small stream at an elevation of 700 m, Tianchi, Hainan Island (Chen et al., 2003b).
Remarks. This species was described from Hainan, China (Chen et al., 2003b).
It is characterized in the male by the number of upper-eye (large) facets in 15 vertical columns and 16 horizontal rows; and in the pupa by the gill with six slender filaments arranged as 2+2+2 from dorsal to ventral, of which the ventral pair has an elongate stalk, and middle and dorsal pairs have a short stalk, and cocoon wall-pocket-shaped, without an anterodorsal projection.
The paramere was noted to have six hooks in the text but it has only three hooks in Fig. 7 in the original description (Chen et al., 2003b).

***Simulium* (*Nevermannia*) *rufithorax* Brunetti, 1911**
Simulium rufithorax Brunetti, 1911: 282–283 (female and male).
Simulium (*Eusimulium*) *rufithorax*: Datta, 1983: 227.
Simulium (*Nevermannia*) *rufithorax*: Datta, 1992a: 45.

Distribution. India (West Bengal and Karnataka).
Bionomics. Unknown.
Remarks. This species was described from females and males collected from West Bengal, and Karnataka, India by Brunetti (1911). Its pupa and larva are unknown. It was placed in the *S. feuerborni* species-group by Datta (1992a).
It is characterized in the female and male by the scutum reddish brown, without three longitudinal vittae; and in the male by the hind basitarsus much enlarged. It is similar to *S. sasai* (Rubtsov) from Japan (Sato et al., 2005) in that the female scutum

lacks three longitudinal vittae, being exceptional among species of the *S. feuerborni* species-group.

Simulium (Nevermannia) subratai Takaoka, Thapa and Henry, 2011

Simulium (*Nevermannia*) *subratai* Takaoka, Thapa and Henry, 2011c: 62–68 (female, pupa, and larva).

Distribution. India (West Bengal).

Bionomics. Aquatic habitat: A moderately flowing trickle (width 12–13 cm, depth 2–3 cm, 15–16 °C, shaded, elevation 2354 m), Dali, Darjeeling (Takaoka et al., 2011c).

Remarks. This species was described from Darjeeling, India (Takaoka et al., 2011c). Its male is unknown.

It is striking that the female thorax is entirely yellow, and larval abdomen lacks colored markings, both characteristics being shared by *S. saitoi* in the same species-group from Japan (Takaoka and Saito, 2000). The pupa is characterized by the gill with six slender filaments arranged as 2+2+2 from dorsal to ventral.

Simulium (Nevermannia) vessabutrae Takaoka and Srisuka, 2010
(Fig. 44B, H)

Simulium (*Nevermannia*) *vessabutrae* Takaoka and Srisuka, 2010c: 97–104 (female, male, pupa, and larva).

Distribution. Thailand.

Bionomics. Aquatic habitat: A stream (width 30 cm, depth 5 cm, flow moderate, 20.0 °C, partially shaded, elevation 1228 m), Khun Sathan National Park, Nan Province (Takaoka and Srisuka, 2010c).

Remarks. This species was described from Thailand by Takaoka and Srisuka (2010c).

It is characterized in the male by the number of upper-eye (large) facets in 17 vertical columns and 18 horizontal rows and paramere with nine hooks; and in the pupa by the gill with six filaments arranged as 2+(2+2) from inside to outside, each pair with a medium-long or long stalk, and arising from a long common basal stalk slightly widened toward its apex (Fig. 44B).

Simulium (Nevermannia) wangxianense Chen, Zhang and Bi, 2004

Simulium (*Nevermannia*) *wangxianense* Chen, Zhang and Bi, 2004: 369–371 (female, male, pupa, and larva).

Distribution. China (Hunan).

Bionomics. Aquatic habitat: A small stream at an elevation of 1250 m, Chenzhou City, Hunan (Chen et al., 2004).

Remarks. This species was described from Hunan, China (Chen et al., 2004).

It is characterized in the male by the number of upper-eye (large) facets in 18 or 19 vertical columns and 13 horizontal rows, and paramere with six or seven hooks; in the pupa by the gill with six slender filaments arranged as 1+2+(1+2) from dorsal to ventral, and cocoon wall-pocket-shaped, without an anterodorsal projection; and in the larva by the labral fan with 30–32 primary rays and posterior circlet with 74 rows of up to 14 hooklets per row.

Simulium (Nevermannia) wichaii Takaoka, 2010

Simulium (*Nevermannia*) *wichaii* Takaoka [in Takaoka and Srisuka, 2010e]: 273–279 (male, pupa, and larva).

Distribution. Thailand.

Bionomics. Aquatic habitat: A small, slow-flowing stream (width 10 cm, depth 2 cm, 13.0 °C, shaded, elevation 1881 m) in Doi Pha Hom Pok National Park, Chiang Mai (Takaoka and Srisuka, 2010e).

Remarks. This species was described from Thailand by Takaoka and Srisuka (2010e). Its female is unknown.

It is characterized in the male by the smaller number of upper-eye (large) facets in 14 vertical columns and 16 or 17 horizontal rows; in the pupa by dark large tubercles on the head and thorax, gill filaments with relatively larger tubercles on annular ridges; and in the larva by the reduced number (16) of the primary rays of the labral fan.

Simulium (Nevermannia) sp. A Takaoka, 1989

Simulium (*Nevermannia*) sp. A Takaoka, 1989b: 248–249 (female).

Distribution. Myanmar.

Bionomics. No information.

Remarks. This species was described from females captured by a Malaise trap in Myanmar (Takaoka, 1989b). Its male, pupa, and larva are unknown.

It is characterized in the female by the slender fore basitarsus ca. 8.5 times as long as its greatest width, tarsal claw with a basal tooth ca. 0.4 times as long as the claw and abdominal segments 7–9 shiny.

Simulium (Nevermannia) sp. "A" Datta, 1973

Simulium (*Eusimulium*) sp. "A" Datta, 1973: 375–376 (female).

Distribution. India (West Bengal).

Bionomics. One female was caught by a light trap, Darjeeling Government College area at an elevation of 2016 m (Datta, 1973).

Remarks. This species was described from a single female captured by a light trap in Darjeeling, North India (Datta, 1973). Its male, pupa, and larva are unknown. It was placed in the subgenus *Eusimulium* (Adler, 2022). However, the female terminalia suggest that it belongs to the *S. feuerborni* species-group in the subgenus *Nevermannia*. It is similar to *S. praelargum* (complex).

Simulium (Nevermannia) sp. D Datta, Dey and Paul, 1975

Simulium (*Eusimulium*) sp. D Datta, Dey and Paul, 1975: 97 (pupa).

Distribution. India (West Bengal).

Bionomics. Aquatic habitat: A stream at Lebon, Darjeeling at an elevation of 2000 m (Datta et al., 1975).

Remarks. This unnamed species was described from pupae collected in Darjeeling, North India (Datta et al., 1975), Its female, male and larva are unknown. It was placed in *Nevermannia* species unplaced to group by Adler (2022). However, it is transferred to the *S. feuerborni* species-group since it is similar to *S. praelargum* (complex) except the cocoon without an anterodorsal projection. It may be the same species as recorded as *S.* (*N.*) sp. 2 (nr. *fruticosum*) by Henry et al. (2010).

Simulium (Nevermannia) sp. Xue, 1991

Titanopteryx kirgisorum Xue, 1991: 93–94 (female, male, pupa, and larva) [junior homonym of *S.* (*M.*) *kirgisorum* Rubtsov, 1956 but no available substitute name (Adler, 2022)].

Distribution. China (Yunnan).

Bionomics. Aquatic habitat: A stream (16–19 °C, elevation 2300 m) (Xue, 1991).

Remarks. This species was described from Yunnan, China by Xue (1991). The new name given by Xue (1991) was a homonym of *S.* (*M.*) *kirgisorum* Rubtsov, 1956 but no substitute name is available (Adler, 2022). It was listed under the subgenus *Montisimulium* in the world list of black flies (Adler, 2022), but it should be transferred to the *S. feuerborni* species-group in the subgenus *Nevermannia*, according to the morphological characteristics illustrated by Xue (1991).

It is characterized in the female by the sensory vesicle long, about half the length of the third palpal segment; in the male by the paramere with five distinct hooks; in the pupa by the gill with six filaments arranged as 4+2 from dorsal to ventral, arising from a medium-long common basal stalk, ventral pair with a medium long stalk, and cocoon with a distinct anterodorsal projection; and in the larva by the labral fan with 26–28 primary rays, rectal organ with four to six short thumb-like secondary lobules per lobe and posterior circlet with ca. 80 rows of ca. 13 hooklets per row.

Simulium (Nevermannia) sp. 2 (nr. *fruticosum*) Henry, Dey, Varma, Thapa and Procunier, 2010

Simulium (*Nevermannia*) sp. 2 (nr. *fruticosum*) Henry, Dey, Varma, Thapa and Procunier, 2010: 436.

Distribution. India (West Bengal).

Bionomics. No information.

Remarks. This species was recorded from India by Henry et al. (2010). However, no information on specimens they collected and their morphological characteristics was given. If it is similar to *S. fruticosum*, it may be the same species as *S.* (*N.*) sp. D recorded from West Bengal, India by Datta et al. (1975).

Simulium (*Nevermannia*) *ruficorne* Species-group

Simulium (*Nevermannia*) *angustitarse* (Lundström, 1911)

Melusina angustitarse Lundström, 1911: 22–23.
Eusimulium angustitarse: Rubtsov, 1989: 592–595 (female, male, pupa, and larva).
Simulium (*Nevermannia*) *angustitarse*: Crosskey, 1987: 458; Chen and An, 2003: 182–183 (female, male, pupa, and larva).

Distribution. This species was widely distributed in Europe, North Africa, Central Asia, and Siberia in the Palearctic Region. In the Oriental Region, it was recorded from Yunnan, China (Chen and An, 2003).

Bionomics. Aquatic habitats: Warmer and small streams overgrown with vegetation. Females are a blood-sucker (Rubtsov, 1956).

Remarks. It is characterized in the pupa by the gill with four slender filaments arranged as 2+2 from dorsal to ventral, of which two filaments of the dorsal pair have a short stalk and those of the ventral pair have no stalk, and cocoon wall-pocket-shaped, somewhat produced anterodorsally; and in the larva by lacking a postgenal cleft.

Simulium (*Nevermannia*) *aureohirtum* Brunetti, 1911

(Figs. 45C–F and 46C)

Simulium aureohirtum Brunetti, 1911: 283–288 (male); Edwards, 1934: 134–137 (female, pupa, and larva).
Simulium (*Eusimulium*) *aureohirtum*: Puri, 1933c: 1–9 (female, male, pupa, and larva); Takaoka, 1979: 382–384 (female, male, pupa, and larva).
Simulium (*Nevermannia*) *aureohirtum*: Takaoka, 2003: 37–40 (female, male, pupa, and larva).
Simulium philippinensis Delfinado, 1962 (from the Philippines), Synonymized by Takaoka, 1983.
Simulium (*Eusimulium*) *tuaranense* Smart and Clifford, 1969 (from Sabah, Malaysia). Synonymized by Crosskey, 1973.
Simulium sp. A Lewis, 1964: 294 (female). Identified as *S.* (*N.*) *aureohirtum* by Shrestha and Takaoka, 2009.

Distribution. This species was widely distributed in the Oriental Region and parts of the Palearctic and Australasian Regions (Adler and Crosskey, 2017).

Bionomics. Aquatic habitats: Small streams and irrigation channels with warm water up to 33.4 °C (Takaoka, 2003). The female of this species undergoes autogenous ovarian development without blood meals at least for the first batch of eggs (Takaoka and Noda, 1979; Takaoka, 1989a).

Remarks. This species is the only member of this species-group that has six slender pupal gill filaments per side (Fig. 46C). The variation in the length and thickness of the pupal gill filaments was reported based on pupae collected from Halmahera, Indonesia (Takaoka, 2003).

In Thailand, this species is divided into two distinct lineages (Thaijarern et al., 2014; Pramual et al., 2021).

***Simulium* (*Nevermannia*) *falcoe* (Shiraki, 1935)**
Eusimulium falcoe Shiraki, 1935: 13–15 (female).
Simulium (*Nevermannia*) *falcoe*: Takaoka, 1979: 384.

Distribution. Taiwan.

Bionomics. Probably ornithophilic, since two females were collected among the feathers of *Falco tinnunculus japonicus* in Kappansan (ca. 300 m in elevation) (Shiraki, 1935). No information of aquatic habitats.

Remarks. This species was described from two females collected in Taiwan (Shiraki, 1935). Its male, pupa, and larva remain unknown.

According to the original description, the female of *S. falcoe* is almost identical to that of *S. aureohirtum* including the color of the antenna (yellow except the first flagellomere darkened) and legs [mid and hind tibiae each with a dark subbasal band clearly seen in Plate I-3 (Shiraki, 1935), though not mentioned in the original description]. Despite a difference in the relative length of the height of the frons against its greatest width, which is ca. 1.5 in this species but almost equal in *S. aureohirtum*, this species is likely to be synonymous with *S. aureohirtum*.

***Simulium* (*Nevermannia*) *geniculare* (Shiraki, 1935)**
Eusimulium geniculare Shiraki, 1935: 19–20 (female).
Simulium (*Nevermannia*) *geniculare*: Takaoka, 1979: 384.

Distribution. Taiwan.

Bionomics. A female of this species was observed to be biting a man (Shiraki, 1935). Aquatic habitats are unknown.

Remarks. This species was described from a single female collected while biting on the ear of Dr. Shiraki in his laboratory (Shiraki, 1935). Its male, pupa, and larva are unknown.

It is characterized by the back of the head covered with dark short hairs, antenna entirely yellow, scutum with three faint greyish longitudinal vittae, and all femora and tibiae yellowish except the apical caps darkened [though the mid tibia has a dark subbasal band in Plate I-5 (Shiraki, 1935)]. There is a possibility that a darkened first flagellomere of the antenna and a dark subbasal band of the hind tibia were overlooked.

This species was once synonymized with *S. aureohirtum* (Takaoka, 2003). However, it is distinguished from the latter species by the dark short hairs on the back of the head, and the presence of three faint longitudinal vittae on the scutum (yellow short hairs on the back of the head and absence of longitudinal vittae on the scutum in *S. aureohirtum*). In addition, *S. geniculare* is a man-biter (Shiraki, 1935), whereas *S. aureohirtum* is autogenous at least for the first batch of eggs and feed probably on birds for the subsequent ovarian development (Takaoka and Noda, 1979).

Simulium (_Nevermannia_) _glatthaari_ Takaoka and Davies, 1995

(Fig. 46A, B, K, I)

Simulium (*Nevermannia*) *glatthaari* Takaoka and Davies, 1995b: 127–132 (female, male, pupa, and larva).

Distribution. Indonesia (Sumatra).

Bionomics. Aquatic habitat: A rice-field drainage brook (width 1 m, depth 0.3 m, elevation 1200 m) at the foot of Mt. Sibayak, North Sumatra (Takaoka and Davies, 1995b).

Remarks. This species was described from Sumatra, Indonesia (Takaoka and Davies, 1995b). It is the first species of this species-group that has eight pupal gill filaments per side (Fig. 46A, B).

It is characterized in the female by the sensory vesicle medium-long, 0.47 times as long as the third palpal segment, and claw with a medium-long tooth 0.35 times as long as the claw; in the male by the large number of upper-eye (large) facets in 22 vertical columns and 22 horizontal rows; in the pupa by the gill with eight filaments arranged as 3+2+3 or 2+3+3 from dorsal to ventral, arising from an extremely short common basal stalk (Fig. 46A, B); and in the larva by the rectal organ simple, lacking secondary lobules.

Apart from the difference in the number of the pupal gill filaments, this species is distinguished from *S. aureohirtum* in the pupa by the prominent terminal hooks and cocoon without an anterodorsal projection, and in the larva by the hypostoma having less produced corner teeth, which are much shorter than the median tooth (Fig. 46K), and mandibular comb-teeth with an enlarged second tooth compared to the third one (Fig. 46I).

Simulium (_Nevermannia_) _karavalliense_ Anbalagan, Rekha, Vijayan, Balachandran, Dinakaran and Krishnan, 2020

Simulium (*Nevermannia*) *karavalliense* Anbalagan, Rekha, Vijayan, Balachandran, Dinakaran and Krishnan, 2020a: 374–380 (female, male, pupa, and larva).

Distribution. India (Tamil Nadu).

Bionomics. Aquatic habitat: A small moderately flowing stream (24.3 °C, exposed to the sun, elevation 229 m), Kollli Hills, Tamil Nadu, southern India (Anbalagan et al., 2020a).

Remarks. This species was described from south India (Anbalagan et al., 2020a). It is the second species of this species-group that has eight pupal gill filaments per side.

It is distinguished from *S. glatthaari* from Sumatra (Takaoka and Davies, 1995b) in the female by the longer sensory vesicle (0.65 as long as the third palpal segment); in the male by the smaller number of male upper-eye (large) facets in 14 vertical columns and 16 horizontal rows; in the pupa by the gill with eight filaments arranged as 3+3+2 from dorsal to ventral, arising from a short common basal stalk; and in the larva by the rectal organ with six to eight secondary lobules per lobe.

It is exceptional that the median sclerite of this species has a bifid apex. In most species of this species-group, the median sclerite is simple rod-shaped.

***Simulium* (*Nevermannia*) *wayani* Takaoka and Chen, 2018**

(Figs. 45A, B, G–M and 46D–H, J, L, M)

Simulium (*Nevermannia*) *wayani* Takaoka and Chen [in Takaoka et al., 2018d]: 964–972 (female, male pupa, and larva).

Distribution. Indonesia (Timor).

Bionomics. Aquatic habitat: A river (width 20–25 m, elevation 287 m) slowly running in Soe, Timor (Takaoka et al., 2018d).

Remarks. This species was described from Timor, Indonesia (Takaoka et al., 2018d).

It is characterized in the female by the spermatheca without a dark sclerotized neck (Fig. 45G); and in the pupa by the gill with four somewhat inflated filaments, which are variable in length (Fig 46D–G). Notably, COI gene sequences of *S. wayani* are not different from those of *S. aureohirtum*, although they are different from those of *S. ornatipes* Skuse, which has four pupal gill filaments, from the Australasian Region (Low et al., unpublished data). The result suggests that this species may have been evolved from *S. aureohirtum.* On the contrary, Adler et al. (2019c) showed by a chromosomal analysis that *S. wayani* is closer to *S. ornatipes* than to *S. aureohirtum.*

Simulium (*Nevermannia*) *vernum* Species-group

***Simulium* (*Nevermannia*) *aberrans* Delfinado, 1969**

Simulium aberrans Delfinado, 1969: 203–206 (female).

Simulium (*Eusimulium*) *aberrans*: Takaoka, 1983: 13–15 (female, male, pupa, and larva).

Distribution. Philippines (Luzon).

Bionomics. Aquatic habitat: A clean mountain stream (0.5–1.0 m wide) (Takaoka, 1983). An adult female of this species was caught by a light trap set at an elevation of 1800–2000 m, Mountain Province, Luzon (Delfinado, 1969).

Remarks. This species was described from a single female from Luzon, the Philippines (Delfinado, 1969). The female was redescribed and the male, pupa and larva were described by Takaoka (1983).

This is the only species of the *S. vernum* species-group recorded from the Philippines.

It is characterized by the simple cocoon and larval abdomen with an accessory sclerite on each side of abdominal segment 9 (similar to Fig. 48P).

***Simulium* (*Nevermannia*) *bumthangense* Takaoka and Somboon, 2008**

Simulium (*Nevermannia*) *bumthangense* Takaoka and Somboon, 2008: 236–241 (male and pupa).

Distribution. Bhutan.

Bionomics. Aquatic habitat: A small stream at an elevation of 2570 m (Takaoka and Somboon, 2008).

Remarks. This species was described from Bhutan (Takaoka and Somboon, 2008). Its female and larva remain unknown.

The male of this species is characterized by the ventral plate with its posterior margin nearly straight. This species is similar in the pupa to *S. thrimshanglaense* from Bhutan but is distinguished in the male from the latter by the smaller number of upper-eye (large) facets in 16 or 17 vertical columns and 17 horizontal rows (20 vertical columns and 23 horizontal rows in the latter species).

***Simulium* (*Nevermannia*) *burmense* Takaoka, 1989**

Simulium (*Nevermannia*) *burmense* Takaoka, 1989b: 244–245 (female).

Distribution. Myanmar.

Bionomics. Females of this species were caught by a Malaise trap at an elevation of ca. 2300 m in Kambaiti, northern Myanmar (Takaoka 1989b). Aquatic habitats are unknown.

Remarks. This species was described from females collected in northern Myanmar (Takaoka 1989b). Its male, pupa, and larva remain unknown.

The female is characterized by the femora yellow except their apical portions darkened and genital fork with a projection directed posteromedially on each arm.

***Simulium* (*Nevermannia*) *cangshanense* (Xue, 1993)**

Eusimulium cangshanense Xue 1993a: 229–231 (female, male, pupa, and larva).

Simulium (*Nevermannia*) *cangshanense*: Chen and An, 2003: 197–199 (female, male, pupa, and larva).

Distribution. China (Yunnan).

Bionomics. Aquatic habitats: Small streams at an elevation of 2300–2500 m, Daili county, Yunnan (Xue 1993a).

Remarks. This species was described from Yunnan, China (Xue 1993a).

It is characterized in the male by the number of upper-eye (large) facets in 17 horizontal rows; and in the pupa by the gill with four slender filaments of the same thickness and cocoon with a long anterodorsal projection.

***Simulium* (*Nevermannia*) *caudisclerum* Takaoka and Davies, 1995**

Simulium (*Nevermannia*) *caudisclerum* Takaoka and Davies, 1995a: 88–94 (female, male, pupa, and larva).

Distribution. Malaysia (Peninsular).

Bionomics. Aquatic habitat: A stream (width 1.5–2.5 m, elevation ca. 2000 m), Cameron's Highlands (Takaoka and Davies, 1995a).

Remarks. This species was described based on reared adults, pupal exuviae, and mature larvae from Peninsular Malaysia (Takaoka and Davies, 1995a).

It is characterized by the number of male upper-eye (large) facets in 14 vertical columns and 15 horizontal rows, pupal gill with four filaments in pairs each with medium-long to long stalk, arising from a medium-long common basal stalk, cocoon with an elongate anterodorsal projection and larval abdomen with accessory sclerites on the last segment.

***Simulium* (*Nevermannia*) *cheni* (Xue, 1993)**
Eusimulium cheni Xue 1993a: 232 (female, male, pupa, and larva).
Simulium (*Nevermannia*) *cheni*: Chen and An, 2003: 199–200 (female, male, pupa, and larva).

Distribution. China (Yunnan).

Bionomics. Aquatic habitats: Small streams at an elevation of 2300–2500 m, Daili county, Yunnan (Xue 1993a).

Remarks. This species was described from Yunnan, China (Xue 1993a).

It is characterized in the pupa by the gill with four slender filaments of the same thickness and cocoon with a long anterodorsal projection.

***Simulium* (*Nevermannia*) *chenzhouense* Chen, Zhang and Bi, 2004**
Simulium (*Nevermannia*) *chenzhouense* Chen, Zhang and Bi, 2004: 365–367 (female, male, pupa, and larva).

Distribution. China (Hunan).

Bionomics. Aquatic habitat: A rivulet in Chenzhou, at an elevation of 1350 m (Chen et al., 2004).

Remarks. This species was described from Hunan, China (Chen et al., 2004).

It is characterized in the male by the number of upper-eye (large) facets in 10 vertical columns and 14 horizontal rows; and in the pupa by the gill with four slender filaments of the same thickness and cocoon with a short anterodorsal projection.

***Simulium* (*Nevermannia*) *chomthongense* Takaoka, Srisuka and Choochote, 2012 (complex)**
Simulium (*Nevermannia*) *chomthongense* Takaoka, Srisuka and Choochote [in Takaoka et al., 2012e]: 382–389 (female, male, pupa, and larva).
Simulium (*Eusimulium*) sp. A Takaoka and Suzuki, 1984: 12–13 (larva).
Simulium (Nevermannia) caudisclerum (*nec* Takaoka and Davies): Kuvangkadilok et al., 1998: 215–230; 1999b: 197–207; Takaoka and Choochote, 2004a: 189.

Distribution. Thailand.

Bionomics. Aquatic habitats: Small streams (10–15 °C, elevation 1700–2200 m) in Doi Inthanon National Parks (Takaoka et al., 2012e).

Remarks. This species was first described as *Simulium* (*Eusimulium*) sp. A based on a larva collected from Doi Inthanon National Parks, Thailand (Takaoka and Suzuki, 1984) and later identified as *S.* (*Nevermannia*) *caudisclerum* from Peninsular Malaysia (Takaoka and Choochote, 2004a). After females and males reared from pupae from the type locality were obtained, this species was described as a new species (Takaoka et al., 2012e). This species is distinguished from *S. caudisclerum* in the male by the number of upper-eye (large) facets in 15 or 16 vertical columns and 17 or 18 horizontal rows (14 vertical columns and 15 horizontal rows in *S. caudisclerum*) and relative width of the hind basitarsus against that of the hind femur (1.3–1.4 in this species versus 1.0 in *S. caudisclerum*); and in the pupa by the dorsal surface of abdominal segments 1–3 dark gray (yellow in *S. caudisclerum*). The pupa and larva of this species are similar to those of *S. caudisclerum* in many characteristics including the cocoon with an anterodorsal projection and presence of an accessory sclerite on each side of larval abdominal segment 9.

Adler et al. (2020a) revealed that this species is composed of two cytoforms (A and B).

***Simulium* (*Nevermannia*) *gandakiense* Adler, Takaoka and Ruiz-Arrondo, 2020**

Simulium (*Nevermannia*) *gandakiense* Adler, Takaoka and Ruiz-Arrondo, 2020b: 738–744 (larva).

Distribution. Nepal.

Bionomics. Aquatic habitat: A forest stream (width 2 m, elevation 2962 m) on a trail of Manasulu before Surki between Bimthang and Chauli Kharka, Dudh Khola River valley, Gandaki Zone, Nepal (Adler et al., 2020b).

Remarks. This species was described from larvae in Nepal (Adler et al., 2020b). Its adults and pupa are unknown.

It is characterized by having a pharate pupal gill with four slender filaments, of which one of two filaments of the dorsal pair is slightly thinner than the three other filaments, and an accessory sclerite on each side of the last abdominal segment.

This species was described together with *S. magnacentrum*, which was also placed in the *S. vernum* species-group. It is morphologically distinguished from *S. magnacentrum* by the yellow head capsule and ocherous body (light to medium brown head capsule and dark gray body in the latter species).

This species was also characterized chromosomally.

***Simulium* (*Nevermannia*) *gracile* Datta, 1973**

Simulium (*Eusimulium*) *gracilis* Datta, 1973: 368–371 (female, male, pupa, and larva).

Simulium (*Nevermannia*) *gracile*: Shrestha and Takaoka, 2009: 255.

Distribution. India (West Bengal), China (Tibet) and Nepal.

Bionomics. Aquatic habitats: Trickles at an elevation of 1735–2025 m, Darjeeling (Datta, 1973).

Remarks. This species was described from Darjeeling, India (Datta, 1973). It was later recorded from Nepal (Shrestha and Takaoka, 2009).

It is characterized in the male by the number of upper-eye (large) facets in 15 vertical columns and ca. 16 horizontal rows; and in the pupa by the gill with four slender filaments of the same length and thickness and cocoon with a long bifid anterodorsal projection.

Simulium (Nevermannia) khunklangense Takaoka and Srisuka, 2013

Simulium (Nevermannia) khunklangense Takaoka and Srisuka, 2013 [in Takaoka et al., 2013d]: 281–287 (female, male, pupa, and larva).

Distribution. Thailand.

Bionomics. Aquatic habitat: A small stream (width 0.5 m, depth 5.5 cm, bottom sandy and muddy, 17.5 °C, partially shaded, elevation 1611 m) moderately flowing in a forest, Doi Inthanon National Parks (Takaoka et al., 2013d).

Remarks. This species was described from Doi Inthanon National Parks, Thailand (Takaoka et al., 2013d).

It is similar to *S. chomthongense* from Thailand in many female and larval characteristics but is distinguished in the male by the number of upper-eye (large) facets in 18 vertical columns and 19 or 20 horizontal rows (15 or 16 vertical columns and 17 or 18 horizontal rows in *S. chomthongense*) and relative width of the hind basitarsus against the hind femur (1.1 in this species versus 1.3–1.4 in *S. chomthongense*). The difference in the sequences of the mitochondrial 16S rRNA gene (516 base pairs) between this species and *S. chomthongense* is only two base pairs (Takaoka et al., 2013d). Adler et al. (2020a) chromosomally proved the difference between *S. khunklangense* and *S. chomthongense*.

Simulium (Nevermannia) kurtaki Takaoka and Davies, 1995

Simulium (Nevermannia) kurtaki Takaoka and Davies, 1995a: 94–97 (female, pupa, and larva).

Distribution. Malaysia (Peninsular).

Bionomics. Aquatic habitats: A stream (width 0.2–0.3 m, 17 °C, elevation ca. 1600 m), just south of Gunong Padang Camp, on the slope of Mt. Tahan, and a stream (width 0.6–1.5 m, 18 °C), just north of Gunong Padang Camp (Takaoka and Davies, 1995a).

Remarks. This species was described from Peninsular Malaysia (Takaoka and Davies, 1995a). Its male is unknown.

It is characterized in the female by the median portion of the frons and clypeus bare; in the pupa by the head and thorax without tubercles, gill with four filaments of the same length and thickness in two pairs each short-stalked and arising from a short common basal stalk and cocoon without an anterolateral projection; and in the larva by the postgenal cleft medium-long, ca. 1.5 times as long as the postgenal bridge, and pointed anteriorly. The gill filaments are as long as the pupal body (ca. 2.6 mm). In the original description (Takaoka and Davies, 1995a), the length of the gill filaments was wrongly described as 4.2–5.5 mm.

Simulium (Nevermannia) laichauense Takaoka and Sofian-Azirun, 2017
(Fig. 48A, B)

Simulium (Nevermannia) laichauense Takaoka and Sofian-Azirun [in Takaoka et al., 2017c]: 68–72 (female and pupa).

Distribution. Vietnam.

Bionomics. Aquatic habitat: A fast-flowing river (width 6 m, elevation 1728 m), Sapa, northern Vietnam (Takaoka et al., 2017c).

Remarks. This species was described from a single female reared from a pupa collected from Sapa, Vietnam (Takaoka et al., 2017c). Its male and larva remain unknown.

The pupa of this species is characterized by the frons and anterior portions of the thorax much darkened (Fig. 48A, B), gill with four slender filaments arising from a short common basal stalk and cocoon without an anterodorsal projection.

***Simulium* (*Nevermannia*) *lekhaniense* Takaoka and Shrestha, 2010**
(Fig. 48K, L)

Simulium (*Nevermannia*) *lekhaniense* Takaoka and Shrestha, 2010: 29–32 (pupa and larva).

Distribution. Nepal.

Bionomics. Aquatic habitats: A small water canal (21.0 °C, exposed to the sun, elevation 1606 m) in Baglung and a stream (width 2 m, 10.0 °C, exposed to the sun, elevation 2532 m) in Mustang (Takaoka and Shrestha, 2010).

Remarks. This species was described from a pupa and a larva in Nepal (Takaoka and Shrestha, 2010). Its female and male remain unknown.

The pupa is characterized by the unique shape of the cocoon, which has an elongated anterodorsal projection with an open space on its basal half (Fig. 48K, L), and gill with four slender filaments, of which two filaments of the dorsal pair are longer than the two other filaments of the ventral pair.

***Simulium* (*Nevermannia*) *letense* Takaoka and Shrestha, 2010**
(Figs. 47E and 48C, D, F, J)

Simulium (*Nevermannia*) *letense* Takaoka and Shrestha, 2010: 32–37 (female, pupa, and larva).

Distribution. Nepal.

Bionomics. Aquatic habitats: A very rapidly-flowing stream (width 2 m, 10.0 °C, exposed to the sun, elevation 2532 m), Lete, in Mustang, and a small water canal (21.0 °C, exposed to the sun, elevation 1606 m) in Baglung (Takaoka and Shrestha, 2010).

Remarks. This species was described from Mustang, Nepal (Takaoka and Shrestha, 2010). The male of this species is unknown.

The female of this species is characterized by the claw with a small basal tooth (Fig. 47E), and the pupa is most remarkable in having the thorax with distinct carinae on the dorsal surface (Fig. 48C, D), gill with four short filaments arising from a black common basal stalk (Fig. 48F), and cocoon round and flat, with an anteroventral collar (Fig. 48J) [patellate according to Crosskey (1990)]. Notably, these pupal characteristics are shared by *S. planipuparium* Rubtsov described from Tadzhikistan (Rubtsov, 1956).

Simulium (_Nevermannia_) _magnacentrum_ Adler, Takaoka and Ruiz-Arrondo, 2020

Simulium (*Nevermannia*) *magnacentrum* Adler, Takaoka and Ruiz-Arrondo, 2020b: 744–745 (larva).

Distribution. Nepal.

Bionomics. Aquatic habitat: A forest stream (width 2m, elevation 2962 m) on a trail of Manasulu before Surki between Bimthang and Chauli Kharka, Dudh Khola River valley, Gandaki Zone, Nepal (Adler et al., 2020b).

Remarks. This species was described from larvae in Nepal (Adler et al., 2020b). Its adults and pupa are unknown.

It is characterized by having a pharate pupal gill with four slender filaments, of which one of two filaments of the dorsal pair is slightly thinner than the three other filaments, and an accessory sclerite on each side of the last abdominal segment.

This species was described together with *S. gandakiense*, which is also placed in the *S. vernum* species-group. It is morphologically distinguished from *S. gandakiense* by the light to medium brown head capsule and dark gray body (yellow head capsule and ocherous body in the latter species).

This species was also characterized chromosomally.

Simulium (_Nevermannia_) _purii_ Datta, 1973

Simulium (*Eusimulium*) *purii* Datta, 1973: 371–373 (female, male, pupa, and larva).
Simulium (*Nevermannia*) *purii*: Chen and An, 2003: 211–212 (female, male, pupa, and larva).

Distribution. India (West Bengal), Nepal and China (Tibet).

Bionomics. Aquatic habitats: Trickles at an elevation of 1750–1900 m, Darjeeling (Datta, 1973).

Remarks. This species was described from Darjeeling, India (Datta, 1973) and was recorded from Tibet (Chen and An, 2003) and Nepal (Shrestha and Takaoka, 2009).

It is characterized in the male by the number of upper-eye (large) facets in 15 vertical columns and ca. 15 horizontal rows [though 16 or 17 vertical columns and 17–19 horizontal rows in males in Nepal (Shrestha and Takaoka, 2009)]; and in the pupa by the gill with four slender filaments of the same length and thickness [though one of two filaments of the dorsal pair is slightly longer and thicker than the three others in the pupae in Nepal (Shrestha and Takaoka, 2009)], and cocoon with a long anterodorsal projection.

Simulium (_Nevermannia_) _qiaolaoense_ Chen, 2001

Simulium (*Nevermannia*) *qiaolaoense* Chen, 2001b: 560–566 (male, pupa, and larva).

Distribution. China (Guizhou).

Bionomics. Aquatic habitat: A small stream at an elevation of 1400 m in Mt. Leigong (Chen, 2001b).

Remarks. This species was described from Guizhou, China (Chen, 2001b). Its female is unknown.

It is characterized in the male by the number of upper-eye (large) facets in 12 vertical columns and 16 horizontal rows, and ventral plate saddle-shaped; and in the pupa by the gill with four slender filaments, of which the dorsal filament is somewhat longer and thicker than the three others, and cocoon with a short anterodorsal projection. The accessory sclerite was noted to be marked and moderately sclerotized but probably the term "accessory sclerite" was mistaken as "rectal scales." In fact, the accessory sclerite is absent in Fig. 20 (Chen, 2001b).

***Simulium* (*Nevermannia*) *qingshuiense* Chen, 2001**

Simulium (*Nevermannia*) *qingshuiense* Chen, 2001b: 560–563 (female, male, pupa, and larva).

Distribution. China (Guizhou).

Bionomics. Aquatic habitats: Rivulets in Mt. Leigong (Chen, 2001b).

Remarks. This species was described from Guizhou, China (Chen, 2001b).

It is characterized in the male by the number of upper-eye (large) facets in 16 vertical columns and 16 horizontal rows; and in the pupa by the gill with four slender filaments, of which the dorsal filament is ca. 1.2 times as long as and 1.3 times as thick as the three others, and cocoon with a short anterodorsal projection. The accessory sclerite was noted to be marked and moderately sclerotized but probably the term "accessory sclerite" was mistaken as "rectal scales." In fact, the accessory sclerite is absent in Fig. 11 (Chen, 2001b).

***Simulium* (*Nevermannia*) *silvestre* (Rubtsov, 1956)**

Eusimulium silvestre Rubtsov, 1956: 433–434 (female, male, pupa, and larva).

Simulium (*Nevermannia*) *silvestre*: Crosskey, 1987: 462.

Distribution. China (Sichuan and Yunnan), Finland, Norway, Sweden, Russia, Ukraine, Mongolia, Siberia, Canada, and USA.

Bionomics. Aquatic habitats: Small streams flowing from marshy areas in forests and containing small quantity of organic residue. Females of this species sometimes attack human (Rubtsov, 1956).

Remarks. This species was described from Russia by Rubtsov (1956). It was recorded from Yunnan, China by Xue (1987). There is a possibility that this is the same species as either *S. cangshanense* or *S. cheni*, both of which were described as new species from Yunnan by Xue (1993a).

It is characterized in the pupa by the gill with four filaments of the same length and thickness and cocoon with an elongate anterodorsal projection.

***Simulium* (*Nevermannia*) *taulingense* Takaoka, 1979**

(Fig. 48G, H)

Simulium (*Eusimulium*) *taulingense* Takaoka: 1979: 373–374 (female, male, pupa, and larva).

Distribution. Taiwan.

Bionomics. Aquatic habitat: A small stream at an elevation of 2600 m in Ta U Lin (Takaoka, 1979).

Remarks. This species was described from Taiwan (Takaoka, 1979).

The pupa of this species is remarkable in having the gill composed of an inflated dumbbell-like structure (Fig. 48G). This is the only species in the *S. vernum* species-group in having such a unique shape of the pupal gill.

***Simulium* (*Nevermannia*) *tayense* Takaoka and Ya'cob, 2017**
(Figs. 47H and 48N)

Simulium (*Nevermannia*) *tayense* Takaoka and Ya'cob [in Takaoka et al., 2017c]: 72–79 (female, male, pupa, and larva).

Distribution. Vietnam.

Bionomics. Aquatic habitats: A river (width 6 m, depth 0.6 m, bottom rocky, exposed to the sun, elevation 1728 m), fast running in a forest, and a small stream (width 0.5 m, depth 0.1 m, bottom rocky, 11.0 °C, exposed to the sun, elevation 1750 m), slowly running from a secondary forest, Sapa, northern Vietnam (Takaoka et al., 2017c).

Remarks. This species was described from Sapa, Vietnam (Takaoka et al., 2017c).

This species is characterized by the number of male upper-eye (large) facets in 17 vertical columns and 19 horizontal rows, pupal gill with four slender filaments arising from a medium-long common basal stalk, cocoon with an anterodorsal projection, and larval abdomen with an accessory sclerite on each side of segment 9.

***Simulium* (*Nevermannia*) *thrimshinglaense* Takaoka and Somboon, 2008**
(Figs. 47A, C, D, F, I–R and 48E, I, M, O)

Simulium (*Nevermannia*) *thrimshinglaense* Takaoka and Somboon, 2008: 229–236 (female, male, pupa, and larva).

Distribution. Bhutan.

Bionomics. Aquatic habitats: Two small streams at an elevation of 3720 and 3596 m, Mongar (Takaoka and Somboon, 2008).

Remarks. This species was described from Bhutan (Takaoka and Somboon, 2008).

It is characterized by the bare pupal frons, gill with four filaments, of which the dorsal filament of the dorsal pair somewhat longer than the three other filaments (Fig. 48E), and cocoon with an anterodorsal projection (Fig. 48I).

***Simulium* (*Nevermannia*) *tjibodense* Edwards, 1934**

Simulium latipes Meigen var. *tjibodense* Edwards, 1934: 134 (female, male, pupa, and larva).

Simulium (*Eusimulium*) *tjibodense*: Crosskey, 1973: 424.

Simulium (*Nevermannia*) *tjibodense*: Crosskey, 1987: 462; Takaoka and Davies, 1996: 8–10 (female, male, pupa, and larva).

Distribution. Indonesia (Java).

Bionomics. Aquatic habitat: A small stream in Mt. Gede (elevation 2600 m), Cibodas (Edwards, 1934).

Remarks. This species was described from Cibodas, West Java, Indonesia (Edwards, 1934).

It is characterized by the number of male upper-eye (large) facets in 17 vertical columns and 18 horizontal rows, pupal gill with four slender filaments in two pairs each having a long stalk, and cocoon with an elongate anterodorsal projection.

***Simulium* (*Nevermannia*) *tosariense* Edwards, 1934**

Simulium latipes Meigen var. *tosariense* Edwards, 1934: 132–134 (female, male, pupa, and larva).

Simulium (*Eusimulium*) *tosariense*: Crosskey, 1973: 424; Datta, 1983: 228.

Simulium (*Nevermannia*) *tosariense*: Crosskey, 1987: 462; Takaoka and Davies, 1996: 6–7 (female, male, pupa, and larva).

Distribution. Indonesia (Java).

Bionomics. Unknown.

Remarks. This species was described from Tosari, East Java, Indonesia (Edwards, 1934).

It is characterized by the number of male upper-eye (large) facets in 19 vertical columns and 22 horizontal rows, pupal gill with four slender filaments in two pairs each with a short stalk, and cocoon with an elongate anterodorsal projection.

***Simulium* (*Nevermannia*) *yushangense* Takaoka, 1979**

(Fig. 48P)

Simulium (*Eusimulium*) *yushangense* Takaoka, 1979: 370–373 (male, pupa, and larva).

Distribution. Taiwan.

Bionomics. Aquatic habitat: A small stream at an elevation of 2600 m in Ta U Lin (Takaoka, 1979).

Remarks. This species was described from Taiwan (Takaoka, 1979). Its female remains unknown.

It is characterized by the pupal gill with four slender filaments in two pairs each with a short stalk, of which the dorsal filament of the dorsal pair is longer and thicker than the three other filaments, cocoon with an anterodorsal projection, and larval abdomen with an accessory sclerite on each side of segment 9 (Fig. 48P).

***Simulium* (*Nevermannia*) *zhangjiajiense* Chen, Zhang and Bi, 2004**

Simulium (*Nevermannia*) *zhangjiajiense* Chen, Zhang and Bi, 2004: 367–369 (male, pupa, and larva).

Distribution. China (Hunan).

Bionomics. Aquatic habitat: A small stream at an elevation of 480 m, Hunan Province (Chen et al., 2004).

Remarks. This species was described from Hunan, China (Chen et al., 2004). Its female is unknown.

It is characterized in the male by the number of upper-eye (large) facets in 17 vertical columns and 13 horizontal rows; and in the pupa by the gill with four slender filaments, of which the dorsal filament is ca. 1.3 times as long as the three others; and in the larva by the mandible with a supernumerary serration, and rectal organ compound, with 10–12 secondary lobules per lobe. The accessory sclerite was noted to be marked and moderately sclerotized but probably the term "accessory sclerite" was mistaken as "rectal scales." In fact, the accessory sclerite is absent in Fig. 28 (Chen et al., 2004).

Simulium (_Nevermannia_) sp. A Datta, 1975

Simulium (*Nevermannia*) sp. A Datta, 1975b: 92–95 (pupa and larva).

Distribution. India (West Bengal).

Bionomics. Aquatic habitats: A trickle flowing down the Hill Cart Road at an elevation of 1600 m, and a trickle at an elevation of 2480 m, both in Darjeeling (Datta, 1975b).

Remarks. This unnamed species was described from pupae and larvae from Darjeeling, north India (Datta, 1975b). Its female and male are unknown. The larval mandibular serration composed of a relatively large tooth and a small tooth, and the presence of a supernumerary serration suggest that this species belongs to the *vernum* species-group in the subgenus *Nevermannia*.

It is characterized by the pupal gill with four slender filaments arranged as 2+2 from dorsal to ventral, cocoon simple, without an anterodorsal projection, and larval mandible with a supernumerary serration.

Simulium (_Nevermannia_) Species Unplaced to Species-group

Simulium (_Nevermannia_) sp. C Zhang, Wen and Chen, 1999

Simulium (*Nevermannia*) sp. C Zhang, Wen and Chen, 1999: 232.

Distribution. China (Guizhou).

Bionomics. Unknown.

Remarks. This unnamed species was reported from Guizhou, China by Zhang et al. (1999). However, no information of specimens and their morphological characteristics was given. This species was not treated in the monographs on Chinese black flies by Chen and An (2003) and Chen (2016).

Subgenus **Simulium** *Latreille s. str.*

Simulium (*Simulium*) *argentipes* Species-group

Simulium (*Simulium*) *arboreum* Takaoka, 2006

(Fig. 49B, P)

Simulium (*Simulium*) *arboreum* Takaoka, 2006a: 99–103 (female and pupa).

Distribution. Philippines (Luzon).

Bionomics. Aquatic habitat: A stream (width 2–5 m, 18 °C, exposed to the sun, elevation 1640 m) flowing from a forest at Sumigor Bridge, along the road from Banaue to Abatan, Ifugao Province (Takaoka, 2006a).

Remarks. This species was described from a female reared from a pupa collected from Luzon, the Philippines (Takaoka, 2006a). Its male and larva are unknown.

It is characterized in the female by the darkened legs; and in the pupa by the gill of arborescent type, with 38 thread-like filaments (Fig. 49P) and the absence of the terminal hooks.

Simulium (*Simulium*) *argentipes* Edwards, 1928

Simulium argentipes Edwards, 1928: 62–63 (female).

Simulium (*Simulium*) *argentipes* Edwards: Crosskey, 1973: 425; Takaoka and Davies, 1995a: 146–147 (female).

Distribution. Malaysia (Peninsular).

Bionomics. Aquatic habitat: A stream (width 6 m, depth 0.5 m, 22 °C, exposed to the sun, flow fast, elevation 499 m) at Sungai Lemoi near Ulu Jelai Power Station, Cameron's Highlands, Pahang (Unpublished data).

Remarks. This species was described from females collected from Cameron's Highlands (Edwards, 1928). The female was redescribed based on the holotype specimen and its terminalia and claw were illustrated (Takaoka and Davies, 1995a).

The female of this species is characterized by the darkened legs.

Simulium (*Simulium*) *beludense* Takaoka, 1996

Simulium (*Simulium*) *beludense* Takaoka, 1996b: 157–160 (female and male).

Simulium (*Simulium*) *nigripilosum* (*nec* Edwards) Smart and Clifford, 1969: 18–21 (female, male, pupa, and larva).

Distribution. Malaysia (Sabah and Sarawak).

Bionomics. Aquatic habitats: Streams and rivers at an elevation of 14–60 m in lowlands (Smart and Clifford, 1969).

Remarks. This species was recorded as *S. nigripilosum* Edwards from Sabah (Smart and Clifford, 1969). Later, it was found that this identification was wrong, and subsequently it was described as *S. beludense* based on adult female and male specimens deposited at British Museum (Natural History) (Takaoka, 1996b).

It is characterized by darkened female legs, pupal gill with eight short filaments and cocoon of corbicular type (similar to Fig. 49R, S).

Simulium (_Simulium_) _bidentatum_ (Shiraki, 1935)
Odagmia bidentatum Shiraki, 1935: 34–37 (female).
Gnus bidentatum: Rubtsov, 1959–1964: 441–442.
Simulium (*Gnus*) *bidentatum*: Takaoka, 1976b: 393–397 (female, male, pupa, and larva); Takaoka and Okazawa, 1988: 100.
Simulium (*Odagmia*) *bidentatum*: Ogata and Sasa, 1954: 331.
Simulium bidentatum: Bentinck, 1955: 11 (illustrations of female, male, and pupa).
Simulium (*Simulium*) *bidentatum*: Takaoka and Saito, 2007: 54.

Distribution. Japan (Honshu, Kyushu, Shikoku, and Nansei Islands), China (Fujian, Guizhou, Liaoning, Shandong, Sichuan, and Yunnan), and Korea (South).

Bionomics. Aquatic habitats: Streams and rivers in lowlands in Japan (Takaoka, 1976b).

This is one of the common species in Japan and is a natural vector of *Onchocerca japonica* Uni, Bain and Takaoka, a causative agent of zoonotic onchocerciasis in Japan (Fukuda et al., 2010). The life history from oviposition to adult emergence of this species was studied in the laboratory by Hadi and Takaoka (1995). Larvae of this species were infected with mermithid, microsporidan, and other fungal pathogens in Kyushu, Japan (Takaoka, 1982).

Remarks. This species was described from Japan (Shiraki, 1935). Its male and pupa were illustrated by Bentinck (1955) and its larva was illustrated by Ogata et al. (1956) and Orii et al. (1969). The female, male, pupa, and larva of this species were fully redescribed and illustrated by Takaoka (1976b). This species was recorded from Yunnan, China (Xue, 1987), where a new species, *S. tanae* Xue in the *S. argentipes* species-group, was described by Xue (1992). It is thus suspected that the species formerly regarded as *S. bidentatum* in Yunnan was the same as *S. tanae*.

It is characterized in the female by yellowish legs; and in the male by the scutum covered with brassy short hairs; in the pupa by the presence of small terminal hooks; and in the larva by abdomen with paired protuberances dorsally (similar to Fig. 49W).

Simulium (_Simulium_) _canlaonense_ Delfinado, 1969
Simulium (*Simulium*) *canlaonense* Delfinado, 1969: 200–202 (female).

Distribution. Philippines (Negros).

Bionomics. Unknown.

Remarks. This species was described from a female collected from Negros, the Philippines (Delfinado, 1969). The female was redescribed and illustrated by Takaoka (1983). Its male, pupa, and larva are unknown.

It is characterized in the female by darkened legs.

Simulium (_Simulium_) _haiduanense_ Takaoka, Low and Huang, 2018
(Fig. 49A, C–O, Q, T–V)
Simulium (*Simulium*) *haiduanense* Takaoka, Low and Huang [in Takaoka et al., 2018b]: 885–891 (female, male, pupa, and larva).

Distribution. Taiwan.

Bionomics. Aquatic habitats: A moderately flowing stream (width 2.5 m, 18.0 °C, partially shaded, elevation 681 m) and a river (width 10 m, 18.5 °C, shaded, elevation 850 m) (Takaoka et al., 2018b).

Remarks. This species was described from Taiwan (Takaoka et al., 2018b).

It is characterized in the female by yellowish legs; in the pupa by having small terminal hooks; and in the larva by the abdomen lacking paired dorsal protuberances.

***Simulium* (*Simulium*) *minangkabaum* Takaoka and Sigit, 1997**
(Fig. 49W)

Simulium (Simulium) minangkabaum Takaoka and Sigit, 1997: 76–80 (female, male, pupa, and larva).

Distribution. Indonesia (Sumatra).

Bionomics. Aquatic habitats: Two streams (width 3–5 m, 20 °C, exposed to the sun, elevation 970–1460 m) in hilly cultivated fields (Takaoka and Sigit, 1997).

Remarks. This species was described from Sumatra, Indonesia (Takaoka and Sigit, 1997).

It is characterized in the female by the darkened legs; in the pupa by the gill with 8–10 filaments and the presence of small terminal hooks; and in the larva by abdominal segments 1–5 each with a pair of conical protuberances dorsally (Fig. 49W).

***Simulium* (*Simulium*) *sansahoense* Takaoka and Chen, 2017**
(Fig. 49R, S)

Simulium (*Simulium*) *sansahoense* Takaoka and Chen [in Takaoka et al., 2017e]: 82–89 (female, male, pupa, and larva).

Distribution. Vietnam.

Bionomics. Aquatic habitat: A river (width 6 m, 10 °C, exposed to the sun, elevation 1194 m) moderately flowing in a natural forest, Sapa, Lao Cai Province, northern Vietnam (Takaoka et al., 2017e).

Remarks. This species was described from Vietnam (Takaoka et al., 2017e).

It is characterized in the female by yellowish legs; in the pupa by the presence of small terminal hooks; and in the larva by abdominal segments 1–5 each with a pair of conical protuberances dorsally (similar to Fig. 49W).

***Simulium* (*Simulium*) *tanae* Xue, 1992**

Simulium (*Simulium*) *tanae* Xue, 1992: 93–96 (female, male, pupa, and larva).

Distribution. China (Yunnan).

Bionomics. Aquatic habitats: Rivers at an elevation of 850–2100 m in Yunnan (Xue, 1992).

Remarks. This species was described from Yunnan, south China (Xue, 1992).

This species is similar to *S. bidentatum* in many characteristics including the female yellowish legs but is barely distinguished by the female mandible with 32 inner teeth (26 inner teeth in *S. bidentatum*) and male abdominal segments 2, 6, and 7 each with a pair of shiny dorsolateral patches (abdominal segments 2, 5, 6, and 7 each with such shiny patches in *S. bidentatum*).

Simulium (*Simulium*) *christophersi* Species-group

Simulium (*Simulium*) *atipornae* Takaoka, Srisuka and Choochote, 2014

(Fig. 50A–S)

Simulium (*Simulium*) *atipornae* Takaoka, Srisuka and Choochote [in Takaoka et al., 2014h]: 725–731 (female, male, pupa, and larva).

Distribution. Thailand and Vietnam.

Bionomics. Aquatic habitat: A fast-flowing stream (width 1.6 m, depth 18 cm, bottom gravelly and sandy, 18.0 °C, pH 4.55, shaded by a dense forest canopy, elevation 1550 m), Man-Daeng, Lom Kao District, Phetchabun Province (Takaoka et al., 2014h).

Remarks. This species was described from Thailand (Takaoka et al., 2014h). It was also recorded from Vietnam (Takaoka et al., 2015b).

It is characterized in the female by the cibarium with 73–85 tiny processes near the posterodorsal margin (Fig. 50A); in the male by the large number of upper-eye (large) facets in 21 or 22 vertical columns and 22 or 23 horizontal rows; in the pupa by the integument of the head and the anterior half of the thorax bare, abdominal segments 7 and 8 with spine-combs, segment 9 lacking terminal hooks and cocoon medium brown, wall-pocket-shaped, with one or two large open spaces and few to several small open spaces anterolaterally on each side (Fig. 50P).

Simulium (*Simulium*) *fuscopilosum* Edwards, 1928

Simulium fuscopilosum Edwards, 1928: 63 (female).
Simulium (*Simulium*) *fuscopilosum*: Takaoka and Davies, 1995a: 148–150 (female).

Distribution. Malaysia (Peninsular).

Bionomics. Aquatic habitat: Unknown. Females of this species were recorded to feed human bloods (Takaoka and Davies, 1995a).

Remarks. This species was described from females collected from Cameron's Highlands by Edwards (1928). The female of this species was redescribed by Takaoka and Davies (1995a). Its male, pupa, and larva remain unknown.

The female of this species is characterized by the scutum without longitudinal vittae.

Simulium (*Simulium*) *howletti* Puri, 1932

Simulium (*Simulium*) *howletti* Puri, 1932d: 505–509 (female and male).

Distribution. India (Bihar) and China (Tibet).

Bionomics. Unknown.

Remarks. This species was described from Bihar, India by Puri (1932d). Its pupa and larva remain unknown.

It is characterized in the female by the scutum with three longitudinal vittae, mid tibia pale yellow except its apical one-fourth black and hind tibia brownish yellow except its apical one-fourth black; and in the male by the fore coxa brown and style with its greatest width medially when viewed ventrolaterally.

Simulium (_Simulium_) _rasuli_ Takaoka, Ya'cob and Sofian-Azirun, 2018

Simulium (*Simulium*) *rasuli* Takaoka, Ya'cob and Sofian-Azirun [in Ya'cob et al., 2018]: 1–3 (female).

Distribution. Malaysia (Peninsular).

Bionomics. Unknown.

Remarks. This species was described from females collected by a Malaise trap from Genting Highland, Peninsular Malaysia by Ya'cob et al. (2018). Its male, pupa, and larva remain unknown.

The female of this species is distinguished from that of *S. fuscopilosum* by the scutum with longitudinal vittae and entirely darkened femora and tibiae.

Simulium (_Simulium_) sp. C Takaoka and Davies, 1995

Simulium (*Simulium*) sp. C Takaoka and Davies, 1995a: 157–158 (pupa).

Distribution. Malaysia (Peninsular).

Bionomics. Aquatic habitat: A stream on the southern slope of Mt. Tahan, Pahang (Takaoka and Davies, 1995a).

Remarks. *Simulium* (*S.*) sp. C was described from a pupal exuviae and its associated cocoon from Peninsular Malaysia (Takaoka and Davies, 1995a). Takaoka et al. (2014h) suggested that *S.* (*S.*) sp. C is the pupa of *S. atipornae* or *S. fuscopilosum* or other related species.

The pupa of this species is characterized by the gill with six thread-like filaments, absence of terminal hooks and cocoon wall-pocket-shaped, with large anterolateral open spaces on each side.

Simulium (_Simulium_) _chungi_ Species-group

Simulium (_Simulium_) _chungi_ Takaoka and Huang, 2006

(Fig. 51A–Q)

Simulium (*Simulium*) *chungi* Takaoka and Huang, 2006: 219–227 (female, male, pupa, and larva); Takaoka and Huang, 2018: 149.

Distribution. Taiwan and Vietnam.

Bionomics. Aquatic habitats: A mountain river (width 8–10 m, 14.3 °C, elevation 50–60 m) and a mountain stream (width 2–4 m, 14 °C, elevation 600 m) in Yilan County, and Yunei River, Yisheng Village, Fuxing District, Taoyuang City (Takaoka and Huang, 2006).

Remarks. This species was described from Taiwan (Takaoka and Huang, 2006). It was recorded from Vietnam based on adult females attracted to a human (Takaoka et al., 2014c). Some morphological features of the male of this species show intraspecific variation (Takaoka and Huang, 2018).

It is characterized in the female by the cibarium without a median projection, scutum with five longitudinal vittae, claw simple, ovipositor valve with a narrow tail-like tip posteromedially (Fig. 51B); in the male by the style with a long basal protuberance (Fig. 51I) and ventral plate with its body rounded posteriorly when viewed ventrally (Fig. 51H), lacking teeth on its posterior surface (Fig. 51K); in the pupa by the gill with six filaments (Fig. 51L) and abdominal segment 9 with a pair of conical terminal hooks.

***Simulium* (*Simulium*) *taipokauense* Takaoka, Davies and Dudgeon, 1995**

Simulium (*Simulium*) *taipokauense* Takaoka, Davies and Dudgeon, 1995: 194–195 (male).

Distribution. China (Hong Kong).

Bionomics. Unknown.

Remarks. This species was described from adult males caught by light traps in Hong Kong (Takaoka et al., 1995). Its female, pupa, and larva are unknown.

The male of this species is similar to that of *S. chungi* in many characteristics including the genitalia but is distinguished by the mid femur and tibia entirely yellow.

Simulium (*Simulium*) *crocinum* Species-group

***Simulium* (*Simulium*) *crocinum* Takaoka and Choochote, 2004**

(Figs. 52F and 53H)

Simulium (*Simulium*) *crocinum* Takaoka and Choochote, 2004c: 319–324 (female, male, pupa, and larva).

Distribution. Thailand.

Bionomics. Aquatic habitat: A mountain stream (width 0.5 m, 18.5 °C, elevation ca. 1500 m) in Doi Inthanon National Parks (Takaoka and Choochote, 2004c).

Remarks. This species was described from Thailand (Takaoka and Choochote, 2004c).

It is characterized by the pupal thorax bare on the anterior half except the small raised portion at the base of the gill with tubercles (Fig. 53H), pupal gill with six filaments, of which two filaments of the ventral pair are much thinner than the four other filaments (Fig. 53H) and cocoon thickly woven, with an anterolateral open space on each side.

***Simulium* (*Simulium*) *ephippioidum* Chen and Wen, 1999**

Simulium (*Simulium*) *ephippioidum* Chen and Wen, 1999: 436–438 (female, male, pupa, and larva).

Distribution. China (Guizhou and Hubei).

Bionomics. Aquatic habitat: A rapid stream in Mt. Fanjing (Chen and Wen, 1999).

Remarks. This species was described from Guizhou, China (Chen and Wen, 1999).

It is characterized in the female by the cibarium with a cluster of minute processes, and basal section of the radius fully haired; in the male by the number of upper-eye (large) facets in 15 vertical columns and 16 horizontal rows and ventral plate with its body saddle-shaped; in the pupa by the gill with six slender filaments of almost the same thickness and last abdominal segment with a pair of terminal hooks.

The cocoon was noted as "boot-shaped" but it is "wall-packet-shaped" in Fig. 1 (Chen and Wen, 1999).

***Simulium* (*Simulium*) *fuzhouense* Zang and Wang, 1991**

Simulium (*Simulium*) *fuzhouense* Zang and Wang, 1991b: 109–112 (female, male, pupa, and larva).

Distribution. China (Fujian and Guizhou).

Bionomics. Aquatic habitat: A small stream in the suburb of Fuzhou City (Zhang and Wang 1991b).

Remarks. This species was described from Fujian, China by Zhang and Wang (1991b).

It is similar to *S. ufengense* from Taiwan (Takaoka, 1979) but is barely distinguished by the presence of spines on the anterior surface of the basal protuberance in the male style and pupal frons covered with tubercles.

***Simulium* (*Simulium*) *laocaiense* Takaoka and Lau, 2017**

(Fig. 53B, K, N)

Simulium (*Simulium*) *laocaiense* Takaoka and Lau [in Takaoka et al., 2017c]: 90–95 (female, pupa, and larva).

Distribution. Vietnam.

Bionomics. Aquatic habitat: A stream (width 0.5–1.0 m, bottom rocky, 9.0 °C, partially shaded, elevation 1680 m) slowly flowing from a secondary forest, Oguy Ho, Sapa, Lao Cai Province (Takaoka et al., 2017c).

Remarks. This species was described from northern Vietnam (Takaoka et al., 2017c). Its male remains unknown.

It is characterized in the pupa by the frons and thorax moderately covered with tubercles, gill with six dark filaments, of which three outer filaments are somewhat longer than their counter inner filaments (Fig. 53B) and cocoon wall-pocket-shaped, with an anterodorsal portion somewhat produced forwardly (Fig. 53K).

***Simulium* (*Simulium*) *ledongense* Yang and Chen, 2001**

Simulium (*Simulium*) *ledongense* Yang and Chen, 2001: 90–93 (female, male, pupa, and larva).

Distribution. China (Hainan).

Bionomics. Aquatic habitat: A small stream at an elevation of 600 m flowing from Mt. Jianfen, Hainan (Yang and Chen, 2001).

Remarks. This species was described from Hainan, China (Yang and Chen, 2001).

It is characterized in the female by the haired basal portion of the radium; in the male by the ventral plate with its body saddle-shaped; and in the pupa by the gill with six slender filaments, of which two filaments of the ventral pair are much thinner than the four other filaments.

***Simulium* (*Simulium*) *mediocoloratum* Takaoka and Choochote, 2004**
(Fig. 53E)

Simulium (*Simulium*) *mediocoloratum* Takaoka and Choochote, 2004c: 314–319 (female, male, pupa, and larva).

Distribution. Thailand.

Bionomics. Aquatic habitats: A cascading stream of Monthathan Waterfall (11.7 °C, elevation ca. 750 m) and a stream of Siribhume Waterfall (width 0.5 m, 18.5 °C, elevation ca. 1500 m) (Takaoka and Choochote, 2004c).

Remarks. This species was described from Thailand (Takaoka and Choochote, 2004c).

It is characterized by the pupal thorax bare on the anterior half except the small raised portion at the base of the gill with tubercles (Fig. 53E), gill with six filaments arranged as 2+2+2 from dorsal to ventral, which are subequal in length and thickness to one another, though all three inner filaments are somewhat shorter than the three outer filaments (Fig. 53E), and cocoon thinly woven, probably with a large anterolateral open space on each side.

***Simulium* (*Simulium*) *parawaterfallum* Zhang, Yang and Chen, 2003**

Simulium (*Simulium*) *parawaterfallum* Zhang, Yang and Chen, 2003: 742–744 (female, male, pupa, and larva).

Distribution. China (Hainan).

Bionomics. Aquatic habitat: A large rock in a waterfall (Zhang et al., 2003).

Remarks. This species was described from Hainan, China (Zhang et al., 2003).

It is similar to *S. waterfallum* from Hainan (Zhang et al., 2003) but is distinguished in the female by the genital fork without a lateral projection on each arm; in the male by the horn-like protuberance of the style lacking spines along its anterior margin; in the pupa by the dorsum of abdominal segment 1 lacking tubercles and cocoon without an anterolateral open space on each side.

***Simulium* (*Simulium*) *ramulosum* Chen, 2000**

Simulium (*Simulium*) *ramulosum* Chen, 2000: 103–105 (female, pupa, and larva).

Distribution. China (Guizhou).

Bionomics. Aquatic habitat: A trickle in Mt. Leigong (elevation 800–1000 m) (Chen, 2000).

Remarks. This species was described from Guizhou, China (Chen, 2000). Its male is unknown. It was tentatively transferred from *Simulium* s. str. species unplaced to species-group to the *S. crocinum* species-group because of the female scutum without any pattern, cibarium unarmed, and pupal abdominal segment 9 without terminal hooks (Takaoka, 2017a). It should be confirmed on whether the ovipositor valves are narrowly transparent along its inner and posterior margins, one of the diagnostic characteristics of the *S. crocinum* species-group.

It is characterized in the female by the sensory vesicle medium-long, 0.4 times as long as the third palpal segment, and basal section of the radius bare; in the pupa by the thoracic integument with trichomes each with 15–22 branches and gill with six slender filaments, of which the outer filament of the dorsal pair is somewhat longer and much thicker than the five other filaments; and in the larva by the rectal organ with about eight finger-like secondary lobules per lobe.

The cocoon was noted to be "boot-shaped" but it is "wall-pocket-shaped" in Fig. 11 (Chen, 2000). The female mandible was noted to lack outer teeth, a characteristic very rarely occurring in the subgenus *Simulium* s. str.

***Simulium* (*Simulium*) *rudnicki* Takaoka and Davies, 1995**

(Figs. 52K and 53A, M)

Simulium (*Simulium*) *rudnicki* Takaoka and Davies, 1995a: 155 (female); Takaoka and Choochote, 2004c: 311–314 (male, pupa, and larva).

Distribution. Malaysia (Peninsular) and Thailand.

Bionomics. Aquatic habitat: A stream of Mae Klang Waterfall (width 5 m, 19.5 °C, elevation ca. 400 m) in Chiang Mai Province (Takaoka and Choochote, 2004c).

Remarks. This species was described from a female caught by an aspirator while flying around a human in Langkawi Island, Peninsular Malaysia (Takaoka and Davies, 1995a). Its male, pupa, and larva were described based on specimens collected from Thailand (Takaoka and Choochote, 2004c).

It is characterized in the pupa by the thorax entirely bare on its anterior half, gill with six filaments arranged as 2+2+2 from dorsal to ventral, which are subequal in thickness and length, though the inner filament of the ventral pair is somewhat thinner than the five other filaments (Fig. 53A), and cocoon transparent, thin, film-like, with a large anterolateral open space on each side.

***Simulium* (*Simulium*) *serenum* Huang and Takaoka, 2009**

(Fig. 53G, L)

Simulium (*Simulium*) *serenum* Huang and Takaoka, 2009a: 33–37 (female and pupa); Takaoka and Huang, 2018: 147–149 (male).

Distribution. Taiwan.

Bionomics. Aquatic habitats: A small stream (width 0.5 m, 21 °C, shaded, elevation 761 m) in Nantou County, and a slow-flowing stream (width 2 m, 20.5 °C, shaded, elevation ca. 40 m) in Hualien County (Huang and Takaoka, 2009a).

Remarks. This species was described from Taiwan and placed in the *S. griseifrons* species-group (Huang and Takaoka, 2009a). It was later transferred to the *S. crocinum* species-group (Takaoka, 2017a). The male of this species was described by Takaoka and Huang (2018).

The pupa of *S. serenum* is characterized by the cocoon bearing a deep anterodorsal cleft medially and a large anterolateral open space on each side (Fig. 53L).

Simulium (Simulium) shangchuanense An and Hao, 1998

Simulium (*Simulium*) *shangchuanense* An and Hao [in An et al., 1998]: 187–190 (female, male, pupa, and larva).

Distribution. China (Guangdong and Shangchuan Island).

Bionomics. Aquatic habitats: Streams (width 2–5 m, depth 0.1–0.5 m, elevation 20 m) (An et al., 1998).

Remarks. This species was described from Guangdong, China (An et al., 1998). This species was placed in the *S. crocinum* species-group by the female scutum without longitudinal vittae, male ventral plate with its body flat, without any teeth, and style with a basal protuberance (Takaoka, 2017a). However, it should be noted that the ovipositor valves, which are short, close together in the middle, and with several short hairs only near their inner margins, are quite different from those of most species of the *S. crocinum* species-group.

It is characterized in the female by the basal portion of the radius with a few hairs and sternite 8 with its posterior margin slightly concave; and in the pupa by the gill with six filaments decreasing in thickness from dorsal to ventral.

Simulium (Simulium) taichungense Takaoka and Huang, 2018

Simulium (*Simulium*) *taichungense* Takaoka and Huang, 2018: 142–147 (female, male, and pupa).

Distribution. Taiwan.

Bionomics. Aquatic habitat: A moderately flowing stream (width 2 m, 20.5 °C, partially shaded, elevation 958 m) in Heping District, Taichung City (Takaoka and Huang, 2018).

Remarks. This species was described from Taiwan (Takaoka and Huang, 2018). Its larva is unknown.

It is characterized by the pupal gill with six unpigmented filaments and pupal abdomen with distinct spine-combs on the dorsal surface of segments 7–9, by which it is easily distinguished from all the 21 species (except *S. vietnamense* from Vietnam, of which the pupa is unknown) in the *S. crocinum* species-group including *S. serenum* and *S. ufengense*, both from Taiwan (Huang and Takaoka, 2009b; Takaoka, 1979).

Simulium (Simulium) tenuatum Chen, 2000

Simulium (*Simulium*) *tenuatum* Chen, 2000: 100–103 (female, male, pupa, and larva).

Distribution. China (Guizhou).

Bionomics. Aquatic habitat: A small, shaded stream in Leigong Mountain Nature Reserve (elevation 600–800 m) (Chen, 2000).

Remarks. This species was described from Guizhou, China (Chen, 2000).

It is characterized in the female by the basal section of the radius fully haired; in the male by the number of upper-eye (large) facets in 14 vertical columns and 16 horizontal rows; and in the pupa by the gill with six slender filaments, of which the upper filament of the dorsal pair is slightly thicker than the five other filaments, and last abdominal segment with a pair of terminal hooks.

The cocoon was noted as "boot-shaped" but it is "wall-packet-shaped" in Fig. 5 (Chen, 2000).

Simulium (_Simulium_) _turgidum_ Takaoka and Pham, 2017
(Fig. 53D, J, O)

Simulium (*Simulium*) *turgidum* Takaoka and Pham [in Takaoka et al., 2017c]: 96–101 (female, pupa, and larva).

Distribution. Vietnam.

Bionomics. Aquatic habitats: A stream (width 0.5–2.5 m, bottom rocky, 10.0 °C, exposed to the sun, elevation 1812 m) moderately flowing in a forest, and a stream (width 0.5 m, bottom rocky, 11.0 °C, exposed to the sun, elevation 1750 m) moderately flowing in a secondary forest, Sapa, Lao Cai Province (Takaoka et al., 2017c).

Remarks. This species was described from Vietnam (Takaoka et al., 2017c). Its male is unknown.

It is characterized by the pupal gill with six dark inflated filaments (Fig. 53D) and cocoon with an anterolateral open space on each side (Fig. 53J).

This species is similar to *S. tumidilfilum* and *S. waterfallum* from China in having the pupal gill with inflated filaments and cocoon with an anterolateral open space on each side but is distinguished by the female ovipositor valve, which is triangular in this species but tongue-like in the latter two species (Takaoka et al., 2017c).

Simulium (_Simulium_) _ufengense_ Takaoka, 1979
(Fig. 53C)

Simulium (*Simulium*) *ufengense* Takaoka, 1979: 390–392 (male and pupa); Huang and Takaoka, 2009b: 91–95 (female and larva).

Distribution. Taiwan.

Bionomics. Aquatic habitat: A shallow stream (width 6 m, shaded) near the basin of a waterfall in a lowland, Wu Feng, Chao Chi (Takaoka, 1979).

Remarks. This species was described from Taiwan (Takaoka, 1979). Its female and larva were later described by Huang and Takaoka (2009b). It was placed in the *S. griseifrons* species-group by Takaoka and Davies (1995a) but was later transferred to the *S. crocinum* species-group (Takaoka, 2017a).

It is characterized by the pupal gill with six short and slender filaments arranged as 2+2+2 from dorsal to ventral, of which two inner filaments of the dorsal and middle pairs are directed inward, whereas the four other filaments are directed forward (Fig. 53C), pupal abdomen with spine-combs on segment 8 and cocoon wall-pocket-shaped, without an anterolateral open space on each side.

***Simulium* (*Simulium*) *vietnamense* Takaoka, Sofian-Azirun and Chen, 2014**
Simulium (*Simulium*) *vietnamense* Takaoka, Sofian-Azirun and Chen [in Takaoka et al., 2014c]: 743–746 (female).

Distribution. Vietnam.
Bionomics. Unknown.
Remarks. This species was described from two females collected by a sweeping net while flying around a human in Tam Dao National Parks, Vinh Phuc Province (Takaoka et al., 2014c). The male, pupa, and larva of this species are unknown.
The female of this species is characterized by the haired basal portion of the radial vein.

***Simulium* (*Simulium*) *waterfallum* Zhang, Yang and Chen, 2003**
Simulium (*Simulium*) *waterfallum* Zhang, Yang and Chen, 2003: 741–742 (female, male, pupa, and larva).

Distribution. China (Hainan).
Bionomics. Aquatic habitat: A large rock adjacent to a waterfall, at an elevation of 600 m (Zhang et al., 2003).
Remarks. This species was described from Hainan, China (Zhang et al., 2003).
It is characterized in the male by the number of upper-eye (large) facets in 13 vertical columns and 15 horizontal rows; in the pupa by the gill with six inflated filaments, dorsum of abdominal segment 1 densely covered with tubercles and cocoon shoe-shaped, with a small anterolateral open space on each side.

***Simulium* (*Simulium*) *wulingense* Zhang and Chen, 2000**
Simulium (*Simulium*) *wulingense* Zhang and Chen, 2000: 221–223 (female, male, pupa, and larva).

Distribution. China (Guizhou).
Bionomics. Aquatic habitat: a stream at an elevation of 700 m (Zhang and Chen 2000).
Remarks. This species was described from Guizhou, China (Zhang and Chen 2000).
This species is similar to *S. fuzhouense* but is distinguished in the pupa by the dorsum of abdominal segment 6 with spine-combs (without spine-combs in the latter species), rectal organ with eight or nine secondary lobules per lobe (15–20 secondary lobules in the latter species), and posterior circlet with 74–78 rows of up to 10–12 hooklets per row (110–115 rows of up to 13 or 14 hooklets per row in the latter species).

***Simulium* (*Simulium*) *yongi* Takaoka and Davies, 1997**
(Figs. 52A–E, G–J, L–Q and 53I)
Simulium (*Simulium*) *yongi* Takaoka and Davies, 1997: 11–16 (female, male, pupa, and larva).

Distribution. Malaysia (Peninsular) and Thailand.

Bionomics. Aquatic habitat: A cascading stream (elevation ca. 600 m) in Cameron's Highlands (Takaoka and Davies, 1997).

Remarks. This species was described from Cameron's Highlands (Takaoka and Davies, 1997). The pupa and larva of this species were wrongly associated with the female of *S. pahangense* (Takaoka and Davies, 1995a). This species was also recorded from Thailand (Kuvangkadilok and Takaoka, 2000).

This species is characterized in the female by the basal portion of the radius haired on its apical one-fourth to two-thirds; and in the pupa by the cocoon formed of an almost transparent thin membrane and with a large anterodorsal incision (Fig. 53I).

***Simulium* (*Simulium*) *yukawai* Takaoka and Pham, 2019**

(Fig. 53F)

Simulium (*Simulium*) *yukawai* Takaoka and Pham [in Takaoka et al., 2019h]: 246–251 (female and pupa).

Distribution. Vietnam.

Bionomics. Aquatic habitat: A small stream (width 0.5 m, bottom rocky, 11.0 °C, exposed to the sun, elevation 1750 m) moderately flowing in a secondary forest, in Sapa (Takaoka et al., 2019h).

Remarks. This species was described from Vietnam (Takaoka et al., 2019h). Its male and larva remain unknown.

It is characterized in the female by having all the femora yellow with their apical caps light brown and the paraproct covered with 48–50 distinct hairs on its ventral and lateral surfaces; and in the pupa by the frons and most of the thorax bare, gill with six dark slender filaments in three pairs closely arising from the base, of which two outer filaments of the dorsal and middle pairs are much longer than the four other filaments (Fig. 53F), abdomen with distinct spine-combs on the dorsal surface of segments 7–9, and cocoon with a large anterolateral open space on each side.

This species is distinguished from all the three other species of the *S. crocinum* species-group recorded from Vietnam, *S. laocaiense*, *S. turgidum* and *S. vietnamense*, by the female fore and mid femora yellow except the apical caps darkened (entirely yellowish in the three known species); and from *S. laocaiense* by the cocoon with a large anterolateral open space on each side (no open space in *S. laocaiense*) and from *S. turgidum* by the pupal gill filaments slender (inflated in *S. turgidum*).

***Simulium* (*Simulium*) *dumogaense* Species-group**

***Simulium* (*Simulium*) *contractum* Takaoka, 2003**

(Figs. 55A, H, I and 56E)

Simulium (*Simulium*) *contractum* Takaoka, 2003: 539–541 (pupa); Takaoka et al., 2020g: 684–689 (female, male, and larva).

Distribution. Indonesia (Sulawesi).

Bionomics. Aquatic habitat: A slow-flowing lowland stream (width 0.5–1.0 m, 27.0 °C, exposed to the sun, elevation 390 m) in South Sulawesi (Takaoka, 2003).

Remarks. This species was described from Sulawesi, Indonesia by Takaoka (2003). Its female, male, and larva were described by Takaoka et al. (2020g).

It is characterized in the pupa by the head densely covered with round tubercles, gill with six filaments each with somewhat narrowed portion basally (Fig. 56E) and abdominal segments 6 and 7 each lacking an outer hook ventrally on each side.

***Simulium* (*Simulium*) *dumogaense* Takaoka and Roberts, 1988**

(Figs. 54B, F, H–J, 55C–G, Q–U and 56A, F)

Simulium (*Simulium*) *dumogaense* Takaoka and Roberts, 1988: 212–214 (female, male, pupa, and larva); Takaoka, 2003: 453–461 (female, male, pupa, and larva).

Distribution. Indonesia (Sulawesi).

Bionomics Aquatic habitats: Small streams (width 0.1–2.0 m) at various elevations ranging from 500 to 1500 m (Takaoka and Roberts, 1988).

Remarks. This species was described from Sulawesi, Indonesia by Takaoka and Roberts (1988) and redescribed by Takaoka (2003).

It is characterized in the female by the claw with a small subbasal tooth (Fig. 54F); in the male by the scutum with polymorphic color patterns (Fig. 55C–G); in the pupa by the frons densely and elaborately covered with round tubercles and two branched trichomes arising close together on each side (Fig. 56A) and gill with six short filaments (Fig. 56F); and in the larva by the postgenal cleft bullet-shaped.

***Simulium* (*Simulium*) *fuscitarse* Takaoka, 2003**

(Fig. 56J, Q)

Simulium (*Simulium*) *fuscitarse* Takaoka, 2003: 461–466 (female, pupa, and larva).

Distribution. Indonesia (Sulawesi).

Bionomics. Aquatic habitats: A mountain stream (width 2.0–5.0 m, bottom with pebbles, flow fast, 19.0 °C, elevation 1000 m) in Central Sulawesi, and a forest river (width 10–15 m, bottom rocky, with many stones, 20.0 °C, flow fast, shaded, elevation ca. 500 m) in South Sulawesi (Takaoka, 2003).

Remarks. This species was described from Sulawesi, Indonesia by Takaoka (2003). Its male is unknown.

It is characterized in the female by the scutum without pattern, basal portion of the radial vein bare, hind basitarsus entirely darkened, claw with a small basal tooth; in the pupa by the head densely covered with round tubercles, abdominal segment 5 with a pair of bifid hooks on each side and abdominal segments 6 and 7 each with a pair of bifid inner and unbranched outer hooks ventrally on each side; and in the larva by the postgenal cleft medium-long and triangular (Fig. 56Q).

***Simulium* (*Simulium*) *landonoense* Takaoka, 2003**

(Fig. 56I)

Simulium (*Simulium*) *landonoense* Takaoka, 2003: 481–484 (female, male, pupa, and larva).

Distribution. Indonesia (Sulawesi).

Bionomics. Aquatic habitats: Various types of water courses from a small ditch ca. 0.5 m wide to a river ca. 20 m wide, in lowland areas with an elevation of 100–220 m in South Sulawesi (Takaoka, 2003).

Remarks. This species was described from Sulawesi, Indonesia by Takaoka (2003).

It is characterized in the pupa by the gill with eight thread-like filaments (Fig. 56I). Apart from the number of gill filaments, this species is indistinguishable from *S. tumpaense*.

***Simulium* (*Simulium*) *mediodentatum* Takaoka, 2003**

(Figs. 54C, N–P and 56C, K, P, N, O)

Simulium (*Simulium*) *mediodentatum* Takaoka, 2003: 448–453 (female, pupa, and larva).

Distribution. Indonesia (Sulawesi).

Bionomics. Aquatic habitat: A slow-flowing mountain stream (width 3 m, bottom with sands, 17.0 °C, partially shaded, elevation 1520 m) in Central Sulawesi (Takaoka, 2003).

Remarks. This species was described from Sulawesi, Indonesia by Takaoka (2003). Its male is unknown.

It is characterized in the female by the sensory vesicle with a large opening, scutum without pattern, basal portion of the radial vein bare, claw with a sharply-pointed medium-sized basal tooth; in the pupa by the frons with three pairs of trichomes having two to four branches (Fig. 56C), and moderately covered with round tubercles bearing tiny secondary projections, gill with six filaments forwarded close together, and cocoon shoe-shaped and compactly woven (Fig. 56K); and in the larva by the postgenal cleft medium-long and triangular.

***Simulium* (*Simulium*) *nudifrons* Takaoka, 2003**

(Fig. 56H, M)

Simulium (*Simulium*) *nudifrons* Takaoka, 2003: 442–448 (female, male, pupa, and larva).

Distribution. Indonesia (Sulawesi).

Bionomics. Aquatic habitats: A forest stream (width 5 m, bottom rocky, 23.0 °C, shaded, elevation 140 m) in southeast Sulawesi, a small stream (width 2–3 m, bottom sandy, shaded, 20.0 °C, elevation 480 m) in South Sulawesi, and a small stream (width 2–3 m, bottom rocky, 20.0 °C, shaded, elevation 980 m) in South Sulawesi (Takaoka, 2003).

Remarks. This species was described from Sulawesi, Indonesia by Takaoka (2003).

It is characterized in the female by the scutum with color patterns, basal portion of the female radial vein partially haired, and claw with a sharply-pointed medium-sized basal tooth; in the pupa by the frons bare and with two unbranched trichomes on each side widely separated from each other, gill with six short filaments (0.6–0.9 mm long) (Fig. 56H), abdomen with stout hooked spines dorsally on segments 2–4, and cocoon shoe-shaped and compactly woven; and in the larva by the postgenal cleft bullet-shaped.

***Simulium* (*Simulium*) *tomohonense* Takaoka, 2003**
(Figs. 54A, D, G, Q–S, 55K, L–P and 56B, D, L)
Simulium (*Simulium*) *tomohonense* Takaoka, 2003: 431–437 (female, male, pupa, and larva).

Distribution. Indonesia (Sulawesi).

Bionomics. Aquatic habitats: A forest stream (width 1 m, 24.0 °C, bottom with sands and pebbles, shaded, elevation ca. 1000 m) and a river (width 5 m, 23.0 °C, bottom rocky, partially shaded, elevation 440 m) in North Sulawesi; and a small stream (width 1–2 m, bottom sandy, 22.0 °C, elevation 860 m) in South Sulawesi (Takaoka, 2003).

Remarks. This species was described from Sulawesi, Indonesia by Takaoka (2003).

It is characterized in the female and male by the hind basitarsi entirely darkened (Fig. 55K); in the female by the basal portion of the radial vein partially haired, and claw with a sharply-pointed medium-sized basal tooth (Fig. 54G); in the pupa by the head with two frontal trichomes widely separated from each other on each side (Fig. 56B), abdomen with stout hooked spines dorsally on segments 2–4 and cocoon boot-shaped and roughly woven (Fig. 56L); and in the larva by the head capsule darkened.

***Simulium* (*Simulium*) *tumpaense* Takaoka and Roberts, 1988**
(Figs. 54E, 55B, J, V–Z and 56G, R)
Simulium (*Simulium*) *tumpaense* Takaoka and Roberts, 1988: 215–218 (female, male, pupa, and larva); Takaoka, 2003: 472–481 (female, male, pupa, and larva).

Distribution. Indonesia (Sulawesi).

Bionomics. Aquatic habitats: Various types of water courses from a stream with width of less than 1.0 m to a river ca. 15 m wide, at various elevations ranging from 100 to 1450 m (Takaoka and Roberts, 1988).

Remarks. This species was described from Sulawesi, Indonesia by Takaoka and Roberts (1988) and redescribed by Takaoka (2003).

It is characterized by the female claw without tooth (Fig. 54E), pupal gill with six filaments widely divergent when viewed laterally (Fig. 56G), pupal abdominal segments 6 and 7 each with an inner hook only and lacking an outer hook on each side and larval postgenal cleft large and rounded (Fig. 56R).

***Simulium* (*Simulium*) *watanabei* Takaoka, 2003**
(Figs. 54K–M and 55A'–D')
Simulium (*Simulium*) *watanabei* Takaoka, 2003: 466–472 (female, male, pupa, and larva).

Distribution. Indonesia (Sulawesi).

Bionomics. Aquatic habitats: Three fast-flowing forest streams (width 1.0–3.0 m, bottom rocky or with sands, 20.0–22.0 °C, elevation 480–980 m) in South Sulawesi, and a moderately-flowing forest stream (width 2.0–3.0 m, bottom with sands, 20.0 °C, shaded, elevation 480 m) in South Sulawesi; and a slow-flowing stream (width 3 m, bottom with sands and pebbles, 17.0 °C, partially shaded, elevation 1520 m) in Central Sulawesi (Takaoka, 2003).

Remarks. This species was described from Sulawesi, Indonesia by Takaoka (2003).

It is characterized in the female by the scutum without pattern, basal portion of the radial vein bare, claw with a small subbasal tooth; in the male by the hind basitarsus nearly parallel-sided; and in the pupa by the head almost bare on its dorsal half.

Simulium (*Simulium*) *seramense* Takaoka, 2003

Simulium (*Simulium*) *seramense* Takaoka, 2003: 437–441 (female, pupa, and larva).

Distribution. Indonesia (Seram).

Bionomics. Aquatic habitat: A river (width ca. 10 m, depth 0.3–0.5 m, 23.0 °C, bottom with stones, shaded, elevation ca. 20 m), just downstream of the waterwall, Rumahkai, near Tifulae, West Seram (Takaoka, 2003).

Remarks. *Simulium seramense* was described from Seram, Indonesia by Takaoka (2003). Its male is unknown. This is the only species of not only the *S. dumogaense* species-group but also the subgenus *Simulium* s. str. recorded from the Australasian Region.

This species is similar to *S. tomohonense* in having the female and male basitarsi entirely darkened, female scutum with patterns formed by grayish-white pruinose and non-pruinose black areas, female claw with a sharply pointed medium-long tooth, basal portion of the female radial vein partially haired and cocoon boot-shaped, coarsely woven, with small open spaces anteriorly. However, it is distinguished from the latter species by the following characteristics (those of *S. tomohonense* in parentheses): in the female by the ovipositor valve almost entirely covered with stout hairs (covered with hairs on the basal half and bare on the apical half); in the pupa by the thorax with cone-shaped and round tubercles (with round tubercles only), frontal trichomes simple or bifid (bifid to quadrifid) and six gill filaments short, 1.2–1.5 mm long (1.5–1.8 mm long); and in the larva by the posterior circlet with 152–162 rows of up to 22 hooklets per row (130 rows of up to 19 hooklets per row).

Simulium (*Simulium*) *eximium* Species-group

Simulium (*Simulium*) *eximium* De Meijere, 1913

(Fig. 57B–D, L–N, Q, T, U)

Simulium eximium De Meijere, 1913: 330 (female and male).

Simulium (*Simulium*) *eximium* De Meijere: Edwards, 1934: 104–106 (female, male, pupa, and larva); Takaoka and Davies, 1996: 48–54 (female, male, pupa, and larva).

Distribution. Indonesia (Bali, Flores, Java, Lombok, and Sumatra).

Bionomics. Aquatic habitats: Rapids of fast-flowing streams (width 0.5–1.0 m, shaded) in natural forests in Cibodas Botanic Garden and in Puncak, West Java (Takaoka and Davies, 1996).

Remarks. This species was described by De Meijere (1913). Its pupa and larva were described by Edwards (1934), and its female, male, pupa, and larva were redescribed by Takaoka and Davies (1996).

It is characterized in the female by the cibarium with ca. 20 tiny processes and paraproct having a ridge directed ventrally (Fig. 57C); in the male by the variable number of upper-eye (large) facets in 14–19 vertical columns and 17–23 horizontal rows; and in the pupa by the gill with nine or 10 short slender filaments (Fig. 57M, N) and cocoon boot-shaped, with small to moderate-sized open spaces anteriorly (Fig. 57Q).

***Simulium* (*Simulium*) *ranauense* Takaoka, Hadi and Sigit, 2006**

(Fig. 57A)

Simulium (*Simulium*) *ranauense* Takaoka, Hadi and Sigit, 2006b: 27–33 (female, male, pupa, and larva).

Simulium (*Simulium*) sp. nr. *eximium* Takaoka et al., 2000: 165.

Distribution. Indonesia (Sumatra).

Bionomics. Aquatic habitats: Rapids in the fast-flowing rivers (width 2–15 m, 20.0–24.0 °C, elevation 340–560 m) in Bengkulu and West Sumatra (Takaoka et al., 2006b).

Remarks. This species was described from Sumatra, Indonesia (Takaoka et al., 2006b).

It is similar to *S. eximium* but is distinguished in the male by abdominal segments 2 and 4–7 each with a pair of silvery dorsolateral patches; in the pupa by the cocoon boot-shaped, with small open spaces anteriorly; and in the larva by the labral fan with 38–42 primary rays. Hadi et al. (1996b) showed that this species is chromosomally distinguished from *S. eximium*.

***Simulium* (*Simulium*) *thienemanni* Edwards, 1934**

(Fig. 57P, S)

Simulium (*Simulium*) *thienemanni* Edwards, 1934: 106–108 (female, male, pupa, and larva); Takaoka and Davies, 1996: 57–61 (female, pupa, and larva).

Distribution. Indonesia (Java).

Bionomics. Aquatic habitat: A rapid of a forest stream (width 2–4 m) in East Java (Takaoka and Davies, 1996).

Remarks. This species was described from Java, Indonesia by Edwards (1934).

It is characterized by the pupal gill with 10 short filaments, of which four filaments of the dorsal and ventral pairs are somewhat inflated (Fig. 57P) and cocoon boot-shaped, with an ear-shaped lateral flap on each anterolateral side (Fig. 57S).

Simulium (_Simulium_) _upikae_ Takaoka and Davies, 1996

(Fig. 57E–K, O, R)

Simulium (*Simulium*) *upikae* Takaoka and Davies, 1996: 54–57 (female, male, pupa, and larva).

Distribution. Indonesia (Java)

Bionomics. Aquatic habitats: Small streams (width 0.5–3.0 m, elevation 1000–1220 m) running in a tea cultivation field and a natural forest in West Java (Takaoka and Davies, 1996).

Remarks. This species was described from West Java, Indonesia by Takaoka and Davies (1996).

It is similar to *S. eximium* but is distinguished in the female by the cibarium with seven tiny processes and paraproct with a ridge directed posterolaterally; in the pupa by the gill with 10 slender filaments of the same length except two filaments of the ventral pair 0.5–0.6 times as long as the eight other filaments (Fig. 57O) and cocoon boot-shaped, with small open spaces anteriorly (Fig. 57R).

Simulium (_Simulium_) _griseifrons_ Species-group

Simulium (_Simulium_) _choochotei_ Takaoka, 2002

(Figs. 58F, H and 60J, S)

Simulium (*Simulium*) *choochotei* Takaoka [in Takaoka and Choochote, 2002]: 120–127 (female, male, pupa, and larva).

Distribution. Thailand.

Bionomics. Aquatic habitat: A cascading waterfall (Monthathan Waterfall) (16–17 °C, shaded, elevation 750 m) in Chiang Mai Province (Takaoka and Choochote, 2002).

Remarks. This species was described from northern Thailand (Takaoka and Choochote, 2002).

It is characterized in the female and male by the haired basal portion of the radius; in the female by the ovipositor valve with numerous stout hairs (Fig. 58H); in the pupa by the gill with six short, somewhat inflated filaments (Fig. 60J), terminal hooks distinct and cocoon boot-shaped (Fig. 60S); and in the larva by the posterior circlet with ca. 210 rows of up to 25 hooklets per row.

Simulium (_Simulium_) _digrammicum_ Edwards, 1928

(Figs. 59A and 60D, R)

Simulium digrammicum Edwards, 1928: 61–62 (female).

Simulium (*Simulium*) *digrammicum*: Takaoka and Suzuki, 1984: 40; Takaoka and Davies, 1995a: 102–105 (female); Takaoka and Choochote, 2002: 127–131 (male, pupa, and larva).

Distribution. Malaysia (Peninsular) and Thailand.

Bionomics. Aquatic habitat: A stream (width 2–5 m, shaded, 18 °C, elevation 1265 m) at Siribhume Waterfall, Chiang Mai Province (Takaoka and Choochote,

2002). Females of this species were reported as a human biter (Takaoka and Davies, 1995a).

Remarks. This species was described from females collected from Cameron's Highlands (Edwards, 1928). It was treated as a junior synonym of *S. griseifrons* Brunetti, 1911 from India by Puri (1932c) but resurrected by Takaoka and Davies (1995a), who also redescribed and illustrated the female based on the lectotype and paralectotype specimens. The male, pupa, and larva were later described based on specimens collected from Thailand (Takaoka and Choochote, 2002).

Females of this species were also collected from Mt. Tahan, Pahang and Gombak Forest Reserve, Selangor (Takaoka and Davies, 1995a). No additional material was recovered from Cameron's Highlands or other areas except for one set of pupal exuviae and cocoon from Simpang Pulai. The pupal exuviae is almost the same as that described from Thailand (Takaoka and Choochote, 2002) except the relative thickness of the six gill filaments (Unpublished data).

It is characterized by the cocoon shoe-shaped with a flap anterolaterally on each side and a small open space on each side (Fig. 60R).

***Simulium* (*Simulium*) *gravelyi* Puri, 1933**

Simulium (*Simulium*) *gravelyi* Puri, 1933a: 803–807 (female, male, and pupa).

Distribution. India (Tamil Nadu) and China (Tibet).

Bionomics. Aquatic habitats: Rocks in cascades, Palni Hills (Puri, 1933a).

Remarks. This species was described from southern India by Puri (1933a). Its larva remains unknown.

This species is characterized in the female by the yellowish femora and tibiae; in the male by the ventral plate with its body nearly quadrate when viewed ventrally; and in the pupa by the cocoon shoe-shaped.

***Simulium* (*Simulium*) *griseifrons* Brunetti, 1911**

Simulium griseifrons Brunetti, 1911: 285 (female).
Simulium (*Simulium*) *griseifrons*: Puri, 1932c: 1137–1141 (female, male, and pupa).

Distribution. India (Himachal Pradesh, Kashmir, and Uttar Pradesh) and China (Tibet).

Bionomics. Aquatic habitats: Torrential streams at elevations of 1950–2100 m, Fagu and Matiana, and Nara Nag, Kashmir (Puri, 1932c).

Remarks. This species was described from a female collected from Kalighad, Kumaon, Western Himalayas by Brunetti (1911) and its female, male, and pupa were described by Puri (1932c). Its larva remains unknown.

It is characterized in the female by the basal portion of the radius with or without hairs; in the male by the basal portion of the radius with hairs; and in the pupa by the cocoon shoe-shaped, with a small open space on each lateral surface.

***Simulium* (*Simulium*) *japonicum* Matsumura, 1931**

Simulium japonicum Matsumura, 1931: 407 (female); Bentinck, 1955: 8 (female, male, and pupa); Rubtsov, 1959–1964: 540–541 (female, male, pupa, and larva).

Simulium (Simulium) japonicum: Ogata and Sasa, 1954: 332; 1955: 10–18; Ogata et al., 1956: 90–91 (female, male, pupa, and larva); Ogata, 1956: 62–65 (female and pupa); 1966: 129; Orii et al., 1969: 1–13 (larva); Takaoka, 1977a: 201–205 (female, male, pupa, and larva).
Simulium oshimanum Shiraki, 1935: 56–59 (female). Synonymized by Ogata (1956).

Distribution. Japan (Hokkaido, Honshu, Kyushu, Shikoku, and Nansei Islands), China (Fujian, Jilin, Liaoning, and Tibet), Korea (South) and Siberia.

Bionomics. This species is one of the common species in Japan. Aquatic habitats: Rapids in mountainous streams. The female of this species is a man-biter in many localities in Japan including Nansei Islands (Takaoka, 1977a). This species is a potential vector of *Onchocerca japonica*, a parasite of wild boar and a causative agent of zoonotic onchocerciasis in Japan (Fukuda et al., 2008).

Remarks. This species was originally described from Hokkaido, Japan (Matsumura, 1931). Its male and pupa were illustrated by Bentinck (1955), and its larva was illustrated by Orii et al. (1969). The full description of the female, male, pupa, and larva was given by Takaoka (1977a).

It is characterized in the male by the number of upper-eye (large) facets in ca. 17 horizontal rows, hind basitarsus 3.5 times as long as its greatest width and nearly as wide as the hind tibia and femur and ventral plate subquadrate (when viewed ventrally), with setae on its ventral surface; in the pupa by the gill with six filaments decreasing in length and thickness from dorsal to ventral and cocoon wall-pocket-shaped or shoe-shaped, with a large anterolateral open space on each side; and in the larva by the postgenal cleft triangular.

***Simulium* (*Simulium*) *maenoi* Takaoka and Choochote, 2002**

(Figs. 58A, 59B–F and 60A, C, I, P)

Simulium (Simulium) maenoi Takaoka and Choochote, 2002: 115–120 (female, male, pupa, and larva); Takaoka et al., 2014b: 746–747.

Distribution. Thailand and Vietnam.

Bionomics. Aquatic habitats: Cascading streams (16–19 °C) at Sai Yoi Waterfall (elevation 830 m) and Monthathan Waterfall (elevation 750 m) in Chiang Mai Province (Takaoka and Choochote, 2002).

Remarks. This species was originally described from Thailand (Takaoka and Choochote, 2002). The record of *S. maenoi* from Vietnam was based on three females captured in Tam Dao National Park, Vinh Phuc Province, while flying around a human (Takaoka et al., 2014b).

It is characterized by the cocoon wall-pocket-shaped, with a small narrow open space on each side (Fig. 60P).

***Simulium* (*Simulium*) *phukaense* Takaoka and Choochote, 2005**

(Figs. 58D, I, J and 60G, N)

Simulium (Simulium) phukaense Takaoka and Choochote, 2005g: 95–101 (female, male, and pupa).

Distribution. Thailand.

Bionomics. Aquatic habitat: A cascading stream (width ca. 1.5 m, depth ca. 1.5 cm, exposed to the sun, 18 °C, elevation 1250 m) in a sparsely forested area, Doi Phu Ka National Park, Nan Province (Takaoka and Choochote, 2005g).

Remarks. This species was described from northern Thailand (Takaoka and Choochote, 2005g). Its larva is unknown.

It is characterized in the pupa by the gill with six filaments decreasing in length and thickness from dorsal to ventral (Fig. 60G), spine-combs only on abdominal segment 8, abdominal segment 9 lacking terminal hooks and cocoon shoe-shaped (Fig. 60N).

***Simulium (Simulium) pukaengense* Takaoka and Choochote, 2005**

(Figs. 58B and 60H, O)

Simulium (Simulium) pukaengense Takaoka and Choochote, 2005f: 227–234 (female, male, pupa, and larva).

Distribution. Thailand.

Bionomics. Aquatic habitat: A fast-flowing cascading stream (width 0.5 m, depth 10–15 cm, 23 °C, elevation 640 m) in a sparsely forested area, Pu Kaeng Waterfall, Chiang Rai Province (Takaoka and Choochote, 2005f).

Remarks. This species was described from northern Thailand (Takaoka and Choochote, 2005f).

It is characterized by the haired basal portion of the radius of the female and male, male abdominal segments 2, 6, and 7 each with a pair of white pruinose dorsolateral or lateral spots, pupal gill with six filaments decreasing in length and diameter from dorsal to ventral (Fig. 60H), terminal hooks present, cocoon shoe-shaped (Fig. 60O) and larval rectal organ with 19–28 finger-like secondary lobules per lobe.

***Simulium* (*Simulium*) *spoonatum* An and Yan, 1998**

Simulium (*Simulium*) *spoonatum* An and Yan [in An et al., 1998]: 190–192 (female, male, and pupa).

Distribution. China (Guangdong).

Bionomics. Aquatic habitats: Streams (width 2–5 m, depth 0.1–0.5 m, elevation 20 m) (An et al., 1998).

Remarks. This species was described from Guangdong, China (An et al., 1998). Its larva is unknown. This species was placed in the *S. crocinum* species-group based on the male ventral plate flat, without any teeth, and style with a long basal protuberance (Takaoka, 2017a). However, the scutum with longitudinal vittae and the shape of the ovipositor valves, which are triangular, close together in the middle and with several short hairs only near the inner margins, are quite different from those of most species of the *S. crocinum* species-group. This species is here transferred to the *S. griseifrons* species-group, though it should be confirmed on whether the female cibarium has a dorsomedian projection, one of the diagnostic characteristics of the *S. griseifrons* species-group.

It is characterized in the female by the basal portion of the radius with 19 hairs, cibarium without minute processes and scutum with five longitudinal vittae; and in the pupa by the gill with six slender filaments of the same thickness and cocoon with several small open spaces along each anterolateral margin.

***Simulium* (*Simulium*) *suchariti* Takaoka and Choochote, 2004**
(Figs. 58C, G, 59G–J and 60B, E, Q)

Simulium (*Simulium*) *suchariti* Takaoka and Choochote, 2004b: 31–33 (female); Takaoka and Choochote, 2004c: 324–326 (pupa and larva); Takaoka et al., 2017k: 27–29 (male).

Distribution. Thailand.

Bionomics. Aquatic habitats: A small stream (width 0.3 m, bottom rocky, 6.5 °C, shaded, elevation 2565 m) slowly flowing in a natural forest, Ang Ka, and a stream (width 1.5 m, depth 13 cm, bottom sandy, 14.5 °C, pH 5.4, partially shaded, elevation 2210 m) fast flowing in a natural forest, Doi Inthanon National Parks, Chiang Mai Province (Takaoka and Choochote, 2004c).

Remarks. This species was described from females collected from Thailand by Takaoka and Choochote (2004b) and its pupa and larva were described by Takaoka and Choochote (2004c). The male of this species was later described by Takaoka et al. (2017k).

It is characterized in the female by the cibarium with tubercles on the dorsomedian process and near its base (Fig. 58C); in the male by abdominal segments 2, 6, and 7 each with a pair of shiny dorsolateral spots and abdominal segment 10 with ca. 15 hairs on each side of the posteroventral surface; in the pupa by the gill with six filaments decreasing in length from dorsal to ventral, frons and thorax moderately or densely covered with tubercles, terminal hooks absent, and cocoon with a large anterolateral open space on each side (Fig. 60Q); and in the larva by the head spots indistinct and sheath of the subesophageal ganglion unpigmented.

***Simulium* (*Simulium*) *tenuitarsus* Puri, 1933**

Simulium (*Simulium*) *tenuitarsus* Puri, 1933a: 809–812 (male and pupa).

Distribution. India (West Bengal).

Bionomics. Aquatic habitat: A large stream north of Marianbarie, Tea Estate, West Bengal (Puri, 1933a).

Remarks. This species was described from India by Puri (1933a). The female and larva of this species are unknown.

This species is characterized in the male by the haired basal portion of the radial vein, and ventral plate with its body quadrate when viewed ventrally; and in the pupa by the cocoon boot-shaped.

***Simulium* (*Simulium*) *thongsahuani* Takaoka and Choochote, 2010**
(Fig. 60K, M)

Simulium (*Simulium*) *thongsahuani* Takaoka and Choochote [in Takaoka et al., 2010b]: 17–25 (female, male, pupa, and larva).

Distribution. Thailand and Malaysia (Peninsular).

Bionomics. Aquatic habitat: A stream (width 1 m, depth 20 cm, 24.0 °C, elevation 200–240 m) at Ton Nga Chang Waterfall, Songkhla Province (Takaoka et al., 2010b).

Remarks. This species was described from southern Thailand by Takaoka et al. (2010b), and subsequently recorded from Peninsular Malaysia by Takaoka et al. (2019a).

It is characterized in the male by abdominal segments 2 and 5–7 each with a pair of shiny dorsolateral spots and style lacking an apical spine; in the pupa by the gill with six filaments, of which three ventral filaments are slender, subequal in thickness to one another and much thinner than three dorsal filaments, and cocoon wall-pocket-shaped; and in the larva by the darkened head capsule.

Simulium (_Simulium_) _visuti_ Takaoka and Choochote, 2006
(Figs. 58E, K, L, 59K–M and 60F, L, T–W)

Simulium (*Simulium*) *visuti* Takaoka and Choochote, 2006d: 115–123 (female, male, pupa, and larva).

Distribution. Thailand.

Bionomics. Aquatic habitat: A moderately flowing stream (width ca. 2 m, depth 20–30 cm, 15 °C, elevation 1050 m) in a sparsely forested area, in Nan Province (Takaoka and Choochote, 2006d).

Remarks. This species was described from northern Thailand by Takaoka and Choochote (2006d).

It is characterized in the female and male by the haired basal portion of the radius; in the female by the paraproct with elongate stout hairs (Fig. 58K, L); in the male by the ventral plate with its body subquadrate when viewed ventrally (Fig. 59K); in the pupa by the cocoon shoe-shaped (Fig. 60T); and in the larva by the body gradually widened to abdominal segment 8, then narrowed to abdominal segment 9, hypostoma with nine or 10 bristles on each side and rectal organ with 17–23 finger-like secondary lobules per lobe.

Simulium (_Simulium_) _grossifilum_ Species-group

Simulium (_Simulium_) _grossifilum_ Takaoka and Davies, 1995

Simulium (*Simulium*) *grossifilum* Takaoka and Davies, 1995a: 105–115 (female, male, pupa, and larva); Takaoka et al., 2014b: 746.

Distribution. Malaysia (Peninsular), Thailand, and Vietnam.

Bionomics. Aquatic habitats: Small to medium-sized streams at low to high elevations (160–1315 m) in Cameron's Highlands (Ya'cob et al., 2016).

Remarks. This species was originally described from Peninsular Malaysia (Takaoka and Davies, 1995a). It was later recorded from Thailand (Kuvangkadilok and Takaoka, 2000) and Vietnam (Takaoka et al., 2014d). The record of this species from Vietnam was based on three females captured in Tam Dao National Park, Vinh

Phuc Province, while flying around a human (Takaoka et al., 2014b). The aquatic stages remain to be collected from Vietnam.

It is characterized in the male by the number of upper-eye (large) facets in 17 vertical columns and 16 horizontal rows; and in the pupa by the gill with six inflated tubular filaments and abdomen with small terminal hooks.

***Simulium* (*Simulium*) *hengshanense* Bi and Chen, 2004**

Simulium (*Simulium*) *hengshanense* Bi and Chen, 2004: 569–571 (female, male, pupa, and larva).

Distribution. China (Hubei and Hunan).

Bionomics. Aquatic habitat: A small, stream (elevation 1080 m), Hengshan, Hunan Province (Bi and Chen, 2004).

Remarks. This species was described from China (Bi and Chen, 2004).

It is characterized by the number of male upper-eye (large) facets in 10 vertical columns and 13 horizontal rows, pupal gill with six thread-like filaments, and abdominal segment 9 with a pair of terminal hooks.

***Simulium* (*Simulium*) *lushanense* Chen, Kang and Zhang, 2007**

Simulium (*Simulium*) *lushanense* Chen, Kang and Zhang, 2007a: 579–580 (male and pupa).

Distribution. China (Jiangxi).

Bionomics. Aquatic habitat: A small, shaded stream (elevation 300 m), Lushan, Jiangxi Province, China (Chen et al., 2007a).

Remarks. This species was described from China (Chen et al., 2007a). Its female and larva remain unknown.

It is characterized by the number of male upper-eye (large) facets in 12 vertical columns and 11 horizontal rows and pupal abdominal segment 9 with a pair of terminal hooks. The pupal gill was described to be composed of two inflated horns with six tube-like filaments [though, the pupal gill has seven inflated tubular filaments according to Fig. 2 by Chen et al. (2007a)].

***Simulium* (*Simulium*) *phayaoense* Takaoka and Choochote, 2005**
(Fig. 61A–T)

Simulium (*Simulium*) *phayaoense* Takaoka and Choochote, 2005f: 219–227 (female, male, pupa, and larva).

Distribution. Thailand.

Bionomics. Aquatic habitat: A fast-flowing cascading stream (width 0.5–1.0 m, depth 10–20 cm, 21–24 °C, elevation 680 m) in a densely forested area at Champa Thong Waterfall, Phayao Province (Takaoka and Choochote, 2005f).

Remarks. This species was described from northern Thailand (Takaoka and Choochote, 2005f).

It is similar to *S. grossifilum* but is distinguished from the latter species in the female by the sensory vesicle 0.41 times as long as the third palpal segment; in the male by abdominal segments 2, 6, and 7 each with a pair of whitish pruinose spots; in the pupa by abdominal segment 9 lacking terminal hooks; and in the larva by the postgenal cleft 2.4–2.8 times as long as the postgenal bridge (Fig. 61T) and posterior circlet with 84–98 rows of hooklets.

Simulium (*Simulium*) *indicum* Species-group

Simulium (_Simulium_) _indicum_ Becher, 1885

Simulium indicum Becher, 1885: 519 (female).
Simulium (*Himalayum*) *indicum*: Lewis, 1973: 460–463 (male, pupa and larva); Lewis, 1974: 25–33 (female).
Simulium (*Simulium*) *indicum*: Otsuka et al., 2003: 113–120.

Distribution. India, Afghanistan, Bangladesh, Bhutan, Myanmar, China (Tibet), Nepal, and Pakistan (Kashmir).

Bionomics. Aquatic habitats: A number of watercourses at elevations from 450 to 2025 m in India (Datta, 1997). This is a well-known man-biting species in many areas of southern slopes of Himalayan mountains (Lewis, 1974).

Remarks. Lewis (1973) newly created the subgenus *Himalayum* for this species, but Otsuka et al. (2003) synonymized *Himalayum* with the subgenus *Simulium* s. str. by using a combination of morphological and DNA analysis. Takaoka (2017a) created the *S. indicum* species-group in the subgenus *Simulium* s. str. for *S. indicum* and a few other related species.

It is characterized in the female by the frons densely covered with fine hairs and basal portion of the radius fully haired; in the pupa by abdominal segments 5–9 lacking spine-combs dorsally and cocoon shoe-shaped; and in the larva by the presence of an accessory sclerite.

Simulium (_Simulium_) _nigrogilvum_ Summers, 1911
(Figs. 62A, C, D, H–M and 63A–J)

Simulium nigrogilvum Summers, 1911: 586–588 (female).
Simulium (*Himalayum*) *nigrogilvum*: Takaoka and Suzuki, 1984: 22–27 (female, male, pupa, and larva).
Simulium (*Simulium*) *nigrogilvum*: Otsuka et al., 2003: 113–120.

Distribution. Myanmar, Thailand, and Vietnam (?).

Bionomics. Aquatic habitat: A stream (width 7 m, depth 35 cm, bottom sandy, 22.5 °C, elevation 704 m) in Mong La, Shan State in Myanmar (Takaoka et al., 2018j). The females of this species are anthropophilic (Choochote et al., 2005) and transmit an unknown filarial parasite in Thailand (Fukuda et al., 2003) and *Onchocerca* type I (Aupalee et al., 2020b).

Remarks. This species was assigned to the subgenus *Himalayum* (Takaoka and Suzuki, 1984) but was later transferred to the subgenus *Simulium* s. str. (Otsuka et al., 2003). This species was recorded from Vietnam by Crosskey and Howard (1997). However, data including the material collected and its locality were not given. This species was recently recorded from Myanmar by Takaoka et al. (2018j).

It is characterized in the female by the frons fully haired, mandible with distinct outer teeth (Fig. 62A), fore basitarsus with a thick hair crest, subcosta haired, basal portion of the radial vein fully haired, claw with a small subbasal tooth (Fig. 62C), abdominal segments 2 and 3 pale, and ovipositor valve protruded posteriorly (Fig. 62D); and in the pupa by the head and thoracic integument with tubercles, gill with six short filaments, of which the ventral filament of the ventral pair is as thick as or thinner than the counter filament, and thinner than the four other filaments when compared basally (Fig. 63A) [though wrongly stated as "all filaments are almost equal in length and thickness" in Takaoka and Suzuki (1984)], dorsum of the abdomen lacking spine-combs and terminal hooks and cocoon shoe-shaped (Fig. 63C); and in the larva by the mandible with a supernumerary serration with numerous teeth and postgenal cleft long but its apex not reaching the posterior margin of the hypostoma.

***Simulium* (*Simulium*) *umphangense* Takaoka, Srisuka and Saeung, 2017**

(Fig. 62B, E–G)

Simulium (*Simulium*) *umphangense* Takaoka, Srisuka and Saeung, 2017j: 945–948 (female).

Distribution. Thailand.

Bionomics. Females of this species take bloods of humans (Takaoka et al., 2017j). Aquatic habitats are unknown.

Remarks. This species was described from females collected on humans in Thailand by Takaoka et al. (2017j). The male, pupa, and larva are unknown.

It is characterized by the fore tibia almost entirely darkened and ovipositor valve not protruded posteriorly (Fig. 62E).

***Simulium* (*Simulium*) *vanellum* Huang, Phasuk, Chanpaisaeng and Adler, 2010**

Simulium (*Simulium*) *vanellum* Huang, Phasuk, Chanpaisaeng and Adler, 2010: 50–57 (female, male, pupa, and larva).

Distribution. Thailand.

Bionomics. Aquatic habitats: Streams (width 2–10 m, depth 2–45 cm, 20–25 °C, pH 7.39–7.91, velocity 0.3–0.8 m/sec), Kanchanaburi Province (Huang et al., 2010).

Remarks. This species was described from Thailand by Huang et al. (2010).

It is characterized in the female by the mandible with six vestigial outer teeth; in the male by having a pair of pruinose dorsolateral spots only on abdominal segment 6; and in the larva by the mandible without a supernumerary serration.

Simulium (*Simulium*) *iridescens* Species-group

***Simulium* (*Simulium*) *iridescens* De Meijere, 1913**

(Fig. 64A–L, N–R)

Simulium iridescens De Meijere, 1913: 333 (female and male); Edwards. 1934: 112–114 (female, male, pupa, and larva).

Simulium (*Simulium*) *iridescens* De Meijere: Takaoka and Davies, 1996: 67–70 (female, male, pupa, and larva).

Distribution. Indonesia (Bali, Java, Lombok, and Sumatra).

Bionomics. Aquatic habitats: Small streams (width 0.2–3.0 m) in lowlands (Takaoka and Davies, 1996).

Remarks. The species was described from East Java, Indonesia by De Meijere (1913). Its pupa and larva were described by Edwards (1934). The redescription of all stages of this species was carried out by Takaoka and Davies (1996). It was recorded from Sumatra (Takaoka et al., 2000) and from Bali and Lombok (Takaoka et al., 2017e).

It was first placed in the *S. melanopus* species-group by Takaoka and Davies (1996) but was transferred to the *S. iridescens* species-group by Takaoka (2017a).

It is characterized by the pupal gill with six thread-like filaments (Fig. 64N) and frons moderately covered with tubercles.

Simulium (_Simulium_) _javaense_ Takaoka and Hadi, 1991

(Fig. 64M)

Simulium (*Simulium*) *javaense* Takaoka and Hadi, 1991: 363–366 (female, male, pupa, and larva).
Simulium iridescens var. ?: Edwards, 1934: 114 (male, pupa, and larva).

Distribution. Indonesia (Bali and Java).

Bionomics. Aquatic habitat: A small stream (width 0.3 m, shaded) fast flowing on a hilly slope, East Java (Takaoka and Hadi, 1991).

Remarks. This species was first reported as *S. iridescens* var. ? by Edwards (1934). It was described as a new species, *S. javaense*, by Takaoka and Hadi (1991). It was recorded from Bali (Takaoka et al., 2017e).

It is similar to *S. iridescens* in many features including the female and male genitalia but is distinguished in the pupa by the bare frons and gill filaments short and somewhat inflated (Fig. 64M); and in the larva by the dorsum of each abdominal segment with a reddish brown band.

Simulium (_Simulium_) _malyschevi_ Species-group

Simulium (_Simulium_) _baimaii_ Kuvangkadilok and Takaoka, 1999

(Fig. 66C)

Simulium (*Simulium*) *baimaii* Kuvangkadilok and Takaoka, 1999 [in Takaoka and Kuvangkadilok, 1999]: 507–508 (pupa and larva).
Simulium (*Simulium*) *baimaii*: Takaoka and Choochote, 2006b: 49–53 (female and male).

Distribution. Thailand.

Bionomics. Aquatic habitat: A small stream (width ca. 2.0 m, depth 10–15 cm, 22.0 °C, partially shaded, elevation 1210 m) moderately running in a sparsely forested area at Wang Kwang Waterfall, Loei Province (Takaoka and Kuvangkadilok, 1999).

Remarks. This species was originally described from a pupa and two larvae from Thailand (Takaoka and Kuvangkadilok, 1999). It was placed in the *S. malyschevi* species-group when its female and male were described by Takaoka and Choochote (2006b).

It is characterized in the female by the ventral surface of abdominal segment 7 lacking a medial sternal plate and branched hairs on it; in the male by the number of upper-eye (large) facets in 21 vertical columns and 21 horizontal rows; in the pupa by the gill composed of two inflated filaments (Fig. 66C), thoracic trichomes unbranched and cocoon wall-pocket-shaped, without an anterolateral open space on each side; and in the larva by abdominal segments lacking any protuberances dorsally and postgenal cleft bullet-shaped, not reaching the posterior margin of the hypostoma.

***Simulium (Simulium) damingense* Chen, Zhang and Zhang, 2007**

Simulium (Simulium) damingense Chen, Zhang and Zhang, 2007b: 779–781 (female, male, pupa, and larva).

Distribution. China (Guangxi).

Bionomics. Aquatic habitat: A fast-flowing stream (elevation 1764 m) in Guangxi (Chen et al., 2007b).

Remarks. This species was described from China by Chen et al. (2007b).

It is characterized in the female by the sternite 7 with branched hairs and simple claw lacking a small subbasal tooth (probably overlooked?); in the male by the number of upper-eye (large) facets in 11 vertical columns and 12 horizontal rows; in the pupa by the head and thorax bare, with unbranched trichomes; and in the larva by abdominal segments lacking any protuberances dorsally, and postgenal cleft long, reaching the posterior margin of the hypostoma.

***Simulium (Simulium) hirtipannus* Puri, 1932**

Simulium (Simulium) hirtipannus Puri, 1932d: 509–513 (female, male, and pupa); Zhang and Wang, 1991a: 489 (larva).

Distribution. India (Assam) and China (Fujian, Guizhou, and Tibet).

Bionomics. Aquatic habitat: A large torrential hill stream in Assam, India (Puri, 1932d).

Remarks. This species was described from India by Puri (1932d). Its mature larva was described based on specimens from Fujian, China by Zhang and Wang (1991a).

This species is similar to *S. damingense* from China (Chen et al., 2007b) except the female claw with a small subbasal tooth and larval postgenal cleft long but not reaching the posterior margin of the hypostoma.

The number of male upper-eye (large) facets of this species is unknown.

Simulium (Simulium) jacuticum Rubtsov, 1940

Simulium (*Gnus*) *jacuticum* Rubtsov, 1940: 365–366 (female, male, pupa, and larva).

Distribution. China (Fujian, Heilongjiang, Inner Mongolia, and Shanxi) and Siberia.

Bionomics. Aquatic habitats: Large rivers (Rubtsov, 1956).

Remarks. This species was described from Siberia by Rubtsov (1940).

The male and pupa of this species seem to be similar to those of *S. damingense* but its larva is distinguished from the latter by the postgenal cleft not reaching the posterior margin of the hypostoma. Information on the presence or absence of the branched hairs on the ventral surface of female abdominal segment 7, the number of male upper-eye (large) facets and most pupal characteristics is not available.

Simulium (Simulium) lomkaoense Takaoka and Srisuka, 2014
(Figs. 65E and 66B, D, E, J)

Simulium (*Simulium*) *lomkaoense* Takaoka and Srisuka [Takaoka et al., 2014j]: 1109–1115 (female, male, pupa, and larva).

Distribution. Thailand.

Bionomics. Aquatic habitat: A fast-flowing stream (width 1.6 m, depth 18 cm, bottom gravelly and sandy, 18.0 °C, pH 4.55, shaded, elevation 1550 m) in Phetchabun Province (Takaoka et al., 2014j).

Remarks. This species was described from Thailand by Takaoka et al. (2014j).

It is similar to *S. baimaii* from Thailand in many characteristics including abdominal segment 7 of the female lacking branched hairs, pupal gill with two inflated filaments (Fig. 66B), simple wall-pocket-shaped cocoon (Fig. 66D), and larval abdomen lacking dorsal protuberances. However, this species is distinguished from the latter species in the female by the bare subcosta; in the male by the number of upper-eye (large) facets in 22 or 23 vertical columns and 23 or 24 horizontal rows, fore basitarsus greatly dilated, 4.46–4.57 times as long as its greatest width and hind basitarsus spindle-shaped, 4.55–4.61 times as long as wide; and in the larva by the sheath of the subesophageal ganglion of the larva dark pigmented (Fig. 66J).

Simulium (Simulium) obliquum Takaoka and Low, 2017
(Figs. 65A–C, F–Q and 66A, F, G, H, I, L, M)

Simulium (*Simulium*) *obliquum* Takaoka and Low [Takaoka et al., 2017c]: 101–108 (female, male, pupa, and larva).

Distribution. Vietnam.

Bionomics. Aquatic habitat: A river (width 8–10 m, flow moderate, bottom of pebbles, 17 °C, elevation 162 m) fast flowing in a natural forest, Pu Mat, Nghe An Province, (Takaoka et al., 2017c).

Remarks. This species was described from northern Vietnam by Takaoka et al. (2017c).

It is characterized in the female by the sensory vesicle 0.2 times as long as the third palpal segment and genital fork with two projections on each arm (Fig. 65F, G);

in the pupa by the cocoon wall-pocket-shaped, with a large anterolateral open space on each side (Fig. 66F); and in the larva by the abdomen with multiple pairs of protuberances (Fig. 66L, M).

***Simulium* (*Simulium*) *phidimense* Takaoka and Shrestha, 2020**
(Figs. 65D and 66G)

Simulium (*Simulium*) *phidimense* Takaoka and Shrestha [in Takaoka et al., 2020c]: 212–217 (female, male, and pupa).

Distribution. Nepal.

Bionomics. Aquatic habitat: A river (flow moderate, bottom of pebbles, 17 °C, elevation 579 m), Phidim (Takaoka et al., 2020c).

Remarks. This species was described from Nepal by Takaoka et al. (2020c). Its larva is unknown.

The female, male, and pupa of this species are similar to those of *S. obliquum* from Vietnam. However, this species is distinguished from the latter species by the number of branched hairs on sternite 7 of the female (22–26 in this species versus 46–51 in *S. obliquum*).

***Simulium* (*Simulium*) *siripoomense* Takaoka and Saito, 1996**
(Fig. 66K)

Simulium (*Byssodon*) *siripoomense* Takaoka and Saito, 1996: 163–165 (larva).
Simulium (*Simulium*) *siripoomense* Takaoka and Mulla, 2000: 25–28 (female and pupa).

Distribution. Thailand.

Bionomics. Aquatic habitats: Streams (width 3–6 m, depth 10–50 cm, 24.0 °C) slowly or rapidly flowing in a forest at Siribhume Waterfall, Chiang Mai Province (Takaoka and Saito, 1996).

Remarks. This species was described from two larvae in Thailand and was assigned to the subgenus *Byssodon* by Takaoka and Saito (1996). However, it was transferred to the *S. malyschevi* species-group of the subgenus *Simulium*, when its female and pupa were described (Takaoka and Mulla, 2000).

It is similar to *S. obliquum* from Vietnam in many characteristics including the female abdominal segment 7 with branched hairs on its ventral surface, and larval abdomen with multiple pairs of prominent protuberances. However, this species is distinguished from the latter species by the relative length of the female sensory vesicle against the third palpal segment (0.4 in this species versus 0.2 in *S. obliquum*) and larval postgenal cleft [reaching the posterior margin of the hypostoma in this species (Fig. 66K) versus not reaching the posterior margin of the hypostoma in *S. obliquum*].

Simulium (*Simulium*) *melanopus* Species-group

Simulium (*Simulium*) *abatanense* Takaoka, 1983

Simulium (*Simulium*) *abatanense* Takaoka, 1983: 118–120 (female and pupa).

Distribution. Philippines (Luzon).

Bionomics. Aquatic habitat: A small stream (width 0.5–1.0 m, shaded) flowing down on a mountain slope, Abatan, northern Luzon (Takaoka, 1983).

Remarks. This species was described from northern Luzon, the Philippines by Takaoka (1983). Its male and larva are unknown.

The female of this species is characterized by the ovipositor valves fused with the eighth sternite directed ventrally, and paraproct much expanded ventrally, with a ventrointernal plate rounded apically.

Simulium (*Simulium*) *acostai* Takaoka, 1983

Simulium (*Simulium*) *acostai* Takaoka, 1983: 113–115 (female, male, pupa, and larva).

Distribution. Philippines (Mindanao).

Bionomics. Aquatic habitat: A small stream (width 2.0–3.0 m) fast flowing in a mountainous area, Davao (Takaoka, 1983).

Remarks. This species was described from Mindanao, the Philippines by Takaoka (1983).

It is characterized by the male hind basitarsus moderately enlarged, nearly parallel-sided, 4.5 times as long as its greatest width, much narrower than the hind tibia and femur, and yellow on the basal two-thirds.

Simulium (*Simulium*) *aquilum* Takaoka, 2006

(Fig. 69H)

Simulium (*Simulium*) *aquilum* Takaoka, 2006c: 317–319 (female and pupa).

Distribution. Philippines (Luzon).

Bionomics. Aquatic habitat: A fast-flowing stream (width 0.5–3.0 m, 15.0 °C, shaded, elevation 1680 m), Banaue, Ifugao Province (Takaoka, 2006c).

Remarks. This species was described from northern Luzon, the Philippines by Takaoka (2006c). Its male and larva are unknown.

It is characterized in the female by the sensory vesicle with an enlarged opening, fore coxa darkened, and the basal portion of the radial vein haired on the apical one-third or one-fourth; and in the pupa by the frons densely covered with round tubercles.

Simulium (*Simulium*) *atrum* Delfinado, 1969

Simulium (*Simulium*) *atrum* Delfinado, 1969: 200 (female): Takaoka, 1983: 99–101 (female).

Distribution. Philippines (Luzon, Mindanao and Tawi Tawi).

Bionomics. Unknown.

Remarks. This species was described from two females caught by a Malaise trap, at Katanglad, Mindanao, the Philippines by Delfinado (1969). The female of this species was redescribed and illustrated by Takaoka (1983). Its male, pupa, and larva are unknown.

The female of this species is characterized by the legs darkened except the basal half of the hind basitarsus and second tarsomere yellow, abdominal segments 2 and 6–8 shiny dorsally and ovipositor valves relatively short, rounded posteromedially.

***Simulium* (*Simulium*) *banahauense* Takaoka, 2007**

(Fig. 68Y)

Simulium (*Simulium*) *banahauense* Takaoka, 2007a: 11–16 (female, male, and pupa).

Distribution. Philippines (Luzon).

Bionomics. Aquatic habitat: A stream (width 6–10 m, flow moderate, 21.0 °C, partially shaded, elevation ca. 600 m), one of tributaries of Cristalino River, Quezon Province (Takaoka, 2007a).

Remarks. This species was described from Luzon, the Philippines (Takaoka, 2007a). Its larva is unknown.

It is characterized by the female and male coxae darkened and male aedeagal membrane bare (Fig. 68Y).

***Simulium* (*Simulium*) *bishopi* Takaoka and Davies, 1995**

Simulium (*Simulium*) *bishopi* Takaoka and Davies, 1995a: 111–115 (female, male, pupa, and larva).

Distribution. Indonesia (Sumatra), Malaysia (Peninsular) and Thailand.

Bionomics. Aquatic habitats: Small to medium-sized streams at low to high elevations (159–1813 m) in Cameron's Highlands and other areas of Peninsular Malaysia (Ya'cob et al., 2016).

Remarks. This species was described from females, males, pupae, and larvae collected from Gombak, Selangor (Takaoka and Davies, 1995a). It was recorded from Sumatra, Indonesia (Takaoka et al., 2000) and southern Thailand by Takaoka et al. (2019g).

It is characterized in the pupa by the frons and the anterior half of the thorax bare.

***Simulium* (*Simulium*) *cheedhangi* Takaoka, Ya'cob and Sofian-Azirun, 2015**

(Figs. 67C and 69C)

Simulium (*Simulium*) *cheedhangi* Takaoka, Ya'cob and Sofian-Azirun [in Takaoka et al., 2015a]: 44–48 (female, pupa, and larva).

Distribution. Malaysia (Sabah and Sarawak).

Bionomics. Aquatic habitat: A stream (width 0.3–0.5 m, bottom muddy, 18.6 °C, partially shaded, elevation 1544 m) slowly flowing in natural forests, crossing a logging road from Para Rabata to Lepa Bunga, Bakalalan (Takaoka et al., 2015a).

Remarks. This species was described from Sarawak by Takaoka et al. (2015a). Its male is unknown.

It is characterized in the female by the claw with a medium-sized tooth; in the pupa by the dorsal pair of gill filaments with a somewhat inflated stalk, which is 1.6 times as thick as the interspiracular trunk (Fig. 69C) and cocoon shoe-shaped, with a low anteroventral collar.

***Simulium* (*Simulium*) *concavum* Takaoka, 2006**

(Figs. 67K, L and 69E, P, Q)

Simulium (*Simulium*) *concavum* Takaoka, 2006b: 287–293 (female, male, and pupa).

Distribution. Philippines (Luzon).

Bionomics. Aquatic habitat: A moderately flowing stream (width ca. 10 m, elevation ca. 1100 m), east of Banaue Town, Ifugao Province (Takaoka, 2006b).

Remarks. This species was described from Luzon, the Philippines by Takaoka (2006b). Its larva is unknown.

It is characterized in the male by the small number of upper-eye (large) facets in 12 or 13 vertical columns and 14 horizontal rows and scutum with a non-pruinose dark median vitta and two non-pruinose dark elliptical lateral spots; and in the pupa by the frons and the anterior half of the thorax widely bare.

***Simulium* (*Simulium*) *conspicuum* Takaoka, 2006**

(Fig. 69T)

Simulium (*Simulium*) *conspicuum* Takaoka, 2006c: 321–325 (female, male, pupa, and larva).

Distribution. Philippines (Luzon).

Bionomics. Aquatic habitats: Small streams (width 0.3–3.0 m, 14.0–15.0 °C, elevation 1400–1730 m) in Mountain and Ifugao Provinces (Takaoka, 2006c).

Remarks. This species was described from northern Luzon, the Philippines by Takaoka (2006c).

It is characterized in the female and male by the darkened fore coxae; in the pupa by the cocoon wall-pocket-shaped, with its ventral margin much shorter than its dorsal margin when viewed laterally (Fig. 69T).

***Simulium* (*Simulium*) *crassimanum* Edwards, 1933**

(Fig. 69F)

Simulium crassimanum Edwards, 1933: 256 (female).

Simulium (*Simulium*) *crassimanum*: Takaoka et al., 2015b: 2–8 (female, male, and pupa).

Simulium (*Simulium*) *liewi* Takaoka, 2007b: 296–300 (female and pupa). Synonymized by Takaoka et al., 2015b: 2.

Distribution. Malaysia (Sabah).

Bionomics. Aquatic habitat: A small stream (width 1.0–1.2 m, depth 7–12 cm, bottom rocky or with pebbles, 16.1–16.3 °C, shaded, elevation 1728–1890 m) slow flowing in a natural forest, above Carson Waterfall, Mt. Kinabalu (Takaoka et al., 2015b).

Remarks. This species was originally described on the basis of females collected at Kamborangah in the Kinabalu National Park, Sabah (Edwards, 1933). Smart and Clifford (1969) redescribed the female of this species based on the holotype specimen. The paratype female of this species was examined and its genitalia were illustrated by Takaoka (1983). Its female was redescribed and its male and pupa were described by Takaoka et al. (2015b).

It is characterized in the female by the antenna entirely dark brown, fore coxa dark brown, fore tibia dark brown with a wide white area on its outer surface, mid tibia entirely dark brown to black, mid basitarsus yellowish on its basal three-fourths, hind tibia dark brown to black with its base yellow, hind basitarsus whitish on its basal three-fifths, basal portion of the radial vein entirely bare and tergite 2 silvery iridescent; and in the pupa by the frons and the dorsal surface of the anterior half of the thorax bare, six gill filaments without annular ridges (Fig. 69F) and cocoon wall-pocket-shaped or shoe-shaped, with a low anteroventral collar.

***Simulium* (*Simulium*) *discrepans* Delfinado, 1969**

Simulium (*Simulium*) *discrepans* Delfinado, 1969: 202 (male); Takaoka, 1983: 96–99 (female, male, pupa, and larva).

Distribution. Philippines (Luzon, Bohol, Cebu, Leyte, Mindanao, Mindoro, Negros, and Panay).

Bionomics. Aquatic habitats: Rapids of rivers (width 5–16 m) in lowlands (Takaoka, 1983).

Remarks. This species was described from a male dissected out from a pupa collected from Laguna, Luzon, the Philippines by Delfinado (1969). The female caught by a light trap at Abatan, Mountain Province was wrongly associated with this species. The female and larva of this species were described by Takaoka (1983).

It is characterized in the female by the simple claw; in the male by the hind basitarsus slender and parallel-sided and ventral plate without setae; and in the pupa by the frons and the anterior half of the thorax almost bare and abdominal segments 5–9 lacking spine-combs dorsally.

***Simulium* (*Simulium*) *forcipatum* Delfinado, 1969**

(Figs. 67E, O–Q and 68A–C, S–X)

Simulium (*Simulium*) *forcipatum* Delfinado, 1969: 202 (male); Takaoka, 1983: 134–135 (male); Takaoka, 2006c: 309–317 (female, male, pupa, and larva).

Distribution. Philippines (Luzon).

Bionomics. Aquatic habitats: Rivers (width 5–30 m, 14.5–18.0 °C, elevation 920–1400 m) in Ifugao Province, and streams (width 0.5–3.0 m, 15.0–23.2 °C, elevation 192–1680 m) in Ifugao and Laguna Provinces (Takaoka, 2006c).

Remarks. This species was described from a single male, collected at Abatan, Mountain Province, northern Luzon, the Philippines by Delfinado (1969). Its female, pupa, and larva were described by Takaoka (2006c).

It is characterized in the female and male by the darkened fore coxae; in the female by the ovipositor valves triangular (Fig. 67O); in the male by the scutum with variable color patterns (Fig. 68A–C), ventral plate Y-shaped, with two vertical rows of teeth on the posterior surface (Fig. 68S, V); in the pupa by the frons and the anterior narrow portion of the thorax bare; and in the larva by the hypostoma with 11–13 hypostomal bristles on each side.

This species was suspected to be synonymous with *S. atrum* Delfinado by Takaoka (2006c).

***Simulium* (*Simulium*) *gatchaliani* Takaoka, 1983**
Simulium (*Simulium*) *gatchaliani* Takaoka, 1983: 115–116 (male and pupa).

Distribution. Philippines (Mindanao).

Bionomics. Aquatic habitat: A cascading small stream (width 0.3–0.8 m) in a mountainous area, Davao, Mindanao (Takaoka, 1983).

Remarks. This species was described from Mindanao, the Philippines (Takaoka, 1983). Its female and larva are unknown.

The male of this species is characterized by the small number of upper-eye (large) facets in 12 vertical columns and 14 horizontal rows.

***Simulium* (*Simulium*) *halconense* Takaoka and Tenedero, 2007**
Simulium (*Simulium*) *halconense* Takaoka and Tenedero, 2007a: 38–42 (female and pupa).

Distribution. Philippines (Mindoro).

Bionomics. Aquatic habitat: A slow-flowing stream (width 1.5–3.0 m, shaded, 25.0 °C, elevation 78 m) near Mt. Halcon, Mindoro Island (Takaoka and Tenedero, 2007a).

Remarks. This species was described from Mindoro Island, the Philippines by Takaoka and Tenedero (2007a). Its male and larva are unknown.

It is characterized in the female by the fore basitarsus greatly dilated, 4.6 times as long as its greatest width; and in the pupa by the dorsum of abdominal segment 1 moderately covered with tubercles on the posterior one-third.

***Simulium* (*Simulium*) *iwahigense* Takaoka, 1983**
Simulium (*Simulium*) *iwahigense* Takaoka, 1983: 116–118 (female, male, pupa, and larva).

Distribution. Philippines (Palawan).

Bionomics. Aquatic habitats: Small streams (width ca. 3.0 m, flow slow, shaded) in lowlands and mountainous areas (Takaoka, 1983).

Remarks. This species was described from Palawan, the Philippines by Takaoka (1983).

It is characterized in the female by the genital fork with strongly angulate lateral ridges and spermatheca relatively enlarged; and in the male by the hind basitarsus narrow, nearly parallel-sided, and yellow on the basal three-fifths.

***Simulium* (*Simulium*) *lardizabalae* Takaoka and Sofian-Azirun, 2015**
(Figs. 68D, F–L and 69B)
Simulium (*Simulium*) *lardizabalae* Takaoka and Sofian-Azirun [in Takaoka et al., 2015b]: 20–23 (male and pupa).

Distribution. Malaysia (Sabah).

Bionomics. Aquatic habitat: A small stream (width 0.5 m, depth 10 cm, 13.9 °C, shaded, elevation 2134 m) slowly flowing in a natural forest, along the trekking route to the peak of Mt. Kinabalu from Mesilau, 2–3 km from Mesilau (Takaoka et al., 2015b).

Remarks. This species was described from Sabah by Takaoka et al. (2015b). Its female and larva are unknown.

The male and pupa of this species are similar to those of *S. nigripilosum* in the male leg color, the arrangement of the pupal gill filaments (Fig. 69B) and the wall-pocket-shaped cocoon. However, it is distinguished from the latter species in the male by the number of upper-eye (large) facets in 25 vertical columns and 26 horizontal rows (17–19 vertical columns and 19–21 horizontal rows in *S. nigripilosum*); and in the pupa by the frons and the dorsal surface of the anterior half of the thorax bare and abdominal segment 7 with spine-combs dorsally.

***Simulium* (*Simulium*) *laterale* Edwards, 1933**

(Fig. 69R, S)

Simulium laterale Edwards, 1933: 256 (female).
Simulium (*Simulium*) *laterale*: Takaoka et al., 2015b: 8–14 (female, male, and pupa).
Simulium kinabaluense Smart and Clifford, 1969: 29–30 (female, male, pupa, and larva). Synonymized by Takaoka et al. (2015b).

Distribution. Malaysia (Sabah).

Bionomics. Aquatic habitats: A small stream (width 0.2 m, depth 2–3 cm, bottom sandy, 16.4 °C, shaded, elevation 1728 m) slowly flowing in a natural forest, Timpohon, Mt. Kinabalu; and a stream (width 10–15 m, depth 0.5 m, exposed to the sun) just below the Park Office of Mesilau (Takaoka et al., 2015b).

Remarks. This species was originally described on the basis of females collected at Lumu in Kinabalu National Park (Edwards, 1933). Smart and Clifford (1969) redescribed the female of this species from reared specimens and described the male, pupa, and larva of this species. The paralectotype female of this species was examined and its terminalia were illustrated by Takaoka (1983). *Simulium kinabaluense* Smart and Clifford was synonymized with this species by Takaoka et al. (2015b).

It is characterized in the female by the hind tibia entirely darkened and the basal portion of the radius bare; and in the pupa by the cocoon shoe-shaped, with a relatively high anterior collar having small depressed portion medially (Fig. 69R, S).

***Simulium* (*Simulium*) *maklarini* Takaoka, 2007**

Simulium (*Simulium*) *maklarini* Takaoka, 2007b: 291–295 (female and pupa).

Distribution. Malaysia (Sabah).

Bionomics. Aquatic habitat: A slow-flowing stream (width ca. 20 cm, 23.0 °C, shaded, elevation 550 m) in Hot Spring Park, Kundasang (Takaoka, 2007b).

Remarks. This species was described from one female reared from a pupa collected in Sabah (Takaoka, 2007b).

It is characterized by the female fore coxa whitish yellow, hind tibia white at its base, six pupal gill filaments divergent at an angle of ca. 90 degrees and cocoon shoe-shaped, with a moderate anteroventral collar.

***Simulium* (*Simulium*) *malaibaense* Takaoka and Tenedero, 2007**

(Fig. 67D)

Simulium (*Simulium*) *malaibaense* Takaoka and Tenedero, 2007b: 89–96 (female, male, pupa, and larva).

Distribution. Philippines (Negros).

Bionomics. Aquatic habitat: A moderately flowing river (width 6–9 m, 21.0 °C, elevation ca. 600 m) in Negros Occidental (Takaoka and Tenedero, 2007b).

Remarks. This species was described from Negros, the Philippines by Takaoka and Tenedero (2007b).

It is characterized in the female by the claw with a tiny rudimentary subbasal tooth (Fig. 67D); and in the pupa by the gill filaments widely divergent when viewed laterally.

***Simulium* (*Simulium*) *mayoyaoense* Takaoka, 2006**

(Fig. 69G)

Simulium (*Simulium*) *mayoyaoense* Takaoka, 2006b: 293–297 (female, male, and pupa).

Distribution. Philippines (Luzon).

Bionomics. Aquatic habitat: A moderately flowing stream (width ca. 5 m, 22.0 °C, elevation ca. 1100 m), east of Banaue Town, Ifugao Province (Takaoka, 2006b).

Remarks. This species was described from northern Luzon, the Philippines by Takaoka (2006b). Its larva is unknown.

It is characterized in the male by the large number of upper-eye (large) facets in 19 or 20 vertical columns and 19 or 20 horizontal rows and hind basitarsus yellowish on its basal half and slightly more; and in the pupa by the gill with six filaments widely divergent when viewed laterally (Fig. 69G) and the dorsum of abdominal segment 1 sparsely covered with tubercles.

***Simulium* (*Simulium*) *melanopus* Edwards, 1929**

Simulium melanopus Edwards, 1929: 13–14 (female and male); Delfinado, 1962: 48–50 (female and male).

Simulium (*Simulium*) *melanopus* Edwards: Delfinado, 1969: 202–203; Takaoka, 1983: 93–96 (female, male, pupa, and larva).

Distribution. Philippines (Luzon, Leyte, Mindanao, and Negros).

Bionomics. Aquatic habitats: Rapids of rivers (width 5–15 m) and streams (width 2–4 m) in lowlands (Takaoka, 1983).

Remarks. This species was described from females and males from Luzon, the Philippines by Edwards (1929). The wrong association between the female and male was pointed out by Takaoka (1983). The female thought as *S. melanopus* was that of *S. discrepans*. The female, pupa, and larva of this species were described by Takaoka (1983).

It is characterized by the pupal gill filaments covered with relatively large tubercles on their surface.

Simulium (Simulium) mirum Takaoka, Sofian-Azirun and Low, 2016
(Fig. 67A)

Simulium (*Simulium*) *laterale* (*nec* Edwards): Smart and Clifford, 1969: 22–26 (female, male, pupa, and larva).
Simulium (*Simulium*) sp.: Takaoka et al., 2015b: 28.
Simulium (*Simulium*) *mirum* Takaoka, Sofian-Azirun and Low [in Takaoka et al., 2016a]: 3–12 (female, male, pupa, and larva).

Distribution. Indonesia (Kalimantan) and Malaysia (Sabah and Sarawak).

Bionomics. Aquatic habitats: A small stream (width 0.2 m, depth 2–3 cm, 16.4 °C, shaded, elevation 1728 m) slowly flowing in a natural forest, a small stream (width 3.5 m, depth 10–12 cm, 17.5 °C, shaded, elevation 1591 m) moderately to fastly flowing in a natural forest, Mt. Kinabalu; a stream (width 6.0 m, depth 0.5 m, 17.0 °C, partially shaded, elevation 1714 m) moderately to fastly flowing in a natural forest, Mesilau, Kundasang, Sabah; and a stream (width 3.0–4.0 m, bottom rocky, 23.0 °C, partially shaded, elevation 928 m) moderately flowing in a pasture, Bakalalan, Sarawak (Takaoka et al., 2016a).

Remarks. This species was described from Sabah and Sarawak, Malaysia by Takaoka et al. (2016a). It was once wrongly identified as *S. laterale* from Sabah, Malaysia by Smart and Clifford (1969).

It is striking that this species has dimorphic male scutal color patterns linked with different numbers of upper-eye (large) facets: type A male scutum white pruinose with a medial vitta and lateral round spots non-pruinose and upper-eye (large) facets in 12–14 vertical columns and 15 or 16 horizontal rows, whereas type B male scutum white pruinose with an inverted-T-like medial large portion non-pruinose and upper-eye (large) facets in 16 or 17 vertical columns and 17 or 18 horizontal rows. Genetic evidence supported the conspecificity of the two types of the males of this species.

Simulium (Simulium) murudense Takaoka, Ya'cob and Sofian-Azirun, 2015
(Fig. 69N, O)

Simulium (*Simulium*) *murudense* Takaoka, Ya'cob and Sofian-Azirun, 2015a: 39–44 (female, male, and pupa).

Distribution. Malaysia (Sarawak).

Bionomics. Aquatic habitat: A stream (width 0.3–0.5 m, bottom muddy, 14.4 °C, partially shaded, elevation 2115 m) slowly flowing in a grassland, Church Camp, Bakalalan (Takaoka et al., 2015a).

Remarks. This species was described from Sarawak by Takaoka et al. (2015a). Its larva is unknown. It is remarkable that the fore coxae of the female and male are darkened and the basal portion of the radial vein of the female is fully haired.

***Simulium (Simulium) neodiscrepans* Takaoka, 2006**

(Figs. 67B, F–J, 68E, M–R and 69U–W)

Simulium (Simulium) neodiscrepans Takaoka, 2006b: 301–306 (female, male, pupa, and larva).

Distribution. Philippines (Luzon).

Bionomics. Aquatic habitat: A moderately flowing river (width 10–15 m, 29.0 °C, elevation ca. 60 m), Ilocos Sur Province (Takaoka, 2006b).

Remarks. This species was described from northern Luzon, the Philippines by Takaoka (2006b).

It is characterized in the female by the femora mostly yellow, claw simple, ovipositor lobes of the eighth sternite short (Fig. 67F); in the male by the femora dark yellow or ocherous, hind basitarsus slender (Fig. 68E), ventral plate without setae (Fig. 68M), and paramere with the wide base covered with setae (Fig. 68R); and in the pupa by the frons almost bare and abdomen without spine-combs on segments 5–9; and in the larva by the rectal organ simple or with one to four finger-like or thumb-like secondary lobules per lobe.

***Simulium (Simulium) nigripilosum* Edwards, 1933**

(Fig. 69D, J–M)

Simulium nigripilosum Edwards, 1933: 256 (female).

Simulium (Simulium) nigripilosum: Takaoka et al., 2015b: 14–20 (female, male, and pupa).

Distribution. Malaysia (Sabah).

Bionomics. Aquatic habitats: A stream (width 0.5 m, depth 2–5 cm, shaded) slow-flowing in a forest, on the right side of the trekking route to the peak of Mt. Kinabalu, 2 km up from Carson Falls, a small stream (width 1.0 m, depth 7–10 cm, bottom rocky, 16.1 °C, shaded, elevation 1890 m) slow-flowing in a natural forest, down-stream of Carson Falls, Mt. Kinabalu, and a stream (width 0.5 m, depth 2 cm, bottom concrete, 16.0 °C, partially shaded, elevation 1934 m) moderately flowing in a cultivated field, Mesilau, Kundasang (Takaoka et al., 2015b).

Remarks. This species was originally described on the basis of females and males collected at Kamborangah in Kinabalu National Park (Edwards, 1933). Smart and Clifford (1969) associated their adult females reared from pupae with this species, but this association was wrong (Takaoka 1996b). The species regarded as *S. nigripilosum* by Smart and Clifford (1969) was described as a new species, *S. beludense* (Takaoka 1996b). It was placed in the *S. argentipes* species-group (Takaoka and Saito, 2007). The terminalia of the holotype female of *S. nigripilosum* were illustrated by Takaoka (1983). The female of *S. nigripilosum* was redescribed based on females reared from pupae, and the male and pupa were described by Takaoka et al. (2015b).

It is characterized in the female by the antenna dark brown, fore coxa dark brown, fore tibia dark brown with a wide white area on its outer surface, mid tibia entirely dark brown to black, mid basitarsus yellowish on its basal two-thirds, hind tibia dark brown to black with its base yellow, hind basitarsus whitish on its basal two-thirds, claw with a small basal tooth and the basal portion of the radial vein entirely bare; and in the pupa by the frons and the dorsal surface of the thorax densely covered with round tubercles, gill with six filaments, of which the upper filament of the dorsal pair is much thicker than the five other filaments (Fig. 69D), and cocoon wall-pocket-shaped (Fig. 69J–M).

***Simulium* (*Simulium*) *palawanense* Delfinado, 1971**

Simulium (*Simulium*) *palawanense* Delfinado, 1971: 132–133 (female); Takaoka, 1983: 101–102 (female).

Distribution. Philippines (Palawan).

Bionomics. Unknown.

Remarks. This species was described from females from Palawan, the Philippines by Delfinado (1971). Its male, pupa, and larva are unknown.

It is characterized by the ovipositor valves membranous and paraproct slightly widened toward its apex when viewed laterally.

***Simulium* (*Simulium*) *pandum* Takaoka, 2006**

(Fig. 67M, N)

Simulium (*Simulium*) *pandum* Takaoka, 2006b: 297–301 (female, male, and pupa).

Distribution. Philippines (Luzon).

Bionomics. Aquatic habitat: A moderately flowing mountain stream (width 2–5 m, 18.0 °C, elevation 1640 m) at Sumigor Bridge, Ifugao Province (Takaoka, 2006b).

Remarks. This species was described from northern Luzon, the Philippines by Takaoka (2006b). Its larva is unknown.

It is characterized in the female by the ovipositor lobes of the eighth sternite acutely bent ventrally (Fig. 67M, N); in the male by the hind basitarsus yellow on its basal half; and in the pupa by the dorsum of abdominal segment 1 moderately covered with tubercles.

***Simulium* (*Simulium*) *retusum* Delfinado, 1971**

Simulium (*Simulium*) *retusum* Delfinado, 1971: 134–135. (female); Takaoka, 1983: 102.

Distribution. Philippines (Palawan).

Bionomics. Unknown.

Remarks. This species was described from females from Palawan, the Philippines by Delfinado (1971). Its male, pupa, and larva are unknown. Its description was so brief, lacking important information on leg color and the presence or absence of a claw tooth, thus making its recognition difficult.

***Simulium* (*Simulium*) *salebrosum* Takaoka, 1983**

Simulium (*Simulium*) *salebrosum* Takaoka, 1983: 109–111 (female, male, pupa, and larva).

Distribution. Philippines (Mindanao).

Bionomics. Aquatic habitats: Rivers (width 25–30 m, elevation ca. 200 m) in Cotabato (Takaoka, 1983).

Remarks. This species was described from Mindanao, the Philippines by Takaoka (1983).

It is characterized in the pupa by the thoracic integument roughened, forming numerous ridges and densely covered with armed tubercles, and gill filaments with relatively large tubercles on the ridges.

***Simulium* (*Simulium*) *simulacrum* Delfinado, 1969**

Simulium (*Simulium*) *simulacrum* Delfinado, 1969: 202 (male); Takaoka, 1983: 102–104 (male and pupa).

Distribution. Philippines (Luzon).

Bionomics. Aquatic habitat: A small stream (width 0.6–1.2 m) in Banaue (Takaoka, 1983).

Remarks. This species was described from a single male captured by a light trap at Abatan, Mountain Province by Delfinado (1969). Its male and pupa were described by Takaoka (1983).

The male of this species is characterized by the hind basitarsus yellow on its basal two-thirds.

***Simulium* (*Simulium*) *subatrum* Takaoka, 1983**

(Fig. 69X)

Simulium (*Simulium*) *subatrum* Takaoka, 1983: 104–107 (female, male, pupa, and larva).

Distribution. Philippines (Luzon and Palawan).

Bionomics. Aquatic habitat: A cascading mountain stream (width 5–8 m) in Palawan (Takaoka, 1983).

Remarks. This species was described from Palawan and Luzon, the Philippines by Takaoka (1983).

It is characterized in the female and male by the darkened fore coxae; in the female by abdominal segments 2 and 6–8 shiny dorsally, and ovipositor valves membranous, nearly triangular and tapered posteromedially; and in the male by the ventral plate subquadrate when viewed ventrally, without teeth on its posterior surface.

***Simulium* (*Simulium*) *sumigorense* Takaoka, 2006**

(Fig. 69I)

Simulium (*Simulium*) *sumigorense* Takaoka, 2006c: 319–321 (female, male, and pupa).

Distribution. Philippines (Luzon).

Bionomics. Aquatic habitat: A fast-flowing stream (width 0.5–3.0 m, 15.0 °C, shaded, elevation 1680 m), Banaue (Takaoka, 2006c).

Remarks. This species was described from northern Luzon, the Philippines by Takaoka (2006c). Its larva is unknown.

It is characterized by the pupal gill with eight thread-like filaments (Fig. 69I).

***Simulium* (*Simulium*) *suplidoi* Takaoka, 1983**
Simulium (*Simulium*) *suplidoi* Takaoka, 1983: 111–113 (female, male, pupa, and larva).

Distribution. Philippines (Negros).

Bionomics. Aquatic habitat: A creek (width 4–6 m, shaded) flowing in a forest (Takaoka, 1983).

Remarks. This species was described from Negros, the Philippines by Takaoka (1983).

It is characterized in the female by the ovipositor valves fused to the sternite 8 pointed apically, with their inner margins concave medially; and in the male by the small number of upper-eye (large) facets in 12 vertical columns and 14 horizontal rows.

***Simulium* (*Simulium*) *taalense* Takaoka, 1983**
Simulium (*Simulium*) *taalense* Takaoka, 1983: 107–109 (female, male, pupa, and larva).

Distribution. Philippines (Luzon and Mindoro).

Bionomics. Aquatic habitat: A small stream (width ca. 2 m, partially shaded) in a hilly area, near the Lake Taal (Takaoka, 1983).

Remarks. This species was described from Luzon, the Philippines by Takaoka (1983). It was later recorded from Mindoro by Takaoka and Tenedero (2007a).

It is characterized in the female by the scutum with yellow short hairs, ovipositor valves fused to the sternite 8 elongate, with its apical half much narrowed and furnished with numerous stout hairs; and in the pupa by the cocoon boot-shaped.

***Simulium* (*Simulium*) *timpohonense* Takaoka and Sofian-Azirun, 2015** (Fig. 69A)
Simulium (*Simulium*) *timpohonense* Takaoka and Sofian-Azirun [in Takaoka et al., 2015b]: 23–27 (male and pupa).

Distribution. Malaysia (Sabah).

Bionomics. Aquatic habitat: A small stream (width 1.0 m, depth 7–10 cm, 16.1 °C, shaded, elevation 1890 m) slowly flowing in a natural forest, just below Carson Falls, Timpohon, Mt. Kinabalu (Takaoka et al., 2015b).

Remarks. This species was described from a male reared from a pupa from Sabah (Takaoka et al., 2015b). Its female and larva are unknown.

It is characterized in the male by the scutum with a characteristic pattern, fore coxa whitish yellow, hind tibia brownish black with its base yellowish white and hind basitarsus yellowish white on its basal three-fifths and dark brown on its apical two-fifths; and in the pupa by the frons and the thorax covered with tubercles, gill with six filaments decreasing in length from dorsal to ventral (Fig. 69A) and cocoon shoe-shaped, with a low anteroventral collar.

Simulium (*Simulium*) *tomentosum* Delfinado, 1969

Simulium (*Simulium*) *tomentosum* Delfinado, 1969: 203 (male); Takaoka, 1983: 135–136 (male).

Distribution. Philippines (Luzon).

Bionomics. Unknown.

Remarks. This species was described from a single male from Mountain Province, Luzon, the Philippines by Delfinado (1969). Its female, pupa, and larva are unknown.

It is characterized by the scutum silvery pruinose except a median vitta and two small round lateral spots, which are non-pruinose.

Simulium (*Simulium*) *multistriatum* Species-group

Simulium (*Simulium*) *barraudi* Puri, 1932

(Fig. 71L–N)

Simulium (Simulium) barraudi Puri, 1932c: 1130–1132 (female, male, and pupa); Takaoka et al., 2020d: 303–306 (male and pupa); Deng et al., 1994: 13 (larva).

Simulium (*Simulium*) sp. B Lewis, 1964: 293 (female). Identified as *S.* (*S.*) *barraudi* by Takaoka et al. (2020d).

Distribution. Bhutan, India (Himachal Pradesh and Kashmir), Myanmar, Nepal, Pakistan, and China (Tibet).

Bionomics. Aquatic habitat: A river (width over 10 m, flow rapid, elevation 651 m), Tamor, Nepal (Takaoka et al., 2020d).

Remarks. This species was originally described from India by Puri (1932c). It was later recorded from Tibet (Deng et al., 1994), Bhutan (Takaoka et al., 2021c), Myanmar (Takaoka et al., 2018j), and Nepal (Takaoka et al., 2020d).

The female of this species is similar in leg color to that of *S. digitatum* Puri from India (Puri, 1932c) but is distinguished by sternite 8, which is bare medially in this species but is haired medially in *S. digitatum*. The female of this species is also distinguished from that of *S. novolineatum* from India (Puri, 1932c, 1933b) by the mid basitarsus, which is yellow on the basal three-fourths in this species but is yellow on the basal half in *S. novolineatum*.

The male of this species is characterized by the ventral plate with its body nearly quadrate when viewed ventrally and without serrated posterolateral margins (Fig. 71L, N) and style with its greatest width at the apical one-third (not near the base) when viewed ventrolaterally.

Simulium (*Simulium*) *bullatum* Takaoka and Choochote, 2005

(Fig. 72H)

Simulium (*Simulium*) *bullatum* Takaoka and Choochote, 2005c: 43–47 (female and pupa); Pramual, 2014: 552–553 (larva).

Distribution. Thailand.

Bionomics. Aquatic habitat: A small stream (width 1.0 m, depth 20 cm, shaded, 20 °C, elevation 1420 m) moderately flowing in Flower Garden, Doi Pui National Park, Chiang Mai Province (Takaoka and Choochote, 2005c).

Remarks. This species was described from northern Thailand by Takaoka and Choochote (2005c). Its larva was described by Pramual (2014). Its male is unknown.

It is characterized in the female by the cibarium with several processes near the base of the dorsomedian projection; and in the pupa by the basal fenestra greatly inflated at the base of the gill (Fig. 72H).

***Simulium* (*Simulium*) *chainarongi* Kuvangkadilok and Takaoka, 1999**
(Fig. 72E, P)

Simulium (*Simulium*) *chainarongi* Kuvangkadilok and Takaoka [in Takaoka and Kuvangkadilok, 1999]: 501–504 (female, male, pupa, and larva).

Distribution. Thailand.

Bionomics. Aquatic habitats: A stream (width 2–10 m, 26.5 °C, elevation 140 m) at Klang Lum Duan, and Huai Luang Waterfall (width 5 m, flow fast, 24.0 °C, shaded, elevation 300 m) (Takaoka and Kuvangkadilok, 1999).

Remarks. This species was described from Thailand by Takaoka and Kuvangkadilok (1999).

It is characterized in the female by the mid and hind femora entirely darkened; in the male by the hind basitarsus dark yellow on its basal two-fifths and brownish black on tits apical three-fifths and style not narrowed medially when viewed ventrolaterally; in the pupa by the cocoon shoe-shaped; and in the larva by the presence of paired dorsal protuberances on abdominal segments 1–6.

***Simulium* (*Simulium*) *chaliowae* Takaoka and Boonkemtong, 1999**

Simulium (*Simulium*) *chaliowae* Takaoka and Boonkemtong [in Takaoka and Kuvangkadilok, 1999]: 497–501 (female, male, and pupa); Pramual, 2014: 549–554 (larva).

Distribution. Thailand.

Bionomics. Aquatic habitat: A stream (width 3 m, 20 °C, elevation 290) at Maekham Waterfall (Takaoka and Kuvangkadilok, 1999).

Remarks. This species was described from Thailand by Takaoka and Kuvangkadilok (1999). Its larva was described by Pramual (2014).

It is characterized in the female by the basal portion of the radius with a few hairs near the apical end; in the male by the hind basitarsus entirely darkened, basal protuberance of the style with many teeth along its anterior margin; in the pupa by the gill with eight filaments widely divergent when viewed laterally and cocoon shoe-shaped; and in the larva by abdominal segments 2–6 each with a pair of protuberances dorsally.

Simulium (Simulium) *chanyae* Takaoka and Choochote, 2007
(Fig. 72I)

Simulium (*Simulium*) *chanyae* Takaoka and Choochote, 2007: 19–22 (female); Takaoka et al., 2018c: 805–815 (female, male, and pupa).

Distribution. Myanmar and Thailand.

Bionomics. Aquatic habitat: A stream (width 2.5–3.0 m, depth 0.1–0.3 m, bottom rocky and muddy, 20.0 °C, shaded, elevation 1279 m) moderately flowing, Wain-Ya Village, Shan State, Central Myanmar (Takaoka et al., 2018c).

Remarks. *Simulium chanyae* was originally described from two females caught on a human in Thailand (Takaoka and Choochote, 2007). The male and pupa of this species were described based on the specimens from Myanmar (Takaoka et al., 2018c). Its larva is unknown.

It is characterized in the female by the femora yellowish except their apical caps darkened; in the male by the subcosta haired; and in the pupa by the gill filaments widely divergent when viewed laterally and cocoon without anterolateral windows (Fig. 72I).

Simulium (Simulium) *daoense* Takaoka and Adler, 2017

Simulium (*Simulium*) *daoense* Takaoka and Adler [in Takaoka et al., 2017c]: 108–116 (female, male, pupa, and larva).

Distribution. Vietnam and Thailand.

Bionomics. Aquatic habitat: A small shallow stream (width 0.5 m, bottom of pebbles, 7.0 °C, partially shaded, elevation 1315 m) moderately flowing from a forest, Sapa (Takaoka et al., 2017c).

Remarks. This was described from northern Vietnam by Takaoka et al. (2017c). It was recorded from Thailand by Thaijarern et al. (2018).

It is characterized in the female by the haired basal portion of the radial vein; in the male by the sensory vesicle ellipsoidal; and in the pupa by the mediolateral and ventrolateral trichomes of the pupal thorax unbranched, gill filaments divergent at an angle of ca. 90 degrees when viewed laterally and cocoon with an enlarged open space on each side.

Simulium (Simulium) *demolaense* Takaoka and Somboon, 2008
(Fig. 72A)

Simulium (*Simulium*) *demolaense* Takaoka and Somboon, 2008: 241–245 (male and pupa).

Distribution. Bhutan.

Bionomics. Aquatic habitat: A stream (width 1.5 m, depth 10 cm, shaded, 24 °C, elevation 501 m) moderately flowing in a densely forested area (Takaoka and Somboon, 2008).

Remarks. This species was described from Bhutan by Takaoka and Somboon (2008). Its female and larva are unknown.

It is characterized in the male by the ventral plate with a few setae on its anterior surface; and in the pupa by the stalks of the dorsal and ventral paired filaments at a right angle when viewed laterally.

***Simulium* (*Simulium*) *dentatum* Puri, 1932**

Simulium (*Simulium*) *dentatum* Puri, 1932c: 1135–1137 (female, male and pupa); Datta, 1974a: 21–22 (larva).

Distribution. India (West Bengal), Bhutan, and China (Tibet).

Bionomics. Aquatic habitats: Streams in Kurseong (elevation, 1800 m), Darjeeling (Puri, 1932c).

Remarks. This species was described from North India by Puri (1932c). It was later recorded from Bhutan by Datta (1991a, b). Its larva was described by Datta (1974a).

It is characterized by the male ventral plate with its narrow body having a flattened keel-like process produced ventrally and two vertical rows of sharp teeth on its posterior surface.

***Simulium* (*Simulium*) *deothangense* Takaoka and Somboon, 2008**

Simulium (*Simulium*) *deothangense* Takaoka and Somboon, 2008: 245–250: (female, male, and pupa).

Distribution. Bhutan.

Bionomics. Aquatic habitat: A stream (width 1.5 m, depth 10 cm, shaded, 24 °C, elevation 501 m) moderately flowing in a densely forested area (Takaoka and Somboon, 2008).

Remarks. This species was described from Bhutan by Takaoka and Somboon (2008). Its larva is unknown.

It is characterized in the pupa by the gill filaments widely divergent when viewed laterally and cocoon with a small anterolateral open space on each side.

***Simulium* (*Simulium*) *fenestratum* Edwards, 1934 (complex)**

Simulium (*Simulium*) *fenestratum* Edwards, 1934: 110–111 (female, male, pupa, and larva).

Distribution. Indonesia (Sumatra), Laos, and Thailand.

Bionomics. This species is common in North and West Sumatra, occurring in various sizes of flowing waters from irrigation ditches (0.3–0.5 m wide) to rivers (6–8 m wide) exposed to the sun or shaded, at an elevation of 300–1450 m (Takaoka et al., 2000).

Larvae of this species were frequently infected with mermithids, microsporidans and other fungi (Takaoka et al., 2000).

Remarks. This species was described from Sumatra, Indonesia by Edwards (1934). It was recorded from Laos (Adler and Crosskey, 2017) and Thailand (Takaoka and Saito, 1996).

The populations of this species in Thailand are genetically different form that in Sumatra, Indonesia (Hew et al., 2023a).

It is characterized in the female by the haired basal portion of the radius; in the male by the hind basitarsus almost entirely darkened though its basal one-third somewhat paler; and in the larva by the rectal organ with 15–20 secondary lobules per lobe. Takaoka et al. (2000) suspected that this species is a species-complex because they found two types of male and larval specimens. The male hind basitarsus was almost entirely darkened in one type but darkened except its basal portion yellow in the other type. The larval abdomen has paired protuberances in one type but has no such protuberances in the other type. Further studies are needed to clarify the species status of this species.

***Simulium* (*Simulium*) *gokarnaense* Takaoka and Shrestha, 2020**
Simulium (*Simulium*) *gokarnaense* Takaoka and Shrestha [in Takaoka et al., 2020d]: 300–303 (male and pupa).

Distribution. Nepal.

Bionomics. Aquatic habitat: A stream (elevation 1317 m), Kathmandu (Takaoka et al., 2020d).

Remarks. This species was described from Nepal by Takaoka et al. (2020d). Its female and larva are unknown.

It is characterized in the male by the number of upper-eye (large) facets in 18 vertical columns and 19 horizontal rows, hind basitarsus moderately enlarged, slightly narrower than the hind tibia and femur, style slightly narrowed medially when viewed ventrolaterally, with a long basal protuberance having no spines on its anterior surface, and ventral plate with its body subquadrate, slightly widened posteriorly when viewed ventrally; and in the pupa by the gill with eight filaments widely divergent basally at an angle of 120 degrees when viewed laterally.

***Simulium* (*Simulium*) *hillycum* Maskey, 1989**
Simulium (*Simulium*) *hillycum* Maskey, 1989: 38–40 (female, male, pupa, and larva).

Distribution. Nepal.

Bionomics. Aquatic habitats: Rivers and streams in the mountain sides, with water temperature 18–23 °C (Maskey, 1989).

Remarks. This species was described from Nepal by Maskey (1989).

It is characterized in the female by the fore basitarsus ca. 4.5 times as long as its greatest width, mid tibia entirely darkened and ovipositor valve without distinct hairs; in the male by the style with a long basal protuberance having several small spines on its anterior surface; in the pupa by the stalks of the dorsal and ventral pairs of gill filaments divergent at ca. 90 degrees when viewed laterally; and in the larva by the rectal organ with 10 secondary lobules per lobe and posterior circlet with 76–78 rows of 14–16 hooklets per row.

No information is available on whether the basal portion of the female radius is haired or bare.

***Simulium* (*Simulium*) *hirtinervis* Edwards, 1928**

(Figs. 70B–D and 72D, N)

Simulium hirtinervis Edwards, 1928: 63–64 (female).
Simulium (*Simulium*) *hirtinervis*: Takaoka and Davies, 1995a: 115–120 (female, male, pupa, and larva).

Distribution. Malaysia (Peninsular) and Vietnam.

Bionomics. Aquatic habitats: Small to large streams at low to high elevations (220–1600 m) in Cameron's Highlands (Ya'cob et al., 2016).

Remarks. This species was described from females from Peninsular Malaysia (Edwards, 1928). The male, pupa, and larva of this species were described by Takaoka and Davies (1995a).

It is characterized in the female by the haired basal portion of the radial vein. The mid tibia was wrongly described to be blackish except its base whitish in Takaoka and Davies (1995a), but it is whitish except its apical one-sixth blackish. Abdominal segment 2 was also wrongly described to be dark brown in Takaoka and Davies (1995a); it is whitish on the basal four-fifths and dark brown on the rest.

***Simulium* (*Simulium*) *kisapense* Takaoka, Sofian-Azirun and Ya'cob, 2012**

Simulium (*Simulium*) *kisapense* Takaoka, Sofian-Azirun and Ya'cob, 2012d: 1198–1205 (female, male, pupa, and larva).

Distribution. Malaysia (Peninsular).

Bionomics. Aquatic habitat: The Kisap river (width ca. 10 m, flow moderate, bottom pebbles, 24.0 °C, elevation 42 m) (Takaoka et al., 2012d).

Remarks. This species was described from Langkawi Island, Peninsular Malaysia (Takaoka et al., 2012d).

It is characterized in the female by the bare basal section of the radial vein; in the male by the ventral plate with setae; and in the pupa by the gill with eight filaments divergent at an angle of 90 degrees or more when viewed laterally and cocoon shoe-shaped, with small open space(s) on each side.

***Simulium* (*Simulium*) *lacduongense* Takaoka and Ya'cob, 2015**

Simulium (*Simulium*) *lacduongense* Takaoka and Ya'cob [in Takaoka et al., 2015c]: 60–67 (female, male, and pupa).

Distribution. Vietnam.

Bionomics. Aquatic habitat: A small stream (width 3–6 m, depth 10–20 cm, bottom sandy, 20.0 °C, elevation 1440 m) slowly flowing in an open land, Lac Duong, Lam Dong Province (Takaoka et al., 2015c).

Remarks. This species was described from southern Vietnam by Takaoka et al. (2015c). Its larva is unknown.

It is characterized in the female by the haired basal portion of the radial vein; in the male by the number of upper-eye (large) facets in 19 vertical columns and

20 horizontal rows and basal protuberance of the style with several tiny spines on its anterior surface; and in the pupa by the gill with eight filaments widely divergent when viewed laterally and cocoon wall-pocket-shaped, with an anterolateral open space on each side.

Simulium (Simulium) lampangense Takaoka and Choochote, 2005
(Figs. 70A, E–I, 71A and 72G, J)

Simulium (*Simulium*) *lampangense* Takaoka and Choochote, 2005i: 319–326 (female, male, and pupa); Thaijarern et al., 2017: 268 (larva).

Distribution. Thailand.

Bionomics. Aquatic habitat: A cascading calcareous stream (width 5–10 m, bottom rocky, shaded, 23.5 °C, elevation 570 m) (Takaoka and Choochote, 2005i).

Remarks. This species was described from northern Thailand by Takaoka and Choochote (2005i). Its larva was described by Thaijarern et al. (2017).

It is characterized in the male by the style much narrowed on its apical two-thirds, and its basal protuberance with several minute spines on its anterior surface; in the pupa by the stalks of the ventral and dorsal pairs of gill filaments widely divergent when viewed laterally (Fig. 72G); and in the larva by abdominal segments 3–7 each with a pair of protuberances dorsally.

Simulium (Simulium) laui Takaoka and Sofian-Azirun, 2015
(Figs. 71B–H and 72O–R)

Simulium (*Simulium*) *laui* Takaoka and Sofian-Azirun [in Takaoka et al., 2015c]: 67–74 (female, male, pupa, and larva).

Distribution. Myanmar and Vietnam.

Bionomics. Aquatic habitat: A small shallow stream (width 2–4 m, 17.0 °C, elevation 1415 m) slowly flowing in an open land, Suoi Vang, Lam Dong Province (Takaoka et al., 2015c).

Remarks. *Simulium laui* was originally described from Vietnam (Takaoka et al., 2015c). It was recorded from Myanmar by Takaoka et al. (2018j).

It is characterized in the female by the bare basal portion of the radius; in the male by the large number of upper-eye (large) facets in 20 or 21 vertical columns and 21 horizontal rows and basal protuberance of the style with several tiny spines on the anterior surface; and in the pupa by the gill with eight filaments widely divergent when viewed laterally and cocoon wall-pocket-shaped, with an anterolateral open space on each side.

Simulium (Simulium) malayense Takaoka and Davies, 1995 (complex)

Simulium (*Simulium*) *malayense* Takaoka and Davies, 1995a: 120–123 (female, pupa, and larva).

Distribution. Malaysia (Peninsular), Thailand, and Vietnam.

Bionomics. Aquatic habitats: Small to medium-sized streams at low to high elevations (220–1400 m) in Cameron's Highlands (Ya'cob et al., 2016).

Remarks. This species was described from females, pupae and larvae collected from Fraser's Hill (Takaoka and Davies, 1995a). This species was also recorded from Thailand (Kuvangkadilok and Takaoka, 2000) and Vietnam (Pham, 1998, 1999).

It is characterized in the female by the bare basal portion of the radial vein; and in the pupa by the frons and the anterodorsal surface of the thorax bare and cocoon wall-pocket-shaped, without open spaces anterolaterally and abdominal segment 1 dark gray dorsally.

***Simulium* (*Simulium*) *maskeyae* Takaoka and Shrestha, 2022**

(Fig. 71I–K)

Simulium (*Simulium*) *maskeyae* Takaoka and Shrestha [in Takaoka et al., 2020d]: 295–300 (female, male, and pupa).

Distribution. Nepal.

Bionomics. Aquatic habitat: A stream (26 °C, elevation 1348 m), Far West (Takaoka et al., 2020d).

Remarks. This species was described from Nepal by Takaoka et al. (2020d). Its larva is unknown.

It is characterized in the female by the sensory vesicle 0.37–0.41 times as long as the third palpal segment, and ovipositor valve with two to five medium-long hairs near its anterior margin; in the male by the number of upper-eye (large) facets in 20 or 21 vertical columns and 19 or 20 horizontal rows, hind basitarsus moderately enlarged, slightly narrower than the hind tibia and femur, style slightly narrowed medially when viewed ventrolaterally, with a long basal protuberance having no or a few spines on its anterior surface, and ventral plate with its body subquadrate, slightly widened posteriorly, having its posterior margin rounded when viewed ventrally; and in the pupa by the gill with eight filaments widely divergent basally at an angle of 130 degrees when viewed laterally.

***Simulium* (*Simulium*) *novolineatum* Puri, 1933**

Simulium (*Simulium*) *lineatum* Puri, 1932c: 1125–1130 (female, male, and pupa).
Simulium (*Simulium*) *novolineatum* Puri, 1933b: 817; Datta, 1992a: 51–52.

Distribution. India (Arunachal Pradesh, Karnataka, Uttar Pradesh, and West Bengal) and Myanmar.

Bionomics. Aquatic habitats: Fast-flowing streams (Datta, 1992a).

Remarks. This species was described under the name of *S. lineatum* by Puri (1932c), but renamed as *S. novolineatum* since the name *S. lineatum* was preoccupied by *S.* (*W.*) *lineatum* (Meigen, 1804) from Europe (Puri, 1933b). Its larva is unknown.

It is characterized in the female by the fore basitarsus about five times as long as its greatest width and ovipositor valve with a few short hairs; in the male by the hind basitarsus enlarged, slightly wider than the hind tibia, style narrowed medially, and ventral plate with its body nearly quadrate when viewed ventrally, having a ventrally produced process, of which the anterior surface is bare and the posterior surface with several ridges; and in the pupa by the gill with eight filaments and cocoon with an open space anterolaterally on each side.

***Simulium* (*Simulium*) *ratekholaense* Takaoka and Shrestha, 2020**

Simulium (*Simulium*) *ratekholaense* Takaoka and Shrestha [in Takaoka et al., 2020d]: 289–295 (female, male, and pupa).

Distribution. Nepal.

Bionomics. Aquatic habitat: A stream (flow moderate, elevation 1787 m) running from a forest, Ilam (Takaoka et al., 2020d).

Remarks. This species was described from Nepal by Takaoka et al. (2020d). Its larva is unknown.

It is characterized in the female by the ovipositor valve with seven to nine short to medium-long hairs; in the male by the number of upper-eye (large) facets in 19 vertical columns and 21 or 22 horizontal rows, hind basitarsus greatly enlarged, slightly wider than the hind tibia and femur, style slightly narrowed medially, with a long basal protuberance having several tiny spines on its anterior surface, ventral plate Y-shaped, with its body narrow, nearly parallel-sided when viewed ventrally; and in the pupa by the gill with eight filaments moderately divergent basally at a right angle when viewed laterally.

***Simulium* (*Simulium*) *rotifilis* Chen and Zhang, 1998**

Simulium (*Simulium*) *rotifilis* Chen and Zhang, 1998: 59–60 (pupa and larva); Zeng et al., 2006: 161–167 (female).

Distribution. China (Guizhou and Jiangxi).

Bionomics. Aquatic habitat: A stream (19 °C, elevation 1400 m) (Chen and Zhang, 1998).

Remarks. This species was described from Guizhou, China (Chen and Zhang, 1998). Its female was described by Zeng et al. (2006). Its male is unknown.

It is characterized by the pupal gill composed of an enlarged inflated basal fenestra and eight slender filaments.

***Simulium* (*Simulium*) *sakishimaense* Takaoka, 1977**

Simulium (*Simulium*) *sakishimaense* Takaoka, 1977a: 197–201 (female, male, pupa, and larva).

Simulium (*Simulium*) *fenestratum* (*nec* Edwards), Ogata, 1966: 126–128.

Simulium (*Simulium*) *sakishimaense fujianense* Zhang and Wang, 1991a: 484–486 (female, male, pupa, and larva).

Distribution. Japan (Nansei Islands), China (Fujian, Guizhou, Hong Kong, Hubei, Jiangxi, Yunnan, and Zhejiang), and Taiwan.

Bionomics. Aquatic habitats: Streams and rivers fast flowing in hilly or mountainous areas in Ishigaki Island and Iriomote Island, Japan (Takaoka 1977a).

Remarks. This species was first recorded from Ishigaki Island, Japan as *S. fenestratum* Edwards from Java by Ogata (1966). It was described as a new species, *S. sakishimaense* by Takaoka (1977a). It was recorded from China (Hong Kong) and Taiwan (Takaoka, 1979; Takaoka et al., 1995).

It is characterized in the female by the basal portion of the radius bare or partially haired; in the male by the basal protuberance of the style with tiny spines on its anterior surface; in the pupa by the gill filaments diverged at a right angle when viewed laterally and cocoon wall-pocket-shaped, with a large anterolateral open space on each side.

***Simulium* (*Simulium*) *takense* Takaoka and Choochote, 2005**
(Figs. 71O, P and 72C, L)

Simulium (*Simulium*) *takense* Takaoka and Choochote, 2005e: 127–133 (female, male, pupa, and larva).

Distribution. Thailand.

Bionomics. Aquatic habitat: A small calcareous stream (width ca. 1 m, depth ca. 5 cm, flow moderate, shaded, 24 °C, elevation 420 m) at Nang Khruang Waterfall, Tak Province (Takaoka and Choochote, 2005e).

Remarks. This species was described from northern Thailand (Takaoka and Choochote 2005e). Its larva is unknown.

It is characterized in the male by the ventral plate with its body having two rows of teeth converging ventrally to each other on its posterior surface (Fig. 71O, P); and in the pupa by the gill filaments divergent when viewed laterally (Fig. 72C), abdominal segments 8 and 9 each with distinct spine-combs dorsally, abdominal segment 9 with a pair of distinct hooks and cocoon shoe-shaped (Fig. 72L).

***Simulium* (*Simulium*) *thimphuense* Takaoka, Namgay and Somboon, 2021**
(Fig. 72F, K)

Simulium (*Simulium*) *thimphuense* Takaoka, Namgay and Somboon, 2021c: 81–82 (pupa).

Distribution. Bhutan.

Bionomics. Aquatic habitat: A fast-flowing stream (partially shaded, elevation 2391 m) in Thimphu district (Takaoka et al., 2021c).

Remarks. This species was described from Bhutan by Takaoka et al. (2021c). Its female, male, and larva are unknown.

It is characterized by the acute angle between the stalks of the dorsal and ventral pairs of the pupal gill filaments when viewed laterally (Fig. 72F) and cocoon wall-pocket-shaped, with several small open spaces anterolaterally on each side (Fig. 72K).

Simulium (_Simulium_) _triglobus_ Takaoka and Kuvangkadilok, 1999

(Figs. 70J, K and 72M)

Simulium (*Simulium*) *triglobus* Takaoka and Kuvangkadilok, 1999: 504–507 (female, pupa, and larva); Pramual et al., 2022a: 60–65 (male).

Distribution. Thailand.

Bionomics. Aquatic habitat: A small stream (width 0.5–1.0 m, 19 °C, elevation 500 m) at Tontong Waterfall, Nan Province (Takaoka and Kuvangkadilok, 1999).

Remarks. This species was described from northern Thailand by Takaoka and Kuvangkadilok (1999). Its male was described by Pramual et al. (2022a).

It is characterized in the female by having three spermathecae (Fig. 70J); in the male by the ventral plate with its body hexagonal when viewed ventrally; and in the pupa by the gill with eight filaments widely divergent when viewed laterally and cocoon shoe-shaped, with small open spaces anteriorly (Fig. 72M).

Simulium (_Simulium_) _ubonae_ Thaijarern, Wongpakam, Kangrang and Pramual, 2019

Simulium (*Simulium*) *ubonae* Thaijarern, Wongpakam, Kangrang and Pramual, 2019b: 463–473 (female, male, pupa, and larva).

Distribution. Thailand.

Bionomics. Aquatic habitat: A stream (width 5 m, depth 8 cm, flow moderate, elevation 1226 m) in an open area, Loei Province (Thaijarern et al., 2019b).

Remarks. This species was described from northern Thailand by Thaijarern et al. (2019b).

It is characterized in the female by the sensory vesicle 0.41 times as long as the third palpal segment and fore basitarsus 4.3 times as long as its greatest width; in the male by the small number of upper-eye (large) facets in 15 vertical and 15 or 16 horizontal rows, hind basitarsus 5.4 times as long as its greatest width and style with a long basal protuberance having small spines on its anterior surface; in the pupa by the gill with eight filaments widely divergent when viewed laterally and cocoon shoe-shaped, with a small open space anterolaterally on each side; and in the larva by abdominal segments 2–6 each with a pair of dorsolateral protuberances.

Simulium (_Simulium_) _uncum_ Zhang and Chen, 2001

Simulium (*Simulium*) *uncum* Zhang and Chen, 2001: 216–218 (female, male, pupa, and larva).

Distribution. China (Guizhou).

Bionomics. Aquatic habitat: A mountain stream at an elevation of 1400 m (Zhang and Chen, 2001).

Remarks. This species was described from Guizhou, China (Zhang and Chen, 2001).

It is characterized in the female by the sternite 8 with several hairs even on its central portion; in the male by the number of upper-eye (large) facets in 16 vertical columns and 13 horizontal rows and the ventral plate without teeth; in the pupa by the thoracic trichomes unbranched and only abdominal segment 8 with spine-combs; and in the larva by the posterior circlet with ca. 40 rows of hooklets (the number of rows of hooklets appears to be too small).

***Simulium* (*Simulium*) *undecimum* Takaoka, Srisuka and Saeung, 2019**
Simulium (*Simulium*) *undecimum* Takaoka, Srisuka and Saeung, 2019e: 87–93 (female, male, pupa, and larva).

Distribution. Thailand.

Bionomics. Aquatic habitat: A small stream (width 1.5 m, bottom sandy, flow fast, 14.5 °C, partially shaded, elevation 2210 m) in a natural forest, Doi Inthanon National Parks, Chiang Mai Province (Takaoka et al., 2019e).

Remarks. This species was described from Thailand by Takaoka et al. (2019e).

It is characterized in the female by the cibarium with 17 tiny processes and ovipositor valve with several hairs; in the male by the number of upper-eye (large) facets in 21 vertical columns and 21 horizontal rows, style nearly parallel-sided when viewed ventrolaterally, with a long basal protuberance having many spines on its anterior surface and ventral plate Y-shaped when viewed ventrally, without setae on its anteroventral surface; in the pupa by the gill with eight filaments divergent at ca. 90 degrees when viewed laterally, abdominal segments 7 and 8 each with spine-combs and cocoon wall-pocket-shaped, with open spaces anterolaterally on each side; and in the larva by the thorax and abdomen lacking paired dorsal protuberances.

***Simulium* sp. nr. *novolineatum* Puri, 1932**
Simulium sp. nr. *novolineatum* Puri, 1932: Datta and Pal, 1975: 157–159 (pupa).

Distribution. India (West Bengal).

Bionomics. Unknown.

Remarks. This species was recorded based on a single pupa collected from Darjeeling, India by Datta and Pal (1975). Its female, male, and larva are unknown.

According to Datta and Pal (1975), it is similar to *S. novolineatum* but is distinguished by the ventral surface of abdominal segment 4 with a strongly chitinized seta on each side and cocoon with a very short neck. The extent of chitinization of a seta and the presence or absence of a short neck on the cocoon probably are intraspecific variation.

Simulium (*Simulium*) *nebulicola* Species-group

Simulium (*Simulium*) *celsum* Takaoka and Davies, 1996

(Fig. 73D, F–K, N, S)

Simulium (*Simulium*) *celsum* Takaoka and Davies, 1996: 75–79 (female, male, pupa, and larva).

Distribution. Indonesia (Java).

Bionomics. Aquatic habitats: Small streams (width 0.3–0.6 m) in the National Cibodas Botanic Garden (Takaoka and Davies, 1996).

Remarks. This species was described from west Java by Takaoka and Davies (1996).

It is characterized in the female and male by the femora yellowish except their apices darkened (Fig. 73D); and in the pupa by the presence of terminal hooks.

Simulium (*Simulium*) *nebulicola* Edwards, 1934

(Fig. 73C, P, T)

Simulium (*Simulium*) *nebulicola* Edwards, 1934: 114–115 (male); Takaoka and Davies, 1996: 71–75 (female, male, pupa, and larva).

Distribution. Indonesia (Java, Flores, and Sumatra).

Bionomics. Aquatic habitat: A small stream (width less than 0.5 m, shaded, elevation ca. 1300 m) running down the slope of a natural forest, Puncak, West Java (Takaoka and Davies, 1996).

Remarks. This species was described from Java by Edwards (1934). Its female, pupa, and larva were described by Takaoka and Davies (1996).

It is characterized in the female and male by the femora dark except their bases yellow (Fig. 73C); and in the pupa by the gill with six short filaments (Fig. 73P) and abdomen lacking terminal hooks.

Simulium (*Simulium*) *sumatraense* Takaoka and Sigit, 1997

(Fig. 73A, B, E, L, M, O, Q, R)

Simulium (*Simulium*) *sumatraense* Takaoka and Sigit, 1997: 72–76 (female, male, pupa, and larva).

Distribution. Indonesia (Sumatra).

Bionomics. Aquatic habitats: Small streams (width 0.3–0.6 m, completely or partially shaded) running from a forest to rice fields (Takaoka and Sigit, 1997).

Remarks. This species was described from Sumatra, Indonesia by Takaoka and Sigit (1997).

It is characterized in the female and male by the mid femora entirely dark; in the female by the hind tibia yellowish on its basal half and with a dark subbasal marking (Fig. 73B); in the pupa by the presence of terminal hooks; and in the larva by the body dark gray.

Simulium (*Simulium*) *nobile* Species-group

Simulium (*Simulium*) *asishi* Datta, 1988

(Fig. 75E)

Simulium (*Simulium*) *asishi* Datta, 1988: 1–8 (female, male, pupa, and larva).

Distribution. India (West Bengal).

Bionomics. Aquatic habitats: Slow-flowing streams (Datta, 1988). Females of this species feed on human blood (Datta, 1992b).

Remarks. This species was described from India by Datta (1988).

It is similar to *S. nodosum* but is distinguished in the female by the claw with a small subbasal tooth and ovipositor valves directed posteriorly; in the male by the ventral plate with its body much constricted basally when viewed ventrally; and in the pupa by the gill with three somewhat inflated short tubular filaments with the same width throughout their length, though their tips tapered (Fig. 75E).

Simulium (*Simulium*) *baltazarae* Delfinado, 1962

(Fig. 75B)

Simulium baltazarae Delfinado, 1962: 50–53 (female, male, and pupa).
Simulium (*Simulium*) *nobile* (*nec* De Meijere): Delfinado, 1969: 203.
Simulium (*Simulium*) *baltazarae*: Delfinado, 1971: 132; Takaoka, 1983: 120–124 (female, male, pupa, and larva).

Distribution. Philippines (Luzon, Cebu, Mindanao, Mindoro, Negros, and Samar).

Bionomics. Aquatic habitats: Rapids in rivers (5–30 m wide). This is one of the commonest species in the Philippines (Takaoka, 1983).

Remarks. This species was described from adult female and male specimens reared from pupae by Delfinado (1962). It was once synonymized with *S. nobile* from Java (Delfinado, 1969) but was soon resurrected by Delfinado (1971). The revised descriptions of the female, male, and pupa of this species, together with the description of the larva, were provided by Takaoka (1983).

It is characterized in the female and male by the antenna yellow except two apical flagellomeres black (similar to Fig. 74A); in the male by the number of upper-eye (large) facets in 13 vertical columns and 13 horizontal rows, abdominal segments 2 and 4–7 each with a pair of shiny dorsolateral patches; in the pupa by the frons and thorax moderately covered with tubercles and gill with 10 filaments (Fig. 75B); and in the larva by abdomen without dorsal protuberances and spatulate setae.

Simulium (*Simulium*) *benquetense* Takaoka, 1983

(Fig. 75C)

Simulium (*Simulium*) *benquetense* Takaoka, 1983: 125–126 (female, male, pupa, and larva).

Distribution. Philippines (Luzon).

Bionomics. Aquatic habitats: Rapids of a river (width ca. 10 m) and a stream (exposed to the sun) in northern highlands (Takaoka, 1983).

Remarks. This species was described from Luzon, the Philippines by Takaoka (1983).

It is characterized in the female by the antenna brown to brownish black except its scape and pedicel yellow; in the male by the antenna yellow to yellowish brown except two apical flagellomeres black and the number of upper-eye (large) facets in 17 vertical columns and 17 horizontal rows; in the pupa by the frons and thorax bare and gill with six slender filaments (Fig. 75C); and in the larva by abdomen without dorsal protuberances and dark spatulate setae.

***Simulium* (*Simulium*) *cotabatoense* Takaoka, 1983**

(Fig. 75A)

Simulium (*Simulium*) *cotabatoense* Takaoka, 1983: 124–125 (female, male, pupa, and larva).

Distribution. Philippines (Mindanao).

Bionomics. Aquatic habitats: Rapids of a river (width 6–30 m) and streams (width 1–2 m) in lowlands (Takaoka, 1983).

Remarks. This species was described from Mindanao, the Philippines by Takaoka (1983).

It is similar to *S. baltazarae* but is distinguished from the latter species by the pupal gill with 12 slender filaments (Fig. 75A).

***Simulium* (*Simulium*) *delfinadoae* Takaoka, 1983**

(Figs. 74S, T and 75D)

Simulium (*Simulium*) sp. A Delfinado, 1969: 203 (pupa).

Simulium (*Simulium*) *delfinadoae* Takaoka, 1983: 129–131 (female, male, pupa, and larva).

Distribution. Philippines (Luzon and Negros).

Bionomics. Aquatic habitats: Rapids of a river (width ca. 15 m) and small streams (width 0.5–2.0 m) in lowlands (Takaoka, 1983).

Remarks. This was recorded as *S.* (*S.*) sp. A from the Philippines by Delfinado (1969). It was later described as a new species, *S. delfinadoae* by Takaoka (1983).

It is characterized in the female and male by the antenna dark yellow except two or three apical flagellomeres black; in the female by the ovipositor valves well sclerotized, with their inner margins concave; in the male by the number of upper-eye (large) facets in 16 vertical columns and 17 horizontal rows; in the pupa by the frons and thorax mostly bare and gill with six inflated short tubular filaments

(Fig. 75D); and in the larva by abdomen without dorsal protuberances and dark spatulate setae.

***Simulium* (*Simulium*) *kiuliense* Smart and Clifford, 1969**

Simulium (*Simulium*) *kiuliense* Smart and Clifford, 1969: 26–29 (female, male, pupa, and larva).

Distribution. Malaysia (Sabah).

Bionomics. Aquatic habitats: Rivers (Smart and Clifford, 1969).

Remarks. This species was described from Sabah, Malaysia by Smart and Clifford (1969). This species was once synonymized with *S. nobile* by Crosskey (1973), but was resurrected by Low et al. (2016a) based on the result of a genetic analysis using concatenated COI, 16S rRNA, and 28S rRNA sequences. A genetic distance between this species and *S. nobile* is 4.20–4.25.

This species is morphologically indistinguishable from *S. nobile.*

***Simulium* (*Simulium*) *latistylum* Takaoka, 1983**

Simulium (*Simulium*) *latistylum* Takaoka, 1983: 128–129 (male, pupa, and larva): Takaoka and Tenedero, 2007c: 223–226 (female).

Distribution. Philippines (Palawan).

Bionomics. Aquatic habitats: Various types of watercourses from small irrigation channels to large rivers with slow to moderate flows in lowlands (Takaoka and Tenedero, 2007c).

Remarks. This species was described from Palawan, the Philippines by Takaoka (1983). Its female was described by Takaoka and Tenedero (2007c).

It is characterized in the female and male by the antenna dark brown except the scape, pedicel, and the base of the first flagellomere yellow; in the male by the number of upper-eye (large) facets in 14 vertical columns and 14 horizontal rows and style much widened when viewed ventrolaterally; in the pupa by the frons and thorax mostly bare and gill with six slender filaments decreasing in length from dorsal to ventral; and in the larva by abdomen without dorsal protuberances and dark spatulate setae.

***Simulium* (*Simulium*) *leytense* Takaoka, 1983**

Simulium (*Simulium*) *leytense* Takaoka, 1983: 126–128 (female, male, pupa, and larva).

Distribution. Philippines (Leyte, Luzon, Negros, and Panay).

Bionomics. Aquatic habitats: Rapids of rivers (10–15 m wide) and streams (1–2 m wide) moderately flowing in lowlands (Takaoka, 1983).

Remarks. This species was described from Leyte, Negros, Panay, and southern parts of Luzon, the Philippines by Takaoka (1983).

It is similar to *S. benquetense* but is distinguished from the latter species by the number of male upper-eye (large) facets in 13 vertical columns and 13 horizontal rows.

Simulium (Simulium) nobile De Meijere, 1907

(Figs. 74A, E, F and 75K, Q, R)

Simulium nobile De Meijere, 1907: 206 (male); Edwards, 1934: 115–117 (female, male, pupa, and larva).
Simulium (*Simulium*) *nobile*: Takaoka and Davies, 1996: 61–66 (female, male, pupa, and larva).

Distribution. Indonesia (Bali, Java, and Sumatra).

Bionomics. Aquatic habitats: Small streams and rivers (width 0.1–9.0 m) in Java (Takaoka and Davies, 1996).

Remarks. This species was described from adult male specimens from central Java by De Meijere (1907). Its female, male, pupa and larva were briefly described by Edwards (1934) and were fully redescribed by Takaoka and Davies (1996).

It is characterized in the female and male by the antenna yellow except two apical flagellomeres black (Fig. 74A); in the male by the number of upper-eye (large) facets in 13 or 14 vertical columns and 13 or 14 horizontal rows; in the pupa by the frons and thorax sparsely or moderately covered with tubercles and gill with six slender filaments of subequal length; and in the larva by thoracic segment 3 and abdominal segments 1–5 each with a pair of conical protuberances dorsally (Fig. 75K), and abdominal segments covered with spatulate setae and branched setae (Fig. 75Q, R).

Simulium (Simulium) nodosum Puri, 1933

(Figs. 74C, H–M, U–A' and 75F, I, J, O)

Simulium (*Simulium*) *nodosum* Puri. 1933b: 813–817 (female, pupa, and larva); Lewis, 1974: 34–37 (female, male, pupa, and larva); Zhang and Wang 1991a: 486–487 (male).
Simulium shirakii Takahasi and Kawano, 1940: 82 (female). Synonymized by Low et al. (2015a).

Distribution. India, Myanmar, China (Fujian, Hainan, Hong Kong, and Yunnan), Taiwan, Thailand, and Vietnam.

Bionomics. Aquatic habitats: Streams and rivers of lowlands. This species is a human-biter and a vector of an *Onchocerca* species in Thailand (Takaoka et al., 2003).

Remarks. *Simulium nodosum* was originally described from India by Puri (1933b) and was redescribed by Lewis (1974). It is one of the simuliid species with a wide distribution in the Oriental Region from India to Taiwan (Adler, 2022). The species identity of specimens from Myanmar, Thailand, Vietnam, and Taiwan was confirmed by a DNA sequenced-based analysis by Low et al. (2015a).

It is characterized in the female by the wide frons (frons:head ratio 1:2.8–3.1), sensory vesicle medium-long (0.26–0.35 times as long as the third palpal segment), claw without tooth (Fig. 74H) and ovipositor valves short, rounded, directed posteriorly, each having an apical transparent projection directed inwardly

(Fig. 74I); in the male by the ventral plate with its body relatively wide when viewed ventrally (Fig. 74U); in the pupa by the gill with three inflated tubular filaments with their tips rounded (Fig. 75F) and cocoon simple, wall-pocket-shaped (Fig. 75I, J).

***Simulium* (*Simulium*) *timorense* Takaoka, Hadi and Sigit, 2006**
(Figs. 74B, D, G, N–R, B'–E' and 75G, H, L–N, P)

Simulium (*Simulium*) *timorense* Takaoka, Hadi and Sigit, 2006a: 18–23 (female, male, pupa, and larva).

Distribution. Indonesia (Flores, Lombok, Sumba, Sumbawa, and West Timor).

Bionomics. Aquatic habitats: Small streams and rivers (width 0.2–20.0 m, 17.0–27.0 °C) at low to high elevations (150–1250 m) in Flores, the Lesser Sunda Archipelago (Takaoka et al., 2006a).

Remarks. This species was described from West Timor and Flores, Indonesia by Takaoka et al. (2006a). It was later recorded from other islands (Lombok, Sumba, and Sumbawa) of the Lesser Sunda Archipelago (Takaoka et al., 2017e).

It is similar to *S. nobile* including the female terminalia and male genitalia (Fig. 74N–R, B'–E'), claw with a small subbasal tooth (Fig. 74G) and cocoon shoe-shaped (Fig. 75G, H), but is distinguished from the latter species by the larval abdomen lacking paired conical protuberances dorsally.

***Simulium* (*Simulium*) *vanluni* Ya'cob, Takaoka and Sofian-Azirun, 2017**

Simulium (*Simulium*) *nobile* (*nec* De Meijere): Takaoka and Davies, 1995a: 123–124.

Simulium (*Simulium*) *vanluni* Ya'cob, Takaoka and Sofian-Azirun, 2017a: 32–38 (female, male, pupa, and larva).

Distribution. Malaysia (Peninsular) and Thailand.

Bionomics. Aquatic habitats: Small streams and rivers (width 3.7–12.0 m, 22–26 °C) in Peninsular Malaysia (Takaoka and Davies, 1995a; Ya'cob et al., 2017a).

Remarks. This species was once regarded as *S. nobile* by Takaoka and Davies (1995a). It was later described as a new species, *S. vanluni*, by Ya'cob et al. (2017a) based on the result of a genetic analysis using concatenated COI, 16S rRNA, and 28S rRNA sequences (Low et al., 2016a). A genetic distance between this species and *S. nobile* is 1.52–1.73.

This species is morphologically indistinguishable from *S. nobile.*

***Simulium* (*Simulium*) *ornatum* Species-group**

***Simulium* (*Simulium*) *ornatum* Meigen, 1818 (complex)**

Simulium ornatum Meigen, 1818: 290–291 (female).

Odagmia ornata (Meigen): Rubtsov, 1989: 778–781 (female, male, pupa, and larva).
Simulium (*Simulium*) *ornatum*: Crosskey, 1987: 476.

Distribution. Many European and Middle Eastern countries and China (Guizhou, Jilin, Liaoning, and Yunnan).

Bionomics. Aquatic habitats: Small streams and sometimes from very small but warm springs with temperatures fluctuating from 10 °C to 20 °C (Rubtsov, 1989). Females are blood-suckers of man and animals (Rubtsov, 1989).

Remarks. This species was described from German by Meigen (1818). It is similar to species of the *S. variegatum* species-group except that the female and male pleural membranes are haired and pupal gill has eight filaments.

***Simulium* (*Simulium*) *pavlovskii* Species-group**

***Simulium* (*Simulium*) *jieyangense* An, Yan, Yang and Hao, 1994**

Simulium (*Simulium*) *jieyangense* An, Yan, Yang and Hao, 1994: 4–6 (female, male, and pupa).

Distribution. China (Guangdong).

Bionomics. Unknown.

Remarks. This species was described from Guangdong, China by An et al. (1994). Its larva is unknown.

The female ovipositor valves are rounded posteromedially, differing from those of most species in the *S. pavlovskii* species-group, of which ovipositor valves are triangular.

***Simulium* (*Simulium*) *striatum* Species-group**

***Simulium* (*Simulium*) *argyrocinctum* De Meijere, 1913**

Simulium argyrocinctum De Meijere, 1913: 332–333 (female and male); Edwards, 1934: 108–109 (female, male, pupa, and larva).
Simulium (*Simulium*) *argyrocinctum*: Takaoka and Davies, 1996: 44–48 (female, male, pupa, and larva).

Distribution. Indonesia (Java and Sumatra).

Bionomics. Aquatic habitats: Small streams and rivers (width 0.2–9.0 m) (Takaoka and Davies, 1996).

Remarks. The female and male of this species were first described from Java, Indonesia by De Meijere (1913) and were redescribed by Edwards (1934), who at the same time described its pupa and larva for the first time. The female, male, pupa, and larva of this species were redescribed and illustrated by Takaoka and Davies (1996).

Simulium angyrocinctum is the only species of the *S. striatum* species-group from Java and Sumatra. It is characterized by the haired basal portion of the female radial vein, pupal gill with 10 filaments arranged in five pairs and larval thoracic segments 1–3 and abdominal segments 1–8 each having a pair of small conical dorsolateral protuberances.

***Simulium* (*Simulium*) *bagmaticum* Maskey, 1987 [1985]**

Simulium (*Simulium*) *bagmaticum* Maskey, 1987 [1985]: 16–22 (female, male, pupa, and larva).

Distribution. Nepal.

Bionomics. Unknown.

Remarks. This species was described from Nepal by Maskey (1987). This was the first member of the *S. striatum* species-group recorded from Nepal.

It is characterized by the pupal gill with 10 filaments arranged as 2+3+3+2 from dorsal to ventral, of which four or five filaments are somewhat inflated basally, and pupal abdomen lacking terminal hooks (Maskey, 1985).

It is difficult to compare it with other related species for lack of information on certain key features including the relative size of the female frons, presence or absence of hairs on the basal portion of the radial vein, number of male upper-eye (large) facets, and presence or absence of paired protuberances on the larval thorax and abdomen.

***Simulium* (*Simulium*) *bahense* Chen, 2003**

Simulium (*Simulium*) *bahense* Chen, 2003: 336–338 (female, male, and pupa).

Distribution. China (Yunnan).

Bionomics. Aquatic habitats: Small streams at an elevation of 650 m (Chen, 2003).

Remarks. This species was described from Yunnan, China by Chen (2003). Its larva is unknown.

This species is similar to *S. chiangmaiense* originally described from Thailand and *S. kapuri* from Assam, India, in having the pupal gill with eight filaments, of which two filaments of the ventral pair are thinner than the others (Datta, 1975a; Takaoka and Suzuki, 1984). However, it differs from *S. chiangmaiense* by the number of male upper-eye (large) facets in 12 vertical columns and 12 horizontal rows (15–17 vertical columns and 16–18 horizontal rows in *S. chiangmaiense*), and from *S. kapuri* by having pupal terminal hooks (no pupal terminal hooks in *S. kapuri*).

Simulium (Simulium) baliense Takaoka and Sofian-Azirun, 2017

(Fig. 77A)

Simulium (Simulium) baliense Takaoka and Sofian-Azirun [in Takaoka et al., 2017e]: 177–182 (female, male, pupa, and larva).

Distribution. Indonesia (Bali).

Bionomics. Aquatic habitats: Small to medium-sized streams and rivers (width 0.2–15.0 m, depth 0.2–1.0 m, 20–23 °C) (Takaoka et al., 2017e).

Remarks. This species was described from Indonesia by Takaoka et al. (2017e).

It is similar to *S. argyrocinctum* in many characteristics, but is distinguished from the latter species in the female by the frontal ratio (1.31:1.00:1.43 in this species versus 1.4–1.5:1.0:1.6–1.7 in *S. argyrocinctum*) and relative length of the fore basitarsus against its greatest width (6.5 in this species versus 4.8 in *S. argyrocinctum*); in the male by the number of upper-eye (large) facets (14 or 15 vertical columns and 15 horizontal rows in this species versus 17 or 18 vertical columns and 17 or 18 horizontal rows in *S. argyrocinctum*) and relative length of the fore basitarsus against its greatest width (5.64 in this species versus 6.5 in *S. argyrocinctum*); and in the larva by lacking paired conical dorsolateral protuberances.

Simulium (Simulium) beniense Takaoka and Shrestha, 2020

Simulium (Simulium) beniense Takaoka and Shrestha [in Takaoka et al., 2020a]: 39–43 (female and pupa).

Distribution. Nepal.

Bionomics. Aquatic habitat: A stream (width 3.0–4.0 m, elevation ca. 900 m) (Takaoka et al., 2020a).

Remarks. This species was described from a single female reared from a pupa in Nepal by Takaoka et al. (2020a). Its male and larva are unknown.

It is characterized in the female by the haired basal portion of the radial vein and yellowish femora; and in the pupa by the arrangement of 10 gill filaments as 2+3+3+2 from dorsal to ventral and relative thickness of 10 gill filaments, i.e., two filaments of the dorsal pair and three or four filaments of the middle triplets are 1.3–1.5 times as thick as the four or five other filaments.

Simulium (Simulium) chiangmaiense Takaoka and Suzuki, 1984

(Fig. 77E)

Simulium (Simulium) chiangmaiense Takaoka and Suzuki, 1984: 38–39 (female, male, and pupa).

Distribution. Bhutan, Myanmar, Thailand, China (Yunnan) (?), and Vietnam.

Bionomics. Aquatic habitats: Streams and rivers (width 3–50 m, depth 20–35 cm, bottom sandy, 22.5–22.8 °C, elevation 659–958 m) in Myanmar (Takaoka et al., 2018j).

Remarks. *Simulium chiangmaiense* was originally described from Thailand (Takaoka and Suzuki, 1984). Its larva is unknown.

It is characterized in the male by the number of upper-eye (large) facets in 15–17 vertical columns and 16–18 horizontal rows; and in the pupa by the frons and the anterior portion of the thorax covered with relatively larger tubercles bearing tiny secondary projections and gill with eight filaments, of which two filaments of the ventral pair are thinner than the others (Fig. 77E) and all filaments are furnished with marked annular ridges and furrows. In Myanmar, pupae of this species were shown to be variable in the number of gill filaments: eight, nine, or 10 (Takaoka et al., 2018j).

The distribution record of this species from Yunnan, China (Chen and An, 2003) should be confirmed, as suggested by Takaoka et al. (2018j), because it is possible that the species identified as *S. chiangmaiense* was *S. bahense* described from Yunnan, China by Chen (2003). Both species are indistinguishable in the pupal stage, although the numbers of male upper-eye (large) facets are clearly different from each other: 12 vertical columns and 12 horizontal rows in *S. bahense* versus 15–17 vertical columns and 16–18 horizontal rows in *S. chiangmaiense*.

***Simulium* (*Simulium*) *chainpurense* Takaoka and Shrestha, 2020**

Simulium (*Simulium*) *chainpurense* Takaoka and Shrestha (in Takaoka et al., 2020a): 46–49 (female, male, pupa, and larva).

Distribution. Nepal.

Bionomics. Aquatic habitat: A stream (19 °C, elevation 1230 m) (Takaoka et al., 2020a).

Remarks. This species was described from Nepal by Takaoka et al. (2020a).

It is characterized in the female by the basal portion of the radial vein bare or partially haired and hind tibia with a dark subbasal spot; in the male by the number of upper-eye (large) facets in 21 vertical columns and 21 horizontal rows and hind basitarsus somewhat enlarged; in the pupa by the gill with 10 filaments, of which the lower filament of the ventral pair is slightly thicker than the nine other filaments, and dorsum of abdominal segment 1 sparsely covered with tubercles; and in the larva by having conical paired dorsolateral protuberances on thoracic segment 3 and abdominal segments 1–7 (Takaoka et al., 2020a).

***Simulium* (*Simulium*) *consimile* Puri, 1932**

Simulium (*Simulium*) *consimilis* Puri, 1932e: 527–529 (female, male, and pupa).

Simulium (*Simulium*) *consimile*: Davies and Györkös, 1992: 1041–1044 (female, male, pupa, and larva).

Distribution. India (Punjab and Himachal Pradesh), Pakistan and Sri Lanka.

Bionomics. Aquatic habitats: Rivers (width 5–16 m) and a small stream (width 20–25 cm) in an elevation of 60–600 m, in Sri Lanka (Davies and Györkös, 1992).

Remarks. This species was described from India by Puri (1932e). Its larva was described, when its female, male and pupa were redescribed based on specimens from Sri Lanka by Davies and Györkös (1992).

It is characterized in the female by the scutum with yellow hairs, basal portion of the radial vein bare, and fore and mid femora yellow to brownish yellow; in the male by the number of upper-eye (large) facets in 16 vertical columns and 15 or 16 horizontal rows and somewhat enlarged hind basitarsus; and in the pupa by the gill with 10 filaments of almost the same thickness, of which two middle triplets partially overlap when viewed laterally. Whether the larva of this species has paired conical dorsolateral protuberances is unknown.

***Simulium* (*Simulium*) *dadagaunense* Takaoka and Shrestha, 2020**

Simulium (*Simulium*) *dadagaunense* Takaoka and Shrestha (in Takaoka et al., 2020b): 91–94 (female and pupa).

Distribution. Nepal.

Bionomics. Aquatic habitat: A stream (elevation 1080 m) (Takaoka et al., 2020b).

Remarks. This species was described from a single female reared from a pupa in Nepal by Takaoka et al. (2020b). Its male and larva are unknown.

The female of this species is characterized by the relatively narrow frons, weakly developed fronto-ocular area, cibarium with four tiny processes, basal portion of the radial vein bare and all femora yellowish or ocherous with the apical caps darkened. The pupa of this species is similar to *S. beniense* from Nepal.

***Simulium* (*Simulium*) *grisescens* Brunetti, 1911**

Simulium grisescens Brunetti, 1911: 283–284 (male)

Simulium (*Simulium*) *grisescens*: Puri, 1932e: 523–526 (female, male, and pupa); Datta, 1974a: 18–19 (larva); Takaoka et al., 2020b: 95–100 (female, male, pupa, and larva).

Distribution. Bhutan, India, Nepal, and Pakistan.

Bionomics. Aquatic habitats: Various streams in Coonoor (elevation 1800 m), Nilgiri Hills, South India, and in other localities, North India (Puri, 1932e).

Remarks. This species was described from a single male collected at Darjeeling area, India by Brunetti (1911). The female and pupa of this species were described by Puri (1932e), and its larva was described by Datta (1974a). The female, male, pupa and larva of this species were redescribed based on specimens collected from Nepal by Takaoka et al. (2020b).

This species is characterized in the female by the scutum with brassy hairs and haired basal portion of the radial vein; in the male by the number of upper-eye (large) facets in 16 or 17 vertical columns and 17 horizontal rows and somewhat enlarged hind basitarsus; in the pupa by the integument of the frons and the anterodorsal portion of the thorax with relatively large tubercles having tiny secondary projections and gill with 10 filaments of almost the same thickness, arranged as 2+3+3+2 from dorsal to ventral, of which two middle triplets overlap when viewed laterally; and in the larva by thoracic segment 3 and abdominal segments 1–8 each with a pair of conical dorsolateral protuberances.

***Simulium* (*Simulium*) *jeffreyi* Takaoka and Davies, 1995**
(Figs. 76B, C and 77L)

Simulium (*Simulium*) *jeffreyi* Takaoka and Davies, 1995a: 124–130 (female, male, pupa, and larva).

Distribution. Malaysia (Peninsular) and Vietnam.

Bionomics. Aquatic habitats: Small to medium-sized streams at low elevations (less than 500 m) in various provinces of Peninsular Malaysia (Ya'cob et al., 2016).

Remarks. This species was described from Peninsular Malaysia (Takaoka and Davies, 1995a). It was later recorded from Vietnam (Pham, 1999).

It is characterized by the pupal gill with 10 filaments arranged as 2+1+2+1+2+2 from dorsal to ventral, of which five lower filaments are much slenderer than the five upper filaments.

The male upper-eye (large) facets were in 14 vertical columns and 14 or 15 horizontal rows in the original description (Takaoka and Davies, 1995a), but the number of upper-eye (large) facets varies by populations: 13 vertical columns and 13 horizontal rows in a population at Kedah, and 11 vertical columns and 13 horizontal rows in a population at Cameron's Highlands (Takaoka et al., 2018m).

***Simulium* (*Simulium*) *kapuri* Datta, 1975**

Simulium (*Simulium*) *kapuri* Datta, 1975a: 31–35 (female, pupa, and larva); Datta, 1992a: 51.

Distribution. India (Assam and West Bengal).

Bionomics. Aquatic habitats: Streams and rivers (Datta, 1992a).

Remarks. *Simulium kapuri* was originally described from Assam, India (Datta, 1975a). Its male is unknown.

It is characterized by the pupal gill with eight filaments, of which two filaments of the ventral pair are thinner than the six other filaments, lack of pupal terminal hooks and presence of a supernumerary serration of the larval mandible (Datta, 1975a). There is a possibility that a supernumerary serration on the larval mandible had been accidentally formed, since it is not unusual even for species in the subgenus *Simulium* s. str. It should be confirmed on whether this occurs always in other larval specimens.

Simulium (Simulium) latistriatum Senior-White, 1922

Simulium latistriatum Senior-White, 1922: 129 (female).
Simulium (Simulium) striatum (*nec* Brunetti): Puri, 1932e: 516–521 (female, male, and pupa).
Simulium (Simulium) sp. nr. *striatum*: Takaoka et al., 2020a: 49–50 (pupa).

Distribution. Bhutan, India (Karnataka, Kerala, Maharashtra, and Tamil Nadu), and Nepal.

Bionomics. Aquatic habitats: Various streams in Coonoor (elevation 1800 m), Nilgiri Hills, and other localities, South India (Puri, 1932e).

Remarks. This species was described from South India by Senior-White (1922). Its female was redescribed and its male and pupa were described under the name, *S. striatum*, by Puri (1932e), who, in the same paper, synonymized *S. latistriatum* with *S. striatum*. As noted by Takaoka et al. (2020a), *S. latistriatum* is easily distinguished from *S. striatum,* redescribed by Davies and Györkös (1992), by the following characteristics (those of *S. striatum* in parentheses): the relative length of the female frons against its greatest width, which is a little less than twice (1.2), relative length of the female fore basitarsus against its greatest width, which is a little less than five (3.8), female hind tibia, which is dark brown to black except its basal one-third to two-thirds yellow (dark brown except its basal tip yellow), male hind femur, which is dark gray to black except its basal one-third yellowish gray (dark brown except its base light to medium brown), relative length of the male fore basitarsus against its greatest width, which is a little more than five (4.2), and pupal terminal hooks, which are present (absent). Thus, *S. latistriatum* is resurrected from a junior synonym of *S. striatum.* The previous distribution records of *S. striatum* from Bhutan (Datta, 1991a) and India (Puri, 1932e) should be interpreted as those of *S. latistriatum*.

There is a possibility that this species is composed of at least two species because its distribution is widely separated latitudinally into two areas: one in Nepal and Bhutan and the other in South India. In fact, the females of this species are different between South India and Nepal by the length of the frons relative to its greatest width, which is a little less than twice (South India) but 0.9 (Nepal).

Simulium (Simulium) lineothorax Puri, 1932

Simulium (Simulium) lineothorax Puri, 1932e: 521–523 (female, male, and pupa).

Distribution. India (Assam and Tamil Nadu).

Bionomics. Aquatic habitats: Large streams (Puri, 1932e).

Remarks. Simulium lineothorax was described from India by Puri (1932e). Its larva is unknown. This species is similar to *S. striatum* but differs from the latter species and most of other related species in the *S. striatum* species-group by the male scutum with five longitudinal vittae.

Simulium (*Simulium*) *maeklongkeense* Takaoka, Srisuka and Saeung, 2020 (Fig. 77H)

Simulium (*Simulium*) *maeklongkeense* Takaoka, Srisuka and Saeung [in Aupalee et al., 2020a]: 9–11 (male and pupa).

Distribution. Thailand.

Bionomics. Aquatic habitat: A stream (width 20 m, depth 50 cm, bottom sandy, elevation 570 m, 22.7 °C, pH 6.47), at Mae Klong River, Tak Province (Aupalee et al., 2020a).

Remarks. This species was described from Thailand by Aupalee et al. (2020a). Its female and larva are unknown.

It is characterized in the male by the number of upper-eye (large) facets in 14 or 15 vertical columns and 15 or 16 horizontal rows, scutum with brassy short hairs and moderately enlarged hind basitarsus; and in the pupa by the frons and thorax moderately covered with round tubercles, gill with 10 thread-like filaments (of different thickness) arranged as 2+2+2+1+1+2 from dorsal to ventral and cocoon shoe-shaped, woven by relatively thick and broad threads and with or without small open spaces.

This species is morphologically distinguished from all other species (including six other Thai species) of the *S. striatum* species-group by the shoe-shaped cocoon woven by relatively thick and broad threads (Fig. 77H) except *S. subpalmatum* from Sri Lanka (Davies and Györkös, 1992) which has a similar shoe-shaped cocoon woven by relatively thick and broad threads and no open spaces. However, this species is distinguished from the latter species by the number of male upper-eye (large) facets in 14 or 15 vertical columns and 15 or 16 horizontal rows (13 vertical columns and 12 horizontal rows in *S. subpalmatum*) and pupal gill filaments arranged as 2+2+2+1+1+2 (arranged as 2+3+1+2+2 in *S. subpalmatum*).

Simulium (*Simulium*) *menglaense* Chen, 2003

Simulium (*Simulium*) *menglaense* Chen, 2003: 338–340 (female, male, and pupa).

Distribution. China (Yunnan and Hainan).

Bionomics. Aquatic habitat: A rapid of a small mountain stream (elevation 650 m) (Chen, 2003).

Remarks. *Simulium menglaense* was originally described from Yunnan, China by Chen (2003). Its larva is unknown.

It is characterized in the male by the number of upper-eye (large) facets in 11 vertical columns and 11 horizontal rows, hind basitarsus parallel-sided and abdominal segments 2, 6, and 7 each with a pair of white dorsolateral spots; and in the pupa by the gill with 10 filaments arranged as 2+3+3+2 from dorsal to ventral, of which two filaments of the ventral pair and one or two filaments of two middle triplets are thinner than the others.

Simulium (Simulium) nakhonense Takaoka and Suzuki, 1984
(Fig. 76L)

Simulium (Simulium) nakhonense Takaoka and Suzuki, 1984: 33–37 (female, male, pupa, and larva).

Distribution. China (Hainan), Myanmar, Thailand, and Vietnam.

Bionomics. Aquatic habitats: Large streams (width 3–7 m, depth 20–35 cm, bottom sandy, 22.5–22.8°C, elevation 659–704 m) in Myanmar (Takaoka et al., 2018j).

Remarks. *Simulium nakhonense* was originally described from Thailand (Takaoka and Suzuki, 1984).

It is characterized by the pupal gill with 10 filaments, usually arranged in five pairs or four pairs plus two individual filaments, of which four or five lower filaments are thinner than the others, all filaments with moderately defined annular ridges and furrows except the basal portions of the ventral paired filaments, which are smooth, and the frons and the anterior portion of the thorax covered with round tubercles without tiny secondary projections.

Simulium (Simulium) nujiangense Xue, 1993

Simulium (Simulium) nujiangense Xue, 1993b: 234–235 (female, pupa, and larva).

Distribution. China (Yunnan).

Bionomics. Aquatic habitat: A stream (25–27 °C, elevation 850–900 m) (Xue 1993b).

Remarks. This species was described from Yunnan, China (Xue 1993b). Its male is unknown.

It is characterized by the pupal gill with nine filaments arranged as 2+2+3+2 from dorsal to ventral.

There is a possibility that this is a variant of *S. chiangmaiense*, which has the pupal gill with eight, nine, or 10 filaments in Myanmar (Takaoka et al., 2018j). Further comparison is impossible because no detailed information on its pupa and larva is available except the pupal gill, cocoon and postgenal cleft of the larval head, all of which were illustrated (Xue 1993b).

The female scutum was noted to have no color pattern in the original description (Xue 1993b). Probably it was overlooked because females of all the other species of the *S. striatum* species-group have five longitudinal vittae. The dorsal and ventral surfaces of pupal abdomen in Fig. 4 (Xue 1993b) show the presence of spine-combs dorsally on segment 7 only, two pairs of hooks ventrally on segments 5 and 6 only, and no terminal hooks on segment 9. Probably, spine-combs on the dorsum of segment 8 (and 9), and paired hooks on the ventral surface of segment 7, as well as terminal hooks, were overlooked. In addition, no information is available on whether paired dorsal protuberances on the larval thorax and abdomen are present.

Simulium (_Simulium_) _pallidum_ Puri, 1932

Simulium (*Simulium*) *pallidum* Puri, 1932e: 529–531 (female, male, and pupa): Takaoka et al., 2020a: 49.

Distribution. India (Himachal Pradesh and Punjab), Nepal, and Pakistan.

Bionomics. Aquatic habitats: Small streams (Puri, 1932e).

Remarks. This species was described from India by Puri (1932e). Its larva is unknown.

It is characterized in the female by the scutum with pale golden hairs, haired basal portion of the radial vein, and mid tibia entirely yellow; in the male by the scutum with golden hairs, mid tibia entirely yellow, and hind basitarsus slightly enlarged, nearly parallel-sided; and in the pupa by the gill filaments of almost the same thickness, arranged as 2+3+3+2, of which two middle triplets completely overlap when viewed laterally.

Simulium (_Simulium_) _palmatum_ Puri, 1932

Simulium (*Simulium*) *grisescens* var. *palmatum* Puri, 1932e: 526–527 (female, male, and pupa).

Distribution. India (Karnataka, Kerala, and Maharashtra).

Bionomics. Aquatic habitats: Streams and rivers (Puri, 1932e).

Remarks. This species was described as a variety of *S. grisescens* from India since its female and male were almost identical to those of *S. grisescens* (Puri, 1932e). Its larva is unknown.

This species is characterized in the pupa by the gill with 10 filaments, of which five upper filaments are half to one-third as long as the five lower filaments and are much dilated in their basal half, and cocoon shoe-shaped, without any open spaces (Puri, 1932e). Puri (1932e), in his original description, noted that tubercles were absent on the frons and the anterior three-fourths of the thorax, but, in fact, tubercles were present according to the observation made by Davies and Györkös (1992).

Simulium (_Simulium_) _perakense_ Takaoka, Ya'cob and Sofian-Azirun, 2018
(Figs. 76N and 77C)

Simulium (*Simulium*) *perakense* Takaoka, Ya'cob and Sofian-Azirun, 2018m: 52–59 (female, male, pupa, and larva).

Distribution. Malaysia (Peninsular).

Bionomics. Aquatic habitats: A moderately flowing river (width 10–15 m, depth 30–50 cm, 20.0 °C, elevation 588 m), near the bridge of the main road, Nineteen Point, Batan Padang Province, Perak State, and a ditch (width 0.2–0.3 m, elevation 450 m), near the entrance of Resort, Janda Baik, Pahang State (Takaoka et al., 2018m).

Remarks. This species was described from Peninsular Malaysia (Takaoka et al., 2018m).

It is characterized in the female by having the haired basal portion of the radial vein, and scutum with brassy fine recumbent short hairs; in the male by the number of upper-eye (large) facets in 16–18 vertical columns and in 17–19 horizontal rows, scutum with brassy recumbent short hairs; in the pupa by the gill with 10 slender filaments of almost the same thickness, of which two middle triplets overlap when viewed laterally (Fig. 77C); and in the larva by having paired dorsolateral protuberances on the thorax and abdomen.

***Simulium* (*Simulium*) *phraense* Takaoka, Srisuka and Saeung, 2017**
(Figs. 76K and 77D)

Simulium (*Simulium*) *phraense* Takaoka, Srisuka and Saeung [in Takaoka et al., 2018e]: 562–568 (female, male, pupa, and larva).

Distribution. Thailand.

Bionomics. Aquatic habitats: The Mae Yom River (width 3.6 m, depth 15 cm, bottom rocky, 29.0 °C, flow fast, elevation 194 m) in Mae Yom National Park, Song District, Phrae Province, and six streams (elevation 199–796 m) in Phrae Province and four other provinces (Lampang, Phayao, Khamphaeng Phet, and Tak) (Takaoka et al., 2018e).

Remarks. This species was described from Thailand by Takaoka et al. (2018e).

It is characterized in the female by the haired basal portion of the radius and scutum covered with yellow short hairs; in the male by the number of upper-eye (large) facets in 16 or 17 vertical columns and 16 or 17 (rarely 15) horizontal rows and scutum covered with brassy short hairs; in the pupa by the gill with 10 short filaments arranged as 2+1+2+1+2+2 (rarely 3+2+1+2+2) from dorsal to ventral, of which five lower filaments are much thinner than the five upper filaments (Fig. 77D), and all filaments lacking annular ridges on their surface; and in the larva by thoracic segment 3 and abdominal segments 1–8 each with a pair of dorsolateral conical protuberances. This species is easily distinguished from all other related species (except *S. poolpholi*) by the pupal gill filaments lacking annular ridges.

***Simulium* (*Simulium*) *pingtungense* Huang and Takaoka, 2008**

Simulium (*Simulium*) *pingtungense* Huang and Takaoka, 2008c: 309–317 (female, male, pupa, and larva).

Distribution. Taiwan.

Bionomics. Aquatic habitats: A small stream (width 0.5–0.7 m), a large stream (width 3–4 m) and a rapid-flowing river (width 15–20 m) in lowlands (elevation 50–241 m) (Huang and Takaoka, 2008c).

Remarks. This species was described from Taiwan (Huang and Takaoka, 2008c).

It is characterized in the female by having the haired basal portion of the radial vein; in the male by the number of upper-eye (large) facets in 19 vertical columns and 20 horizontal rows) and scutum with copper-colored short hairs; in the pupa by the gill with 10 slender filaments of almost the same thickness, of which two middle triplets overlap when viewed laterally; and in the larva by lacking paired dorsolateral protuberances on the thorax and abdomen.

This was the second species of the *S. striatum* species-group recorded from Taiwan. It is distinguished from *S. quinquestriatum* from Taiwan by the male scutum with copper-colored short hairs, and larval thorax and abdomen without paired dorsolateral protuberances.

***Simulium* (*Simulium*) *poolpholi* Takaoka, Srisuka and Saeung, 2023**
Simulium (*Simulium*) *poolpholi* Takaoka, Srisuka and Saeung [in Srisuka et al., 2023a]: 3–10 (female, male, pupa, and larva).

Distribution. Thailand.

Bionomics. Aquatic habitat: A large lowland stream (width 30 m, depth 40 cm, bottom rocky, flow fast, elevation 110 m) in Ubon Ratchathani Province (Srisuka et al., 2023a).

Remarks. This species was described from Thailand by Srisuka et al. (2023a).

It is similar to *S. phraense* but is distinguished from the latter species by the arrangement of pupal gill filaments: 2+(2+1)+(2+1)+2 or 2+2+1+2+1+2 or 2+3+(2+1)+2 or 2+2+1+3+2 or 2+3+3+2 or 2+2+1+(2+1)+2 or 2+(1+2)+3+2 from dorsal to ventral.

***Simulium* (*Simulium*) *pothigaiense* Anbalagan, 2018**
Simulium (*Simulium*) *pothigaiense* Anbalagan [in Anbalagan et al., 2018]: 1407–1410 (female, male, pupa, and larva).

Distribution. India (Tamil Nadu).

Bionomics. Aquatic habitat: A moderately-flowing stream (width 3–4 m, depth 10–14 cm, 19.2 °C, elevation 1400 m, pH 6.7), Agasthyamalai Biosphere Reserve, Tamil Nadu State (Anbalagan et al., 2018).

Remarks. This species was described from southern India (Anbalagan et al., 2018). It is exceptional that the cocoon is wall-pocket-shaped, and the posterior circlet of the larva has 46 rows of hooklets, which are much fewer than those of other species in the *S. striatum* species-group.

Apart from these two characteristics, it is characterized in the female by the basal section of the radius fully haired; in the male by the number of upper-eye (large) facets in 18 vertical columns and 20 horizontal rows, scutum with golden yellow short hairs, and abdominal segments 2–7 each with a pair of dorsolateral shiny patches; in the pupa by the head and thoracic trichomes all unbranched and gill with 10 filaments of the same thickness, arranged as 2+3+3+2 from dorsal to ventral, dorsum of abdominal segments 1 and 2 darkened and with tubercles; and in the larva by lacking paired protuberances on the dorsum of the thorax and the abdomen.

The postgenal cleft was noted to be large, rounded and twice as long as the postgenal bridge but it is narrow and about four times the length of the postgenal bridge in Fig. 3D (Anbalagan et al., 2018).

***Simulium* (*Simulium*) *quinquestriatum* (Shiraki, 1935)**
(Fig. 77B, I, J, M)
Stilboplax 5-striatum Shiraki, 1935: 27–33 (female).
Simulium (*Simulium*) *quinquestriatum*: Simuliid Research Group of the Japan Society of Sanitary Zoology, 1972: 192; Takaoka, 1979: 396–399 (female, male, pupa and larva); Takaoka et al., 2022a: 11–12 (female, male, pupa, and larva).

Distribution. Taiwan, China (Fujian, Guangdong, Guizhou, Hainan, Hong Kong, Hubei, Jiangxi, Liaoning, Sichuan, Tibet and Yunnan) (?), Korea (South) (?), and Vietnam (?).

Bionomics. Aquatic habitats: Rivers (width 5–12 m) and small streams (width 0.3–1.0 m) in lowlands (Takaoka, 1979).

Remarks. Simulium quinquestriatum was described as *Stilboplax 5-striatum* from eight females and one male collected from Taiwan (Shiraki, 1935). It was transferred to the subgenus *Simulium* s. str. of the genus *Simulium* s. l. when a list of black flies in Japan was constructed (Simuliid Research Group of the Japan Society of Sanitary Zoology, 1974).

Takaoka (1979) redescribed the female, male, pupa and larva of this species based on specimens collected from Taiwan. The specimens previously examined are likely to have included those of not only *S. quinquestriatum* but also *S. yanpingense*, since the number of the horizontal rows of male upper-eye (large) facets was noted by Takaoka (1979) to be ca. 16, which is close to 17 or 18 in *S. yanpingense*.

Simulium quinquestriatum was later recorded from China, Japan, Korea, Thailand, and Vietnam. However, the species regarded as *S. quinquestriatum* in Thailand and Japan was described as new species, *S. wangkwaiense* (Aupalee et al., 2020a), and *S. tanegashimaense* (Takaoka et al., 2022a). The other distribution records from China, Korea, and Vietnam should be reconfirmed.

Simulium quinquestriatum, redescribed by Takaoka et al. (2022a), is characterized in the female by having the haired basal portion of the radial vein; in the male by the number of upper-eye (large) facets in 19 or 20 vertical columns and 19 or 20 horizontal rows and scutum with yellow short hairs; in the pupa by the gill with 10 slender filaments widely spread like a fan, of which two filaments of the dorsal pair have their stalk directed posteriorly (Fig. 77B); and in the larva by the dorsum of the posterior abdominal segments densely covered with dark chisel-like spinous setae (Fig. 77M).

Simulium (_Simulium_) _salyanense_ Takaoka and Shrestha, 2020

Simulium (*Simulium*) *salyanense* Takaoka and Shrestha (in Takaoka et al., 2020a): 43–46 (female, male, and pupa).

Distribution. Nepal.

Bionomics. Aquatic habitat: A stream (20.5 °C, elevation 1106 m) (Takaoka et al., 2020a).

Remarks. This species was described from females and males reared from pupae in Nepal by Takaoka et al. (2020a). Its larva is unknown.

The female and pupa of this species are similar to those of *S. beniense* from Nepal but are distinguished by the darkened female hind femora, the pupal gill with 10 filaments of different thickness, of which the lower filament of the ventral pair is much thinner than the nine other filaments (Takaoka et al., 2020a).

***Simulium* (*Simulium*) *striatum* Brunetti, 1912**
Simulium striatum Brunetti, 1912: 90–91 (female).
Simulium (*Simulium*) *striatum*: Davies and Györkös, 1992: 1041–1044 (female, male, and pupa).

Distribution. Sri Lanka.

Bionomics. Aquatic habitat: An outlet draining Imudurawa Tank near Amparai Airport in eastern Sri Lanka (Davies and Györkös, 1992).

Remarks. This species was described from females collected from Sri Lanka by Brunetti (1912). The male and pupa of this species were described, when its female was redescribed based on specimens from Sri Lanka by Davies and Györkös (1992). The male, which was described as *S. striatum* by Senior-White (1922), was identified as probably *S.* (*G.*) *pattoni* Senior-White by Davies and Györkös (1992).

This species is characterized in the female by the frons narrow, with its narrowest one-twelfths of the width of the head, scutum with gray hairs, hind tibia dark brown with its basal tip yellow, fore basitarsus flattened, 3.8 times as long as its greatest width and basal portion of the radial vein bare; in the male by the number of upper-eye (large) facets in 18 vertical columns and 17 or 18 horizontal rows, scutum with coppery hairs and abdominal segments 5–9 semi-shiny but without silvery dorsolateral spots, and fore and hind basitarsi enlarged, 4.2 and 3.5 times as long as their greatest widths, respectively; and in the pupa by the head and thorax covered with tubercles, gill with 10 filaments of almost the same thickness, of which two middle triplets completely overlap basally when viewed laterally and terminal hooks absent. The extremely narrow frons relative to the head in the female of *S. striatum* was thought to be erroneous because the female frons is usually wide in the subgenus *Simulium* s. str. (Takaoka et al., 2020a). Whether spine-combs on the dorsum of pupal abdomen is present or not was not noted in Davies and Györkös (1992).

***Simulium* (*Simulium*) *subpalmatum* Davies and Györkös, 1992**
Simulium (*Simulium*) *subpalmatum* Davies and Györkös, 1992: 1035–1041 (female, male, pupa, and larva).

Distribution. Sri Lanka.

Bionomics. Aquatic habitats: Streams and Rivers (width 1–15 m) (Davies and Györkös, 1992).

Remarks. This species was described from Sri Lanka by Davies and Györkös (1992).

It is similar to *S. palmatum* from India in sharing the gill with 10 filaments, of which five upper filaments are much dilated in their basal half, and cocoon shoe-shaped, without open spaces (Puri, 1932e). However, this species is distinguished from *S. palmatum* by the following characteristics (those of *S. palmatum* in parentheses): the relative length of the female hind basitarsus against its greatest width, which is 5.0 (5.8–5.9), number of the outer teeth of the mandible, which is 10 or 11 (14), relative length of the five upper gill filaments against the five lower filaments, which is 1.0 (0.5–0.7), and relative thickness of the five upper gill filaments against the five lower filaments is about twice (four times).

Simulium (Simulium) tadtonense Takaoka, Srisuka and Saeung, 2020

(Fig. 77F)

Simulium (Simulium) tadtonense Takaoka, Srisuka and Saeung [in Aupalee et al., 2020a]: 7–9 (female, male, and pupa).

Distribution. Thailand.

Bionomics. Aquatic habitat: A stream (width 30 m, depth 30 cm, bottom rocky, 19.3 °C, pH 6.5, elevation 181 m) at Tad Ton Waterfall, Chaiyaphum Province (Aupalee et al., 2020a).

Remarks. This species was described from Thailand by Aupalee et al. (2020a). Its larva is unknown.

It is characterized in the female by the narrow frons and haired basal portion of the radial vein; in the male by the number of upper-eye (large) facets in 18 vertical columns and 18 horizontal rows, scutum with brassy short hairs and hind basitarsus yellowish white on the basal half or little more; in the pupa by the frons densely covered with round tubercles, two short frontal trichomes on each side and gill with 10 filaments arranged as 2+(2+1)+(2+1)+2 or 2+(1+2)+(2+1)+2 or 2+3+3+2 from dorsal to ventral, and cocoon shoe-shaped, with a low anterior wall (Fig. 77F).

This species is similar to *S. wangkwaiense* from Thailand (Aupalee et al., 2020a) in many characteristics including the haired basal portion of the female radius, the number of the male upper-eye (large) facets and the arrangement of pupal gill filaments. However, there are differences in the relative size of the female frons, color of short hairs on the male scutum, relative thickness of 10 pupal gill filaments, and presence or absence of tubercles on the dorsum of pupal abdominal segment 1.

Simulium (Simulium) tavanense Takaoka and Sofian-Azirun, 2017

Simulium (Simulium) tavanense Takaoka and Sofian-Azirun [in Takaoka et al., 2017c]: 118–125 (female, male, pupa, and larva).

Distribution. Vietnam.

Bionomics. Aquatic habitat: A small stream (width 1.0 m, depth 8 cm, bottom pebbles, 13 °C, elevation 1048 m) (Takaoka et al., 2017c).

Remarks. Simulium tavanense was described from Vietnam (Takaoka et al., 2017c).

It is characterized in the female by the haired basal portion of the radial vein; in the male by the scutum with yellow short hairs, the number of upper-eye (large) facets in 18 vertical columns and 19 horizontal rows and the subcosta and the basal portion of the radial vein haired; in the pupa by the frons and the anterior portion of the thorax covered with tubercles with tiny secondary projections and gill with 10 filaments of the subequal thickness, arranged as 2+3+3+2 from dorsal to ventral, of which two middle triplets partially overlap basally when viewed laterally; and in the larva by abdominal segments 1–7 each with a pair of cone-like dorsolateral protuberances.

***Simulium* (*Simulium*) *taythienense* Takaoka, Sofian-Azirun and Ya'cob, 2014**
Simulium (*Simulium*) *taythienense* Takaoka, Sofian-Azirun and Ya'cob [in Takaoka et al., 2014d]: 353–361 (female, male, pupa, and larva).

Remarks. Vietnam.

Bionomics. Aquatic habitat: An irrigation channel (width 0.5–1.0 m, bottom muddy, 21 °C, elevation 283 m) (Takaoka et al., 2014d).

Remarks. *Simulium taythienense* was described from Vietnam (Takaoka et al., 2014d).

It is characterized in the female by the enlarged opening of the sensory vesicle and the haired basal portion of the radial vein; in the male by the scutum with brassy short hairs and the number of upper-eye (large) facets in 17 or 18 vertical columns and 17 or 18 horizontal rows; in the pupa by the frons and the anterior portion of the thorax covered with tubercles without tiny secondary projections and gill with 10 filaments of the subequal thickness, arranged as 2+3+3+2 from dorsal to ventral, of which two middle triplets are not overlapping basally when viewed laterally; and in the larva by thoracic segments 1–3 and abdominal segments 1–8 each with a pair of cone-like dorsolateral protuberances.

***Simulium* (*Simulium*) *thailandicum* Takaoka and Suzuki, 1984**
Simulium (*Simulium*) *thailandicum* Takaoka and Suzuki, 1984: 37–38 (male and pupa).

Distribution. Thailand, China (Yunnan), and Vietnam.

Bionomics. Unknown.

Remarks. *Simulium thailandicum* was originally described from a single male reared from a pupa collected from Thailand (Takaoka and Suzuki, 1984). Its female and larva are unknown.

It is characterized in the pupa by the frons and the anterior portion of the thorax covered with tubercles having tiny secondary projections and gill with 10 filaments arranged as 2+2+2+2+2 from dorsal to ventral, of which six filaments of the second, fourth, and fifth pairs are somewhat thinner than the four other filaments.

***Simulium* (*Simulium*) *thilorsuense* Takaoka, Srisuka and Saeung, 2022**
Simulium (*Simulium*) *thilorsuense* Takaoka, Srisuka and Saeung [in Aupalee et al., 2022]: 2–8 (female, male, pupa, and larva).

Distribution. Thailand.

Bionomics. Aquatic habitat: A fast-flowing calcareous streamlet originating from Thi Lor Su Waterfall (80–100 m wide), Umphang Wildlife Sanctuary, Tak Province (Aupalee et al., 2022).

Remarks. This species was described from Thailand by Aupalee et al. (2022).

It is characterized in the female and male by darkened legs; in the female by the basal portion of radius haired; in the male by the high number of upper-eye (large) facets in 20 vertical columns and 19 or 20 horizontal rows and scutum with brassy short hairs; in the pupa by the gill with 10 filaments of almost the same thickness;

and in the larva by abdominal segments 1–8 each with a pair of conical dorsolateral protuberances.

It is distinguished from all the seven other species of this species-group recorded from Thailand in the male by the greater number of upper-eye (large) facets in 20 vertical columns and 19 or 20 horizontal rows; and in the pupa by the combination of the arrangement of 10 filaments (2+3+3+2 from dorsal to ventral), subequal thickness of gill filaments and tubercles on the frons without secondary projections.

***Simulium* (*Simulium*) *valparaiense* Anbalagan, 2018**

Simulium (*Simulium*) *valparaiense* Anbalagan [in Anbalagan et al., 2018]: 1410–1415 (female, male, pupa, and larva).

Distribution. India (Tamil Nadu).

Bionomics. Aquatic habitat: A medium-flowing stream (width 8–10 m, depth 15–20 cm, 20 °C, elevation 1200 m, pH 6.4), Srikundra, Tamil Nadu State (Anbalagan et al., 2018).

Remarks. This species was described together with *S. pothigaiense* from South India (Anbalagan et al., 2018). It is exceptional that the cocoon is wall-pocket-shaped, and the posterior circlet of the larva has 57 rows of hooklets, which are much fewer than those of other species in the *S. striatum* species-group.

Apart from these two characteristics, it is characterized in the female by the cibarium without minute processes and the basal section of the radium bare; in the male by the number of upper-eye (large) facets in 16 vertical columns and 18 horizontal rows and abdominal segments 2–7 each with a pair of shiny dorsolateral patches; in the pupa by the head and thoracic trichomes all unbranched, gill with 10 filaments of the same thickness, arranged as 2+3+3+2 from dorsal to ventral, dorsum of abdominal segments 1 and 2 with tubercles and cocoon truncate and wall-pocket-shaped; and in the larva by the dorsum of abdomen with paired protuberances.

The postgenal cleft was noted to be large, round, and 1.3 times as long as the postgenal bridge, but it is narrow and about four times the length of the postgenal bridge in Fig. 6D (Anbalagan et al., 2018).

***Simulium* (*Simulium*) *wangkwaiense* Takaoka, Srisuka and Saeung, 2020 (Fig. 76O)**

Simulium (*Simulium*) *wangkwaiense* Takaoka, Srisuka and Saeung [in Aupalee et al., 2020a]: 2–7 (female, male, pupa, and larva).

Simulium (*Simulium*) *quinquestriatum* (*nec* Shiraki, 1935): Takaoka and Saito, 1996: 167–168.

Distribution. Thailand.

Bionomics. Aquatic habitat: A stream (width 4 m, depth 40 cm, bed rocky, expose to the sun, 19.2 °C, pH 6.15, elevation 413 m), Wang Kwai Waterfall, Chomthong district, Chiang Mai Province (Aupalee et al., 2020a).

Remarks. *Simulium wangkwaiense* was the same species that was misidentified in Thailand by Takaoka and Saito (1996) as *S. quinquestriatum* (Shiraki), originally described from Taiwan (Shiraki, 1935) and redescribed by Takaoka (1979).

It is characterized in the female by the haired basal portion of the radius; in the male by the number of upper-eye (large) facets in 18 or 19 vertical columns and 18 or 19 horizontal rows, scutum with yellowish short hairs and moderately enlarged hind basitarsus; in the pupa by the frons and thorax with relatively large tubercles having tiny secondary projections, gill with 10 thread-like filaments of almost the same thickness, arranged as 2+3+3+2 from dorsal to ventral, of which two middle triplets overlap basally when viewed laterally, and dorsum of abdominal segment 1 sparsely covered with tiny tubercles; and in the larva by thoracic segments 1 and 3 and abdominal segments 1–8 each with a pair of cone-like dorsolateral protuberances.

This species is similar to *S. quinquestriatum* from Taiwan (Shiraki, 1935; Takaoka et al., 2022b), and *S. tavanense* from Vietnam (Takaoka et al., 2017c) by sharing a combination of three characteristics: the haired basal portion of the female radial vein, male scutum with yellow short hairs, and 10 pupal gill filaments of almost the same thickness. However, this species is distinguished from *S. quinquestriatum* by the pupal gill with 10 slender filaments, of which stalks of the two middle triplets overlap basally when viewed laterally (not overlap in *S. quinquestriatum*) and the dorsum of larval posterior abdominal segments lacking dark chisel-like spinous setae (densely covered with dark chisel-like spinous setae in *S. quinquestriatum*); from *S. tavanense* in the male by the subcosta and the basal portion of the radial vein bare (haired in *S. tavanense*), and in the pupa by the dorsum of abdominal segment 1 with tubercles (dorsum of abdominal segment 1 lacking tubercles in *S. tavanense*).

This species is morphologically distinguished from all the other Thai species of the *S. striatum* species-group by the pupal gill with 10 filaments of the same thickness.

***Simulium* (*Simulium*) *wuzhishanense* Chen, 2003**

Simulium (*Simulium*) *wuzhishanense* Chen, 2003: 340–341 (female, male, pupa, and larva).

Distribution. China (Hainan).

Bionomics. Aquatic habitat: A small shaded stream in Hainan (Chen, 2003).

Remarks. This species was described from Hainan, China by Chen (2003).

It is characterized in the female by the basal portion of the radius haired; in the male by the low number of upper-eye (large) facets in 12 vertical columns and 14 horizontal rows and abdominal segments 2, 6, and 7 each with a pair of white dorsolateral spots; in the pupa by the gill with 10 filaments of almost the same thickness and arranged as 2+3+3+2 from dorsal to ventral; and in the larva by thoracic and abdominal segments lacking paired dorsolateral protuberances.

***Simulium* (*Simulium*) *xuandai* Takaoka, Sofian-Azirun and Ya'cob, 2014**

Simulium (*Simulium*) *xuandai* Takaoka, Sofian-Azirun and Ya'cob [in Takaoka et al., 2014d]: 361–364 (male and pupa).

Distribution. Myanmar and Vietnam.

Bionomics. Aquatic habitat: A moderately flowing forest stream (width 4–5 m, depth 10–15 cm, bottom gravelly, stony, rocky, 18 °C, shaded, elevation 700 m) (Takaoka et al., 2014d).

Remarks. *Simulium xuandai* was originally described from Vietnam (Takaoka et al., 2014d). Its female and larva are unknown. It was recorded from Myanmar by Takaoka et al., 2018j).

It is characterized in the male by the large number of upper-eye (large) facets in 20 vertical columns and 20 horizontal rows and scutum with yellow short hairs; in the pupa by the frons and the anterior portion of the thorax covered with tubercles having tiny secondary projections and the gill with 10 filaments of almost the same thickness, arranged as 2+3+3+2 from dorsal to ventral, of which two middle triplets overlap basally when viewed laterally.

Simulium (_Simulium_) _yanaense_ Anbalagan, 2019

Simulium (*Simulium*) *yanaense* Anbalagan [in Anbalagan et al., 2019a]: 13574–13578 (male, pupa, and larva).

Distribution. India (Karnataka).

Bionomics. Aquatic habitats: Streams (width 2.0–3.5 m, depth 5–30 cm, elevation 55 m) in Karnataka (Anbalagan et al., 2019a).

Remarks. This species was described from southern India (Anbalagan et al., 2019a). Its female is unknown.

It is characterized in the male by the number of upper-eye (large) facets in 12 vertical columns and 16 horizontal rows, the basal section of the radius bare and abdominal segments 2–7 each with a pair of shiny dorsolateral patches; in the pupa by the thoracic integument bare, with trichomes all unbranched and cocoon shoe-shaped, without open spaces anteriorly; and in the larva by the dorsum of abdomen without paired dorsolateral protuberances.

Five lower filaments are about half as thick as the five upper filaments, according to Fig. 2A of Anbalagan et al. (2019a), though this pupal characteristic was not mentioned.

The coxite was noted to be 1.2 times as long as its greatest width, but it is 0.65 times as long as its greatest width in Fig. 1D of Anbalagan et al. (2019a), and the postgenal cleft was described to be arrow-head-shaped and 3.1 times as long as the postgenal bridge, but it is rounded and 5.0 times as long as the postgenal bridge in Fig. 2D of Anbalagan et al. (2019a).

Simulium (_Simulium_) _yanpingense_ Takaoka and Huang, 2022
(Fig. 76A, D–J, M, P–W)

Simulium (*Simulium*) *yanpingense* Takaoka and Huang [in Takaoka et al., 2022a]: 3–6 (female, male, and pupa).

Distribution. Taiwan.

Bionomics. Aquatic habitat: A river (elevation 320 m), Yanping Village, Taitung County (Takaoka et al., 2022a).

Remarks. This species was described from Taiwan (Takaoka et al., 2022a). Its larva is unknown.

This species is similar to *S. pingtungense* and *S. quinquestriatum*, both from Taiwan (Shiraki, 1935; Takaoka, 1979; Huang and Takaoka, 2008c), in many characteristics including the haired basal portion of the female radius and dark female and male leg color, but it is distinguished from both species by the number of male upper-eye (large) facets in 17 (rarely 18) vertical columns and 17 or 18 horizontal rows (19 vertical columns and 20 horizontal rows in *S. pingtungense* and 19 or 20 vertical columns and 19 or 20 horizontal rows in *S. quinquestriatum*); from *S. pingtungense* by the greatest width relative to the narrowest width of the female frons (1.25–1.31 in this species versus 1.40–1.45 in *S. pingtungense*) and relative thickness of the lowermost filament of the pupal gill (not thickest in this species versus thickest in *S. pingtungense*); and from *S. quinquestriatum* by the male scutum covered with brassy short hairs (yellow short hairs in *S. quinquestriatum*).

Simulium (_Simulium_) sp. B Takaoka and Davies, 1995
Simulium (*Simulium*) sp. B Takaoka and Davies, 1995a: 131–132 (pupa).

Distribution. Malaysia (Peninsular).

Bionomics. Aquatic habitat: A river (width 8–12 m) in Gombak, Peninsular Malaysia.

Remarks. This unnamed species was described from a pupa collected from Gombak, Selangor (Takaoka and Davies, 1995a).

This pupa differs from those of *S. jeffreyi* and *S. perakense* by lacking spine-combs on abdominal segment 8. The pupal gill filaments are almost the same in arrangement as those of *S. perakense* but differ from the latter by the relative length of the five upper filaments, which are much shorter than the five lower filaments.

Simulium (_Simulium_) sp. nr. _grisescens_ Datta, 1975
Simulium (*Simulium*) sp. nr. *grisescens* Datta and Pal, 1975: 159 (larva).

Distribution. India (West Bengal).

Bionomics. Unknown.

Remarks. This species was reported from a single larva from Darjeeling by Datta and Pal (1975). Its female, male, and pupa are unknown.

The larva of this species is distinguished from *S. grisescens* by the second antennal article with an unpigmented annulation medially (Datta and Pal, 1975).

Simulium (_Simulium_) Diju sp. Lewis, 1974
Simulium (*Simulium*) Diju sp. Lewis, 1974: 38 (pupa).

Distribution. India (Assam).

Bionomics. Unknown.

Remarks. This unnamed species was reported from Assam, India by Lewis (1974). Its female, male, and larva are unknown.

Morphological characteristics of its pupa are not available except that the gill is composed of 10 filaments arranged as 2+3+3+2 from dorsal to ventral. It is provisionally placed in the *S. striatum* species-group of the subgenus *Simulium* s. str. judging from the number and arrangement of 10 gill filaments.

Simulium (_Simulium_) Doigroon sp. Lewis, 1974

Simulium (*Simulium*) Doigroon sp. Lewis, 1974: 38 (pupa).

Distribution. India (Assam).

Bionomics. Unknown.

Remarks. This unnamed species was reported from Assam, India by Lewis (1974). Its female, male, and larva are unknown.

Morphological characteristics of its pupa are not available except that the pupal gill is composed of eight filaments arranged as 2+2+2+2 from dorsal to ventral, of which two filaments of the ventral pair are much thinner than the six other filaments and the cocoon is shoe-shaped. It is similar to *S. kapuri* in the *S. striatum* species-group judging from the number, arrangement, and relative thickness of eight gill filaments and shape of the cocoon.

Simulium (_Simulium_) _tuberosum_ Species-group

Simulium (_Simulium_) _brevipar_ Subgroup

Simulium (_Simulium_) _brevipar_ Takaoka and Davies, 1995 (complex)

Simulium (*Simulium*) *brevipar* Takaoka and Davies, 1995a: 132–137 (female, male, pupa, and larva).

Distribution. Malaysia (Peninsular) and Thailand.

Bionomics. Aquatic habitats: Small to medium-sized streams at low to high elevations (159–1813 m) in various provinces of Peninsular Malaysia (Ya'cob et al., 2016).

Remarks. This species was described from females, males, pupae, and larvae collected from Fraser's Hill, Pahang (Takaoka and Davies, 1995a).

It is characterized in the pupa by the pit-like organ adjacent to the base of the gill and gill with six slender filaments of almost the same thickness except the uppermost filament which is much thicker than the five other filaments when compared basally.

Simulium (_Simulium_) _huense_ Takaoka and Ya'cob, 2017

(Figs. 78G and 79B, H, J–M)

Simulium (*Simulium*) *cavum* Takaoka and Ya'cob [in Takaoka et al., 2015c] (pre-occupied name): 74–79 (male, pupa, and larva).

Simulium (*Simulium*) *huense* Takaoka and Ya'cob [in Takaoka et al., 2017c]: 139.

Distribution. Vietnam.

Bionomics. Aquatic habitats: A small stream (width 1–2 cm, 14.0 °C, shaded, elevation 1273 m) and a small stream (width 0.5–1.0 m, depth 5–30 cm, 15.0 °C, shaded, elevation 1186 m) moderately flowing from a natural forest, near the top of Bach Ma National Park, Thua Thien Hue Province (Takaoka et al., 2015c).

Remarks. This species was described as *S. cavum* Takaoka and Ya'cob from central Vietnam (Takaoka et al., 2015c). The substitute name *huense* was given for this species, since the name *cavum* was preoccupied by an African species *Simulium* (*Metomphalus*) *cavum* Gibbins, 1938 (Adler and Crosskey, 2016). Thus, *cavum* Takaoka and Ya'cob, 2015 is a junior primary homonym of *cavum* Gibbins, 1938. The female of this species remains unknown.

This species is distinguished from *S. brevipar* (complex), *S. tiomanense*, *S. tianchi*, and *S. yuphae* (complex) in the male by the number of upper-eye (large) facets in 21 vertical columns [c.f., in 17 vertical columns in *S. brevipar* (complex), 16 vertical columns in *S. tiomanense*, 15 vertical columns in *S. tianchi* and 17 or 18 vertical columns in *S. yuphae* (complex)]; from *S. brevipar* (complex), *S. sigiti*, and *S. yuphae* (complex) in the pupa by the relatively large tubercles on the frons without tiny secondary projections (c.f., large tubercles with tiny secondary projections in the latter three known species); and also from *S. tianchi* and *S. sigiti* in the pupa by the gill with six filaments of different length and thickness (Fig. 79B) (c.f., six gill filaments are subequal in length and thickness to one another in the latter two known species).

Simulium (Simulium) lowi Takaoka and Adler, 2017

(Fig. 79A)

Simulium (*Simulium*) *brevipar* (*nec* Takaoka and Davies): Takaoka et al., 2014d: 365.
Simulium (*Simulium*) *lowi* Takaoka and Adler [in Takaoka et al., 2017c]: 145–147 (male and pupa).

Distribution. Vietnam.

Bionomics. Aquatic habitat: A small stream (width 0.5 m, bottom rocky, 17 °C, shaded, elevation 975 m) slow-flowing in bushes along the road near the Tam Dao National Park, Vinh Phuc Province (Takaoka et al., 2014d).

Remarks. This species was first reported as *S. brevipar* from northern Vietnam (Takaoka et al., 2014d) and later was described as a new species, *S. lowi*, using morphological and COI gene-sequence analyses (Takaoka et al., 2017c).

Simulium (Simulium) sapaense Takaoka and Low, 2017

(Figs. 78H and 79C)

Simulium (*Simulium*) *sapaense* Takaoka and Low [in Takaoka et al., 2017c]: 138–145 (female, male, and pupa).

Distribution. Vietnam.

Bionomics. Aquatic habitat: A small stream (width 0.5–1.0 m, depth 7 cm, bottom rocky, 9.0 °C, partially shaded, elevation 1680 m) slowly flowing from a secondary forest, Sapa, Lao Cai Province (Takaoka et al., 2017c).

Remarks. This species was described from northern Vietnam (Takaoka et al., 2017c). Its larva is unknown.

It is characterized in the male by the number of upper-eye (large) facets in 17 or 18 vertical columns and 19 or 20 horizontal rows; and in the pupa by the vicinity of the pit-like organ covered with tubercles.

Simulium (*Simulium*) *sigiti* Takaoka and Hadi, 1991

(Fig. 79G)

Simulium (*Simulium*) *sigiti* Takaoka and Hadi, 1991: 357–363 (female, male, pupa, and larva).

Distribution. Indonesia (Java).

Bionomics. Aquatic habitat: A small stream (width 0.5 m, shaded, elevation 1250 m) running down through a natural forest near the upper border of an open field cultivated for tea plantation in Puncak, West Java (Takaoka and Hadi, 1991).

Remarks. This species was described from Java, Indonesia (Takaoka and Hadi, 1991). This was the first species of the *S. tuberosum* species-group recorded from Indonesia, and it was also the first species having a pit-like organ at the base of the pupal gill beside species of the *S. clathrinum* species-group of the subgenus *Morops*.

It is characterized in the male by the number of upper-eye (large) facets in 20 horizontal rows and fore and hind basitarsi 4.4 and 4.5 times as long as their greatest widths, respectively; in the pupa by the head and thoracic integument covered with relatively large tubercles having tiny secondary projections, gill with six slender filaments of the same length and thickness (Fig. 79G), and abdominal segments 6–9 or 7–9 each with spine-combs; and in the larva by the postgenal cleft twice as long as the postgenal bridge and rectal organ with two or three secondary lobules per lobe.

Simulium (*Simulium*) *tatopaniense* Takaoka and shrestha, 2020

(Fig. 79F, I)

Simulium (*Simulium*) *tatopaniense* Takaoka and Shrestha [in Takaoka et al., 2020e]: 307–313 (female, male, and pupa).

Distribution. Nepal.

Bionomics. Aquatic habitat: A small stream (elevation 1576 m) at Tatopani (Takaoka et al., 2020e).

Remarks. This species was described from Nepal (Takaoka et al., 2020e). Its larva is unknown.

It is similar to *S. huense* from Vietnam in sharing a combination of two characteristics: the number of male upper-eye (large) facets in 20 or 21 vertical columns and 20 or 21 horizontal rows and the pupal gill with six slender filaments, of which the uppermost filament is much thicker than the five other filaments. However, this species is distinguished from the latter species by the relative length of the male fore basitarsus against its greatest width (4.2 in this species versus 5.4 in *S. huense*), shape of the male hind basitarsus (wedge-shaped in this species versus spindle-shaped in *S. huense*), and tiny secondary projections on relatively large tubercles on the pupal frons (present in this species versus absent in *S. huense*).

***Simulium* (*Simulium*) *tianchi* Chen, Zhang and Yang, 2003**

Simulium (*Simulium*) *tianchi* Chen, Zhang and Yang, 2003b: 746–749 (female, male, pupa, and larva).

Distribution. China (Hainan).

Bionomics. Aquatic habitat: A fast-flowing stream at an elevation of 650 m in Hainan (Chen et al., 2003b).

Remarks. This species was described from Hainan, China by Chen et al. (2003b).

It is characterized in the female by the fore basitarsus 3.6 times as long as its greatest width; in the male by the number of upper-eye (large) facets in 15 vertical columns and 12 horizontal rows and hind basitarsus 3.4 times as long as its greatest width; in the pupa by the gill with six slender short filaments of the same length and thickness; and in the larva by the postgenal cleft 4.5 times as long as the postgenal bridge.

***Simulium* (*Simulium*) *tiomanense* Takaoka, Sofian-Azirun and Belabut, 2012** (Fig. 79E)

Simulium (*Simulium*) *tiomanense* Takaoka, Sofian-Azirun and Belabut, 2012a: 408–412 (male, pupa, and larva).

Distribution. Malaysia (Peninsular).

Bionomics. Aquatic habitat: A small stream (width 0.5–2.0 m, 25.0 °C, shaded) moderately flowing in a forest, Tioman Island (Takaoka et al., 2012a).

Remarks. This species was described from a pharate male, its associated pupal exuviae and a larva collected from Tioman Island (Takaoka et al., 2012a). The female remains unknown.

It is characterized in the male by the dichoptic head (in place of the usual holoptic head), which is divided by a very narrow frons, as well as the round enlarged facets (in place of the usual hexagonal facets); and in the pupa by the small pit-like organ and the gill filaments divergent vertically at an angle of 90 degrees or slightly more (Fig. 79E).

***Simulium* (*Simulium*) *yuphae* Takaoka and Choochote, 2005 (complex)** (Figs. 78A–F, I–N and 79D)

Simulium (*Simulium*) *yuphae* Takaoka and Choochote, 2005i: 326–333 (female, male, pupa, and larva).

Simulium (*Simulium*) *brevipar* (*nec* Takaoka and Davies): Takaoka and Saito, 1996: 168.

Distribution. Thailand and Vietnam.

Bionomics. Aquatic habitats: Small to medium-sized streams (16–26 °C, elevation 830–1420 m) in mountainous areas in Chiang Mai Province (Takaoka and Choochote, 2005i).

Remarks. *Simulium yuphae* was described from Thailand by Takaoka and Choochote (2005i). This species is the same species previously misidentified as *S. brevipar* by Takaoka and Saito (1996).

This species is similar to *S. brevipar* (complex) but is distinguished from the latter species by the configuration of the pupal gill filaments, i.e., one of two filaments of the ventral pair is always directed posteriorly (Fig. 79D) in this species, whereas that is directed forward in *S. brevipar* (complex).

Simulium (*Simulium*) *nitidithorax* Subgroup

Simulium (*Simulium*) *aeneifacies* Edwards, 1933

(Figs. 80O and 81E)

Simulium (*Simulium*) *aeneifacies* Edwards, 1933: 256 (female); Takaoka, 2008b: 65–68 (female, male, pupa, and larva).

Distribution. Malaysia (Sabah).

Bionomics. Aquatic habitats: A stream (width 0.3–0.5 m, 12.0 °C, shaded, elevation 2250 m), and a stream (width 0.4–1.5 m, 13.0 °C, shaded, elevation 1860–1970 m), both moderately flowing in a natural forest near the Carson Falls, Kinabalu Parks (Takaoka, 2008b).

Remarks. This species was described based on a single female collected at Kamborangah (elevation 2160 m), Mt. Kinabalu, Sabah (Edwards, 1933). Its male, pupa, and larva were described by Takaoka (2008b).

It is characterized by the short female sensory vesicle.

Simulium (*Simulium*) *alberti* Takaoka, 2008

(Figs. 80C–G, J, L, M, R–U and 81A, C, I)

Simulium (*Simulium*) *alberti* Takaoka, 2008b: 56–62 (female, male, pupa, and larva).

Distribution. Malaysia (Sabah).

Bionomics. Aquatic habitat: A stream (width 0.5–2.0 m, 14.0 °C, shaded, elevation 1940 m) moderately flowing in a natural forest in Mesilau, Kinabalu National Park (Takaoka, 2008b).

Remarks. This species was described from Sabah, Malaysia (Takaoka, 2008b)

It is characterized by the elongate female sensory vesicle and pupal gill composed of six filaments in three pairs with relatively longer stalks (Fig. 81I).

Simulium (*Simulium*) *congi* Takaoka and Sofian-Azirun, 2015

(Fig. 81J, M–P)

Simulium (*Simulium*) *congi* Takaoka and Sofian-Azirun [in Takaoka et al., 2015c]: 80–84 (male, pupa, and larva).

Distribution. Vietnam.

Bionomics. Aquatic habitat: A small stream (width 0.4 m, depth 10 cm, bottom muddy, 15.0 °C, partially shaded, elevation 1722 m) slowly flowing in a swampy area, Dinh Kno, Lac Duong, Lam Dong Province (Takaoka et al., 2015c).

Remarks. This species was described from one male reared from a pupa, one pharate male and one mature larva collected from Lam Dong Province, southern Vietnam (Takaoka et al., 2015c). The female of this species remains unknown.

This species is distinguished from all other species of the *S. tuberosum* species-group by having the swollen basal fenestra at the base of the pupal gill (Fig. 81J), a characteristic not reported in species of the *S. tuberosum* species-group.

***Simulium* (*Simulium*) *jasmoni* Takaoka, Sofian-Azirun and Belabut, 2012**

Simulium (*Simulium*) *jasmoni* Takaoka, Sofian-Azirun and Belabut, 2012a: 404–408 (female, male, pupa, and larva).

Distribution. Malaysia (Peninsular).

Bionomics. Aquatic habitat: A small stream (width 0.2–0.3 m, 25.0 °C, shaded) slow-flowing in a forest, near Sungai Tedan, Tioman Island (Takaoka et al., 2012a).

Remarks. This species was described from Tioman Island (Takaoka et al., 2012a).

It is characterized in the female by the claw with a minute subbasal swelling; and in the pupa by the gill filaments divergent basally at an angle of 90 degrees or slightly more, when viewed laterally.

***Simulium* (*Simulium*) *jianshiense* Takaoka, Otsuka and Adler, 2021**
(Fig. 81H)

Simulium (*Simulium*) *suzukii* (*nec* Rubtsov): Takaoka, 1979: 395–396.
Simulium (*Simulium*) *jianshiense* Takaoka, Otsuka and Adler [in Takaoka et al., 2021b]: 3–8 (female, male, and pupa).

Distribution. Taiwan.

Bionomics. Aquatic habitats: Lowland streams (width 0.5–3.0 m) and rivers (width ca. 10 m) (Takaoka, 1979).

Remarks. This species was first recorded as *S. suzukii* from Taiwan (Takaoka, 1979). However, it was designated as cytoform "L" of the *S. tani* (complex) by Huang and Adler (2011), who showed that this species is chromosomally more similar to *S.* (*S.*) *tani* originally described from Peninsular Malaysia than to *S. suzukii*. Later, based on morphological, chromosomal and genetic analyses, this species was described as a new species, *S. jianshiense*, by Takaoka et al. (2021b). Its larva remains undescribed.

Simulium jianshiense is similar to *S. suzukii* from Hokkaido and *S. ryukyuense* from Okinawa Island and Amamioshima Island in the arrangement of six gill filaments (Fig. 81H), but is distinguished from *S. suzukii* in the female by the relative length of the labrum to the clypeus (0.51–0.52 versus 0.61–0.63) and the number of tiny processes on the cibarium (60–76 versus 103–115), and in the male by the color of the fore coxa (yellow versus light brown); and from *S. ryukyuense* in the female by the height of the frons relative to its narrowest width (1.19–1.23 versus 1.31–1.57) and the sensory vesicle's relative length to the third palpal segment (long versus medium-long) and its opening size (moderate versus large).

Simulium (_Simulium_) _keningauense_ Takaoka, 2008

(Figs. 80B and 81B, D, G)

Simulium (*Simulium*) *keningauense* Takaoka, 2008b: 72–75 (female, male, pupa, and larva).

Distribution. Malaysia (Sabah and Sarawak) and Indonesia (Kalimantan).

Bionomics. Aquatic habitats: Small streams (width 0.5–6.0 m) to rivers (width 8–20 m) at medium to high elevations (400–1880 m) in Sabah and Sarawak. Water temperature ranges from 14.0 °C to 25.0 °C (Takaoka, 2008b).

Remarks. This species was described from Sabah and Sarawak (Takaoka, 2008b).

It is characterized by the female claw with a tiny subbasal swelling (Fig. 80B) and pupal thorax covered with sharply-pointed conical tubercles (Fig. 81B).

Simulium (_Simulium_) _lunduense_ Takaoka, 2008

(Fig. 80K,Q)

Simulium (*Simulium*) *lunduense* Takaoka, 2008b: 75–79 (female, male, pupa, and larva).

Distribution. Malaysia (Sarawak).

Bionomics. Aquatic habitats: Streams and rivers (width 2–20 m, 24.0–26.0 °C) in a lowland in Sarawak (Takaoka, 2008b).

Remarks. This species was described from Sarawak (Takaoka, 2008b).

It is similar to *S. keningauense* in having the pupal thorax covered with sharply pointed conical tubercles. However, it is distinguished from *S. keningauense* in the male by the narrower hind basitarsus (Fig. 80K) and style tapered toward the apex when viewed ventrolaterally (Fig. 80Q); and in the pupa by the short anterodorsal trichomes on the thorax much shorter than the frontal trichomes and gill filaments in three pairs all short to long stalked.

Simulium (_Simulium_) _mainitense_ Takaoka and Tenedero, 2019

Simulium (*Simulium*) *mainitense* Takaoka and Tenedero, 2019: 384–389 (female, male, and pupa).

Distribution. Philippines (Palawan).

Bionomics. Aquatic habitat: A river (width 5–6 m, 23.0 °C, shaded, elevation 200 m), moderately flowing in a forest, Mainit, 10 km north of Brook's Point (Takaoka and Tenedero, 2019).

Remarks. This species was described from Palawan, the Philippines (Takaoka and Tenedero, 2019). Its larva remains unknown.

It is characterized by the medium-long female sensory vesicle, pupal thorax covered with round tubercles and pupal gill filaments, of which the longest is 2.0–2.2 mm.

***Simulium* (*Simulium*) *masilauense* Takaoka, 2008**
(Fig. 80N)

Simulium (*Simulium*) *masilauense* Takaoka, 2008b: 62–65 (female, male, pupa, and larva).

Distribution. Malaysia (Sabah).

Bionomics. Aquatic habitat: An irrigation stream (width 1 m, 14 °C, elevation 1880 m) (Takaoka, 2008b).

Remarks. This species was described from Sabah, Malaysia by Takaoka (2008b).

It is characterized in the female by the sensory vesicle, 0.59 times as long as the third palpal segment, and scutum with brown short hairs; in the male by the number of upper-eye (large) facets in 17 vertical columns and 19 horizontal rows and hind basitarsus enlarged, 3.4–3.9 times as long as its greatest width; in the pupa by the gill with six slender filaments in three pairs each with a short stalk, abdominal segments 7, 8, and 9 each with spine-combs and last abdominal segment with triangular terminal hooks; and in the larva by the postgenal cleft long, 3.9–5.5 times as long as the postgenal bridge, and last abdominal segment with a pair of small ventral papillae.

***Simulium* (*Simulium*) *nigrifacies* Datta, 1974**

Simulium (*Simulium*) *nigrifacies* Datta, 1974a: 17–18 (male and pupa); Deng et al., 1994: 13 (larva).

Distribution. India, Myanmar, and China (Tibet).

Bionomics. Aquatic habitat: A small stream (width 30–50 cm, depth 2–5 cm, bottom muddy, 23.0 °C, shaded, elevation 623 m) slowly flowing, Namp-San Waterfall, Nar-Sam Village, Lashio, Shan State, central Myanmar (Takaoka et al., 2018c).

Remarks. This species was described from India by Datta (1974a). Its larva was briefly described and illustrated by Deng et al. (1994). The female of this species is unknown.

It is characterized by the pupal gill with six short filaments, of which the outer filament of the dorsal pair is much thicker than the five other filaments. The absence of a pit-like organ near the base of the gill was confirmed by Takaoka et al. (2018c).

***Simulium* (*Simulium*) *nitidithorax* Puri, 1932**

Simulium (*Simulium*) *nitidithorax* Puri, 1932b: 909–912 (female, male, and pupa); Chen and An, 2003: 342 (larva).

Distribution. India (Arunachal Pradesh and West Bengal) and China (Fujian, Hainan, and Tibet).

Bionomics. Aquatic habitat: A small stream in Mari Bengal Terai, India (Puri, 1932b).

Remarks. This species was described from India by Puri (1932b). It was once recorded from Thailand by Takaoka and Suzuki (1984) but later the species considered *S. nitidithorax* was identified as *S. tani* described from Peninsular Malaysia (Takaoka and Saito, 1996).

It is characterized by the male fore basitarsus greatly dilated (about four times as long as its greatest width), pupal gill with six short filaments in three sessile pairs and pupal abdomen with terminal hooks.

***Simulium* (*Simulium*) *prominentum* Chen and Zhang, 2002**

Simulium (*Simulium*) *prominentum* Chen and Zhang, 2002: 77–78 (pupa and larva).

Distribution. China (Guizhou).

Bionomics. Aquatic habitat: A fast-flowing mountainous stream at an elevation of 1400 m (Chen and Zhang, 2002).

Remarks. This species was described from Guizhou Province, China by Chen and Zhang (2002). Its adults are unknown.

It is characterized in the pupa by the head and thoracic integument covered with rather large tubercles having tiny secondary projections, gill with six short filaments, of which the outer filament of the dorsal pair is thicker than the five other filaments, abdominal segments 7 and 8 each with spine-combs and terminal hooks absent; and in the larva by the postgenal cleft twice as long as the postgenal bridge and posterior circlet with 76 rows of 16–20 hooklets per row.

***Simulium* (*Simulium*) *puliense* Takaoka, 1979**

(Fig. 81K)

Simulium (*Simulium*) *puliense* Takaoka, 1979: 392–395 (male, pupa, and shed larval skin); Huang and Takaoka, 2008b: 305–308 (female).

Distribution. Taiwan.

Bionomics. Aquatic habitat: A small stream (width 0.3 m, flow slow) (Takaoka, 1979).

Remarks. This species was described from Taiwan by Takaoka (1979). Its female was described by Huang and Takaoka (2008b).

It is characterized in the female by the sensory vesicle 0.3 times as long as the third palpal segment, scutum with brown short hairs, and fore basitarsus dilated, 4.1–4.2 times as long as its greatest width; in the male by the number of upper-eye (large) facets in 21 horizontal rows and abdominal segments 2, 6, and 7 each with a pair of shiny dorsolateral patches; in the pupa by the head and thoracic integument with somewhat large tubercles, gill with six short slender filaments directed forward close together, of which the outer filament of the dorsal pair 1.3 times as thick as the five other filaments (Fig. 81K), dorsum of abdominal segments 7, 8, and 9 each with a transverse row of spine-combs on each side and abdominal segment 9 without terminal hooks; and in the larva by the postgenal cleft long, ca. 4.2 times as long as the postgenal bridge, and posterior circlet with 76 rows of ca. 14 hooklets per row.

Chen (2016) wrongly illustrated the pupal gill with five filaments of *S. puliense*.

Simulium (_Simulium_) _quasifrenum_ Delfinado, 1971

Simulium (*Simulium*) *quasifrenum* Delfinado, 1971: 133–134 (male); Takaoka, 1983: 131–134 (female, male, pupa, and larva).

Distribution. Philippines (Palawan).

Bionomics. Aquatic habitats: Streams at low to medium elevations (Takaoka, 1983).

Remarks. This species was described based on adult males by Delfinado (1971). Its female, pupa, and larva were described by Takaoka (1983).

This species is similar to *S. keningauense* and *S. lunduense* in having the pupal thorax with sharply pointed cone-like tubercles (similar to Fig. 81B). However, it is distinguished from *S. keningauense* in the female by the relative length of the sensory vesicle against the third palpal segment (0.29–0.35 in *S. quasifrenum* versus 0.42–0.48 in *S. keningauense*) and claw without a subbasal swelling [with a tiny subbasal swelling (Fig. 80B) in *S. keningauense*]; and in the pupa by the angle formed basally by the upper filament of the dorsal pair and the lower filament of the ventral pair when viewed laterally (90 degrees or greater in *S. quasifrenum* versus 30–40 degrees in *S. keningauense*).

Simulium (_Simulium_) _rangjungense_ Takaoka and Somboon, 2008

(Figs. 80A, H, I and 81F)

Simulium (*Simulium*) *rangjungense* Takaoka and Somboon, 2008: 250–256 (female, male, and pupa).

Distribution. Bhutan.

Bionomics. Aquatic habitat: An irrigation channel (width 30 cm, depth 10 cm, 24.0 °C, exposed to the sun, elevation 1051 m) in a sparsely forested area in Rangjung (Takaoka and Somboon, 2008).

Remarks. This species was described from Bhutan by Takaoka and Somboon (2008). Its larva is unknown.

It is similar to *S. tani* (complex) from Peninsular Malaysia in having the female claw with a tiny subbasal swelling (similar to Fig. 80B) but is distinguished from the latter species in the pupa by the gill with six filaments, of which three outer filaments are subequal in thickness [slightly becoming thinner from dorsal to ventral in *S. tani* (complex)], and dorsomedial trichome on each side of the thorax, which is extremely short (Fig. 81F) [short (similar to Fig. 81E) in *S. tani* (complex)].

Simulium (_Simulium_) _ryukyuense_ Ogata, 1966

Simulium (*Simulium*) *ryukyuense* Ogata, 1966: 123–130 (female, male, pupa, and larva); Takaoka et al., 2021d: 8.

Simulium (*Simulium*) *suzukii* (*nec* Rubtsov): Takaoka, 1977a: 209–213 (female, male, pupa, and larva).

Simulium (*Simulium*) *suzukii* cytoform 'D': Adler et al., 2013: 6–8.

Distribution. Japan (Nansei Islands).

Bionomics. Aquatic habitats: Streams at low to medium elevations in Amami-oshima Island and Okinawa Island (Takaoka, 1977a).

Remarks. This species was described from Okinawa Island, Nansei Islands, Japan by Ogata (1966). This species was synonymized with *S. suzukii* by Takaoka (1977a), but was resurrected based on morphological, chromosomal and genetic analyses by Takaoka et al. (2021d).

This species is distinguished from *S. suzukii* cytoform "C" from Hokkaido by the color of the male fore coxa and relative length of the female sensory vesicle to the third palpal segment (Takaoka et al., 2021d), as shown later under *S. suzukii*.

***Simulium* (*Simulium*) *sabahense* Smart and Clifford, 1969**

(Figs. 80P and 81L)

Simulium (*Simulium*) *sabahense* Smart and Clifford, 1969: 31–35 (female, male, pupa, and larva); Takaoka, 2008b: 68–72 (female, male, pupa, and Larva).

Distribution. Malaysia (Sabah and Sarawak).

Bionomics. Aquatic habitats: A rapidly flowing river (width 7–10 m, 22.0 °C, shaded, elevation 545–565 m) and a small stream (width 0.2 m, flow moderate, 23.0 °C, shaded, elevation 550 m) in a natural forest, in Hot Spring Park, Poring, Sabah; and a river (width 5–10 m, 24.0 °C) in a lowland in Pueh, Sarawak (Takaoka, 2008b).

Remarks. This species was originally described from Sabah (Smart and Clifford, 1969). The female, male, pupa, and larva of this species were redescribed by Takaoka (2008b).

It is characterized by the medium-long female sensory vesicle, the number of male upper-eye (large) facets in 12–14 vertical columns and 13–15 horizontal rows, the short anterodorsal and mediolateral trichomes of the pupal thorax and the pupal gill with six filaments decreasing in length and thickness from dorsal to ventral (Fig. 81L).

***Simulium* (*Simulium*) *sanyaense* Cai, Liu and An, 2016**

Simulium (*Simulium*) *sanyaense* Cai, Liu and An, 2016: 214–216 (female and pupa).

Distribution. China (Hainan).

Bionomics. Aquatic habitats: Streams (width 2–3 m, shaded, elevation 200 m) (Cai et al., 2016).

Remarks. This species was described from Hainan, China by Cai et al. (2016). Its male and larva are unknown.

It is characterized in the female by the sensory vesicle long, 0.5–0.6 times as long as the third palpal segment, scutum covered with black fine hairs, hind femur and tibia brownish black except their bases brownish yellow and tergites of abdominal segments 5–7 large and shiny (probably those of abdominal segments 6–8 large and shiny); and in the pupa by the head and thoracic integument densely covered with tubercles, gill with six slender filaments arranged as 2+2+2 from dorsal to ventral, all lying vertically.

This species is similar to *S. suzukii* in having the long female sensory vesicle and the same arrangement of the pupal gill.

In the original description (Cai et al., 2016), the presence or absence of the pupal terminal hooks was not mentioned.

Simulium (Simulium) suoivangense Takaoka and Pham, 2017
Simulium (*Simulium*) *tani* (complex): morphoform "b": Takaoka et al., 2015c: 85–86.
Simulium (*Simulium*) *suoivangene* Takaoka and Pham [in Takaoka et al., 2017c]: 148–153 (female, male, pupa, and larva).

Distribution. Vietnam.

Bionomics. Aquatic habitats: A river (width 4–6 m, 18.0 °C, elevation 1452 m), moderately flowing in an open land, Suoi Vang, Lam Dong Province, and a river (width 4–5 m, depth 20–30 cm, bottom of pebbles, 15.0 °C, partially shaded, elevation 223 m), rapidly flowing in a natural forest, Pu Mat National Park, Nghe An Province (Takaoka et al., 2017c).

Remarks. This species was described from Vietnam by Takaoka et al. (2017c). It is the same species previously recognized as morphoform "b" of *S. tani* (complex) from Vietnam by Takaoka et al. (2015c). The recognition of morphoform "b" as a distinct species was supported by DNA sequence-based and chromosomal analyses (Adler et al., 2016; Low et al., 2016b).

It is characterized by lacking pupal terminal hooks, differing from all 11 cytoforms and morphoform "a" among *S. tani* (complex), all of which have pupal terminal hooks (Takaoka et al., 2015c).

Simulium (Simulium) suzukii Rubtsov, 1963 (complex)
Simulium suzukii Rubtsov, 1959–1964: 525–526 (female, male, pupa, and larva).
Simulium (*Simulium*) *suzukii*: Orii et al., 1969: 1–13 (larva).
Simulium tuberosum (*nec* Lundström): Bentinck, 1955: 9 (illustrations of female, male, and pupa).

Distribution. Japan, Korea (South), Siberia and China (Guizhou, Hong Kong, Jilin, Jiangxi, Sichuan, and Yunnan).

Bionomics. Aquatic habitats: Medium-sized to large mountainous streams, and lowland streams in Japan (Bentinck, 1955).

Remarks. This species was first recognized as *S. tuberosum* in Japan by Ogata et al. (1956). It was described as a new species, *S. suzukii,* by Rubtsov (1959–1964). Adler et al. (2013) showed by a chromosome analysis that it is a species complex consisting of four cytoforms: cytoforms "A" and "B" from South Korea, cytoform "C" from Hokkaido, Japan, and cytoform "D" from Okinawa, Japan. Takaoka et al. (2021d) reported that cytoform "D" was *S. ryukyuense* Ogata, 1966, which was once sunk as a junior synonym of *S. suzukii* by Takaoka (1977a).

This species is distinguished from *S. ryukyuense* by the light brown male fore coxae and long female sensory vesicle (0.51–0.63 times as long as the third palpal segment). In *S. ryukyuense*, the male forecoxae are yellow, and the female sensory vesicle is medium-long (0.33–0.42 times as long as the third palpal segment).

Simulium (Simulium) tani Takaoka and Davies, 1995 (complex)
Simulium (*Simulium*) *tani* Takaoka and Davies, 1995a: 137–141 (female, male, pupa, and larva).

Distribution. Indonesia (Sumatra), Malaysia (Peninsular), Myanmar, Thailand, and Vietnam.

Bionomics. Aquatic habitats: Streams at low to high elevations (224–1813 m) in Cameron's Highlands and other areas in Peninsular Malaysia (Ya'cob et al., 2016).

Remarks. Simulium tani was described from Peninsular Malaysia (Takaoka and Davies, 1995a). This species is a species complex consisting of 10 cytoforms in Thailand (Tangkawanit et al., 2009b). The type of *S. tani* from Peninsular Malaysia was shown to be different from any cytoforms reported from Thailand and Taiwan and was designated as cytoform K of *S. tani* (complex) (Adler et al., 2013). It was recorded from Sumatra, Indonesia by Takaoka et al. (2000), Myanmar by Takaoka et al. (2018j), and Vietnam by Takaoka et al. (2017c).

The population of *S. tani* in Sumatra, Indonesia was genetically different from those of *S. tani* from Thailand (Hew et al., 2023a).

It is characterized in the female by the sensory vesicle elongate and claw with a minute subbasal swelling (similar to Fig. 80B); in the pupa by the gill with six slender filaments directed forward close together (similar to Fig. 81G); and in the larva by the postgenal cleft, of which both lateral margins are rounded, and last abdominal segment with a pair of small ventral papillae.

***Simulium* (*Simulium*) *tukurongense* Takaoka and Tenedero, 2019**

Simulium (*Simulium*) *tukurongense* Takaoka and Tenedero, 2019: 389–392 (female and pupa).

Distribution. Philippines (Palawan).

Bionomics. Aquatic habitats: The Tukurong River (width 30 m, 23.0 °C, shaded, elevation 160 m), moderately to rapidly flowing in a forest, Montible, Puerto Princesa City, and the Estrella River (width 20 m, 22.0 °C, elevation 110 m), near Estrella Waterfall, Nara (Takaoka and Tenedero, 2019).

Remarks. This species was described from Palawan, the Philippines (Takaoka and Tenedero, 2019). Its male and larva remains unknown.

It is similar to *S. mainitense* but is barely distinguished from the latter species in the female by the relative length of the labrum against the clypeus (0.5–0.6 in this species versus 0.7 in *S. mainitense*); and in the pupa by the length of the longest gill filament (1.3–1.6 mm in this species versus 2.0–2.2 mm in *S. mainitense*).

***Simulium* (*Simulium*) *xuandei* Takaoka and Pham, 2015**

Simulium (*Simulium*) *xuandei* Takaoka and Pham [in Takaoka et al., 2015c]: 86–93 (female, male, pupa and larva).

Distribution. Vietnam.

Bionomics. Aquatic habitats: Two small streams (width 1–6 m, exposed to the sun, 16.0 and 18.0 °C, elevation 1452 and 1492 m) in Lam Dong Province (Takaoka et al., 2015c).

Remarks. This species is characterized in the female by the sensory vesicle 0.33–0.37 times as long as the third palpal segment; in the male by the number of upper-eye (large) facets in 18 vertical columns and 19 horizontal rows; and in the pupa by the frons and anterodorsal surface of the thorax with tubercles having tiny secondary projections, and gill with six slender filaments divergent basally at about 90 degrees when viewed laterally.

Simulium (*Simulium*) *rufibasis* Subgroup

Simulium (*Simulium*) *arisanum* Shiraki, 1935

Simulium arisanum Shiraki, 1935; 80–82 (female).
Simulium (Simulium) arisanum: Takaoka, 1979: 400.

Distribution. Taiwan.

Bionomics. Unknown.

Remarks. This species was described from Taiwan by Shiraki (1935). Its male, pupa, and larva are unknown. It is tentatively placed in the *S. rufibasis* subgroup, although the presence of a pair of stout hairs on the sternite 7 is not confirmed.

The female of this species is characterized by the scutum with yellow short hairs and hind basitarsus yellowish on the basal half. The pupa tentatively considered to be that of *S. arisanum* is characterized by the head and thoracic integument almost bare (Low et al., 2018).

Simulium (*Simulium*) *dailekhense* Takaoka and Shrestha, 2020

(Fig. 82F–J)

Simulium (*Simulium*) *dailekhense* Takaoka and Shrestha [in Takaoka et al., 2020e]: 313–318 (female, male, and pupa).
Simulium (*Simulium*) *ramosum* (*nec* Puri): Shrestha and Takaoka, 2009: 256.

Distribution. Nepal.

Bionomics. Aquatic habitats: Three streams at high elevations (1600–2167 m) (Takaoka et al., 2020e).

Remarks. This species is the same species as that previously recorded as *S. ramosum* Puri, based on two females and their pupal exuviae from Nepal (Shrestha and Takaoka, 2009). Takaoka et al. (2020e), after reexamining these specimens, described this species as a new species, *S. dailekhense*, based on the difference of the male fore coxa (light or medium brown in this species versus yellowish in *S. ramosum*) and male hind basitarsus (yellowish on the basal four-ninths in this species versus yellowish on the basal half in *S. ramosum*). The female and pupa of this species are indistinguishable from those of *S. ramosum*.

The pupa of this species is characterized by the branched frontal and thoracic trichomes, as in *S. ramosum*.

Simulium (*Simulium*) *doipuiense* Takaoka and Choochote, 2005 (complex)

Simulium (*Simulium*) sp. E Takaoka and Suzuki, 1984: 42–44 (female and pupa).
Simulium (*Simulium*) *doipuiense* Takaoka and Choochote, 2005d: 106–110 (female, male, pupa, and larva).

Distribution. Thailand and Vietnam.

Bionomics. Aquatic habitats: Streams (width 0.2–5.0 m, 18.5–20.0 °C, shaded, elevation 1360 m) in Chiang Mai Province (Takaoka and Choochote, 2005d).

Remarks. *Simulium doipuiense* is the species formerly recorded as *S.* sp. E from Thailand (Takaoka and Suzuki, 1984). It is a species complex consisting of two cytoforms (A and B) (Tangkawanit et al., 2009b).

According to the original description (Takaoka and Choochote 2005d), it is characterized in the female by having the antenna dark brown except the scape, pedicel and first flagellomere yellow, scutum covered with yellow short hairs, hind tibia yellowish white except the apical two-fifths medium to dark brown and stem of the genital fork without a round apex; in the male by the number of upper-eye (large) facets in 17 vertical columns and 16 or 17 horizontal rows, hind basitarsus yellow to dark yellow on the basal one-third to slightly more than the basal two-fifths and dark brown on the rest; in the pupa by the gill with six filaments arranged as 2+2+2 from dorsal to ventral, decreasing in length and thickness from dorsal to ventral, with the longest 1.5–1.8 mm and the shortest 0.8–1.0 mm; in the larva by the head capsule yellow, with head spots indistinct or faintly negative.

***Simulium* (*Simulium*) *fuscicoxae* Takaoka and Ya'cob, 2017**

Simulium (*Simulium*) *rufibasis* (*nec* Brunetti): Takaoka et al., 2015c: 85

Simulium (*Simulium*) *fuscicoxae* Takaoka and Ya'cob [in Takaoka et al., 2017c]: 127–130 (female and pupa).

Distribution. Vietnam.

Bionomics. Aquatic habitat: A small stream (width 1–2 m, 16.0 °C, elevation 1492 m) slowly flowing in an open land, 45 km from Dalat, Lam Dong Province (Takaoka et al., 2015b).

Remarks. This species was first reported as *S. rufibasis* based on a female and its associated pupal exuviae and cocoon from Lam Dong Province, southern Vietnam (Takaoka et al., 2015b). Subsequently it was described as a new species (Takaoka et al., 2017c). Its male and larva are unknown.

It is characterized in the female by the sensory vesicle having a large opening, scutum with dark short hairs and dark fore coxae; and in the pupa by lacking spine-combs on the dorsal surface of abdominal segment 7.

***Simulium* (*Simulium*) *giayense* Takaoka and Lau, 2017**

(Fig. 83C)

Simulium (*Simulium*) *giayense* Takaoka and Lau [in Takaoka et al., 2017c]: 137–138 (female, male, and pupa).

Distribution. Vietnam.

Bionomics. Aquatic habitat: A small stream (width 0.5 m, bottom rocky, 11.0 °C, elevation 1750 m) moderately flowing in a secondary forest, Sapa, Lao Cai Province (Takaoka et al., 2017c).

Remarks. This species was described from northern Vietnam (Takaoka et al., 2017c). Its larva is unknown.

It is similar in the female to *S. rosliramlii* but is distinguished in the male by the scutum covered with yellow short hairs; and in the pupa by the frons and the dorsal surface of the anterior half of the thorax densely covered with tubercles (Fig. 83C).

***Simulium* (*Simulium*) *hehuanense* Takaoka, 2021**
Simulium (*Simulium*) *rufibasis* (*nec* Brunetti): Takaoka, 1979: 395.
Simulium (*Simulium*) *hehuanense* Takaoka [in Takaoka et al., 2021b]: 404–409 (female, male, and pupa).

Distribution. Taiwan.

Bionomics. Aquatic habitat: A stream at an elevation of ca. 2600 m at Dayuling (Takaoka et al., 2021b).

Remarks. This was described from Taiwan by Takaoka et al. (2021b). It was first recorded as *S. rufibasis* from Taiwan (Takaoka, 1979). Its larva is unknown.

It is characterized in the female by the height of the frons nearly as long as its narrowest width, hind tibia yellow on the basal half and dark on the rest, and clustered stout hairs on sternite 7 nearly twice as long as other hairs on its surrounding area; in the male by the number of upper-eye (large) facets in 19 vertical columns and 19 horizontal rows, fore basitarsus 5.52 times as long as its greatest width, and abdominal segments 2, 6, and 7 each with a pair of shiny, white-pruinose dorsolateral spots; and in the pupa by the frons moderately covered with tubercles and gill with six filaments, of which the longest dorsal filament is 2.3–2.5 mm long.

This species is distinguished from *S. rufibasis* from India (Puri, 1932b) by the following characteristics (those of *S. rufibasis* in parentheses): the female hind basitarsus yellowish white on the basal three-fifths (yellow on the basal half to one-third), absence of a pair of shiny dorsolateral spots on the male abdominal segment 5 (presence), and absence of pupal terminal hooks (presence).

***Simulium* (*Simulium*) *langtangense* Takaoka and Shrestha, 2020**
Simulium (*Simulium*) *langtangense* Takaoka and Shrestha [in Takaoka et al., 2020e]: 322–326 (female, male, and pupa).
Simulium (*Simulium*) *rufibasis* (*nec* Brunetti): Shrestha and Takaoka, 2009: 256–257.

Distribution. Nepal.

Bionomics. Aquatic habitat: A stream (elevation 1552 m) on the way to Langtang (Takaoka et al., 2020e).

Remarks. This species was previously recorded as *S. rufibasis*, based on a female and its pupal exuviae from Nepal (Shrestha and Takaoka, 2009). Takaoka et al. (2020e) described this species as a new species, *S. langtangense*, based on the difference of the position of the shiny dorsolateral spots on the male abdomen (present on abdominal segments 2, 6, and 7 in this species versus present on abdominal segments 2 and 5–7 in *S. rufibasis*).

The darkened male fore coxa of this species, combined with the pupal frons and thorax moderately covered with tubercles, separates it from all the other members of the *S. rufibasis* subgroup.

Simulium (_Simulium_) _manooni_ Takaoka and Choochote, 2005
(Figs. 82A, C and 83B, G, H, I)

Simulium (*Simulium*) *manooni* Takaoka and Choochote, 2005d: 99–106 (female, male, pupa, and larva).

Distribution. Thailand.

Bionomics. Aquatic habitat: A small stream (width 1 m, depth 0.2 m, shaded, 18.5 °C, elevation 1360 m) in Doi Pui National Park (Takaoka and Choochote, 2005d).

Remarks. This species is similar to *S. giayense* from Vietnam, *S. dailekhense* from Nepal and *S. ramosum* from India in having the same arrangement of the pupal gill filaments, i.e., an acute angle between the stalks of the dorsal and ventral pairs of filaments when viewed laterally (Fig. 83G) (Puri, 1932b; Takaoka et al., 2017c, 2020e). However, this species is distinguished from *S. giayense* by the integument of the frons moderately covered with tubercles (Fig. 83B) [densely covered with tubercles (Fig. 83C) in *S. giayense*], and from the latter two species by the unbranched trichomes on the frons and thorax (branched trichomes in the latter two species).

This species is characterized in the female by the dark short hairs on the scutum; in the male by the number of upper-eye (large) facets in 17 or 18 vertical columns and 18 horizontal rows and the fore coxa light brown on its outer surface; and in the larva by the dark spots on the head capsule.

Simulium (_Simulium_) _nakanoi_ Takaoka and Shrestha, 2020
(Figs. 82D and 83D)

Simulium (*Simulium*) *nakanoi* Takaoka and Shrestha [in Takaoka et al., 2020e]: 318–322 (female, male, and pupa).

Distribution. Nepal.

Bionomics. Aquatic habitat: A stream (elevation (1600 m) in Shivapuri National Park (Takaoka et al., 2020e).

Remarks. This species was described from Nepal by Takaoka et al. (2020e). Its larva is unknown.

This species is similar in the pupa to *S. pradyai* from Bhutan and *S. rosliramlii* from Vietnam in sharing the pupal frons and most of the thorax bare, without tubercles (Takaoka and Somboon, 1998; Takaoka et al., 2017c). However, this species is distinguished from *S. pradyai* by the relative length of the female fore basitarsus against its greatest width (6.0–6.2 in this species versus 5.36 in *S. pradyai*) and the number of male upper-eye (large) facets (13 or 14 vertical columns in this species versus 17 vertical columns in *S. pradyai*), and from *S. rosliramlii* by the color of the male fore coxa (whitish yellow in this species versus medium brown in *S. rosliramlii*) and of the male hind tibia (dark brown except the basal half or little less yellowish white in this species versus dark brown except the basal tip dark yellow to light brown in *S. rosliramlii*).

***Simulium* (*Simulium*) *pradyai* Takaoka, 2008**

Simulium (*Simulium*) *pradyai* Takaoka, 2008 [in Takaoka and Somboon, 2008]: 256–261 (female and pupa).

Distribution. Bhutan and Nepal.

Bionomics. Aquatic habitats: A small stream (width 2 m, depth 0.1–0.2 m, flow moderate, 22 °C, elevation 2570 m) in a sparsely forested area, Dungmithang, Bumthang, and a small stream (width ca. 1 m, depth 0.1 m, flow slow, 5 °C, elevation 3720 m) in a densely forested area, both Thrimshingla, Mongar (Takaoka and Somboon, 2008).

Remarks. This species was described from Bhutan (Takaoka and Somboon, 2008). Its male and larvae are unknown. This species was recorded from Nepal by Shrestha and Takaoka (2009).

It is characterized by the pupal frons and the dorsal surface of the anterior half of the thorax bare, gill with six slender filaments, of which two filaments of the ventral pair have a long stalk, and last abdominal segment with a pair of weakly developed terminal hooks.

***Simulium* (*Simulium*) *ramosum* Puri, 1932**

Simulium (*Simulium*) *ramosum* Puri, 1932b: 905–906 (female, male, and pupa); Datta, 1974b: 466–467 (larva).

Distribution. India (Himachal Pradesh and West Bengal).

Bionomics. Aquatic habitats: Various streams in the Shimla Hills, Himachal Pradesh (Puri, 1932b).

Remarks. This species was described by Puri (1932b). Its larva was described by Datta (1974b).

It is similar to *S. rufibasis* but is easily distinguished by the pupal head and thorax with branched trichomes.

***Simulium* (*Simulium*) *rosliramlii* Takaoka and Chen, 2017**
(Fig. 83A)

Simulium (*Simulium*) *rosliramlii* Takaoka and Chen [in Takaoka et al., 2017c]: 130–136 (female, male, and pupa).

Distribution. Vietnam.

Bionomics. Aquatic habitat: A river (width 6 m, depth 0.6 m, bottom rocky, 11.0 °C, elevation 1728 m) fast-flowing in a forest, Lai Chau, Sapa, Lao Cai Province (Takaoka et al., 2017c).

Remarks. This species was described from northern Vietnam (Takaoka et al., 2017c). Its larva is unknown.

It is characterized by the pupal frons and most of the thorax bare, without tubercles (Fig. 83A).

Simulium (Simulium) rufibasis Brunetti, 1911

Simulium rufibasis Brunetti, 1911: 282–288 (female).
Simulium (*Simulium*) *rufibasis*: Puri, 1932b: 899–903 (female, male, and pupa); Datta, 1974a: 19–20 (larva).

Distribution. India, Pakistan, Myanmar and China (Anhui, Fujian, Guizhou, Hubei, Liaoning, Sichuan, and Yunnan).

Bionomics. Aquatic habitats: Small to large streams at high elevations from 1600 to 2500 m in West Bengal (Puri, 1932b).

Remarks. This species was described from India by Brunetti (1911). Its female was redescribed and its male and pupa were first described by Puri (1932b). Its larva was described by Datta (1974a). This species was once recorded from various countries outside India, such as Japan, Korea, Taiwan, Thailand, Nepal, and Vietnam, but later these previous records were found to be incorrect based on the reexaminations by Adler et al. (2019a) and Takaoka et al. (2017c, 2020e, 2021b, 2018i). Thus, the records of this species from China, Myanmar and Pakistan should be reevaluated. For example, the species recognized as *S. rufibasis* in China has no pupal terminal hooks (Chen, 2016), suggesting that it is different from *S. rufibasis*, which has pupal terminal hooks (Puri, 1932b).

It is characterized in the female by the scutum with brassy short hairs, fore coxa yellow, hind tibia pale yellow on the basal two-fifths to one-third, and hind basitarsus pale yellow on the basal half to one-third; in the male by the scutum with copper-colored short hairs, fore coxa black and hind basitarsus slightly yellow on the basal one-third and much enlarged, and abdominal segments 2 and 5–7 each with a pair of shiny dorsolateral patches; in the pupa by the head and thoracic integument with tubercles, gill with six filaments decreasing in length and thickness from dorsal to ventral and terminal hooks poorly developed.

Simulium (Simulium) setsukoae Takaoka and Choochote, 2004
(Fig. 82B)

Simulium (*Simulium*) sp. D Takaoka and Suzuki, 1984: 42 (female).
Simulium (*Simulium*) *setsukoae* Takaoka and Choochote, 2004: 33–35 (female); Tangkawanit et al., 2009a: 37–38 (male, pupa, and larva).

Distribution. Thailand.

Bionomics. Six females of this species were collected while biting a human at an elevation of ca. 2400 m in Doi Inthanon National Parks (Takaoka and Suzuki, 1984).

Remarks. The female of this species was first recorded as *S*. sp. D from Thailand by Takaoka and Suzuki (1984). It was later described as a new species by Takaoka and Choochote (2004b). Its male, pupa, and larva were briefly described by Tangkawanit et al. (2009a).

It is characterized in the female by the sternite 7 with a pair of clustered short hairs, which are nearly as long as those on the surrounding area (Fig. 82B); and in the male by the upper-eye (large) facets in ca. 20 vertical columns; and in the pupa by the frons with round tubercles and the upper filament of the dorsal pair of the gill about twice as thick as the counter lower filament.

***Simulium* (*Simulium*) *tenebrosum* Takaoka, Srisuka and Saeung, 2018 (complex)**

(Figs. 82K, M–S and 83J, K)

Simulium (*Simulium*) *rufibasis* (*nec* Brunetti), Takaoka and Suzuki, 1984: 41–42 (female legs).
Simulium (*Simulium*) *tenebrosum* Takaoka, Srisuka and Saeung [in Takaoka et al., 2018i]: 409–415 (female, male, pupa, and larva).

Distribution. Thailand.

Bionomics. Aquatic habitat: A stream at an elevation of 2210 m in Doi Inthanon National Park, Chiang Mai, where many adult females of this species were captured on humans (Choochote et al., 2005).

Remarks. This species was recorded as *S. rufibasis* from Thailand, based on 28 females caught by a sweeping net while biting a human, and a male reared from a pupa collected in Doi Inthanon National Parks (Takaoka and Suzuki, 1984).

Takaoka et al. (2018i) described this species as a new species, *S. tenebrosum*, based on the difference of the position of the shiny dorsolateral spots on the male abdomen (present on abdominal segments 2, 6, and 7 in this species versus present on abdominal segments 2 and 5–7 in *S. rufibasis*), relative length of the male fore basitarsus against its greatest width (7.2–8.2 in this species versus 6.0 in *S. rufibasis*) and pupal terminal hooks (absent in this species versus present in *S. rufibasis*). The female of this species is almost indistinguishable from not only *S. rufibasis* but also *S. ramosum* from India (Puri, 1932b).

***Simulium* (*Simulium*) *weji* Takaoka, 2001**

(Figs. 82E, L and 83F)

Simulium (*Simulium*) *weji* Takaoka, 2001c: 349–354 (female, male, pupa, and larva); Tangkawanit et al., 2009a: 39 (pupa and larva).

Distribution. Thailand.

Bionomics. Aquatic habitat: A calcareous stream (width 1.0–3.0 m, 23.0 °C, elevation 580 m), Tharn Thong Waterfall, Lampang Province (Takaoka, 2001c).

Remarks. This species was described from Thailand (Takaoka, 2001c). Its pupa and larva were morphologically studied by Tangkawanit et al. (2009a).

It is characterized in the female by the hind tibia mostly darkened except the small basal portion and paired clustered long hairs on the sternite 7, which are two to three times as long as those on the surrounding area (Fig. 82E); in the male by the number of upper-eye (large) facets in 16 vertical columns and 17 horizontal rows and ventral plate with its body having rounded lateral margins when viewed ventrally (Fig. 82L); in the pupa by the frons and thorax covered with cone-like tubercles and antennal sheath with tubercles; and in the larva by the body dark greenish on thoracic segment 1 and abdominal segments 5–8 and light brown or light green on the rest.

Simulium (_Simulium_) _xiulinense_ Takaoka, 2021

Simulium (*Simulium*) *xiulinense* Takaoka [in Takaoka et al., 2021b]: 409 (male and pupa).

Distribution. Taiwan.

Bionomics. Aquatic habitat: A stream at an elevation of ca. 2600 m at Dayuling (Takaoka et al., 2021b).

Remarks. This species was described from Taiwan (Takaoka et al., 2021b). Its female and larva are unknown.

It is characterized in the male by the large number of upper-eye (large) facets in 20 or 21 vertical columns and 22 horizontal rows, fore basitarsus 6.64 times as long as its greatest width and abdominal segments 2 and 5–7 each with a pair of shiny dorsolateral patches; and in the pupa by the gill with six filaments, of which the longest uppermost filament is ca. 3.0 mm long.

This species is distinguished from *S. rufibasis* from India (Puri, 1932b) by the following characteristics (those of *S. rufibasis* in parentheses): the male hind basitarsus yellowish white on the basal three-fifths (yellow on the basal half to one-third) and absence of pupal terminal hooks (presence).

Simulium (_Simulium_) _variegatum_ Species-group

Simulium (_Simulium_) _barnesi_ Takaoka and Suzuki, 1984

Simulium (*Simulium*) *barnesi* Takaoka and Suzuki, 1984: 31–33 (female).

Distribution. Thailand and India (Arunachal Pradesh).

Bionomics. Unknown.

Remarks. This species was described from a female collected from Doi Chom Chang, near Chiang Mai (Takaoka and Suzuki, 1984). Later, one female was recorded from India (Borah et al., 2012). However, its male, pupa and larva are unknown.

It is characterized by the dark legs, and paraproct with its ventral surface flat, lacking a shallow groove, and its anteromedial surface dark pigmented.

Simulium (_Simulium_) _bulla_ Davies and Györkös, 1987

Simulium (*Simulium*) *bulla* Davies and Györkös, 1987b: 2737–2742 (female, male, pupa, and larva).

Distribution. Sri Lanka.

Bionomics. Aquatic habitats: Rapid trickles on rock surfaces and in streams (width 0.1–10 m, bottom rocky, elevation 400–1800 m) (Davies and Györkös, 1987b).

Remarks. This species was described from Sri Lanka (Davies and Györkös, 1987b).

It is characterized in the female and male by the scutum with yellow hairs and yellowish legs including fore coxae; in the male by the number of upper-eye (large) facets in 19 vertical columns and 19 horizontal rows; and in the pupa by the thorax without tubercles.

Simulium (Simulium) caohaiense Chen and Zhang, 1997
Simulium (*Simulium*) *caohaiense* Chen and Zhang, 1997a: 194–196 (female, male, pupa, and larva).

Distribution. China (Guizhou).

Bionomics. Aquatic habitat: A small stream (20 °C, elevation 2200 m) (Chen and Zhang, 1997a).

Remarks. This species was described from Guizhou, China (Chen and Zhang, 1997a).

It is characterized in the female by the yellowish legs; in the male by the number of upper-eye (large) facets in 18 horizontal rows and yellowish legs; and in the larva by the rectal organ compound, each lobe with six or seven secondary lobules.

Simulium (Simulium) chamlongi Takaoka and Suzuki, 1984
Simulium (*Simulium*) *chamlongi* Takaoka and Suzuki, 1984: 27–30 (female, male, pupa, and larva); Crosskey and Howard, 1997; Pham, 1999: 335; Takaoka *et al.*, 2015c: 93; Takaoka *et al.*, 2017c: 154.

Distribution. China (Yunnan and Tibet), Myanmar, Thailand, and Vietnam.

Bionomics. Aquatic habitat: A small stream (width 1–2 m, 16.0 °C, exposed to the sun, elevation 1492 m) slowly flowing in an open land, 45 km from Dalat, Lam Dong Province (Takaoka et al., 2015c).

Remarks. This species was originally described from Thailand by Takaoka and Suzuki (1984). It was later recorded from Vietnam by Crosskey and Howard (1997), Pham (1999), and Takaoka et al. (2015c).

The pupa of this species is characterized by the frons lacking tubercles. The male of this species has the large number of upper-eye (large) facets in 21–23 vertical columns and 22 or 23 horizontal rows.

Simulium (Simulium) cruszi Davies and Györkös, 1992
Simulium (*Simulium*) *cruszi* Davies and Györkös, 1992: 1031–1035 (female, male, pupa, and larva).

Distribution. Sri Lanka.

Bionomics. Aquatic habitats: A stream (width 0.3 m, 16 °C) draining a small pond, and a stream (width 0.6–0.8 m, 15 °C), just northeast of Horton Plains (Davies and Györkös, 1992).

Remarks. This species was described from Sri Lanka by Davies and Györkös (1992).

It is characterized in the female by the sensory vesicle long, half as long as the third palpal segment, scutum with coppery short hairs, fore basitarsus 10.5 times as long as its greatest width, and legs dark brown except the fore coxa and trochanter, the basal one-fifth of the tibia and the basal three-fifths of the basitarsus of the hind leg yellow; in the male by the number of upper-eye (large) facets in 22 or 23 vertical columns and 23 or 24 horizontal rows, scutum with coppery short hairs, legs darkened, fore and hind basitarsi 10 and 3.1–3.4 times as long as their greatest widths, respectively; and in the pupa by lacking terminal hooks.

***Simulium* (*Simulium*) *elongatum* Takaoka, 2003**

(Figs. 87G, H and 89K, L, W)

Simulium (*Simulium*) *elongatum* Takaoka, 2003: 501–505 (female, male, pupa, and larva).

Distribution. Indonesia (Sulawesi).

Bionomics. Aquatic habitats: Two moderately-flowing mountain streams (width 0.5–3.0 m, 16.0–17.0 °C, partially shaded, elevation 1520–1600 m) in Lore Utara (Takaoka, 2003).

Remarks. This species was described from Central Sulawesi by Takaoka (2003).

It is characterized in the female by the tibiae yellowish except apical caps darkened; in the male by abdominal segments 2 and 4–7 each with a pair of dorsolateral shiny spots; in the pupa by the gill composed of a somewhat inflated elongate common basal stalk with five finger-like projections and six slender thread-like filaments arising from the apices of the five finger-like projections (six slender apical filaments about two-thirds the total length of the gill) (Fig. 87G, H) and cocoon wall-pocket-shaped with a long anterodorsal projection (Fig. 89K, L); and in the larva by the postgenal cleft short (0.5–0.6 times as long as the postgenal bridge) and round.

***Simulium* (*Simulium*) *flavum* Takaoka, 2003**

(Figs. 88A, B and 89H)

Simulium (*Simulium*) *flavum* Takaoka, 2003: 505–511 (female, male, pupa, and larva).

Distribution. Indonesia (Sulawesi).

Bionomics. Aquatic habitats: Two roadside ditches (width 0.3–0.5 m, depth 0.2 m, 19.0 °C, elevation 1120–1200 m), a river (width 6.0–8.0 m, bottom with stones, 19.0 °C, elevation 1500 m), and a stream (width 3.0–4.0 m, 21.0 °C, elevation 1450 m), all between Malino and Sinjai (Takaoka, 2003).

Remarks. This species was described from South Sulawesi by Takaoka (2003).

It is characterized in the female and male by the yellow scutum; in the pupa by the bare frons and gill with six filaments (at least the basal half of each filament is somewhat inflated and the apical half is tapered to become a slender apical filament), of which two are directed upward and four are forward (Fig. 88A, B), and cocoon simple wall-pocket-shaped (Fig. 89H); and in the larva by the postgenal cleft bullet-shaped and nearly as long as the postgenal bridge.

Simulium (Simulium) globosum Takaoka, 2003

(Figs. 88K, L and 89S, T)

Simulium (*Simulium*) *globosum* Takaoka, 2003: 528–534 (female, male, pupa, and larva).

Distribution. Indonesia (Sulawesi).

Bionomics. Aquatic habitats: Various types of water courses ranging from a roadside ditch (0.4 m wide) to a river (6–8 m wide), with temperature of 18.0–22.0 °C and an elevation of 1120–1450 m, in Malino and Loka (Takaoka, 2003).

Remarks. This species was described from South Sulawesi by Takaoka (2003).

This species is similar to *S. flavum*, *S. lorense*, *S. malinoense*, and *S. quinquefiliatum* by having the female and male scuta yellow and female cibarium furnished with numerous tiny processes. However, it is distinguished by the pupal gill composed of a single ellipsoidal structure (though both ends depressed) with six slender filaments (Fig. 88K, L).

Simulium (Simulium) gurneyae Senior-White, 1922

Simulium (*Simulium*) *gurneyae* Senior-White, 1922: 130 (female).
Simulium (*Simulium*) *gurneyae*: Puri, 1932a: 890–894 (female, male, and pupa).

Distribution. India (Karnataka, Maharashtra, and Tamil Nadu).

Bionomics. Aquatic habitats: Streams in and around Coonoor (elevation 1800 m) in Ootacumundo (elevation 2100 m) and near Old and New Mahableshwar (elevation 1350) (Puri, 1932a).

Remarks. This species was described from females by Senior-White (1922). Its male and pupa were described by Puri (1932a). Its larva is unknown.

It is characterized in the pupa by the anterior part of the thorax without tubercles, gill with six slender filaments, of which the uppermost filament is directed forward, then downward, and thickest of all, about twice as thick as the lowermost filament. The gill filaments spread out slightly at their base when viewed laterally.

Simulium (Simulium) hackeri Edwards, 1928

Simulium hackeri Edwards, 1928: 60–61 (female).
Simulium (*Simulium*) *hackeri*: Takaoka and Suzuki, 1984: 32 (female genitalia); Takaoka and Davies, 1995a: 142–145 (female and male); Ya'cob et al. 2018a: 111–113 (pupa and larva).

Distribution. Malaysia (Peninsular).

Bionomics. Aquatic habitat: A small stream (ca. 4 m wide) at an elevation of 1315 m in Cameron's Highlands (Ya'cob et al., 2018a).

Remarks. This species was described from females collected from Cameron's Highlands by Edwards (1928). The female terminalia were illustrated by Takaoka and Suzuki (1984). Its female and male were described by Takaoka and Davies (1995a), and its pupa and larva were described by Ya'cob et al. (2018).

It is characterized by the darkened legs of the female and male and bare pupal frons.

***Simulium* (*Simulium*) *himalayense* Puri, 1932**

Simulium (*Simulium*) *himalayense* Puri, 1932a: 885–890 (female, male, and pupa).

Distribution. India, Myanmar, Nepal, Pakistan, and China (Tibet).

Bionomics. Aquatic habitats: Streams in and around Shimla (elevation 1640–1968 m), in Kasauli (elevation 1312–1640 m), and near Fago (elevation 2624 m), Theog (elevation 2296 m), and Matiana (elevation 2624 m) (Puri, 1932a).

Remarks. This species was originally described from India (Puri, 1932a).

It is similar to *S. chamlongi* from Thailand (Takaoka and Suzuki, 1984) but is distinguished by the pupal frons and thorax covered with minute tubercles. It differs from *S. gurneyae* and *S. nilgiricum* from India (Puri, 1932a) by the pupal gill filaments, which are of almost the same thickness and spread out widely at an angle of over 90 degrees when viewed laterally.

***Simulium* (*Simulium*) *karenkoense* (Shiraki, 1935)**

Odagmia karenkoensis Shiraki, 1935: 42–45 (female).

Simulium (*Simulium*) *karenkoense*: Takaoka, 1979: 400.

Distribution. Taiwan.

Bionomics. Aquatic habitats are unknown.

Remarks. This species was described from a single female collected from Taiwan by Shiraki (1935). Its male, pupa, and larva are unknown.

The female of this species is characterized by the dark short hairs on the scutum and dark legs.

***Simulium* (*Simulium*) *kiewmaepanense* Takaoka, Srisuka and Saeung, 2017**

Simulium (*Simulium*) *kiewmaepanense* Takaoka, Srisuka and Saeung, 2017l: 1221–1222 (female and pupa).

Distribution. Thailand.

Bionomics. Aquatic habitat: A stream (width 1.5 m, depth 13 cm, bottom sandy, 14.5 °C, pH 5.4, partially shaded, elevation 2210 m) fast flowing in a natural forest, Doi Inthanon National Parks (Takaoka et al., 2017l).

Remarks. This species was described from Thailand by Takaoka et al. (2017l). Its male and larva are unknown.

It is characterized in the female by the femora yellow except the apical caps darkened and tibiae almost entirely darkened; and in the pupa by the integument of the head and the anterior half of the thorax densely or moderately covered with round tubercles, gill with six thread-like filaments and abdomen with spine-combs dorsally on segments 7 and 8.

***Simulium* (*Simulium*) *laxum* Takaoka, 2003**
(Figs. 87C, D and 89G)

Simulium (*Simulium*) *laxum* Takaoka, 2003: 492–496 (female, male, pupa, and larva).

Distribution. Indonesia (Sulawesi).

Bionomics. Aquatic habitats: Several slow to fast-flowing streams (width 1.0–5.0 m, 17.0–23.0 °C, elevation 970–1520 m) in Napu and Mamasa (Takaoka, 2003).

Remarks. This species was described from Central and South Sulawesi by Takaoka (2003).

It is characterized in the female by the haired basal portion of the radius; in the male by abdominal segments 2–7 each with a pair of dorsolateral shiny spots; and in the pupa by the gill with six slender thread-like filaments and a relatively large round basal fenestra (Fig. 87C, D), and cocoon thinly and roughly woven leaving small open spaces, covering the entire body and most of gill filaments even anteriorly so that its anterior opening is not discernible, though individual threads are visible (Fig. 89G).

***Simulium* (*Simulium*) *linduense* Takaoka, 2003**
(Figs. 87E, F and 89I, J)

Simulium (*Simulium*) *linduense* Takaoka, 2003: 496–501 (female, male, pupa, and larva).

Distribution. Indonesia (Sulawesi).

Bionomics. Aquatic habitats: Two moderately-flowing mountain streams (width 0.6–3.0 m, 17.0–18.0 °C, shaded, elevation 1480–1520 m) in Lore Utara (Takaoka, 2003).

Remarks. This species was described from Central Sulawesi by Takaoka (2003).

It is similar to *S. elongatum* in having the essentially similar arrangement of the pupal gill filaments. However, this species is distinguished from the latter species by all six slender thread-like apical filaments about half the total length of the gill (Fig. 87E, F), and cocoon wall-pocket-shaped, with a short anterodorsal projection (Fig. 89I, J).

***Simulium* (*Simulium*) *lorense* Takaoka, 2003**
(Figs. 88C, D and 89M, N)

Simulium (*Simulium*) *lorense* Takaoka, 2003: 511–516 (female, male, pupa, and larva).

Distribution. Indonesia (Sulawesi).

Bionomics. Aquatic habitats: A moderately flowing river (Saluledo River) (width ca. 5 m, depth 0.3–0.5 m, 21.0 °C, elevation 970 m), in Pormas, and a forest stream (width 0.3 m, 17.0 °C, partially shaded, elevation 1520 m), in Lore Utara (Takaoka, 2003).

Remarks. This species was described from Central and South Sulawesi by Takaoka (2003).

This species is similar to *S. flavum* by having the yellow female and male scutum and pupal gill with six filaments (at lease their basal halves are somewhat inflated and their apical halves are tapered toward apical slender portions), of which two are directed upward and four are forward (Fig. 88C, D), and larval postgenal cleft medium-long (nearly as long as the postgenal bridge) and bullet-shaped. However, this species is distinguished by male abdominal segments 2–7 each with a pair of dorsolateral pale spots, pupal frons covered with tubercles, and cocoon wall-pocket-shaped, with a medium-long anterodorsal projection (Fig. 89M, N).

***Simulium* (*Simulium*) *luculentum* Takaoka, Srisula and Saeung, 2019**

Simulium (*Simulium*) *luculentum* Takaoka, Srisuka and Saeung, 2019f: 155–160 (female, male, pupa, and larva).

Distribution. Thailand.

Bionomics. Aquatic habitat: A stream (width 1.6 m, depth 20 cm, bottom sandy, 20.3 °C, partially shaded, elevation 1109 m) moderately flowing in a natural forest of Doi Inthanon National Parks (Takaoka et al., 2019f).

Remarks. This species was described from Thailand by Takaoka et al. (2019f).

It is characterized in the female by the mid femur and tibia entirely whitish yellow to yellow and the fore and hind tibiae yellow with the apical caps medium brown; in the male by the number of upper-eye (large) facets in 22 or 23 vertical columns and 22 or 23 horizontal rows; and in the pupa by the head and thorax moderately covered with tubercles and gill with six filaments somewhat inflated basally.

***Simulium* (*Simulium*) *luridum* Takaoka, 2003**

(Figs. 86B, G–I, K, R–V and 89A–E)

Simulium (*Simulium*) *luridum* Takaoka, 2003: 534–539 (female, male, pupa, and larva).

Distribution. Indonesia (Sulawesi).

Bionomics. Aquatic habitats: Several slow-flowing streams or ditches (width 0.1–1.0 m, 18.0–25.0 °C, elevation 820–1050 m) in Tomohon (Takaoka, 2003).

Remarks. This species was described from North Sulawesi by Takaoka (2003). This is the only species found in North Sulawesi among 11 species of the *S. variegatum* species-group in Sulawesi.

It is distinguished from the 10 other species in the female by the unarmed cibarium (Fig. 86B) and ellipsoidal spermatheca; in the male by the ventral plate with its body narrow when viewed ventrally (Fig. 86S); in the pupa by the frons and thorax covered with spine-like processes (Fig. 89A), gill with six filaments, of which five are directed forward and one is upward, basal fenestra extended posteriorly (Fig. 86B, C) and abdomen without terminal hooks.

***Simulium* (*Simulium*) *malinoense* Takaoka, 2003**

(Figs. 88E, F and 89Y)

Simulium (*Simulium*) *malinoense* Takaoka, 2003: 516–520 (female, male, pupa, and larva).

Distribution. Indonesia (Sulawesi).

Bionomics. Aquatic habitats: A moderately-flowing river (width 6–8 m, bottom with stones, 19.0 °C, elevation 1500 m) in Kanreiapia, and a forest stream (width 0.3 m, 17.0 °C, partially shaded, elevation 1520 m) in Lore Utara (Takaoka, 2003).

Remarks. This species was described from Central and South Sulawesi by Takaoka (2003).

It is similar to *S. flavum* and *S. lorense* by having the yellow female and male scuta and pupal gill with six filaments (at least their basal halves are somewhat inflated and their apical halves tapered toward apical slender portions), of which two are directed upward and four are forward (Fig. 88E, F), and larval postgenal cleft medium-long and bullet-shaped. However, this species is distinguished from *S. flavum* by the pupal frons covered with tubercles, and from *S. lorense* by the cocoon without an anterodorsal projection.

***Simulium* (*Simulium*) *molliculum* Takaoka, 2003**

(Figs. 86A, C–F, J, L–Q, 87A, B and 89F, U, V, Z)

Simulium (*Simulium*) *molliculum* Takaoka, 2003: 484–491 (female, male, pupa, and larva).

Distribution. Indonesia (Sulawesi).

Bionomics. Aquatic habitats: Two small mountainous streams (width 0.5–3.0 m, 16.0–17.0 °C, partially or entirely shaded, elevation 1520–1600 m) in Napu (Takaoka, 2003).

Remarks. This species was described from Central Sulawesi by Takaoka (2003).

It is characterized in the female by the haired basal portion of the radius; in the male by abdominal segments 2–7 each with a pair of dorsolateral shiny spots; and in the pupa by the gill with six slender thread-like filaments directed forward close together and a relatively large round basal fenestra (Fig. 87A, B), and cocoon thinly woven and soft, covering the entire body and most of gill filaments even anteriorly so that the anterior opening is not discernible, though there appears to be a long anterodorsal projection bending down like a hood (Fig. 89F).

***Simulium* (*Simulium*) *nilgiricum* Puri, 1932**

Simulium (*Simulium*) *nilgiricum* Puri, 1932a: 894–897 (female, male, and pupa).

Distribution. India (Karnataka, Maharashtra, and Tamil Nadu) and Sri Lanka.

Bionomics. Aquatic habitats are unknown.

Remarks. This species was described from India by Puri (1932a). It was later recorded from Sri Lanka by Datta (1983). Its larva is unknown.

It is characterized in the pupa by the unique shape of the cocoon, of which the anterior border curved backward so that the length of the cocoon is much longer at its top than at its base (similar to Fig. 69T) and gill with six slender filaments of almost the same thickness.

***Simulium* (*Simulium*) *nubis* Davies and Györkös, 1987**
Simulium (*Simulium*) *nubis* Davies and Györkös, 1987b: 2742–2745 (female, male, pupa, and larva).

Distribution. Sri Lanka.
Bionomics. Aquatic habitats: Streams with fast to moderate flows, 0.3–6.0 m wide, and more often at an elevation of 1000–2000 m (Davies and Györkös, 1987b). One female was caught while feeding on a human on a high ground (ca. 2000 m in elevation) in Nuwara Eliya (Davies and Györkös, 1987b).
Remarks. This species was described from Sri Lanka (Davies and Györkös, 1987b).
It is characterized in the female and male by the scutum with dark hairs and darkened legs including fore coxae; in the male by the number of upper-eye (large) facets in 17 vertical columns and 17 horizontal rows; and in the pupa by the thorax without tubercles.
This species is distinguished in the female and male from *S. bulla* from Sri Lanka by the darker hairs on the scutum and darker legs.

***Simulium* (*Simulium*) *oitanum* (Shiraki, 1935)**
Odagmia oitana Shiraki, 1935: 37–40 (female).
Simulium oitana: Bentinck, 1955: 10 (female, male, and pupa).
Odagmia aokii Takahasi, 1941: Synonymized by Uemoto, 1978.
Simulium (*Odagmia*) *aokii*: Takaoka, 1976b: 390–393 (female, male, pupa, and larva).
Simulium (*Simulium*) *oitanum*: Takaoka and Saito, 2007: 58.

Distribution. Japan (Kyushu, Hokkaido, Honshu, Shikoku, and Nansei Islands), China (Guizhou, Jilin, Liaoning, and Yunnan), Korea (South), Nepal (?), and Siberia.
Bionomics. According to Bentinck (1955), this species is one of the four most common lowland black fly species in Japan. It is widespread and found throughout the year and most commonly breeds in lowland streams and ditches. Females have been captured while biting man and cattle. Takaoka et al. (1992) reported that in northeast Japan, this species is a natural vector of *Onchocerca* type A [=*Onchocerca* type I, a bovine parasite, designated by Takaoka and Bain (1990)]. Fukuda et al. (2008) reported that this is an experimental vector of *Onchocerca japonica* Uni, Bain and Takaoka, a parasite of wild boar and a causative agent of zoonotic onchocerciasis in Japan, and two other *Onchocerca* species, *O. skrjabini* Rukhlyadev and *O. eberhardi* Uni and Bain, parasites of sika deer.

Remarks. This species was described under the genus *Odagmia* from females collected from Oita, Japan (Shiraki, 1935). It was then treated under the subgenus *Odagmia* in the genus *Simulium* s. l. (Takaoka and Okazawa, 1988), and finally transferred to the subgenus *Simulium* s. str. in the same genus (Takaoka and Saito, 2007). Illustrations of its female cibarium, legs, and terminalia and male genitalia were provided by Bentinck (1955). The female, male, pupa, and larva of this species were fully redescribed under the name of *S.* (*Odagmia*) *aokii* by Takaoka (1976b).

According to the original description by Shiraki (1935), the middle and hind femora of the female of *S. oitanum* are mostly darkened, a characteristic separating it from *S. aokii* described by Takahasi (1941). The female of *S. aokii* has the mid and hind femora yellowish except their apical caps blackish brown (Takaoka, 1976b). Later, the hind femora and tibiae of the female of *S. aokii* in Oita (type locality of *S. oitanum*) were shown to be changeable in color by months: females with the dark femora and tibiae began to appear in the middle of autumn (November), increased in relative abundance up to the middle of winder (January), and then decreased gradually until the mid-spring (April) (Hadi et al., 1996a). *Simulium aokii* was synonymized by Uemoto (1978). However, a question remains to be solved because there is a gap of the time for the occurrence of adult females in Oita between the two species: i.e., females of *S. oinatum* were captured in August (Shiraki, 1935) whereas females of *S. aokii* were captured inside a cattle-shed only from October to May (Takaoka, 1994). Further trials to collect pupae of *S. oitanum* in Oita in August had failed (Unpublished data).

It is characterized in the female by the frontal ratio 1.2:1.0:0.9, and cibarium with ca. 15 processes near the posterodorsal margin; in the male by the number of upper-eye (large) facets in ca. 16 horizontal rows, and fore and hind basitarsi ca. 6.6 and ca. 4.0 times as long as their greatest widths, respectively, and in the pupa by the frons and dorsal surface of the thorax densely covered with tubercles, gill with six slender filaments arranged as 2+2+2 from dorsal to ventral, stalks of the dorsal and ventral pairs widely divergent at ca. 130 degrees and dorsum of abdominal segment 1 sparsely covered with tubercles; in the larva by the postgenal cleft long, mitre-shaped, moderately pointed apically, ca. 3.6 times as long as the postgenal bridge, and rectal organ simple or with a few secondary lobules per lobe.

***Simulium* (*Simulium*) *palopoense* Takaoka, 2003**

(Figs. 88I, J and 89Q, R, X)

Simulium (*Simulium*) *palopoense* Takaoka, 2003: 523–528 (female, male, pupa, and larva).

Distribution. Indonesia (Sulawesi).

Bionomics. Aquatic habitats: Two small mountainous streams (width 0.5–3.0 m, 16.0–17.0 °C, partially or entirely shaded, elevation 1520–1600 m) in Napu (Takaoka, 2003).

Remarks. This species was described from Central and South Sulawesi by Takaoka (2003).

It is distinctive among species of the *S. variegatum* species-group in having the female tergites 5–8 shiny and pupal gill composed of three globes (one anteriorly, one dorsally and one inwardly) and two short finger-like inflated filaments directed outwardly (Fig. 88I, J).

***Simulium* (*Simulium*) *paranubis* Davies and Györkös, 1992**

Simulium (*Simulium*) *paranubis* Davies and Györkös, 1992: 1035 (male and pupa).

Distribution. Sri Lanka.

Bionomics. Aquatic habitat: A small stream (width 0.3–6 m, 16 °C) draining a small pond at an elevation of 1000–2000 m (Davies and Györkös, 1992).

Remarks. This species was described from a single male reared from a pupa collected from Sri Lanka (Davies and Györkös, 1992).

This species is similar to *S. nubis* from Sri Lanka, but is barely distinguished from the latter species in the male by the number of upper-eye (large) facets in 19 or 20 vertical columns and 18 or 19 horizontal rows and abdominal segments 2, 6, and 7 each with a pair of shiny dorsolateral spots; and in the pupa by the frons and thorax without tubercles.

This species is distinguished in the male from *S. bulla* from Sri Lanka by the darker hairs on the scutum and darker legs.

***Simulium* (*Simulium*) *phuluense* Takaoka and Sofian-Azirun, 2017**
(Fig. 84B–O, Q, S–U)

Simulium (*Simulium*) *phuluense* Takaoka and Sofian-Azirun [in Takaoka et al., 2017c]: 154–161 (female, male, pupa, and larva).

Distribution. Vietnam.

Bionomics. Aquatic habitat: A small shallow stream (width 2.0 m, depth 15 cm, bottom of pebbles, 9.0 °C, partially shaded, elevation 1192 m) moderately flowing in a natural forest, Sapa, Lao Cai Province (Takaoka et al., 2017c).

Remarks. This species was described from northern Vietnam by Takaoka et al. (2017c).

It is characterized in the female by the hind tibia yellow on little less than the basal half and gradually darkened toward the apex, with its apical cap dark brown on the outer and posterior surface and widely yellow except the apical cap dark brown on the inner surface; and in the male by the number of upper-eye (large) facets in 23 or 24 vertical columns and 23 or 24 horizontal rows and subcosta with hairs.

***Simulium* (*Simulium*) *qianense* Chen and Chen, 2001**

Simulium (*Simulium*) *qianense* Chen and Chen, 2001: 208–212 (female, male, pupa, and larva).

Distribution. China (Guizhou and Hubei).

Bionomics. Aquatic habitats: Small streams in Leigongshan Nature Reserve (elevation 600–1300 m) (Chen and Chen, 2001).

Remarks. This species was described from Guizhou, China (Chen and Chen, 2001).

It is characterized in the male by the small number of upper-eye (large) facets in 11 vertical columns and 12 horizontal rows. The term "accessory sclerite" was wrongly used for "rectal scales" in the original description of the larva (Chen and Chen, 2001).

***Simulium* (*Simulium*) *quinquefiliatum* Takaoka, 2003**

(Figs. 88G, H and 89O, P)

Simulium (*Simulium*) *quinquefiliatum* Takaoka, 2003: 520–523 (female and pupa).

Distribution. Indonesia (Sulawesi).

Bionomics. Aquatic habitat: A moderately flowing river (width 6–8 m, bottom with stones, 19.0 °C, elevation 1500 m) in Kanreiapia (Takaoka, 2003).

Remarks. This species was described from South Sulawesi by Takaoka (2003). Its male and larva are unknown.

It is similar to *S. flavum*, *S. lorense* and *S. malinoense* in having the yellow female and male scuta. However, it is distinguished in the pupa by having five inflated filaments, of which one inflated filament has two apical slender filaments (Fig. 88G, H).

***Simulium* (*Simulium*) *srisukai* Takaoka and Saeung, 2017**

(Fig. 84A, P, R)

Simulium (*Simulium*) *srisukai* Takaoka and Saeung [in Takaoka et al., 2017l]: 1213–1221 (female, male, pupa, and larva).

Distribution. Thailand.

Bionomics. Aquatic habitat: A stream (width 45 cm, depth 10 cm, bottom sandy, 17.8 °C, partially shaded, elevation 1241 m) moderately flowing in a natural forest, Mae Klong Yai Village, Tak Province (Takaoka et al., 2017l).

Remarks. This species was described from northern Thailand by Takaoka et al. (2017l).

It is characterized in the female by the femora and tibiae darkened; in the male by the large number of upper-eye (large) facets in 24–26 vertical columns and 24–26 horizontal rows; and in the pupa by the frons and anterior half of the thorax bare and pupal gill with six filaments, each inflated at least on its basal one-fourth (Fig. 84P).

***Simulium* (*Simulium*) *taiwanicum* Takaoka, 1979**

Simulium (*Simulium*) *taiwanicum* Takaoka, 1979: 388–390 (female, male, pupa, and larva).

Distribution. Taiwan and China (Jiangxi and Sichuan).

Bionomics. Aquatic habitat: A mountain stream (width 0.2–1.0 m, shaded, elevation ca. 1000 m) in Shi Tou (Takaoka, 1979).

Remarks. This species was described from Taiwan by Takaoka (1979).

It is characterized in the female by the cibarium with ca. 40 tiny processes, femora and tibiae yellowish with dark apices; in the male by the large number of upper-eye (large) facets in 22 horizontal rows; in the pupa by the frons densely covered with round tubercles and gill with six slender filaments; and in the larva by the rectal organ with ca. 10 finger-like secondary lobules per lobe. It is distinguished from *S. karenkoense* from Taiwan by the yellowish legs.

***Simulium* (*Simulium*) *tulshii* Takaoka and Shrestha, 2010**
(Fig. 85A–X)

Simulium (*Simulium*) *tulshii* Takaoka and Shrestha, 2010: 49–56 (female, male, pupa, and larva).

Distribution. Nepal.

Bionomics. Aquatic habitats: Three moderately or rapidly flowing small streams (8.0–15.5 °C, elevation 2745–3743 m) in Mustang (Takaoka and Shrestha, 2010).

Remarks. This species was described from Nepal by Takaoka and Shrestha (2010).

It is characterized in the female by the cibarium with numerous sharply pointed tiny processes along the posterodorsal margin (Fig. 85A) and ovipositor valve produced posteromedially, with its apical one-third bare (Fig. 85C); in the pupa by the frons covered with sharply-pointed tubercles (Fig. 85P), gill with six filaments, all inflated at least on their basal one-third to basal half (Fig. 85Q, R) and covered with fine spinous processes (Fig. 85S), and cocoon shoe-shaped (Fig. 85T); and in the larva by the rectal organ without secondary lobules and ventral papillae absent.

***Simulium* (*Simulium*) sp. A Maskey, 1998**

Simulium (*Simulium*) sp. A Maskey, 1998.

Distribution. Nepal.

Bionomics. Unknown.

Remarks. Information for morphological characteristics of this species is not available.

***Simulium* (*Simulium*) sp. B Maskey, 1998**

Simulium (*Simulium*) sp. B Maskey, 1998.

Distribution. Nepal.

Bionomics. Unknown.

Remarks. Information for morphological characteristics of this species is not available.

Simulium (*Simulium*) *yadongense* Species-group

Simulium (*Simulium*) *zunyiense* Chen, Yiu and Zhang, 2012

Simulium (*Simulium*) *zunyiense* Chen, Yiu and Zhang, 2012: 2–4 (female, male, pupa, and larva).

Distribution. China (Guizhou).

Bionomics. Aquatic habitat: A trickle (elevation 1483 m) in Dadong, Kuankuoshui, Guizhou (Chen et al., 2012).

Remarks. This species was described by Chen et al. (2012). This species was transferred from the *S. variegatum* species-group to the *S. yadongense* species-group (Takaoka, 2017a).

It is characterized in the female by the cibarium with ca. 30 minute processes, scutum with golden yellow short hairs, fore basitarsus 7.2 times as long as its greatest width, claw with a small subbasal tooth, ovipositor valve triangular and pointed posteromedially, with its inner margin widely concave; in the male by the number of upper-eye (large) facets in 22 vertical columns and 26 horizontal rows, scutum with golden yellow short hairs, hind basitarsus somewhat enlarged, style with a short subbasal protuberance and ventral plate Y-shaped, with a ventrally produced process, and with teeth along posterolateral margins; in the pupa by the frons and the dorsum of the thorax uniformly covered with tubercles, gill with six thread-like filaments, abdominal segments 7–9 each with spine-combs dorsally, last abdominal segment with a pair of terminal hooks, and cocoon wall-pocket-shaped, with many small perforations near the anterior margin; and in the larva by the postgenal cleft triangular, 1.5 times as long as the postgenal bridge [though it is nearly as long as the postgenal bridge, according to the illustration in Chen et al. (2012)].

Simulium (*Simulium*) s. str. Species Unplaced to Species-group

Simulium (*Simulium*) *ambiguum* Shiraki, 1935

Simulium (*Simulium*) *ambiguum* Shiraki, 1935: 71–73 (female and male).

Distribution. Taiwan.

Bionomics. Unknown.

Remarks. This species was described based on a single female and three males collected from Taiwan by Shiraki (1935). Its pupa and larva are unknown.

This species is characterized in the female by having the striated scutum with dark hairs, dark legs, and simple claw; and in the male by the scutum with dark hairs, dark legs, and abdominal segments 4 and 5 (probably 6 and 7) each with a pair of shiny spots. The characteristics suggest that the female is assignable to the *S. striatum* species-group, but the male is placed in either the *S. tuberosum* species-group or the *S. venustum* species-group. The female of this species is distinguished from *S. pingtungense, S. quinquestriatum* and *S. yanpingense*, all in the *S. striatum* species-group from Taiwan by the scutum with dark short hairs.

***Simulium* (*Simulium*) *biforaminiferum* Datta, 1974**
Simulium (*Simulium*) *biforaminiferum* Datta, 1974b: 463–466 (pupa and larva); Datta 1991a: 31 (female).

Distribution. India (West Bengal) and Bhutan.

Bionomics. Aquatic habitat: A trickle, at an elevation of 2031–2050 m, West Bengal (Datta, 1974b).

Remarks. This species was described from four pupae and a larva from West Bengal, India (Datta, 1974b). Its male is unknown. The female of this species dissected out of a pupa was briefly described by Datta (1991a), who suggested that it belongs to the *S. tuberosum* species-group based on the pharate female with the shiny frons, unpatterned scutum, and simple claw without any tooth. However, it is unlikely because almost all species in the *S. tuberosum* species-group have a simple wall-pocket-shaped cocoon without an open space on each anterolateral surface.

It had been long treated as a species belonging to *Simulium* s. str. species unplaced to species-group until it was tentatively transferred to the *S. griseifrons* species-group, redefined by Takaoka (2017a).

This species is here returned to *Simulium* s. str. species unplaced to species-group, since it is unlikely to be a member the *S. griseifrons* species-group due to the scutum of the pharate female not patterned, as reported by Datta (1991a), which was overlooked by Takaoka (2017a).

It is characterized in the female by the scutum not patterned, and claw without any tooth; in the pupa by the gill with six filaments, of which the dorsal filament is stoutest of all, and the five other filaments arise from the base of the dorsal filament, dorsum of abdominal segments 7, 8, and 9 with spine-combs, last abdominal segment with a pair of terminal hooks, and cocoon wall-pocket-shaped, with a large anterolateral window on each side (Datta, 1974b, 1991a).

***Simulium* (*Simulium*) *cremnosi* Davies and Györkös, 1987**
Simulium (*Simulium*) *cremnosi* Davies and Györkös, 1987b: 2734–2737 (female, male, pupa, and larva).

Distribution. Sri Lanka.

Bionomics. Aquatic habitats: Moderate to fast flowing rocky streams, often in cascades in mountainous areas at an elevation of 250–1700 m. Two females were collected while biting a human (Davies and Györkös, 1987b).

Remarks. This species was described from Sri Lanka by Davies and Györkös (1987b).

It is characterized in the female by the large body size (4.0 mm long), sensory vesicle with a large opening, scutum with variable patterns of gray pruinosity, dark legs, simple claw, and tergites of abdominal segments 3–9 shiny; in the male by the large body size (3.7 mm long), the number of upper-eye (large) facets in 14 vertical columns and 14 horizontal rows, dark legs, abdominal segments 2–6 each with a pair of shiny dorsolateral spots and ventral plate with its body rectangular when viewed ventrally; and in the pupa by the gill with six short slender filaments, abdomen without terminal hooks and cocoon corbicular-shaped. No similar species was recorded from other countries in the Oriental Region.

***Simulium* (*Simulium*) *katoi* Shiraki, 1935**
Simulium (*Simulium*) *katoi* Shiraki, 1935: 53–56 (female).

Distribution. Taiwan.
Bionomics. Unknown.
Remarks. This species was described based on a single female collected from Taiwan by Shiraki (1935). Its male, pupa, and larva are unknown.

The female of this species is characterized by having the striated scutum with yellow hairs, yellowish legs, and simple claw. These characteristics suggest that it is similar to *S. chungi* in the *S. chungi* species-group and *S. sakishimaense* in the *S. multistriatum* species-group, thought there is a difference in the color of the femora.

***Simulium* (*Simulium*) *multifurcatum* Zhang and Wang, 1991**
Simulium (*Simulium*) *multifurcatum* Zhang and Wang, 1991a: 483–484 (pupa and larva).

Distribution. China (Fujian).
Bionomics. Aquatic habitats: Small streams (Zhang and Wang, 1991a).
Remarks. This species was described from pupae and larvae collected from Fujian, China by Zhang and Wang (1991a). Its female and male remain unknown.

This species is similar to *S. biforaminiferum* from India in having the cocoon wall-pocket-shaped, with a large open space on each anterolateral surface, but is distinguished by lacking pupal terminal hooks.

***Simulium* (*Simulium*) *neoacontum* Chen, Xiu and Zhang, 2012**
Simulium (*Simulium*) *neoacontum* Chen, Xiu and Zhang, 2012: 7–9 (male and pupa).

Distribution. China (Guizhou).
Bionomics. Aquatic habitat: A shaded stream at an elevation of 682 m in Rangshui, Kuankuoshui, Guizhou (Chen et al., 2012).
Remarks. This species was described from a single male reared from a pupa by Chen et al. (2012). Its female and larva are unknown. According to the original description and illustrations by Chen et al. (2012), the male has upper-eye (large) facets in 12 horizontal rows, scutum with golden yellow short hairs, fore basitarsus 6.5 times as long as its greatest width, style without a basal protuberance and a terminal spine, and ventral plate Y-shaped with a ventrally produced process, and the pupa has the gill with six slender filaments in three pairs and cocoon wall-pocket-shaped, with a few small spaces near anterior margin on each side.

***Simulium* (*Simulium*) *palniense* Puri, 1933**
Simulium (*Simulium*) *palniense* Puri, 1933a: 807–809 (female, male, and pupa).

Distribution. India (Tamil Nadu).

Bionomics. Aquatic habitat: A large stream below Fairy Falls, Palni Hills, South India (Puri, 1933a).

Remarks. This species was described from South India by Puri (1933a). Its larva is unknown.

It is characterized in the female by the striated scutum with yellow hairs, darkened legs, simple claw, and ovipositor valves triangular, with their inner margins nearly straight, moderately darkened and close to each other; in the male by the ventral plate with its body rectangular, having a ventrally produced process, of which its posterolateral margins are serrated along the basal half, and its anterior surface is setose; and in the pupa by the gill with six slender filaments and boot-shaped cocoon. The female and pupal characteristics agree well with those of the *S. griseifrons* species-group, redefined by Takaoka (2017a), but the male characteristic of the ventral plate with serrated posterolateral margins departs the definition of this species-group.

***Simulium* (*Simulium*) *parvulum* Takaoka, 2003**
Simulium (*Simulium*) *parvulum* Takaoka, 2003: 541–542 (larva).

Distribution. Indonesia (Sulawesi).

Bionomics. Aquatic habitat: A cascading stream (width 0.3–0.5 m, 20.0 °C, partially shaded, elevation 1050 m) in a natural forest (Takaoka, 2003).

Remarks. This species was described based on a single mature larva collected from Sulawesi, Indonesia by Takaoka (2003). Its female, male and pupa are unknown.

The larva of this species is characterized by the small body size (3.8 mm long), triangular postgenal cleft, smaller number of main rays on the labral fan (28 main rays), and pharate pupal gill with six dark slender short filaments. The shape of the postgenal cleft suggests that this species is closer to species of the *S. dumogaense* species-group than to those of the *S. variegatum* species-group, both of which are recorded from Sulawesi (Takaoka, 2003).

***Simulium* (*Simulium*) *shennongjiaense* Yang, Luo and Chen, 2005**
Simulium (*Simulium*) *shennongjiaense* Yang, Luo and Chen, 2005: 839–841 (female, male, pupa, and larva).

Distribution. China (Guangxi and Hubei).

Bionomics. Aquatic habitat: A rapid-flowing, mountainous stream at an elevation of 1680 m (Yang et al., 2005).

Remarks. This species was described from Hubei, China (Yang et al., 2005), and was placed in the *S. tuberosum* species-group by Chen (2016). It may be placed in the *S. griseifrons* species-group by having the female cibarium with a median projection, basal section of the radius bare, ovipositor valve triangular, male style with a short basal protuberance, ventral plate flat, without any teeth, and cocoon with an anterolateral open space on each side, according to Yang et al. (2005). However, the female scutum has no longitudinal vittae, departing from the definition of the *S. griseifrons* species-group.

***Simulium* (*Simulium*) *singtamense* Datta and Pal, 1975**
Simulium (*Simulium*) *singtamense* Datta and Pal, 1975: 154–157 (pupa and larva).

Distribution. India (West Bengal) and Nepal (?).

Bionomics. Aquatic habitat: A stream exposed to the sun, in Singtam, Darjeeling (Datta and Pal, 1975).

Remarks. This species was described from five pupae and four larvae in Singtam, Darjeeling, India by Datta and Pal (1975). Its female and male are unknown.

It is characterized in the pupa by the gill with six slender filaments arranged as 2+2+2 from dorsal to ventral, decreasing in thickness from dorsal to ventral, widely divergent vertically, dorsum of abdominal segments 7 and 8 each with spine-combs, abdominal segment 9 with a pair of terminal hooks, and cocoon shoe-shaped, with a low anteroventral neck; and in the larva by the postgenal cleft elongate-subcordate.

***Simulium* (*Simulium*) *tamorense* Takaoka and Shrestha, 2010**
(Figs. 90A–P and 91J–P)
Simulium (*Simulium*) *tamorense* Takaoka and Shrestha, 2010: 41–48 (female, male, pupa, and larva).

Distribution. Nepal.

Bionomics. Aquatic habitat: A rapid-flowing river (width over 10 m, 17.0 °C, elevation 651 m), Tamor, Taplejung (Takaoka and Shrestha, 2010).

Remarks. This species was described as a species in the *S. multistriatum* species-group from Nepal (Takaoka and Shrestha, 2010). It is here transferred to *Simulium* s. str. species unplaced to species-group, since it departs from the definition of the *S. multistriatum* species-group by having the cibarium lacking a dorsomedian projection (Fig. 90A), ovipositor valve with its inner margin well sclerotized (Fig. 90B), and style with a blunt basal protuberance (Fig. 90I, J), and ventral plate with a median process produced ventrally from the anterior portion of its body (Fig. 90K), and comb-teeth of the larval mandible increasing in length from the first to third (Fig. 91N).

Beside the above-mentioned characteristics, it is characterized in the pupa by the gill with eight short filaments, all ca. 0.5 mm long (Fig. 91J), abdominal segments 5–9 lacking spine-combs dorsally, and cocoon shoe-shaped (Fig. 91K, L); and in the larva by abdominal segments 1–7 each with paired protuberances dorsally (Fig. 91M).

***Simulium* (*Simulium*) *titarense* Takaoka and Shrestha, 2010**
(Fig. 91A–I)
Simulium (*Simulium*) *titarense* Takaoka and Shrestha, 2010: 37–41 (male and pupa).

Distribution. Nepal.

Bionomics. Aquatic habitat: A stream (width 2 m, flow moderate, 18.0 °C, elevation 1700 m), Mustang (Takaoka and Shrestha, 2010).

Remarks. This species was described from Nepal by Takaoka and Shrestha (2010). Its female and larva are unknown. It was placed in the *S. griseifrons* species-group by Takaoka and Shrestha (2010), before this species-group was split into six species-groups by Takaoka (2017a). However, this species is here moved to *Simulium* s. str. species unplaced to species-group since it is difficult to place this species in any one of the six species-groups derived from the former *S. griseifrons* species-group.

It is characterized in the male by the style with a triangular basal protuberance (Fig. 91C) and ventral plate with its body quadrate when viewed ventrally, having a ventrally produced cone-like process, of which the posterior surface is bare and has no teeth (Fig. 91B, D, E); and in the pupa by the gill with six filaments (Fig. 91H), abdomen lacking terminal hooks, and cocoon wall-pocket-shaped, with lateral open spaces anterolaterally on each side (Fig. 91I).

Simulium (_Simulium_) sp. E Zhang, Wen and Chen, 1999

Simulium (*Simulium*) sp. E Zhang, Wen and Chen, 1999: 234.

Distribution. China (Guizhou).

Bionomics. Unknown.

Remarks. This unnamed species was reported from Guizhou, China by Zhang et al. (1999). However, its morphological characteristics were not noted. This species was not treated in two monographs on Chinese black flies by Chen and An (2003) and Chen (2016).

Simulium (_Simulium_) sp. F Zhang, Wen and Chen, 1999

Simulium (*Simulium*) sp. F Zhang, Wen and Chen, 1999: 234.

Distribution. China (Guizhou).

Bionomics. Unknown.

Remarks. This unnamed species was reported from Guizhou, China by Zhang et al. (1999). However, its morphological characteristics were not noted. This species was not treated in two monographs on Chinese black flies by Chen and An (2003) and Chen (2016).

Simulium (_Simulium_) sp. G Zhang, Wen and Chen, 1999

Simulium (*Simulium*) sp. G Zhang, Wen and Chen, 1999: 234.

Distribution. China (Guizhou).

Bionomics. Unknown.

Remarks. This unnamed species was reported from Guizhou, China by Zhang et al. (1999). However, its morphological characteristics were not noted. This species was not treated in two monographs on Chinese black flies by Chen and An (2003) and Chen (2016).

***Simulium* (*Simulium*) sp. H Zhang, Wen and Chen, 1999**

Simulium (*Simulium*) sp. H Zhang, Wen and Chen, 1999: 234.

Distribution. China (Guizhou).

Bionomics. Unknown.

Remarks. This unnamed species was reported from Guizhou, China by Zhang et al. (1999). However, its morphological characteristics were not described. This species was not treated in two monographs on Chinese black flies by Chen and An (2003) and Chen (2016).

***Simulium* (*Simulium*) Kamti sp. Lewis, 1974**

Simulium (*Simulium*) Kamti sp. Lewis, 1974: 38 (pupa).

Distribution. India (Uttar Pradesh).

Bionomics. Unknown.

Remarks. This unnamed species was reported from Uttar Pradesh, India, by Lewis (1974). Its female, male, and larva are unknown. Morphological characteristics of its pupa are not available except that the pupal gill is composed of six filaments and pupal thorax covered with rather large tubercles and branched trichomes. It is similar to *S*. (*S*.) *ramosum* in the *S. tuberosum* species-group in having the branched trichomes on the pupal thorax. However, detailed information for its morphology is needed for its assignment to species-group.

***Simulium* (*Simulium*) Rengma sp. Lewis, 1974**

Simulium (*Simulium*) Rengma sp. Lewis, 1974: 38 (pupa).

Distribution. India (Assam).

Bionomics. Unknown.

Remarks. This unnamed species was reported from Assam, India, by Lewis (1974). Its female, male, and larva are unknown.

Morphological characteristics of its pupa are not available except that the pupal gill is composed of six filaments and pupal thorax covered with small tubercles and unbranched trichomes. It is difficult to place this species in any species-group for lack of detailed information for its morphology.

***Simulium* (*Simulium*) Assam sp. Lewis, 1974**

Simulium (*Simulium*?) Assam sp. Lewis, 1974: 38 (pupa).

Distribution. India (Assam).

Bionomics. Unknown.

Remarks. This unnamed species was reported from Assam, India, by Lewis (1974). Its female, male, and larva are unknown.

Morphological characteristics of its pupa are not available except that the pupal gill is composed of six slender filaments and the cocoon is wall-pocket-shaped. It is difficult to place this species in any species-group for lack of detailed information for its morphology.

***Simulium* (*Simulium*) Subansiri sp. Lewis, 1974**

Simulium (*Simulium*) Subansiri sp. Lewis, 1974: 38 (pupa).

Distribution. India (Assam).

Bionomics. Unknown.

Remarks. This unnamed species was reported from Assam, India, by Lewis (1974). Its female, male, and larva are unknown.

Morphological characteristics of its pupa are not available except that the pupal gill is composed of eight filaments of the same thickness, arranged as 2+2+2+2 from dorsal to ventral, and the cocoon is wall-pocket-shaped. The arrangement of the gill filaments and the shape of the cocoon are similar to those of species in the *S. multistriatum* species-group. However, detailed information for its morphology is needed for its assignment to species-group.

Subgenus Wallacellum *Takaoka*

***Simulium* (*Wallacellum*) *amplum* Takaoka, 2009**
(Figs. 92U, 93B, Q–T and 95J, K)

Simulium (*Wallacellum*) *amplum* Takaoka, 2009a: 56–63 (female, male, and pupa).

Distribution. Philippines (Luzon).

Bionomics. Aquatic habitats: A mountain stream (width 3–5 m, shaded, 23.8 °C, elevation 169 m), at Maria Makiling Falls, Laguna Province, a small forest stream (width 1–2 m, shaded, 20 °C, elevation 1110 m) in Banaue, Ifugao Province, and a moderately flowing stream (width 6–10 m, shaded, 21 °C, elevation 600 m), a tributary of Cristalino River, Quezon Province (Takaoka, 2009a).

Remarks. This species was described from the Philippines (Takaoka, 2009a). Its larva is unknown.

It is characterized in the female and male by the calcipala extended up to the middle of the third tarsomere; in the female by the small sensory vesicle with a large opening, ovipositor valve somewhat produced posteriorly (similar to Fig. 92O), and genital fork with arms each extended posteriorly, then curved posterolaterally (Fig. 92U); in the male by the number of upper-eye (large) facets in 11–13 vertical columns and 11–13 horizontal rows, hind basitarsus spindle-shaped (Fig. 93B) and ventral plate much narrowed posteriorly, with a notch posteromedially when viewed ventrally (Fig. 93Q); and in the pupa by the gill with four slender filaments, of which the dorsal filament is directed upward, then curved inward, and much shorter than the three other filaments and cocoon simple, wall-pocket-shaped (Fig. 95J, K).

***Simulium* (*Wallacellum*) *cabrerai* Takaoka, 1983**

(Fig. 94G)

Simulium (*Wallacellum*) *cabrerai* Takaoka, 1983: 28–30 (female, male, pupa, and larva).

Distribution. Philippines (Luzon).

Bionomics. Aquatic habitat: A slow-flowing shaded stream on the semi-cultivated mountain slope at an elevation of ca. 2000 m (Takaoka, 1983).

Remarks. This species was described from Luzon, the Philippines (Takaoka, 1983).

It is characterized by the elongate female sensory vesicle, ovipositor valve produced posteriorly (similar to Fig. 92O), and genital fork with arms each extended posteriorly along the outer margin of the paraproct (similar to Fig. 92U); in the male by the large number of upper-eye (large) facets in 16 vertical columns and 16 horizontal rows and ventral plate slightly narrowed posteriorly, with its posterior margin rounded posteriorly when viewed ventrally; in the pupa by the gill with four short tubular filaments, which lack tiny protuberances on their surface (Fig. 94G); and in the larva by the presence of the accessory sclerites (similar to Fig. 95Z).

***Simulium* (*Wallacellum*) *carinatum* Delfinado, 1969**

(Figs. 94F and 95E, F, Z)

Simulium (*Eusimulium*) *carinatum* Delfinado, 1969: 206–207 (female, male, and pupa).

Simulium (*Wallacellum*) *carinatum*: Takaoka, 1983: 25–28 (female, male, pupa, and larva).

Distribution. Philippines (Luzon and Negros).

Bionomics. Aquatic habitats: Shaded creeks or rivers (with 5–12 m), and small slow-flowing streams in lowlands (Takaoka, 1983).

Remarks. This species was described under the subgenus *Eusimulium* from the Philippines (Delfinado, 1969). Its female, male, and pupa were redescribed, and the larva was described by Takaoka (1983).

It is characterized in the female by the small sensory vesicle, ovipositor valve produced posteriorly (similar to Fig. 92O), and genital fork with arms each extended posteriorly along the outer margin of the paraproct (similar to Fig. 92U); in the male by the number of upper-eye (large) facets in 12 vertical columns and 13 horizontal rows, and ventral plate somewhat narrowed posteriorly, and with its posterior margin produced medially when viewed ventrally; in the pupa by the gill with four inflated short tubular filaments, which are covered with tiny protuberances (Fig. 94F); and in the larva by the presence of the accessory sclerites (Fig. 95Z).

***Simulium* (*Wallacellum*) *celebesense* Takaoka, 2003**

Simulium (*Wallacellum*) *celebesense* Takaoka, 2003: 23–25 (female, male, pupa, and larva).

Distribution. Indonesia (Sulawesi).

Bionomics. Aquatic habitats: Small streams (width 0.2–2.0 m, shaded, 21.0–25.5 °C) at middle to high elevations (650–1040 m) in North and South Sulawesi, and a stream (width ca. 5.0 m, 23.0 °C, shaded) in a lowland (elevation 140 m), at Air Terjun, Southeast Sulawesi (Takaoka, 2003).

Remarks. This species was described from Sulawesi, Indonesia (Takaoka, 2003).

It is similar to *S. alfurense* from Seram and Biak but is barely distinguished by the relative length of the dorsal filament of the inner pair of the pupal gill against its counter ventral filament (0.8–1.0 in this species versus 0.4–0.6 in *S. alfurense*), and larval posterior circlet with 84–90 rows of up to 16 hooklets per row (102–106 rows of up to 18 hooklets in *S. alfurense*).

***Simulium* (*Wallacellum*) *claveriaense* Takaoka, 2006**

(Figs. 92J, V, 93I–L, 94C and 95B, H, I, A')

Simulium (*Wallacellum*) *claveriaense* Takaoka, 2006d: 335–342 (female, male, pupa, and larva).

Distribution. Philippines (Luzon).

Bionomics. Aquatic habitats: A small cascading stream (width 0.1–0.5 m, depth less than 1 cm, bottom rocky, 24.0 °C, shaded, elevation ca. 40 m) at Claveria, Cagayan Province, a small slow-flowing shallow stream (width 0.5–1.0 m, bottom rocky, partially shaded, elevation 490 m), Palina, Benquet Province, and a slow-flowing stream (width 0.3–1.0 m, depth 2–5 cm, bottom rocky, exposed to the sun, 25.5 °C, elevation 250 m) in Ilocos Sur Province (Takaoka, 2006d).

Remarks. This species was described from Luzon, the Philippines (Takaoka, 2006d).

It is characterized in the female and male by the calcipala extended up to the apex of the third tarsomere (Fig. 92J); in the female by the small sensory vesicle (similar to Fig. 92C), ovipositor valve somewhat produced posteriorly (similar to Fig. 92O) and genital fork with a small projection directed posteromedially as well as a distinct projection directed anteriorly on each arm (similar to Fig. 92R); in the male by the number of upper-eye (large) facets in 14 vertical columns and 13 or 14 horizontal rows and ventral plate wide, with its posterior margin nearly straight when viewed ventrally (Fig. 93I); in the pupa by the frons with three long trichomes each with four to six branches on each side (Fig. 95B), gill with four short slender filaments arranged as 2+2 from dorsal to ventral (Fig. 94C) and cocoon wall-pocket-shaped, produced anterodorsally in the form of a broad triangular projection, much protruded beyond the level of the anterior tips of the ventrolateral margins (Fig. 95H, I); and in the larva by the presence of the accessory sclerites (Fig. 95A').

***Simulium* (*Wallacellum*) *makilingense* Takaoka, 1983**

Simulium (*Wallacellum*) *makilingense* Takaoka, 1983: 34–36 (female, male, and pupa).

Distribution. Philippines (Luzon).

Bionomics. Aquatic habitat: A small shaded stream (width 0.2–0.6 m) flowing down a banana plantation at a foothill of Mt. Makiling (Takaoka, 1983).

Remarks. This species was described from Luzon, the Philippines (Takaoka, 1983). Its larva is unknown.

It is characterized in the female and male by the calcipala extended up to the base of the fourth tarsomere; in the female by the small sensory vesicle, and genital fork with a small projection directed posteromedially as well as a distinct projection directed anteriorly on each arm (similar to Fig. 91R); in the male by the number of upper-eye (large) facets in 13 vertical columns and 13 horizontal rows and ventral plate wide and with its lateral margins rounded and its posterior margin nearly straight when viewed ventrally; and in the pupa by the frons with three bifid or trifid long trichomes on each side, gill with four long slender filaments arranged as 2+2 from dorsal to ventral and cocoon wall-pocket-shaped, produced anterodorsally in the form of a broad triangular projection, slightly protruded beyond the level of the anterior tips of the ventrolateral margins.

***Simulium* (*Wallacellum*) *marilogense* Takaoka, 2009**
(Figs. 92A, B, G–I, K, L–N, R, 93U and 95A, C, G, P, Q)

Simulium (*Wallacellum*) *marilogense* Takaoka, 2009a: 39–47 (female, male, pupa, and larva).

Distribution. Philippines (Mindanao).

Bionomics. Aquatic habitat: A small shallow stream (width 1–2 m, bottom rocky and sandy, shaded, elevation 1000–1040 m) moderately flowing in a natural forest at Epol Waterfalls, Davao City (Takaoka, 2009a).

Remarks. This species was described from Mindanao, the Philippines (Takaoka, 2009a).

It is characterized in the female and male by the calcipala elongate, almost reaching the end of the second tarsomere; in the female by the elongate sensory vesicle, ovipositor valve triangular (Fig. 92L), genital fork with arms each having a distinct projection directed posteromedially as well as a distinct projection directed anteriorly (Fig. 92R), and paraproct moderately produced ventrally (Fig. 92N); in the male by the number of upper-eye (large) facets in 13 vertical columns and 13 or 14 horizontal rows and aedeagal membrane with spinous setae (Fig. 93U); and in the pupa by the gill with four short slender filaments arranged as 2+2 from dorsal to ventral and cocoon shoe-shaped.

***Simulium* (*Wallacellum*) *molawinense* Takaoka, 2006**
(Figs. 93M–P and 95N, O)

Simulium (*Wallacellum*) *molawinense* Takaoka, 2006d: 342–346 (male and pupa).

Distribution. Philippines (Luzon).

Bionomics. Aquatic habitats: Rapidly or moderately flowing forest streams (width 3–7 m, depth 2–5 cm, 22–24 °C, shaded, elevation 169–192 m), Los Baños (Takaoka, 2006d).

Remarks. This species was described from Luzon, the Philippines (Takaoka, 2006d). Its female and larva are unknown.

It is characterized in the male by the number of upper-eye (large) facets in 15 vertical columns and 15 horizontal rows, calcipala extended up to the middle of the third tarsomere and ventral plate narrowed posteriorly without a notch posteromedially when viewed ventrally (Fig. 93M); and in the pupa by the gill with four short slender filaments arranged as 2+2 from dorsal to ventral and cocoon shoe-shaped (Fig. 95N, O).

***Simulium* (*Wallacellum*) *ogonukii* Takaoka, 1983**

Simulium (*Wallacellum*) *ogonukii* Takaoka, 1983: 36–37 (female, pupa, and larva); Takaoka, 2009a: 47–49 (male).

Distribution. Philippines (Mindanao).

Bionomics. Aquatic habitats: A mountain stream (width 2–3 m, elevation over 1000 m) flowing down the slope of Mt. Apo, Davao City (Takaoka, 1983), and a moderately flowing stream (width 1–3 m, bottom stony, 19 °C, elevation 1360 m), at Mainit, Davao del Sur (Takaoka, 2009a).

Remarks. This species was described from Mindanao, the Philippines (Takaoka, 1983). Its male was described by Takaoka (2009a).

It is characterized in the female and male by the calcipala elongate, extended up to the base of the third tarsomere; in the female by the elongate sensory vesicle, ovipositor valve triangular (similar to Fig. 92L) and genital fork with its arms each having a narrow projection directed posteromedially in addition to a distinct projection directed anteriorly (similar to Fig. 92T); in the male by the number of upper-eye (large) facets in 13 vertical columns and 13 or 14 horizontal rows; and in the pupa by the antennal sheath with distinct spinous projections (similar to Fig. 95D), gill with four short slender filaments arranged as 2+2 from dorsal to ventral and cocoon shoe-shaped.

***Simulium* (*Wallacellum*) *resimum* Takaoka, 1983**

(Fig. 94H)

Simulium (*Wallacellum*) *resimum* Takaoka, 1983: 33–34 (female, male, pupa, and larva).

Distribution. Philippines (Mindanao).

Bionomics. Aquatic habitat: A fast-flowing stream (width 2–3 m, elevation ca. 1000 m) on the slope of Mt. Apo (Takaoka, 1983).

Remarks. This species was described from Mindanao, the Philippines (Takaoka, 1983).

It is characterized in the female by the small sensory vesicle, ovipositor valve produced posteriorly (similar to Fig. 92O), and genital fork with lateral arms each extended posteriorly along the outer margin of the paraproct (similar to Fig. 92U); in the male by the number of upper-eye (large) facets in 15 vertical columns and 16 horizontal rows, and ventral plate much narrowed posteriorly and with its posterior margin depressed medially (similar to Fig. 93Q); in the pupa by the gill with four somewhat inflated filaments, of which the dorsal filament is directed upward, and curved inward and backward, and the inner middle filament is much shorter than the half of the outer middle filament (Fig. 94H); and in the larva by the presence of the accessory sclerites (similar to Fig. 95Z).

***Simulium* (*Wallacellum*) *recurvum* Takaoka, 1983**

(Figs. 93A and 94E)

Simulium (*Wallacellum*) *recurvum* Takaoka, 1983: 32–33 (female, male, pupa, and larva).

Distribution. Philippines (Luzon and Negros).

Bionomics. Aquatic habitats: Shaded streams (width 0.3–1.5 m) on the slope of the rice terrace, Banaue, Luzon, and rivers (width 5–7 m) in a lowland, Negros (Takaoka, 1983).

Remarks. This species was described from the Philippines (Takaoka, 1983).

It is characterized in the female by the small sensory vesicle, ovipositor valve produced posteriorly (similar to Fig. 92O), and genital fork with lateral arms each extended posteriorly along the outer margin of the paraproct (similar to Fig. 92U); in the male by the number of upper-eye (large) facets in 14 vertical columns and 14 horizontal rows, and ventral plate much narrowed posteriorly and with its posterior margin depressed medially (similar to Fig. 93Q); in the pupa by the gill with four slender filaments, of which the dorsal filament is directed upward, and curved inward and backward, and the inner middle filament is much longer than the half of the outer middle filament (Fig. 94E); and in the larva by the presence of the accessory sclerites (similar to Fig. 95Z).

***Simulium* (*Wallacellum*) *spinosibranchium* Takaoka, 1983**

(Fig. 94D)

Simulium (*Wallacellum*) *spinosibranchium* Takaoka, 1983: 30–32 (female, male, and pupa).

Distribution. Philippines (Luzon).

Bionomics. Aquatic habitat: A slow-flowing shaded stream on the semi-cultivated mountain slope at an elevation of ca. 2000 m in Luzon (Takaoka, 1983).

Remarks. This species was described from, Luzon, the Philippines (Takaoka, 1983). Its larva remains unknown.

It is characterized in the female by the small sensory vesicle, ovipositor valve produced posteriorly (similar to Fig. 92O), and genital fork with lateral arms each extended posteriorly along the outer margin of the paraproct (similar to Fig. 92U); in the male by the number of upper-eye (large) facets in 14 vertical columns and 14 horizontal rows, and ventral plate slightly narrowed posteriorly and with its posterior margin of W-shape when viewed ventrally; and in the pupa by the frons densely covered with tubercles, antennal sheath with eight groups of one to three spinous projections (similar to Fig. 95D), and gill with four short inflated tubular filaments, which are covered with large raised protuberances each capped with a black point, lying in a subparallel row on the outer surface from the base to the apex (Fig. 94D).

Simulium (_Wallacellum_) _suyoense_ Takaoka, 2006

(Figs. 92F, S, 93C–H, 94A and 95T–W)

Simulium (*Wallacellum*) *suyoense* Takaoka, 2006d: 327–335 (female, male, pupa, and larva).

Distribution. Philippines (Luzon and Mindoro).

Bionomics. Aquatic habitat: A slow-flowing stream (width 0.3–1.0 m, depth 2–5 cm, bottom rocky, 25.5 °C, elevation 250 m) in Ilocos Sur Province, Luzon (Takaoka, 2006d).

Remarks. This species was described from the Philippines (Takaoka, 2006d).

It is characterized in the female by the elongate sensory vesicle, cibarium with its dorsal margin deeply depressed (Fig. 92F), and genital fork with a projection directed anteriorly but without a projection directed posteomedially on each arm (Fig. 92S); in the male by the number of upper-eye (large) facets in 14 or 15 vertical columns and 15 or 16 horizontal rows, and ventral plate slightly narrowed posteriorly and with its posterior margin rounded posteriorly when viewed ventrally (Fig. 93C); in the pupa by the antennal sheaths with spinous projections (similar to Fig. 95D), and gill with four slender filaments arranged as (2+1)+1 from dorsal to ventral, arising from a bulbous basal common stalk (Fig. 94A); and in the larva by the yellowish white head capsule with indistinct or faintly positive spots.

Simulium (_Wallacellum_) _tenederoi_ Takaoka, 2009

(Figs. 92C, O–Q, T, W and 95L, M, R, S, Y)

Simulium (*Wallacellum*) *tenederoi* Takaoka, 2009a: 46–56 (female, male, pupa, and larva).

Distribution. Philippines (Samar).

Bionomics. Aquatic habitat: A small shallow stream (width 0.2–0.5 m, bottom rocky, shaded, 25 °C, elevation 110–130 m) slowly to moderately flowing down a grassy hill with coconut trees, in Western Samar Province (Takaoka, 2009a).

Remarks. This species was described from Samar, the Philippines (Takaoka, 2009a).

It is characterized in the female and male by the calcipala extended beyond the apex of the third tarsomere; in the female by the small sensory vesicle (Fig. 92C), ovipositor valve produced posteriorly (Fig. 92O), genital fork with a small projection directed posteomedially in addition to a moderately developed projection directed anteriorly on each arm (Fig. 92T), and paraproct much produced anteroventrally (Fig. 92Q); in the male by the number of upper-eye (large) facets in 14 or 15 vertical columns and 14 or 15 horizontal rows; in the pupa by the gill with four slender filaments arranged as 2+2 from dorsal to ventral, lying vertically, and cocoon shoe-shaped (Fig. 95L, M); and in the larva by the head capsule with a dark pigmented sheath of the subesophageal ganglion ventrally (Fig. 95Y), and abdomen with characteristic reddish-brown markings (Fig. 95R, S).

Simulium (_Wallacellum_) _tuyense_ Takaoka, 1983

(Fig. 94B)

Simulium (*Wallacellum*) *tuyense* Takaoka, 1983: 38–39 (male, pupa, and larva); Takaoka and Tenedero, 2022b: 452–454 (female).

Distribution. Philippines (Luzon, Samar, and Palawan).

Bionomics. Aquatic habitat: A slow-flowing trickle (width 20 cm, depth 1 cm, shaded) in a lowland coconut forest in Batangus, Luzon (Takaoka, 1983).

Remarks. This species was described from the Philippines (Takaoka, 1983). Its female was described by Takaoka and Tenedero (2022b).

It is characterized in the female by the ovipositor valve triangular (similar to Fig. 92L); in the male by the number of upper-eye (large) facets in 10 vertical columns and 12 horizontal rows; in the pupa by the antennal sheath with spinous projections (similar to Fig. 95D), and gill with four short slender filaments arranged as 2+2 from dorsal to ventral (Fig. 94B).

This species is similar to *S. yonakuniense* Takaoka, described from the Nansei Islands, Japan (Takaoka, 1972).

***Simulium* (*Wallacellum*) *yonakuniense* Takaoka, 1972**

Simulium (*Morops*) *yonakuniense* Takaoka, 1972: 521–523 (female, male, pupa, and larva).

Simulium (*Wallacellum*) *yonakuniense* Takaoka, 1983: 21.

Distribution. Japan (Yonakuni Island in Nansei Islands) and Taiwan (Lan Yu Island).

Bionomics. Aquatic habitat: A short shaded slow stream (ca. 5 m long) flowing to a pond, in Yonakuni Island (Takaoka, 1972).

Remarks. This species was described from Yonakuni Island, the Nansei Islands, Japan (Takaoka, 1972) and later recorded from Lan Yu Island, Taiwan (Chung, 1986). It was first assigned to the subgenus *Morops* but later transferred to the subgenus *Wallacellum* (Takaoka, 1983).

It is characterized in the female and male by the calcipala elongate, almost reaching the end of the second tarsomere; in the female by the elongate sensory vesicle, ovipositor valve triangular (similar to Fig. 92L) and genital fork with arms each having a distinct projection directed posteromedially in addition to a small projection directed anteriorly (similar to Fig. 92R); in the male by the number of upper-eye (large) facets in ca. 13 horizontal rows; in the pupa by the frons sparsely covered with tubercles, antennal sheath with distinct spinous projections (similar to Fig. 95D), and gill with four short slender filaments arranged as 2+2 from dorsal to ventral.

This species is the only member of the subgenus *Wallacellum* recorded from Japan and Taiwan, and marks the most northern distribution of the subgenus.

***Simulium* (*Wallacellum*) sp. Takaoka, Hadi and Sigit, 2006**

Simulium (*Wallacellum*) sp. Takaoka, Hadi and Sigit, 2006a: 25 (larva).

Distribution. Indonesia (Java, Timor, and Flores).

Bionomics. Aquatic habitat: A river (width 10–20 m, 22 °C, elevation 890 m) in Ruteng (Takaoka et al., 2006a).

Remarks. This species was recorded from Flores, Indonesia by Takaoka et al. (2006a). Its female, male and pupa are unknown. It was recorded from Timor by Almet et al. (2016) and Java (Hadi, pers. com.).

The larva of this species is characterized by the pharate pupal gill with four slender filaments, postgenal cleft medium-long, hanging-bell-shaped, with a narrow anterior extension, labral fan with 39 primary rays, rectal organ without secondary lobules and posterior circlet with 108 rows of up to 18 hooklets per row.

Simulium (*Wallacellum*) *alfurense* Takaoka, 2003

(Figs. 92D, E and 95D, X)

Simulium (*Wallacellum*) *alfurense* Takaoka, 2003: 14–23 (female, male, pupa, and larva).

Distribution. Indonesia (Biak and Seram).

Bionomics. Aquatic habitats: A trickle (width 2–10 cm, depth less than 0.2 cm, shaded, 25.5 °C) to rivers (width 10–15 m, depth 0.3–0.5 m, shaded, 23.0 °C) in lowland areas (elevation 3–50 m) near the sea, in Seram; a slow-flowing stream (width ca. 5 m, 26.0 °C, elevation ca. 50 m), crossing the road between the town of Biak and Wardo Waterfall in Biak (Takaoka, 2003).

Remarks. This species was described from Seram and Biak, Indonesia (Takaoka, 2003).

It is characterized in the female and male by the hind tibia with a narrowly constricted low process on its inner surface along the anterior margin from the base to a little before the midpoint (similar to Fig. 92G) and calcipala elongate, almost reaching the end of the second tarsomere (similar to Fig. 92I); in the female by the elongate sensory vesicle (similar to Fig. 92B), ovipositor valve triangular (similar to Fig. 92L) and genital fork with arms each having a distinct projection directed posteromedially as well as a distinct projection directed anteriorly (similar to Fig. 92R); in the male by the small number of upper-eye (large) facets in 11 vertical columns and 13 horizontal rows and ventral plate wide when viewed ventrally; and in the pupa by the bare frons, antennal sheath with eight raised portions each with one to three spinous projections apically (Fig. 95D) and gill with four short slender filaments arranged as 2+2 from dorsal to ventral and variable in their relative lengths.

Subgenus Wilhelmia *Rubtsov*

Simulium (*Wilhelmia*) *takahasii* (Rubtsov, 1962)

(Figs. 96A–S and 97A–I)

Wilhelmia takahasii Rubtsov, 1962 (1959–1964): 411–412 (female, male, and pupa).

Simulium (*Wilhelmia*) *takahasii*: Orii *et al.*, 1969: 11 (illustration of head capsule of larva); Takaoka *et al.* (2023b): 2–7 (female, male, pupa, and larva).
Stilboplax equinum (*nec* Linnaeus): Shiraki, 1935: 23–27 (female and male).
Simulium equinum (*nec* Linnaeus): Kani, 1944: 5 (pupa).
Simulium salopiense (*nec* Edwards): Bentinck, 1955: 7–8 (illustrations of female, male, and pupa).
Simulium (*Wilhelmia*) *salopiense* (*nec* Edwards): Ogata *et al.*, 1956: 79–80 (female, male, pupa, and larva).

Distribution. Japan (Honshu, Shikoku, and Kyushu), China (Beijing, Guizhou, Qinghai, Shangxi, Xinjiang, and Yunnan), and Korea (South).

Bionomics. Aquatic habitat: A small stream (width 1–2 m, depth 10–20 cm, elevation 500) running in a rice field in Yufuin, Oita, Japan (Takaoka, 1985). Females of this species are severe biters on cattle and horses, and rarely on humans (Bentinck, 1955; Ogata et al., 1956). This species is unique in mating, blood-feeding and ovipositing in captivity (Takaoka, 1985) and is an experimental vector of *Dirofilaria immitis* (Leidy) and *Brugia pahangi* (Buckley and Edeson) (Takaoka et al., 1986; Takaoka and Baba, 1987).

Remarks. This species was formerly recorded from Japan as *Simulium equinum* Linnaeus or *S. salopiense* Edwards (Shiraki, 1935; Kani, 1944; Bentinck, 1955; Ogata et al., 1956), and later was described as a new species, *Wilhelmia takahasii*, by Rubtsov (1959–1964) based on the illustrations given by Bentinck (1955). The larval salivary gland chromosomes of this species were studied by Adler et al. (2015). The revised descriptions of the female, male, pupa, and mature larva of this species were provided by Takaoka et al. (2023b).

It is characterized in the female and male by the pleural membrane and part of the upper mesopleuron region haired, basal portion of the radius fully haired and hind basitarsi slender, nearly parallel-sided (Fig. 96I); in the female by the sensory vesicle medium-long (0.49 times as long as the third palpal segment) (Fig. 96A), cibarium unarmed (Fig. 96B), scutum with three dark vittae and spermatheca enlarged, egg-plant-shaped (Fig. 96H); in the male by the number of upper-eye (large) facets in 18 or 19 vertical columns and 19 horizontal rows, abdominal segments 2 and 5–7 each with a pair of shiny dorsolateral patches, ventral plate subtriangular, flat, with long basal arms divergent (Fig. 96J), median sclerite rounded, with numerous short hooks and spines (Fig. 96N), paramere with numerous distinct hooks (Fig. 96O) and cercus enlarged and with numerous hairs (Fig. 96R, S); in the pupa by the frons bare, gill composed of two inflated tubular filaments oriented dorso-ventrally, then inwardly, and six short tubular filaments in between (Fig. 97A, B), abdomen lacking spine-combs dorsally, abdominal segments 5–7 each with a pair of unbranched hooks ventrally on each side, last abdominal segment with a pair of conical terminal hooks (Fig. 97C) and cocoon shoe-shaped (Fig. 97D, E); and in the mature larva by the postgenal cleft medium-long, 1.8–2.0 times as long as the postgenal bridge (Fig. 97H), abdomen covered with dark tiny bifid or trifid setae dorsally (Fig. 97I), ventral papillae absent, rectal organ simple, and posterior circlet with 94–96 rows of up to 28–30 hooklets per row.

Simulium (Wilhelmia) xingyiense Chen and Zhang, 1998

Simulium (Wilhelmia) xingyiense Chen and Zhang, 1998: 57–59 (female, male, pupa, and larva).

Distribution. China (Guizhou).

Bionomics. Aquatic habitats: Streams (elevation 400–2200 m) (Chen and Zhang, 1998).

Remarks. This species was described from Guizhou, China (Chen and Zhang, 1998).

It is characterized in the male by the number of upper-eye (large) facets in 16 horizontal rows; and in the pupa by the gill with eight short filaments, of which the dorsal and the ventral filaments are two to three times as thick as the six other filaments, according to Fig. 6 in Chen and Zhang (1998).

Chapter 4
The Fauna of Black Flies in 15 Countries or Areas in the Oriental Region

In this Chapter, black fly species in the Oriental Region are treated by countries or areas. Among 21 countries involved, Bangladesh, Maldives, Brunei, Singapore, Cambodia, Timor (East), Laos, and Palau are not included because no records of black flies are available in Bangladesh, Maldives, Brunei, Singapore, Cambodia, and Timor (East), and only one and two species of black flies were recorded in Palau and Laos, respectively.

The fauna of black flies in Malaysia is divided into two: that of Peninsular and that of Sabah and Sarawak. The fauna of black flies in Indonesia is divided into three: that of the Sunda Archipelago, that of East Indonesia (Sulawesi, Maluku, and Irian Jaya), and that of Kalimantan, the last being combined that of Sabah and Sarawak and treated as the fauna of black flies in Borneo.

The black fly fauna of each of 15 countries or areas is briefly overviewed and its species list and identification keys are provided.

1 The Fauna of Black Flies in India

Overview

Becher (1885) described *Simulium* (*Simulium*) *indicum* from Assam, India. Brunetti (1911) described seven new species based on adult specimens collected from the Himalayas, India: *S.* (*Nevermannia*) *rufithorax*, *S.* (*N.*) *aureohirtum*, *S.* (*N.*) *senile*, *S.* (*Gomphostilbia*) *metatarsale*, *S.* (*S.*) *griseifrons*, *S.* (*S.*) *grisescens*, and *S.* (*S.*) *rufibasis*. Senior-White (1922) described two new species: *S.* (*S.*) *gurneyae* and *S.* (*S.*) *latistriatum*. Edwards (1927) described two new species: *S.* (*S.*) *kashmiricum* and *S.* (*S.*) *stevensoni* (at that time, under the subgenus *Tetisimulium*), from Kashmir, North-West India.

Puri (1932a) described two new species: *S.* (*S.*) *himalayense* and *S.* (*S.*) *nilgiricum*, and described the male and pupa of *S.* (*S.*) *gurneyae* from the Himalayas.

H. Takaoka, *The Black Flies of Subtropical and Tropical Asia*,
https://doi.org/10.1007/978-981-99-5219-9_4

Puri (1932b) described three new species: *S.* (*S.*) *christophersi*, *S.* (*S.*) *nitidithorax*, and *S.* (*S.*) *ramosum*, and described the male and pupa of *S.* (*S.*) *rufibasis*. Puri (1932c) described four new species: *S.* (*S.*) *novolineatum* (at that time, as *S.* (*S.*) *lineatum*), *S.* (*S.*) *barraudi*, *S.* (*S.*) *dentatum*, and *S.* (*S.*) *digitatum*, and described the male and pupa of *S.* (*S.*) *griseifrons*. Puri (1932d) described two new species: *S.* (*S.*) *howletti* from Pusa, Bihar and *S.* (*S.*) *hirtipannus* from Assam. Puri (1932e) described four new species: *S.* (*S.*) *consimile*, *S.* (*S.*) *lineothorax*, and *S.* (*S.*) *palmatum* (at that time, as a variety of *S.* (*S.*) *grisescens*) and *S.* (*S.*) *pallidum*, and described the male and pupa of *S.* (*S.*) *grisescens*. In the same paper, Puri (1932e) synonymized *S.* (*S.*) *latistriatum* with *S.* (*S.*) *striatum* described from Sri Lanka (Brunetti, 1912). However, *S.* (*S.*) *latistriatum* is here resurrected from a junior synonym of *S. striatum*. Puri (1933a) described three new species: *S.* (*S.*) *gravelyi*, *S.* (*S.*) *tenuitarsus*, and *S.* (*S.*) *palniense*. Puri (1933b) described *S.* (*S.*) *nodosum*. Puri (1933c) described the pupa and larva of *S.* (*N.*) *aureohirtum* and recorded *S.* (*E.*) *aureum* from Himachal Pradesh, West-North India. Puri (1933d) reported *S.* (*Wilhelmia*) *pseudequinum* (at that time, as *S.* (*W.*) *equinum* var. *mediterraneum*) from Kashmir and described its female and male.

Lewis (1973) described a new species, *S.* (*S.*) *rashidi*, based on pupae and larvae collected on nymphs of mayflies in Himachal Pradesh, India, and Bahrain and Kolalai, Pakistan, and recorded *S.* (*S.*) sp. nr. *ephemerophilum* based on a larva collected on a nymph of a mayfly in Himachal Pradesh, India. Lewis (1974) redescribed *S. indicum* and *S. nodosum*, and recorded nine unnamed species from Assam and Uttar Pradesh.

Datta (1973) described six new species and one unnamed species from Darjeeling, West Bengal: *S.* (*N.*) *praelargum*, *S.* (*N.*) *gracile*, *S.* (*N.*) *purii*, *S.* (*N.*) sp. "A," *S.* (*M.*) *nemorivagum*, *S.* (*G.*) *tenuistylum*, and *S.* (*G.*) *darjeelingense*. Datta (1974a) described *S.* (*S.*) *nigrifacies* and described the larvae of *S.* (*S.*) *grisescens*, *S.* (*S.*) *dentatum*, *S.* (*S.*) *himalayense*, and *S.* (*S.*) *rufibasis*. Datta (1974b) described two new species, *S.* (*M.*) *dasguptai* and *S.* (*S.*) *biforaminiferum*, and described the female of *S.* (*M.*) *nemorivagum* and the larva of *S.* (*S.*) *ramosum*. Datta (1975a) described five new species and reported two unnamed species from Assam: *S.* (*G.*) *bucolicum*, *S.* (*G.*) *litoreum*, *S.* (*G.*) *unum*, *S.* (*G.*) *fidum*, and *S.* (*S.*) *kapuri*, *S.* (*G.*) sp. nr. *varicorne* and *S.* (*G.*) sp. nr. *litoreum*. Datta (1975b) described a new species, *S.* (*M.*) *ghoomense*, from Darjeeling. Datta et al. (1975a) recorded four unnamed species from Darjeeling: *S.* (*N.*) sp. A, *S.* (*M.*) sp. B, *S.* (*M.*) sp. C, and *S.* (*N.*) sp. D. Datta and Pal (1975) described *S.* (*S.*) *singtamense* based on pupal and larval specimens from Darjeeling and described the pupa of *S.* (*S.*) sp. nr. *novolineatum* and the larva of *S.* (*S.*) sp. nr *grisescens*. Datta (1983) synonymized *S.* (*S.*) *kashmiricum* with *S.* (*S.*) *indicum*. Datta (1985a) described *S.* (*S.*) *adventicium* from Himachal Pradesh. Datta (1985b) recorded *S.* (*E.*) *aureum* from Kashmir. Datta (1988) described *S.* (*S.*) *asishi* from West Bengal. Datta (1991b) reported *S.* (*G.*) ? *sundaicum* from West Bengal. Datta (1992a) reported *S.* (*M.*) sp. nr. *ghoomense* and *S.* (*N.*) sp., based on one larva and one female reared from a pupa, respectively, collected from Himachal Pradesh. Stauder and Kiel (2004) reported the phoretic association of *S. rashidi* and *S. ephemerophilum* with Heptageniidae in Himachal Pradesh.

Takaoka et al. (2009c) described the male of *S.* (*M.*) *dasguptai*. Takaoka et al. (2010e) described two new species: *S.* (*G.*) *sachini* and *S.* (*G.*) *williei*, based on two pupae, and one pupa and two larvae, respectively, from Darjeeling. Henry et al. (2010) recorded *S.* (*M.*) *taktsangense* and *S.* (*N.*) sp. 2 (nr. *fruticosum*), which are similar to *S.* (*M.*) sp. C and *S.* (*N.*) sp. D, respectively. Takaoka et al. (2011c) described a new species, *S.* (*N.*) *subratai*. Borah et al. (2012) recorded three species: *S.* (*G.*) *decuplum*, *S.* (*S.*) *barnesi*, and *S.* (*S.*) *pradyai*, synonymized *S.* (*G.*) *unum* with *S.* (*G.*) *darjeelingense*, and identified the larva, which was formerly reported as *S.* (*G.*) sp. nr. *varicorne* by Datta (1975a), as *S.* (*G.*) *parahiyangum*, from Assam. Takaoka and Somboon (2008) described a new species, *S.* (*M.*) *dattai* [the species formerly regarded as *S.* (*M.*) sp. B in India by Datta et al. (1975)]. Takaoka et al. (2020a) pointed out that the species regarded as *S.* (*S.*) *striatum* from India and Nepal is different from that from Sri Lanka (the type locality); thus, an unnamed species, *S.* (*S.*) sp. nr. *striatum*, was proposed for populations of India and Nepal. Takaoka et al. (2022b) described the females and males of *S.* (*G.*) *williei* and *S.* (*G.*) *sachini* based on Indian specimens.

From South India, Datta (1985b) recorded *S.* (*G.*) *pattoni*. Anbalagan et al. (2014) described two new species: *S.* (*G.*) *peteri* and *S.* (*G.*) *takaokai*. Anbalagan et al. (2015a,b) described three new species: *S.* (*G.*) *cauveryense*, *S.* (*G.*) *panagudiense*, and *S.* (*G.*) *kottoorense*. Anbalagan et al. (2018) described two new species: *S.* (*S.*) *pothigainense* and *S.* (*S.*) *valparaiense*. Anbalagan et al. (2019a) described a new species, *S.* (*S.*) *yanaense*. Anbalagan et al. (2019b) described a new species, *S.* (*G.*) *kumbakkaraiense*. Vijayan et al. (2019) described a new species, *S.* (*G.*) *agasthyamalaiense*. Anbalagan et al. (2020a) described a new species, *S.* (*N.*) *karavalliense*. Anbalagan et al. (2020b) described two new species: *S.* (*G.*) *dinakarani* and *S.* (*G.*) *krishnani*.

In total, the fauna of black flies in India are represented by 91 species, of which 76 species are recorded only from the Oriental areas of India, 13 species are recorded only from Kashmir and Himachal Pradesh in the Palearctic Region and two species are common to both regions in India.

List of 91 Species of Black Flies in India

Genus *Simulium* s. l.
Subgenus *Eusimulium*
*S. aureum**
Subgenus *Gomphostilbia*
S. batoense species-group
S. agasthyamalaiense, S. decuplum, S. dinakarani, S. kottoorense, S. krishnani, S. kumbakkaraiense, S. panagudiense, S. parahiyangum, S. pattoni, S. peteri, S. takaokai, and *S.* Pabho sp. (Lewis, 1974)
S. darjeelingense species-group
S. darjeelingense

Gomphostilbia species unplaced to species-group
S. bucolicum, S. cauveryense, S. fidum, S. litoreum, S. metatarsale, S. tenuistylum, S. sp. nr. *litoreum, S.* ? *sundaicum*, *S.* Ranga sp. (Lewis, 1974), and *S.* Lakhimpur sp. (Lewis, 1974)
Subgenus *Montisimulium*
S. dasguptai, S. dattai, S. ghoomense, S. nemorivagum, S. taktsangense, S. sp. nr. *ghoomense**, and *S.* sp. C (Datta et al., 1975)
Subgenus *Nevermannia*
S. feuerborni species-group
S. praelargum (complex), *S. rufithorax, S. senile**, *S. subratai, S.* sp. "A" (Datta, 1973), *S.* sp. D (Datta et al., 1975), and *S.* sp. 2 (nr. *fruticosum*) (Henry et al., 2010)
S. ruficorne species-group
*S. aureohirtum*** and *S. karavalliense*
S. vernum species-group
S. gracile, *S. purii*, *S.* sp. A (Datta et al., 1975), and *S.* sp. (Datta, 1992a)*
Subgenus *Simulium* s. str.
S. bezzii species-group
*S. stevensoni**
S. christophersi species-group
*S. christophersi** and *S. howletti**
S. ephemerophilum species-group
S. ephemerophilum (probably same as *S.* sp. nr. *ephemerophilum*)* and *S. rashidi**
S. griseifrons species-group
S. gravelyi, *S. griseifrons*, and *S. tenuitarsus*
S. indicum species-group
*S. indicum***
S. malyschevi species-group
S. hirtipannus
S. multistriatum species-group
*S. barraudi**, *S. dentatum*, *S. digitatum**, *S. novolineatum*, and *S.* sp. nr. *novolineatum*
S. nobile species-group
S. asishi and *S. nodosum*
S. ornatum species-group
*S. adventicium**
S. striatum species-group
S. consimile, *S. grisescens*, *S. kapuri*, *S. latistriatum, S. lineothorax, S. pallidum*, *S. pothigainense*, *S. valparaiense, S. yanaense*, *S.* sp. nr. *grisescens*, *S.* Diju sp. (Lewis, 1974), and *S.* Doigroon sp. (Lewis, 1974),
S. tuberosum species-group
S. nigrifacies, *S. nitidithorax*, *S. pradyai*, *S. ramosum*, and *S. rufibasis*
S. variegatum species-group
S. barnesi, *S. gurnyae*, *S. himalayense*, and *S. nilgiricum*

Simulium s. sp. species unplaced to species-group
S. biforaminiferum, *S. palniense*, *S. singtamense*, *S.* Assam sp. (Lewis, 1974),
S. Kamti sp. (Lewis, 1974), *S.* Rengma sp. (Lewis, 1974), and
S. Subansiri sp. (Lewis, 1974),
Subgenus *Wilhelmia*
*S. pseudequinum**

*These 13 species were recorded from Kashmir and Himachal Pradesh in the Palearctic Region, though *S.* (*S.*) *barraudi* was recorded also from Bhutan, Myanmar, and Nepal in the Oriental Region.

**These two species were recorded from both the Oriental Region and Kashmir and Himachal Pradesh in the Palearctic Region.

*Keys to Identify 82 Species of Black Flies in India**

Females**

1. Postnotum with hairs .... *S.* (*E.*) *aureum*
   Postnotum without hairs 2
2. Katepisternum haired 3
   Katepisternum bare 15
3. Claw with small basal tooth .... *S.* (*G.*) *pattoni*
   Claw with medium-long or long basal tooth 4
4. Fore coxa yellow 5
   Fore coxa brown 14
5. Base of radius with tuft of yellow hairs 6
   Base of radius with tuft of black hairs 7
6. Hind tibia yellowish white on basal half, and light to dark brown on rest. Claw with medium-long tooth 0.38 times as long as claw (Fig. 36E) .... *S.* (*G.*) *sachini*
   Hind tibia light brown except basal one-fourth to one-third whitish yellow and apical cap dark brown. Claw with long basal tooth 0.52 as long as claw (Fig. 36F) .... *S.* (*G.*) *williei*
7. Scutum without longitudinal vittae .... *S.* (*G.*) *panagudiense*
   Scutum with three longitudinal vittae 8
8. Hind tibia yellowish except apical cap black .... *S.* (*G.*) *tenuistylum*
   Hind tibia light to dark brown except basal one-six to one-fourth white 9
9. Mandible with 32 inner teeth and 12 outer teeth .... *S.* (*G.*) *kottoorense*
   Mandible with 12–24 inner teeth and 0–13 outer teeth 10
10. Mandible with 12 or 24 inner teeth and lacking outer teeth 11
    Mandible with 16–21 inner teeth and 7–13 outer teeth 12
11. Mandible with 12 inner teeth .... *S.* (*G.*) *dinakarani*
    Mandible with 24 inner teeth .... *S.* (*G.*) *krishnani*
12. Fore basitarsus 5.8–6.3 times as long as its greatest width .... *S.* (*G.*) *agasthyamalaiense* and *S.* (*G.*) *peteri*

Fore basitarsus 4.2–5.0 times as long as its greatest width 13
13. Fore basitarsus 4.2 times as long as its greatest width .... *S.* (*G.*) *takaokai*
Fore basitarsus 5.0 times as long as its greatest width *S.* (*G.*) *kunbakkaraiense*
14. Base of radius with tuft of yellow hairs .... *S.* (*G.*) *bucolicum*
Base of radius with tuft of black hairs ....*S.* (*G.*) *darjeelingense*
15. Claw with medium-long or long basal tooth 16
Claw simple or with small subbasal tooth 23
16. Antenna yellow except flagellomere 1 darkened, or gray to white except scape, pedicel and base of flagellomere 1 dark yellow 17
Antenna mostly darkened 18
17. Mid tibia brown ....*S.* (*N.*) *karavalliense*
Mid tibia yellow on basal half or little more, with dark subbasal marking .... *S.* (*N.*) *aureohirtum*
18. Scutum yellow or orange red 19
Scutum reddish brown or grayish black or black 20
19. Scutum yellow, without three dark longitudinal vittae ....*S.* (*N.*) *subratai*
Scutum orange red, with three dark longitudinal vittae .... *S.* (*N.*) *praelargum* (complex) and *S.* (*N.*) sp. "A"
20. Scutum reddish brown ....*S.* (*N.*) *rufithorax*
Scutum grayish black or black 21
21. Claw with long basal tooth 0.43–0.49 times as long as claw ....*S.* (*N.*) *gracile* and *S.* (*N.*) *purii*
Claw with medium-long basal tooth 0.27–0.29 times as long as claw (similar to Fig. 42C) 22
22. Sensory vesicle long, 0.65 times as long as third palpal segment .... *S.* (*M.*) *nemorivagum*
Sensory vesicle long, 0.52 times as long as third palpal segment ....*S.* (*M.*) *ghoomense*
23. Pleural membrane haired 24
Pleural membrane bare 26
24. Claw simple, without subbasal tooth .... *S.* (*W.*) *pseudequinum*
Claw with subbasal tooth 25
25. Scutum without longitudinal vittae. Abdominal segments 6–8 shiny .... *S.* (*S.*) *adventicium*
Scutum with three longitudinal vittae. Abdominal segments 6–8 not shiny .... *S.* (*S.*) *stevensoni*
26. Claw with subbasal tooth (similar to Fig. 62C) 27
Claw simple, without tooth 34
27. Basal portion of radius haired ....*S.* (*S.*) *indicum*
Basal portion of radius bare 28
28. Abdominal segment 7 with a pair of clustered branched hairs ventrally (similar to Fig. 65C) .... *S.* (*S.*) *hirtipannus*
Abdominal segment 7 without a pair of clustered branched hairs ventrally 29

29. Ovipositor valve well sclerotized and rounded posteriorly, with stout hairs apically .... *S.* (*S.*) *asishi*
    Ovipositor valve membranous and not rounded posteromedially, without stout hairs apically 30
30. Ovipositor valve with inner margin nearly straight .... *S.* (*S.*) *christophersi*
    Ovipositor valve with inner margin moderately to greatly concave (similar to Fig. 84C) 31
31. Mid femur entirely yellow or yellow gradually darkened at apex .... *S.* (*S.*) *himalayense* and *S.* (*S.*) *gurneyae*
    Mid femur dark brown 32
32. Mid tibia yellow except apical one-fourth black .... *S.* (*S.*) *howletti*
    Mid tibia dark brown or dark brown except basal one-third yellowish brown 33
33. Mid basitarsus whitish yellow on basal three-fourths .... *S.* (*S.*) *barnesi*
    Mid basitarsus yellowish brown on basal one-third .... *S.* (*S.*) *nilgiricum*
34. Scutum with longitudinal vittae (similar to Fig. 76B) 35
    Scutum without longitudinal vittae 41
35. Ovipositor valve with ventrally produced process along inner margin (similar to Fig. 76D, E) 36
    Ovipositor valve without ventrally produced process along inner margin 39
36. Basal portion of radius haired ....*S.* (*S.*) *grisescens*, *S.* (*S.*) *kapuri*, *S.* (*S.*) *palmatum*, and *S.* (*S.*) *pothigaiense*
    Bas al portion of radius bare 37
37. Hind tibia black except basal one-fourth yellow or dark brown except basal one-third to two-thirds yellow .... *S.* (*S.*) *consimile* and *S.* (*S.*) *latistriatum*
    Hind tibia dark brown to brownish black or black except base pale yellow 38
38. Hind tibia dark brown to brownish black .... *S.* (*S.*) *valparaiense*
    Hind tibia black except base yellow .... *S.* (*S.*) *lineothorax*
39. Mid tibia pale yellow except apical one-fifth dark brown .... *S.* (*S.*) *dentatum*
    Mid tibia pale yellow except apical tip gray to dark gray 40
40. Ovipositor valve with eight or nine short hairs .... *S.* (*S.*) *digitatum*
    Ovipositor valve with two or three short hairs .... *S.* (*S.*) *novolineatum* and *S.* (*S.*) *barraudi*
41. Abdominal segment 7 with a pair of clustered stout hairs ventrally (similar to Fig. 82C) .... *S.* (*S.*) *rufibasis* and *S.* (*S.*) *ramosum*
    Abdominal segment 7 without a pair of clustered stout hairs ventrally .... *S.* (*S.*) *nitidithorax*

Males***

1. Coxite longer than style 2
   Coxite shorter than style 22
2. Postnotum with hairs .... *S.* (*E.*) *aureum*
   Postnotum without hairs 3

3. Katepisternum haired 4
   Katepisternum bare 14
4. Hind basitarsus brownish except apical one-fifth black .. *S.* (*G.*) *darjeelingense*
   Hind basitarsus medium to dark brownish except basal half yellow to light brown 5
5. Hind basitarsus enlarged, nearly as wide as or slightly wider than hind tibia ....*S. (G.) metatarsale* and *S. (G.) tenuistylum*
   Hind basitarsus slender or somewhat enlarged, less than 0.82 times as wide as hind tibia 6
6. Mid and hind tibiae with dark subbasal marking ....*S.* (*G.*) *fidum*
   Mid and hind tibiae without dark subbasal marking 7
7. Mid tibia medium to dark brown 8
   Mid tibia whitish yellow on basal one-third and light to dark brown on rest...13
8. Fore coxa brown ....*S.* (*G.*) *takaokai*
   Fore coxa yellow 9
9. Abdominal segments 2–7, each with a pair of shiny dorsolateral spots 10
   Abdominal segments 2 and 5–7, each with a pair of shiny dorsolateral spots 11
10. Upper-eye (large) facets in 12 vertical columns .... *S.* (*G.*) *peteri*
    Upper-eye (large) facets in 20 vertical columns .... *S.* (*G.*) *kottoorense*
11. Upper-eye (large) facets in 20 vertical columns ...... *S.* (*G.*) *agasthyamalaiense*
    Upper-eye (large) facets in 13 vertical columns 12
12. Fore basitarsus 6.6 times as long as its greatest width ....*S.* (*G.*) *cauveryense*
    Fore basitarsus 6.0–6.2 times as long as its greatest width .... *S.* (*G.*) *dinakarani*, *S.* (*G.*) *krishnani*, and *S.* (*G.*) *kumbakkaraiense*
13. Ventral plate with ventral margin deeply concave medially when viewed caudally (similar to Fig. 36Y) ....*S.* (*G.*) *sachini*
    Ventral plate with ventral margin rounded when viewed caudally (Fig. 36V) ... .*S.* (*G.*) *williei*
14. Pleural membrane haired .... *S.* (*W.*) *pseudequinum*
    Pleural membrane bare 15
15. Style lacking terminal spine ....*S.* (*N.*) *senile*
    Style with terminal spine 16
16. Antenna yellow except flagellomere 1 darkened ....*S.* (*N.*) *aureohirtum*
    Antenna mostly darkened 17
17. Paramere with one distinct hook 18
    Paramere with several hooks..19
18. Ventral plate with median keel (similar to Fig. 45K) ....*S.* (*N.*) *karavalliense*
    Ventral plate without median keel (similar to Fig. 47P) ....*S.* (*N.*) *gracile* and *S.* (*N.*) *purii*
19. Scutum velvet black .... *S.* (*M.*) *ghoomense*
    Scutum reddish gray or reddish brown 20

20. Scutum reddish brown .... *S.* (*N.*) *rufithorax*
    Scutum reddish gray 21
21. Scutum without dark longitudinal lines. Upper-eye (large) facets in 12 vertical columns and 14 horizontal rows .... *S.* (*M.*) *nemorivagum*
    Scutum with dark longitudinal lines. Upper-eye (large) facets in 16 vertical columns and 17 horizontal rows ....*S.* (*N.*) *praelargum* (complex)
22. Pleural membrane haired .... *S.* (*S.*) *adventicium*
    Pleural membrane bare 23
23. Basal portion of radius haired ....*S.* (*S.*) *indicum*
    Basal portion of radius bare 24
24. Scutum white pruinose with inverted-V-shaped, non-pruinose black area (similar to Fig. 74S, T) 25
    Scutum otherwise 26
25. Ventral plate greatly constricted basally when viewed ventrally, with ratio of narrowest to widest 0.4:1.0 .... *S.* (*S.*) *asishi*
    Ventral plate moderately constricted basally when viewed ventrally, with ratio of narrowest to widest 0.8:1.0 (Fig. 74U) .... *S.* (*S.*) *nodosum*
26. Style without spines basally 27
    Style with spines basally or elongate basal protuberance with pointed apex ...31
27. Scutum with copper-colored short hairs. Hind basitarsus slender, 0.6 times as wide as hind tibia .... *S.* (*S.*) *hirtipannus*
    Scutum with golden yellow short hairs. Hind basitarsus enlarge, 0.7–1.0 times as wide as hind tibia 28
28. Hind basitarsus as wide as hind tibia 29
    Hind basitarsus 0.70–0.77 times as wide as hind tibia 30
29. Hind basitarsus grayish black on basal half and black on rest ....
    ....*S.* (*S.*) *christophersi*
    Hind basitarsus brownish yellow on basal half and black on rest ....
    ....*S.* (*S.*) *howletti*
30. Hind femur slightly yellowish gray ....*S.* (*S.*) *gurneyae*
    Hind femur black ....*S.* (*S.*) *himalayense* and *S.* (*S.*) *nilgiricum*
31. Style with elongate basal protuberance with apex pointed (similar to Fig. 71C)
    32
    Style with short round basal protuberance 35
32. Ventral plate with teeth on posterior surface 33
    Ventral plate without teeth on posterior surface 34
33. Ventral plate with teeth in two vertical rows on posterior surface ....
    ....*S.* (*S.*) *dentatum*
    Ventral plate with teeth randomly on posterior surface ....
    .... *S.* (*S.*) *novolineatum*
34. Ventral plate nearly quadrate, 0.7 times as long as wide when viewed ventrally (Fig. 71L) .... *S.* (*S.*) *barraudi*
    Ventral plate rectangular, 1.4 times as long as wide when viewed ventrally .....
    ....*S.* (*S.*) *digitatum*

35. Ventral plate nearly quadrate or rectangular when viewed ventrally, with teeth on each posterolateral margin (similar to Fig. 80L) 36
Ventral plate saddle-shaped (similar to Fig. 76P) or rectangular when viewed ventrally, without teeth on each posterolateral margin 40
36. Ventral plate rectangular, much longer than wide when viewed ventrally .... *S.* (*S.*) *palniense*
Ventral plate nearly quadrate, as long as or shorter than wide when viewed ventrally 37
37. Hind basitarsus 0.7 times as wide as hind tibia ....*S.* (*S.*) *nitidithorax*
Hind basitarsus nearly as wide as hind tibia 38
38. Abdominal segments 2, 6, and 7 each with a pair of shiny dorsolateral spots... ....*S.* (*S.*) *nigrifacies*
Abdominal segments 2 and 5–7 each with a pair of shiny dorsolateral spots 39
39. Fore coxa black ....*S.* (*S.*) *rufibasis*
Fore coxa yellow ....*S.* (*S.*) *ramosum*
40. Ventral plate quadrate or rectangular when viewed ventrally 41
Ventral plate saddle-shaped when viewed ventrally (similar to Fig. 76P) 43
41. Hind basitarsus as wide as hind tibia. Ventral plate bare ...... *S.* (*S.*) *griseifrons*
Hind basitarsus 0.7 or 0.9 times as wide as hind tibia. Ventral plate with setae on ventral surface 42
42. Hind basitarsus 0.7 times as wide as hind tibia ....*S.* (*S.*) *gravelyi*
Hind basitarsus 0.9 times as wide as hind tibia .... *S.* (*S.*) *tenuitarsus*
43. Scutum with longitudinal vittae ....S. (S.) *lineothorax*
Scutum without longitudinal vittae 44
44. Scutum with golden yellow short hairs 45
Scutum with copper-colored short hairs 48
45. Abdominal segments 2 and 5–7 each with a pair of shiny dorsolateral spots ... . *S.* (*S.*) *pallidum*
Abdominal segments 2–7 each with a pair of shiny dorsolateral spots 46
46. Upper-eye (large) facets in 12 vertical columns and 16 horizontal rows ....*S.* (*S.*) *yanaense*
Upper-eye (large) facets in 16 or 18 vertical columns and 18 or 20 horizontal rows 47
47. Upper-eye (large) facets in 16 vertical columns and 18 horizontal rows ....*S.* (*S.*) *valparaiense*
Upper-eye (large) facets in 18 vertical columns and 20 horizontal rows .... *S.* (*S.*) *pothigaiense*
48. Hind basitarsus somewhat enlarged, 0.6 times as wide as hind tibia .... *S.* (*S.*) *consimile*
Hind basitarsus greatly enlarged, as wide as hind tibia 49

49. Mid femur grayish yellow .... *S.* (*S.*) *latistriatum*
Mid femur black except base yellow ....
....*S.* (*S.*) *grisescens* and *S.* (*S.*) *palmatum*

Pupae****

1. Gill with three filaments 2
Gill with four, six, eight, 10, 12, or 14 filaments 3
2. Gill with three inflated tubular filaments with rounded apices (Fig. 75F) ....
.... *S.* (*S.*) *nodosum*
Gill with three filaments with tapered apices (Fig. 75E) ....*S.* (*S.*) *asishi*
3. Gill with four filaments 4
Gill with six, eight, 10, 12, or 14 filaments 7
4. Cocoon without anterodorsal projection 5
Cocoon with anterodorsal projection 6
5. Frons with two trichomes on each side. Abdominal segment 5 without spine-combs .... *S. (E.) aureum*
Frons with three trichomes on each side. Abdominal segment 5 with spine-combs ....*S. (N.)* sp. A
6. Cocoon with bifid anterodorsal projection ....*S.* (*N.*) *gracile*
Cocoon with single anterodorsal projection .... *S. (N.) puri* and *S. (N.)* sp.
7. Gill with six filaments 8
Gill with eight, 10, 12, or 14 filaments 26
8. Gill filaments as long as or longer than pupal body 9
Gill filaments much shorter than pupal body 10
9. Cocoon with distinct anterodorsal projection .....*S.* (*N.*) *praelargum* (complex)
Cocoon simple, without anterodorsal projection ....
....*S. (N.)* sp. D and *S.* sp. 2 (nr. *fruticosum*)
10. Cocoon wall-pocket-shaped, with short anterodorsal projection ....
.... *S.* (*N.*) *aureohirtum*
Cocoon wall-pocket-shaped, without anterodorsal projection, or shoe-shaped, or boot-shaped 11
11. Cocoon wall-pocket-shaped 12
Cocoon shoe-shaped or boot-shaped 22
12. Cocoon with dorsal margin much longer than ventral margin when viewed laterally (similar to Fig. 95H, I) .... *S.* (*S.*) *nilgiricum*
Cocoon with dorsal margin much shorter than ventral margin when viewed laterally 13
13. Cocoon with large anterolateral window on each side (similar to Fig. 66F) 14
Cocoon without anterolateral window on each side 15
14. Frons with micro-tubercles. Terminal hooks present ....*S.* (*S.*) *biforminiferum*
Frons bare, without micro-tubercles. Terminal hooks absent ....
.... *S.* (*S.*) *hirtipannus*
15. Terminal hooks absent ....*S.* (*S.*) *nigrifacies*
Terminal hooks present 16

16. Frons bare, without tubercles 17
 Frons with tubercles 18
17. Cocoon deeply concave anteromedially when viewed dorsally, with dorsal margin 0.63 times as long as ventral margin ....*S.* (*S.*) *gurneyae*
 Cocoon moderately concave anteromedially when viewed dorsally, with dorsal margin over 0.8 times as long as ventral margin when viewed laterally ....*S.* (*S.*) *christophersi*
18. Thoracic trichomes branched ....*S.* (*S.*) *ramosum*
 Thoracic trichomes unbranched 19
19. Gill filaments moderately divergent at right angle or less formed at base between dorsal filament of dorsal pair and ventral filament of ventral pair when viewed laterally 20
 Gill filaments widely divergent at angle of 130–150 degrees formed at base between dorsal filament of dorsal pair and ventral filament of ventral pair when viewed laterally 21
20. Dorsal, middle and ventral pairs of gill filaments all sessile ....*S.* (*S.*) *nitidithorax*
 At least dorsal and ventral pairs of gill filaments short-stalked ....*S.* (*S.*) *rufibasis*
21. Abdominal segment 6 with spine-combs dorsally ....*S.* (*S.*) *himalayense*
 Abdominal segment 6 without spine-combs dorsally .... *S.* (*S.*) *singtamense*
22. Abdominal segments 5–9 without spine-combs dorsally. Cocoon shoe-shaped .... *S.* (*S.*) *indicum*
 At least abdominal segment 8 with spine-combs dorsally. Cocoon boot-shaped 23
23. Dorsal, middle, and ventral pairs of gill filaments all sessile .....*S.* (*S.*) *gravelyi*
 At least one of the three pairs of gill filaments short-stalked 24
24. Middle pair of gill filaments short-stalked .... *S.* (*S.*) *palniense*
 Middle pair of gill filaments sessile 25
25. Ventral pair of gill filaments short-stalked ....*S.* (*S.*) *tenuitarsus*
 Ventral pair of gill filaments sessile ....*S.* (*S.*) *griseifrons*
26. Gill with eight filaments 27
 Gill with 10, 12, or 14 filaments 43
27. Eight gill filaments arising from inflated bulbous structure 28
 Eight gill filaments arising from common basal stalk 29
28. Gill with one horn-like structure having eight slender filaments (Fig. 37C) ....*S.* (*G.*) *williei*
 Gill with one bifid structure having six finger-like projections and eight slender filaments (Fig. 37I) .... *S.* (*G.*) *sachini*
29. Gill composed of two somewhat inflated basal arms each directed dorsally and ventrally and six short filaments in-between, each with constriction basally (similar to Fig. 97A) .... *S.* (*W.*) *pseudequinum*
 Gill composed of eight thread-like filaments arranged otherwise 30

30. Frons with three trichomes on each side. Abdominal segment 9 with grapnel-like hooklets on each ventrolateral surface 31
    Frons with two trichomes on each side. Abdominal segment 9 without grapnel-like hooklets 38
31. Eight gill filaments arranged as 2+(2+2)+2 from dorsal to ventral ....
    .... *S.* (*G.*) *darjeelingense*
    Eight gill filaments arranged as 3+3+2 or (3+3)+2 from dorsal to ventral ....
   32
32. Eight gill filaments arranged as 3+3+2 from dorsal to ventral (dorsal and middle triplets not sharing stalk) 33
    Eight gill filaments arranged as (3+3)+2 from dorsal to ventral (dorsal and middle triplets sharing stalk) 36
33. Two filaments of ventral pair of gill about twice as long as six other filaments
    .... *S.* (*G.*) ? *sundaicum*
    Two filaments of ventral pair of gill slightly to somewhat longer than six other filaments 34
34. Stalk of ventral pair of gill over half as long as interspiracular trunk ....
    .... *S.* (*G.*) *agasthyamalaiense*, *S.* (*G.*) *dinakarani*,
    S. (*G.*) *panagudiense*, and *S.* (*G.*) *tenuistylum*
    Stalk of ventral pair of gill extremely short, less than one-third as long as interspiracular trunk 35
35. Common basal stalk of gill short ....
    ....*S.* (*G.*) *kumbakkaraiense* and *S.* (*G.*) *kottoorense*
    Common basal stalk of gill absent ....*S.* (*G.*) *krishnani* and *S.* (*G.*) *peteri*
36. Two filaments of ventral pair of gill subequal in thickness to each other ....
    ....*S.* (*G.*) *bucolicum* and *S.* (*G.*) *takaokai*
    Two filaments of ventral pair of gill different in thickness from each other 37
37. Upper filament of ventral pair of gill thicker than lower counter filament ....
    .... *S.* (*G.*) *litoreum*
    Lower filament of ventral pair of gill thicker than upper counter filament ....
    .... *S.* (*G.*) *fidum*
38. Eight gill filaments arranged as 3+3+2 from dorsal to ventral ....
    ....*S.* (*N.*) *karavalliense*
    Eight gill filaments arranged as 2+2+2+2 from dorsal to ventral 39
39. Cocoon wall-pocket-shaped, without open spaces (attached on mayfly nymphs).
    .... *S.* (*S.*) *rashidi*
    Cocoon wall-pocket-shaped, with open space(s) anterolaterally on each side, or shoe-shaped, with several open spaces anterolaterally on each side (attached on plants, rocks or stones or artificial substrates like plastic sheet) 40
40. Cocoon wall-pocket-shaped, usually with one open space anterolaterally on each side (similar to Fig. 72J) ....*S.* (*S.*) *barraudi*, *S.* (*S.*) *dentatum*,
    S. (*S.*) *digitatum*, and *S.* (*S.*) *novolineatum*
    Cocoon shoe-shaped, with several open spaces anterolaterally on each side 41

41. Two filaments of ventral pair of gill much thinner than six other filaments (similar to Fig. 77E) .... *S.* (*S.*) *kapuri*
    All eight filaments of gill subequal in thickness to one another 42
42. Terminal hooks present .... *S.* (*S.*) *adventicium*
    Terminal hooks absent .... *S.* (*S.*) sp. nr. *novolineatum*
43. Gill with 10 filaments 44
    Gill with 12 or 14 filaments 48
44. Abdominal segment 9 with grapnel-like hooklets 45
    Abdominal segment 9 without grapnel-like hooklets 46
45. Gill with 10 filaments arranged as 4+4+2 or 3+1+4+2 from dorsal to ventral. Cocoon wall-pocket-shaped .... *S.* (*G.*) *decuplum*
    Gill with 10 filaments arranged as 2+3+3+2 from dorsal to ventral. Cocoon shoe-shaped .... *S.* (*G.*) *cauveryense*
46. Five ventral filaments much thinner than five dorsal filaments. Cocoon shoe-shaped, without open spaces anteriorly on each side .... *S.* (*S.*) *palmatum* and *S.* (*S.*) *yanaense*
    All ten filaments subequal in thickness to one another. Cocoon wall-pocket-shaped, without open space(s) anterolaterally on each side or shoe-shaped with several open spaces anteriorly on each side 47
47. Cocoon wall-pocket-shaped, without open space(s) anterolaterally on each side .... *S.* (*S.*) *pothigaiense* and *S.* (*S.*) *valparaiense*
    Cocoon shoe-shaped, with several open spaces anteriorly on each side (similar to Fig. 77G) .... *S.* (*S.*) *consimile*, *S.* (*S.*) *grisescens*, *S.* (*S.*) *lineothorax*, *S.* (*S.*) *pallidum*, and *S.* (*S.*) *latistriatum*
48. Gill with 12 filaments 49
    Gill with 14 filaments 50
49. Cocoon with anterodorsal projection .... *S.* (*M.*) *ghoomense*
    Cocoon without anterodorsal projection ....*S.* (*M.*) *nemorivagum*
50. Gill composed of one horn-like structure with 14 slender filaments ....*S.* (*M.*) *dasguptai*
    Gill composed of 14 thread-like filaments arising from medium-long or long common basal stalk 51
51. Fourteen gill filaments arising from medium-long common basal stalk .... *S.* (*M.*) *dattai*
    Fourteen gill filaments arising from long common basal stalk ...... *S.* (*M.*) sp. C

Mature larvae*****

1. Ventral papillae present 2
   Ventral papillae absent 17
2. Postgenal cleft absent or vestigial (Fig. 42X) 3
   Postgenal cleft present, over 0.3 times as long as postgenal bridge 5

3. Pharate pupal gill with 12 filaments arising from short common basal stalk .... *S.* (*M.*) *ghoomense*, *S.* (*M.*) *nemorivagum*, and *S.* (*M.*) sp. nr. *ghoomense*
   Pharate pupal gill with 14 filaments 4
4. Pharate pupal gill with long common basal stalk ....*S.* (*M.*) *taktsangense*
   Pharate pupal gill with inflated tubular structure ....*S.* (*M.*) *dasguptai*
5. Hypostoma with lateral margins smooth. Postgenal cleft much longer than postgenal bridge 6
   Hypostoma with lateral margins serrated. Postgenal cleft short, less than 0.8 times as long as postgenal bridge 10
6. Postgenal cleft widely reaching posterior margin of hypostoma. Abdominal segments 1–5 each with a pair of cone-like protuberances dorsally (Fig. 23T, U). Posterior abdominal segments with dark, spinous large setae dorsally .... *S.* (*G.*) *parahiyangum*
   Postgenal cleft medium-long to long. Abdominal segments 1–5 without a pair of cone-like protuberances dorsally. Posterior abdominal segments without dark spinous large setae 7
7. Postgenal cleft almost reaching posterior margin of hypostoma 8
   Postgenal cleft 2.9–6.5 times as long as postgenal bridge 9
8. Pharate pupal gill with eight filaments arising from common basal stalk ....*S.* (*G.*) *fidum*
   Pharate pupal gill with eight filaments arising from horn-like inflated structure ....*S.* (*G.*) *williei*
9. Postgenal cleft 2.9 times as long as postgenal bridge .... *S.* (*G.*) *bucolicum*
   Postgenal cleft 6.5 times as long as postgenal bridge .... *S.* (*G.*) *litoreum* and *S.* (*G.*) sp. nr. *litoreum*
10. Postgenal cleft short, 0.32 times as long as postgenal bridge ....*S.* (*N.*) *praelargum* (complex)
    Postgenal cleft medium-long, 0.7–0.8 times as long as postgenal bridge 11
11. Mandible with supernumerary serration 12
    Mandible without supernumerary serration 14
12. Pharate pupal gill with four filaments arising from medium-long common basal stalk .... *S.* (*N.*) *purii*
    Pharate pupal gill with four filaments arising from short common basal stalk 13
13. Stalk of ventral pair of pharate pupal gill extremely short, much shorter than that of dorsal pair ....*S.* (*N.*) *gracile*
    Stalk of ventral pair of pharate pupal gill short, nearly as long as that of dorsal pair ....*S.* (*N.*) sp. A
14. Pharate pupal gill with four filaments ....*S.* (*E.*) *aureum*
    Pharate pupal gill with six or eight filaments 15
15. Pharate pupal gill with eight filaments ....*S.* (*N.*) *karavalliense*
    Pharate pupal gill with six filaments 16
16. Head spots markedly positive. Mandibular serration composed of two teeth of same size. Hypostoma with lateral margins markedly serrated .... *S.* (*N.*) *aureohirtum*

Head spots faintly positive. Mandibular serration composted of two teeth of different size. Hypostoma with lateral margins weakly serrated .... *S.* (*N.*) *subratai*

17. Larvae occurring on nymphs of mayfly 18
Larvae occurring on other substrates 19
18. Hypostoma with median tooth longest of all. Apical tooth of mandible dentate .... *S.* (*S.*) *ephemerophilum*
Hypostoma with median tooth shorter than each corner tooth. Apical tooth of mandible not dentate ....*S.* (*S.*) *rashidi*
19. Accessory sclerite present (similar to Fig. 63J) ....*S.* (*S.*) *indicum*
Accessory sclerite absent 20
20. Postgenal cleft medium-long, 0.8 times as long as postgenal bridge .... *S.* (*S.*) *adventicium*
Postgenal cleft long, much longer than postgenal bridge 21
21. Rectal organ simple, without secondary lobules .... *S.* (*W.*) *pseudequinum*
Rectal organ compound, with secondary lobules 22
22. Postgenal cleft reaching posterior margin of hypostoma 23
Postgenal cleft not reaching posterior margin of hypostoma 24
23. Pharate pupal gill with three short filaments .... *S.* (*S.*) *asishi*
Pharate pupal gill with three bulbous filaments .... *S.* (*S.*) *nodosum*
24. Postgenal cleft longer than wide 25
Postgenal cleft nearly as long as or shorter than its greatest width 28
25. Postgenal cleft triangular, with its greatest width 0.6 times its length .... *S.* (*S.*) *rufibasis* and *S.* (*S.*) *ramosum*
Postgenal cleft spear-head-shaped, with its greatest width 0.7–0.9 times its length 26
26. Head without positive spots. Postgenal cleft 2.5 times as long as postgenal bridge ....*S.* (*S.*) *biforaminiferum*
Head with positive spots. Postgenal cleft 5.5–7.0 times as long as postgenal bridge 27
27. Postgenal cleft 7.0 times as long as postgenal bridge. Labral fan with ca. 35 primary rays ....*S.* (*S.*) *singtamense*
Postgenal cleft 5.5 times as long as postgenal bridge. Labral fan with ca. 50 primary rays ....*S.* (*S.*) *himalayense*
28. Pharate pupal gill with eight filaments ....*S.* (*S.*) *kapuri* and *S.* (*S.*) *dentatum*
Pharate pupal gill with 10 filaments 29
29. Five lower filaments of pharate pupal gill about half as thick as five upper filaments ....*S.* (*S.*) *yanaense*
All 10 filaments of pharate pupal gill subequal in thickness to one another ....*S.* (*S.*) *grisescens* and *S.* (*S.*) sp. nr. *grisescens*

* Nine unnamed species recorded from Assam by Lewis (1974) are not included because no detailed information for their morphological characteristics is available.

** *Simulium* (*G.*) *fidum*, *S.* (*G.*) *litoreum*, *S.* (*G.*) *metatarsale*, *S.* (*G.*) sp. nr. *litoreum*, *S.* (*G.*) ? *sundaicum*, *S.* (*M.*) *dattai*, *S.* (*M.*) *taktsangense*, *S.* (*M.*) sp. B, *S.* (*M.*) sp. nr. *ghoomense*, *S.* (*N.*) sp. A, *S.* (*N.*) sp. D, *S.* (*N.*) *senile*, *S.* sp. 2 (nr. *fruticosum*), *S.* (*N.*) sp., *S.* (*S.*) *biforaminiferum*, *S.* (*S.*) *nigrifacies*, *S.* (*S.*) *rashidi*, *S.* (*S.*) *singtamense*, *S.* (*S.*) *yanaense*, *S.* (*S.*) sp. nr. *grisescens*, *S.* (*S.*) sp. nr. *novolineatum*, and *S.* (*S.*) *ephemerophilum* are not included because their females are unknown. *Simulium* (*G.*) *decuplum* and *S.* (*G.*) *parahiyangum* are not included because their females are not known from India.

*** *S.* (*G.*) *bucolium*, *S.* (*G.*) *litoreum*, *S.* (*G.*) sp. nr. *litoreum*, *S.* (*M.*) *dattai*, *S.* (*M.*) *taktsangense*, *S.* (*M.*) sp. B, *S.* (*M.*) sp. nr. *ghoomense*, *S.* (*N.*) *subratai*, *S.* (*N.*) sp. A, *S.* (*N.*) sp. "A," *S.* (*N.*) sp. D, *S.* sp. 2 (nr. *fruticosum*), *S.* (*N.*) sp., *S.* (*S.*) *barnesi*, *S.* (*S.*) *biforaminiferum*, *S.* (*S.*) *kapuri*, *S.* (*S.*) *rashidi*, *S.* (*S.*) *singtamense*, *S.* (*S.*) sp. nr. *grisescens*, *S.* (*S.*) sp. nr. *novolineatum*, *S.* (*S.*) *ephemerophilum*, and *S.* (*S.*) *stevensoni* are not included because their males are unknown. *Simulium* (*G.*) ? *sundaicum* is not included because its male is briefly described based on the pharate male dissected out of a pupa. *Simulium* (*G.*) *decuplum*, *S.* (*G.*) *parahiyangum*, and *S.* (*G.*) *pattoni* are not included because their males are not known from India.

**** *Simulium* (*G.*) *metatarsale*, *S.* (*G.*) sp. nr. *litoreum*, *S.* (*M.*) *taktsangense*, *S.* (*N.*) *senile*, *S.* (*N.*) *rufithorax*, *S.* (*N.*) sp. "A", *S.* (*S.*) *barnesi*, *S.* (*S.*) *howletti*, *S.* (*S.*) sp. nr. *grisescens*, *S.* (*S.*) *ephemerophilum*, and *S.* (*S.*) *stevensoni* are not included because their pupae are unknown. *Simulium* (*G.*) *parahiyangum* and *S.* (*G.*) *pattoni* are not included because their pupae are not known from India.

***** *Simulium* (*G.*) *darjeelingense*, *S.* (*G.*) *metatarsale*, *S.* (*M.*) sp. C, *S.* (*N.*) *senile*, *S.* (*N.*) *rufithorax*, *S.* (*N.*) sp. "A," *S.* (*N.*) sp. D, *S.* sp. 2 (nr. *fruticosum*), *S.* (*N.*) sp., *S.* (*S.*) *christophersi*, *S.* (*S.*) *barnesi*, *S.* (*S.*) *barraudi*, *S.* (*S.*) *consimile*, *S.* (*S.*) *digitatum*, *S.* (*S.*) *gravelyi*. *S.* (*S.*) *griseifrons*, *S.* (*S.*) *hirtipannus*, *S.* (*S.*) *howletti*, *S.* (*S.*) *lineothorax*, *S.* (*S.*) *nigrifacies*, *S.* (*S.*) *nitidithorax*, *S.* (*S.*) *novolineatum*, *S.* (*S.*) *palniense*, *S.* (*S.*) *palmatum*, *S.* (*S.*) *stevensoni*, *S.* (*S.*) *tenuitarsus*, *S.* (*S.*) sp. nr. *novolineatum*, and *S.* (*S.*) *latistriatum* are not included because their larvae are unknown. *Simulium* (*G.*) ? *sundaicum* is not included because no detailed information is available for its larva. *Simulium* (*G.*) *decuplum*, *S.* (*G.*) *pattoni*, *S.* (*G.*) *sachini*, and *S.* (*M.*) *dattai* are not included because their larvae are not known from India. *Simulium* (*S.*) *pothigaiense* and *S.* (*S.*) *valparaiense* run up to couplet 23 but cannot run further due to the lack of information for the accurate shape and size of their postgenal clefts. *Simulium* (*G.*) *agasthyamalaiense*, *S.* (*G.*) *dinakarani*, *S.* (*G.*) *kottoorense*, *S.* (*G.*) *krishnani*, *S.* (*G.*) *kumbakkaraiense*, *S.* (*G.*) *panagudiense*, *S.* (*G.*) *peteri*, and *S.* (*G.*) *takaokai* run up to couplet 6 but cannot go further because of unreliable information on the size of their postgenal clefts.

2 The Fauna of Black Flies in Nepal

Overview

In Nepal, Lewis (1964) recorded three black fly species consisting of one new species, *Simulium* (*Gomphostilbia*) *nepalense*, and two unnamed species, *S.* sp. A and *S.* sp. B.

Maskey (1987, 1989, 1998) described three new species: *S.* (*Simulium*) *hillycum*, *S.* (*S.*) *bagmaticum*, and *S.* (*S.*) *vega* and recorded two unnamed species, *S.* (*S.*) sp. A and *S.* (*S.*) sp. B. Adler and Crosskey (2009) listed another nine known species: *S.* (*G.*) *bucolicum*, *S.* (*G.*) *siamense*, *S.* (*G.*) *varicorne*, *S.* (*Nevermannia*) *aureohirtum*, *S.* (*N.*) *praelargum*, *S.* (*S.*) *indicum*, *S.* (*S.*) *himalayense*, *S.* (*S.*) *oitanum*, and *S.* (*S.*) *singtamense*. Shrestha and Takaoka (2009) added 10 more species: *S.* (*G.*) *bhutanense*, *S.* (*G.*) *darjeelingense*, *S.* (*G.*) *gombakense*, *S.* (*Montisimulium*) *dattai*, *S.* (*M.*) *nemorivagum*, *S.* (*N.*) *gracile*, *S.* (*N.*) *purii*, *S.* (*S.*) *pradyai*, *S.* (*S.*) *rufibasis*, and *S.* (*S.*) *ramosum*, and identified *S.* sp. A recorded by Lewis (1964) as *S.* (*N.*) *aureohirtum*. Takaoka and Shrestha (2010) described 10 new species: *S.* (*A.*) *suchitrae*, *S.* (*Eusimulium*) *jomsomense*, *S.* (*G.*) *baglungense*, *S.* (*G.*) *butwalense*, *S.* (*G.*) *nuwakotense*, *S.* (*N.*) *lekhaniense*, *S.* (*N.*) *letense*, *S.* (*S.*) *titarense*, *S.* (*S.*) *tamorense*, *S.* (*S.*) *tulshii*, and recorded *S.* (*G.*) *sachini* [the species formerly misidentified as *S.* (*G.*) *gombakense* by Shrestha and Takaoka (2009)]. Takaoka et al. (2020a,b) described four new species of the *S. striatum* species-group: *S.* (*S.*) *beniense*, *S.* (*S.*) *chainpurense*, *S.* (*S.*) *dadagaunense*, and *S.* (*S.*) *salyanense*, and recorded three species of the same species-group: *S.* (*S.*) *grisescens*, *S.* (*S.*) *pallidum*, and *S.* (*S.*) *latistriatum* (at that time, as *S.* (*S.*) sp. nr. *striatum*). Takaoka et al. (2020f) described six new species: *S.* (*G.*) *dadeldhuraense*, *S.* (*G.*) *lamjungense*, *S.* (*G.*) *nareshi*, *S.* (*G.*) *shresthae*, *S.* (*G.*) *shivapuriense* [the species formerly regarded as *S.* (*G.*) *darjeelingense* by Shrestha and Takaoka (2009)], and *S.* (*M.*) *jumlaense*. Takaoka et al. (2020c) described three new species: *S.* (*G.*) *purandharaense*, *S.* (*N.*) *nivritiae*, and *S.* (*S.*) *phidimense*. Takaoka et al. (2020d) described three new species of the *S. multistriatum* species-group: *S.* (*S.*) *ratekholaense*, *S.* (*S.*) *maskeyae*, and *S.* (*S.*) *gokarnaense*, and identified *S.* sp. B recorded by Lewis (1964) as *S.* (*S.*) *barraudi*. Takaoka et al. (2020e) described four new species of the *S. tuberosum* species-group: *S.* (*S.*) *nakanoi*, *S.* (*S.*) *tatopaniense*, *S.* (*S.*) *dailekhense* [the species formerly misidentified as *S.* (*S.*) *ramosum* by Shrestha and Takaoka (2009)], and *S.* (*S.*) *langtangense* [the species formerly regarded as *S.* (*S.*) *rufibasis* by Shrestha and Takaoka (2009)]. *Simulium* (*S.*) *vega* was synonymized with *S.* (*S.*) *himalayense* in the 2008 version of World list of black flies (Adler, 2020). Adler et al. (2020b) described two new species of the *S. vernum* species-group: *S.* (*N.*) *gandakiense* and *S.* (*N.*) *magnacentrum*.

The fauna of black flies in Nepal is represented by 57 species including two unnamed species and five recorded species, of which seven species should be reconfirmed for their distribution records from Nepal. All the species are placed in six subgenera of the genus *Simulium* s. l., as shown in the following list.

List of 57 Species of Black Flies in Nepal

Subgenus *Asiosimulium*
S. suchitrae
Subgenus *Eusimulium*
S. jomsomense
Subgenus *Gomphostilbia*
S. asakoae species-group
S. lamjungense, *S. nareshi*, and *S. nepalense*
S. batoense species-group
S. baglungense, *S. bhutanense*, *S. bucolicum**, *S. butwalense*, *S. dadeldhuraense*, *S. purandharaense*, and *S. siamense* (complex)*
S. darjeelingense species-group
S. shresthae and *S. shivapuriense*
S. gombakense species-group
S. nuwakotense and *S. sachini*
S. varicorne species-group
*S. varicorne**
Subgenus *Montisimulium*
S. dattai, *S. jumlaense*, and *S. nemorivagum*
Subgenus *Nevermannia*
S. feuerborni species-group
S. praelargum (complex) and *S. nivritiae*
S. ruficorne species-group
S. aureohirtum
S. vernum species-group
S. gandakiense, *S. gracile*, *S. lekhaniense*, *S. letense*, *S. magnacentrum*, and *S. purii*
Subgenus *Simulium* s. str.
S. indicum species-group
S. indicum
S. malyschevi species-group
S. phidimense
S. multistriatum species-group
S. barraudi, *S. gokarnaense*, *S. hillycum*, *S. maskeyae*, and *S. ratekholaense*
S. striatum species-group
S. bagmaticum, *S. beniense*, *S. chainpurense*, *S. dadagaunense*, *S. grisescens*, *S. latistriatum*, *S. pallidum*, and *S. salyanense*
S. tuberosum species-group
S. dailekhense, *S. langtangense*, *S. nakanoi*, *S. pradyai*, and *S. tatopaniense*
S. variegatum species-group
S. himalayense, *S. oitanum**, and *S. tulshii*, *S.* sp. A, *S.* sp. B
Simulium s. str. species unplaced to species-group
*S. singtamense**, *S. tamorense*, and *S. titarense*

* The records of these five species from Nepal should be confirmed (Takaoka et al., 2020e).

Keys to Identify 50 Species of Black Flies in Nepal

Females**

1. Postnotum haired .... *S.* (*E.*) *jomsomense*
   Postnotum bare 2
2. Katepisternum haired 3
   Katepisternum bare 10
3. Claw with small basal tooth .... *S.* (*G.*) *bhutanense*
   Claw with large basal tooth 4
4. Hind tibia whitish yellow on little less than basal half or more 5
   Hind tibia whitish yellow on basal one-third or less 8
5. Hind tibia without dark subbasal marking 6
   Hind tibia with dark subbasal marking 7
6. Hind tibia whitish yellow on basal half. Claw with medium-long tooth, 0.4 times as long as claw (Fig. 36E) .... *S.* (*G.*) *sachini*
   Hind tibia whitish yellow on basal two-thirds. Claw with long tooth, 0.5 times as long as claw .... *S.* (*G.*) *nepalense*
7. Sensory vesicle short, 0.2 times as long as third palpal segment .... *S.* (*G.*) *lamjungense*
   Sensory vesicle long, 0.5 times as long as third palpal segment .... *S.* (*G.*) *nareshi*
8. Base of radius with tuft of yellow hairs .... *S.* (*G.*) *purandharaense*
   Base of radius with tuft of dark hairs 9
9. Sensory vesicle medium-long, 0.33–0.35 times as long as third palpal segment .... *S.* (*G.*) *dadeldhuraense*
   Sensory vesicle elongate, 0.5 times as long as third palpal segment .... *S.* (*G.*) *baglungense*
10. Frons moderately or densely covered with fine hairs 11
    Frons with several hairs along both lateral margins and near antennal base 19
11. Scutum ochreous 12
    Scutum dark brown to black 13
12. Cibarium with 11 tiny conical processes near ventral margin. Height of frons 2.1 times as long as its narrowest width .... *S.* (*N.*) *nivritiae*
    Cibarium with 73 tiny conical processes near ventral margin. Height of frons 2.6 times as long as its narrowest width .... *S.* (*N.*) *praelargum* (complex)

13. Fore tarsus with thick dorsal hair crest (similar to Fig. 62H) ..... *S.* (*S.*) *indicum*
    Fore tarsus with moderate dorsal hair crest 14
14. Claw with tiny to medium-long tooth 0.03–0.29 times as long as claw 15
    Claw with large tooth, 0.4–0.5 times as long as claw 17
15. Claw with tiny tooth 0.03 times as long as claw (similar to Fig. 42D) ....
    .... *S.* (*M.*) *jumlaense*
    Claw with medium-long tooth 0.25–0.29 times as long as claw (Figs. 42C and 47E) 16
16. Sensory vesicle medium-long, 0.41 times as long as third palpal segment (Fig. 47B) .... *S.* (*N.*) *letense*
    Sensory vesicle long, 0.65 times as long as third palpal segment ....
    .... *S.* (*M.*) *nemorivagum*
17. Fore coxa yellow .... *S.* (*N.*) *aureohirtum*
    Fore coxa brown 18
18. Third palpal segment not enlarged, with sensory vesicle short, 0.3 times as long as third palpal segment (Fig. 4C) .... *S.* (*A.*) *suchitrae*
    Third palpal segment enlarged, with sensory vesicle long, half as long as third palpal segment (similar to Fig. 47A) .... *S.* (*N.*) *gracile* and *S.* (*N.*) *purii*
19. Claw with small subbasal tooth (similar to Fig. 65B) 20
    Claw without tooth 22
20. Hind tibia darkened except base whitish. Sternite 7 with branched hairs (similar to Fig. 65C) .... *S.* (*S.*) *phidimense*
    Hind tibia yellow except apical cap darkened. Sternite 7 without branched hairs 21
21. Ovipositor valve extended posteromedially, longer than sternite 8, and with rounded apical tip (Fig. 85C) .... *S.* (*S.*) *tulshii*
    Ovipositor valve shorter than sternite 8, and with pointed apical tip (similar to Fig. 84C) .... *S.* (*S.*) *himalayense*
22. Scutum without longitudinal vittae 23
    Scutum with longitudinal vittae 27
23. Fore basitarsus greatly dilated, 3.7 times as long as its greatest width. Sternite 7 without cluster of stout hairs .... *S.* (*S.*) *tatopaniense*
    Fore basitarsus moderately dilated, 4.7–6.2 times as long as its greatest width. Sternite 7 with cluster of stout hairs (similar to Fig. 82C) 24
24. Cluster of stout hairs on sternite 7 slightly longer than those on surrounding area.
    .... *S.* (*S.*) *dailekhense*
    Cluster of stout hairs on sternite 7 2.0–2.5 times as long as those on surrounding area 25
25. Fore basitarsus greatly dilated, 4.7 times as long as its greatest width ....
    .... *S.* (*S.*) *langtangense*
    Fore basitarsus greatly dilated, 5.4–6.2 times as long as its greatest width 26
26. Fore basitarsus greatly dilated, 5.4 times as long as its greatest width ....
    .... *S.* (*S.*) *pradyai*

Fore basitarsus greatly dilated, 6.0–6.2 times as long as its greatest width .... *S.* (*S.*) *nakanoi*

27. Ovipositor valve without short process produced ventrally along inner margin (similar to Fig. 70E) 28
Ovipositor valve with short process produced ventrally along inner margin (Fig. 76D, E) 32
28. Mid tibia black .... *S.* (*S.*) *hillycum*
Mid tibia entirely yellow or yellow except apical cap darkened 29
29. Mid femur dark brown except extreme basal tip whitish yellow .... *S.* (*S.*) *ratekholaense*
Mid femur entirely yellow or yellow with apical tip greyish 30
30. Cibarium without median projection and with 10–20 tiny processes (Fig. 90A) .... *S.* (*S.*) *tamorense*
Cibarium with median projection and without tiny processes (similar to Fig. 70A) 31
31. Fore basitarsus 4.0 times as long as its greatest width .... *S.* (*S.*) *barraudi*
Fore basitarsus 5.4 times as long as its greatest width .... *S.* (*S.*) *maskeyae*
32. Fronto-ocular area poorly developed (similar to Fig. 74C) .... *S.* (*S.*) *dadagaunense*
Fronto-ocular area well developed 33
33. Hind tibia yellowish on basal one-third to two-thirds 34
Hind tibia light to dark brown except base yellowish white 36
34. Hind tibia with dark subbasal marking .... *S.* (*S.*) *chainpurense*
Hind tibia without dark subbasal marking 35
35. Middle tibia yellow .... *S.* (*S.*) *pallidum*
Middle tibia yellow to brownish yellow on basal half and brown to dark brown on rest .... *S.* (*S.*) *latistriatum*
36. Mid femur medium to dark brown .... *S.* (*S.*) *grisescens*
Mid femur entirely yellow or dark yellow to light brown except apical cap dark brown 37
37. Mid femur entirely yellow .... *S.* (*S.*) *beniense*
Mid femur dark yellow to light brown except apical cap dark brown .... *S.* (*S.*) *salyanense*

Males***

1. Style shorter than coxite (Figs. 19M and 45I) 2
Style longer than coxite (Fig. 90H) 16
2. Katepisternum bare 3
Katepisternum haired 8
3. Hind basitarsus not enlarged, dark yellow on basal half and medium to dark brown on rest (similar to Fig. 45H) .... *S.* (*N.*) *aureohirtum*

Hind basitarsus moderately enlarged, entirely darkened (Fig. 10H) 4
4. Paramere with one hook (Fig. 10M) 5
Paramere with several hooks 6
5. Ventral plate Y-shaped, with median keel when viewed ventrally (Fig. 10I) ....
....*S.* (*E.*) *jomsomense*
Ventral plate lamellate, without median keel when viewed ventrally (similar to Fig. 47M) .... *S.* (*N.*) *gracile* and *S.* (*N.*) *purii*
6. Upper-eye (large) facets in 12 vertical columns and 14 horizontal rows ....
.... *S.* (*M.*) *nemorivagum*
Upper-eye (large) facets in 16–18 vertical columns and 18–20 horizontal rows 7
7. Paramere with five hooks ....*S.* (*N.*) *praelargum* (complex)
Paramere with six hooks ....*S.* (*N.*) *nivritiae*
8. Fore coxa dark yellow to medium brown 9
Fore coxa whitish yellow 13
9. Hind basitarsus yellow on basal two-thirds and medium brown on rest ....
....*S.* (*G.*) *butwalense*
Hind basitarsus entirely darkened 10
10. Upper-eye (large) facets in 12 vertical columns and 14 horizontal rows ....
.... *S.* (*G.*) *shresthae*
Upper-eye (large) facets in 14–17 vertical columns and 15–18 horizontal rows 11
11. Upper-eye (large) facets in 14 vertical columns and 15 horizontal rows ....
....*S.* (*G.*) *shivapuriense*
Upper-eye (large) facets in 17 vertical columns and 18 horizontal rows 12
12. Fore basitarsus 7.2 times as long as its greatest width ...... *S.* (*G.*) *nuwakotense*
Fore basitarsus 7.6–7.8 times as long as its greatest width .... *S.* (*G.*) *sachini*
13. Hind tibia not enlarged, nearly parallel-sided ....*S.* (*G.*) *bhutanense*
Hind tibia enlarged, wedge-shaped 14
14. Hind tibia without dark subbasal marking. Ventral plate with ventral margin somewhat concave medially when viewed caudally ....*S.* (*G.*) *nepalense*
Hind tibia with dark subbasal marking. Ventral plate with ventral margin nearly straight or rounded medially when viewed caudally 15
15. Upper-eye (large) facets in 11 vertical columns and 13 horizontal rows ....
....*S.* (*G.*) *lamjungense*
Upper-eye (large) facets in 15 vertical columns and 16 horizontal rows ....
....*S.* (*G.*) *nareshi*
16. Basal portion of radius haired 17
Basal portion of radius bare 18
17. Fore tarsus with thick dorsal hair crest (similar to Fig. 62H) .....*S.* (*S.*) *indicum*
Fore tarsus with moderate dorsal hair crest ....*S.* (*S.*) *titarense*
18. Style without basal protuberance (Fig. 85J) 19
Style with short to long basal protuberance 20

19. Hind tibia whitish yellow except apical cap medium brown ....*S.* (*S.*) *tulshii*
Hind tibia medium to dark brown ....*S.* (*S.*) *himalayense*
20. Style with long basal protuberance with pointed apex (similar to Fig. 71C) 21
Style with short basal protuberance without pointed apex 25
21. Fore tibia whitish except apical cap darkened ....*S.* (*S.*) *hillycum*
Fore tibia light to dark brown with outer surface white medially 22
22. Hind basitarsus medium to dark brown with basal one-third light brown .... *S.* (*S.*) *ratekholaense*
Hind basitarsus yellow on basal half and medium to dark brown on rest 23
23. Style with its greatest width at apical one-third when viewed ventrolaterally. Ventral plate without teeth along posterior margin when viewed laterally .... *S.* (*S.*) *barraudi*
Style with its greatest width at base when viewed ventrolaterally. Ventral plater with teeth along posterior margin 24
24. Ventral plate with lateral margins constricted medially when viewed caudally .... *S.* (*S.*) *maskeyae*
Ventral plate with lateral margins nearly parallel-sided, without constriction medially when viewed caudally .... *S.* (*S.*) *gokarnaense*
25. Ventral plate saddle-shaped when viewed ventrally (Fig. 76P) 26
Ventral plate nearly quadrate when viewed ventrally (Fig. 78I) 30
26. Scutum covered with brassy short hairs 27
Scutum covered with yellow short hairs 28
27. Mid tibia medium to dark brown except basal tip yellow ...... *S.* (*S.*) *grisescens*
Mid tibia grayish yellow except apical tip dark gray .... *S.* (*S.*) *latistriatum*
28. Mid tibia yellow .... *S.* (*S.*) *pallidum*
Mid tibia light to dark brown or medium to dark brown except basal tip yellowish white 29
29. Fore basitarsus 6.3–6.5 times as long as its greatest width ... *S.* (*S.*) *salyanense*
Fore basitarsus 5.3–5.7 times as long as its greatest width .... *S.* (*S.*) *chainpurense*
30. Hind tibia yellowish white on basal half and medium to dark brown on rest. Upper-eye (large) facets in 13 or 14 vertical columns and 16 or 17 horizontal rows ....*S.* (*S.*) *nakanoi*
Hind tibia medium to dark brown except base yellowish white. Upper-eye (large) facets in 17–20 vertical columns and 18–20 horizontal rows 31
31. Fore basitarsus 5.4–6.2 times as long as its greatest width ....*S.* (*S.*) *dailekhense*
Fore basitarsus 3.2–4.2 times as long as its greatest width 32
32. Hind tibia medium to dark brown except basal two-fifths whitish yellow. Fore basitarsus 3.2–3.5 times as long as its greatest width .... *S.* (*S.*) *langtangense*

Hind tibia medium to dark brown except basal half whitish yellow. Fore basitarsus 4.2 times as long as its greatest width .... *S.* (*S.*) *tatopaniense*

Pupae****

1. Gill composed of one Y-shaped inflated structure with eight slender filaments (Fig. 37H, I) 2
   Gill composed of 4–28 slender filaments, or six filaments which are somewhat inflated at least on basal one-third 3
2. Dorsal arm of inflated structure of gill with one constriction (Fig. 37I) .... *S.* (*G.*) *sachini*
   Dorsal arm of inflated structure of gill with two constrictions (Fig. 37H) .... *S.* (*G.*) *nuwakotense*
3. Gill with four filaments (Fig. 10P) 4
   Gill with 6–28 filaments 8
4. Cocoon wall-pocket-shaped (Fig. 10Q) .... *S.* (*E.*) *jomsomense*
   Cocoon wall-pocket-shaped, with anterodorsal projection(s), or shoe-shaped, with anteroventral collar 5
5. Gill with black common basal stalk (Fig. 48F). Cocoon rounded, with anteroventral collar (Fig. 48J) .... *S.* (*N.*) *letense*
   Gill with brown common basal stalk. Cocoon with anterodorsal projection(s) 6
6. Cocoon with two anterodorsal projections .... *S.* (*N.*) *gracile*
   Cocoon with one anterodorsal projection 7
7. Anterodorsal projection of cocoon with large open space on basal half (Fig. 48K) .... *S.* (*N.*) *lekhaniense*
   Anterodorsal projection of cocoon without open space .... *S.* (*N.*) *purii*
8. Gill with six filaments 9
   Gill with 8–28 filaments 21
9. Cocoon shoe-shaped (similar to Fig. 63C) 10
   Cocoon wall-pocket-shaped 11
10. All gill filaments directed forward (similar to Fig. 63A). Cuticle of gill filaments without spinous processes .... *S.* (*S.*) *indicum*
    Four gill filaments directed forward and two filaments directed upward, then backward (Fig. 85Q, R). Cuticle of gill filaments with fine spinous processes (Fig. 85S) .... *S.* (*S.*) *tulshii*
11. Cocoon with anterodorsal projection 12
    Cocoon without anterodorsal projection 13
12. Six gill filaments arranged as 2+2+2 from dorsal to ventral and widely divergent (Fig. 46C) .... *S.* (*N.*) *aureohirtum*
    Six gill filaments arranged as 4+2 from dorsal to ventral and directed forward close together (similar to Fig. 44C) .... *S.* (*N.*) *praelargum* (complex)
13. Cocoon with large open space(s) anterolaterally on each side (Fig. 91I) 14
    Cocoon without open space anterolaterally on each side 15

14. Frons bare, without tubercles .... *S.* (*S.*) *phidimense*
    Frons densely covered with tubercles ....*S.* (*S.*) *titarense*
15. All gill filaments longer than pupal body and arranged as 4+2 from dorsal to ventral .... *S.* (*N.*) *nivritiae*
    All gill filaments shorter than pupal body and arranged as 2+2+2 from dorsal to ventral 16
16. Thorax with pit-like organ at base of gill (similar to Fig. 79H) .... *S.* (*S.*) *tatopaniense*
    Thorax without pit-like organ at base of gill 17
17. Frons and dorsal surface of anterior half of thorax bare (similar to Fig. 83A) 18
    Frons and dorsal surface of anterior half of thorax moderately or densely covered with tubercles (similar to Fig. 83B, C) 19
18. Stalk of ventral pair of gill longer than those of dorsal and middle pairs .... *S.* (*S.*) *pradyai*
    Stalk of ventral pair of gill nearly as long as those of dorsal and middle pairs (Fig. 83D)... *S.* (*S.*) *nakanoi*
19. Dorsomedial trichomes on thorax with three to five branches ....*S.* (*S.*) *dailekhense*
    Dorsomedial trichomes on thorax unbranched 20
20. Last abdominal segment without terminal hooks .... *S.* (*S.*) *langtangense*
    Last abdominal segment with terminal hooks ....*S.* (*S.*) *himalayense*
21. Gill with eight filaments 22
    Gill with 10–28 filaments 36
22. Eight gill filaments arranged as 3+3+2 or (3+3)+2 or 2+1+3+2 from dorsal to ventral. Last abdominal segment with grapnel-like hooklets 23
    Eight gill filaments arranged as 2+2+2+2 from dorsal to ventral. Last abdominal segment without grapnel-like hooklets 31
23. Gill with round inflated basal fenestra 24
    Gill without inflated basal fenestra 25
24. Eight gill filaments arranged as 3+3+2 from dorsal to ventral. Round inflated basal fenestra twice as thick as interspiracular trunk ....*S.* (*G.*) *baglungense*
    Eight gill filaments arranged as 2+1+3+2 from dorsal to ventral. Round inflated basal fenestra three times as thick as interspiracular trunk .... *S.* (*G.*) *dadeldhuraense*
25. Two filaments of ventral pair of gill subequal in thickness to each other 26
    Two filaments of ventral pair of gill different in thickness from each other 29
26. Eight gill filaments arranged as 3+3+2 from dorsal to ventral (dorsal and middle triplets not sharing stalk) ....*S.* (*G.*) *bhutanense*
    Eight gill filaments arranged as (3+3)+2 from dorsal to ventral (dorsal and middle triplets sharing stalk) 27

27. Terminal hooks triangular, with outer margins not crenulated .... *S.* (*G.*) *purandharaense*
    Terminal hooks wide, plate-like, with outer margins crenulated 28
28. Terminal hooks each with outer margin 1.4 times as long as inner margin .... *S.* (*G.*) *nareshi*
    Terminal hooks each with outer margin 2.7–3.0 times as long as inner margin .... *S.* (*G.*) *lamjungense*
29. Stalk of ventral pair of gill shorter than common basal stalk .... *S.* (*G.*) *shivapuriense*
    Stalk of ventral pair of gill longer than common basal stalk 30
30. Eight gill filaments arranged as (3+3)+2 from dorsal to ventral (Fig. 29E) .... *S.* (*G.*) *shresthae*
    Eight gill filaments arranged as (2+1+3)+2 from dorsal to ventral .... *S.* (*G.*) *butwalense*
31. Cocoon shoe-shaped (Fig. 91K, L) .... *S.* (*S.*) *tamorense*
    Cocoon wall-pocket-shaped with small to large open space(s) anterolaterally on each side 32
32. Cocoon with few small open spaces along anterolateral margin on each side .... *S.* (*S.*) *hillycum*
    Cocoon with at least one large open space anterolaterally on each side 33
33. Stalk of dorsal pair and that of ventral pair of gill divergent basally at right angle when viewed laterally 34
    Stalk of dorsal pair and that of ventral pair of gill divergent basally at obtuse angle when viewed laterally 35
34. Frontal trichomes subequal in length to each other .... *S.* (*S.*) *ratekholaense*
    Frontal trichomes different in length from each other ....*S.* (*S.*) *barraudi*
35. Frontal trichomes subequal in length to each other .... *S.* (*S.*) *maskeyae*
    Frontal trichomes different in length from each other .... *S.* (*S.*) *gokarnaense*
36. Gill with 10 filaments 37
    Gill with 12–28 filaments 40
37. Last abdominal segment without terminal hooks. Four or five filaments of gill somewhat inflated basally, 2.0–3.0 times as thick as other slender filaments when compared basally .... *S.* (*S.*) *bagmaticum*
    Last abdominal segment with terminal hooks. All filaments of gill subequal in thickness to one another, or some filaments 1.5–2.0 times as thick as other filaments when compared basally 38
38. Two filaments of ventral pair of gill 0.5–0.7 times as thick as two filaments of dorsal pair .... *S.* (*S.*) *beniense*, *S.* (*S.*) *dadagaunense*, and *S.* (*S.*) *salyanense*
    Two filaments of ventral pair of gill as thick as or slightly thicker than two filaments of dorsal pair 39
39. Two middle triplets of gill filaments almost entirely overlapping basally .... *S.* (*S.*) *pallidum*

Two middle triplets of gill filaments not (or partially) overlapping basally ....
..........................*S.* (*S.*) *chainpurense, S.* (*S.*) *latistriatum*, and *S.* (*S.*) *grisescens*

40. Gill with 12 filaments (similar to Fig 42S) .... *S.* (*M.*) *nemorivagum*
Gill with 14 or 28 filaments 41

41. Gill with 14 filaments, without enlarged basal fenestra .... *S.* (*M.*) *dattai*
Gill with 28 filaments, with enlarged basal fenestra (Fig. 6B) ....
....*S.* (*A.*) *suchitrae*

Mature larvae*****

1. Ventral papillae present 2
Ventral papillae absent 13

2. Postgenal cleft extremely small, vestigial (similar to Fig. 42X) ....
.... *S.* (*M.*) *nemorivagum*
Postgenal cleft over 0.29 times as long as postgenal bridge 3

3. Postgenal cleft 0.29–0.32 times as long as postgenal bridge (Fig. 42Y) 4
Postgenal cleft over 0.5 times as long as postgenal bridge 5

4. Abdomen with reddish brown markings dorsally ....
....*S.* (*N.*) *praelargum* (complex)
Abdomen without reddish brown markings dorsally .... *S.* (*M.*) *dattai*

5. Postgenal cleft 0.5–0.7 times as long as postgenal bridge 6
Postgenal cleft over 2.6 times as long as postgenal bridge 12

6. Accessory sclerite present (similar to Fig. 48P) 7
Accessory sclerite absent 8

7. Head capsule yellowish. Body ocherous .... *S.* (*N.*) *gandakiense*
Head capsule light to medium brown. Body dark gray ....
.... *S.* (*N.*) *magnacentrum*

8. Mandibular serration composed of two large teeth (similar to Fig. 46I) ....
....*S.* (*N.*) *aureohirtum*
Mandibular serration composed of one large tooth and one small tooth 9

9. Postgenal cleft nearly as long as postgenal bridge 10
Postgenal cleft 0.7 times as long as postgenal bridge 11

10. Common basal stalk of pharate pupal gill black ....*S.* (*N.*) *letense*
Common basal stalk of pharate pupal gill not black .... *S.* (*N.*) *lekhaniense*

11. Common basal stalk of pharate pupal gill short ....*S.* (*N.*) *gracile*
Common basal stalk of pharate pupal gill medium-long .... *S.* (*N.*) *puri*

12. Postgenal cleft 2.6 times as long as postgenal bridge .... *S.* (*G.*) *sachini*
Postgenal cleft reaching posterior margin of hypostoma ...... *S.* (*G.*) *butwalense*

13. Postgenal cleft reaching posterior margin of hypostoma .... *S.* (*A.*) *suchitrae*
Postgenal cleft not reaching posterior margin of hypostoma 14

14. Accessory sclerite present (similar to Fig. 63J) ....*S.* (*S.*) *indicum*
Accessory sclerite absent 15

15. Rectal organ simple, without secondary lobules ....*S.* (*S.*) *tulshii*
    Rectal organ compound, with secondary lobules 16
16. Pharate pupal gill with six filaments ....
    .... *S.* (*S.*) *himalayense* and *S.* (*S.*) *singtamense*
    Pharate pupal gill with eight or 10 filaments 17
17. Pharate pupal gill with 10 filaments 18
    Pharate pupal gill with eight filaments 19
18. Four of five filaments of pharate pupal gill much thicker than other filaments ....
    .... *S.* (*S.*) *bagmaticum*
    All filaments of pharate pupal gill subequal in thickness to one another ....
    .... *S.* (*S.*) *chainpurense* and *S.* (*S.*) *grisescens*
19. Abdominal segments 1–7 each with a pair of round protuberances dorsally (Fig. 91M) .... *S.* (*S.*) *tamorense*
    Abdominal segments1–7 without such protuberances dorsally ....
    .... *S.* (*S.*) *hillycum*

*Keys provided by Takaoka and Shrestha (2010) are here revised. Five of nine species recorded by Adler and Crosskey (2009): *S.* (*G.*) *bucolicum*, *S.* (*G.*) *siamense* (complex), *S.* (*G.*) *varicorne*, *S.* (*S.*) *oitanum*, and *S.* (*S.*) *singtamense*, are not included in the present keys because their records from Nepal are not validated (Takaoka et al., 2020e). The two unnamed species, (*S.*) sp. A and *S.* (*S.*) sp. B, reported by Maskey (1998), are also not included because of the lack of information for their morphological characteristics.

*******Simulium* (*G.*) *butwalense*, *S.* (*M.*) *dattai*, *S.* (*N.*) *gandakiense*, *S.* (*S.*) *gokarnaense*, *S.* (*N.*) *lekhaniense*, *S.* (*N.*) *magnacentrum*, *S.* (*G.*) *nuwakotense*, *S.* (*G.*) *shresthae*, and *S.* (*G.*) *shivapuriense* are not included because their females are unknown.

******Simulium* (*A.*) *suchitrae*, *S.* (*G.*) *baglungense*, *S.* (*G.*) *dadeldhuraense*, *S.* (*G.*) *purandharaense*, *S.* (*M.*) *dattai*, *S.* (*M.*) *jumlaense*, *S.* (*N.*) *gandakiense*, *S.* (*N.*) *lekhaniense*, *S.* (*N.*) *letense*, *S.* (*N.*) *magnacentrum*, *S.* (*S.*) *beniense*, and *S.* (*S.*) *dadagaunense* are not included because their males are unknown.

********Simulium* (*G.*) *nepalense*, *S.* (*M.*) *jumlaense*, *S.* (*N.*) *gandakiense*, and *S.* (*N.*) *magnacentrum* are not included because their pupae are unknown.

**********Simulium* (*E.*) *jomsomense*, *S.* (*G.*) *baglungense*, *S.* (*G.*) *bhutanense*, *S.* (*G.*) *dadeldhuraense*, *S.* (*G.*) *nareshi*, *S.* (*G.*) *nepalense*, *S.* (*G.*) *nuwakotense*, *S.* (*G.*) *purandharaense*, *S.* (*G.*) *shivapuriense*, *S.* (*G.*) *shresthae*, *S.* (*M.*) *jumlaense*, *S.* (*N.*) *nivritiae*, *S.* (*S.*) *barraudi*, *S.* (*S.*) *beniense*, *S.* (*S.*) *dadagaunense*, *S.* (*S.*) *dailekhense*, *S.* (*S.*) *gokarnaense*, *S.* (*S.*) *langtangense*, *S.* (*S.*) *maskeyae*, *S.* (*S.*) *nakanoi*, *S.* (*S.*) *pallidum*, *S.* (*S.*) *phidimense*, *S.* (*S.*) *pradyai*, *S.* (*S.*) *ratekholaense*, *S.* (*S.*) *salyanense*, *S.* (*S.*) *latistriatum*, *S.* (*S.*) *tatopaniense*, and *S.* (*S.*) *titarense* are not included because their larvae are unknown.

3 The Fauna of Black Flies in Bhutan

Overview

Lewis (1974) recorded *Simulium* (*Simulium*) *indicum* (at that time, under the subgenus *Himalayum*) from Bhutan. Datta (1991a) recorded five species: *S.* (*Nevermannia*) *aureohirtum*, *S.* (*S.*) *biforaminiferum*, *S.* (*S.*) *dentatum*, *S.* (*S.*) *grisescens*, and *S.* (*S.*) *striatum.*

Takaoka and Somboon (2008) described 11 new species including *S.* (*Montisimulium*) *dattai* [the species formerly reported as *S.* (*Eusimulium*) sp. C by Datta et al. (1975)] and recorded *S.* (*M.*) *nemorivagum*. Takaoka et al. (2021c) described a new species, *S. thimphuense* and recorded four species: *S.* (*S.*) *barraudi*, *S.* (*S.*) *chiangmaiense*, *S.* (*S.*) *nodosum*, and *S.* (*S.*) *himalayense*.

Takaoka et al. (2020a) pointed out that the species formerly regarded as *S.* (*S.*) *striatum* in India by Puri (1932e) is different from *S.* (*S.*) *striatum* from Sri Lanka (type locality). For this reason, the species regarded as *S.* (*S.*) *striatum* in Bhutan by Datta (1991a) was treated as *S.* (*S.*) sp. nr. *striatum*, which is here regarded as *S. latistriatum*.

The fauna of black flies in Bhutan is represented by 23 species, of which two are placed in the subgenus *Gomphostilbia*, three are in the subgenus *Montisimulium*, four are in the subgenus *Nevermannia*, and 14 are in the subgenus *Simulium* s. str. (Takaoka and Somboon, 2008; Takaoka et al., 2021c).

List of 23 Species of Black Flies in Bhutan

Subgenus *Gomphostilbia*
S. batoense species-group
S. bhutanense, and *S. chuzargangense*
Subgenus *Montisimulium*
S. dattai, *S. nemorivagum*, and *S. taktsangense*
Subgenus *Nevermannia*
S. feuerborni species-group
S. mongarense
S. ruficorne species-group
S. aureohirtum
S. vernum species-group
S. bumthangense and *S. thrimshinglaense*
Subgenus *Simulium* s. str.
S. indicum species-group
S. indicum
S. multistriatum species-group
S. barraudi, *S. demolaense*, *S. dentatum*, *S. deothangense,* and *S. thimphuense*

S. nobile species-group
S. nodosum
S. striatum species-group
S. chiangmaiense, *S. grisescens*, and *S. latistriatum*
S. tuberosum species-group
S. pradyai, and *S. rangjungense*
S. variegatum species-group
S. himalayense
Simulium s. str. species unplaced to species-group
S. biforaminiferum

Keys to Identify 23 Species of Black Flies in Bhutan

Females*

1. Katepisternum haired 2
   Katepisternum bare 3
2. Claw with medium-long tooth 0.31 times as long as claw .... *S.* (*G.*) *bhutanense*
   Claw with long tooth 0.52 times as long as claw .... *S.* (*G.*) *chuzargangense*
3. Claw with medium-long or long tooth 4
   Claw simple or with small tooth 6
4. Claw with medium-long tooth 0.29 times length of claw (similar to Fig. 42C). .... *S.* (*M.*) *nemorivagum*
   Claw with long tooth half as long as claw 5
5. Antenna yellow except first flagellomere darkened .... *S.* (*N.*) *aureohirtum*
   Antenna entirely dark brown .... *S.* (*N.*) *thrimshinglaense*
6. Claw with small tooth (similar to Fig. 62C) 7
   Claw without tooth 8
7. Frons densely covered with fine hairs. Basal portion of radius haired .... *S.* (*S.*) *indicum*
   Frons almost bare. Basal portion of radius bare .... *S.* (*S.*) *himalayense*
8. Scutum with longitudinal vittae 9
   Scutum without longitudinal vittae 13
9. Ovipositor valve without ventrally produced lobe near its inner margin 10
   Ovipositor valve with ventrally produced lobe near its inner margin (similar to Fig. 76D, E) 12
10. Ovipositor valve with five to nine hairs .... *S.* (*S.*) *dentatum*
    Ovipositor valve with two or three hairs 11
11. Mid femur whitish yellow on basal half or little more and light to medium brown on rest .... *S.* (*S.*) *deothangense*
    Mid femur yellow except apical tip grayish .... *S.* (*S.*) *barraudi*

12. Basal portion of radius haired ...... *S.* (*S.*) *chiangmaiense* and *S.* (*S.*) *grisescens*
    Basal portion of radium bare .... *S.* (*S.*) *latistriatum*
13. Ovipositor valve well sclerotized, with several stout long hairs (Fig. 74I) ....
    .... *S.* (*S.*) *nodosum*
    Ovipositor valve membranous, with few slender short hairs 14
14. Scutum with whitish yellow short hairs. Sternite 7 with a pair of stout clustered hairs (similar to Fig. 82C) .... *S.* (*S.*) *pradyai*
    Scutum with dark brown short hairs. Sternite 7 without a pair of stout clustered hairs .... *S.* (*S.*) *rangjungense*

Males**

1. Katepistenum with hairs ....*S.* (*G.*) *bhutanense*
   Katepisternum without hairs 2
2. Basal portion of radius haired 3
   Basal portion of radium bare 8
3. Hind basitarsus slender (similar to Fig. 62I) 4
   Hind basitarsus enlarged (similar to Fig. 43J) 5
4. Ventral plate with median keel (similar to Fig. 45K) ....*S.* (*N.*) *aureohirtum*
   Ventral plate without median keel (similar to Fig. 62M) ....*S.* (*S.*) *indicum*
5. Paramere with one hook (Fig. 47Q) 6
   Paramere with five or six hooks (similar to Fig. 43P) 7
6. Upper-eye (large) facets in 20 vertical columns and 23 horizontal rows ....
   ....*S.* (*N.*) *thrimshinglaense*
   Upper-eye (large) facets in 16 or 17 vertical columns and 17 horizontal rows ....
   ....*S.* (*N.*) *bumthangense*
7. Upper-eye (large) facets in 12 vertical columns and 14 horizontal rows ....
   .... *S.* (*M.*) *nemorivagum*
   Upper-eye (large) facets in 18 vertical columns and 20 horizontal rows ....
   ....*S.* (*N.*) *mongarense*
8. Scutum white pruinose with inverted V-shaped non-pruinose marking (similar to Fig. 74S, T) .... *S.* (*S.*) *nodosum*
   Scutum white pruinose on shoulders, along lateral margins and on prescutellar area 9
9. Scutum with brassy short hairs 10
   Scutum with whitish yellow or yellow short hairs 12
10. Ventral plate quadrate when viewed ventrally (similar to Fig. 80L), with teeth on each posteolateral margin ....*S.* (*S.*) *rangjungense*
    Ventral plate saddle-shaped when viewed ventrally (similar to Fig. 76P), without teeth 11
11. Mid femur grayish yellow except apical tip dark gray ....*S.* (*S.*) *latistriatum*
    Mid femur medium to dark brown ....
    ....*S.* (*S.*) *chiangmaiense* and *S.* (*S.*) *grisescens*
12. Style without basal protuberance ....*S.* (*S.*) *himalayense*
    Style with long basal protuberance (similar to Fig. 71C) 13

13. Style with its greatest width at apical one-third .... *S.* (*S.*) *barraudi*
Style with its greatest width at or near base 14
14. Ventral plate with many setae on ventral surface ....*S.* (*S.*) *deothangense*
Ventral plate almost bare 15
15. Hind basitarsus little wider than hind tibia. Ventral plate with two vertical rows of distinct teeth on posterior surface ....*S.* (*S.*) *dentatum*
Hind basitarsus 0.9 times as wide as hind tibia. Ventral plate with weak teeth irregularly on posterior surface ....*S.* (*S.*) *demolaense*

Pupae***

1. Gill with three inflated tubular filaments (Fig. 75F) ....*S.* (*S.*) *nodosum*
Gill with four, six, eight, 10, 12, or 14 filaments 2
2. Gill with four slender filaments (Fig. 48E) 3
Gill with six, eight, 10, 12, or 14 filaments 4
3. Gill filaments shorter than pupal body (Fig. 48E) ....*S.* (*N.*) *thrimshinglaense*
Gill filaments longer than pupal body ....*S.* (*N.*) *bumthangense*
4. Gill with six filaments 5
Gill with eight, 10, 12, or 14 filaments 11
5. Cocoon shoe-shaped (similar to Fig. 63C) ....*S.* (*S.*) *indicum*
Cocoon wall-pocket-shaped 6
6. Cocoon with short to long anterolateral projection 7
Cocoon without anterolateral projection 8
7. Gill filaments arranged as 2+2+2 from dorsal to ventral (Fig. 46C) ....
....*S.* (*N.*) *aureohirtum*
Gill filaments arranged as 4+2 from dorsal to ventral ....*S.* (*N.*) *mongarense*
8. Cocoon with anterolateral open space on each side ......*S.* (*S.*) *biforaminiferum*
Cocoon without anterolateral open space on each side 9
9. Frons almost bare .... *S.* (*S.*) *pradyai*
Frons with tubercles 10
10. Stalks of dorsal and ventral pairs of gill divergent at ca. 90 degrees when viewed laterally.. *S.* (*S.*) *himalayense*
Stalks of dorsal and ventral pairs of gill divergent at ca. 50 degrees when viewed laterally ....*S.* (*S.*) *rangjungense*
11. Gill with eight filaments 12
Gill with 10, 12, or 14 filaments 16
12. Gill filaments arranged as 3+3+2 from dorsal to ventral. Cocoon without anterolateral open space on each side ....*S.* (*G.*) *bhutanense*
Gill filaments arranged as 2+2+2+2 from dorsal to ventral. Cocoon wall-pocket-shaped, with anterolateral open space(s) on each side, or shoe-shaped, with small open spaces anteriorly on each side 13
13. Two filaments of ventral pair of gill much thinner than six other filaments (Fig. 77E) ....*S.* (*S.*) *chiangmaiense*
All eight filaments subequal in thickness to one another 14

14. Stalks of dorsal and ventral pairs of gill divergent at angle of 40–80 degrees when viewed laterally (Fig. 72F) .... S. (S.) *thimphuense*
    Stalks of dorsal and ventral pairs divergent at angle of 90 degrees or more when viewed when viewed laterally 15
15. Stalks of dorsal and ventral pairs divergent at angle of ca. 90 degrees when viewed laterally (Fig. 72A) ....S. (S.) *barraudi* and S. (S.) *demolaense*
    Stalks of dorsal and ventral pairs divergent at angle of ca. 120 degrees.............
    .... S. (S.) *dentatum* and S. (S.) *deothangense*
16. Gill with 10 filaments ....S. (S.) *grisescens* and S. (S.) *latistriatum*
    Gill with 12 or 14 filaments 17
17. Gill with 12 filaments .... S. (M.) *nemorivagum*
    Gill with 14 filaments 18
18. Gill with long common basal stalk ....S. (M.) *taktsangense*
    Gill with medium-long common basal stalk .... S. (M.) *dattai*

Mature larvae****

1. Ventral papillae present (similar to Fig 48P) 2
   Ventral papillae absent 7
2. Pharate pupal gill with four filaments ....S. (N.) *thrimsinglaense*
   Pharate pupal gill with six, 12, or 14 filaments 3
3. Pharate pupal gill with six filaments 4
   Pharate pupal gill with 12 or 14 filaments 5
4. Abdomen with distinct reddish-brown markings on dorsal surface (similar to Fig. 44K, L) .... S. (N.) *mongarense*
   Abdomen without reddish-brown markings on dorsal surface ....
   .... S. (N.) *aureohirtum*
5. Pharate pupal gill with 12 filaments .... S. (M.) *nemorivagum*
   Pharate pupal gill with 14 filaments 6
6. Postgenal cleft small, quadrate or M-shaped, 0.29–0.34 times as long as postgenal bridge (Fig. 42Y) ....S. (M.) *dattai*
   Postgenal cleft extremely small and vestigial (similar to Fig. 42X) ....
   .... S. (M.) *taktsangense*
7. Last abdominal segment with accessory sclerite (similar to Fig. 63J) ....
   ....S. (S.) *indicum*
   Last abdominal segment without accessory sclerite 8
8. Postgenal cleft long, reaching posterior margin of hypostoma (similar to Fig. 75N) .... S. (S.) *nodosum*
   Postgenal cleft long, but not reaching posterior margin of hypostoma 9
9. Pharate pupal gill with six filaments 10
   Pharate pupal gill with eight or 10 filaments 11
10. Postgenal cleft subtriangular ....S. (S.) *biforaminiferum*
    Postgenal cleft rounded...S. (S.) *himalayense*

11. Pharate pupal gill with 10 filaments. Abdominal segments 1–8 each with a pair of cone-shaped dorsolateral protuberances (similar to Fig. 77L)....................... .... *S.* (*S.*) *grisescens*
    Pharate pupal gill with eight filaments. Abdominal segments 1–8 each without a pair of cone-shaped dorsolateral protuberances 12
12. Body length 4.5 mm. Labral fan with 48 primary rays ...... *S.* (*S.*) *thimphuense*
    Body length ca. 6.5 mm. Labral fan with 56 primary rays .... *S.* (*S.*) *dentatum*

**Simulium* (*M.*) *dattai*, *S.* (*M.*) *taktsangense*, *S.* (*N.*) *mongarense*, *S.* (*N.*) *bumthangense*, *S.* (*S.*) *biforaminiferum*, *S.* (*S.*) *demolaense*, and *S.* (*S.*) *thimphuense* are not included because their females are unknown.

** *Simulium* (*G.*) *chuzargangense*, *S.* (*M.*) *dattai*, *S.* (*M.*) *taktsangense*, *S.* (*S.*) *biforaminiferum*, *S.* (*S.*) *thimphuense*, and *S.* (*S.*) *pradyai* are not included because their males are unknown.

*** *Simulium* (*G.*) *chuzargangense* is not included because its pupa is unknown.

*****Simulium* (*G.*) *bhutanense*, *S.* (*G.*) *chuzargangense*, *S.* (*N.*) *bumthangense*, *S.* (*S.*) *barraudi*, *S.* (*S.*) *chiangmaiense*, *S.* (*S.*) *demolaense*, *S.* (*S.*) *deothangense*, *S.* (*S.*) *rangjungense*, *S.* (*S.*) *latistriatum*, and *S.* (*S.*) *pradyai* are not included because their larvae are unknown.

4 The Fauna of Black Flies in Myanmar

Overview

From Myanmar, Lewis (1974) first recorded *Simulium* (*Simulium*) *indicum*. Takaoka (1989b) described three new species: *S.* (*Gomphostilbia*) *namense*, *S.* (*Montisimulium*) *kambaitense* and *S.* (*Nevermannia*) *burmense*, and recorded four species: *S.* (*N.*) sp. A, *S.* (*S.*) *chamlongi*, *S.* (*S.*) *novolineatum*, and *S.* (*S.*) *rufibasis*, based on adult specimens preserved at British Museum Natural History.

Takaoka et al. (2017g, 2018j) described two new species: *S.* (*G.*) *myanmarense* and *S.* (*G.*) *monglaense*, and recorded 14 species: *S.* (*Asiosimulium*) sp., *S.* (*G.*) *asakoae*, *S.* (*G.*) *decuplum*, *S.* (*G.*) *dentistylum*, *S.* (*G.*) *lurauense*, *S.* (*N.*) *aureohirtum*, *S.* (*S.*) *nigrogilvum*, *S.* (*S.*) *laui*, *S.* (*S.*) *barraudi*, *S.* (*S.*) *nodosum*, *S.* (*S.*) *chiangmaiense*, *S.* (*S.*) *nakhonense*, *S.* (*S.*) *xuandai*, and *S.* (*S.*) *himalayense*. Takaoka et al. (2017h) described a new species, *S.* (*Asiosimulium*) *shanense* [the species formerly reported as *S.* (*A.*) sp.]. Takaoka et al. (2018c) recorded five more species: *S.* (*G.*) *gombakense*, *S.* (*G.*) *siamense* (complex), *S.* (*S.*) *chanyae*, *S.* (*S.*) *nigrifacies*, and *S.* (*S.*) *tani* (complex).

The fauna of black flies in Myanmar is represented by 29 species, of which one species is placed in the subgenus *Asiosimulium*, nine species are in the subgenus *Gomphostilbia*, one species is in the subgenus *Montisimulium*, three species are in the subgenus *Nevermannia*, and 15 species are in the subgenus *Simulium* s. str.

List of 29 Species of Black Flies in Myanmar

Subgenus *Asiosimulium*
S. shanense
Subgenus *Gomphostilbia*
S. asakoae species-group
S. asakoae, *S. lurauense*, *S. myanmarense*, and *S. monglaense*
S. batoense species-group
S. decuplum, *S. dentistylum*, and *S. siamense* (complex)
S. ceylonicum species-group
S. namense
S. gombakense species-group
S. gombakense
Subgenus *Montisimulium*
S. kambaitense
Subgenus *Nevermannia*
S. feuerborni species-group
S. sp. A
S. ruficorne species-group
S. aureohirtum
S. vernum species-group
S. burmense
Subgenus *Simulium* s. str.
S. indicum species-group
S. indicum and *S. nigrogilvum*
S. multistriatum species-group
S. barraudi, *S. chanyae*, *S. laui*, and *S. novolineatum*
S. nobile species-group
S. nodosum
S. striatum species-group
S. chiangmaiense, *S. nakhonense*, and *S. xuandai*
S. tuberosum species-group
S. nigrifacies, *S. rufibasis*, and *S. tani* (complex)
S. variegatum species-group
S. chamlongi and *S. himalayense*

Keys to Identify 29 Species of Black Flies in Myanmar

Females*

1. Katepisternum haired 2
   Katepisternum bare 9

2. Base of radius with tuft of dark hairs 3
   Base of radius with tuft of yellow hairs 4
3. Sensory vesicle long, 0.54 times as long as third palpal segment ....
   .... *S.* (*G.*) *decuplum*
   Sensory vesicle medium-long, 0.37 times as long as third palpal segment.........
   .... *S.* (*G.*) *siamense* (complex)
4. Sensory vesicle long, 0.50–0.61 times as long as third palpal segment 5
   Sensory vesicle medium-long, 0.15–0.40 times as third palpal segment 6
5. Frons:head ratio 1.0:4.2 .... *S.* (*G.*) *lurauense*
   Frons:head ratio 1.0:5.2–5.3 .... *S.* (*G.*) *gombakense*
6. Mandible with three to six outer teeth 7
   Mandible without outer teeth 8
7. Sensory vesicle 0.3–0.4 times as long as third palpal segment ....
   ....*S.* (*G.*) *asakoae*
   Sensory vesicle 0.25 times as long as third palpal segment ..... *S.* (*G.*) *namense*
8. Frontal ratio 1.5:1.0:1.9 ....*S.* (*G.*) *myanmarense*
   Frontal ratio 1.9:1.0:2.3–2.6 ....*S.* (*G.*) *monglaense*
9. Claw with medium-sized or large basal tooth 10
   Claw with small subbasal tooth or without any tooth 13
10. Hind tibia yellow on basal half and darkened on rest, and with dark subbasal marking (Fig. 45B)... *S.* (*N.*) *aureohirtum*
    Hind tibia dark brown though medial portion light brown 11
11. Mandible with 13 outer teeth. Claw with long basal tooth half as long as claw
    .... *S.* (*N.*) *burmense*
    Mandible without outer teeth. Claw with medium-sized basal tooth 0.37–0.40 times as long as claw 12
12. Fore basitarsus 7.4 times as long as its greatest width. Ovipositor valves tongue-like, widely separated basally from each other, and with inner margins converged apically .... *S.* (*M.*) *kambaitense*
    Fore basitarsus 8.5 times as long as its greatest width. Ovipositor valves triangular, narrowly separated from each other, and with inner margins parallel-sided ....*S.* (*N.*) sp. A
13. Claw with small subbasal tooth 14
    Claw without any tooth 15
14. Frons densely covered with fine hairs. Basal portion of radius haired..............
    ....*S.* (*S.*) *indicum* and *S.* (*S.*) *nigrogilvum*
    Frons with several stout hairs along each lateral margin. Basal portion of radius bare...*S.* (*S.*) *himalayense*
15. Scutum without longitudinal vittae 16
    Scutum with longitudinal vittae 18
16. Sternite 7 with a pair of tuft hairs (similar to Fig. 82C) ....*S.* (*S.*) *rufibasis*
    Sternite 7 without a pair of tuft hairs 17

17. Ovipositor valves strongly sclerotized, each with several stout hairs apically (Fig. 74I) .... *S.* (*S.*) *nodosum*
    Ovipositor valves triangular, membranous, each without stout hairs apically .... *S.* (*S.*) *tani* (complex)
18. Ovipositor valves each with ventrally produced lobe along inner margin (Fig. 76D, E) .... *S.* (*S.*) *chiangmaiense* and *S.* (*S.*) *nakhonense*
    Ovipositor valves each without ventrally produced lobe 19
19. Hind tibia dark brown except base yellowish .... *S.* (*S.*) *laui*
    Hind tibia yellow on basal three-fourths 20
20. Mid basitarsus yellow on basal half .... *S.* (*S.*) *novolineatum*
    Mid basitarsus yellow on basal three-fourths or two-thirds 21
21. Mid femur yellow except apical tip grayish .... *S.* (*S.*) *barraudi*
    Mid femur yellow on basal one-third or little more and light brown on rest .... *S.* (*S.*) *chanyae*

Males**

1. Katepisternum haired 2
   Katepisternum bare 9
2. Hind basitarsus slender, nearly parallel-sided 3
   Hind basitarsus enlarged, spindle-shaped or wedge-shaped 6
3. Fore coxa yellow. Ventral plate nearly parallel-sided or widened posteriorly when viewed ventrally 4
   Fore coxa dark yellow to medium brown. Ventral plater gradually narrowed posteriorly when viewed ventrally 5
4. Ventral plate nearly parallel-sided when viewed ventrally .... *S.* (*G.*) *siamense* (complex)
   Ventral plate widened posteriorly when viewed ventrally (Fig. 36P) .... *S.* (*G.*) *gombakense*
5. Ventral plate with ventral margin straight or slightly concave medially when viewed caudally (similar to Fig. 20N) .... *S.* (*G.*) *decuplum*
   Ventral plate produced ventrally with ventral margin convex medially when viewed caudally .... *S.* (*G.*) *dentistylum*
6. Upper-eye (large) facets in 10 or 11 vertical columns and 13 or 14 horizontal rows .... *S.* (*G.*) *asakoae*
   Upper-eye (large) facets in 14 or 15 vertical columns and 15 or 16 horizontal rows 7
7. Hind tibia with dark subbasal marking. Hind basitarsus slightly wider than hind femur .... *S.* (*G.*) *namense*
   Hind tibia without dark subbasal marking. Hind basitarsus slightly narrower than hind femur 8
8. Fore basitarsus 6.6 times as long as its greatest width .... *S.* (*G.*) *lurauense*
   Fore basitarsus 7.4–8.8 times as long as its greatest width .... *S.* (*G.*) *monglaense* and *S.* (*G.*) *myanmarense*

9. Basal portion of radius haired 10
   Basal portion radius bare 12
10. Style longer than coxite (Fig. 62J) ....*S.* (*S.*) *indicum* and *S.* (*S.*) *nigrogilvum*
   Style shorter than coxite 11
11. Hind basitarsus enlarged, spindle-shaped (similar to Fig. 42J)..........................
   .... *S.* (*M.*) *kambaitense*
   Hind basitarsus slender, nearly parallel-sided (similar to Fig. 45H) ....
   .... *S.* (*N.*) *aureohirtum*
12. Scutum white pruinose with inverted-V-shaped, non-pruinose black area (similar to Fig. 74S, T)... *S.* (*S.*) *nodosum*
   Scutum white pruinose on each shoulder, along each lateral margin and on prescutellar area 13
13. Style without basal protuberance ..... *S.* (*S.*) *chamlongi* and *S.* (*S.*) *himalayense*
   Style with basal protuberance 14
14. Abdominal segments 2, 6, and 7 each with a pair of shiny dorsolateral spots...
   ....*S.* (*S.*) *nigrifacies*
   Abdominal segments 2, 5, 6, and 7 each with a pair of shiny dorsolateral spots 15
15. Style with long basal protuberance, with pointed apex (similar to Fig. 71C) ....
   16
   Style with short round protuberance (similar to Figs. 76Q and 80M) 19
16. Mid femur light to dark brown 17
   Mid femur whitish yellow or grayish yellow 18
17. Upper-eye (large) facets in 17 or 18 vertical columns and in 18 or 19 horizontal rows .... *S.* (*S.*) *chanyae*
   Upper-eye (large) facets in 20 or 21 vertical columns and in 20 horizontal rows .... *S.* (*S.*) *laui*
18. Style with greatest width at apical one-third when viewed ventrolaterally ....
   .... *S.* (*S.*) *barraudi*
   Style with greatest width near base when viewed ventrolaterally ....
   .... *S.* (*S.*) *novolineatum*
19. Ventral plate quadrate when viewed ventrally (similar to Fig. 80L), and with teeth on each lateral margin when viewed caudally (similar to Fig. 80S) 20
   Ventral plate rounded when viewed ventrally (similar to Fig. 76P), and without teeth on each lateral margin when viewed caudally (Fig. 76T) 21
20. Fore coxa black. Hind basitarsus black except basal one-third yellow ....
   ....*S.* (*S.*) *rufibasis*
   Fore coxa yellow. Hind basitarsus white on basal half and brown on rest ....
   .... *S.* (*S.*) *tani* (complex)
21. Scutum with whitish yellow short hairs. Upper-eye (large) facets in 20 vertical columns ....*S.* (*S.*) *xuandai*
   Scutum with brassy short hairs. Upper-eye (large) facets in 15–17 vertical columns 22

22. Hind basitarsus whitish yellow on basal half or little more .... *S.* (*S.*) *chiangmaiense*
Hind basitarsus whitish yellow on basal two-thirds or little less .... *S.* (*S.*) *nakhonense*

Pupae***

1. Gill with three inflated tubular filaments (Fig. 75F) .... *S.* (*S.*) *nodosum*
Gill otherwise 2
2. Gill composed of one inflated structure and eight slender filaments (Fig. 37E) .... *S.* (*G.*) *gombakense*
Gill composed of six, eight, nine, or 10 slender or somewhat inflated filaments ...3
3. Gill with six slender or somewhat inflated filaments 4
Gill with eight, nine, or 10 slender filaments 11
4. Cocoon shoe-shaped (Fig. 63C) 5
Cocoon wall-pocket-shaped..6
5. Integument of thorax bare .... *S.* (*S.*) *indicum*
Integument of thorax moderately covered with tubercles .... *S.* (*S.*) *nigrogilvum*
6. Cocoon with short anterodorsal projection (similar to Fig. 46H) .... *S.* (*N.*) *aureohirtum*
Cocoon without anterodorsal projection 7
7. Outer (or dorsal) filament of dorsal pair of gill much thicker than outer filament of middle pair. Terminal hooks absent .... *S.* (*S.*) *nigrifacies*
Outer (or dorsal) filament of dorsal pair of gill subequal in thickness to outer filament of middle pair. Terminal hooks present 8
8. Six gill filaments without short common basal stalk (similar to Fig. 81G)........ .... *S.* (*S.*) *tani* (complex)
Six gill filaments with short common basal stalk (similar to Fig. 83D, E) 9
9. Upper filament of dorsal pair of gill and lower filament of ventral pair moderately divergent basally at ca. 90 degrees when viewed laterally .... *S.* (*S.*) *rufibasis*
Upper filament of dorsal pair of gill and lower filament of ventral pair widely divergent basally at ca. 120 degrees when viewed laterally 10
10. Integument of head and thorax sparsely covered with tubercles .... *S.* (*S.*) *himalayense*
Integument of head and thorax bare .... *S.* (*S.*) *chamlongi*
11. Gill with eight filaments 12
Gill with nine or 10 filaments or eight filaments on one side and 10 filaments on other side or eight filaments on one side and nine filaments on other side, or nine filaments on one side and 10 filaments on other side 21

12. Eight gill filaments arranged as 2+2+2+2 from dorsal to ventral (Figs. 72A and 77E) 13
    Eight gill filaments arranged as (3+3)+2 or 2+1+3+2 from dorsal to ventral 16
13. Cocoon shoe-shaped. Terminal hooks present .... *S.* (*S.*) *chiangmaiense*
    Cocoon wall-pocket-shaped. Terminal hooks absent 14
14. Cocoon without open space anterolaterally on each side (Fig. 72I) .... *S.* (*S.*) *chanyae*
    Cocoon with large open space anterolaterally on each side (similar to Fig. 72J) 15
15. Stalks of dorsal and ventral pairs of gill divergent basally at 90 degrees when viewed laterally (similar to Fig. 72D) .... *S.* (*S.*) *barraudi* and *S.* (*S.*) *novolineatum*
    Stalks of dorsal and ventral pairs of gill divergent basally at ca. 120 degrees when viewed laterally (similar to Fig. 72G) .... *S.* (*S.*) *laui*
16. Cocoon with anterodorsal projection .... *S.* (*G.*) *myanmarense*
    Cocoon without anterodorsal projection 17
17. Two filaments of ventral pair of gill 1.5–2.0 times as long as six other filaments .... *S.* (*G.*) *siamense* (complex)
    Two filaments of ventral pair of gill as long as or slightly longer than six other filaments 18
18. Eight gill filaments arranged as 2+1+3+2 from dorsal to ventral and all short .... *S.* (*G.*) *dentistylum*
    Eight gill filaments arranged as (3+3)+2 from dorsal to ventral and all long 19
19. Dorsum of abdominal segments 1 and 2 darkened .... *S.* (*G.*) *asakoae*
    Dorsum of abdominal segments 1 and 2 yellowish 20
20. Terminal hooks wide, plate-like, with outer margins crenulated .... *S.* (*G.*) *monglaense*
    Terminal hooks short, blunt, with outer margins smooth .... *S.* (*G.*) *lurauense*
21. Gill with nine filaments on each side, or eight filaments on one side and 10 filaments on other side, or eight filaments on one side and nine filaments on other side, or nine filaments on one side and 10 filaments on other side .... *S.* (*S.*) *chiangmaiense*
    Gill with 10 filaments on each side 22
22. Cocoon wall-pocket-shaped (similar to Fig. 20S) .... *S.* (*G.*) *decuplum*
    Cocoon shoe-shaped (similar to Fig. 77G) 23
23. Ten gill filaments arranged as 2+2+2+2+2 or 2+2+1+2+1+2 from dorsal to ventral .... *S.* (*S.*) *nakhonense*
    Ten gill filaments arranged as 2+3+3+2 from dorsal to ventral 24
24. All 10 gill filaments subequal in thickness to one another .... *S.* (*S.*) *xuandai*

At least two filaments of ventral pair of gill much thinner than other filaments.. *S.* (*S.*) *chiangmaiense*

Mature larvae****

1. Ventral papillae present (Fig. 3E) 2
   Ventral papillae absent 10
2. Mandibular serration composed of two large teeth (similar to Fig. 46I) ....
   ....*S.* (*N.*) *aureohirtum*
   Mandibular serration composed of one large tooth and one small tooth 3
3. Postgenal cleft reaching posterior margin of hypostoma (similar to Fig. 23V) .
   .... *S.* (*G.*) *dentistylum*
   Postgenal cleft not reaching posterior margin of hypostoma 4
4. Pharate pupal gill composed of one inflated structure and eight slender filaments
   .... *S.* (*G.*) *gombakense*
   Pharate pupal gill composed of eight or 10 slender filaments 5
5. Postgenal cleft long, approaching posterior margin of hypostoma. Abdomen with dark branched setae dorsally at least on posterior segments 6
   Postgenal cleft medium-long, 1.0–3.0 times as long as postgenal bridge. Abdomen with unbranched setae dorsally at least on posterior segments 7
6. Body grayish yellow. Pharate pupal gill with 10 slender filaments ....
   .... *S.* (*G.*) *decuplum*
   Body yellow with grayish brown transverse markings. Pharate pupal gill with eight slender filaments ....*S.* (*G.*) *siamense* (complex)
7. Postgenal cleft pointed apically and 2.2–2.8 times as long as postgenal bridge
   8
   Postgenal cleft rounded apically and 1.1–1.9 times as long as postgenal bridge 9
8. Labral fan with 30 primary rays ....*S.* (*G.*) *lurauense*
   Labral fan with 36–48 primary rays ....*S.* (*G.*) *asakoae*
9. Abdominal segments 1 and 2 with dark broad band ....*S.* (*G.*) *myanmarense*
   Abdominal segments 1 and 4 with dark broad band ....*S.* (*G.*) *monglaense*
10. Accessory sclerite present (Fig. 63J) ....*S.* (*S.*) *indicum* and *S.* (*S.*) *nigrogilvum*
    Accessory sclerite absent 11
11. Postgenal cleft reaching posterior margin of hypostoma (similar to Fig. 75N)
    .... *S.* (*S.*) *nodosum*
    Postgenal cleft not reaching posterior margin of hypostoma 12
12. Pharate pupal gill with six slender filaments 13
    Pharate pupal gill with eight or 10 slender filaments 15
13. Postgenal cleft subcordate, rounded laterally ....
    ....*S.* (*S.*) *chamlongi* and *S.* (*S.*) *himalayense*
    Postgenal cleft bullet-shaped, not (or slightly) rounded laterally 14
14. Cephalic apotome with faint positive head spots ....*S.* (*S.*) *tani* (complex)
    Cephalic apotome without positive head spots ....*S.* (*S.*) *rufibasis*

15. Pharate pupal gill with 10 filaments. Abdominal segments 1–8 each with a pair of protuberances dorsally .... *S.* (*S.*) *nakhonense*
    Pharate pupal gill with eight filaments. Abdominal segments 1–8 without protuberances dorsally .... *S.* (*S.*) *laui*

**Simulium* (*A.*) *shanense*, *S.* (*G.*) *dentistylum*, *S.* (*S.*) *nigrifacies*, and *S.* (*S.*) *xuandai* are not included because their females are unknown.

** *Simulium* (*N.*) *burmense* and *S.* (*N.*) sp. A are not included because their males are unknown.

****Simulium* (*A.*) *shanense*, *S.* (*G.*) *namense*, *S.* (*M.*) *kambaitense*, *S.* (*N.*) *burmense*, and *S.* (*N.*) sp. A are not included because their pupae are unknown.

*****Simulium* (*A.*) *shanense*, *S.* (*G.*) *namense*, *S.* (*M.*) *kambaitense*, *S.* (*N.*) *burmense*, *S.* (*N.*) sp. A, *S.* (*S.*) *nigrifacies*, *S.* (*S.*) *chanyae*, *S.* (*S.*) *chiangmaiense*, and *S.* (*S.*) *xuandai* are not included because their larvae are unknown. *Simulium* (*S.*) *barraudi* is not included because its larva is unknown from Myanmar.

5 The Fauna of Black Flies in Sri Lanka

Overview

From Sri Lanka, only four species of black flies were known: *Simulium* (*Simulium*) *striatum* (Brunetti, 1912), *S.* (*Gomphostilbia*) *ceylonicum* (Enderlein, 1921), *S.* (*G.*) *pattoni* (Senior-White 1922), and *S.* (*S.*) *consimile* (Puri,1932e), until Datta (1983) recorded *S.* (*S.*) *nilgiricum*. Davies and Györkös (1987a, b, 1988a, b, 1992) described 11 new species including *S.* (*Byssodon*) *languidum* (the only species of the subgenus *Byssodon* in the Oriental Region) and *S.* (*G.*) *trirugosum* (at that time, under the subgenus *Morops*) and recorded *S.* (*Nevermannia*) *aureohirtum*.

The fauna of black flies in Sri Lanka is represented by 17 species in the genus *Simulium* s. l., of which one species is placed in the subgenus *Byssodon*, six species are in the subgenus *Gomphostilbia*, one species is in the subgenus *Nevermannia*, and nine species are in the subgenus *Simulium* s. str. (Davies and Györkös, 1992).

List of 17 Species of Black Flies in Sri Lanka

Subgenus *Byssodon*
S. languidum
Subgenus *Gomphostilbia*
S. batoense species-group
S. dola, *S. krombeini*, and *S. pattoni*

S. ceylonicum species-group
S. ceylonicum and *S. ela*
S. varicorne species-group
S. trirugosum
Subgenus *Nevermannia*
S. ruficorne species-group
S. aureohirtum
Subgenus *Simulium* s. str.
S. striatum species-group
S. consimile, *S. striatum*, and *S. subpalmatum*
S. variegatum species-group
S. bulla, *S. cruszi*, *S. nilgiricum*, *S. nubis*, and *S. paranubis*
Simulium s. str. species unplaced to species-group
S. cremnosi

Keys to Identify 17 Species of Black Flies of Sri Lanka

Females**

1. Katepisternum haired 2
   Katepisternum bare 7
2. Pleural membrane haired .... *S. (G.) trirugosum*
   Pleural membrane bare 3
3. Claw with small tooth .... *S. (G.) pattoni*
   Claw with large tooth 4
4. Sensory vesicle long, 0.7 times as long as third palpal segment .... *S. (G.) ela*
   Sensory vesicle 0.3–0.5 times as long as third palpal segment 5
5. Sensory vesicle 0.5 times as long as third palpal segment .... *S. (G.) krombeini*
   Sensory vesicle 0.30–0.36 times as long as third palpal segment 6
6. Height of frons relative to its narrowest width 3.0 .... *S. (G.) ceylonicum*
   Height of frons relative to its narrowest width 2.1 .... *S. (G.) dola*
7. Claw with large basal tooth 8
   Claw simple or with small subbasal tooth 9
8. Basal portion of radius haired .... *S. (N.) aureohirtum*
   Basal portion of radius bare .... *S. (B.) languidum*
9. Claw with small subbasal tooth 10
   Claw simple, without tooth 12
10. Fore basitarsus slender, 10.5 times as long as its greatest width. Ovipositor valve rounded posteromedially .... *S. (S.) cruszi*
    Fore basitarsus slender, 4.5–6.0 times as long as its greatest width. Ovipositor valve pointed posteromedially 11

11. Hind tibia yellow on basal three-fourths and dark brown on rest .... *S.* (*S.*) *bulla*
    Hind tibia yellow on basal one-fifth or one-third and dark brown on rest .... *S.* (*S.*) *nilgiricum* and *S.* (*S.*) *nubis*
12. Ovipositor valve without ventrally produced process along inner margin. Paraproct slightly produced ventrally beyond ventral tip of cercus .... *S.* (*S.*) *cremnosi*
    Ovipositor valve with ventrally produced process along inner margin. Paraproct greatly produced ventrally beyond ventral tip of cercus 13
13. Basal portion of radius with 6–13 hairs .... *S.* (*S.*) *subpalmatum*
    Basal portion of radius bare or with one or two hairs 14
14. Frons:head ratio 1:12. Frontal ratio 2.4:1.0:2.8. Sensory vesicle 0.4 times as long as third palpal segment. Hind tibia dark brown to black except basal tip yellow .... *S.* (*S.*) *striatum*
    Frons:head ratio 1:5. Frontal ratio 1.3:1.0:1.5. Sensory vesicle 0.5 times as long as third palpal segment. Hind tibia dark brown to black except basal one-fifth yellow .... *S.* (*S.*) *consimile*

Males***

1. Katepisternum haired. 2
   Katepisternum bare 7
2. Pleural membrane haired .... *S.* (*G.*) *trirugosum*
   Pleural membrane bare 3
3. Hind basitarsus pale brown on basal one-third to two-fifths, and enlarged, 2.7 times as long as its greatest width and subequal in width to hind tibia............. .... *S.* (*G.*) *ela*
   Hind basitarsus yellow on basal three-fifths, and slender or somewhat enlarged, 3.4–5.0 times as long as its greatest width, and much narrower than hind tibia 4
4. Hind basitarsus slender, nearly parallel-sided, 5.0 times as long as its greatest width 5
   Hind basitarsus somewhat enlarged, spindle-shaped, 3.4–3.8 times as long as its greatest width 6
5. Upper-eye (large) facets in 11 vertical columns and 11 horizontal rows. Fore basitarsus 5.2 times as long as its greatest width ....*S.* (*G.*) *dola*
   Upper-eye (large) facets in 12 or 13 vertical columns and 12 horizontal rows. Fore basitarsus 6.3 times as long as its greatest width ....*S.* (*G.*) *krombeini*
6. Upper-eye (large) facets in 17 vertical columns and 17 horizontal rows .... *S.* (*G.*) *pattoni*
   Upper-eye (large) facets in 13 vertical columns and 14 horizontal rows ....*S.* (*G.*) *ceylonicum*
7. Coxite longer than style 8
   Coxite shorter than style 9
8. Basal portion of radius haired .... *S.* (*N.*) *aureohirtum*

Basal portion of radius bare .... *S.* (*B.*) *languidum*

9. Ventral plate wider than long when viewed ventrally and without teeth on posterior surface 10
   Ventral plate much longer than wide when viewed ventrally (similar to Fig. 84I) and with teeth on posterior surface 13
10. Ventral plate subquadrate when viewed ventrally and not produced ventrally (similar to Fig. 61H) ....*S.* (*S.*) *cremnosi*
    Ventral plate rounded posteriorly when viewed ventrally (similar to Fig. 76P) and much produced ventrally 11
11. Upper-eye (large) facets in 13 vertical columns and 12 horizontal rows ....*S.* (*S.*) *subpalmatum*
    Upper-eye (large) facets in 16–18 vertical columns and 15–18 horizontal rows 12
12. Upper-eye (large) facets in 16 vertical columns and 15 or 16 horizontal rows.. .... *S.* (*S.*) *consimile*
    Upper-eye (large) facets in 18 vertical columns and 17 or18 horizontal rows ... .*S.* (*S.*) *striatum*
13. Upper-eye (large) facets in 22 or 23 vertical columns and 23 or 24 horizontal rows ....*S.* (*S.*) *cruszi*
    Upper-eye (large) facets in 17–20 vertical columns and 17–19 horizontal rows 14
14. Upper-eye (large) facets in 17 vertical columns and 17 horizontal rows ....*S.* (*S.*) *nubis*
    Upper-eye (large) facets in 19 or 20 vertical columns and 18 or19 horizontal rows 15
15. Mid tibia yellow .... *S.* (*S.*) *bulla*
    Mid tibia dark brown ....*S.* (*S.*) *paranubis*

Pupae

1. Gill with three filaments 2
   Gill with six, eight, and 10 filaments 3
2. Gill filaments 1.5 times as long as pupal body. Abdominal segments 5–8 each with spine-combs dorsally. Cocoon wall-pocket-shaped ...... *S.* (*G.*) *trirugosum*
   Gill filaments half as long as pupal body. Abdominal segments 5–8 each without spine-combs dorsally. Cocoon shoe-shaped .... *S.* (*B.*) *languidum*
3. Gill with six filaments 4
   Gill with eight or 10 filaments 10
4. Cocoon shoe-shaped, with open spaces anteriorly ....*S.* (*S.*) *cremnosi*
   Cocoon wall-pocket-shaped, without open spaces anteriorly 5
5. Cocoon wall-pocket-shaped, with short anterodorsal projection (similar to Fig. 46H)...*S.* (*N.*) *aureohirtum*
   Cocoon wall-pocket-shaped, without anterodorsal projection 6
6. Cocoon longer at top than at base (similar to Fig. 69T) .... *S.* (*S.*) *nilgiricum*

Cocoon longer at base than at top 7

7. Thorax bare without tubercles 8
   Thorax with tubercles 9
8. Gill with round transparent swollen basal fenestra at base .... *S.* (*S.*) *bulla*
   Gill without swollen basal fenestra at base ....*S.* (*S.*) *paranubis*
9. Face with two trichomes ....*S.* (*S.*) *cruszi*
   Face without trichomes ....*S.* (*S.*) *nubis*
10. Gill with eight filaments 11
    Gill with 10 filaments 15
11. Gill with eight filaments arranged as 1+2+(1+2)+2 from dorsal to ventral .... *S.* (*G.*) *pattoni*
    Gill with eight filaments arranged as (1+2)+(1+2)+2 from dorsal to ventral 12
12. Stalks of dorsal triplet and ventral pair of gill widely divergent at angle of 140 degrees when viewed laterally ....*S.* (*G.*) *dola*
    Stalks of dorsal triplet and ventral pair of gill moderately divergent at angle of 60–90 degrees when viewed laterally 13
13. Stalk of ventral pair of gill much longer than primary and secondary stalks of middle triplet combined ....*S.* (*G.*) *krombeini*
    Stalk of ventral pair of gill nearly as long as or slightly shorter than primary and secondary stalks of middle triplet combined 14
14. Stalk of ventral pair of gill as thick as primary stalk of middle triplet ....*S.* (*G.*) *ceylonicum*
    Stalk of ventral pair of gill 0.75 times as thick as primary stalk of middle triplet .... *S.* (*G.*) *ela*
15. Five upper filaments of gill about twice as thick as five lower filaments. Cocoon shoe-shaped, without open spaces anteriorly .... *S.* (*S.*) *subpalmatum*
    All 10 filaments of gill subequal in thickness to one another. Cocoon shoe-shaped, with open spaces anteriorly 16
16. Two middle triplets of gill filaments almost completely overlapping when viewed laterally .... *S.* (*S.*) *striatum*
    Two middle triplets of gill filaments not or partially overlapping when viewed laterally ....*S.* (*S.*) *consimile*

Mature larvae****

1. Ventral papillae present 2
   Ventral papillae absent 7
2. Postgenal cleft reaching posterior margin of hypostoma 3
   Postgenal cleft medium-long to long, not reaching posterior margin of hypostoma 4
3. Rectal organ simple, without secondary lobules .... *S.* (*B.*) *languidum*
   Rectal organ compound, with 8–10 secondary lobules per lobe .... *S.* (*G.*) *trirugosum*

4. Postgenal cleft medium-long, ca. 0.8 times as long as postgenal bridge .... *S. (N.) aureohirtum*
   Postgenal cleft long, much longer than postgenal bridge 5
5. Rectal organ simple, without secondary lobules.. .... *S. (G.) ceylonicum*, *S. (G.) krombeini*, and *S. (G.) ela*
   Rectal organ compound, with 15–34 secondary lobules in three lobes 6
6. Postgenal cleft 2.5–3.0 times as long as postgenal bridge. Posterior circlet with ca. 105 rows of 11 or 12 hooklets per row ....*S. (G.) dola*
   Postgenal cleft 8.0 times as long as postgenal bridge. Posterior circlet with ca. 60 rows of nine or 10 hooklets per row ....*S. (G.) pattoni*
7. Postgenal cleft rounded, with its widest in middle 0.38 times as wide as width of head capsule 8
   Postgenal cleft deltoid, with its widest in middle 0.27–0.31 times as wide as width of head capsule 9
8. Posterior circlet with ca. 86 rows of up to 14 hooklets per row .... *S. (S.) subpalmatum*
   Posterior circlet with 102–104 rows of up to 16 hooklets per row ....*S. (S.) consimile*
9. Posterior circlet with ca. 180 rows of up to 20 hooklets per row ....*S. (S.) cremnosi*
   Posterior circlet with ca. 80 rows or ca. 108 rows of up to 12–14 hooklets per row 10
10. Posterior circlet with ca. 80 rows of up to 12 hooklets per row ..... *S. (S.) bulla*
   Posterior circlet with ca. 108 rows of up to 14 hooklets per row........................ .... *S. (S.) nubis*, *S. (S.) paranubis*, and *S. (S.) cruszi*

*Keys to identify 16 species of black flies in Sri Lanka presented by Davies and Györkös (1987a, 1992) are revised.

****Simulium (S.) paranubis*** is not included because its female is unknown.

***Simulium (S.) nilgiricum* goes to the couplet 13 but cannot go further because of insufficient information on the male.

****Simulium (S.) striatum* and *S. (S.) nilgiricum* are not included because their larvae are unknown.

6 The Fauna of Black Flies in Thailand

Overview

In Thailand, Summers (1911) was the first to describe a new species, *S. (S.) nigrogilvum*. Edwards (1928) recorded *S. (S.) digrammicum* and *S. (S.) hackeri* described from Peninsular Malaysia, and synonymized *S. (S.) nigrogilvum* with *S. (S.) indicum*. Puri (1932c) synonymized *S. (S.) digrammicum* with *S. (S.) griseifrons*.

Takaoka and Suzuki (1984) added 17 taxa consisting of seven new species including *S.* (*S.*) *barnesi* (the species previously misidentified as *S.* (*S.*) *hackeri*), five newly recorded species, *S.* (*N.*) *aureohirtum*, *S.* (*S.*) *nodosum*, *S.* (*S.*) *nitidithorax*, *S.* (*S.*) *sakishimaense*, *S.* (*S.*) *rufibasis*, and five unnamed species, *S.* (*N.*) sp. A, *S.* (*N.*) sp. B, *S.* (*G.*) sp. C, *S.* (*S.*) sp. D and *S.* (*S.*) sp. E; and resurrected *S.* (*S.*) *digrammicum* and *S.* (*S.*) *nigrogilvum* from synonyms of *S.* (*S.*) *griseifrons* and *S.* (*S.*) *indicum*, respectively. Takaoka and Davies (1995a) suggested that *S.* (*N.*) sp. A is *S.* (*N.*) *caudiscreum* described from Peninsular Malaysia.

Takaoka and Saito (1996) described a new species, *S.* (*Byssodon*) *siripoomense*, recorded four species, *S.* (*G.*) *decuplum*, *S.* (G.) *parahiyangum*, *S.* (*S.*) *brevipar*, and *S.* (*S.*) *quinquestriatum*, and identified *S.* (*G.*) sp. C as *S.* (*G.*) *asakoae*. Takaoka and Adler (1997) described a new species, *S.* (*D.*) *courtneyi*, recorded *S.* (*D.*) *pahangense*, and proposed a new subgenus *Daviesellum*. Takaoka and Kuvangkadilok (1999) described three new species. Kuvangkadilok et al. (1998) reported C-banding of polytene chromosomes of five Thai species of black flies including *S.* (*Montisimulium*) sp. G. Kuvangkadilok et al. (1999a) recorded *S.* (*N.*) *feuerborni* [the species formerly regarded as *S.* (*N.*) sp. B]. Kuvangkadilok and Takaoka (2000) described a new species, *S.* (*G.*) *chumpornense,* and recorded nine species: S. (*G.*) *angulistylum*, *S.* (*G.*) *sheilae*, *S.* (*G.*) *dentistylum*, *S.* (*G.*) *gombakense*, *S.* (*S.*) *grossifilum*, *S.* (*S.*) *malayense*, *S.* (*S.*) *nobile*, *S.* (*S.*) *rudnicki*, and *S.* (*S.*) *yongi*. Takaoka and Mulla (2000) described the female and pupa of *S.* (*S.*) *siripoomense*, with its subgeneric transfer from *Byssodon* to *Simulium* s. str. Takaoka and Choochote (2004a) recorded *S.* (*G.*) *burtoni*.

Phasuk et al. (2005) studied chromosomally and morphologically 12 species of the subgenus *Gomphostilbia* including *S.* (*G.*) *dentistylum* and *S.* (*G.*) *novemarticulatum,* both newly recorded from Thailand and four unnamed species, *S.* sp. nr. *sheilae*, *S.* sp. A, *S.* sp. B and *S.* sp. C. Jitklang and Kuvangkadilok (2008) described a new species, *S.* (*G.*) *adleri*. Jitklang et al. (2008) described three new species: *S.* (*G.*) *curtatum*, *S.* (*G.*) *doisaketense*, and *S.* (*G.*) *trangense* [the same species as *S.* sp. nr. *sheilae* reported by Phasuk et al. (2005)], reported four unnamed species, *S.* (*G.*) nr. *asakoae* 2, *S.* (*G.*) nr. *asakoae* 3, *S.* (*G.*) nr. *asakoae* 4, and *S.* (*G.*) nr. *sheilae* 3. Pramual and Tangkawanit (2008) described a new species, *S.* (*G.*) *kuvangkadilokae*.

Takaoka (2001c) and Takaoka and Choochote (2002, 2004b,c, 2005a-j, 2006a,c-e, 2007) described 30 new species including *S.* (*S.*) *setsukoae* (the species formerly reported as *S.* (*S.*) sp. D), *S.* (*S.*) *doipuiense* (the species formerly reported as *S.* (*S.*) sp. E) and *S.* (*S.*) *yuphae* [the species formerly recognized as *S.* (*S.*) *brevipar* by Takaoka and Saito (1996)]. Takaoka and Choochote (2005b) proposed a new subgenus *Asiosimulium*. Takaoka and Choochote (2006b) described the female and male of *S.* (*S.*) *baimaii*. Takaoka and Choochote (2009) described the female and male of *S.* (*M.*) *angkaense*.

Takaoka and Srisuka (2010a) described the female of *S.* (*G.*) *kuvangkadilokae* and redescribed its male, pupa and larva. Takaoka et al. (2009a, 2010a, b) described three new species and recorded *S.* (*G.*) *cheongi* from southern Thailand. Takaoka et al. (2010c) described the female of *S.* (*G.*) *gombakense*. Takaoka et al. (2010d) described a new species, *S.* (*M.*) *nanense* and described the female, male, and pupa of *S.* (*M.*) *laoleense*. Takaoka and Srisuka (2009, 2010b-e, 2011) described six new species.

Tangkawanit et al. (2018) described a new species, *S.* (*A.*) *phurueaense*. Thaijerern et al. (2018) described a new species, *S.* (*S.*) *ubonae*.

Takaoka et al. (2012e, 2013c, d, 2014h, i, j, 2017f-n, 2018e-i, k, 2019e, 2020h, 2021e) described 46 new species including *S.* (*N.*) *chomthongense* (the species formerly regarded as *S.* (*N.*) *caudisclerum*), *S.* (*S.*) *tenebrosum* (the species formerly regarded as *S.* (*S.*) *rufibasis*), *S.* (*G.*) *chiangraiense, S.* (*G.*) *pamiangense, S.* (*G.*) *huaikaeoense, S.* (*G.*) *khaokhoense*, *S.* (*G.*) *maeklangense* (all these five new species derived from the species formerly regarded as *S.* (*G.*) *decuplum*), and *S.* (*G.*) *ngaoense* [the species formerly regarded as *S.* (*G.*) *dentistylum* and *S.* (*G.*) *parahiyangum*] and recorded two species, *S.* (*S.*) *bishopi* and *S.* (*G.*) *thuathienense*. Takaoka et al. (2018l) described the female of *S.* (*G.*) *chayamaritae*. Srisuka et al. (2015a) described the male, pupa and larva of *S.* (*A.*) *wanchaii*. Srisuka et al. (2015b) recorded *S.* (*S.*) *brevipar* (complex) from Pha Hom Pok National Park.

Srisuka et al. (2019, 2021, 2022a,b, 2023a,b) described six new species. Aupalee et al. (2020a, b, 2022) described four new species of the *S. striatum* species-group including *S.* (*S.*) *wangkwaiense* (the species formerly regarded as *S.* (*S.*) *quinquestriatum*). Pramual et al. (2022a) described the male of *S.* (S.) *triglobus*. Pramual et al. (2022b) described a new species, *S.* (*G.*) *prayooki* [the species formerly designated as cytoform C of *S.* (*G.*) *angulistylum* (complex)].

Chromosomal and molecular analyses revealed the cryptic diversity by disclosing two or more cytospecies, cytoforms or molecular lineages in Thai populations of the following 10 morphospecies: *S. angulistylum* (Pramual and Kuvangkadilok, 2012), *S. asakoae* (Jitklang et al., 2008), *S. siamense* (Kuvangkadilok et al., 2008; Pramual and Wongpakam, 2011), *S. feuerborni* (Pramual et al., 2015; Ya'cob et al., 2017b), *S.* (*N.*) *chomthongense* (Adler et al., 2020a), *S. malayense*, *S. fenestratum* (Thaijarern et al., 2018), *S. tani* (Tangkawanit et al., 2009b), *S. tenebrosum* (Adler et al., 2019a), and *S. doipuiense* (Tangkawanit et al., 2009b; Adler et al., 2016). Each of these 10 morphospecies is thus considered a species complex except *S.* (*G.*) *asakoae*, which was re-analyzed by Low et al. (2020).

Phylogenetic relationships of Thai black fly species, those of species in the subgenus *Gomphostilbia*, and those of three small subgenera (*Asiosimulium*, *Daviesellum*, and *Wallacellum*) were molecularly studied by Thanwisai et al. (2006), Phayuhasena et al. (2010), Pramual et al. (2011), and Otsuka et al. (2007), respectively.

Ecological studies of adults and immature stages of black flies in Thailand were carried out, revealing different seasonal and elevational patterns of species composition and abundance (Choochote et al., 2005; Ishii et al., 2008; Srisuka et al., 2015b, 2017).

Seven species are known to bite humans: *S.* (*G.*) *asakoae*, *S.* (*S.*) *chamlongi*, *S.* (*S.*) *doipuiense* (complex), *S.* (*S.*) *nigrogilvum* and *S.* (*S.*) *nodosum*, *S.* (*S.*) *tenebrosum* (complex) (the species formerly reported as *S.* (*S.*) *rufibasis*), and *S.* (*S.*) *umphangense* (Choochote et al., 2005; Pramual et al., 2016; Takaoka et al.,

2017i). *Simulium* (*G.*) *asakoae*, *S.* (*G.*) *khelangense*, *S.* (*S.*) *nigrogilvum*, and *S.* (*S.*) *nodosum* were incriminated as natural vectors of four different kinds of filariae including an *Onchocerca* species (Fukuda et al., 2003; Ishii et al., 2008; Takaoka et al., 2003; Aupalee et al., 2020b). *Simulium* (*G.*) *chumpornense* is natural vectors of *Trypanosoma* and *Leucocytozoon* (Jumpato et al., 2018; Thaijarern et al., 2019a; Pramual et al., 2020).

Natural infections of black flies with mermithids, microsporidans and other fungi were also investigated by Aupalee et al. (2020b), Huang et al. (2021), and Jitklang et al. (2012).

All of these species are classified in six of the 11 subgenera of the genus *Simulium* s. l. in the Oriental Region: five species in *Asiosimulium*, two species in *Daviesellum*, 69 species in *Gomphostilbia*, seven species in *Montisimulium*, 10 species in *Nevermannia*, and 58 species in *Simulium* s. str. The species in three of these subgenera are further placed in species-groups: seven species-groups of *Gomphostilbia*, three species-groups of *Nevermannia* and 12 species-groups of *Simulium* s. str. (Adler, 2022; Takaoka, 2012, 2017a).

*List of 153 Species of Black Flies in Thailand**

Subgenus *Asiosimulium*
S. furvum, *S. oblongum*, *S. phurueaense*, *S. saeungae*, and *S. wanchaii*
Subgenus *Daviesellum*
S. courtneyi, and *S. pahangense*
Subgenus *Gomphostilbia*
S. asakoae species-group
S. asakoae, *S. banluangense*, *S. chaowaense*, *S. chiangduoense*, *S. huaimorense*, *S. inthanonense*, *S. junkumae*, *S. kiewfinense*, *S. kiewlomense*, *S. klonglanense*, *S. loeiense*, *S. maehongsonense*, *S. maelanoiense*, *S. maewongense*, *S. mokroense*, *S. monglaense*, *S. muangpanense*, *S. myanmarense*, *S. namdanense*, *S. nanoiesne*, *S. nanthaburiense*, *S. phapeungense*, *S. pitasawatae*, *S. puaense*, *S. rampae*, *S. songense*, *S. sutheppuiense*, *S. teerachanense*, *S. thungchangense*, and *S. udomi*
S. batoense species-group
S. chiangraiense, *S. dentistylum*, *S. fukudae*, *S. huaikaeoense*, *S. khaokhoense*, *S. maeklangense*, *S. ngaoense*, *S. pamiangense*, and *S. siamense* (complex)
S. ceylonicum species-group
S. curtatum, *S. pangsidaense*, *S. sheilae*, *S. trangense*, and *S. wijiti*
S. darjeelingense species-group
S. chayamaritae
S. epistum species-group
S. adleri, *S. angulistylum* (complex), *S. cheongi*, *S. datfaense*, *S. isanense*, *S. otsukai*, and *S. prayooki*
S. gombakense species-group

S. gombakense, *S. maleewongae*, *S. paiense* (complex), *S. prayongi*, and *S. thuathienense*

S. varicorne species-group

S. burtoni, *S. chumpornense*, *S. khelangense*, *S. kuvangkadilokae*, *S. novemarticulatum*, and *S. piroonae*

Gomphostilbia species unplaced to species-group

S. sp. A, *S.* sp. B, *S.* sp. C, *S.* nr. *asakoae* 2, *S.* nr. *asakoae* 3, *S.* nr. *asakoae* 4, and *S.* nr. *sheilae* 3

Subgenus *Montisimulium*

S. angkaense, *S. phahompokense*, *S. laoleense*, *S. merga*, *S. nanense*, *S. surachaii, and S.* sp. G

Subgenus *Nevermannia*

S. feuerborni species-group

S. chiangklangense, *S. fangense*, *S. feuerborni* (complex), *S. fruticosum*, *S. maeaiense*, *S. vessabutrae*, and *S. wichaii*

S. ruficorne species-group

S. aureohirtum

S. vernum species-group

S. chomthongense (complex) and *S. khunklangense*

Subgenus *Simulium* s. str.

S. christophersi species-group

S. atipornae

S. crocinum species-group

S. crocinum, *S. mediocoloratum*, *S. rudnicki*, and *S. yongi*

S. griseifrons species-group

S. choochotei, *S. digrammicum*, *S. maenoi*, *S. phukaense*, *S. pukaengense*, *S. suchariti*, *S. thongsahuani*, and *S. visuti*

S. grossifilum species-group

S. grossifilum and *S. phayaoense*

S. indicum species-group

S. nigrogilvum, *S. umphangense*, and *S. vanellum*

S. malyschevi species-group

S. baimaii, *S. lomkaoense*, and *S. siripoomense*

S. melanopus species-group

S. bishopi

S. multistriatum species-group

S. bullatum, *S. chainarongi*, *S. chaliowae*, *S. chanyae*, *S. daoense*, *S. fenestratum* (complex), *S. lampangense*, *S.malayense* (complex), *S. takense*, *S. triglobus*, *S. ubonae*, and *S. undecimum*

S. nobile species-group

S. nodosum and *S vanluni*

S. striatum species-group

S. chiangmaiense, *S. maeklongkeense*, *S. nakhonense*, *S. phraense*, *S. poolpholi*, *S. tadtonense*, *S. thailandicum*, .*S. thilorsuense*, and *S. wangkwaiense*

S. tuberosum species-group

S. brevipar (complex), *S. doipuiense* (complex), *S. manooni*, *S. setsukoae*, *S. tani* (complex), *S. tenebrosum* (complex), *S. weji*, and *S. yuphae* (complex)
S. variegatum species-group
S. barnesi, *S. chamlongi*, *S. kiewmaepanense*, *S. luculentum*, and *S. srisukai*

* *Simulium* (*G.*) *duolongum* and *S.* (*G.*) *tahanense*, both originally described from Peninsular Malaysia, are not included in the new list, because these two species from Thailand were shown to be different from those from Peninsular Malaysia by a DNA analysis (Takaoka et al., 2019c), although they were reported from Thailand with their DNA sequences (Pramual et al., 2011). The identities of these two species reported from Thailand remain unclear.

Keys to Identify 145 Species of Black Flies in Thailand

Earlier keys were provided first for 19 taxa of black flies in Thailand (Takaoka and Suzuki, 1984) and subsequently for 55 species (Takaoka and Choochote, 2004b). Updated keys to identify 145 Thai black fly species are here provided. Due to the large number of the species, keys are provided to identify species of each subgenus following the keys to identify six subgenera.

Keys to Identify Six Subgenera of the Genus Simulium *s. l. in Thailand*

Females

1. Katepisternum haired. .... *Gomphostilbia*
   Katepisternum bare. 2
2. Cibarium with distinct group of spinous processes medially (Fig. 4D)................
   .... *Asiosimulium*
   Cibarium without such processes medially.. 3
3. Ovipositor valve tongue-shaped (Figs. 8C and 42E). 4
   Ovipositor valve otherwise. 5
4. Basal section of radius haired. .... *Montisimulium*
   Basal section of radius bare .... *Daviesellum*
5. Claw with large basal tooth (Fig. 43C). .... *Nevermannia*
   Claw simple or with small subbasal tooth .... *Simulium* s. str.

Males

1. Katepisternum haired. .... *Gomphostilbia*
   Katepisternum bare 2
2. Coxite longer than style (e.g., Fig. 5D) 3

Coxite shorter than style (e.g., Fig. 49I) 5
3. Paramere without hooks (Fig. 5J) ....*Asiosimulium*
Paramere with hooks (e.g., Fig. 11O)..4
4. Paramere with one hook (e.g., Fig. 47Q)..
.............*Nevermannia* (*S. vernum* species-group and *S. ruficorne* species-group)
Paramere with three or more hooks (e.g., Fig. 43P)..
.........................*Montisimulium* and *Nevermannia* (*S. feuerborni* species-group)
5. Coxite much longer than wide (Fig. 8J, M). ....*Daviesellum*
Coxite as long as, or slightly shorter than, wide (e.g., Fig. 49I) ....
....*Simulium* s. str.

Pupae

1. Gill with 19–56 filaments (Fig. 6A) ....*Asiosimulium*
Gill with less than 15 filaments. 2
2. Gill with 12–14 filaments (Fig. 42R, S). Gill filaments with numerous tiny black spots (Fig. 42T).. *Montisimulium*
Gill with less than 10 filaments. Gill filaments without black spots 3
3. Dorsal surface of abdominal segment 2 with stout hooks (Fig. 9C), similar in size to those on abdominal segments 3 and 4 (Fig. 9D). Dorsum of abdominal segments 5–9 without spine-combs ....*Daviesellum*
Dorsal surface of abdominal segment 2 with spines (in place of such distinct hooks), much smaller than hooks on abdominal segments 3 and 4. Spine-combs present at least on dorsum of abdominal segment 8 4
4. Abdominal segment 9 with grapnel-like hooklets on each side (Fig. 3D) ....
.... *Gomphostilbia*
Abdominal segment 9 without grapnel-like hooklets on each side 5
5. Gill with two, three, eight or 10 filaments ....*Simulium* s. str.
Gill with four or six filaments 6
6. Gill with four filaments (similar to Fig. 48E) ....
....*Nevermannia* (*S. vernum* species-group)
Gill with six filaments 7
7. Gill filaments longer than pupal body (Fig. 44A, B)..
.... *Nevermannia* (*S. feuerborni* species-group)
Gill filaments shorter than pupal body (Fig. 46C)...8
8. Cocoon with short anterodorsal projection (Fig. 46H) ....
.... *Nevermannia* (*S. ruficorne* species-group)
Cocoon wall-pocket shaped, without anterodosal projection, shoe-shaped, or boot-shaped ....*Simulium* s. str.

Mature larvae

1. Hypostoma wide, with 12 or 13 apical teeth (Fig. 9L, N)................*Daviesellum*
Hypostoma of moderate width, with nine apical teeth (e.g., Fig. 6H) 2
2. Postgenal cleft vestigial or extremely small (Fig. 42X) .... *Montisimulium*

Postgenal cleft well defined, small to large. 3
3. Ventral papillae present, medium-sized to large (Fig. 3I). 4
Ventral papillae absent or small 5
4. Lateral margins of hypostoma serrated (e.g., Fig. 46K) ....*Nevermannia*
Lateral margins of hypostoma smooth (e.g., Fig. 32U). ....*Gomphostilbia*
5. Pharate pupal gill with 19–56 filaments. ....*Asiosimulium*
Pharate pupal gill with 2–10 filaments. ....*Simulium* s. str.

Keys to Identify Five Species of the Subgenus Asiosimulium

Females

1. Cercus much elongated, about twice as long as its basal width (Fig. 4P)...........
....*S. oblongum* and *S. phurueaense*
Cercus normal or somewhat elongated, less than 1.3 times as long as its basal width (Fig. 4L, N) 2
2. Abdominal segment 5 shiny dorsally ....*S. wanchaii*
Abdominal segment 5 dull dorsally 3
3. Spermatheca globular (Fig. 4Q) .... *S. saeungae*
Spermatheca pear-shaped (Fig. 4S)... *S. furvum*

Males

1. Hind basitarsus slightly widened, 0.75 times as wide as hind tibia (Fig. 5A) .....
....*S. oblongum* and *S. phurueaense*
Hind basitarsus much widened, 0.94–0.96 times as wide as hind tibia (Fig. 5B).
....2
2. Abdominal segments 5–7 without shiny patches ....*S. wanchaii*
Abdominal segments 5–7 each with a pair of shiny dorsolateral patches 3
3. Antenna yellow except apical flagellomere light brown................... *S. saeungae*
Antenna dark brown except scape, pedicel, and base of first flagellomere yellow ....*S. furvum*

Pupae

1. Head and thoracic integument bare .... *S. saeungae*
Head and thoracic integument with tubercles 2
2. Gill with 31–33 filaments ....*S. oblongum* and *S. phurueaense*
Gill with 19 or 22 filaments 3
3. Gill with 19 filaments (Fig. 6A) ....*S. wanchaii*
Gill with 22 filaments .... *S. furvum*

Mature larvae

1. Postgenal cleft with sheath of subesophageal ganglion dark pigmented (Fig. 6I). 2
   Postgenal cleft with sheath of subesophageal ganglion unpigmented 3
2. Pharate pupal gill with 22 filaments.......... *S. furvum*
   Pharate pupal gill with 31–33 filaments .... *S. phurueaense*
3. Labral fan with 30–33 primary rays.......... *S. saeungae*
   Labral fan with 38–45 primary rays 4
4. Labral fan with 38–40 primary rays. Mandible with supernumerary serration .... *S. wanchaii*
   Labral fan with 43–45 primary rays. Mandible without supernumerary serration (similar to Fig. 6G) ....*S. oblongum*

Keys to Identify Two Species of the Subgenus **Daviesellum**

Females

1. Arms of genital fork with moderately developed projection directed anteriorly (Fig. 8D) .... *S. courtneyi*
   Arms of genital fork without such projection (Fig. 8C) .... *S. pahangense*

Males

2. Ventral plate much longer than wide when viewed ventrally (Fig. 8M)........... .... *S. courtneyi*
   Ventral plate somewhat shorter than wide when viewed ventrally (Fig. 8J)....... ....*S. pahangense*

Pupae

1. Frons densely covered with tubercles. .... *S. courtneyi*
   Frons mostly bare except narrow portion near lower margin (Fig. 9A)................ .... *S. pahangense*

Mature larvae

1. Postgenal cleft long, but its apex not reaching posterior margin of hypostoma (Fig. 9O).......... *S. courtneyi*
   Postgenal cleft long, its apex reaching posterior margin of hypostoma (Fig. 9M) ....*S. pahangense*

Keys to Identify 58 Species of the Subgenus Gomphostilbia*

Females**

1. Antenna with seven or eight flagellomeres (Fig. 39A–C) 2
   Antenna with nine flagellomeres 7
2. Antenna with seven flagellomeres (Fig. 39C) .... *S. novemarticulatum*
   Antenna with eight flagellomeres (Fig. 39A, B) 3
3. Tergite 5 shiny ....*S. burtoni*
   Tergite 5 not shiny 4
4. Mandible with three to five teeth on outer margin ....*S. piroonae*
   Mandible without teeth on outer margin 5
5. Hind tibia yellowish white on basal two-thirds ....*S. chumpornense*
   Hind tibia yellowish white on basal half 6
6. Frons:head ratio 1.00:4.21–4.66. Labrum 0.65–0.69 times as long as clypeus. Fore basitarsus 6.29–6.38 times as long as its greatest width ....*S. khelangense*
   Frons:head ratio 1.00:3.78–4.05. Labrum 0.57–0.59 times as long as clypeus. Fore basitarsus 5.56 times as long as its greatest width ......*S. kuvangkadilokae*
7. Posterior surface of hind tibia mostly darkened except basal tip 8
   Posterior surface of hind tibia white or yellow on basal two-fifths or more 12
8. Sensory vesicle over half as long as third palpal segment 9
   Sensory vesicle less than 0.4 times as long as third palpal segment 11
9. Fore basitarsus 7.5 times as long as its greatest width............. *S. chayamaritae*
   Fore basitarsus 5.3–6.8 times as long as its greatest width 10
10. Arms of genital fork each with short projection directed anteriorly (Fig. 20F) .... *S. chiangraiense, S. huaikaeoense, S. khaokhoense, S. maeklangense*, and *S. pamiangense*
    Arms of genital fork without such projection (similar to Fig. 23F) .... *S. ngaoense*
11. Frons:head ratio 1.0:4.5 .... *S. siamense* (complex)
    Frons:head ratio 1.0:5.4–6.4 ....*S. fukudae*
12. Hind tibia with dark subbasal marking (similar to Fig. 33B) 13
    Hind tibia without dark subbasal marking 18
13. Sensory vesicle long, 0.65 times as long as third palpal segment .... *S. adleri*
    Sensory vesicle medium-long, 0.3 times as long as third palpal segment 14
14. Abdominal segment 2 pale .... *S. cheongi* and *S. datfaense*
    Abdominal segment 2 darkened 15
15. Hind tibia whitish yellow on basal half .... *S. maleewongae*
    Hind tibia whitish yellow on basal two-thirds 16
16. Fore basitarsus 7.3–7.4 times as long as its greatest width ....*S. isanense*
    Fore basitarsus 6.5 or 6.6 times as long as its greatest width 17

17. Hind basitarsus 6.5 times as long as its greatest width .... *S. angulistylum* (complex)
Hind basitarsus 7.5–8.0 times as long as its greatest width .... *S. prayooki*
18. Base of radius with tuft of dark hairs or dark hairs mixed with few yellow hairs 19
Base of radius with tuft of yellow hairs or yellow hairs mixed with few dark hairs 23
19. Sensory vesicle 0.25–0.32 times as long as third palpal segment .... *S. wijiti*
Sensory vesicle 0.50–0.68 times as long as third palpal segment 20
20. Frons:head ratio 1.0:5.7. Sensory vesicle enlarged, 0.68 times as long as third palpal segment ....*S. sheilae*
Frons:head ratio 1.0:4.2–5.1. Sensory vesicle enlarged, 0.55–0.63 times as long as third palpal segment 21
21. Fore basitarsus 6.1 times as long as its greatest width .... *S. inthanonense*
Fore basitarsus 5.0–5.2 times as long as its greatest width 22
22. Claw with basal tooth 0.42–0.45 times length of claw ....*S. trangense*
Claw with basal tooth 0.53 times length of claw ....*S. pangsidaense*
23. Genital fork with projection directed posteromedially on each arm (Fig. 36H, I) 24
Genital fork without projection directed posteromedially on each arm 26
24. Sensory vesicle medium-long, 0.4 times as long as third palpal segment .... *S. thuathienense*
Sensory vesicle long, 0.6–0.8 times as long as third palpal segment 25
25. Fore basitarsus 5.8–6.0 times as long as its greatest width ....*S. gombakense*
Fore basitarsus 6.5–7.2 times as long as its greatest width .... *S. paiense* (complex)
26. Base of radial with tuft of yellow hairs mixed with a few dark hairs ....*S. mokroense*
Base of radial with tuft of yellow hairs 27
27. Sensory vesicle elongate, over half as long as third palpal segment 28
Sensory vesicle short to medium-long, less than 0.4 times as long as third palpal segment 30
28. Labrum 0.70–0.71 times as long as clypeus ....*S. kiewlomense*
Labrum 0.59–0.63 times as long as clypeus 29
29. Wing length 1.6 mm .... *S. klonglanense*
Wing length 2.0 mm .... S. *namdanense*
30. Mandible with distinct teeth on outer margin 31
Mandible without distinct teeth on outer margin 42
31. Body length 1.9–2.2 mm 32
Body length 2.4–2.6 mm 40
32. Claw tooth 0.50–0.53 times length of claw 33
Claw tooth 0.43–0.47 times length of claw 37
33. Claw tooth 0.53 times length of claw .... *S. udomi*
Claw tooth 0.50 times length of claw 34
34. Sensory vesicle 0.22–0.24 times length of third palpal segment ... *S. nanoiense*

Sensory vesicle 0.26–0.32 times length of third palpal segment 35
35. Labrum 0.6 times as long as clypeus ....*S. sutheppuiense*
Labrum 0.7 times as long as clypeus 36
36. Fifth palpal segment 2.6 times as long as third .... *S. loeiense*
Fifth palpal segment 2.1 times as long as third ....*S. maelanoiense*
37. Sensory vesicle 0.24 times as long as third palpal segment .... *S. puaense*
Sensory vesicle 0.26–0.35 times as long as third palpal segment 38
38. Sensory vesicle 0.32–0.35 times as long as third palpal segment ....
....*S. pitasawatae*
Sensory vesicle 0.26–0.29 times as long as third palpal segment 39
39. Spermatheca 1.3 times as long as wide ....*S. huaimorense*
Spermatheca 1.6 times as long as wide ....*S. nanthaburiense*
40. Labrum 0.54–0.56 times as long as clypeus .... *S. asakoae*
Labrum 0.61–0.68 times as long as clypeus 41
41. Claw tooth 0.50 times as long as claw .... *S. thungchangense*
Claw tooth 0.47 times as long as claw ....*S. chiangdaoense*
42. Sensory vesicle 0.36 times as long as third palpal segment. Claw tooth 0.56 times as long as claw ....*S. maewongense*
Sensory vesicle 0.25–0.32 times as long as third palpal segment. Claw tooth 0.46–0.50 times as long as claw 43
43. Fore basitarsus 5.5 times as long as its greatest width 44
Fore basitarsus 5.9–6.5 times as long as its greatest width 45
44. Labrum 0.52 times as long as clypeus ....*S. maehongsonense*
Labrum 0.65 times as long as clypeus ....*S. songense*
45. Hind basitarsus 6.5 times as long as its greatest width .... *S. myanmarense*
Hind basitarsus 5.9–6.2 times as long as its greatest width 46
46. Frons:head ratio 1.0:4.7–5.5 ....*S. junkumae*
Frons:head ratio 1.0:4.0 ....*S. muangpanense*

Males***

1. Antenna with seven flagellomeres (similar to Fig. 39C) ....*S. novemarticulatum*
Antenna with eight or nine flagellomeres 2
2. Antenna with eight flagellomeres (similar to Fig. 39A, B) 3
Antenna with nine flagellomeres 5
3. Ventral plate produced ventrally (similar to Fig. 39V) ....*S. burtoni*
Ventral plate nearly flat (Fig. 39P, S) 4
4. Upper-eye (large) facets in 15 vertical columns and 15 horizontal rows ....
.... *S. chumpornense*
Upper-eye (large) facets in 11 or 12 vertical columns and 12 or 13 horizontal rows ....*S. kuvangkadilokae* and *S. piroonae*
5. Hind basitarsus slender, much narrower than hind tibia 6
Hind basitarsus enlarged 19
6. Base of radius with tuft of dark hairs 7
Base of radius with tuft of yellow hairs 13

7. Scutum with brassy fine hairs. Abdominal segments 2 and 5–7 each with dorsolateral pair of shiny whitish-pruinose patches 8
   Scutum with yellow fine hairs. Abdominal segments 2, 6 and 7 or 2 and 5–7 each with dorsolateral pair of shiny whitish-pruinose patches 9
8. Paramere with fine setae on outer surface (Fig. 24P) ....*S. fukudae*
   Paramere without fine setae on outer surface ....*S. siamense* (complex)
9. Abdominal segments 2 and 5–7 each with dorsolateral pair of shiny white-pruinose patches ....*S. dentistylum*
   Abdominal segments 2, 6, and 7 each with dorsolateral pair of shiny whitish-pruinose patches 10
10. Upper-eye (large) facets in 13–15 vertical columns ....*S. khaokhoense, S. huaikaeoense*, and *S. ngaoense*
    Upper-eye (large) facets in 16 or 17 vertical columns 11
11. Fore basitarsus 6.6 times as long as its greatest width .... *S. pamiangense*
    Fore basitarsus 7.6–8.6 times as long as its greatest width 12
12. Fore basitarsus 7.6–7.8 times as long as its greatest width .....*S. chiangraiense*
    Fore basitarsus 7.9–8.6 times as long as its greatest width .....*S. maeklangense*
13. Style of same width from base to apical one-third when viewed ventrolaterally (Fig. 33K) 14
    Style gradually tapered when viewed ventrolaterally 15
14. Fore basitarsus 5.5 times as long as its greatest width .... *S. angulistylum* (complex)
    Fore basitarsus 7.5–7.9 times as long as its greatest width ....*S. isanense*, *S. otsukai*, and *S. prayooki*
15. Ventral plate concave ventromedially when viewed posteriorly (similar to Fig. 36Y) ....*S. maleewongae* and *S. thuathienense*
    Ventral plate rounded ventrally when viewed posteriorly (Fig. 36R, V) 16
16. Upper-eye (large) facets in nine or 10 vertical columns and 10 or 11 horizontal rows 17
    Upper-eye (large) facets in 14 or 15 vertical columns and 14–16 horizontal rows 18
17. Sensory vesicle small, globular, 0.15 times as long as third palpal segment ....*S. cheongi*
    Sensory vesicle small, ellipsoidal, 0.25 times as long as third palpal segment .... *S. adleri*
18. Fore basitarsus 6.8–7.1 times as long as its greatest width ....*S. gombakense*, *S. prayongi*, and part of *S. paiense* (genoform B)
    Fore basitarsus 7.6–8.5 times as long as its greatest width ....*S. paiense* (complex)
19. Fore coxae darkened .... *S. chayamaritae*
    Fore coxae yellow 20
20. Base of radius with tuft of dark hairs 21
    Base of radius with tuft of yellow hairs or yellow hairs with few dark hairs 26

21. Upper-eye (large) facets in 13 horizontal rows 22
Upper-eye (large) facets in 14–17 horizontal rows 23
22. Hind basitarsus whitish-yellow on basal one-third ....*S. trangense*
Hind basitarsus somewhat darkened on basal one-third, though its border not well defined .... *S. sheilae*
23. Upper-eye (large) facets in 15 or 17 vertical columns 24
Upper-eye (large) facets in 13 vertical columns 25
24. Upper-eye (large) facets in 17 vertical columns .... *S. inthanonense*
Upper-eye (large) facets in 15 vertical columns .... *S. wijiti*
25. Upper-eye (large) facets in 16 or 17 horizontal rows ....*S. curtatum*
Upper-eye (large) facets in 14 or 15 horizontal rows ....*S. pangsidaense*
26. Base of radius with tuft of yellow hairs mixed with few dark hairs ....
....*S. mokroense*
Base of radius with tuft of yellow hairs 27
27. Upper-eye (large) facets vermilion 28
Upper-eye (large) facets medium to dark brown 29
28. Upper-eye (large) facets in nine or 10 vertical columns and 12 horizontal rows. Fore basitarsus 8.4–8.7 times as long as its greatest width .... *S. thungchangense*
Upper-eye (large) facets in 11 or 12 vertical columns and 13 or 14 horizontal rows. Fore basitarsus 7.1–7.4 times as long as its greatest width .... *S. asakoae*
29. Antenna yellowish except few apical flagellomeres slightly to somewhat grayish ....*S. teerachanense*
Antenna brownish except scape, pedicel and base of first flagellomere yellowish 30
30. Hind basitarsus 0.9 time as wide as hind femur 31
Hind basitarsus 1.0–1.6 times as wide as hind femur 36
31. Ventral plate with ventral margin rounded when viewed posteriorly (Fig. 13S, T) 32
Ventral plate with ventral margin nearly straight medially when viewed posteriorly (Fig. 13U, V) 34
32. Fore basitarsus 6.4 times as long as its greatest width .... *S. banluangense*
Fore basitarsus 7.4–8.8 times as long as its greatest width 33
33. Wing length 1.6–1.7 mm ....*S. chaowaense*
Wing length 2.0 mm .... *S. myanmarense*
34. Upper-eye (large) facets in 13 vertical columns .... *S. muangpanense*
Upper-eye (large) facets in 14–16 vertical columns 35
35. Upper-eye (large) facets in 14 vertical columns .... *S. klonglanense*
Upper-eye (large) facets in 15 or 16 vertical columns ....*S. namdangense*
36. Upper-eye (large) facets in nine or 10 vertical columns .... *S. puaense*
Upper-eye (large) facets in 11 or more vertical columns 37
37. Upper-eye (large) facets in 11 vertical columns 38
Upper-eye (large) facets in 12–17 vertical columns 41

38. Head nearly as wide as thorax .... *S. loeiense*
    Head slightly wider than thorax 39
39. First antennal flagellomere 1.8–1.9 times as long as second one ....
    .... *S. sutheppuiense* and *S. maelanoiense*
    First antennal flagellomere 1.5–1.6 times as long as second one 40
40. Fore basitarsus 7.7 times as long as its greatest width .... *S. maewongense*
    Fore basitarsus 6.9 times as long as its greatest width .... *S. phapeungense*
41. Upper-eye (large) facets in 16 or 17 vertical columns 42
    Upper-eye (large) facets in 12–14 vertical columns 44
42. Subcosta bare or with two hairs. Wing length 1.9–2.0 mm ....
    .... *S. huaimorense* and *S. kiewfinense*
    Subcosta with 7–12 hairs. Wing length 2.2–2.5 mm 43
43. Ventral plate with ventral margin nearly straight when viewed posteriorly ....
    .... *S. junkumae*
    Ventral plate with ventral margin rounded when viewed posteriorly ....
    .... *S. rampae*
44. Upper-eye (large) facets in 13 or 14 horizontal rows 45
    Upper-eye (large) facets in 15 or 16 horizontal rows 47
45. Ventral plate with ventral margin rounded when viewed posteriorly ....
    .... *S. nanoiense*
    Ventral plate with ventral margin nearly straight when viewed posteriorly ....
   46
46. Upper-eye large facets in 12 vertical columns .... *S. nanthaburiense*
    Upper-eye large facets in 13 vertical columns .... *S. maehongsonense*
47. Ventral plate with ventral margin nearly straight when viewed posteriorly ....
    .... *S. pitasawatae*
    Ventral plate with ventral margin rounded when viewed posteriorly 48
48. Hind basitarsus 1.5 times as wide as hind femur .... *S. kiewlomense*
    Hind basitarsus 1.2–1.3 times as wide as hind femur ....
    .... *S. chiangdaoense* and *S. udomi*

Pupae

1. Gill of inflated structure (e.g., Fig. 40D) 2
   Gill filamentous 7
2. Gill with 12 slender filaments (Fig. 40D) .... *S. kuvangkadilokae*
   Gill with eight slender filaments 3
3. Gill without finger-like, thumb-like, or nipple-like projections (Fig. 37A) ....
   .... *S. maleewongae*
   Gill with four to six finger-like, thumb-like, or nipple-like projections (e.g., Fig. 37E) 4
4. Terminal hooks cone-like .... *S. thuathienense*
   Terminal hooks plate-like with crenulated outer margins 5

5. Inflated structure of gill bifurcated when viewed laterally, with four to six thumb-like or nipple-like projections (Fig. 37G) .... *S. prayongi*
   Inflated structure of gill nearly straight when viewed laterally, with six finger-like or thumb-like projections (Fig. 37E) 6
6. Relative width of second constriction of inflated structure of gill against that of first one 0.8–0.9 .... *S. gombakense*
   Relative width of second constriction of inflated structure of gill against that of first one 1.0–1.4 .... *S. paiense* (complex)
7. Gill with six filaments (e.g., Fig. 40B) 8
   Gill with seven, eight, or 10 filaments 10
8. Cocoon without anterodorsal projection .... *S. piroonae*
   Cocoon with anterodorsal projection 9
9. Cocoon with elongate anterodorsal projection extended beyond anteroventral tips of cocoon (similar to Fig. 14N) .... *S. udomi*
   Cocoon with short anterodorsal projection not extended beyond anteroventral tips of cocoon (similar to Fig. 24T, U) .... *S. fukudae*
10. Gill with 10 filaments (similar to Fig. 20Q) 11
    Gill with seven or eight filaments 12
11. Dorsal surface of abdominal segments 1 and 2 darkened.................................
    .... *S. chiangraiense* and *S. pamiangense*
    Dorsal surface of abdominal segments 1 and 2 not darkened ....
    .... *S. huaikaeoense, S. khaokhoense*, and *S. maeklangense*
12. Frons and anterodorsal part of thorax bare .... *S. adleri*
    Frons and anterodorsal part of thorax with tubercles 13
13. Face with two trichomes on each side (similar to Fig. 29A) ... *S. chayamaritae*
    Face with one trichome on each side 14
14. Antennal sheath with tubercles 15
    Antennal sheath without tubercles 16
15. Gill with eight short filaments arranged as 3+3+2 from dorsal to ventral..........
    .... *S. ngaoense*
    Gill with eight short filaments arranged as 2+1+3+2 from dorsal to ventral ......
    .... *S. dentistylum*
16. Common basal stalk of gill long, 1.3–1.5 times as long as interspiracular trunk. Gill with seven or eight filaments arranged as 3+2+(2+1) or 2+2+(2+1) from dorsal to ventral .... *S. rampae*
    Common basal stalk of gill short to medium-long, shorter than interspiracular trunk. Gill with eight filaments arranged otherwise 17
17. Gill filaments arranged as 1+1+2+4 from dorsal to ventral .... *S. otsukai*
    Gill filaments arranged as 3+3+2 or 6+2 from dorsal to ventral 18
18. Gill filaments arranged in two groups, i.e. one dorsal (four individual and two paired filaments) and one ventral (two paired filaments) .... *S. chumpornense*
    Gill filaments arranged in three groups (3, 3, and 2 filaments from dorsal to ventral) 19
19. Gill with short common basal stalk 20

Gill with medium-long common basal stalk 26

20. All gill filaments nearly of same thickness 21
One or two filaments of ventral pair of gill much thicker than other filaments 23

21. Dorsal surface of abdominal segments 1–5 not darkened .... *S. angulistylum* (complex)
Dorsal surface of abdominal segments 1 and 2 or 1–5 darkened 22

22. Dorsal surface of abdominal segments 1–5 darkened ....*S. isanense*
Dorsal surface of abdominal segments 1 and 2 darkened .... *S. prayooki*

23. One of two filaments of ventral pair much longer and thicker than seven other filaments. Terminal hooks wide, with crenulated outer margins 24
Two filaments of ventral pair of same length and thickness and much longer and thicker than six other filaments. Terminal hooks simple, triangular, with smooth outer margins 25

24. Basal portions of six filaments of two triplets of gill abruptly tapered .... *S. datfaense*
Basal portions of six filaments of two triplets of gill not tapered ......*S cheongi*

25. Dorsal surface of abdominal segment 9 with spine-combs. Terminal hooks triangular, with bases extended laterally when viewed posteriorly .... *S. siamense* (complex)
Dorsal surface of abdominal segment 9 without spine-combs. Terminal hooks cone-like, with bases not extended laterally when viewed posteriorly............... ....*S. novemarticulatum*

26. Dorsal and middle triples of gill consisting of one individual filament and two paired filaments with long stalk ....*S. burtoni*
Dorsal and middle triplets consisting of three filaments, or one individual filament and two paired filaments with short or medium-long stalk 27

27. Cocoon with distinct anterodorsal projection 28
Cocoon with or without anterodorsal bulge, but without anterodorsal projection 32

28. Cocoon with elongate anterodorsal projection extended much beyond anteroventral tips of cocoon (Fig. 14N) .... *S. chiangdaoense* and *S. mokroense*
Cocoon with short anterodorsal projection extended up to anteroventral tips of cocoon 29

29. Dorsum of abdominal segments 1–5 darkened .....*S. inthanonense* and *S. wijiti*
Dorsum of abdominal segments1–5 unpigmented or light yellowish 30

30. Primary stalks of dorsal and middle triplets of gill filaments nearly as long as their common stalk ....*S. maewongense* and *S. myanmarense*
Primary stalks of dorsal and middle triplets of gill filaments much shorter than their common stalk 31

31. Terminal hooks wide, each with outer margin 2.9 times as long as inner margin ....*S. huaimorense*

Terminal hooks triangular, each with outer margin 1.5–1.6 times as long as inner margin .... *S. kiewlomense*

32. Dorsum of abdominal segments 1 and 2 or 1–3 darkened .... *S. asakoae*
Dorsum of abdominal segments 1 and 2 unpigmented or light yellowish 33
33. Dorsum of abdominal segment 1 sparsely covered with minute tubercles 34
Dorsum of abdominal segment 1 without minute tubercles 35
34. Outer margin of terminal hook 2.2 times length of inner margin ..... *S. loeiense*
Outer margin of terminal hook 2.7 times length of inner margin ....
....*S. thungchangense*
35. Terminal hooks conical, each with outer margin as long as or slightly longer than inner margin and not crenulated (similar to Fig. 14G) ....
.... *S. klonglanense* and *S. namdanense*
Terminal hooks triangular or widened and plate-like, each with outer margin over 1.5 times length of inner margin, and undulate or crenulated 36
36. Terminal hooks triangular, each with outer margin 1.5–2.0 times as long as inner margin, and smooth or undulate (Fig. 27H, I) .... *S. sheilae*
Terminal hooks widened and plate-like, each with outer margin over 2.0 times length of inner margin, and crenulated 37
37. Stalk of ventral pair of gill thicker than interspiracular trunk ....
.... *S. banluangense*
Stalk of ventral pair of gill thinner than interspiracular trunk 38
38. Outer margin of terminal hook 3.3–4.0 times as long as inner margin ....
....*S. maehongsonense*, *S. pitasawatae*, and *S. trangense*
Outer margin of terminal hook 2.0–2.9 times as long as inner margin ....
...........................*S. chaowaense, S. junkumae, S. kiewfinense, S. maelanoiense, S. muangpanense, S. nanoiesne, S. phapeungense, S. puaense, S. songense*, *S. sutheppuiense*, and *S. teerachanense*

Mature larvae****

1. Abdomen with protuberances dorsally 2
Abdomen without protuberances dorsally 6
2. Abdominal segments 1 and 2 each with four pairs of protuberances (Fig. 40L)
.... *S. kuvangkadilokae*
Abdominal segments 1 and 2 each with one to three pairs of protuberances 3
3. Abdominal segments 1 and 2 each with three pairs of protuberances ....
.... *S. piroonae*
Abdominal segments 1 and 2 each with one or two pairs of protuberances
....4
4. Thoracic segment 3 and abdominal segments 1–5 each with one pair of dorsolateral protuberances (Fig. 40M, N) .... *S. novemarticulatum*
Thoracic segment 3 and abdominal segments 1–5 each with one dorsal pair and one dorsolateral pair of protuberances 5

5. Postgenal cleft long, reaching posterior margin of hypostoma. Thorax and abdomen densely covered with dark spines of various size (Fig. 23X, Z) .... *S. ngaoense*
   Postgenal cleft long, but not reaching posterior margin of hypostoma. Thorax and abdomen sparsely or moderately covered with unpigmented or dark minute setae .... *S. chumpornense*
6. Postgenal cleft long, with its apex close to or completely reaching posterior margin of hypostoma 7
   Postgenal cleft short or medium-long, less than 3.5 times as long as postgenal bridge 20
7. Abdomen markedly constricted between segments 4 and 5 when viewed dorsally (similar to Fig. 23T) .... *S. dentistylum*
   Abdomen not markedly constricted between segments 4 and 5 when viewed dorsally 8
8. Abdomen moderately covered with unbranched minute dark or unpigmented setae dorsally on segments 5–8 9
   Abdomen moderately or densely covered with branched dark spinous setae or spinules dorsally on segments 5–8 12
9. Postgenal cleft reaching posterior margin of hypostoma (Fig. 27M) ......*S. sheilae*
   Postgenal cleft with its apex close to posterior margin of hypostoma (Fig. 27L) 10
10. Posterior circlet with 71 rows of hooklets .... *S. pangsidaense*
    Posterior circlet with 90–92 rows of hooklets 11
11. Abdomen with reddish brown markings dorsally .... *S. trangense*
    Abdomen without reddish brown markings dorsally .... *S. wijiti*
12. Pharate pupal gill with six filaments .... *S. fukudae*
    Pharate pupal gill with eight or 10 filaments 13
13. Thoracic cuticle moderately covered with minute dark spinules with 7–11 branches dorsally. Pharate pupal gill with 10 filaments........ .... *S. chiangraiense, S. huaikaeoense, S. khaokhoense, S. maeklangense*, and *S. pamiangense*
    Thoracic cuticle almost bare or sparsely covered with minute spinules dorsally. Pharate pupal gill with eight filaments 14
14. Minute dark spinous setae on abdominal segments 5–8 somewhat flat and stout basally, with short branches apically .... *S. siamense* (complex)
    Minute dark spinous setae on abdominal segments 5–8 slender and hair-like basally, with long branches 15
15. Labral fan with 26 primary rays .... *S. chayamaritae*
    Labral fan with 32–41 primary rays 16
16. Rectal organ with 16–18 secondary lobules per lobe .... *S. prayooki*
    Rectal organ with 6–14 secondary lobules per lobe 17
17. Minute dark spinous setae on abdominal segments 5–8 with 5–12 (mostly 8–10) branches .... *S. angulistylum* (complex)

Minute dark spinous setae on abdominal segments 5–8 with 0–5 branches 18
18. Posterior circlet with 94–98 rows of hooklets ....*S. isanense*
Posterior circlet with less than 90 rows of hooklets 19
19. Posterior circlet with ca. 76 rows of hooklets ....*S. burtoni*
Posterior circlet with ca. 86 rows of hooklets ....*S. cheongi*
20. Pharate pupal gill inflated 21
Pharate pupal gill filamentous 23
21. Abdomen with reddish-brown markings on segments 1–8 ...... *S. maleewongae*
Abdomen with ocherous or reddish-brown markings on segments 6–8 22
22. Abdominal segments 1–4 yellow .... *S. thuathienense*
Abdominal segments 1–4 gray .... *S. gombakense* and *S. paiense* (complex)
23. Pharate pupal gill with six filaments .... *S. udomi*
Pharate pupal gill with eight filaments 24
24. Abdominal segments 1–4 unicolor 25
Abdominal segments 1–4 not unicolor 37
25. Abdominal segments 1–4 light ocherous (Fig. 15A–C) 26
Abdominal segments 1–4 grayish or greenish gray or greenish 28
26. Postgenal cleft 2.8–3.3 times as long as postgenal bridge ...... *S. teerachanense*
Postgenal cleft 1.2–2.4 times as long as postgenal bridge 27
27. Labral fan with 23 or 24 primary rays .... *S. pitasawatae*
Labral fan with 29 or 30 primary rays .... *S. sutheppuiense*
28. Pharate pupal gill with elongate common basal stalk .... *S. rampae*
Pharate pupal gill with medium-long common basal stalk 29
29. Postgenal cleft 1.5–4.0 times as long as postgenal bridge 30
Postgenal cleft 0.8–1.4 times as long as postgenal bridge 33
30. Head spots moderately positive 31
Head spots indistinct or faintly positive 32
31. Postgenal cleft 1.8–2.0 times as long as postgenal bridge .... *S. doisaketense*
Postgenal cleft 2.2–2.7 times as long as postgenal bridge .... *S. asakoae*
32. Postgenal cleft 2.5 times as long as postgenal bridge .... *S. namdanense*
Postgenal cleft 3.7–4.0 times as long as postgenal bridge ....*S. klonglanense*
33. Abdomen dark gray except segment 6 unpigmented ....*S. adleri*
Abdominal color pattern otherwise 34
34. Abdominal segments 5–8 pinkish (Fig. 15Q) .... *S. thungchangense*
Abdominal color pattern otherwise 35
35. Abdominal segments 7 and 8 light gray but abdominal segments 5 and 6 not darkened (Fig. 15I) ....*S. huaimorense*
Abdominal color pattern otherwise 36
36. Abdominal segments 5, 7, and 8 gray but abdominal segment 6 not darkened (Fig.15H) .... *S. loeiense*
Abdominal segments 5–8 grayish (Fig. 15G) .... *S. chaowaense*
37. Abdominal segment 1 gray but abdominal segments 2–4 ocherous (Fig. 15P) .... *S. maehongsonense*
Abdominal color pattern otherwise 38

38. Abdominal segments 1 and 2 gray or green but abdominal segments 3 and 4 not darkened 39
    Abdominal color pattern otherwise 40
39. Abdominal segments 7 and 8 gray (Fig. 15L) .... *S. myanmarense*
    Abdominal segments 7 and 8 not gray (Fig. 15M) .... *S. maewongense*
40. Abdominal segments 1–3 gray or greenish gray but abdominal segment 4 not darkened 41
    Abdominal color pattern otherwise 44
41. Abdominal segments 7 and 8 gray with reddish-brown pigments (Fig. 15J)..... .... *S. nanoiense*
    Abdominal segments 7 and 8 not gray with reddish-brown pigments (Fig. 15K, S) 42
42. Labral fan with 40 primary rays .... *S. inthanonense* (morphoform 1)
    Labral fan with 21–25 primary rays 43
43. Posterior circlet with 94–99 rows of hooklets .... *S. kiewlomense*
    Posterior circlet with 81–84 rows of hooklets .... *S. junkumae*
44. Abdominal segments 1 and 3 greenish gray (Fig. 15T)................................... .... *S. inthanonense* (morphoform 2)
    Abdominal color pattern otherwise 45
45. Abdominal segments 1, 3, and 4 gray but abdominal segment 2 not darkened (Fig. 15N) .... *S. phapeungense*
    Abdominal color pattern otherwise 46
46. Abdominal segments 1 and 4 grayish but abdominal segments 2 and 3 not darkened (Fig. 15O) .... *S. muangpanense*
    Abdominal segments 1–3 light green but abdominal segment 4 not green (Fig. 15R) .... *S. chiangdaoense* and *S. mokroense*

* The following seven unnamed species are not included because of insufficient information on their morphological characteristics: *S.* sp. A, *S.* sp. B, *S.* sp. C, *S.* nr. *asakoae* 2, *S.* nr. *asakoae* 3, *S.* nr. *asakoae* 4, and *S.* nr. *sheilae* 3.

***Simulium dentistylum*, *S. doisaketense*, *S. otsukai*, *S. phapeungense* and *S. prayongi* are not included because their females are unknown. *Simulium curtatum* runs to couplet 21 but cannot be further compared with *S. inthanonense* and *S. sheilae* due to the brevity of the original description of its female.

*** *Simulium datfaense*, *S. doisaketense*, and *S. khelangense* are not included because their males are unknown.

**** *Simulium khelangense* is not included because its pupa is unknown. *Simulium doisaketense* and *S. curtatum* run to couplet 31 but cannot run further.

**** *Simulium chiangdaoense*, *S. datfaense*, *S. khelangense, S. otsukai*, and *S. prayongi* are not included because their larvae are unknown. *Simulium curtatum* runs to couplet 38 but cannot run further.

Keys to Identify Six Species of the Subgenus **Montisimulium***

Females

1. Claw with small subbasal tooth (Fig. 42D) .... *S. phahompokense*
   Claw with medium-sized basal tooth (Fig. 42C) 2
2. Claw tooth 0.28 times as long as claw .... *S. surachaii*
   Claw tooth 0.32–0.38 times as long as claw 3
3. Hind basitarsus 7.3 times as long as its greatest width .... *S. angkaense*
   Hind basitarsus 6.3–6.6 times as long as its greatest width 4
4. Sensory vesicle 0.43–0.50 times as long as third palpal segment ...... *S. nanense*
   Sensory vesicle 0.6 times as long as third palpal segment (Fig. 42A) 5
5. Fifth palpal segment 1.1 times length of third .... *S. laoleense*
   Fifth palpal segment 1.9 times length of third .... *S. merga*

Males**

1. Hind basitarsus 5.4–5.9 times as long as its greatest width .... *S. angkaense*
   Hind basitarsus 4.5–4.9 times as long as its greatest width 2
2. Hind basitarsus subequal in width to hind femur (Fig. 42J) .... *S. merga*
   Hind basitarsus 0.8 or 0.9 times as wide as hind femur 3
3. Hind basitarsus 0.9 times as wide as hind femur .... *S. laoleense*
   Hind basitarsus 0.8 times as wide as hind femur .... *S. nanense*

Pupae***

1. Gill composed of two inflated tubular trunks with 14 slender filaments (Fig. 42R) .... *S. merga*
   Gill composed of 12 or 14 slender filaments 2
2. Gill with 12 slender filaments (Fig. 42S) .... *S. angkaense*
   Gill with 14 slender filaments 3
3. Outer group of four gill filaments arising at same level as dorsal and inner groups .... *S. laoleense*
   Outer group of four gill filaments arising more basally than dorsal and inner groups.. *S. nanense*

Mature larvae****

1. Antenna with hyaline bands on segment 2 (Fig. 42W). Abdomen with reddish-brown markings dorsally on segments 5–9 .... *S. angkaense*
   Antenna without hyaline bands on segment 2. Abdomen without reddish-brown markings dorsally on segments 5–9 2
2. Eyebrow well defined, dark brown .... *S. merga*
   Eyebrow not well defined, only somewhat darkened at most near both ends 3

3. Labral fan with 28–32 primary rays .... *S. laoleense*
   Labral fan with 22 primary rays .... *S. nanense*

**Simulium* sp. G is not included because its female, male, and pupa are unknown, and its larval morphological characteristics are unknown.

** *Simulium phahompokense* and *S. surachaii* are not included because their males are unknown.

*** *Simulium phahompokense* and *S. surachaii* are not included because their pupae are unknown.

**** *Simulium phahompokense* and *S. surachaii* are not included because their larvae are unknown.

Keys to Identify 10 Species of the Subgenus **Nevermannia**

Females*

1. Antenna yellow with at least first flagellomere darkened. Hind tibia yellow on basal half, dark on apical half, and with subbasal dark ring (similar to Fig. 45B) ....*S. aureohirtum*
   Antenna almost all darkened or mostly so except few basal segments pale. Hind tibia brownish 2
2. Scutum with three dark longitudinal vittae. 3
   Scutum without longitudinal vittae 6
3. Genital fork with strongly sclerotized horizontal bar at base of anterodorsal projection of each arm (Fig. 43F) ....*S. maeaiense*
   Genital fork without strongly sclerotized horizontal bar at base of anterodorsal projection of each arm (Fig. 43E) 4
4. Mandible with 17 or 18 inner teeth .... *S. vessabutrae*
   Mandible with 20–23 inner teeth 5
5. Frons:head ratio 1.0:5.4 .... *S. feuerborni* (complex)
   Frons:head ratio 1.0:4.6–5.2 ....
   .... *chiangklangense*, *S. fangense*, and *S. fruticosum*
6. Hind basitarsus 7.2 times as long as its greatest width ....
   .... *S. chomthongense* (complex)
   Hind basitarsus 6.8–6.9 times as long as its greatest width ......*S. khunklangense*

Males**

1. Antenna yellow or yellowish brown except first flagellomere darkened. Hind basitarsus slender, parallel-sided, much narrower than hind tibia (similar to Fig. 45H) .... *S. aureohirtum*

Antenna almost all darkened. Hind basitarsus inflated, nearly as wide as hind tibia (Fig. 43J) 2

2. Paramere with three to seven hooks (Fig. 43P) 3
   Paramere with one hook (Fig. 47Q) 7
3. Upper-eye (large) facets in 21 vertical columns .... *S. fangense*
   Upper-eye (large) facets in 14–19 vertical columns 4
4. Upper-eye (large) facets in 19 vertical columns ....*S. maeaiense*
   Upper-eye (large) facets in 14–17 vertical columns 5
5. Upper-eye (large) facets in 17 vertical columns. Paramere with nine hooks........ .... *S. vessabutrae*
   Upper-eye (large) facets in 14–16 vertical columns. Paramere with three to eight hooks 6
6. Upper-eye (large) facets in 16 vertical columns. Paramere with three or four hooks .... *S. feuerborni* (complex)
   Upper-eye (large) facets in 14 or 15 vertical columns. Paramere with six to eight hooks .... *S. fruticosum* and *S. wichaii*
7. Upper-eye (large) facets in 15 or 16 vertical columns .... *S. chomthongense* (complex)
   Upper-eye (large) facets in 18 vertical columns ....*S. khunklangense*

Pupae

1. Gill with four filaments (Fig. 48E) 2
   Gill with six filaments (Figs. 44A, B and 46C) 3
2. Length of common basal stalk of gill relative to interspiracular trunk 0.76–0.79 .... *S. chomthongense* (complex)
   Length of common basal stalk of gill relative to interspiracular trunk 0.50–0.65...*S. khunklangense*
3. All gill filaments divergent widely from base (Fig. 46C) .... *S. aureohirtum*
   All gill filaments directed forward close together (Fig. 44A, B) 4
4. Cocoon with anterodorsal projection (Fig. 44I, J) 5
   Cocoon without anterodorsal projection 7
5. Gill filaments arranged as 2+1+2+1 from inside to outside, lying horizontally (Fig. 44E) ....*S. maeaiense*
   Gill filaments arranged otherwise, not lying horizontally 6
6. Gill filaments arranged as 4+2 from dorsal to ventral ... *S. feuerborni* (complex)
   Gill filaments arranged as 2+1+1+2 or 3+1+2 from dorsal to ventral .... *S. fangense*
7. Common basal stalk longer than interspiracular trunk (Fig. 44B) .... *S. vessabutrae*
   Common basal stalk much shorter than interspiracular trunk 8
8. Frons and thorax covered with dark brown rather large tubercles. Facial trichome short and slender, slightly longer than frontal ones. Gill filaments covered with rather large tubercles on ridges ....*S. wichaii*

Frons and thorax covered with light brown small tubercles. Facial trichome medium-long and somewhat stout, 2.5 times as long as frontal ones. Gill filaments covered with small tubercles on ridge 9
9. Stalk of ventral paired filaments medium-long to long (Fig. 44C, D) .... *S. fruticosum*
Stalk of ventral paired filaments short (Fig. 44A) ....*S. chiangklangense*

Mature larvae***

1. Abdomen with characteristic reddish-brown markings dorsally (Fig. 44K, L). . 2
Abdomen without colored markings dorsally..7
2. Thoracic segment 1 with tawny transverse band dorsally 3
Thoracic segment 1 without colored band dorsally 4
3. Ventral surface of head capsule darkened medially .... *S. feuerborni* (complex)
Ventral surface of head capsule not darkened medially ....*S. vessabutrae*
4. Labral fan with 16 primary rays ....*S. wichaii*
Labral fan with 23 or more primary rays 5
5. Postgenal cleft 0.3 times as long as postgenal bridge (Fig. 44O) ....*S. maeaiense*
Postgenal cleft 0.5–0.6 times as long as postgenal bridge 6
6. Rectal organ with 17–21 secondary lobules per lobe .... *S. fangense*
Rectal organ with 11 or 12 secondary lobules per lobe ....*S. fruticosum*
7. Accessory sclerites present (similar to Fig. 48P)... .... *S. chomthongense* (complex) and *S. khunklangense*
Accessory sclerites absent ....*S. aureohirtum*

* *Simulium wichaii* is not included because its female is unknown.
** *Simulium chiangklangense* is not included because its male is unknown.
*** *Simulium chiangklangense* is not included because its larva is unknown.

Keys to Identify 53 Species of the Subgenus **Simulium** *s. str.*

Females*

1. Claw with small subbasal tooth (Fig. 62C) 2
Claw without tooth 15
2. Basal section of radius fully haired. Fore basitarsus with thick dorsal hair crest (similar to Fig. 62H) 3
Basal section of radius bare. Fore basitarsus with moderate dorsal hair crest 5
3. Fore tibia entirely darkened .... *S. umphangense*
Fore tibia yellow on basal half and darkened on apical half 4
4. Mandible with five or six distinct outer teeth (Fig. 62A) ....*S. nigrogilvum*
Mandible with six vestigial outer teeth ....*S. vanellum*
5. Mid and hind femora mostly yellowish. 6

Mid and hind femora mostly brownish 8
6. Hind tibia mostly darkened except base yellow .... *S. kiewmaepanense*
Hind tibia mostly yellowish except apical cap darkened 7
7. Mid tibia entirely yellowish ....*S. luculentum*
Mid tibia yellowish except posterior surface mostly darkened .....*S. chamlongi*
8. Scutum densely covered with yellow hairs.. 9
Scutum moderately covered with brassy hairs 11
9. Ovipositor valve with inner margin nearly straight (Fig. 50C) ......*S. atipornae*
Ovipositor valve with inner margin concave (similar to Fig. 84C).................10
10. Anteromedian surface of paraproct unpigmented (similar to Fig. 84E) ....*S. srisukai*
Anteromedian surface of paraproct darkened ....*S. barnesi*
11. Antenna yellow or tawny with two apical flagellomeres blackish (similar to Fig. 74A) .... *S. vanluni*
Antenna brownish-black with scape, pedicel and base of first flagellomere yellow..12
12. Ovipositor valve elongated, tapered apically (similar to Fig. 67M) ...*S. bishopi*
Ovipositor valve short, obliquely slanted posterolaterally (similar to Fig. 65D) 13
13. Abdominal segment 7 with branched hairs ventrally (similar to Fig. 65C) .... *S. siripoomense*
Abdominal segment 7 with unbranched hairs ventrally 14
14. Subcosta haired .... *S. baimaii*
Subcosta bare ....*S. lomkaoense*
15. Scutum with five distinct longitudinal vittae 16
Scutum without longitudinal vittae 37
16. Scutum with one median, two submedian, and two lateral dark vittae when illuminated anteriorly and viewed dorsally (Fig. 76B, C) 17
Scutum with one median, two submedian, and two sublateral dark vittae when illuminated anteriorly and viewed dorsally (Fig. 70C, D) 27
17. Fore tibia medium to dark brown, without white area on outer surface. Hind tibia dark brown to brownish black with base yellow. Ovipositor valve with ventrally produced round lobe along inner margin (similar to 76D, E) ....*S. tadtonense*, *S. wangkwaiense*, *S. nakhonense*, *S. chiangmaiense*, *S. phraense*, *S. poolpholi*, and *S. thilorsuense*
Fore tibia with white shiny area on outer surface. Hind tibia whitish or yellowish on basal two-thirds or more, dark brown to brownish black on rest. Ovipositor valve without such lobe along inner margin 18
18. Basal section of radius fully haired 19
Basal section of radius bare or partially haired 24
19. Ovipositor valve with elongate internal projection (Fig. 61B) 20
Ovipositor valve without such internal projection 21

20. Sensory vesicle 0.41 times as long as third palpal segment .... *S. grossifilum*
    Sensory vesicle 0.34 times as long as third palpal segment .... *S. phayaoense*
21. Ovipositor valve with 22 or 23 stout hairs, as long as or longer than those on sternite 8 (Fig. 58H) .... *S. choochotei*
    Ovipositor valve with 12–16 hairs, much shorter than those on sternite 8 22
22. Paraproct with ca. 60 medium-long to long stout hairs, several of which are much longer than those on sternite 8 (Fig. 58L) .... *S. visuti*
    Paraproct with 28–30 short to medium-long hairs, all of which are much shorter than those on sternite 8. 23
23. Cibarium with minute spinous processes near base of dorsomedian projection (Fig. 58B) .... *S. pukaengense*
    Cibarium without such spinous processes near base of dorsomedian projection .... *S. digrammicum*
24. Ovipositor valve narrowly darkened along inner margin (Fig. 58G) .... *S. suchariti*
    Ovipositor valve unpigmented along inner margin 25
25. Cibarium with medial projection shorter than its basal width (Fig. 58D). Abdominal segment 2 entirely dark brown .... *S. phukaense*
    Cibarium with medial projection much longer than its basal width (Fig. 58A). Abdominal segment 2 pale yellow on anterior half to three-fourths 26
26. Fore basitarsus 5.5 times as long as its greatest width .... *S. maenoi*
    Fore basitarsus 4.5 times as long as its greatest width .... *S. thongsahuani*
27. Mid femur entirely yellow. Spermathecae three in number (Fig. 70J) .... *S. triglobus*
    Mid femur with darkened area. Spermatheca one in number..........28
28. Hind femur yellowish almost entirely .... *S. chanyae*
    Hind femur darkened at least on more than apical half 29
29. Paraproct tapered ventrally, when viewed laterally 30
    Paraproct rounded ventrally, when viewed laterally 32
30. Cibarium without any processes dorsomedially (Fig. 70A) .... *S. malayense* (complex)
    Cibarium with minute spinous processes dorsomedially 31
31. Fore basitarsus 5.0 times as long as its greatest width .... *S. undecimum*
    Fore basitarsus 4.1 times as long as its greatest width .... *S. bullatum*
32. Ovipositor valve with 6–10 short to medium-long hairs 33
    Ovipositor valve with one to five short to medium-long hairs 34
33. Sternite 8 with ca. 40 medium-long to long hairs on each side .... *S. takense*
    Sternite 8 with 21–24 short to long hairs on each side .... *S. ubonae*
34. Basal section of radius with few hairs near apical end .... *S. chaliowae* and *S. fenestratum* (complex)
    Basal section of radius entirely bare 35
35. Fore basitarsus 5.5–5.9 times as long as its greatest width .... *S. daoense*
    Fore basitarsus 4.7–4.9 times as long as its greatest width 36
36. Sensory vesicle 0.35 times as long as third palpal segment ....*S. chainarongi*
    Sensory vesicle 0.44 times as long as third palpal segment ..... *S. lampangense*

37. Mid tarsomeres 2 and 3 entirely yellow ....*S. nodosum*
Mid tarsomeres 2 and 3 light brown to brownish black 38
38. Mid femur entirely yellow. 39
Mid femur almost dark brown 41
39. Mid tibia yellowish white with apical cap light to medium brown ....
.... *S. mediocoloratum*
Mid tibia entirely yellowish 40
40. Dorsal surface of anterior half of abdominal segment 2 medium brown...........
.... *S. crocinum*
Dorsal surface of anterior half of abdominal segment 2 white or whitish yellow .... *S. rudnicki*
41. Abdominal segment 7 with a pair of clustered hairs ventrally. 42
Abdominal segment 7 without such clustered hairs ventrally 46
42. Pair of clustered hairs on abdominal segment 7 short, subequal in length to those on surrounding area (Fig. 82B). ....*S. setsukoae*
Pair of clustered hairs on abdominal segment 7 much longer than those on surrounding area (Fig. 82C–E) 43
43. Mid tibia light brown with base somewhat yellow ....*S. weji*
Mid tibia white on basal two-fifths or more, medium brown on rest 44
44. Hind basitarsus yellow on basal half, dark brown on rest ....
.... *S. tenebrosum* (complex)
Hind basitarsus yellowish on basal three-fifths, dark brown on rest 45
45. Hind tibia yellowish white except apical two-fifths medium to dark brown......
.... *S. doipuiense* (complex)
Hind tibia brownish black except little more than basal two-fifths yellowish white ....*S. manooni*
46. Hind basitarsus white on little more than basal half, and brownish black on rest. Scutum covered with yellowish hairs as well as dark ones. ....*S. yongi*
Hind basitarsus white on basal three-fifths or more, and brownish black on rest. Scutum covered with dark brown hairs only 47
47. Sensory vesicle long, 0.6–0.7 times as long as third palpal segment ....
.... *S. tani* (complex)
Sensory vesicle medium-long, 0.3 times as long as third palpal segment..........
.... *S. brevipar* (complex) and *S. yuphae* (complex)

Males**

1. Basal section of radius fully haired.. 2
Basal section of radius bare or with few hairs apically. 7
2. Abdominal segments 2 and 5–7 each with a pair of shiny whitish pruinose spots dorsally or dorsolaterally... 3
Abdominal segments 2, 6, and 7 each with a pair of shiny white pruinose spots dorsally or dorsolaterally...4

3. Upper-eye (large) facets in 17 vertical columns. Hind basitarsus nearly parallel-sided .... *S. choochotei*
   Upper-eye (large) facets in 20 vertical columns. Hind basitarsus spindle-shaped........ *S. visuti*
4. Fore basitarsus with thick dorsal hair crest (Fig. 62H). Hind basitarsus nearly parallel-sided (Fig. 62I)........ 5
   Fore basitarsus with moderate dorsal hair crest. Hind basitarsus spindle-shaped........ 6
5. Upper-eye (large) facets in 20 vertical columns .... *S. nigrogilvum*
   Upper-eye (large) facets in 17 vertical columns .... *S. vanellum*
6. Upper-eye (large) facets in 19 vertical columns. Ventral plate rectangular, nearly parallel-sided, when viewed ventrally .... *S. pukaengense*
   Upper-eye (large) facets in 21 or 22 vertical columns. Ventral plate gradually narrowed posteriorly, when viewed ventrally. .... *S. digrammicum*
7. Scutum broadly silvery pruinose with transverse, inverted-V shaped, black band (similar to Fig. 74S, T) 8
   Scutum otherwise 9
8. Ventral plate with body narrow, parallel-sided, when viewed ventrally (similar to Fig. 74B') .... *S. vanluni*
   Ventral plate with body wide, widened medially when viewed ventrally (Fig. 74U)........ *S. nodosum*
9. Mid femur and tibia almost yellow 10
   Mid femur and tibia otherwise........ 11
10. Hind tibia brownish black except base narrowly yellow........ *S. rudnicki*
    Hind tibia dark brown to brownish black except base widely yellow........ *S. crocinum*
11. Mid tibia mostly dark (including posterior surface) 12
    Mid tibia whitish or yellowish (at least on posterior surface) on basal two-fifths or more and dark brown on rest 31
12. Fore tibia medium brown, with white shiny area on outer surface........ 13
    Fore tibia medium to dark brown, without white area on outer surface 27
13. Hind basitarsus white or whitish yellow on basal half or less and dark brown to brownish black on rest. 14
    Hind basitarsus white or whitish yellow on more than basal half and light to dark brown on rest 23
14. Hind basitarsus somewhat enlarged, 4.0–5.0 times as long as its greatest width and much narrower than greatest width of hind tibia 15
    Hind basitarsus much enlarged, 3.2–3.5 times as long as its greatest width and subequal to, or slightly wider than, greatest width of hind tibia. 18
15. Abdominal segments 2, 6, and 7 each with a pair of silvery dorsolateral spots. .... *S. atipornae*
    Abdominal segments 2 and 5–7 each with a pair of silvery dorsolateral spots 16
16. Upper-eye (large) facets in 19 vertical columns .... *S. chamlongi*
    Upper-eye (large) facets in 22–26 vertical columns 17

17. Upper-eye (large) facets in 22 or 23 vertical columns .... *S. luculentum*
    Upper-eye (large) facets in 24–26 vertical columns .... *S. srisukai*
18. Abdominal segments, 2 and 4–7 each with a pair of silvery dorsolateral spots .... *S. bishopi*
    Abdominal segments 2 and 5–7 or 2, 6, and 7 each with a pair of silvery dorsolateral spots..19
19. Abdominal segments 2 and 5–7 each with a pair of silvery dorsolateral spots .... *S. tani* (complex)
    Abdominal segments 2, 6, and 7 each with a pair of silvery dorsolateral spots 20
20. Upper-eye (large) facets in 20 vertical columns .... *S. setsukoae*
    Upper-eye (large) facets in 15–18 vertical columns. 21
21. Hind basitarsus spindle-shaped. .... *S. manooni*
    Hind basitarsus wedge-shaped (Fig. 78H) 22
22. Style with well-produced basal projection densely covered with spines (Fig. 78J) ....*S. brevipar* (complex) and *S. yuphae* (complex)
    Style with basal area somewhat produced, sparsely covered with several spines 23
23. Fore basitarsus 7.2–8.2 times as long as its greatest width................................ .... *S. tenebrosum* (complex)
    Fore basitarsus 5.9–6.1 times as long as its greatest width .... *S. doipuiense* (complex)
24. Hind basitarsus enlarged, whitish yellow on basal three-fifths or little less and dark brown on rest 25
    Hind basitarsus narrow (similar to that of female), parallel-sided, white except apical one-sixth light brown (Fig. 65K) 26
25. Hind basitarsus spindle-shaped .... *S. lomkaoense*
    Hind basitarsus wedge-shaped .... *S. weji*
26. Hind tibia entirely brownish black. .... *S. siripoomense*
    Hind tibia brownish black except base narrowly yellow. .... *S. baimaii*
27. Abdominal segments 2, 6, and 7 each with a pair of silvery dorsolateral spots. .... *S. phukaense*
    Abdominal segments 2 and 5–7 each with a pair of silvery dorsolateral spots 28
28. Hind basitarsus whitish yellow on basal three-fifths and brownish black on apical two-fifths............ *S. chiangmaiense*, *S. maeklongkeense*, *S. nakhonense*, *S. phraense*, *S. tadtonense*, and *S. thailandicum*
    Hind basitarsus whitish yellow on basal half or slightly more, brownish black on rest..29
29. Scutum with yellow short hairs .... *S. wangkawiense*
    Scutum with brassy short hairs 30
30. Upper-eye (large) facets in 16 or 17 vertical columns .... *S. poolpholi*
    Upper-eye (large) facets in 20 vertical columns .... *S. thilorsuense*
31. Hind basitarsus whitish on basal half or little more and brownish black on rest. Body of ventral plate much wider than long 32

Hind basitarsus entirely light to dark brown, or so with basal two-fifths to one-third whitish yellow or dark yellow. Body of ventral plate as long as or longer than wide 35

32. Scutum covered with dark brown recumbent hairs only.......... *S. yongi*
Scutum covered with golden-yellow recumbent hairs interspersed with brown hairs anteriorly 33
33. Ventral plate with body nearly quadrate, shorter than arms .... *S. mediocoloratum*
Ventral plate with body rectangular, longer than arms (Fig. 61H) 34
34. Abdominal segments 2 and 5–7 each with a pair of silvery dorsolateral spots .... *S. grossifilum*
Abdominal segments 2, 6, and 7 each with a pair of silvery dorsolateral spots .... *S. phayaoense*
35. Hind basitarsus entirely light to dark brown. 36
Hind basitarsus brownish black with basal two-fifths to one-third whitish yellow or dark yellow 38
36. Ventral plate subquadrate when viewed ventrally, and with distinct teeth on posterior surface. .... *S. chaliowae*
Ventral plate gradually narrowed posteriorly when viewed ventrally, and without teeth on posterior surface (Fig. 59B) 37
37. Fore basitarsus 6.8 times as long as its greatest width.......... *S. maenoi*
Fore basitarsus 5.4 times as long as its greatest width .... *S. thongsahuani*
38. Upper-eye (large) facets in 20 or 21 vertical columns and 20 or 21 horizontal rows 39
Upper-eye (large) facets in 15–18 vertical columns and 16–19 horizontal rows 42
39. Ventral plate without setae .... *S. undecimum*
Ventral plate with setae 40
40. Ventral plate without teeth (Fig. 59G) .... *S. suchariti*
Ventral plate with teeth (Fig. 71P)..........41
41. Ventral plate with posterior margin rounded posteriorly when viewed ventrally .... *S. chainarongi*
Ventral plate with posterior margin concave when viewed ventrally.......... *S. takense*
42. Upper-eye (large) facets in 15 vertical columns and 16 horizontal rows. Ventral plate with two vertical rows of teeth nearly parallel-sided on posterior surface. 43
Upper-eye (large) facets in 17 or 18 vertical columns and 17–19 horizontal rows. Ventral plate with teeth irregularly situated on posterior surface. 44
43. Ventral plate nearly as long as wide .... *S. ubonae*
Ventral plate 1.7 times as long as wide .... *S. malayense* (complex)
44. Body of ventral plate hexagonal when viewed ventrally.......... *S. triglobus*
Body of ventral plate quadrate or subquadrate when viewed ventrally 45

45. Sensory vesicle 0.3–0.4 times as long as third palpal segment .... *S. daoense*
    Sensory vesicle 0.2 times as long as third palpal segment 46
46. Basal projection of style with one spine .... *S. fenestratum* (complex)
    Basal projection of style with several spines 47
47. Mid tibia whitish except apical cap darkened .... *S. chanyae*
    Mid tibia whitish on basal two-fifths, then gradually darkened toward apex......
    .... *S. lampangense*

Pupae***

1. Gill with two or three inflated tubes 2
   Gill with 6–10 filaments 4
2. Gill with two inflated tubes 3
   Gill with three inflated tubes with rounded apex (Fig. 75F). .... *S. nodosum*
3. Gill with long inflated common basal stalk (Fig. 66C) .... *S. baimaii*
   Gill with short common basal stalk (Fig. 66B) ....*S. lomkaoense*
4. Gill with six filaments 5
   Gill with 8–10 filaments 35
5. Gill with six inflated tubular filaments with minute spines (Fig. 61M, N)....... 6
   Gill with six thread-like filaments (with basal portions somewhat inflated or not) without minute spines 7
6. Terminal hooks present ....*S. grossifilum*
   Terminal hooks absent. ....*S. phayaoense*
7. Cocoon wall-pocket shaped. 8
   Cocoon shoe-shaped or boot-shaped. 27
8. Cocoon with lateral windows anteriorly 9
   Cocoon without lateral window. 15
9. Gill filaments subequal in length and thickness to one another 10
   Gill filaments decreasing in thickness from dorsal to ventral 13
10. Gill filaments short-stalked. Spine-combs on abdominal segments 7and 8 or 7–9 11
    Gill filaments almost sessile. Spine-combs present only on abdominal segment 8 12
11. Spine-combs on abdominal segments 7 and 8 ....*S. atipornae*
    Spine-combs on abdominal segments 7–9 .... *S. siripoomense*
12. Small raised area near base of gill with tubercles................. *S. mediocoloratum*
    Small raised area near base of gill bare .... *S. rudnicki*
13. Cocoon with small anterolateral window on each side (Fig. 60P). Terminal hooks present...*S. maenoi*
    Cocoon with moderate-sized lateral window on each side (Fig. 60Q). Terminal hooks absent...14
14. Integument of head densely covered with tubercles .... *S. suchariti*
    Integument of head bare.. *S. crocinum*
15. Integument of head and thorax bare (except posterior half of thorax with minute tubercles) (Fig. 83A) 16

Integument of head and thorax moderately or densely covered with tubercles 18

16. Cocoon very thin, transparent, and its anterior margin often not well defined (Fig. 53I). Inner filament of ventral pair of gill narrowed basally. Spine-combs only on dorsum of abdominal segment 8. Terminal hooks absent .... *S. yongi*
Cocoon thickly woven, not transparent, and its anterior margin well defined. Inner filament of ventral pair of gill not narrowed basally. Spine-combs on dorsum of abdominal segments 7–9. Terminal hooks present 17
17. Upper filament of dorsal pair of gill more than twice as thick as interspiracular trunk (Fig. 84P) ....*S. srisukai*
Upper filament of dorsal pair of gill less than twice as thick as interspiracular trunk .... *S. chamlongi*
18. Frontal trichomes each with 8–12 slender branches (similar to Fig. 60B) ....*S. thongsahuani*
Frontal trichomes unbranched 19
19. Gill with short common basal stalk (similar to Fig. 83D) 20
Gill without common basal stalk (Fig. 79A) 23
20. Terminal hooks present 21
Terminal hooks absent 22
21. Basal portion of upper filament of dorsal pair of gill nearly as thick as interspiracular trunk .... *S. kiewmaepanense*
Basal portion of upper filament of dorsal pair of gill 1.6–1.8 times as thick as interspiracular trunk ....*S. luculentum*
22. Frons densely covered with tubercles (similar to Fig. 83C) ....*S. tenebrosum* (complex)
Frons moderately covered with tubercles.(similar to Fig. 83B) ....*S. doipuiense* (complex)
23. Thoracic integument with pit-like organ at base of gill (Fig. 79H) 24
Thoracic integument without pit-like organ at base of gill 25
24. Lower filament of ventral pair of gill directed forward ... *S. brevipar* (complex)
Lower filament of ventral pair of gill directed downward and backward (Fig. 79D).. . *S. yuphae* (complex)
25. Abdominal segment 7 without spine-combs dorsally ....*S. weji*
Abdominal segment 7 with spine-combs dorsally 26
26. Inner filament of ventral pair of gill much thinner than outer filament. Terminal hooks present ....*S. tani* (complex)
Two filaments of ventral pair of gill subequal in thickness (Fig. 83G). Terminal hooks absent ....*S. manooni* and *S. setsukoae*
27. Cocoon loosely woven, with many small open spaces (Fig. 60S) 28
Cocoon tightly woven 29
28. Gill filaments somewhat inflated, larger in diameter than interspiracular trunk (Fig. 60J). Terminal hooks large, longer than their basal width ....*S. choochotei*

Gill filaments not so inflated, smaller in diameter than interspiracular trunk (Fig. 60F). Terminal hooks of moderate size, shorter than their basal width (Fig. 60L) .... *S. visuti*

29. Cocoon with anterolateral flap and small anterolateral open space on each side (Fig. 60R) .... *S. digrammicum*
    Cocoon without such flap and open space 30
30. Gill filaments decreasing in length and thickness from dorsal to ventral. Facial and frontal trichomes branched. 31
    Gill filaments subequal in length and thickness to one another. Facial and frontal trichomes unbranched. 32
31. Gill filaments light yellowish-white to yellowish-brown. Terminal hooks present .... *S. pukaengense*
    Gill filaments dark brown. Terminal hooks absent .... *S. phukaense*
32. Abdominal segment 8 with spine-combs dorsally 33
    Abdominal segment 8 without spine-combs dorsally. 34
33. Frons bare .... *S. bishopi*
    Frons sparsely or moderately covered with tubercles .... *S. vanluni*
34. Two filaments of ventral pair of gill thicker than interspiracular trunk (Fig. 63A) .... *S. nigrogilvum*
    Two filaments of ventral pair of gill thinner than interspiracular trunk .... *S. vanellum*
35. Gill with eight filaments 36
    Gill with 10 filaments. 47
36. Cocoon wall-pocket shaped. 37
    Cocoon shoe-shaped. 42
37. Cocoon without lateral windows 38
    Cocoon with lateral windows 39
38. Stalks of dorsal and ventral pairs widely divergent at 180 degrees when viewed laterally .... *S. chanyae*
    Stalks of dorsal and ventral pairs of gill divergent at 90 degrees when viewed laterally .... *S. malayense* (complex)
39. Gill with large unpigmented basal fenestra (Fig. 72H) .... *S. bullatum*
    Gill without large basal fenestra 40
40. Gill filaments widely divergent, angle between stalks of dorsal and ventral pairs more than 90 degrees when viewed laterally (Fig. 72G). .... *S. lampangense*
    Gill filaments moderately divergent, angle between stalks of dorsal and ventral pairs 90 degrees or less when viewed laterally. 41
41. Spine-combs only on dorsum of abdominal segment 8 .... *S. daoense*
    Spine-combs on dorsum of abdominal segments 7 and 8 .... *S. fenestratum* (complex) and *S. undecimum*
42. Cocoon roughly woven anteriorly, leaving some large open spaces in anterior collar and many small open spaces near anterior margin (Fig. 72M) 43
    Cocoon without open spaces 44
43. All gill filaments subequal in thickness to one another. .... *S. triglobus*

Two filaments of ventral pair of gill much thinner than others (Fig. 77E).......... ....*S. chiangmaiense*

44. Cocoon with open space anterolaterally on each side ....*S. ubonae*
Cocoon without open space anterolaterally on each side (Fig. 72L) 45
45. At least two gill filaments directed posteriorly (Fig. 72C). Terminal hooks present .... *S. takense*
Gill filaments not directed posteriorly. Terminal hooks absent 46
46. Basal portion of upper filament of dorsal pair of gill ca. 1.4 times as thick as lower filament of ventral pair .... *S. chaliowae*
Basal portion of upper filament of dorsal pair of gill about twice as thick as lower filament of ventral pair .... *S. chainarongi*
47. Gill filaments lacking distinct ridges and furrows (Fig. 77D) 48
Gill filaments with ridges and furrows (Fig. 77A)...49
48. Five lower filaments of gill subequal in thickness to one another and about half as thick as five upper filaments, which are subequal in thickness to one another (Fig. 77D)...*S. phraense*
Two lower filaments of gill subequal in thickness to each other and less than half as thick as eight other filaments, which are subequal in thickness to one another .... *S. poolpholi*
49. Gill filaments arranged as 2+3+3+2 from dorsal to ventral. All gill filaments subequal in thickness to one another. 50
Gill filaments arranged as 2+2+2+2+2 or 2+1+2+1+2+2 from dorsal to ventral...52
50. Two filaments of ventral pair of gill always thinner than other filaments .... *S. tadtonense*
All gill filaments subequal in length to one another.......................................51
51. Tubercles on frons with tiny secondary projections .... *S. wangkwaiense*
Tubercles on frons without tiny secondary projections .... *S. thilorsuense*
52. Head integument with round tubercles .... *S. nakhonense*
Head integument with angulated tubercles .... *S. thailandicum*

Mature larvae****

1. Pharate pupal gill with two filaments 2
Pharate pupal gill with three, six, eight, or 10 filaments 3
2. Pharate pupal gill with filaments arising from short stalk .... *S. baimaii*
Pharate pupal gill with filaments lacking stalk .... *S. lomkaoense*
3. Pharate pupal gill with three filaments .... *S. nodosum*
Pharate pupal gill with 6–10 filaments 4
4. Pharate pupal gill with six filaments... 5
Pharate pupal gill with eight or 10 filaments 32
5. Last abdominal segment with accessory sclerite ventrally (Fig. 63J)............... 6
Last abdominal segment without accessory sclerite ventrally 7
6. Mandible with supernumerary serration (Fig. 63F) .... *S. nigrogilvum*
Mandible without supernumerary serration .... *S. vanellum*

7. Abdominal segments 1–5 (or up to 8) each with one or more pairs of protuberances dorsally or dorsolaterally 8
   Abdominal segments lacking protuberances...9
8. Abdominal segments 1–8 each with one to six pairs of protuberances dorsally and dorsolaterally (similar to Fig. 66L) .... *S. siripoomense*
   Abdominal segments 1–5 each with one pair of protuberances dorsally (similar to Fig. 75K)..*S. vanluni*
9. Abdomen with its greatest width on segment 8... 10
   Abdomen with its greatest width on segment 6 or 7 13
10. Cephalic apotome mostly pale yellow. Body length 6.5–8.0 mm .... *S. yongi*
    Cephalic apotome mostly light to medium brown. Body length 8.1–11.2 mm...... 11
11. Posterior circlet with ca. 210 rows of hooklets ....*S. choochotei*
    Posterior circlet with 136–160 rows of hooklets 12
12. Posterior circlet with 144–160 rows of hooklets ....*S. rudnicki*
    Posterior circlet with 136–138 rows of hooklets ....*S. visuti*
13. Postgenal cleft widely rounded apically 14
    Postgenal cleft nearly pointed apically 22
14. Body longer than 7.0 mm 15
    Body shorter than 6.9 mm 16
15. Posterior circlet with 87–93 rows of hooklets ....*S. srisukai*
    Posterior circlet with ca. 120 rows of hooklets .... *S. suchariti*
16. Postgenal cleft moderately widened medially, its greatest width much longer than width at base 17
    Postgenal cleft not widened or slightly so medially, its greatest width as long as or slightly longer than width at base..18
17. Body length 5.2–5.6 mm. Rectal organ with 6–10 secondary lobules per lobe.... .... *S. chamlongi*
    Body length 5.9–6.3 mm. Rectal organ with 11–16 secondary lobules per lobe .... *S. luculentum*
18. Rectal organ with 19–28 secondary lobules per lobe ....*S. pukaengense*
    Rectal organ with 7–17 secondary lobules per lobe.......................................19
19. Labral fan with 30–36 primary rays..*S. maenoi*
    Labral fan with 38–50 primary rays 20
20. Labral fan with 38–40 primary rays... *S. digrammicum*
    Labral fan with 46–50 primary rays 21
21. Posterior circlet with 140–143 rows of hooklets ....*S. thongsahuani*
    Posterior circlet with ca. 104 rows of hooklets ....*S. bishopi*
22. Body length 5.8–8.0 mm 23
    Body length shorter than 5.6 mm 27
23. Pharate pupal gill with inflated filaments 24
    Pharate pupal gill with slender filaments 25
24. Postgenal cleft 2.4–2.8 times as long as postgenal bridge (Fig. 61T). Posterior circlet with 84–98 rows of hooklets... *S. phayaoense*

Postgenal cleft 1.8–2.0 times as long as postgenal bridge. Posterior circlet with ca. 120 rows of hooklets .... *S. grossifilum*
25. Sheath of subesophageal ganglion unpigmented (Fig. 50S) .... *S. atipornae*
Sheath of subesophageal ganglion dark pigmented 26
26. Eyebrow distinct .... *S. mediocoloratum*
Eyebrow indistinct .... *S. crocinum*
27. Body color reddish brown .... *S. tani* (complex)
Body color otherwise 28
28. Anterior half of cephalic apotome broadly dark medially .... *S. weji*
Anterior half of cephalic apotome entirely pale 29
29. Eye-spot region surrounded by darkened area .... *S. brevipar* (complex) and *S. yuphae* (complex)
Eye-spot region not surrounded by darkened area 30
30. Eyebrow well defined .... *S. manooni*
Eyebrow weakly defined or indistinct 31
31. Eyebrow indistinct .... *S. doipuiense* (complex)
Eyebrow weakly defined .... *S. tenebrosum* (complex)
32. Pharate pupal gill with eight filaments 33
Pharate pupal gill with 10 filaments 43
33. Pharate pupal gill with enlarged basal fenestra .... *S. bullatum*
Pharate pupal gill with small basal fenestra 34
34. Abdomen with protuberances dorsally 35
Abdomen without protuberances dorsally 40
35. Abdominal segments 1–8 each with a pair of protuberances dorsally .... *S. chiangmaiense*
Abdominal segments 1–6 or 2–7 each with a pair of protuberances dorsally 36
36. Rectal organ with 15–20 secondary lobules per lobe .... *S. fenestratum* (morphoform A)
Rectal organ with 7–14 secondary lobules per lobe 37
37. Head spots positive 38
Head spots indistinct 39
38. Rectal organ with 13 or 14 secondary lobules per lobe .... *S. ubonae*
Rectal organ with seven to nine secondary lobules per lobe .... *S. chainarongi*
39. Abdominal segments 3–7 each with a pair of protuberances dorsally .... *S. lampangense*
Abdominal segments 2–6 each with a pair of protuberances dorsally .... *S. chaliowae*
40. Rectal organ with 15–22 secondary lobules per lobe .... *S. daoense* and *S. fenestratum* (morphoform B)
Rectal organ with less than 15 secondary lobules per lobe 41
41. Labral fan with 56–60 primary rays .... *S. triglobus*
Labral fan with 40–45 primary rays 42
42. Rectal organ with 8–10 secondary lobules per lobe .... *S. malayense* (complex)

Rectal organ with 13 or 14 secondary lobules per lobe................ *S. undecimum*

43. Pharate pupal gill with 10 filaments arranged as 2+3+3+2 from dorsal to ventral 44
Pharate pupal gill with 10 filaments arranged otherwise (e.g., 2+2+2+2+2 or 2+1+2+1+2+2) from dorsal to ventral 45
44. Labral fan with 39–42 primary rays. Posterior circlet with 127–134 rows of hooklets .... *S. thilorsuense*
Labral fan with 48–51 primary rays. Posterior circlet with 115–122 rows of hooklets .... *S. wangkwaiense*
45. Pharate pupal gill with 10 filaments with ridges and furrows .... *S. nakhonense*
Pharate pupal gill with 10 filaments without distinct ridges and furrows 46
46. Labral fan with 42–46 primary rays ....*S. phraense*
Labral fan with 36–38 primary rays .... *S. poolpholi*

* *Simulium thailandicum* is not included because its female is unknown.

** *Simulium barnesi*, *S. bullatum*, *S. kiewmaepanense*, and *S. umphangense* are not included because their males are unknown.

****Simulium barnesi* and *S. umphangense* are not included because their pupae are unknown.

**** *Simulium barnesi*, *S. chiangmaiense*, *S. chanyae*, *S. kiewmaepanense*, *S. maeklongkeense*, *S. phukaense*, *S. setsukoae*, *S. tadtonense*, *S. takense*, *S. thailandicum*, and *S. umphangense* are not included because their larvae are unknown. Morphoform A and morphoform B of *S. fenestratum* herein designated are based on Takaoka et al. (2000) and Thaijarern et al. (2018), in which two types of larvae are noted in populations of *S. fenestratum* from Sumatra and Thailand: one with protuberances on the dorsum of the abdomen and the other without protuberances. Whether the difference between these two morphoforms is intraspecific should be confirmed.

7 The Fauna of Black Flies in Vietnam

Overview

In Vietnam, Crosskey and Howard (1997) were the first to record seven species of black flies: *Simulium* (*Gomphostilbia*) *siamense* (complex), *S.* (*G.*) *montiblense*, *S.* (*S.*) *chamlongi*, *S.* (*Simulium*) *nigrogilvum*, *S.* (*S.*) *nodosum*. *S.* (*S.*) *nakhonense*, and *S.* (*S.*) *thailandicum*. Pham (1998, 1999) recorded nine species of black flies: *S.* (*G.*) *asakoae*, *S.* (*G.*) *duolongum*, *S.* (*G.*) *inthanonense*, *S.* (*S.*) *chiangmaiense*, *S.* (*S.*) *hirtinervis*, *S.* (*S.*) *jeffreyi*, *S.* (*S.*) *malayense* (complex), *S.* (*S.*) *quinquestriatum*, and *S.* (*S.*) *tani* (complex).

Takaoka et al. (2014c) described a new species, *S.* (*S.*) *vietnamense*, and recorded four species: *S.* (*S.*) *chungi*, *S.* (*S.*) *grossifilum*, *S.* (*S.*) *maenoi*, and *S.* (*S.*) *rufibasis*, based on adult female specimens collected on humans in Tam Dao National Park, Vin Phuc Province. Takaoka et al. (2014d) described four new species and recorded three species: *S.* (*G.*) *brinchangense*, *S.* (*Nevermannia*) *aureohirtum*, and *S.* (*S.*) *brevipar* (complex), from Vin Phuc Province.

Takaoka et al. (2014e, 2015b) described 17 new species and recorded one species, *S.* (*S.*) *atipornae*, from Thua Thien Hue and Lam Dong Province. Takaoka et al. (2017d) published a monograph, in which 22 new species including *S.* (*G.*) *sanchayense* (the species formerly recognized as *S.* (*G.*) *brinchangense*), *S.* (*S.*) *lowi* (the species formerly recognized as *S.* (*S.*) *brevipar*), *S.* (*S.*) *fuscicoxae* (the species formerly regarded as *S.* (*S.*) *rufibasis*), and *S.* (*S.*) *suoivangense* [the species formerly recognized as morphoform "b" of *S.* (*S.*) *tani* (complex)] were described; three species, *S.* (*G.*) *parahiyangum*, *S.* (*Nevermannia*) *maeaiense*, and *S.* (*S.*) *doipuiense* (complex) [the species formerly regarded as *S.* (*S.*) *rufibasis* (in part)] were newly recorded; and the substitute name, *S.* (*S.*) *huense*, was given for the species described under the name of *S.* (*S.*) *cavum*. Adler et al. (2016) recorded *S.* (*S.*) *brevipar* (complex) and *S.* (*S.*) *yuphae* (complex) in Vietnam as cytoform "C" and cytoform "N," respectively. After that, three new species, *S.* (*G.*) *yvonneae*, *S.* (*N.*) *pumatense*, and *S.* (*S.*) *yukawai*, were described from Vietnam by Takaoka et al. (2019b, c, h).

Putt et al. (2023) established DNA barcoding of 45 species of black flies in Vietnam.

The fauna of black flies in Vietnam is represented by 74 species, although nine of 16 species recorded by Crosskey and Howard (1997) and Pham (1998, 1999) should be reconfirmed for their distribution records in Vietnam by future studies. All are assigned in the genus *Simulium* s. l. and further placed to four subgenera: 25 species in the subgenus *Gomphostilbia*, one species in the subgenus *Montisimulium*, eight species in the subgenus *Nevermannia,* and 40 species in the subgenus *Simulium* s. str. The rate of endemic species is 59.5% (44/74 species).

List of 74 Species of Black Flies in Vietnam

Subgenus *Gomphostilbia*

S. asakoae species-group

S. asakoae, *S. chaudinhense*, *S. confertum*, *S. fuscidorsum*, *S. hmongense*, *S. hongthaii*, *S. inthanonense**, *S. phulocense*, *S. quychauense*, *S. sanchayense*, *S. tamdaoense*, *S. thituyenae*, and *S. vinhphucense*

S. batoense species-group

S. dachaisense, *S. duolongum**, *S. lamdongense*, *S. longlanhense*, *S. parahiyangum*, *S. siamense* (complex)*, and *S. yvonneae*

S. ceylonicum species-group

S. unii

S. darjeelingense species-group

S. eshimai

S. epistum species-group

*S. montiblense**
S. gombakense species-group
S. thuathienense
S. varicorne species-group
S. breviflagellum
Subgenus *Montisimulium*
S. nigrofilum
Subgenus *Nevermannia*
S. feuerborni species-group
S. bachmaense, *S. langbiangense*, *S. maeaiense*, *S. phami*, and *S. pumatense*
S. ruficorne species-group
S. aureohirtum
S. vernum species-group
S. laichauense and *S. tayense*
Subgenus *Simulium* s. str.
S. argentipes species-group
S. sansahoense
S. christophersi species-group
S. atipornae
S. chungi species-group
S. chungi
S. crocinum species-group
S. laocaiense, *S. turgidum*, *S. vietnamense*, and *S. yukawai*
S. griseifrons species-group
S. maenoi
S. grossifilum species-group
S. grossifilum
S. indicum species-group
S. nigrogilvum
S. malyschevi species-group
S. obliquum
S. multistriatum species-group
S. daoense, *S. hirtinervis**, *S. lacduongense*, *S. laui*, and *S. malayense* (complex)
S. nobile species-group
S. nodosum
S. striatum species-group
S. chiangmaiense, *S. jeffreyi**, *S. nakhonense*, *S. quinquestriatum**, *S. tavanense*, *S. taythienense*, *S. thailandicum**, and *S. xuandai*
S. tuberosum species-group
S. brevipar (complex), *S. congi*, *S. doipuiense* (complex), *S. fuscicoxae*, *S. giayense*, *S. huense*, *S. lowi*, *S. rosliramlii*, *S. sapaense*, *S. suoivangense*, *S. tani* (complex), *S. xuandei*, and *S. yuphae* (complex)
S. variegatum species-group
*S. chamlongi** and *S. phuluense*

* These nine species should be reconfirmed for their distributions in Vietnam because of the lack of detailed information for the localities and specimens examined, or because of the lack of information for adult morphological characteristics.

*Keys to Identify 74 Species of Black Flies in Vietnam**

Females**

1. Katepisternum haired 2
   Katepisternum bare 19
2. Claw with small subbasal tooth (Fig. 21I) .... *S.* (*G.*) *dachaisense*
   Claw with large basal tooth 3
3. Hind tibia darkened except base yellow 4
   Hind tibia darkened on apical one-third to one-half and yellow on rest 11
4. Sensory vesicle elongate, over 0.43 times length of third palpal segment 5
   Sensory vesicle medium-long, less than 0.40 times length of third palpal segment 7
5. Sensory vesicle 0.6 times length of third palpal segment. Mid and hind femora yellow except apical cap darkened .... *S.* (*G.*) *parahiyangum*
   Sensory vesicle 0.43–0.50 times length of third palpal segment. Mid and hind femora light to medium brown except base yellow 6
6. Labrum 0.6 times length of clypeus. Sensory vesicle long, 0.43–0.45 times length of third palpal segment .... *S.* (*G.*) *lamdongense*
   Labrum 0.8 times length of clypeus. Sensory vesicle long, 0.50 times length of third palpal segment .... *S.* (*G.*) *eshimai*
7. Fore coxa light brown .... *S.* (*G.*) *longlanhense*
   Fore coxa yellowish 8
8. Sternite 8 with 24–28 stout hairs on each side .... *S.* (*G.*) *montiblense*
   Sternite 8 with 9–13 stout hairs on each side 9
9. Fore basitarsus 6.3–6.4 times as long as its greatest width .....*S.* (*G.*) *yvonneae*
   Fore basitarsus 5.4–5.8 times as long as its greatest width 10
10. Frons:head ratio 1.0:5.0 .... *S.* (*G.*) *duolongum*
    Frons:head ratio 1.0:4.5 .... *S.* (*G.*) *siamense* (complex)
11. Sensory vesicle elongate, 0.6 times length of third palpal segment 12
    Sensory vesicle short to medium-long, less than 0.4 times length of third palpal segment 13
12. Frons:head ratio 1.0:5.2 .... *S.* (*G.*) *fuscidorsum*
    Frons:head ratio 1.0:4.3 .... *S.* (*G.*) *thituyenae*
13. Hind tibia darkened on apical one-third 14
    Hind tibia darkened on apical half 18
14. Mandible with 15 outer teeth .... *S.* (*G.*) *thuathienense*
    Mandible lacking outer teeth, or with one to six outer teeth 15

15. Outer margin of mandible without tooth .... *S.* (*G.*) *hongthaii*
    Outer margin of mandible with teeth 16
16. Outer margin of mandible with two teeth ....*S.* (*G.*) *unii*
    Outer margin of mandible with three to six teeth 17
17. Outer margin of mandible with five or six teeth ....*S.* (*G.*) *asakoae*
    Outer margin of mandible with three or four teeth .... *S.* (*G.*) *chaudinhense*
18. Base of radius with tuft of brown hairs............................ *S.* (*G.*) *inthanonense*
    Base of radius with tuft of yellow hairs..
    .... *S.* (*G.*) *confertum* and *S.* (*G.*) *phulocense*
19. Claw with large basal tooth 20
    Claw simple or with small- or medium-sized tooth 27
20. Fore coxa darkened 21
    Fore coxa yellow 22
21. Labrum 0.7 times as long as clypeus. Sensory vesicle half as long as third palpal segment .... *S.* (*N.*) *laichauense*
    Labrum 0.9 times as long as clypeus. Sensory vesicle 0.6–0.7 times as long as third palpal segment .... *S.* (*N.*) *tayense*
22. Scutum without dark longitudinal vittae .... *S.* (*N.*) *aureohirtum*
    Scutum with three dark longitudinal vittae 23
23. Cibarium with 16 tiny conical processes ....*S.* (*N.*) *pumatense*
    Cibarium with 24–32 tiny conical processes (Fig. 43B) 24
24. Sensory vesicle with large opening (Fig. 43A) ....*S.* (*N.*) *langbiangense*
    Sensory vesicle with opening of moderate size 25
25. Genital fork with strongly sclerotized horizontal bar at base of anteriorly directed projection on each arm (Fig. 43F) .... *S.* (*N.*) *maeaiense*
    Genital fork without horizontal bar at base of anteriorly directed projection on each arm 26
26. Genital fork with rounded posteromedial projection on each arm ....
    ....*S.* (*N.*) *phami*
    Genital fork with pointed posteromedial projection on each arm.......................
    .... *S.* (*N.*) *bachmaense*
27. Claw with medium-sized tooth (similar to Fig. 42C) ....*S.* (*M.*) *nigrofilum*
    Claw simple or with small subbasal tooth 28
28. Claws simple (without basal or subbasal tooth) 29
    Claws with subbasal tooth 46
29. Basal section of radius haired 30
    Basal section of radius bare 34
30. Hind tibia darkened except base whitish 31
    Hind tibia white on basal two-thirds or more and dark on rest 32
31. Scutum with copper-colored short hairs .... *S.* (*S.*) *jeffreyi*
    Scutum with yellow or whitish yellow short hairs ....
    .... *S.* (*S.*) *chiangmaiense, S.* (*S.*) *nakhonense, S.* (*S.*) *quinquestriatum, S.* (*S.*) *tavanense*, and *S.* (*S.*) *taythienense*
32. Scutum without longitudinal vittae ....*S.* (*S.*) *vietnamense*
    Scutum with five longitudinal vittae (Fig. 70C) 33

33. Ovipositor valve tapered posteromedially (Fig. 51B) .... *S.* (*S.*) *chungi*
Ovipositor valve rounded posteromedially (similar to Fig. 70E) ....
.... *S.* (*S.*) *lacduongense* and *S.* (*S.*) *hirtinervis*
34. Ovipositor valve with elongate projection on inner surface (similar to Fig. 61B)
.... *S.* (*S.*) *grossifilum*
Ovipositor valve without projection on inner surface 35
35. Scutum with distinct longitudinal vittae. Cibarium with dorsomedian projection
.... 36
Scutum with faint longitudinal vittae or absent. Cibarium without dorsomedian projection 38
36. Dorsomedian projection of cibarium covered with numerous spinous processes (Fig. 58A)... *S.* (*S.*) *maenoi*
Dorsomedian projection of cibarium bare, without spinous processes 37
37. Labrum 0.8 times as long as clypeus ....*S.* (*S.*) *daoense*
Labrum 0.65–0.68 as long as clypeus.. *S.* (*S.*) *laui*
38. Abdominal segment 7 with a pair of clustered hairs ventrally (similar to Fig. 82C) 39
Abdominal segment 7 without a pair of clustered hairs ventrally 41
39. Sensory vesicle with large opening ....*S.* (*S.*) *fuscicoxae*
Sensory vesicle with opening of moderate size 40
40. Antenna darkened except scape, pedicel and base of first flagellomere yellow
.... *S.* (*S.*) *doipuiense* (complex)
Antenna almost entirely darkened .... *S.* (*S.*) *giayense* and *S.* (*S.*) *rosliramlii*
41. Ovipositor valves with inner margins widely concave (Fig. 74I).....................
.... *S.* (*S.*) *nodosum*
Ovipositor valves with inner margins nearly straight 42
42. Hind tibia medium to dark brown except base yellow 43
Hind tibia yellow except apical cap medium brown 44
43. Sensory vesicle over half as long as third palpal segment................................
.... *S.* (*S.*) *suoivangense* and *S.* (*S.*) *tani* (complex)
Sensory vesicle less than 0.4 times as long as third palpal segment ....
.... *S.* (*S.*) *brevipar* (complex), *S.* (*S.*) *lowi*, *S.* (*S.*) *sapaense*, *S.* (*S.*) *xuandei*, and *S.* (*S.*) *yuphae* (complex)
44. Fore femur yellow except apical cap light brown .... *S.* (*S.*) *yukawai*
Fore femur entirely yellow 45
45. Frons:head ratio 1.0:3.6 ....*S.* (*S.*) *laocaiense*
Frons:head ratio 1.0:4.3 .... *S.* (*S.*) *turgidum*
46. Frons densely covered with yellow short hairs .... *S.* (*S.*) *nigrogilvum*
Frons sparsely covered with dark hairs 47
47. Subcosta bare or rarely with one or two hairs ....*S.* (*S.*) *obliquum*
Subcosta with many hairs 48
48. Ovipositor valves with inner margins nearly straight or slightly sinuous (Fig. 50C).. *S.* (*S.*) *atipornae*
Ovipositor valves with inner margins widely concave 49
49. Ovipositor valves rounded posteromedially (Fig. 49C) ...... *S.* (*S.*) *sansahoense*

Ovipositor valves pointed posteromedially (Fig. 84C) 50
50. Fore basitarsus 5.6–5.8 times as long as its greatest width .... *S.* (*S.*) *phuluense*
Fore basitarsus 5.1 times as long as its greatest width .... *S.* (*S.*) *chamlongi*

Males***

1. Katepisternum haired 2
Katepisternum bare 21
2. Antenna with eight flagellomeres ....*S.* (*G.*) *breviflagellum*
Antenna with nine flagellomeres 3
3. Hind basitarsus enlarged 4
Hind basitarsus slender, parallel-sided 15
4. Fore coxa darkened ....*S.* (*G.*) *eshimai*
Fore coxa yellowish 5
5. Upper-eye (large) facets in 16 vertical columns 6
Upper-eye (large) facets in 9–14 vertical columns 7
6. Base of radius with tuft of dark hairs .... *S.* (*G.*) *inthanonense*
Base of radius with tuft of yellow hairs ....*S.* (*G.*) *hongthaii*
7. Antenna yellow except apical four or five or six flagellomeres darkened 8
Antenna darkened except scape, pedicel, and base of first flagellomere 11
8. Upper-eye (large) facets in nine or 10 vertical columns and 11 or 12 horizontal rows 9
Upper-eye (large) facets in 11–13 vertical columns and 13 or 14 horizontal rows 10
9. Upper-eye (large) facets in nine vertical columns and 11 or 12 horizontal rows ....*S.* (*G.*) *vinhphucense*
Upper-eye (large) facets in 10 vertical columns and 12 horizontal rows ....*S.* (*G.*.) *chaudinhense*
10. Upper-eye (large) facets in 11 vertical columns and 13 horizontal rows .... S. (*G.*) *asakoae*
Upper-eye (large) facets in 13 or 14 vertical columns and 14 horizontal rows .. ..*S.* (*G.*) *unii*
11. Upper-eye (large) facets in 14 vertical columns and 14 or 15 horizontal rows 12
Upper-eye (large) facets in 11–13 vertical columns and 13 or 14 horizontal rows 14
12. Paramere with setae ....*S.* (*G.*) *tamdaoense*
Paramere without setae 13
13. Hind basitarsus as wide as hind femur....................................*S.* (*G.*) *confertum*
Hind basitarsus 0.8 times as wide as hind femur .... *S.* (*G.*) *quychauense*
14. Hind basitarsus yellowish on basal two-fifths and dark on rest ....*S.* (*G.*) *sanchayense*
Hind basitarsus white on little more than basal half and dark on rest ....*S.* (*G.*) *phulocense*

15. Abdominal segments 2, 6, and 7 each with a pair of shiny dorsolateral patches 16
Abdominal segments 2 and 5–7 each with a pair of shiny dorsolateral patches 18
16. Style without terminal spine.. S. (G.) *parahiyangum*
Style with terminal spine 17
17. Ventral plate with body tapered from base to middle, abruptly narrowed, then tapered posteriorly when viewed ventrally (Fig. 20K) ....S. (G.) *lamdongense*
Ventral plate with body nearly parallel-sided, though posterolateral corners rounded when viewed posteriorly ....S. (G.) *longlanhense*
18. Upper-eye (large) facets in 10 vertical columns and 10 horizontal rows ....S. (G.) *montiblense*
Upper-eye (large) facets in 15–18 vertical columns and 15–18 horizontal rows 19
19. Upper-eye (large) facets in 18 vertical columns and 18 horizontal rows .... S. (G.) *dachaisense*
Upper-eye (large) facets in 15 or 16 vertical columns and 15 or 16 horizontal rows 20
20. Scutum with golden-yellow short hairs ....S. (G.) *siamense* (complex)
Scutum with brassy short hairs .... S. (G.) *duolongum* and S. (G.) *yvonneae*
21. Coxite longer than style 22
Coxite shorter than style 29
22. Hind basitarsus narrow, parallel-sided (similar to Fig. 45H). Ventral plate with median keel (similar to Fig. 45K) ....S. (N.) *aureohirtum*
Hind basitarsus enlarged, spindle-shaped (similar to Fig. 47L). Ventral plate without median keel (similar to Fig. 47M) 23
23. Fore coxa darkened. Paramere with one hook (similar to Fig. 47Q) ....S. (N.) *tayense*
Fore coxa yellow. Paramere with four to seven hooks (similar to Fig. 43P) 24
24. Ventral plate with posterior margin deeply concave when viewed ventrally (similar to Fig. 42K). Paramere with four hooks ....S. (M.) *nigrofilum*
Ventral plate with posterior margin slightly concave when viewed ventrally (similar to Fig. 43K). Paramere with five to seven hooks 25
25. Antenna yellow except pedicel and little less than apical half of first flagellomere somewhat dark yellow to light brown and base of first flagellomere whitish yellow .... S. (N.) *bachmaense*
Antenna otherwise..26
26. Upper-eye (large) facets in 19 vertical columns and 20 horizontal rows ....S. (N.) *maeaiense*
Upper-eye (large) facets in 15–18 vertical columns and 17–19 horizontal rows 27
27. Upper-eye (large) facets in 15 vertical columns and 17 horizontal rows ....S. (N.) *pumatense*

Upper-eye (large) facets in 17 or 18 vertical columns and 18 or 19 horizontal rows 28
28. Fifth palpal segment 1.7 times as long as third segment .... *S. (N.) langbiangense*
Fifth palpal segment 1.9 times as long as third segment .... *S. (N.) phami*
29. Scutum white pruinose with inverted-V-shaped, non-pruinose black area (similar to Fig. 74S, T) .... *S. (S.) nodosum*
Scutum white pruinose on each shoulder, along each lateral margin and on prescutellar area 30
30. Fore basitarsus with thick hair crest (Fig. 62H) .... *S. (S.) nigrogilvum*
Fore basitarsus with moderate hair crest 31
31. Basal section of radius with four or five hairs .... *S. (S.) tavanense*
Basal section of radius bare 32
32. Style without terminal spine (Fig. 51I) .... *S. (S.) chungi*
Style with terminal spine 33
33. Ventral plate without teeth posteriorly 34
Ventral plate with teeth posteriorly 39
34. Style with long horn-like protuberance (similar to Fig. 61J) .... *S. (S.) grossifilum*
Style with short basal protuberance (Fig. 59C) 35
35. Ventral plate rectangular, longer than wide when viewed ventrally (Fig. 59B). .... *S. (S.) maenoi*
Ventral plate saddle-shaped, shorter than wide when viewed ventrally (Fig. 76P) 36
36. Scutum with yellowish short hairs 37
Scutum with brassy or copper-colored short hairs 38
37. First antennal flagellomere 1.8 times length of second flagellomere .... *S. (S.) quinquestriatum*
First antennal flagellomere 1.6 times length of second flagellomere .... *S. (S.) xuandai*
38. Upper-eye (large) facets in 17 or 18 horizontal rows ....*S. (S.) taythienense*
Upper-eye (large) facets in 16 horizontal rows.. *S. (S.) chiangmaiense*, *S. (S.) nakhonense*, and *S. (S.) thailandicum*
39. Style with numerous short spines on dorsal surface of basal portion 40
Style without short spines on dorsal surface of basal portion or with one or two on tip of long basal protuberance 48
40. Abdominal segments 2, 6 and 7 each with a pair of shiny dorsolateral patches 41
Abdominal segments 2 and 5–7 each with a pair of shiny dorsolateral patches 47
41. Fore coxa darkened 42
Fore coxa yellow 43
42. Scutum with yellow short hairs .... *S. (S.) giayense*
Scutum with brassy short hairs .... *S. (S.) rosliramlii*
43. Upper-eye (large) facets in 21 horizontal rows .... *S. (S.) huense*

Upper-eye (large) facets in 16–19 horizontal rows 44
44. Upper-eye (large) facets in 16 or 17 horizontal rows. Fore basitarsus 5.9–6.1 times as long as its greatest width .... *S.* (*S.*) *doipuiense* (complex)
Upper-eye (large) facets in 18–20 horizontal rows. Fore basitarsus 4.5–5.2 times as long as its greatest width 45
45. Upper-eye (large) facets in 18 horizontal rows ....*S.* (*S.*) *brevipar* (complex) and *S.* (*S.*) *lowi*
Upper-eye (large) facets in 19 or 20 horizontal rows 46
46. Fore basitarsus 4.2–4.5 times as long as its greatest width ....*S.* (*S.*) *yuphae* (complex)
Fore basitarsus 5.2 times as long as its greatest width .... *S.* (*S.*) *sapaense*
47. Upper-eye (large) facets in 20 vertical columns and 21 horizontal rows。 Fore coxa dark. .... *S.* (*S.*) *congi*
Upper-eye (large) facets in 18 vertical columns and 18 or 19 horizontal rows. Fore coxa yellow ....*S.* (*S.*) *suoivangense*, *S.* (*S.*) *tani* (complex), and *S.* (*S.*) *xuandei*
48. Style without any basal protuberance 49
Style with horn-like or round basal protuberance 50
49. Scutum with dark short hairs. Hind basitarsus slender, nearly parallel-sided ... .*S.* (*S.*) *obliquum*
Scutum with yellow short hairs. Hind basitarsus enlarged, spindle-shaped........ .... *S.* (*S.*) *chamlongi, S.* (*S.*) *phuluense*, and *S.* (*S.*) *sansahoense*
50. Style with short, round basal protuberance (Fig. 50K) .... *S.* (*S.*) *atipornae*
Style with elongate, horn-like basal protuberance with pointed apex (Fig. 71C) 51
51. Upper-eye (large) facets in 14 or 15 vertical columns and 15 horizontal rows .... *S.* (*S.*) *hirtinervis*
Upper-eye (large) facets in 17–21 vertical columns and 17–21 horizontal rows 52
52. Upper-eye (large) facets in 20 or 21 vertical columns and 21 horizontal rows .... *S.* (*S.*) *laui*
Upper-eye(large) facets in 17–19 vertical columns and 17–20 horizontal rows 53
53. Upper-eye (large) facets in 20 horizontal rows ....*S.* (*S.*) *lacduongense*
Upper-eye (large) facets in 17–19 horizontal rows 54
54. Hind basitarsus whitish yellow on basal half or little less and dark on rest ....*S.* (*S.*) *daoense*
Hind basitarsus whitish yellow on basal two-fifths and dark on rest .... *S.* (*S.*) *malayense* (complex)

Pupae****

1. Gill with 12 filaments. Gill filaments with numerous tiny black spots (Fig. 42T) .... *S.* (*M.*) *nigrofilum*
Gill with less than 10 filaments. Gill filaments without black spots 2

2. Abdominal segment 9 with grapnel-like hooklets on each side (Fig. 3D) 3
   Abdominal segment 9 without grapnel-like hooklets on each side 26
3. Gill composed of inflated structure with eight slender filaments ....
   ....*S.* (*G.*) *thuathienense*
   Gill composed of eight or 10 slender filaments 4
4. Gill with 10 filaments (Fig. 20Q) ....*S.* (*G.*) *lamdongense*
   Gill with eight filaments 5
5. All gill filaments short (ca. 1.0 mm) 6
   All or some gill filaments medium-long to long (more than 1.5 mm) 9
6. Antennal sheath with tubercles (Fig. 23P) 7
   Antennal sheath without tubercles 8
7. Gill filaments arranged as 3+3+2 from dorsal to ventral (Fig. 23R) ....
   .... *S.* (*G.*) *parahiyangum*
   Gill filaments arranged as 2+1+3+2 from dorsal to ventral (Fig. 23Q) ....
   ....*S.* (*G.*) *longlanhense*
8. Face with additional trichome mediolaterally on each side (Fig. 29A) ....
   .... *S.* (*G.*) *eshimai*
   Face without additional trichome mediolaterally on each side ....
   ....*S.* (*G.*) *montiblense*
9. Common basal stalk of gill longer than interspiracular trunk ....
   .... *S.* (*G.*) *fuscidorsum*
   Common basal stalk of gill shorter than interspiracular trunk 10
10. Dorsal and middle triplets of gill not sharing common stalk 11
    Dorsal and middle triplets of gill sharing common stalk 13
11. Dorsal surface of abdominal segments 1 and 2 pale yellow.............................
    .... *S.* (*G.*) *duolongum* and *S.* (*G.*) *siamense* (complex)
    Dorsal surface of abdominal segments 1 and 2 dark yellow to light brown or grayish 12
12. Inner filament of ventral pair of gill slightly longer than outer filament ....
    .... *S.* (*G.*) *dachaisense*
    Inner filament of ventral pair of gill slightly shorter than outer filament ....
    ....*S.* (*G.*) *yvonneae*
13. Dorsal surface of abdominal segments 1 and 2 (or also 3 and 4) darkened .. 14
    Dorsal surface of abdominal segments 1 and 2 unpigmented 16
14. Cocoon with long anterodorsal projection (Fig. 14M) ....*S.* (*G.*) *hmongense*
    Cocoon without anterodorsal projection 15
15. Dorsal surface of abdominal segments 1 and 2 with tubercles ....
    ....*S.* (*G.*) *chaudinhense*
    Dorsal surface of abdominal segments 1 and 2 lacking tubercles ....
    ....*S.* (*G.*) *asakoae*
16. Cocoon with anterodorsal margin produced anteromedially, appearing as letter W or with short projection (Fig. 14L) 17
    Cocoon with anterodorsal margin somewhat produced anteromedially or not (similar to Fig. 14K) 20

17. Stalk of dorsal triplet of gill against that of ventral pair at angle of 90 degrees or more when viewed laterally (Fig. 14A) 18
Stalk of dorsal triplet of gill against that of ventral pair at acute angle when viewed laterally (Fig. 14D, E) 19
18. Dorsal triplet of gill filaments with secondary stalk longer than primary stalk. Terminal hooks broad, plate-like, with outer margins serrate and three times length of inner margins (similar to Fig. 14I).................... *S.* (*G.*) *breviflagellum*
Dorsal triplet of gill filaments with secondary stalk much shorter than primary stalk. Terminal hooks narrow, plate-like, with outer margins crenulated and 1.8 times length of inner margins (similar to Fig. 14H) ....*S.* (*G.*) *unii*
19. Stalk of ventral pair of gill twice length of primary and secondary stalks of middle triplet combined ....*S.* (*G.*) *phulocense*
Stalk of ventral pair of gill as long as or shorter than length of primary and secondary stalks of middle triplet combined (Fig. 14A) ...*S.* (*G.*) *inthanonense*
20. Abdominal segment 9 without spine-combs dorsally .... *S.* (*G.*) *quychauense*
Abdominal segment 9 with spine-combs dorsally 21
21. Dorsal and middle triplets of gill with short common stalk shorter than their primary stalks ....*S.* (*G.*) *confertum*
Dorsal and middle triplets of gill with short common stalk nearly as long as or longer than their primary stalks 22
22. Terminal hooks flat and narrow, longer than wide (similar to Fig. 14G) .... *S.* (*G.*) *thituyenae*
Terminal hooks flat and wide, shorter than wide 23
23. Stalk of ventral pair of gill as long as or shorter than common basal stalk ....*S.* (*G.*) *vinhphucense*
Stalk of ventral pair of gill longer than common basal stalk 24
24. Stalk of ventral pair of gill shorter than primary and secondary stalks of middle triplet combined. Terminal hooks with outer margins 1.8–1.9 times length of inner margins ....*S.* (*G.*) *tamdaoense*
Stalk of ventral pair of gill longer than primary and secondary stalks of middle triplet combined. Terminal hooks with outer margins 2.6–3.6 times length of inner margins 25
25. Two filaments of ventral pair of gill 2.6 mm long .... *S.* (*G.*) *hongthaii*
Two filaments of ventral pair of gill 3.0 mm long ....*S.* (*G.*) *sanchayense*
26. Gill with three inflated tubular filaments (Fig. 75F) .... *S.* (*S.*) *nodosum*
Gill with four, five, six, eight, or 10 filaments 27
27. Gill with four filaments 28
Gill with five, six, eight, or 10 filaments 29
28. Cocoon with anterodorsal projection. Frons and thorax light ochreous ....*S.* (*N.*) *tayense*
Cocoon without anterodorsal projection. Frons and anterolateral surface of thorax dark brown (Fig. 48A) .... *S.* (*N.*) *laichauense*
29. Gill with five filaments (Fig. 44F) ....*S.* (*N.*) *phami*
Gill with six, eight, or 10 filaments 30
30. Gill with six filaments 31

Gill with eight or 10 filaments 59
31. Cocoon wall-pocket-shaped, extended ventrolaterally 32
Cocoon wall-pocket-shaped, not extended ventrolaterally, with or without anterodorsal projection or with or without anterolateral open space(s) on each side, or shoe-shaped 36
32. Six gill filaments arranged 2+2+2 from dorsal to ventral and widely divergent basally (Fig. 46C). Abdominal segments 7 and 8 each with spine-combs dorsally .... *S.* (*N.*) *aureohirtum*
Six gill filaments arranged differently and arising close together. Abdominal segments 5–9 each with spine-combs dorsally 33
33. Six gill filaments lying horizontally (Fig. 44E) .... *S.* (*N.*) *maeaiense*
Six gill filaments arranged as 4+2 from dorsal to ventral 34
34. Cocoon with short anterodorsal projection .... *S.* (*N.*) *pumatense*
Cocoon without anterodorsal projection 35
35. Four gill filaments in dorsal group arranged as 2+2 from inside to outside....... .... *S.* (*N.*) *langbiangense*
Four gill filaments in dorsal group arranged as 1+2+1 or 1+1+2 from inside to outside... *S.* (*N.*) *bachmaense*
36. Cocoon shoe-shaped (Fig. 63C) .... *S.* (*S.*) *nigrogilvum*
Cocoon wall-pocket-shaped 37
37. Cocoon with anterolateral open space (s) on each side 38
Cocoon without anterolateral open space on each side 44
38. Gill with inflated filaments 39
Gill with thread-like filaments 40
39. Gill filaments pale, divergent, with minute fine projections on their surface (similar to Fig. 61M, N) .... *S.* (*S.*) *grossifilum*
Gill filaments dark brown, not divergent, without minute projection on their surface (Fig. 53D) .... *S.* (*S.*) *turgidum*
40. Cocoon with small anterolateral open space on each side (Fig. 60P) .... *S.* (*S.*) *maenoi*
Cocoon with large anterolateral open spaces on each side (Fig. 51M) 41
41. Terminal hook cone-shaped .... *S.* (*S.*) *chungi*
Terminal hook absent 42
42. Abdominal segment 9 with spine-combs dorsally ....*S.* (*S.*) *obliquum*
Abdominal segment 9 without spine-combs dorsally 43
43. Gill filaments divergent basally at obtuse angle when viewed laterally (Fig. 50O).. *S.* (*S.*) *atipornae*
Gill filaments arising close together (Fig. 53F) ....*S.* (*S.*) *yukawai*
44. Cocoon with short broad anterodorsal projection (Fig. 53K)....*S.* (*S.*) *laocaiense*
Cocoon without anterodorsal projection 45
45. Thoracic integument with pit-like organ at base of gill (Fig. 79H) 46
Thoracic integument without pit-like organ at base of gill 49

46. Thoracic integument with tubercles near pit-like organ .... *S.* (*S.*) *sapaense*
Thoracic integument without tubercles near pit-like organ 47
47. One of two filaments of ventral pair of gill directed posteriorly (Fig. 79D) .... *S.* (*S.*) *yuphae* (complex)
Two filaments of ventral pair of gill directed anteriorly 48
48. Upper filament of dorsal pair of gill abruptly tapered basally, then gradually tapered .... *S.* (*S.*) *brevipar* (complex)
Upper filament of dorsal pair of gill gradually tapered from base to apex (Fig. 79A) .... *S.* (*S.*) *lowi*
49. Relatively large tubercles on frons without tiny secondary projections .... *S.* (*S.*) *huense*
Relatively large tubercles on frons with tiny secondary projections (similar to Fig. 79I) 50
50. Gill with basal fenestra inflated (Fig. 81J) .... *S.* (*S.*) *congi*
Gill with basal fenestra not inflated 51
51. Frons bare (Fig. 83A) 52
Frons covered with tubercles (Fig. 83B, C) 53
52. Stalks of dorsal and ventral pairs of gill at angle of 90 degrees when viewed laterally .... *S.* (*S.*) *rosliramlii*
Stalks of dorsal and ventral pairs of gill at angle of 120 degrees when viewed laterally.. *S.* (*S.*) *chamlongi*
53. Gill filaments directed forward in bundle (similar to Fig. 81G) 54
Gill filaments lying in vertical plane, and stalks of dorsal and ventral pairs of gill lying at angle of 60–120 degrees when viewed laterally 55
54. Terminal hooks present .... *S.* (*S.*) *tani* (complex)
Terminal hooks absent .... *S.* (*S.*) *suoivangense*
55. Abdominal segment 9 without spine-combs dorsally .... *S.* (*S.*) *phuluense*
Abdominal segment 9 with spine-combs dorsally 56
56. Frons densely covered with tubercles (Fig. 83C) .... *S.* (*S.*) *giayense*
Frons moderately covered with tubercles (similar to Fig. 83B) 57
57. Terminal hooks present .... *S.* (*S.*) *xuandei*
Terminal hooks absent 58
58. Abdominal segment 7 with spine-combs dorsally .... *S.* (*S.*) *doipuiense* (complex)
Abdominal segment 7 without spine-combs dorsally .... *S.* (*S.*) *fuscicoxae*
59. Gill with eight filaments (similar to Fig. 49Q) 60
Gill with 10 filaments 64
60. Cocoon shoe-shaped or corbicular-shaped (Fig. 49R, S) 61
Cocoon wall-pocket-shaped 62
61. All filaments of gill of same thickness (similar to Fig. 49Q)............................. .... *S.* (*S.*) *sansahoense*
Two filaments of ventral pair of gill more slender than six other filaments (Fig. 77E) .... *S.* (*S.*) *chiangmaiense*

62. Cocoon without anterolateral open space(s) on each side .... *S.* (*S.*) *malayense* (complex)
    Cocoon with anterolateral open space on each side (Fig. 72J) 63
63. Stalks of dorsal and ventral pairs of gill lying at angle of ca. 120 degrees when viewed laterally (similar to Fig. 72G) ....*S.* (*S.*) *lacduongense* and *S.* (*S.*) *laui*
    Stalks of dorsal and ventral pairs of gill lying at angle of ca. 90 degrees when viewed laterally (Fig. 72D) .... *S.* (*S.*) *daoense* and *S.* (*S.*) *hirtinervis*
64. Ten gill filaments arranged as 2+3+3+2 from dorsal to ventral (Fig. 77B).... 65
    Ten gill filaments arranged as 2+2+2+2+2 or 2+1+2+1+2+2 from dorsal to ventral 67
65. Frons with relatively large tubercles having minute secondary projections .... *S.* (*S.*) *tavanense* and *S.* (*S.*) *xuandai*
    Frons with relatively large tubercles without secondary projections..............66
66. Stalk of dorsal pair of gill directed posteriorly (Fig. 77B)................................ ....*S.* (*S.*) *quinquestriatum*
    Stalk of dorsal pair of gill not directed posteriorly ....*S.* (*S.*) *taythienense*
67. Ten gill filaments arranged as 2+1+2+1+2+2 from dorsal to ventral .... *S.* (*S.*) *jeffreyi*
    Ten gill filaments arranged as 2+2+2+2+2 from dorsal to ventral.................68
68. Frons covered with tubercles of various shapes ....*S.* (*S.*) *thailandicum*
    Frons covered with round tubercles ....*S.* (*S.*) *nakhonense*

Mature larvae*****

1. Lateral margins of hypostoma smooth (Fig. 32U) 2
   Lateral margins of hypostoma serrated 18
2. Postgenal cleft reaching posterior margin of hypostoma (Fig. 23V) 3
   Postgenal cleft not reaching posterior margin of hypostoma 5
3. Abdominal segments 1–5 each with two dorsal and two dorsolateral protuberances (Fig. 23T, U) .... *S.* (*G.*) *parahiyangum*
   Abdominal segments 1–5 without protuberances 4
4. Pharate pupal gill with eight filaments. Anterolateral surfaces of abdominal segment 5 covered with conspicuous dark spinous setae longer than those on dorsal surface (Fig. 23Y)... *S.* (*G.*) *longlanhense*
   Pharate pupal gill with 10 filaments. Anterolateral surfaces of abdominal segment 5 lacking conspicuous dark spinous setae ....*S.* (*G.*) *lamdongense*
5. Postgenal cleft shorter than postgenal bridge 6
   Postgenal cleft slightly to greatly longer than postgenal bridge 7
6. Posterior circlet with 88 rows of hooklets .... *S.* (*G.*) *inthanonense*
   Posterior circlet with 70 rows of hooklets ....*S.* (*G.*) *hmongense*

7. Postgenal cleft medium-long, 1.1–3.0 times as long as postgenal bridge 8
   Postgenal cleft long, more than 3.4 times as long as postgenal bridge 13
8. Pharate pupal gill with inflated structure having eight slender filaments ....
   ....*S.* (*G.*) *thuathienense*
   Pharate pupal gill with eight slender filaments 9
9. Abdominal segments 1–4 ocherous, without grayish markings (Fig. 15D)........
   ....*S.* (*G.*) *asakoae*
   Abdominal segments 1 and 2, or 1–3, or 1–4 grayish, or grayish green.........10
10. Abdominal segments 1–4 grayish green .... *S.* (*G.*) *hongthaii*
    Abdominal segments 1 and 2, or 1–3 grayish 11
11. Abdominal segments 1–3 grayish....................................... *S.* (*G.*) *fuscidorsum*
    Abdominal segments 1 and 2 grayish 12
12. Abdominal segment 4 with grayish markings ventrally ....
    ....*S.* (*G.*) *chaudinhense*
    Abdominal segment 4 without grayish markings ventrally ....*S.* (*G.*) *unii*
13. Pharate pupal gill with eight slender short filaments of same length. Posterior circlet with 80–88 rows of hooklets 14
    Pharate pupal gill with eight slender short filaments, of which two filaments of ventral pair much longer than six other filaments. Posterior circlet with 69–76 rows of hooklets 15
14. Posterior circlet with 80 rows of hooklets ....*S.* (*G.*) *montiblense*
    Posterior circlet with 85–88 rows of hooklets ....*S.* (*G.*) *eshimai*
15. Head spots indistinct or faintly negative .... *S.* (*G.*) *dachaisense*
    Head spots faintly positive 16
16. Postgenal cleft 3.4–4.0 times as long as postgenal bridge ....*S.* (*G.*) *yvonneae*
    Postgenal cleft 5.0–7.6 times as long as postgenal bridge 17
17. Posterior abdominal segments with dark spinous setae each having distinct dark branches arising basally, medially, and apically .....*S.* (*G.*) *siamense* (complex)
    Posterior abdominal segments with dark flat setae each having fine transparent branches only apically ....*S.* (*G.*) *duolongum*
18. Postgenal cleft vestigial, extremely small (similar to Fig. 42X) ....
    ....*S.* (*M.*) *nigrofilum*
    Postgenal cleft well defined, small to large 19
19. Ventral papillae present, medium-sized to large 20
    Ventral papillae absent or small 26
20. Accessory sclerites present (similar to Fig. 48P) .... *S.* (*N.*) *tayense*
    Accessory sclerites absent 21
21. Mandibular serration composed of two large teeth (similar to Fig. 46I). Body without reddish-brown markings .... *S.* (*N.*) *aureohirtum*
    Mandibular serration composed of one large tooth and one small tooth (similar to Fig. 44M). Body with reddish-brown markings (similar to Fig. 44K) 22

22. Postgenal cleft 0.6–0.7 length of postgenal bridge .... *S.* (*N.*) *langbiangense*
    Postgenal cleft 0.4–0.5 length of postgenal bridge 23
23. Labral fan with 23–25 primary rays .... *S.* (*N.*) *maeaiense*
    Labral fan with 30–44 primary rays 24
24. Labral fan with 30 or 31 primary rays .... *S.* (*N.*) *pumatense*
    Labral fan with 34–44 primary rays 25
25. Second antennal article 1.1 times length of first article. Posterior circlet with 74–76 rows of hooklets .... *S.* (*N.*) *phami*
    Second antennal article 0.6 times length of first article. Posterior circlet with 88 rows of hooklets .... *S.* (*N.*) *bachmaense*
26. Accessory sclerite present (Fig. 63J) .... *S.* (*S.*) *nigrogilvum*
    Accessory sclerite absent 27
27. Abdominal segments 1–4 each with six pairs of protuberances dorsally and dorsolaterally (Fig. 66L) .... *S.* (*S.*) *obliquum*
    Abdominal segments 1–4 each with or without paired dorsal protuberances 28
28. Abdomen with paired dorsal protuberances 29
    Abdomen without paired dorsal protuberances 34
29. Postgenal cleft deep, reaching posterior margin of hypostoma (similar to Fig. 75M) .... *S.* (*S.*) *nodosum*
    Postgenal cleft deep but not reaching posterior margin of hypostoma 30
30. Pharate pupal gill with eight filaments 31
    Pharate pupal gill with 10 filaments 32
31. Abdominal segments 1–8 each with a pair of dorsolateral protuberances (Fig. 72N) .... *S.* (*S.*) *hirtinervis*
    Abdominal segments 1–5 each with a pair of dorsolateral protuberances (similar to Fig. 49W) .... *S.* (*S.*) *sansahoense*
32. Pharate pupal gill with 10 filaments arranged as 2+3+3+2 from dorsal to ventral .... *S.* (*S.*) *quinquestriatum*, *S.* (*S.*) *taythienense*, and *S.* (*S.*) *tavanense*
    Pharate pupal gill with 10 filaments arranged as 2+2+2+2+2 or 2+1+2+1+2+2 from dorsal to ventral 33
33. Pharate pupal gill with 10 filaments arranged as 2+1+2+1+2+2 from dorsal to ventral .... *S.* (*S.*) *jeffreyi*
    Pharate pupal gill with 10 filaments arranged as 2+2+2+2+2 from dorsal to ventral .... *S.* (*S.*) *nakhonense*
34. Abdominal segments 1–4 each with dark transverse band .... *S.* (*S.*) *tani* (complex), *S.* (*S.*) *suoivangense*, and *S.* (*S.*) *xuandei*
    Abdominal segments 1–4 each without dark transverse band........................35
35. Postgenal cleft triangular 36
    Postgenal cleft spear-head-shaped or mitre-shaped or bullet-shaped 40
36. Pharate pupal gill with enlarged basal fenestra .... *S.* (*S.*) *congi*

Pharate pupal gill without enlarged basal fenestra 37
37. Body length 5.8–6.6 mm .... *S.* (*S.*) *atipornae*
Body length 4.0–5.6 mm 8
38. Cephalic apotome with indistinct head-spots .... *S.* (*S.*) *doipuiense* (complex)
Cephalic apotome with positive head-spots 39
39. Second antennal article 1.0–1.1 times as long as first article ....
.... *S.* (*S.*) *huense* and *S.* (*S.*) *yuphae* (complex)
Second antennal article 0.8 times as long as first article ....
.... *S.* (*S.*) *brevipar* (complex)
40. Postgenal cleft spear-head-shaped or bullet-shaped, longer than wide (Fig. 84U) 41
Postgenal cleft mitre-shaped or rounded, nearly as long as wide (Fig. 72R) 43
41. Rectal organ with 18–20 secondary lobules per lobe .... *S.* (*S.*) *phuluense*
Rectal organ with 6–10 secondary lobules per lobe 42
42. Posterior circlet with 80 rows of hooklets ....*S.* (*S.*) *chamlongi*
Posterior circlet with 92 rows of hooklets ....*S.* (*S.*) *malayense* (complex)
43. Rectal organ with 17–22 secondary lobules per lobe ....*S.* (*S.*) *daoense*
Rectal organ with 7–12 secondary lobules per lobe ....*S.* (*S.*) *laui*

*Keys provided by Takaoka et al. (2017c) are modified.

***Simulium* (*G.*) *breviflagellum*, *S.* (*G.*) *hmongense*, *S.* (*G.*) *quychauense*, *S.* (*G.*) *sanchayense*, *S.* (*G.*) *tamdaoense*, *S.* (*G.*) *vinhphucense*, *S.* (*N.*) *laichauense*, *S.* (*S.*) *huense*, *S.* (*S.*) *congi*, *S.* (*S.*) *thailandicum*, and *S.* (*S.*) *xuandai* are not included because their females are unknown.

*** *Simulium* (*G.*) *fuscidorsum*, *S.* (*G.*) *hmongense*, *S.* (*G.*) *thituyenae*, *S.* (*G.*) *thuathienense*, *S.* (*S.*) *laocaiense*, *S.* (*S.*) *turgidum*, *S.* (*S.*) *vietnamense*, and *S.* (*S.*) *yukawai* are not included because their males are unknown.

**** *Simulium* (*S.*) *vietnamense* is not included because its pupa is unknown.

******Simulium* (*G.*) *breviflagellum*, *S.* (*G.*) *confertum*, *S.* (*G.*) *phulocense*, *S.* (*G.*) *quychauense*, *S.* (*G.*) *sanchayense*, *S.* (*G.*) *tamdaoense*, *S.* (*G.*) *thituyenae*, *S.* (*G.*) *vinhphucense*, *S.* (*N.*) *laichauense*, *S.* (*S.*) *chiangmaiense*, *S.* (*S.*) *lacduongense*, *S.* (*S.*) *thailandicum*, *S.* (*S.*) *fuscicoxae*, *S.* (*S.*) *lowi*, *S.* (*S.*) *rosliramlii*, *S.* (*S.*) *giayense*, *S.* (*S.*) *vietnamense*, and *S.* (*S.*) *yukawai* are not included because their larvae are unknown.

8 The Fauna of Black Flies in Peninsular Malaysia

Overview

Edwards (1928) described five new species of black flies from Peninsular Malaysia: *Simulium* (*Simulium*) *argentipes*, *S.* (*S.*) *digrammicum*, *S.* (*S.*) *fuscopilosum*, *S.* (*S.*) *hackeri*, and *S.* (*S.*) *hirtinervis*. Puri (1932c) synonymized *S.* (*S.*) *digrammicum* with

S. (*S.*) *griseifrons* from India. Crosskey (1973) recorded seven known species: *S.* (*Nevermannia*) *aureohirtum*, *S.* (*Gomphostilbia*) *metatarsale*, *S.* (*G.*) *atratum* (at that time, as *S.* (*G.*) *sundaicum*), *S.* (*G.*) *varicorne*, *S.* (*G.*) *zonatum*, *S.* (*S.*) *nobile*, and *S.* (*S.*) *nitidithorax*.

Takaoka and Davies (1995a) published a monograph, in which 25 new species were described: *Sulcicnephia unidens*, *S.* (*G.*) *duolongum*, *S.* (*G.*) *tahanense*, *S.* (*G.*) *sextuplum*, *S.* (*G.*) *whartoni* [the species formerly regarded as *S.* (*G.*) *zonatum* by Crosskey (1973)], *S.* (*G.*) *cheongi*, *S.* (*G.*) *angulistylum*, *S.* (*G.*) *decuplum*, *S.* (*G.*) *dentistylum*, *S.* (*G.*) *asakoae*, *S.* (*G.*) *sheilae* [the species formerly regarded as *S.* (*G.*) *metatarsale* by Crosskey (1973)], *S.* (*G.*) *longitruncum*, *S.* (*G.*) *burtoni*, *S.* (*G.*) *novemarticulatum*, *S.* (*G.*) *gombakense* (at that time, under the subgenus *Morops*), *S.* (*N.*) *caudisclerum*, *S.* (*N.*) *kurtaki*, *S.* (*S.*) *grossifilum*, *S.* (*S.*) *bishopi*, *S.* (*S.*) *malayense*, *S.* (*S.*) *jeffreyi*, *S.* (*S.*) *brevipar*, *S.* (*S.*) *tani* [the species formerly regarded as *S.* (*S.*) *nitidithorax* by Crosskey (1973)], *S.* (*Daviesellum*) *pahangense* (at that time, as *S.* (*S.*) *pahangense*), and *S.* (*S.*) *rudnicki*; two named species and three unnamed species: *S.* (*G.*) *parahiyangum*, *S.* (*N.*) *feuerborni*, *S.* (*G.*) sp. A, *S.* (*S.*) sp. B and *S.* (*S.*) sp. C, were added; and *S.* (*S.*) *digrammicum* was resurrected from a junior synonym of *S.* (*S.*) *griseifrons*. A new subgenus *Daviesellum* was established under the genus *Simulium* s. l. (Takaoka and Adler, 1997). *Sulcicnephia unidens*, which was described from pupae and larvae, was synonymized with *S.* (*D.*) *pahangense* (Takaoka and Adler, 1997). Two more new species: *S.* (*G.*) *hoiseni* and *S.* (*S.*) *yongi*, were described (Takaoka and Davies, 1997; Takaoka, 2008a). Takaoka (2000a) described the male and pupa of *S.* (*G.*) *gombakense*, with its subgeneric transfer from *Morops* to *Gomphostilbia*. Takaoka et al. (2010a) described the male, pupa, and larva of *S.* (*G.*) *novemarticulatum*.

Takaoka et al. (2011a,b) described three new species: *S.* (*G.*) *sofiani*, *S.* (*G.*) *roslihashimi*, and *S.* (*G.*) *lurauense*. Takaoka (2012) recorded *S.* (*G.*) *charlesi*. Takaoka et al. (2012a,b,c) described five new species: *S.* (*G.*) *terengganuense*, *S.* (*G.*) *aziruni*, *S.* (*S.*) *kisapense*, *S.* (S.) *jasmoni*, and *S.* (*S.*) *tiomanense*. Takaoka et al. (2013a,b) described two new species: *S.* (*G.*) *izuae* and *S.* (*G.*) *langkawiense*. Takaoka et al. (2014a,b,f,g) described seven new species: *S.* (*G.*) *tanahrataense*, *S.* (*G.*) *brinchangense*, *S.* (*G.*) *azhari*, *S.* (*G.*) *tekamense*, *S.* (*G.*) *johorense*, *S.* (*G.*) *jerantutense*, and *S.* (*G.*) *leparense*. Ya'cob et al. (2014) described a new species, S. (*N.*) *ledangense*. Ya'cob et al. (2017a,b) described two new species: *S.* (*N.*) *pairoti* and *S.* (*S.*) *vanluni*, which are the species previously recognized in Peninsular Malaysia as *S.* (*N.*) *feuerborni* and *S.* (*S.*) *nobile*, respectively. Takaoka et al. (2018a) described a new species, *S.* (*G.*) *sazalyi* [the species formerly regarded as *S.* (*G.*) *parahiyangum*]. Takaoka et al. (2018m) described a new species, *S.* (*S.*) *perakense*, recorded *S.* (*G.*) *adleri* and *S.* (*G.*) *trangense*; and identified *S.* (*G.*) sp. A as *S.* (*G.*) *pegalanense* originally described from Sabah; the females of *S.* (*G.*) *adleri* and *S.* (*G.*) *varicorne*, and the males of *S.* (*G.*) *decuplum*, (*G.*) *tahanense*, and *S.* (*S.*) *malayense* were also described. Ya'cob et al. (2018) described a new species, *S.* (*S.*) *rasuli*. Takaoka et al. (2019a) recorded *S.* (*S.*) *thongsahuani* originally described from southern Thailand. Ya'cob et al. (2019) described the female of *S.* (*G.*) *aziruni*.

Genetic analyses of species of the *S. asakoae* species-group and the *S. ceylonicum* species-group, *S.* (*G.*) *parahiyangum*, *S.* (*S.*) *nobile*, and *S.* (*S.*) *tani* (complex) were carried out by Low et al. (2014, 2015b, 2016) and Takaoka et al. (2018a).

Ecological aspects of black flies including distribution patterns of their aquatic habitats related to elevations have been studied by Ya'cob et al. (2016).

Sex mosaics of adult *S.* (*G.*) *trangense* infected with mermithids were reported by Ya'cob et al. (2021). Natural infections of black fly larvae with mermithids were investigated by Lourdes et al. (2022).

Izwan-Anas et al. (2023) summarized the results of DNA barcoding of 52 Malaysian species of black flies, of which those of 14 species were newly added.

The fauna of black flies in Peninsular Malaysia is thus represented by 64 species including two unnamed species and one dubious species, *S.* (*G.*) *atratum*, of which the distribution record should be reconfirmed. All species are placed in four subgenera of the genus *Simulium* s. l., and in six, three, and 11 species-groups in the subgenera *Gomphostilbia*, *Nevermannia*, and *Simulium* s. str., respectively.

List of 64 Species of Black Flies in Peninsular Malaysia

Subgenus *Daviesellum*
S. pahangense
Subgenus *Gomphostilbia*
S. asakoae species-group
S. asakoae, *S. brinchangense*, *S. hoiseni*, *S. izuae*, *S. lurauense*, *S. roslihashimi*, *S. sofiani*, and *S. tanahrataense*
S. batoense species-group
S. decuplum, *S. dentistylum*, *S. duolongum*, *S. jerantutense*, *S. johorense*, *S. pegalanense*, *S. sazalyi*, *S. sextuplum*, *S. tahanense*, *S. tekamense*, and *S. terengganuense*
S. ceylonicum species-group
S. leparense, *S. longitruncum*, *S. sheilae*, and *S. trangense*
S. epistum species-group
S. adleri, *S. angulistylum* (complex), *S. atratum**, *S. cheongi*, and *S. whartoni*
S. gombakense species-group
S. aziruni, *S. gombakense*, and *S. langkawiense*
S. varicorne species-group
S. burtoni, *S. charlesi*, *S. novemarticulatum*, and *S. varicorne*
Gomphostilbia species unplaced to species-group
S. azhari
Subgenus *Nevermannia*
S. feuerborni species-group
S. ledangense and *S. pairoti*
S. ruficorne species-group
S. aureohirtum

S. vernum species-group
S. caudisclerum and *S. kurtaki*
Subgenus *Simulium* s. str.
S. argentipes species-group
S. argentipes
S. christophersi species-group
S. fuscopilosum, *S. rasuli*, and *S.* sp. C
S. crocinum species-group
S. rudnicki and *S. yongi*
S. griseifrons species-group
S. digrammicum and *S. thongsahuani*
S. grossifilum species-group
S. grossifilum
S. melanopus species-group
S. bishopi
S. multistriatum species-group
S. hirtinervis, *S. kisapense*, and *S. malayense* (complex)
S. nobile species-group
S. vanluni
S. striatum species-group
S. jeffreyi, *S. perakense*, and *S.* sp. B
S. tuberosum species-group
S. brevipar (complex), *S. jasmoni*, *S. tani* (complex), and *S. tiomanense*
S. variegatum species-group
S. hackeri

* The record of *S. atratum* from Peninsular Malaysia should be reconfirmed.

*Keys to Identify 63 Species of Black Flies in Peninsular Malaysia**

Females**

1. Katepisternum haired 2
   Katepisternum bare 32
2. Antenna with seven or eight flagellomeres (Fig. 39A–C) 3
   Antenna with nine flagellomeres 5
3. Antenna with seven flagellomeres (Fig. 39C) .... *S.* (*G.*) *charlesi* and *S.* (*G.*) *novemarticulatum*
   Antenna with eight flagellomeres (Fig. 39A, B) 4
4. Subcosta bare. Tergite 5 shiny .... *S.* (*G.*) *burtoni*
   Subcosta with several hairs. Tergite 5 dull .... *S.* (*G.*) *varicorne*

5. Posterior surface of hind tibia mostly darkened except basal tip 6
   Posterior surface of hind tibia white or yellow on basal two-fifths or more ... 15
6. Basal tooth of claw much shorter than half length of claw (Fig. 21H) 7
   Basal tooth of claw about half length of claw 8
7. Sensory vesicle 0.2 times length of third palpal segment........ *S.* (*G.*) *tahanense*
   Sensory vesicle 0.38–0.41 times length of third palpal segment ....
   .... *S.* (*G.*) *jerantutense*
8. Subcosta with 0–2 hairs 9
   Subcosta with several hairs 10
9. Sensory vesicle 0.56 times length of third palpal segment ....
   .... *S.* (*G.*) *pegalanense*
   Sensory vesicle 0.46 times length of third palpal segment ....
   .... *S.* (*G.*) *tekamense*
10. Sensory vesicle 0.5 or more times length of third palpal segment 11
    Sensory vesicle 0.41 or less times length of third palpal segment 12
11. Mandible without teeth on outer margin .... *S.* (*G.*) *sazalyi*
    Mandible with three teeth on outer margin at some distance from apex............
    .... *S.* (*G.*) *decuplum*
12. Frons widely bare except several hairs along both lateral margins and above lower margin .... *S.* (*G.*) *sextuplum*
    Frons moderately covered with hairs except central portion bare to varying extent 13
13. Frons:head ratio 1.0:7.1 .... *S.* (*G.*) *azhari*
    Frons:head ratio 1.0:4.8–5.0 14
14. Sensory vesicle 0.4 times length of third palpal segment .... *S.* (*G.*) *johorense*
    Sensory vesicle 0.3 times length of third palpal segment...... *S.* (*G.*) *duolongum*
15. Hair tuft of base of radial vein black 16
    Hair tuft of base of radial vein yellow 19
16. Frons:head ratio 1.0:8.0–8.3 .... *S.* (*G.*) *terengganuense*
    Frons:head ratio 1.0:4.2–5.7 17
17. Frons:head ratio 1.0:5.7 .... *S.* (*G.*) *sheilae*
    Frons:head ratio 1.0:4.2–5.0 18
18. Scutum covered with yellowish short hairs .... *S.* (*G.*) *trangense*
    Scutum covered with brownish short hairs .... *S.* (*G.*) *leparense*
19. Hind tibia whitish on basal half or little more or less, with dark subbasal spot 20
    Hind tibia whitish on basal half to two-thirds, without dark subbasal spot 24
20. Abdominal segment 2 dark 21
    Abdominal segment 2 pale 22
21. Sensory vesicle long, 0.65 times length of third palpal segment ....
    .... *S.* (*G.*) *adleri*

Sensory vesicle medium-long, 0.3 times length of third palpal segment .... *S.* (*G.*) *angulistylum* (complex)
22. Ratio of width of head against widest part of frons 5.7 .... *S.* (*G.*) *aziruni*
Ratio of width of head against widest part of frons 6.3–6.5 23
23. Mandible with 20 inner teeth .... *S.* (*G.*) *cheongi*
Mandible with 25 inner teeth ....*S.* (*G.*) *whartoni*
24. Sensory vesicle less than 0.4 times length of third palpal segment 25
Sensory vesicle more than 0.5 times length of third palpal segment 29
25. Hind tibia yellowish on basal half and dark on rest 26
Hind tibia yellowish on basal three-fourths and dark on rest 28
26. Mandible with inner and outer teeth .... *S.* (*G.*) *longitruncum*
Mandible with inner teeth but lacking outer teeth 27
27. Hind basitarsus 0.65 times as wide as hind femur .... *S.* (*G.*) *brinchangense*
Hind basitarsus 0.53 times as wide as hind femur .... *S.* (*G.*) *roslihashimi*
28. Mandible with inner and outer teeth ....*S.* (*G.*) *asakoae*
Mandible with inner teeth but lacking outer teeth ....*S.* (*G.*) *izuae*
29. Claw tooth 0.46 times length of claw .... *S.* (*G.*) *langkawiense*
Claw tooth 0.53–0.56 times length of claw 30
30. Genital fork with distinct lobes directed posteromedially (similar to Fig. 36I) ....*S.* (*G.*) *gombakense*
Genital fork without lobes directed posteromedially 31
31. Base of each arm of genital fork narrow, nearly as wide as apical portion .... *S.* (*G.*) *sofiani*
Base of each arm of genital fork much broader than apical portion .... *S.* (*G.*) *lurauense*
32. Claw with large basal tooth (Fig. 47D) 33
Claw without tooth or with small subbasal tooth (Fig. 84B) or tiny subbasal swelling (Fig. 80B) 37
33. Fore coxa darkened 34
Fore coxa yellow 35
34. Frons densely covered with hairs ....*S.* (*N.*) *caudisclerum*
Frons moderately covered with hairs except mediolongitudinal portion bare .... *S.* (*N.*) *kurtaki*
35. Scutum without dark longitudinal vittae .... *S.* (*N.*) *aureohirtum*
Scutum with three dark longitudinal vittae 36
36. Abdominal segments 2 and 6–8 shiny .... *S.* (*N.*) *pairoti*
Abdominal segments 2 and 5–8 shiny .... *S.* (*N.*) *ledangense*
37. Paraproct with cluster of dark spines anteriorly (Fig. 8E, G) .... *S.* (*D.*) *pahangense*
Paraproct without cluster of dark spines anteriorly 38
38. Claw with small subbasal tooth with pointed apex (Fig. 84B) 39
Claw simple or with tiny subbasal swelling (Fig. 80B) 44
39. Fronto-ocular area undeveloped (similar to Fig. 74B). Antenna mostly yellow except a few apical flagellomeres (similar to Fig. 74A) ....*S.* (*S.*) *vanluni*

Fronto-ocular area developed. Antenna mostly brownish except scape, pedicel, and base of first flagellomere 40

40. Inner margins of ovipositor valves nearly parallel-sided or slightly divergent posteriorly (similar to Fig. 50C) 41
Inner margins of ovipositor valves concave medially (Figs. 49C and 84C) 42

41. Mid and hind tibiae yellow on basal half and dark on rest ....*S.* (*S.*) *fuscopilosum*
Mid and hind tibiae darkened except basal tip yellow ....*S.* (*S.*) *rasuli*

42. Ovipositor valves each with rounded posteromedial corner (similar to Fig. 49C). ....*S.* (*S.*) *argentipes*
Ovipositor valves each with pointed tip (Figs. 84C) 43

43. Ovipositor valves each with tip directed inward (Fig. 84C). Genital fork with projection directed anterodorsally on each arm (similar to Fig. 84D) ....*S.* (*S.*) *hackeri*
Ovipositor valves each with tip directed posteriorly (similar to Fig. 67M). Genital fork without projection directed anterodorsally on each arm (similar to Fig. 67G) ....*S.* (*S.*) *bishopi*

44. Scutum with longitudinal vittae (Fig. 76B, C) 45
Scutum without longitudinal vittae 52

45. Ovipositor valve with round flap produced ventrally along inner margin (Fig. 76D, E) 46
Ovipositor valve without round flap produced ventrally along inner margin 47

46. Fore femur light brown .... *S.* (*S.*) *perakense*
Fore femur yellow, though apical tip slightly darkened ....*S.* (*S.*) *jeffreyi*

47. Ovipositor valves covered with setae along inner margins and each with elongate internal projection (similar to Fig. 61B) ....*S.* (*S.*) *grossifilum*
Ovipositor valves bare along inner margins and without internal projection 48

48. Ovipositor valves each with 8–14 short hairs 49
Ovipositor valves each with few short hairs 50

49. Hind basitarsus yellowish white on little less than basal half and brownish black on apical half .... *S.* (*S.*) *digrammicum*
Hind basitarsus yellowish white on basal two-thirds and medium brown on apical one-third ....*S.* (*S.*) *thongsahuani*

50. Basal portion of radial vein with hairs ....*S.* (*S.*) *hirtinervis*
Basal portion of radial vein without hairs 51

51. Paraproct rounded ventrally ....*S.* (*S.*) *kisapense*
Paraproct tapered ventrally ....*S.* (*S.*) *malayense* (complex)

52. Hind tibia whitish-yellow on basal two-thirds on posterior surface 53
Hind tibia mostly dark brown except base yellowish 54

53. Hind femur yellowish except apical cap brownish .... *S.* (*S.*) *rudnicki*
Hind femur brownish except base yellowish ....*S.* (*S.*) *yongi*

54. Sensory vesicle 0.34 times length of third palpal segment ....*S.* (*S.*) *brevipar* (complex)

Sensory vesicle 0.6–0.7 times length of third palpal segment ....*S.* (*S.*) *jasmoni* and *S.* (*S.*) *tani* (complex)

Males***

1. Katepisternum haired 2
   Katepisternum bare 31
2. Antenna with seven or eight flagellomeres 3
   Antenna with nine flagellomeres 6
3. Antenna with seven flagellomeres 4
   Antenna with eight flagellomeres 5
4. Style widened, slightly narrowed from base to apical one-third, then tapered toward apex ....*S.* (*G.*) *charlesi*
   Style not so widened, tapered from base to apex ......*S.* (*G.*) *novemarticulatum*
5. Ventral plate moderately produced ventrally (similar to Fig. 39T) ....*S.* (*G.*) *burtoni*
   Ventral plate flat, not produced ventrally (similar to Fig. 39Q) .... *S.* (*G.*) *varicorne*
6. Hind basitarsus not widened, nearly parallel-sided 7
   Hind basitarsus widened, spindle-shaped or wedge-shaped 20
7. Posterior surface of hind tibia mostly darkened except basal tip pale 8
   Posterior surface of hind tibia whitish on basal two-fifths or more 14
8. Abdominal segments 2, 6, and 7 each with a pair of shiny dorsolateral patches 9
   Abdominal segments 2, 5, 6, and 7 (or and 8) each with a pair of shiny dorsolateral or lateral patches 10
9. Fore coxa light to medium brown .... *S.* (*G.*) *decuplum*
   Fore coxa yellow ....*S.* (*G.*) *sazalyi*
10. Style broad from base to middle or more (Fig. 41E) ....*S.* (*G.*) *azhari*
    Style tapered from base to apex 11
11. Upper-eye (large) facets in 15 vertical columns and 15 or 16 horizontal rows.. .... *S.* (*G.*) *duolongum* and *S.* (*G.*). *pegalanense*
    Upper-eye (large) facets in 11–14 vertical columns and 11–14 horizontal rows..12
12. Upper-eye (large) facets in 11 vertical columns and 11 horizontal rows ....*S.* (*G.*) *tahanense*
    Upper-eye (large) facets in 13 or 14 vertical columns and 14 horizontal rows 13
13. Dorsal surface of abdomen dark except abdominal segments 2 and 3 yellow .. .. *S.* (*G.*) *dentistylum*
    Dorsal surface of abdomen dark except basal one-third of abdominal segment 2 yellow .... *S.* (*G.*) *johorense*
14. Hair tuft of base of radial vein dark brown ....*S.* (*G.*) *terengganuense*
    Hair tuft of base of radial vein yellowish 15

15. Style angulate when viewed ventrally (similar to Fig. 33J) .... *S. (G.) angulistylum* (complex)
    Style gently curved inward when viewed ventrally 16
16. Hind tibia with dark subbasal spot (Fig. 33C) 17
    Hind tibia without dark subbasal spot 18
17. Upper eye (large) facets in 13 vertical columns and 13 horizontal rows .... *S. (G.) whartoni*
    Upper eye (large) facets in 10 vertical columns and 11 horizontal rows .... *S. (G.) cheongi*
18. Ventral plate nearly parallel-sided when viewed ventrally .... *S. (G.) adleri*
    Ventral plate widened posteriorly when viewed ventrally (Fig. 36P) 19
19. Upper eye (large) facets in 15 vertical columns and 15 or 16 horizontal rows .. .. *S. (G.) gombakense*
    Upper eye (large) facets in 11 or 12 vertical columns and 12 or 13 horizontal rows .... *S. (G.) langkawiense*
20. Hair tuft of base of radial vein dark brown 21
    Hair tuft of base of radial vein yellowish 23
21. Abdominal segments 2–8 each with a pair of dorsolateral or lateral shiny patches .... *S. (G.) leparense*
    Abdominal segments 2 and 5–7 (or 8) with a pair of dorsolateral or lateral shiny patches 22
22. Basal one-third of hind basitarsus whitish-yellow .... *S. (G.) trangense*
    Basal one-third of hind basitarsus somewhat darkened, its border not well defined .... *S. (G.) sheilae*
23. Abdominal segments 2–9 each with a pair of shiny dorsolateral patches .... *S. (G.) hoiseni*
    Abdominal segments 2 and 5–7(or 8) with a pair of shiny dorsolateral patches 24
24. Upper-eye (large) facets in 15 or 16 horizontal rows 25
    Upper-eye (large) facets in 12–14 horizontal rows 27
25. Hind basitarsus yellow on basal half or little less .... *S. (G.) tanahrataense*
    Hind basitarsus almost entirely darkened, though basal half somewhat lighter 26
26. Sensory vesicle short, 0.15 times length of third palpal segment .... *S. (G.) sofiani*
    Sensory vesicle medium-long, 0.25–0.29 times length of third palpal segment .... *S. (G.) lurauense*
27. Upper-eye (large) facets vermilion .... *S. (G.) asakoae*
    Upper-eye (large) facets medium brown 28
28. Antenna almost entirely yellowish .... *S. (G.) roslihashimi*
    Antenna medium to dark brown except scape, pedicel, and base of first flagellomere whitish yellow 29
29. Hind basitarsus yellowish on basal two-fifths and medium to dark brown on rest .... *S. (G.) brinchangense*
    Hind basitarsus whitish on basal half or little more or less 30

30. Upper eye (large)facets in 11 or 12 vertical columns and 12 or 13 horizontal rows .... *S.* (*G.*) *izuae*
Upper eye (large) facets in 13 vertical columns and 14 horizontal rows .... *S.* (*G.*) *longitruncum*
31. Style shorter than coxite 32
Style longer than coxite 35
32. Ventral plate with median keel (similar to Fig. 45I) .... *S.* (*N.*) *aureohirtum*
Ventral plate without median keel 33
33. Fore coxa darkened. Paramere with one hook .... *S.* (*N.*) *caudisclerum*
Fore coxa yellow. Paramere with several hooks 34
34. Upper-eye (large) facets in 13 or 14 vertical columns and 15 horizontal rows .... *S.* (*N.*) *ledangense*
Upper-eye (large) facets in 15 or 16 vertical columns and 16 or 17 horizontal rows .... *S.* (*N.*) *pairoti*
35. Coxite much longer than wide (Fig. 8J) .... *S.* (*D.*) *pahangense*
Coxite as long as or slightly shorter than wide 36
36. Scutum broadly silvery-pruinose with transverse, inverted V-shaped, black band (similar to Fig. 74S, T) .... *S.* (*S.*) *vanluni*
Scutum otherwise 37
37. Ventral plate not serrated posteriorly 38
Ventral plate serrated posteriorly 42
38. Ventral plate saddle-shaped (Fig. 76P) 39
Ventral plate rectangular 40
39. Upper-eye (large) facets in 11–14 vertical columns and 13–15 horizontal rows .... *S.* (*S.*) *jeffreyi*
Upper-eye (large) facets in 16–18 vertical columns and 18 or 19 horizontal rows .... *S.* (*S.*) *perakense*
40. Hind basitarsus entirely darkened 41
Hind basitarsus yellowish on little less than basal half 43
41. Upper-eye (large) facets in 21 vertical columns and 21 or 22 horizontal rows. Style narrowed at basal one-third when viewed ventrolaterally. Style with terminal spine .... *S.* (*S.*) *digrammicum*
Upper-eye (large) facets in 18 or 19 vertical columns and 19 horizontal rows. Style nearly parallel-sided throughout its length when viewed ventrolaterally. Style without terminal spine .... *S.* (*S.*) *thongsahuani*
42. Length of body of ventral plate longer than its arms (similar to Fig. 61H) .... *S.* (*S.*) *grossifilum*
Length of body of ventral plate shorter than its arms (Fig. 52M) .... *S.* (*S.*) *yongi*
43. Style with basal protuberance (Fig. 71C) 44
Style without basal protuberance 49
44. Style with elongate basal protuberance pointed apically (Fig. 71C) 45
Style with short basal protuberance rounded apically (Fig. 80M) 47
45. Hind basitarsus yellow on basal two-fifths and dark on rest .... *S.* (*S.*) *malayense* (complex)

Hind basitarsus yellow on basal half and dark on rest 46
46. Hind tibia black except basal one-third white .... *S.* (*S.*) *hirtinervis*
Hind tibia brownish black except basal tip whitish yellow ...... *S.* (*S.*) *kisapense*
47. Abdomen with a pair of silvery dorsolateral spots on segments 2, 6, and 7 48
Abdomen with a pair of silvery dorsolateral spots on segments 2 and 5–7 ....*S.* (*S.*) *jasmoni* and *S.* (*S.*) *tani* (complex)
48. Head holoptic (frons absent) ....*S.* (*S.*) *brevipar* (complex)
Head dichoptic (frons narrowly ....*S.* (*S.*) *tiomanense*
49. Abdomen with a pair of silvery dorsolateral spots on segments 2 and 4–7 .... *S.* (*S.*) *bishopi*
Abdomen with a pair of silvery dorsolateral spots on segments 2 and 5–7 50
50. Ventral plate with two vertical rows of teeth from base to apical tip when viewed posteriorly (similar to Fig. 49M) ....*S.* (*S.*) *argentipes*
Ventral plate with two vertical rows of teeth from base to basal two-thirds when viewed posteriorly (similar to Fig. 84L) ....*S.* (*S.*) *hackeri*

Pupae****

1. Abdominal segment 9 with grapnel-like hooklets 2
Abdominal segment 9 without grapnel-like hooklets 36
2. Gill composed of inflated structure and eight slender filaments (Fig. 37B, E, F) 3
Gill composed of six, eight, or 10 slender filaments 5
3. Inflated structure simple, horn-like (Fig. 37B) .... *S.* (*G.*) *aziruni*
Inflated structure divided into two round portions each with three finger-like projections 4
4. Ratio of width of middle inflated portion against length of inflated structure 0.3 (Fig. 37E) ....*S.* (*G.*) *gombakense*
Ratio of width of middle inflated portion against length of inflated structure 0.24 (Fig. 37F) .... *S.* (*G.*) *langkawiense*
5. Gill with six filaments (Figs. 14F and 24R) 6
Gill with eight or 10 filaments 7
6. Common basal stalk of gill longer than interspiracular trunk (Fig. 14F). Terminal hooks each with crenulated outer margin .... *S.* (*G.*) *hoiseni*
Common basal stalk of gill shorter than interspiracular trunk. Terminal hooks each with smooth outer margin ....*S.* (*G.*) *sextuplum*
7. Gill with 10 short filaments (similar to Fig. 20Q) .... *S.* (*G.*) *decuplum*
Gill with eight short to long filaments 8
8. All gill filaments (length less than 1.0 mm) much shorter than pupal body. Antennal sheath with tubercles (Fig. 23P) 9
All gill filaments (length of ventral paired filaments more than 1.7 mm) subequal to or longer than pupal body. Antennal sheath bare..10

9. Antennal sheath with nine ridges of tubercles .... *S. (G.) sazalyi*
   Antennal sheath with 10 ridges of tubercles .... *S. (G.) dentistylum*
10. Common basal stalk of gill longer than interspiracular trunk 11
    Common basal stalk of gill shorter than interspiracular trunk 12
11. Gill filaments arranged as (3+3)+2 filaments from dorsal to ventral .... *S. (G.) sofiani*
    Gill filaments arranged as 3+[1+(3+1)] or 3+(3+2) or 3+(3+1+1) filaments from dorsal to ventral .... *S. (G.) longitruncum*
12. Terminal hooks not widened, cone-like or mushroom-like 13
    Terminal hooks flattened, plate-like, each with smooth or crenulated outer margin 23
13. Terminal hook with round, flat apex, appearing mushroom when viewed laterally (Fig. 25V). .... *S. (G.) terengganuense*
    Terminal hook cone-like, tapered apically 14
14. Frons and thoracic integument bare .... *S. (G.) adleri*
    Frons and thoracic integument with tubercles 15
15. Dorsal and middle triplets sharing common stalk 16
    Dorsal and middle triplets not sharing common stalk 19
16. One individual and two paired filaments of dorsal group of gill sharing common stalk 17
    One individual and two paired filaments of dorsal group of gill not sharing common stalk 18
17. Dorsal and middle triplets of gill with medium-long to long common stalk .... *S. (G.) varicorne*
    Dorsal and middle triplets of gill with short common stalk .... *S. (G.) tekamense*
18. Frons with elongate tubercles, longer than width (Fig. 22E) .... *S. (G.) tahanense*
    Frons with round tubercles, shorter than width .... *S. (G.) jerantutense*
19. Two filaments of ventral pair of gill more than 2.5 times length of other filaments .... *S. (G.) duolongum*
    Two filaments of ventral pair of gill less than twice length of other filaments 20
20. Two filaments of ventral pair of gill slightly longer than three filaments of middle triplet .... *S. (G.) azhari*
    Two filaments of ventral pair of gill 1.5–2.0 times length of three filaments of middle triplet 21
21. Middle triplet and ventral pair of gill sharing short common stalk .... *S. (G.) charlesi*
    Middle triplet and ventral pair of gill not sharing common stalk 22
22. Dorsal surface of abdominal segments 1–4 gray .... *S. (G.) johorense*
    Dorsal surface of abdominal segments 1 and 2 yellow .... *S. (G.) novemarticulatum*

23. Terminal hook with outer margin smooth or slightly undulate, less than twice length of inner margin 24
Terminal hook much widened, with outer margin crenulated, more than twice length of inner margin 27
24. Stalk of ventral pair of gill twice length of interspiracular trunk .... *S.* (*G.*) *pegalanense*
Stalk of ventral pair of gill subequal to or little longer than interspiracular trunk 25
25. Secondary stalk of middle triplet of gill long .... *S.* (*G.*) *burtoni*
Secondary stalk of middle triplet of gill short 26
26. Common stalk of dorsal and middle triplets of gill medium-long .... *S.* (*G.*) *lurauense*
Common stalk of dorsal and middle triplets of gill very short .... *S.* (*G.*) *sheilae*
27. One of the two filaments of ventral pair of gill much longer than counter filament 28
Two filaments of ventral pair of gill subequal in length to each other 29
28. One of the two filaments of ventral pair of gill about three times length of counter filament .... *S.* (*G.*) *whartoni*
One of the two filaments of ventral pair of gill about twice length of counter filament .... *S.* (*G.*) *cheongi*
29. Dorsal and middle triplets of gill not sharing common stalk .... *S.* (*G.*) *angulistylum* (complex)
Dorsal and middle triplets of gill sharing common stalk 30
30. Abdominal segments 1 and 2 dark gray, with tubercles on dorsal surface .... *S.* (*G.*) *asakoae*
Abdominal segments 1 and 2 unpigmented, without tubercles 31
31. Stalk of dorsal triplet of gill lying against stalk of ventral pair at angle of 80–100 degreeswhen viewed laterally .... *S.* (*G.*) *trangense*
Stalk of dorsal triplet of gill lying against stalk of ventral pair at angle of 60–70 degrees when viewed laterally 32
32. Anterodorsal margin of cocoon with short median bulge 33
Anterodorsal margin of cocoon without median bulge 34
33. Stalk of ventral pair of gill shorter than common basal stalk .... *S.* (*G.*) *izuae*
Stalk of ventral pair of gill longer than common basal stalk .... *S.* (*G.*) *brinchangense*
34. Cocoon 3.0–3.5 mm .... *S.* (*G.*) *roslihashimi*
Cocoon 3.6–3.9 mm .... *S.* (*G.*) *tanahrataense*
35. Gill with four filaments 36
Gill with six, eight, or 10 filaments 37
36. Cocoon with anterodorsal projection .... *S.* (*N.*) *caudisclerum*
Cocoon without anterodorsal projection .... *S.* (*N.*) *kurtaki*
37. Gill with six filaments 38
Gill with eight or 10 filaments 54
38. Dorsum of abdominal segments without spine-combs .... *S.* (*D.*) *pahangense*

Dorsum of abdominal segments (at least posterior segments) with spine-combs 39

39. Gill filaments longer than pupal body 40
Gill filaments shorter than pupal body 41
40. Stalk of ventral pair of gill long .... *S.* (*N.*) *ledangense*
Stalk of ventral pair of gill short to medium-long .... *S.* (*N.*) *pairoti*
41. Cocoon with anterodorsal projection .... *S.* (*N.*) *aureohirtum*
Cocoon without anterodorsal projection 42
42. Gill filaments inflated (similar to Fig. 61N) .... *S.* (*S.*) *grossifilum*
Gill filaments slender, thread-like 43
43. Cocoon shoe-shaped 44
Cocoon wall-pocket-shaped 46
44. Cocoon with anterolateral window on each side (Fig. 60R) .... *S.* (*S.*) *digrammicum*
Cocoon without anterolateral window on each side 45
45. Head integument bare .... *S.* (*S.*) *bishopi*
Head integument sparsely or moderately covered with tubercles .... *S.* (*S.*) *vanluni*
46. Cocoon with anterolateral window on each side 47
Cocoon without anterolateral window on each side 48
47. Dorsal surface of abdomen with spine-combs on segments 7 and 8 .... *S.* (*S.*) sp. C
Dorsal surface of abdomen with spine-combs on segment 8 .... *S.* (*S.*) *rudnicki*
48. Pit-like organ present at base of gill (Fig. 79H) 49
Pit-like organ absent at base of gill 50
49. Thoracic tubercles smooth .... *S.* (*S.*) *tiomanense*
Thoracic tubercles with tiny secondary projections .... *S.* (*S.*) *brevipar* (complex)
50. Cocoon thin, colorless with deep anteromedian incision (Fig. 53I) .... *S.* (*S.*) *yongi*
Cocoon thick, dark yellow to dark brown, without anteromedian incision 51
51. Frontal and facial trichomes each with 8–12 branches .... *S.* (*S.*) *thongsahuani*
Frontal and facial trichomes unbranched 52
52. Upper filament of dorsal pair of gill divergent at acute angle against lower filament of ventral pair (similar to Fig. 81G) .... *S.* (*S.*) *tani* (complex)
Upper filament of dorsal pair of gill divergent at obtuse angle against lower filament of ventral pair 53
53. Head and anterior half of thorax bare .... *S.* (*S.*) *hackeri*
Head and anterior half of thorax moderately or sparsely covered with tubercles .... *S.* (*S.*) *jasmoni*
54. Gill with eight filaments 55
Gill with 10 filaments 59
55. Eight gill filaments arranged as 3+3+2 from dorsal to ventral .... *S.* (*G.*) *leparense*
Eight gill filaments arranged as 2+2+2+2 from dorsal to ventral 56

56. Cocoon of corbicular shape (similar to Fig. 49R, S) .... *S.* (*S.*) *argentipes*
    Cocoon wall-pocket-shaped 57
57. Cocoon without anterolateral window on each side ....
    .... *S.* (*S.*) *malayense* (complex)
    Cocoon with anterolateral window on each side 58
58. Upper filament of dorsal pair of gill divergent at obtuse angle against lower filament of ventral pair. Cocoon with small anterolateral window on each side .... *S.* (*S.*) *kisapense*
    Upper filament of dorsal pair of gill divergent at angle of 90 degrees or less against lower filament of ventral pair (Fig. 72D). Cocoon with large anterolateral window on each side .... *S.* (*S.*) *hirtinervis*
59. Five lower filaments of gill much slenderer than five upper filaments (similar to Fig. 77D) .... *S.* (*S.*) *jeffreyi*
    Gill filaments subequal in thickness to one another (Fig. 77C) 60
60. Dorsal surface of abdominal segment 8 with spine-combs ..... *S.* (*S.*) *perakense*
    Dorsal surface of abdominal segment 8 without spine-combs .... *S.* (*S.*) sp. B

Mature larvae*****

1. Hypostoma very wide, with prominent median tooth accompanied by six small teeth on each side (Fig. 9L). Posterior circlet with over 400 rows of hooklets .... *S.* (*D.*) *pahangense*
   Hypostoma otherwise, with nine anterior teeth. Posterior circlet with less than 250 rows of hooklets 2
2. Lateral margins of hypostoma smooth, without teeth (Fig. 32U). Main tooth of mandibular serration at acute angle to mandible apically (Fig. 32T) 3
   Lateral margins of hypostoma serrated. Main tooth of mandibular serration at obtuse angle to mandible apically 32
3. Body length 5.0–6.0 mm 4
   Body length less than 5.0 mm 8
4. Head capsule dark brown. Postgenal cleft little less than half length of postgenal bridge .... *S.* (*G.*) *adleri*
   Head capsule yellow to light brown. Postgenal cleft 1.2–3.0 times length of postgenal bridge 5
5. Abdominal segments 2–4 each with three reddish-brown spots on each side (Fig. 25W, X) ....*S.* (*G.*) *terengganuense*
   Abdominal segments 2–4 without colored spots on each side 6
6. Ventral surface of head capsule darkened medially ....*S.* (*G.*) *asakoae*
   Ventral surface of head capsule not darkened medially 7
7. Abdominal segments 5 and 6 each with reddish-brown markings ....
   .... *S.* (*G.*) *brinchangense*
   Abdominal segments 5 and 6 without such markings ...... *S.* (*G.*) *longitruncum*

8. Abdominal segments 1–5 each with dorsal protuberances (Fig. 40M, N) 9
   Abdominal segments 1–5 without such dorsal protuberances 12
9. Pharate pupal gill with eight slender filaments, all less than 1.0 mm long ....*S.* (*G.*) *sazalyi*
   Pharate pupal gill with eight slender filaments, of which two filaments of ventral pair longer than 1.5 mm 10
10. Two filaments of ventral pair of pharate pupal gill subequal in length to six other filaments .... *S.* (*G.*) *varicorne*
    Two filaments of ventral pair of pharate pupal gill much longer than six other filaments 11
11. Stout spinous setae on dorsal surface of abdominal segments 5–8 unbranched (Fig. 40Q) ....*S.* (*G.*) *charlesi*
    Stout spinous setae on dorsal surface of abdominal segments 5–8 bifid (Fig. 40O) ....*S.* (*G.*) *novemarticulatum*
12. Abdomen markedly constricted between segments 4 and 5 .... *S.* (*G.*) *dentistylum*
    Abdomen not so markedly constricted between segments 4 and 5 13
13. Pharate pupal gill composed of inflated structure with eight slender filaments 14
    Pharate pupal gill composed of six, eight, or 10 slender filaments 16
14. Inflated structure of pharate pupal gill horn-like, without finger-like projections .... *S.* (*G.*) *aziruni*
    Inflated structure of pharate pupal gill divided into two round portions, each with three finger-like projections 15
15. Posterior circlet with 74–82 rows of hooklets ....*S.* (*G.*) *gombakense*
    Posterior circlet with 96–100 rows of hooklets .... *S.* (*G.*) *langkawiense*
16. Pharate pupal gill with six slender filaments ....*S.* (*G.*) *sextuplum*
    Pharate pupal gill with eight or 10 slender filaments 17
17. Pharate pupal gill with 10 slender filaments .... *S.* (*G.*) *decuplum*
    Pharate pupal gill with eight slender filaments 18
18. Pharate pupal gill with eight slender filaments, of which one of two ventral paired filaments more than 1.5 times length of other filaments 19
    Pharate pupal gill with eight slender filaments, of which two ventral paired filaments subequal in length to each other 20
19. Longest filament of pharate pupal gill more than twice length of other filaments ....*S.* (*G.*) *whartoni*
    Longest filament of pharate pupal gill ca. 1.4 times length of other filaments .. .. *S.* (*G.*) *cheongi*
20. Pharate pupal gill with eight short slender filaments, 1.2–1.5 mm long............ ....*S.* (*G.*) *angulistylum* (complex)
    Pharate gill with eight slender filaments, of which at least two ventral paired filaments longer than 2.0 mm 21

21. Two filaments of ventral pair of pharate pupal gill 2.0–2.5 times length of other filaments. Abdominal segments 5–8 covered with various sizes of dark stout spinous setae dorsally .... *S.* (*G.*) *duolongum*
    Two filaments of ventral pair of pharate pupal gill less than 1.5 times length of other filaments. Abdominal segments 5–8 bare or covered with pale or dark fine setae 22
22. Body 3.5–3.9 mm long 23
    Body 4.0–5.0 mm long 27
23. Postgenal cleft almost reaching posterior margin of hypostoma ....
    .... *S.* (*G.*) *johorense*
    Postgenal cleft long but not reaching posterior margin of hypostoma (7.5–8.0 times length of postgenal bridge) 24
24. Thoracic segment 1 and abdominal segment 1 each encircled with black transverse band (Fig. 22L) .... *S.* (*G.*) *tahanense*
    Thoracic segment 1 and abdominal segment 1 (together with segments 2–5) each encircled with grayish or reddish-brown transverse band 25
25. Abdominal segments 1–4 each encircled with grayish broad band, though abdominal segment 3 also with sparse reddish-purplish band ....
    .... *S.* (*G.*) *lurauense*
    Abdominal segments 1–4 each encircled with reddish-brown band 26
26. Rectal organ with 8–12 secondary lobules per lobe .... *S.* (*G.*) *burtoni*
    Rectal organ with four or five secondary lobules per lobe ....
    .... *S.* (*G.*) *pegalanense*
27. Postgenal cleft long, with its apex close to or reaching posterior margin of hypostoma, leaving no or very short postgenal bridge (Fig. 27L, M) 28
    Postgenal cleft less than 6.0 times length of postgenal bridge 29
28. Body 4.0–4.4 mm long. Rectal organ with five or six secondary lobules per lobe
    .... *S.* (*G.*) *trangense*
    Body 4.6–5.0 mm long. Rectal organ with 8–11 secondary lobules per lobe ....
    .... *S.* (*G.*) *sheilae*
29. Postgenal cleft 5.9 times length of postgenal bridge .... *S.* (*G.*) *azhari*
    Postgenal cleft less than 3.1 times length of postgenal bridge 30
30. Postgenal cleft 3.1 times length of postgenal bridge. Abdominal segments 1–8 with reddish-brown markings .... *S.* (*G.*) *roslihashimi*
    Postgenal cleft less than 2.6 times length of postgenal bridge. Abdominal segments 1–8 without reddish-brown markings 31
31. Postgenal cleft 2.3 times length of postgenal bridge. Rectal organ with five to seven secondary lobules per lobe .... *S.* (*G.*) *sofiani*
    Postgenal cleft 1.5–1.6 times length of postgenal bridge. Rectal organ with eight or ninesecondary lobules per lobe .... *S.* (*G.*) *izuae*
32. Abdominal segment 9 with large ventral papillae 33
    Abdominal segment 9 without ventral papillae or with small ventral papillae (similar to Fig. 88Z) 37
33. Body with reddish-brown markings (similar to Fig. 44K, L) 34

Body without reddish-brown markings 35
34. Labral fan with 19 primary rays .... *S.* (*N.*) *ledangense*
Labral fan with 34–36 primary rays .... *S.* (*N.*) *pairoti*
35. Accessory sclerites present (similar to Fig. 48P) ....*S.* (*N.*) *caudisclerum*
Accessory sclerites absent 36
36. Head spots boldly positive. Mandibular serration composed of two large teeth. Postgenal cleft not pointed apically (similar to Fig. 46M) ....
.... *S.* (*N.*) *aureohirtum*
Head spots moderately positive. Mandibular serration composed of one large tooth and one small tooth (similar to Fig. 48M). Postgenal cleft pointed apically
....*S.* (*N.*) *kurtaki*
37. Abdominal segments 1–5 or 1–8 each with a pair of dorsal protuberances (Figs. 72N and 75K) 38
Abdominal segments lacking dorsal protuberances 42
38. Postgenal cleft long, reaching posterior margin of hypostoma (similar to Fig. 75N)...*S.* (*S.*) *vanluni*
Postgenal cleft long, but not reaching posterior margin of hypostoma 39
39. Pharate pupal gill with eight filaments 40
Pharate pupal gill with 10 filaments 41
40. Abdominal segments 1–8 each with a pair of dorsolateral protuberances (Fig. 72N) .... *S.* (*S.*) *hirtinervis*
Abdominal segments 1–5 each with a pair of dorsolateral protuberances..........
.... *S.* (*S.*) *argentipes*
41. All filaments of pharate pupal gill subequal in thickness to one another ....
.... *S.* (*S.*) *perakense*
Filaments of pharate pupal gill differing in thickness between five upper filaments and five lower filaments ....*S.* (*S.*) *jeffreyi*
42. Pharate pupal gill with eight filaments 43
Pharate pupal gill with six filaments 44
43. Body length 4.8–5.3 mm. Cephalic apotome dark yellow to medium brown at least on posterior three-fifths .... *S.* (*S.*) *kisapense*
Body length 5.6–6.0 mm. Cephalic apotome entirely pale yellow ....
....*S.* (*S.*) *malayense* (complex)
44. Body creamy .... *S.* (*S.*) *digrammicum*
Body grayish to dark gray or reddish brown 45
45. Body length over 5.5 mm 46
Body length less than 5.2 mm 50
46. Labral fan with 60–64 main rays .... *S.* (*S.*) *rudnicki*
Labral fan with 46–54 main rays..47
47. Posterior circlet with 89–91 rows of up to 13–16 hooks per row ....
....*S.* (*S.*) *hackeri*
Posterior circlet with 120–170 rows of up to 18–22 rows per row 48
48. Posterior circlet with 154–170 rows of hooklets ....*S.* (*S.*) *yongi*
Posterior circlet with 120–143 rows of hooklets 49
49. Posterior circlet with ca. 120 rows of hooklets ....*S.* (*S.*) *grossifilum*
Posterior circlet with 140–143 rows of hooklets ....*S.* (*S.*) *thongsahuani*

50. Postgenal cleft rounded apically .... *S.* (*S.*) *bishopi*
    Postgenal cleft pointed apically 51
51. Postgenal cleft triangular (similar to Fig. 79M) 52
    Postgenal cleft nearly parallel-sided on basal half, then tapered apically (similar to Fig. 81P) 53
52. Posterior circlet with 84 rows of hooklets .... *S.* (*S.*) *tiomanense*
    Posterior circlet with ca. 74 rows of hooklets ....*S.* (*S.*) *brevipar* (complex)
53. Body blackish brown .... *S.* (*S.*) *jasmoni*
    Body reddish brown .... *S.* (*S.*) *tani* (complex)

*Keys provided by Takaoka et al. (2018m) are revised. *Simulium* (*G.*) *atratum* originally described from Java by De Meigen (1913) is not included because morphological characteristics in specimens from Peninsular Malaysia are not available.

** *Simulium* (*G.*) *hoiseni*, *S.* (*G.*) *tanahrataense*, *S.* (*S.*) sp. B, and *S.* (*S.*) sp. C are not included because their females are unknown.

*** *Simulium* (*G.*) *aziruni*, *S.* (*G.*) *jerantutense*, *S.* (*G.*) *sextuplum*, *S.* (*N.*) *kurtaki*, *S.* (*S.*) *fuscopilosum*, *S.* (*S.*) *rasuli*, *S.* (*S.*) sp. B, and *S.* (*S.*) sp. C are not included because their males are unknown.

**** *Simulium* (*S.*) *fuscopilosum* and *S.* (*S.*) *rasuli* are not included because their pupae are unknown.

***** *Simulium* (*G.*) *tekamense*, *S.* (*G.*) *jerantutense*, *S.* (*G.*) *hoiseni*, *S.* (*G.*) *tanahrataense*, *S.* (*S.*) *fuscopilosum*, *S.* (*S.*) *rasuli*, *S.* (*S.*) sp. B, and *S.* (*S.*) sp. C are not included because their larvae are unknown.

9 The Fauna of Black Flies in Borneo (Sabah and Sarawak in Malaysia and Kalimantan in Indonesia)

Overview

Edwards (1933) described five new species from Sabah: *Simulium* (*Nevermannia*) *fuscinervis*, *S.* (*Simulium*) *crassimanum*, *S.* (*S.*) *laterale*, *S.* (*S.*) *nigripilosum*, and *S.* (*S.*) *aeneifacies*. Smart and Clifford (1969) described six new species: *S.* (*N.*) *tuaranense*, *S.* (*Gomphostilbia*) *pegalanense*, *S.* (*G.*) *rayohense* *S.* (*S.*) *kinabaluense*, *S.* (*S.*) *kiuliense*, and *S.* (*S.*) *sabahense*; and described the female, male, and pupa of *S.* (*S.*) *nigripilosum*. Crosskey (1973) synonymized *S.* (*S.*) *kiuliense* and *S.* (*N.*) *tuaranense* with *S.* (*S.*) *nobile* and *S.* (*S.*) *aureohirtum*, respectively. Takaoka (1996b) described a new species, *S.* (*S.*) *beludense* [the species misidentified as *S.* (*S.*) *nigripilosum* by Smart and Clifford (1969)].

Takaoka (2001a) described two new species from Sabah: *S.* (*N.*) *borneoense* and *S.* (*G.*) *guniki*, and recorded three species: *S.* (*G.*) *dentistylum*, *S.* (*G.*) *parahiyangum*, and *S.* (*G.*) *sheilae*. Takaoka (2001b) described two new species from Sarawak:

S. (*G.*) *lehi* and *S.* (*G.*) *sarawakense*. Takaoka (2007b) described two new species from Sabah: *S.* (*S.*) *liewi* and *S.* (*S.*) *maklarini*. Takaoka (2008b) described four new species from Sabah and Sarawak: *S.* (*S.*) *alberti*, *S.* (*S.*) *keningauense*, *S.* (*S.*) *masilauense*, and *S.* (*S.*) *lunduense*. Takaoka (2008c) described four new species from Sarawak: *S.* (*G.*) *barioense*, *S.* (*G.*) *charlesi*, *S.* (*G.*) *kelabitense*, and *S.* (*G.*) *paukatense*. Takaoka (2009c) described three new species from Sarawak: *S.* (*G.*) *auratum*, *S.* (*G.*) *fulgidum*, and *S.* (*G.*) *capillatum* [the species formerly misidentified as *S.* (*G.*) *dentistylum* by Takaoka (2001a)]. Takaoka et al. (2015a) described two new species from Sarawak: *S.* (*S.*) *murudense* and *S.* (*S.*) *cheedhangi*. Ya'cob et al. (2015a,b) described two new species from Sarawak: *S.* (*G.*) *hiroyukii* and *S.* (*G.*) *bakalalanense*, and recorded *S.* (*G.*) *terengganuense* (originally described from Peninsular Malaysia). Takaoka et al. (2015b) described two new species, *S.* (*S.*) *lardizabalae* and *S.* (*S.*) *timpohonense*; synonymized *S.* (*S.*) *liewi* and *S.* (*S.*) *kinabaluense* with *S.* (*S.*) *crassimanum* and *S.* (*S.*) *laterale*, respectively; and described the males and pupae of *S.* (*S.*) *crassimanum*, *S.* (*S.*) *laterale*, and *S.* (*S.*) *nigripilosum*, from Sabah. Low et al. (2016a) resurrected *S.* (*S.*) *kiuliense* from a junior synonym of *S.* (*S.*) *nobile*. Takaoka et al. (2016a) described a new species, *S.* (*S.*) *mirum*, of which the male shows dimorphism for the scutal pattern linked with the different number of male upper-eye (large) facets. Takaoka et al. (2016b, 2017d) described two new species, *S.* (*G.*) *kalimantanense* and *S.* (*G.*) *paserense*, and recorded seven species: *S.* (*G.*) *cheongi*, *S.* (*G.*) *tahanense*, *S.* (*G.*) *lehi*, *S.* (*G.*) *sarawakense*, *S.* (*S.*) *mirum*, *S.* (*S.*) *keningauense*, and *S.* (*S.*) *sabahense*, from Kalimantan, Indonesia. Ya'cob et al. (2020) recorded *S.* (*G.*) *kalimantanense* from Sabah.

The fauna of black flies in Borneo is represented by 41 species of the genus *Simulium* s. l., of which 21 species are placed in the subgenus *Gomphostilbia*, two species are in the subgenus *Nevermannia* and 17 species are in the subgenus *Simulium* s. str. Of 41 species, 32 species are from Sabah and Sarawak, one species, *S.* (*G.*) *paserense*, is from Kalimantan, and eight species: *S.* (*G.*) *kalimantanense*, *S.* (*G.*) *lehi*, *S.* (*G.*) *sarawakense*, *S.* (*G.*) *tahanense*, *S.* (*N.*) *aureohirtum*, *S.* (*S.*) *keningauense*, *S.* (*S.*) *mirum*, and *S.* (*S.*) *sabahense*, are from both areas.

The fauna is characterized by a high endemic species rate (33/41 or 80.5%). Only eight species are common to surrounding islands: six species: *S.* (*G.*) *charlesi*, *S.* (*G.*) *cheongi*, *S.* (*G.*) *pegalanense*, *S.* (*G.*) *sheilae*, *S.* (*G.*) *tahanense*, and *S.* (*G.*) *terengganuense*, are common to Peninsular Malaysia, and one species, *S.* (*G.*) *parahiyangum*, is common to Java, whereas one species, *S.* (*N.*) *aureohirtum*, is common to Peninsular Malaysia, Palawan, Java, and Sulawesi.

List of 41 Species of Black Flies in Borneo

Subgenus *Gomphostilbia*
S. alienigenum species-group

S. kalimantanense
S. batoense species-group
S. bakalalanense, *S. parahiyangum*, *S. paukatense*, *S. pegalanense*, *S. tahanense*, and *S. terengganuense*
S. ceylonicum species-group
S. capillatum and *S. sheilae*
S. epistum species-group
S. auratum, S. barioense, *S. cheongi*, *S. fulgidum*, *S. kelabitense*, *S. lehi*, *S. paserense*, *S. rayohense*, and *S. sarawakense*
S. varicorne species-group
S. charlesi
Gomphostilbia species unplaced to species-group
S. guniki and *S. hiroyukii*
Subgenus *Nevermannia*
S. feuerborni species-group
S. borneoense and *S. fuscinervis*
S. ruficorne species-group
S. aureohirtum
Subgenus *Simulium* s. str.
S. argentipes species-group
S. beludense
S. melanopus species-group
S. cheedhangi, *S. crassimanum*, *S. lardizabalae*, *S. laterale*, *S. maklarini*, *S. mirum*, *S. murudense*, *S. nigripilosum*, and *S. timpohonense*
S. nobile species-group
S. kiuliense
S. tuberosum species-group
S. aeneifacies, *S. alberti*, *S. keningauense*, *S. lunduense*, *S. masilauense*, and *S. sabahense*

*Keys to Identify 41 Species of Black Flies from Borneo**

Female**

1. Katepisternum haired 2
   Katepisternum bare 17
2. Antenna composed of seven flagellomeres (Fig. 39C) .... *S.* (*G.*) *charlesi*
   Antenna composed of nine flagellomeres 3
3. Base of radius with tuft of yellow hairs 4
   Base of radius with tuft of dark hairs 11
4. Hind tibia whitish yellow on basal two-thirds or three-fourths and dark on rest, without dark subbasal marking 5

Hind tibia whitish yellow on little more than basal two-fifths to basal two-thirds, with dark subbasal marking 6

5. Antenna brownish-black except scape, pedicel, second to fourth flagellomeres yellow and first flagellomere light to medium brown though base yellow (Fig. 11A) ....*S.* (*G.*) *kalimantanense*

Antenna medium brown except scape, pedicel and base of first flagellomere pale yellow ....*S.* (*G.*) *lehi*

6. Mid basitarsus brown ....*S.* (*G.*) *rayohense*

Mid basitarsus dark brown except basal half or little more whitish yellow or dark yellow 7

7. Mandible without outer teeth ....*S.* (*G.*) *barioense*

Mandible with outer teeth 8

8. Sensory vesicle 0.30–0.34 times as long as third palpal segment 9

Sensory vesicle 0.23–0.26 times as long as third palpal segment 10

9. Fore basitarsus 5.4 times as long as its greatest width ....*S.* (*G.*) *fulgidum*

Fore basitarsus 6.5 times as long as its greatest width .... *S.* (*G.*) *cheongi*

10. Wing 2.5 mm long ....*S.* (*G.*) *kelabitense*

Wing 1.7 mm long .... *S.* (*G.*) *sarawakense*

11. Fore coxa darkened 12

Fore coxa not darkened 13

12. Seventh sternite developed (Fig. 41C) ....*S.* (*G.*) *guniki*

Seventh sternite undeveloped ....*S.* (*G.*) *hiroyukii*

13. Sensory vesicle over half as long as third palpal segment 14

Sensory vesicle 0.2–0.35 times as long as third palpal segment 15

14. Hind tibia darkened except base pale .... *S.* (*G.*) *parahiyangum*

Hind tibia whitish yellow on basal half and darkened on rest......*S.* (*G.*) *sheilae*

15. Frons with narrowest width in middle and moderately covered with short hairs ....*S.* (*G.*) *terengganuense*

Frons with narrowest width near lower margin and largely bare except several hairs along each lateral margin and above lower margin 16

16. Basal tooth of claw 0.4 times as long as claw ....*S.* (*G.*) *tahanense*

Basal tooth of claw half as long as claw ....*S.* (*G.*) *paukatense*

17. Claw with large basal tooth ....*S.* (*N.*) *aureohirtum*

Claw with or without small subbasal tooth 18

18. Basal portion of radius haired 19

Basal portion of radius bare 20

19. Fore coxa whitish yellow ....*S.* (*S.*) *mirum*

Fore coxa darkened ....*S.* (*S.*) *murudense*

20. Claw with small subbasal tooth (Fig. 74G) 21

Claw without tooth 27

21. Antenna yellow except two apical flagellomeres darkened (similar to Fig. 74A) ....*S.* (*S.*) *kiuliense*

Antenna mostly darkened 22

22. Ovipositor valve short, rounded posteromedially (similar to Fig. 49C) .... *S.* (*S.*) *beludense*
    Ovipositor valve elongate, pointed apically (similar to Fig. 67K) 23
23. Fore coxa whitish yellow to yellow 24
    Fore coxa darkened 25
24. Claw with small tooth 0.26 times length of claw .... *S.* (*S.*) *maklarini*
    Claw with medium-long tooth 0.43 times length of claw (Fig. 67C) .... *S.* (*S.*) *cheedhangi*
25. Hind basitarsus whitish on basal half and darkened on apical half .... *S.* (*S.*) *laterale*
    Hind basitarsus whitish on basal three-fifths and darkened on apical two-fifths 26
26. Mid basitarsus yellowish except apical tip darkened .... *S.* (*S.*) *crassimanum*
    Mid basitarsus whitish yellow on basal two-thirds and darkened on apical one-third .... *S.* (*S.*) *nigripilosum*
27. Sensory vesicle long, 0.59 times as long as third palpal segment 28
    Sensory vesicle medium-long, 0.31–0.49 times as long as third palpal segment 29
28. Mandible with 12 outer teeth .... *S.* (*S.*) *alberti*
    Mandible with 16 outer teeth .... *S.* (*S.*) *masilauense*
29. Fore basitarsus 5.6–5.7 times as long as its greatest width .... *S.* (*S.*) *lunduense*
    Fore basitarsus 4.4–4.9 times as long as its greatest width 30
30. Sensory vesicle 0.31 times as long as third palpal segment .... *S.* (*S.*) *aeneifacies*
    Sensory vesicle 0.39–0.49 times as long as third palpal segment 31
31. Wing 1.5 mm long .... *S.* (*S.*) *sabahense*
    Wing 1.7–1.8 mm long .... *S.* (*S.*) *keningauense*

Males***

1. Katepisternum haired 2
   Katepisternum bare 21
2. Antenna composed of seven flagellomeres .... *S.* (*G.*) *charlesi*
   Antenna composed of nine flagellomeres 3
3. Hind basitarsus enlarged 4
   Hind basitarsus slender, nearly parallel-sided (Fig. 11J) 7
4. Fore coxa darkened 5
   Fore coxa not darkened 6
5. Ventral plate with lateral margins concave when viewed ventrally................... .... *S.* (*G.*) *hiroyukii*
   Ventral plate with lateral margins nearly straight when viewed ventrally .... *S.* (*G.*) *guniki*
6. Upper-eye (large) facets in 13 horizontal rows .... *S.* (*G.*) *sheilae*
   Upper-eye (large) facets in 16 or 17 horizontal rows ....*S.* (*G.*) *capillatum*

7. Base of radius with tuft of yellow hairs 8
   Base of radius with tuft of dark hairs 16
8. Antenna brownish-black except scape, pedicel, second to fourth flagellomeres yellow and first flagellomere light to medium brown though base yellow ....*S.* (*G.*) *kalimantanense*
   Antenna otherwise 9
9. Hind tibia yellowish white on basal three-fourths and medium brown on rest, without dark subbasal marking ....*S.* (*G.*) *lehi*
   Hind tibia whitish yellow to yellow on little more than basal half and medium to dark brown on rest, with light brown subbasal marking 10
10. Upper-eye (large) facets in 15 or 16 vertical columns and 15 horizontal rows . ...*S.* (*G.*) *auratum*
    Upper-eye (large) facets in 9–12 vertical columns and 10–13 horizontal rows 11
11. Antenna entirely yellow or yellow except tip of apical flagellomere somewhat darkened 12
    Antenna yellow, though two to five apical flagellomeres somewhat darkened 13
12. Upper-eye (large) facets in 10 vertical columns and 11 horizontal rows ....*S.* (*G.*) *cheongi*
    Upper-eye (large) facets in 12 vertical columns and 13 horizontal rows ....*S.* (*G.*) *kelabitense*
13. Abdomen dark brown dorsally except tergite 2 yellow .... *S.* (*G.*) *rayohense*
    Abdomen dark brown to brownish black dorsally except segments 2 and 3 yellow or dark yellow, though median area light to dark brown 14
14. Antenna yellow, though two apical flagellomeres somewhat darkened ....*S.* (*G.*) *fulgidum*
    Antenna yellow, though three to five apical flagellomeres somewhat darkened 15
15. Cercus with 9–12 hairs .... *S.* (*G.*) *sarawakense*
    Cercus with 15 hairs .... *S.* (*G.*) *barioense*
16. Abdominal segments 2, 6, and 7 each with a pair of dorsolateral shiny spots .... *S.* (*G.*) *parahiyangum*
    Abdominal segments 2 and 5–7 each with a pair of dorsolateral shiny spots 17
17. Antenna grayish yellow except scape, pedicel, and first flagellomere somewhat darkened and ninth flagellomere grayish ....*S.* (*G.*) *terengganuense*
    Antenna otherwise 18
18. Upper-eye (large) facets in 11 vertical columns 19
    Upper-eye (large) facets in 15–18 vertical columns 20
19. Upper-eye (large) facets in 11 horizontal rows ....*S.* (*G.*) *tahanense*
    Upper-eye (large) facets in 13 horizontal rows .... *S.* (*G.*) *bakalalanense*

20. Upper-eye (large) facets in 15 or 16 vertical columns and in 15 horizontal rows .... *S.* (*G.*) *pegalanense*
Upper-eye (large) facets in 17 or 18 vertical columns and in 16 or 17 horizontal rows.. *S.* (*G.*) *paukatense*
21. Basal portion of radius haired 22
Basal portion of radius bare 24
22. Ventral plate with median keel (similar to Fig. 45I, K) ...... *S.* (*N.*) *aureohirtum*
Ventral plate without median keel 23
23. Paramere with 10 hooks .... *S.* (*N.*) *fuscinervis*
Paramere with eight hooks .... *S.* (*N.*) *borneoense*
24. Scutum white-pruinose with non-pruinose inverted V-shaped transverse band (similar to Fig. 74S, T) .... *S.* (*S.*) *kiuliense*
Scutum otherwise 25
25. Style without basal protuberance 26
Style with basal protuberance having spines (Fig. 80M) 33
26. Fore coxa whitish yellow to yellow 27
Fore coxa darkened 29
27. Hind basitarsus slightly enlarged, 4.7 times as long as its greatest width .... *S.* (*S.*) *beludense*
Hind basitarsus moderately enlarged, 4.0–4.1 times as long as its greatest width 28
28. Fore basitarsus 5.4 times as long as its greatest width ..... *S.* (*S.*) *timpohonense*
Fore basitarsus 4.3 times as long as its greatest width .... *S.* (*S.*) *mirum*
29. Hind basitarsus almost entirely darkened .... *S.* (*S.*) *laterale*
Hind basitarsus yellowish white on little less than basal half and darkened on rest 30
30. Upper-eye (large) facets in 25 vertical columns and 26 horizontal rows .... *S.* (*S.*) *lardizabalae*
Upper-eye (large) facets in 17–20 vertical columns and 19–21 horizontal rows 31
31. Fore basitarsus moderately dilated, 6.4–6.5 times as long as its greatest width .... *S.* (*S.*) *nigripilosum*
Fore basitarsus greatly dilated, 4.0–4.5 times as long as its greatest width ...32
32. Hind basitarsus 3.9–4.0 times as long as its greatest width ... *S.* (*S.*) *murudense*
Hind basitarsus 3.0 times as long as its greatest width ....*S.* (*S.*) *crassimanum*
33. Hind basitarsus 4.5–4.6 times as long as its greatest width (Fig. 80K).............. .... *S.* (*S.*) *lunduense*
Hind basitarsus 3.3–3.9 times as long as its greatest width (Fig. 80J) 34
34. Fore basitarsus 5.5–5.7 times as long as its greatest width 35
Fore basitarsus 6.0–6.8 times as long as its greatest width 36
35. Hind basitarsus 3.3 times as long as its greatest width, and as wide as hind femur .... *S.* (*S.*) *sabahense*
Hind basitarsus 3.8–3.9 times as long as its greatest width, and 0.8–0.9 times as wide as hind femur .... *S.* (*S.*) *keningauense*

36. Middle width of style 0.64 times as wide as its greatest width at base when viewed ventrolaterally (Fig. 80O) .... *S.* (*S.*) *aeneifacies*
Middle width of style 0.76 times as wide as its greatest width at base when viewed ventrolaterally (Fig. 80N) 37
37. Upper-eye (large) facets in 17 vertical columns and 19 horizontal rows .... *S.* (*S.*) *masilauense*
Upper-eye (large) facets in 18–20 vertical columns and 20–22 horizontal rows .... *S.* (*S.*) *alberti*

Pupae****

1. Gill with four filaments (Figs. 11Q and 44G) 2
Gill with 6–16 filaments 3
2. Abdominal segment 9 with grapnel-like hooklets ....*S.* (*G.*) *kalimantanense*
Abdominal segment 9 without grapnel-like hooklets ....*S.* (*N.*) *borneoense*
3. Gill with six filaments (Figs. 24R and 46C) 4
Gill with 8–16 filaments 21
4. Abdominal segment 9 with grapnel-like hooklets ....*S.* (*G.*) *paukatense*
Abdominal segment 9 without grapnel-like hooklets 5
5. Cocoon wall-pocket-shaped, with short anterodorsal projection (similar to Fig. 46H) ....*S.* (*N.*) *aureohirtum*
Cocoon wall-pocket-shaped, without anterodorsal projection, or shoe-shaped..6
6. Terminal hooks absent 7
Terminal hooks present 16
7. Frons densely covered with tubercles 8
Frons bare (without tubercles) or sparsely covered with tubercles ventrally 10
8. Cocoon shoc-shapcd, with high antcrovcntral collar (Fig. 69R, S) .... *S.* (*S.*) *laterale*
Cocoon wall-pocket-shaped, or shoe-shaped, with low anteroventral collar (Fig. 69K–M) 9
9. Upper filament of dorsal pair of gill 1.5–1.6 times as thick as counter filament .... *S.* (*S.*) *nigripilosum*
10. Upper filament of dorsal pair of gill 1.2 times as thick as counter filament .... *S.* (*S.*) *timpohonense*
Cocoon wall-pocket-shaped, or shoe-shaped, with low anteroventral collar 11
Cocoon shoe-shaped, with moderate anteroventral collar 14
11. Cocoon wall-pocket-shaped .... *S.* (*S.*) *lardizabalae*
Cocoon shoe-shaped, with low anteroventral collar 12
12. Gill filaments without annular ridges (Fig. 69F) .... *S.* (*S.*) *crassimanum*
Gill filaments with annular ridges 13

13. Cocoon with small nipple-like projections around opening (Fig. 69N, O) .... *S.* (*S.*) *murudense*
Cocoon without nipple-like projections around opening....... *S.* (*S.*) *cheedhangi*
14. Frons sparsely to moderately covered with tubercles ....*S.* (*S.*) *kiuliense*
Frons almost bare 15
15. Area surrounding facial trichomes bare ....*S.* (*S.*) *maklarini*
Area surrounding facial trichomes moderately covered with tubercles .... *S.* (*S.*) *mirum*
16. Dorsomedial surface of thorax with cone-like tubercles having pointed apex (Fig. 81B) 17
Dorsomedial surface of thorax with round tubercles (Fig. 81A) 18
17. Gill with six thread-like filaments arranged as 1+(1+2)+2 from dorsal to ventral (Fig. 81G) ....*S.* (*S.*) *keningauense*
Gill with six thread-like filaments arranged as 2+2+2 from dorsal to ventral ....*S.* (*S.*) *lunduense*
18. Dorsomedial trichomes twice as long as frontal trichomes (Fig. 81C). Longest gill filament 2.2–2.4 mm (Fig. 81I) ....*S.* (*S.*) *alberti*
Dorsomedial trichomes slightly longer or shorter than frontal trichomes. Longest filament 1.4–mm 19
19. Shortest filament of gill 1.3–1.5 mm long ....*S.* (*S.*) *masilauense*
Shortest filament of gill 0.7–1.1 mm long (Fig. 81L) 20
20. Antennal sheath moderately covered with cone-like tubercles on basal one-third .... *S.* (*S.*) *sabahense*
Antennal sheath very sparsely covered with tubercles near base .... *S.* (*S.*) *aeneifacies*
21. Gill with 13 or 16 filaments 22
Gill with eight filaments 23
22. Gill with 13 slender short filaments arising close together at the same level from base ....*S.* (*G.*) *paserense*
Gill with 13 or 16 slender short filaments arranged as 3+6+4 or 6+6+4 in three groups (Fig. 35D–F) .... *S.* (*G.*) *barioense*
23. Gill with swollen basal fenestra ....*S.* (*G.*) *bakalalanense*
Gill without swollen basal fenestra 24
24. Gill filaments short, less than 1.0 mm 25
Gill filaments 1.6 mm or longer 26
25. Gill filaments arising at same level from common basal stalk (Fig. 35B) ....*S.* (*G.*) *fulgidum*
Gill filaments arranged as 3+3+2 from dorsal to ventral (Fig. 23R).................. .... *S.* (*G.*) *parahiyangum*
26. Gill filaments 1.6 mm long, closely arising at same level from short common stalk (Fig. 41J) ....*S.* (*G.*) *guniki*

Gill filaments over 2.0 mm long, arranged differently 27
27. Inner filament of ventral pair of gill longer and thicker than outer filament ....*S.* (*G.*) *pegalanense*
Inner filament of ventral pair of gill as thick as or thinner than outer filament....28
28. Inner filament of ventral pair of gill nearly as thick as outer filament (Fig. 40E) 29
Inner filament of ventral pair of gill thinner and shorter than outer filament.....35
29. Stalks of dorsal and middle triplets and ventral pair of gill arising at same level from common basal stalk (Fig. 40E) 30
Stalks of dorsal and middle triplets of gill sharing short stalk (Fig. 25S) 31
30. Grapnel-like hooklets absent ....*S.* (*G.*) *lehi*
Grapnel-like hooklets present .... *S.* (*G.*) *charlesi*
31. Filaments of ventral pair of gill much longer than six other filaments ....*S.* (*G.*) *tahanense*
All filaments of gill subequal in length (Fig. 25S) 32
32. Terminal hooks mushroom-shaped (Fig. 25T–V) ....*S.* (*G.*) *terengganuense*
Terminal hooks triangular or widened and plate-like 33
33. Both primary stalk of dorsal triplet and secondary stalk of middle triplet of gill extraordinarily long, being half to two-thirds of entire length of filaments .... *S.* (*G.*) *hiroyukii*
Both primary stalk of dorsal triplet and secondary stalk of middle triplet of gill short (similar to Fig. 27A) 34
34. Frons and thorax almost bare ....*S.* (*G.*) *capillatum*
Frons and thorax moderately covered with tubercles ....*S.* (*G.*) *sheilae*
35. All filaments of gill except outer filament of ventral pair 2.0–2.3 mm long ....*S.* (*G.*) *kelabitense* and *S.* (*G.*) *rayohense*
All filaments of gill except outer filament of ventral pair 1.0–1.6 mm long (Fig. 31B) 36
36. Outer filament of ventral pair of gill 3.0 mm long .... *S.* (*G.*) *sarawakense*
Outer filament of ventral pair of gill 1.8–2.4 mm long 37
37. Three filaments of middle triplet of gill composed of two paired filaments and one individual filament (Fig. 31B) .... *S.* (*G.*) *auratum*
Three filaments of middle triplet of gill arising together at same level .... *S.* (*G.*) *cheongi*

Mature larvae*****

1. Ventral papillae large 2
Ventral papillae absent or small 23
2. Hypostoma with lateral margins smooth, without teeth (Fig. 32U) 3
Hypostoma with lateral margins serrated 22
3. Pharate pupal gill with 13 or 16 filaments 4
Pharate pupal gill with four, six, or eight filaments 5

4. Pharate pupal gill with 13 short slender filaments arising close together at same level from base .... *S.* (*G.*) *paserense*
   Pharate pupal gill with 13 or 16 short slender filaments in three groups .... *S.* (*G.*) *barioense*
5. Pharate pupal gill with four filaments .... *S.* (*G.*) *kalimantanense*
   Pharate pupal gill with six or eight filaments 6
6. Pharate pupal gill with six filaments .... *S.* (*G.*) *paukatense*
   Pharate pupal gill with eight filaments 7
7. Thoracic proleg covered with unbranched and branched setae (Fig. 27N) .... *S.* (*G.*) *capillatum*
   Thoracic proleg without setae 8
8. Abdominal segments 1–5 each with a pair of dorsolateral protuberances 9
   Abdominal segments 1–5 without dorsolateral protuberances 10
9. Thoracic segments 2 and 3 and abdominal segments 1–5 each with a pair of dorsolateral protuberances (similar to Fig. 40M, N) .... *S.* (*G.*) *charlesi*
   Abdominal segments 1–5 each with two pairs of dorsolateral protuberances (Fig. 23T, U) .... *S.* (*G.*) *parahiyangum*
10. Thoracic segment 1 and abdominal segment 1 each with grayish-black transverse band (Fig. 22L) .... *S.* (*G.*) *tahanense*
    Thoracic segment 1 and abdominal segment 1without grayish-black transverse band 11
11. Postgenal cleft reaching posterior border of hypostoma (Fig. 27M) 12
    Postgenal cleft not reaching posterior border of hypostoma 13
12. Labral fan with 31–34 primary rays .... *S.* (*G.*) *sheilae*
    Labral fan with 38 primary rays .... *S.* (*G.*) *pegalanense*
13. Postgenal cleft short, 0.4 times as long as postgenal bridge (Fig. 41N) .... *S.* (*G.*) *guniki*
    Postgenal cleft medium-long to long, longer than postgenal bridge (Fig. 32V) 14
14. Body entirely grayish green or dark gray 15
    Body color otherwise 17
15. Body grayish green .... *S.* (*G.*) *lehi*
    Body dark gray 16
16. Pharate pupal gill with enlarged basal fenestra .... *S.* (*G.*) *bakalalanense*
    Pharate pupal gill with moderate basal fenestra .... *S.* (*G.*) *cheongi*
17. Abdominal cuticle almost bare .... *S.* (*G.*) *hiroyukii*
    Abdominal cuticle with branched dark spinous setae 18
18. Pharate pupal gill with eight filaments arising together from short common basal stalk .... *S.* (*G.*) *fulgidum*
    Pharate pupal gill arranged otherwise 19
19. Body length 5.0–6.0 mm 20
    Body length 4.5 mm or shorter 21

20. Labral fan with 40–47 primary rays ....*S.* (*G.*) *terengganuense*
    Labral fan with 30–32 primary rays .... *S.* (*G.*) *kelabitense*
21. Abdominal segments 1–4 almost bare .... *S.* (*G.*) *sarawakense*
    Abdominal segments 1–4 covered with branched setae ....*S.* (*G.*) *rayohense*
22. Body with reddish-brown markings. Postgenal cleft much shorter than postgenal bridge (similar to Fig. 44O) ....*S.* (*N.*) *borneoense*
    Body without reddish-brown markings. Postgenal cleft nearly as long as postgenal bridge (similar to Fig. 46M) .... *S.* (*N.*) *aureohirtum*
23. Pharate pupal gill with eight filaments ....*S.* (*S.*) *beludense*
    Pharate pupal gill with six filaments 24
24. Postgenal cleft reaching posterior margin of hypostoma (similar to Fig. 75N) . ...*S.* (*S.*) *kiuliense*
    Postgenal cleft not reaching posterior margin of hypostoma 25
25. Body length 6–7 mm. Posterior circlet with 155–170 rows of hooklets .... *S.* (*S.*) *laterale*
    Body length 4.5–5.3 mm. Posterior circlet with 66–100 rows of hooklets 26
26. Posterior circlet with ca. 100 rows of hooklets .... *S.* (*S.*) *mirum*
    Posterior circlet with 66–90 rows of hooklets 27
27. Posterior circlet with 84–90 rows of hooklets 28
    Posterior circlet with 66–78 rows of hooklets 29
28. Postgenal cleft bullet-shaped (similar to Fig. 69X). Rectal organ with 8–13 secondary lobules per lobe .... *S.* (*S.*) *cheedhangi*
    Postgenal cleft triangular (similar to Fig. 81P). Rectal organ with 15–18 secondary lobules per lobe ....*S. alberti*
29. Head capsule almost entirely darkened .... *S.* (*S.*) *aeneifacies* and *S.* (*S.*) *masilauense*
    Head capsule entirely or partially pale 30
30. Posterior circlet with up to 16 hooklets per row .... *S.* (*S.*) *lunduense*
    Posterior circlet with up to 13 hooklets per row 31
31. Cephalic apotome whitish yellow ....*S.* (*S.*) *keningauense*
    Cephalic apotome light brown ....*S.* (*S.*) *sabahense*

*Keys to identify 19 species of the subgenus *Gomphostilbia* from Borneo (Takaoka et al., 2016b) and keys to identify eight species of the *Simulium melanopus* species-group from Sabah (Takaoka et al., 2015c) are here combined and modified.

**_Simulium_ (*G.*) *auratum*, *S.* (*G.*) *bakalalanense*, *S.* (*G.*) *capillatum*, *S.* (*G.*) *paserense*, *S.* (*G.*) *pegalanense*, *S.* (*N.*) *borneoense*, *S.* (*N.*) *fuscinervis*, *S.* (*S.*) *lardizabalae*, and *S.* (*S.*) *timpohonense* are not included because their females are unknown.

***_Simulium_ (*G.*) *paserense* and *S.* (*S.*) *cheedhangi* are not included because their males are unknown.

**** *Simulium* (*G.*) *fuscinervis* is not included because its pupa is unknown.

******Simulium* (*G.*) *auratum, S.* (*N.*) *fuscinervis, S.* (*S.*) *crassimanum, S.* (*S.*) *lardizabalae, S.* (*S.*) *murudense, S.* (*S.*) *maklarini, S.* (*S.*) *nigripilosum*, and *S.* (*S.*) *timpohonense* are not included because their larvae are unknown.

10 The Fauna of Black Flies in the Sunda Archipelago, Indonesia

Overview

De Meijere (1907, 1913) recorded *Simulium* (*Nevermannia*) *aureohirtum* Brunetti and described five species: *S.* (*Gomphostilbia*) *atratum, S.* (*Simulium*) *argyrocinctum, S.* (*S.*) *eximium, S.* (*S.*) *nobile*, and *S.* (*S.*) *iridescens*, all based on adult specimens from Java. Edwards (1925) described a new species, *S.* (*G.*) *varicorne*, from a male specimen collected from Sumatra. Edwards (1934) described six new species: *S.* (*N.*) *feuerborni, S.* (*G.*) *sundaicum, S.* (*G.*) *batoense, S.* (*G.*) *friederichsi, S.* (*S.*) *thienemanni*, and *S.* (*S.*) *nebulicola*, two new subspecies: *S.* (*N.*) *latipes* var. *tosariense* and *S.* (*N.*) *latipes* var. *tjibodense* and recorded two varieties, *S.* (*S.*) *iridescens* var. and *S.* (*G.*) *metatarsale* var., from Java and Sumatra. Most of the species described by De Meijere (1907, 1913) were redescribed and their pupae and larvae were also described for the first time, together with those of *S.* (*G.*) *varicorne*, in the same paper (Edwards, 1934).

Crosskey (1973) raised two subspecies of *S.* (*N.*) *latipes* recorded by Edwards (1934) to the species status as *S.* (*N.*) *tjibodense* and *S.* (*N.*) *tosariense*.

Takaoka and Hadi (1991) described two new species: *S.* (*S.*) *sigiti* and *S.* (*S.*) *javaense* [the species formerly regarded as *S.* (*S.*) *iridescens* var.], from Java. Takaoka and Sigit (1992) described a new species, *S.* (*G.*) *parahiyangum* [its pupa and larva were wrongly associated with the adult of *S.* (*G.*) *varicorne* by Edwards (1934)], from Java. Takaoka and Davies (1996) described four new species: *S.* (*G.*) *atratoides* [its larva wrongly associated with *S.* (*G.*) *atratum* by Edwards (1934)], *S.* (*G.*) *gyorkosae* [the species formerly recorded as *S.* (*G.*) *metatarsale* var. by Edwards (1934)], *S.* (*S.*) *celsum*, and *S.* (*S.*) *upikae*, from Java. Takaoka (2012) synonymized *S.* (*G.*) *sundaicum* with *S.* (*G.*) *atratum*.

Takaoka and Sigit (1997) described three new species from Sumatra: *S.* (*G.*) *padangense, S.* (*S.*) *sumatraense*, and *S.* (*S.*) *minangkabaum*. Takaoka et al. (2000) newly recorded 12 species of black flies including one unnamed species, *S.* (*S.*) sp. nr. *eximium* from Sumatra.

Nie et al. (2006) described a new species, *S.* (*G.*) *sumatraense* based on a female caught inside the ship arriving in China from Sumatra. Adler (2022) reported that the species described by Nie et al. (2006) is a junior homonym of *S.* (*S.*) *sumatraense* Takaoka and Sigit, 1997. Takaoka et al. (2006a) reported nine species of black flies including four new species, four known species, and one unidentified species, *S.* (*Wallacellum*) sp., from Flores. Takaoka et al. (2006b) described a new species,

S. (*S.*) *ranauense* [the species formerly recorded as *S.* (*S.*) sp. nr. *eximium* by Takaoka et al.(2000)], from Sumatra. Takaoka et al. (2016c) described a new species, *S.* (*G.*) *merapiense*, from Java. Takaoka et al. (2017e) described four new species and newly recorded five species of black flies from the Lesser Sunda Archipelago. The previous record of *S.* (*S.*) *upikae* from Flores by Takaoka et al. (2006a) was corrected as that of *S.* (*S.*) *eximium*. Takaoka et al. (2017a) described two new species: *S.* (*G.*) *rangatense* and *S.* (*G.*) *sunapii*, from Flores. Takaoka et al. (2017b) described a new species, *S.* (*G.*) *lemborense*, from Flores. Takaoka et al. (2018d) described three new species: *S.* (*G.*) *dhangi* from Sumba and Timor, *S.* (*G.*) *sumbaense* from Sumba, and *S.* (*N.*) *wayani* from Timor. Adler et al. (2019b) chromosomally studied evolutionary and biogeographic history of *S.* (*N.*) *wayani.*

Hew et al. (2023a) presented the results of DNA barcoding of 57 Indonesian species of black flies. Hew et al. (2023b) showed genetically that *S. atratum* is represented by two lineages.

Friederichs (1925) reported that *S.* (*G.*) *atratum* was feeding on blood of domestic fowl, and *S.* (*S.*) *argyrocinctum* and *S.* (*S.*) *iridescens* were pests of cattle.

Natural infections of black fly larvae in Sumatra with mermithids, microsporidans, and other fungi were reported by Takaoka et al. (2000).

The fauna of black flies in the Sunda Archipelago is represented by 49 species including two unnamed species. The number of species recorded from each island and that of indigenous species (in parentheses) are 27 (8) from Sumatra, 22 (7) from Java, and 22 (11) from the Lessor Sunda Archipelago [8 (0) from Bali, 8 (0) from Lombok, 2 (0) from Sumbawa, 2 (1) from Sumba, 15 (6) from Flores, and 6 (2) from Timor].

List of 49 Species of Black Flies in the Sunda Archipelago, Indonesia

Subgenus *Gomphostilbia*

S. asakoae species-group

S. brevilabrum, *S. gyorkosae*, *S. rutengense*, and *S. sunapii*

S. batoense species-group

S. batoense, *S. duolongum*, *S. friederichsi*, *S. lemborense*, and *S. parahiyangum*

S. ceylonicum species-group

S. rangatense and *S. sheilae*

S. epistum species-group

S. atratum, *S. cheongi*, *S. flavocinctum*, *S. floresense*, *S. merapiense*, *S. padangense*, *S. whartoni*, and *S. zonatum*

S. varicorne species-group

S. sumbaense and *S. varicorne*

Gomphostilbia species unplaced to species-group

S. dhangi, *S.* sp. [*S. sumatraense* An, Nie and Chen, 2006 (junior primary homonym of *S. sumatraense* Takaoka and Sigit, 1997, but no available substitute name)]*
Subgenus *Nevermannia*
S. feuerborni species-group
S. feuerborni (complex)
S. ruficorne species-group
S. aureohirtum, *S. glatthaari*, and *S. wayani*
S. vernum species-group
S. tjibodense and *S. tosariense*
Subgenus *Simulium* s. str.
S. argentipes species-group
S. minangkabaum
S. eximium species-group
S. eximium, *S. ranauense*, *S. thienemanni*, and *S. upikae*
S. iridescens species-group
S. iridescens and *S. javaense*
S. melanopus species-group
S. bishopi
S. multistriatum species-group
S. fenestratum (complex)
S. nebulicola species-group
S. celsum, *S. nebulicola*, and *S. sumatraense*
S. nobile species-group
S. nobile and *S. timorense*
S. striatum species-group
S. argyrocinctum and *S. baliense*
S. tuberosum species-group
S. brevipar (complex), *S. sigiti*, and *S. tani* (complex)
Subgenus *Wallacellum*
S. sp.

*This unnamed species is dubious for its distribution in Sumatra.

*Keys to Identify 48 Species of Black Flies in the Sunda Archipelago**

Females**

1. Katepisternum haired 2
   Katepisternum bare 19
2. Antenna with eight flagellomeres 3
   Antenna with nine flagellomeres 4

3. Frons:head ratio 1.0:3.9 .... *S.* (*G.*) *sumbaense*
   Frons:head ratio 1.0:4.7–5.2 .... *S.* (*G.*) *varicorne*
4. Claw simple, without tooth .... *S.* (*G.*) sp.
   Claw with small or large tooth 5
5. Claw with small subbasal tooth .... *S.* (*G.*) *batoense*
   Claw with large basal tooth 6
6. Hind tibia darkened except base yellow 7
   Hind tibia yellowish on basal two-fifths or more 9
7. Scutum covered with brassy short hairs .... *S.* (*G.*) *duolongum*
   Scutum covered with white or whitish yellow short hairs 8
8. Frons:head ratio 1.0:3.9. Sensory vesicle long, 0.6 times as long as third palpal segment .... *S.* (*G.*) *parahiyangum*
   Frons:head ratio 1.0:6.2–6.3. Sensory vesicle medium-long, 0.3–0.4 times as long as third palpal segment .... *S.* (*G.*) *lemborense*
9. Claw with basal tooth 0.62 times as long as claw .... *S.* (*G.*) *zonatum*
   Claw with basal tooth 0.48–0.55 times as long as claw 10
10. Hind tibia with subbasal dark spot 11
    Hind tibia without subbasal dark spot 15
11. Frons:head ratio 1.0:4.4 .... *S.* (*G.*) *atratum*
    Frons:head ratio 1.0:6.3–6.7 12
12. Sensory vesicle 0.4–0.5 times as long as third palpal segment .... *S.* (*G.*) *floresense*
    Sensory vesicle 0.3 times as long as third palpal segment 13
13. Mandible lacking outer teeth .... *S.* (*G.*) *padangense*
    Mandible with five outer teeth 14
14. Mandible with 20 inner teeth .... *S.* (*G.*) *cheongi*
    Mandible with 25 inner teeth .... *S.* (*G.*) *whartoni*
15. Hind tibia yellowish on basal half or basal two-fifths 16
    Hind tibia yellowish on basal three-fifths 17
16. Sensory vesicle elongate, 0.68 times as long as third palpal segment .... *S.* (*G.*) *sheilae*
    Sensory vesicle medium-long, 0.32 times as long as third palpal segment .... *S.* (*G.*) *brevilabrum*
17. Sensory vesicle enlarged, 0.65 times as long as third palpal segment .... *S.* (*G.*) *merapiense*
    Sensory vesicle medium-sized, 0.25–0.34 times as long as third palpal segment 18
18. Frons:head ratio 1.0:5.8 .... *S.* (*G.*) *sunapii*
    Frons:head ratio 1.0:4.9 .... *S.* (*G.*) *gyorkosae*
19. Claw with medium-long or long basal tooth 20
    Claw simple or with small subbasal tooth 24
20. Spermatheca ellipsoidal 21
    Spermatheca globular 23

21. Claw with basal tooth 0.35 times as long as claw .... *S.* (*N.*) *glatthaari*
    Claw with basal tooth 0.4–0.5 times as long as claw 22
22. Claw with basal tooth 0.4 times as long as claw. Spermatheca with unpigmented neck (Fig. 45G) .... *S.* (*N.*) *wayani*
    Claw with basal tooth 0.5 times as long as claw. Spermatheca with dark neck (Fig. 45F) .... *S.* (*N.*) *aureohirtum*
23. Scutum reddish brown, with three dark longitudinal vittae ....
    .... *S.* (*N.*) *feuerborni* (complex)
    Scutum brownish black, without dark longitudinal vittae ....
    .... *S.* (*N.*) *tosariense* and *S.* (*N.*) *tjibodense*
24. Claw with small basal tooth 25
    Claw without tooth 31
25. Antenna yellow with two apical flagellomeres darkened (Fig. 74A) ....
    .... *S.* (*S.*) *nobile* and *S.* (*S.*) *timorense*
    Antenna dark brown except scape, pedicel, and base of first flagellomere yellow 26
26. Tergites 5–8 shiny .... *S.* (*S.*) *iridescens* and *S.* (*S.*) *javaense*
    Tergites 6–8 shiny 27
27. Ovipositor valves with inner margins greatly concave 28
    Ovipositor valves with inner margins nearly straight 29
28. Ovipositor valves pointed apically .... *S.* (*S.*) *bishopi*
    Ovipositor valves rounded posteromedially (similar to Fig. 49C) ....
    .... *S.* (*S.*) *minangkabaum*
29. Hind tibia with dark subbasal marking (Fig. 73B) .... *S.* (*S.*) *sumatraense*
    Hind tibia without subbasal marking (Fig. 73C, D) 30
30. Hind femur darkened except basal one-fourth yellow (Fig. 73C) ....
    .... *S.* (*S.*) *nebulicola*
    Hind femur yellow except apical one-fifth darkened (Fig. 73D) ....
    .... *S.* (*S.*) *celsum*
31. Basal portion of radial vein bare 32
    Basal portion of radial vein haired 34
32. Sensory vesicle long, 0.7 times as long as third palpal segment ....
    .... *S.* (*S.*) *tani* (complex)
    Sensory vesicle medium-long, 0.34 times as long as third palpal segment ....
   33
33. Frons:head ratio 1.0:4.1. Hind basitarsus 5.0 times as long as its greatest width
    .... *S.* (*S.*) *sigiti*
    Frons:head ratio 1.0:3.6. Hind basitarsus 5.7 times as long as its greatest width... *S.* (*S.*) *brevipar* (complex)
34. Hind tibia whitish yellow on basal two-thirds .....*S.* (*S.*) *fenestratum* (complex)
    Hind tibia darkened except base or basal half yellow 35

35. Ovipositor valves rounded posteromedially, each with round lobe raised ventrally along inner margin (similar to Fig. 76D, E) .... *S.* (*S.*) *argyrocinctum* and *S.* (*S.*) *baliense*
Ovipositor valves tapered medially, each without round lobe raised ventrally along inner margin (Fig. 57C) 36
36. Hind tibia darkened except basal half yellow ....*S.* (*S.*) *thienemanni*
Hind tibia darkened except base yellow 37
37. Cibarium with about six tiny processes. Apex of ventral projection of paraproct directed posterolaterally ....*S.* (*S.*) *upikae*
Cibarium with 20–26 tiny processes. Apex of ventral projection of paraproct directed ventrally (Fig. 57C) ....*S.* (*S.*) *eximium* and *S.* (*S.*) *ranauense*

Males***

1. Katepisternum haired 2
Katepisternum bare 22
2. Antenna with eight flagellomeres 3
Antenna with nine flagellomeres 4
3. Upper-eye (large) facets in 11 or 12 vertical columns and 12 horizontal rows .... *S.* (*G.*) *varicorne*
Upper-eye (large) facets in 13 vertical columns and 13 horizontal rows ....*S.* (*G.*) *sumbaense*
4. Hind basitarsus slender, nearly parallel-sided 5
Hind basitarsus enlarged 17
5. Hind tibia darkened except base yellow 6
Hind tibia yellowish on basal two-fifths or more 10
6. Scutum white-pruinose with three non-pruinose dark vittae (one median long and two submedian elliptical) .... *S.* (*G.*) *friederichsi*
Scutum white-pruinose on shoulders, along lateral margins and on prescutellar area, and remaining medial portion non-pruinose 7
7. Abdomen with a pair of shiny dorsolateral spots on segments 2, 6, and 7. Style without terminal spine .... *S.* (*G.*) *parahiyangum*
Abdomen with a pair of shiny dorsolateral spots on segments 2 and 5–7 (or 5–8). Style with terminal spine 8
8. Upper-eye (large) facets in 13 vertical columns and 13 horizontal rows .... *S.* (*G.*) *lemborense*
Upper-eye (large) facets in 15–17 vertical columns and 15–17 horizontal rows 9
9. Upper-eye (large) facets in 15 vertical columns and 15 horizontal rows ....*S.* (*G.*) *duolongum*
Upper-eye (large) facets in 17 vertical columns and 17 horizontal rows ....*S.* (*G.*) *batoense*
10. Hind tibia without subbasal dark marking 11
Hind tibia with subbasal dark marking 12
11. Antenna entirely yellow .... *S.* (*G.*) *flavocinctum*

Antennal yellow except four or five apical flagellomeres darkened .... *S.* (*G.*) *merapiense*
12. Upper-eye (large) facets in 10 or 11 horizontal rows 13
Upper-eye (large)facets in 12–16 horizontal rows 14
13. Antenna entirely yellow .... *S.* (*G.*) *cheongi*
Antennal yellow except three or four apical flagellomeres darkened .... *S.* (*G.*) *zonatum*
14. Upper-eye (large) facets in 14 vertical columns and 16 horizontal rows .... *S.* (*G.*) *atratum*
Upper-eye (large) facets in 11–13 vertical columns and 13 or 14 horizontal rows 15
15. Fore basitarsus 4.9 times as long as its greatest width ....*S.* (*G.*) *floresense*
Fore basitarsus 6.7–8.4 times as long as its greatest width 16
16. Fore basitarsus 6.7 times as long as its greatest width. Hind basitarsus 6.3 times as long as its greatest width .... *S.* (*G.*) *whartoni*
Fore basitarsus 8.4 times as long as its greatest width. Hind basitarsus 4.7 times as long as its greatest width ....*S.* (*G.*) *padangense*
17. Hair tuft on base of radial vein yellow 18
Hair tuft on base of radial vein darkened 20
18. Upper-eye (large) facets in eight or nine vertical columns and 12 horizontal rows .... *S.* (*G.*) *sunapii*
Upper-eye (large) facets in 10 or 11 vertical columns and 12 or 13 horizontal rows 19
19. Fore basitarsus 5.8 times as long as its greatest width .... *S.* (*G.*) *gyorkosae*
Fore basitarsus 7.0 times as long as its greatest width .... *S.* (*G.*) *brevilabrum*
20. Fore basitarsus ca. 6.2 times as long as its greatest width. Hind basitarsus dark brown with basal one-third whitish .... *S.* (*G.*) *sheilae*
Fore basitarsus 6.9–7.2 times as long as its greatest width. Hind basitarsus dark brown with basal half or little less whitish 21
21. Subcosta bare .... *S.* (*G.*) *rangatense*
Subcosta haired .... *S.* (*G.*) *rutengense*
22. Coxite longer than style 23
Coxite shorter than style 25
23. Hind basitarsus slender, parallel-sided, much narrower than hind tibia (Fig. 45H). Ventral plate with median keel (Fig. 45K) .... *S.* (*N.*) *aureohirtum, S.* (*N.*) *glatthaari*, and *S.* (*N.*) *wayani*
Hind basitarsus somewhat enlarged, spindle-shaped, nearly as wide as hind tibia (Fig. 43J). Ventral plate without median keel (Fig. 43N) 24
24. Paramere with three or four hooks ....*S.* (*N.*) *feuerborni* (complex)
Paramere with one hook ....*S.* (*N.*) *tosariense* and *S.* (*N.*) *tjibodense*
25. Hind basitarsus slender, much narrower than hind tibia (Fig. 74F) .... *S.* (*S.*) *nobile* and *S.* (*S.*) *timorense*
Hind basitarsus enlarged, nearly as wide as or slightly narrower than hind tibia 26
26. Style without basal protuberance 27

Style with basal protuberance 33

27. Hind femur yellow with apical one-fifth darkened (Fig. 73D) .....*S.* (*S.*) *celsum*
Hind femur brownish black except basal one-fourth or less yellow 28
28. Abdominal segments 2 and 5–7 each with a pair of dorsolateral white shiny patches ....*S.* (*S.*) *minangkabaum*
Abdominal segments 2 and 4–7, each with a pair of dorsolateral white shiny patches 29
29. Ventral plate with teeth on posterolateral margins in two vertical rows (Fig. 73K) 30
Ventral plate with teeth on ventral margin in horizontal row (Fig. 64K) 32
30. Fore femur brownish black ....*S.* (*S.*) *bishopi*
Fore femur yellow on basal half or three-fourths and rest darkened 31
31. Fore basitarsus 5.0 times as long as its greatest width ....*S.* (*S.*) *nebulicola*
Fore basitarsus 5.7 times as long as its greatest width ....*S.* (*S.*) *sumatraense*
32. Anterolateral white-pruinose spots on scutum large, with their inner margins convergent anteriorly ....*S.* (*S.*) *iridescens*
Anterolateral white-pruinose spots on scutum medium in size, with their inner margins nearly parallel-sided ....*S.* (*S.*) *javaense*
33. Scutum covered with yellow short hairs 34
Scutum covered with dark short hairs 40
34. Ventral plate with body subquadrate when viewed ventrally ....*S.* (*S.*) *fenestratum* (complex)
Ventral plate with body saddle-shaped or pear-shaped when viewed ventrally 35
35. Ventral plate saddle-shaped when viewed ventrally (Fig. 76P) 36
Ventral plate pear-shaped when viewed ventrally (Fig. 57E) 37
36. Upper-eye (large) facets in 17 or 18 vertical columns and 17 or 18 horizontal rows .... *S.* (*S.*) *argyrocinctum*
Upper-eye (large) facets in 14 or 15 vertical columns and 15 horizontal rows .. ..*S.* (*S.*) *baliense*
37. Hind tibia yellow on basal one-third and darkened on rest ....*S.* (*S.*) *thienemanni*
Hind tibia darkened except base yellow or dark yellow or light brown..........38
38. Abdominal segments 2 and 4–7 each with a pair of white pruinose dorsomedial patches ....*S.* (*S.*) *ranauense*
Abdominal segments 2 and 5–7 each with a pair of white pruinose dorsomedial patches 39
39. Hind basitarsus as wide as hind tibia ....*S.* (*S.*) *eximium*
Hind basitarsus 0.8 times as wide as hind tibia ....*S.* (*S.*) *upikae*

40. Abdominal segments 2, 6 and 7 each with a pair of white pruinose dorsomedial patches ....*S.* (*S.*) *brevipar* (complex)
Abdominal segments 2 and 5–7 each with a pair of white pruinose dorsomedial patches 41
41. Upper-eye (large) facets in 20 vertical columns and 20 horizontal rows ....*S.* (*S.*) *sigiti*
Upper-eye (large) facets in 15 vertical columns and 15 horizontal rows .... *S.* (*S.*) *tani* (complex)

Pupae****

1. Abdominal segment 9 with grapnel-like hooklets on each side (Fig. 3D) 2
Abdominal segment 9 without grapnel-like hooklets on each side 23
2. Gill with four filaments .... *S.* (*G.*) *brevilabrum*
Gill with six, eight, or 10 filaments 3
3. Gill with six filaments 4
Gill with eight or 10 filaments 5
4. Gill with two inflated tubes and four slender filaments (Fig. 41I) ....*S.* (*G.*) *dhangi*
Gill with six slender filaments (Fig. 27C) .... *S.* (*G.*) *rangatense*
5. Gill with 10 filaments (Fig. 27E) ....*S.* (*G.*) *atratoides*
Gill with eight filaments 6
6. Gill with one inflated tubular filament and seven slender filaments (Fig. 31F).. ....*S.* (*G.*) *padangense*
Gill with eight slender filaments 7
7. Cocoon with short anterodorsal projection .... *S.* (*G.*) *gyorkosae*
Cocoon without anterodorsal projection 8
8. All gill filaments short and of same thickness, and arising close together from base 9
All gill filaments otherwise 10
9. Antennal sheath with tuberculate protuberances (similar to Fig. 23P). Terminal hooks simple, cone-like .... *S.* (*G.*) *parahiyangum*
Antennal sheath without such protuberances. Terminal hooks wide, plate-like ....*S.* (*G.*) *merapiense*
10. Gill with eight slender filaments arranged as 6+2 from dorsal to ventral. Outer filament of ventral pair of gill much thinner than inner filament .... *S.* (*G.*) *sumbaense*
Gill with eight slender filaments arranged as 3+3+2 or 2+4+2 from dorsal to ventral. Outer filament of ventral pair of gill as thick as or thicker than inner filament 11
11. Dorsal and middle triplets of filaments of gill sharing medium-long stalk slightly longer than interspiracular trunk (Fig. 40C) .... *S.* (*G.*) *varicorne*
Dorsal and middle triplets of filaments of gill sharing short stalk much shorter than interspiracular trunk 12

12. Outer filament of ventral pair of gill thicker than inner filament when compared basally 13
    Outer filament of ventral pair of gill subequal in thickness to inner filament when compared basally...17
13. Dorsal and middle triplets of filaments of gill sharing short stalk ....
    .... *S. (G.) flavocinctum*
    Dorsal and middle triplets of filaments of gill not sharing stalk or sharing extremely short stalk 14
14. Inner filament of ventral pair of gill somewhat longer and thicker than six filaments of dorsal and middle triplets (Fig. 31D) ....*S. (G.) floresense*
    Inner filament of ventral pair of gill subequal in length and thickness to six filaments of dorsal and middle triplets 15
15. Terminal hooks simple, cone-like (similar to Fig. 31G) .... *S. (G.) zonatum*
    Terminal hooks wide and plate-like, each with outer margin crenulated (similar to Fig. 31H) 16
16. Outer filament of ventral pair of gill 1.8–2.4 mm long and slightly less than twice as thick as inner filament .... *S. (G.) cheongi*
    Outer filament of ventral pair of gill ca. 4.0 mm long and three times as thick as inner filament ....*S. (G.) whartoni*
17. Two filaments of ventral pair of gill more than 1.5 times as long as other filaments 18
    Two filaments of ventral pair of gill slightly longer than other filaments 21
18. Terminal hooks simple, cone-like 19
    Terminal hooks widened, each with outer margin undulate or crenulated 20
19. Two filaments of ventral pair of gill ca. 2.5 times as long as six other filaments. Dorsum of abdominal segments 1 and 2 light yellow ....*S. (G.) duolongum*
    Two filaments of ventral pair of gill little more than three times as long as six other filaments (Fig. 22C). Dorsum of abdominal segments 1 and 2 dark gray....
    .... *S. (G.) lemborense*
20. Gill with eight slender filaments arranged as 2+4+2 from dorsal to ventral .....
    ....*S. (G.) batoense*
    Gill with eight slender filaments arranged as 3+3+2 from dorsal to ventral ....
    .... *S. (G.) atratum*
21. Terminal hooks triangular, each with outer margin smooth or slightly undulate, and 1.5–1.8 times as long as inner margin (Fig. 27H, I) .... *S. (G.) sheilae*
    Terminal hooks wide, plate-like, each with outer margin crenulated, and 2.5–4.0 times as long as inner margin 22
22. Frons densely covered with tubercles .... *S. (G.) rutengense*
    Frons moderately covered with tubercles .... *S. (G.) sunapii*
23. Gill with four filaments (Fig. 46D–G) 24
    Gill with six, eight, nine, or 10 filaments 26

24. Cocoon with short anterodorsal projection (Fig. 46H) .... *S.* (*N.*) *wayani*
 Cocoon simple, without anterodorsal projection 25
25. Four gill filaments arranged as 2+2, each arising from relatively long stalk 1.5–2.0 times as long as common basal stalk ....*S.* (*N.*) *tjibodense*
 Four gill filaments arranged as 2+2, each arising from short stalk nearly as long as common basal stalk ....*S.* (*N.*) *tosariense*
26. Gill with six filaments 27
 Gill with eight, nine, or 10 filaments 37
27. Cocoon with short anterodorsal projection (similar to Fig. 46H) ....
 ....*S.* (*N.*) *aureohirtum*
 Cocoon simple, without anterodorsal projection, or shoe-shaped 28
28. Cocoon shoe-shaped (Fig. 75G) 29
 Cocoon wall-pocket-shaped 30
29. Frons bare ....*S.* (*S.*) *bishopi*
 Frons sparsely or moderately covered with tubercles ....
 .... *S.* (*S.*) *nobile* and *S.* (*S.*) *timorense*
30. Gill with six inflated filaments (Fig. 64M) ....*S.* (*S.*) *javaense*
 Gill with six slender thread-like filaments 31
31. Thorax with pit-like organ near base of gill (similar to Fig. 79H) 32
 Thorax without pit-like organ near base of gill 33
32. Six gill filaments all short and of same length and thickness (Fig. 79G) ....
 ....*S.* (*S.*) *sigiti*
 Six gill filaments decreasing in length from dorsal to ventral and of same thickness except dorsal filament of dorsal pair somewhat thicker than five other filaments ....*S.* (*S.*) *brevipar* (complex)
33. Terminal hooks present 34
 Terminal hooks absent 36
34. Abdominal segments 7–9 each with spine-combs dorsally ....
 ....*S.* (*S.*) *tani* (complex)
 Abdominal segments 7 and 8 each with spine-combs dorsally 35
35. Frons bare ....*S.* (*S.*) *celsum*
 Frons sparsely covered with small tubercles ....*S.* (*S.*) *sumatraense*
36. Upper filament of dorsal pair of gill much longer than other filaments (Fig. 73P).
 ....*S.* (*S.*) *nebulicola*
 Upper filament of dorsal pair of gill as long as or slightly longer than other filaments (Fig. 64N) ....*S.* (*S.*) *iridescens*
37. Gill with eight filaments 38
 Gill with nine or 10 filaments 40
38. Cocoon corbicular-shaped (similar to Fig. 49R, S) ....*S.* (*S.*) *minangkabaum*
 Cocoon wall-pocket-shaped 39

39. Cocoon with anterolateral open space on each side (similar to Fig. 72J) ....S. (S.) *fenestratum* (complex)
    Cocoon without anterolateral open space on each side .... S. (N.) *glatthaari*
40. Thoracic integument without rugose swellings anterodorsally. Abdominal segment 8 with spine-combs dorsally 41
    Thoracic integument with rugose swellings anterodorsally (Fig. 57L). Abdominal segment 8 without spine-combs dorsally 42
41. Facial trichome as long as or slightly longer than frontal trichomes .... S. (S.) *baliense*
    Facial trichome twice as long as frontal trichomes .... S. (S.) *argyrocinctum*
42. Gill with 10 inflated short filaments (Fig. 57P) ....S. (S.) *thienemanni*
    Gill with nine or 10 slender thread-like filaments (Fig. 57M–O) 43
43. Two filaments of ventral pair of gill 0.5–0.6 times as long as eight other filaments (Fig. 57O) ....S. (S.) *upikae*
    All filaments subequal in length to one another (Fig. 57M, N) ....S. (S.) *eximium* and S. (S.) *ranauense*

Mature larvae*****

1. Ventral papillae large or medium in size 2
   Ventral papillae absent or small 24
2. Hypostoma with lateral margins smooth 3
   Hypostoma with lateral margins serrated...20
3. Third article of antenna 1.4–1.5 times as long as first article. Pharate pupal gill composed of two inflated tubular filaments and four slender filaments ....S. (G.) *dhangi*
   Third article of antenna as long as or shorter than first article. Pharate pupal gill composed of eight slender filaments, or one inflated filament and seven or 10 slender filaments 4
4. Abdominal segments 1–4 each with protuberances dorsally (Fig. 23T, U) 5
   Abdominal segments 1–4 each without protuberances dorsally 6
5. Cephalic apotome whitish yellow, with indistinct head spots. Abdominal segments 6–8 without dark stout setae dorsally .... S. (G.) *varicorne*
   Cephalic apotome darkened, with negative head spots. Abdominal segments 6–8 with several dark stout setae dorsally (similar to Fig. 23X) ....S. (G.) *parahiyangum*
6. Postgenal cleft short, shorter than postgenal bridge 7
   Postgenal cleft medium-long to long, longer than postgenal bridge 8
7. Postgenal cleft 0.4 times as long as postgenal bridge .... S. (G.) *merapiense*
   Postgenal cleft 0.8 times as long as postgenal bridge (Fig. 31I) .... S. (G.) *zonatum*
8. Postgenal cleft 1.3 times as long as postgenal bridge (Fig. 27K). Pharate pupal gill with 10 slender filaments ....S. (G.) *atratoides*

Postgenal cleft medium-long to long, over twice as long as postgenal bridge. Pharate pupal gill with eight slender filaments, or one inflated filament and seven slender filaments 9
9. Postgenal cleft medium-long, about twice as long as postgenal bridge 10
Postgenal cleft long, more than 2.5 times as long as postgenal bridge 11
10. Rectal organ with one to three nipple-like secondary lobules per lobe .... *S.* (*G.*) *rutengense*
Rectal organ with four or five finger-like secondary lobules per lobe ....*S.* (*G.*) *gyorkosae*
11. First and second antennal articles subequal in length to each other ....*S.* (*G.*) *floresense*
First antennal article longer than second article 12
12. Abdominal segments 6–8 without dark setae dorsally .... *S.* (*G.*) *atratum*
Abdominal segment 6–8 with dark setae dorsally 13
13. First antennal article as long as third article. Abdominal segments 5–8 with unbranched setae dorsally .... *S.* (*G.*) *sunapii*
First antennal article longer than third article. Abdominal segments 5–8 with branched dark setae 14
14. Abdominal segments 5–8 with various sizes of dark setae, of which larger ones are flat, spinous, with several fine, comb-like, unpigmented branches apically .... *S.* (*G.*) *duolongum*
Abdominal segments 5–8 with dark slender setae branched near base 15
15. Pharate pupal gill composed of one inflated tubular filament and seven slender filaments ....*S.* (*G.*) *padangense*
Pharate pupal gill composed of eight slender filaments 16
16. Pharate pupal gill with eight filaments, of which one of two filaments of ventral pair is much longer and thicker than seven other filaments 17
Pharate pupal gill with eight filaments, of which two filaments of ventral pair are equal in length and thickness 18
17. Postgenal cleft long, 6–8 times as long as postgenal bridge. Posterior circlet with 92 rows of 15 hooklets per row .... *S.* (*G.*) *whartoni*
Postgenal cleft long, almost approaching posterior margin of hypostoma. Posterior circlet with 86 rows of 14 hooklets per row .... *S.* (*G.*) *cheongi*
18. Postgenal cleft long, almost reaching posterior margin of hypostoma (Fig. 27M) .... *S.* (*G.*) *sheilae*
Postgenal cleft long, not reaching posterior margin of hypostoma 19
19. Postgenal cleft 2.6–3.0 times length of postgenal bridge. Abdominal segments 5–8 with unbranched or bifid dark setae .... *S.* (*G.*) *lemborense*
Postgenal cleft 6.5–10 times length of postgenal bridge. Abdominal segments 5–8 with dark setae each with four to nine branches (Fig. 22P) ....*S.* (*G.*) *batoense*
20. Abdomen with reddish-brown markings dorsally ....*S.* (*G.*) *feuerborni* (complex)
Abdomen without reddish-brown markings dorsally 21

21. Mandibular serration composed of one well-developed tooth and one small tooth, and with supernumerary serration (similar to Fig. 48M) .... *S.* (*G.*) *tosariense*
    Mandibular serration composed of two well-developed teeth, and without supernumerary serration 22
22. Hypostoma with median tooth much longer than corner teeth (Fig. 46K). Posterior circlet with ca. 84 rows of up to 22 hooklets. Pharate pupal gill with eight filaments .... *S.* (*N.*) *glatthaari*
    Hypostoma with median tooth much shorter than corner teeth (Fig. 46L). Posterior circlet with 63–71 rows of up to 13–15 hooklets. Pharate pupal gill with four or six filaments 23
23. Pharate pupal gill with six filaments .... *S.* (*N.*) *aureohirtum*
    Pharate pupal gill with four filaments ....*S.* (*N.*) *wayani*
24. Pectination of uppermost primary ray of labral fan different from that of others (Fig. 95V, W). Pharate pupal gill with four filaments ....*S.* (*W.*) sp.
    Pectination of uppermost primary ray of labral fan same as that of others. Pharate pupal gill with six, eight, nine, or 10 filaments 25
25. Postgenal cleft long, reaching widely posterior margin of hypostoma (Fig. 75N) 26
    Postgenal cleft, medium-long to long, but not reaching posterior margin of hypostoma 27
26. Abdominal segments 1–5 each with a pair of dorsal protuberances (Fig. 75K). ....*S.* (*S.*) *nobile*
    Abdominal segments 1–5 without dorsal protuberances .... *S.* (*S.*) *timorense*
27. Postgenal cleft rounded apically (Figs. 49V and 77K) 28
    Postgenal cleft pointed apically (Fig. 73R–T) 35
28. Abdominal segments 1–5 or 1–8 each with a pair of protuberances dorsally 29
    Abdominal segments1–8 without protuberances 30
29. Abdominal segments 1–5 each with a pair of protuberances dorsally (Fig. 49W) .... *S.* (*S.*) *minangkabaum*
    Abdominal segments 1–8 each with a pair of protuberances dorsally (similar to Fig. 77L) .... *S.* (*S.*) *argyrocinctum*
30. Body length 4.5–5.5 mm 31
    Body length over 7.0 mm 32
31. Hypostoma with eight bristles on each side ....*S.* (*S.*) *bishopi*
    Hypostoma with five or six bristles on each side ....*S.* (*S.*) *baliense*
32. Pharate pupal gill with inflated filaments ....*S.* (*S.*) *thienemanni*
    Pharate pupal gill with slender thread-like filaments 33
33. Posterior circlet with ca. 210 rows of hooklets .... *S.* (*S.*) *eximium*

Posterior circlet with 164–192 rows of hooklets 34

34. Labral fan with ca. 50 primary rays .... *S.* (*S.*) *upikae*
Labral fan with 38–42 primary rays .... *S.* (*S.*) *ranauense*

35. Hypostoma with four to six bristles on each side 36
Hypostoma with seven to nine bristles on each side 40

36. Rectal organ with one to three secondary lobules per lobe 37
Rectal organ with 6–12 secondary lobules per lobe 38

37. Bristles on hypostoma lying subparallel to lateral margins ....*S.* (*S.*) *sigiti*
Bristles on hypostoma lying divergent posteriorly from lateral margins (Fig. 64R) ....*S.* (*S.*) *iridescens*

38. Posterior circlet with ca. 88 rows of hooklets .... *S.* (*S.*) *sumatraense*
Posterior circlet with 70–74 rows of hooklets 39

39. Body reddish brown .... *S.* (*S.*) *tani* (complex)
Body dark gray ....*S.* (*S.*) *brevipar* (complex)

40. Rectal organ with one to three secondary lobules per lobe ....*S.* (*S.*) *javaense*
Rectal organ with 6–10 secondary lobules per lobe 41

41. Cephalic apotome much darkened on both sides and behind posterior mediolongitudinal spots. Lateral surface of head capsule widely darkened around eye-spot region, with negative spots near posterior margin ....
....*S.* (*S.*) *nebulicola*
Cephalic apotome somewhat darkened only behind posterior mediolongitudinal spots. Lateral surface of head capsule widely pale with somewhat darkened areas above and behind eye-spot region ....*S.* (*S.*) *celsum*

*The keys given for 22 species of black flies in Java by Takaoka and Davies (1996), and those for 19 species of black flies in the Lesser Sunda Archipelago are combined and updated.

***Simulium* (*G.*) *dhangi*, *S.* (*G.*) *rutengense*, *S.* (*G.*) *rangatense*, *S.* (*G.*) *friederichsi*, and *S.* (*W.*) sp. are not included because their females are unknown. *Simulium* (*G.*) *flavocinctum* is also not included because detailed information for its female is lacking, although it runs up to couplet 9.

****Simulium* (*G.*) *dhangi* and *S.* (*W.*) sp. are not included because their males are unknown.

**** *Simulium* (*G.*) *friederichsi* and *S.* (*W.*) sp. are not included because their pupae are unknown.

******Simulium* (*G.*) *brevilabrum*, *S.* (*G.*) *friederichsi*, and *S.* (*G.*) *rangatense* are not included because their larvae are not known. *Simulium* (*G.*) *flavocinctum* is also not included because detailed information for its larva is lacking, although it was briefly described by Edwards (1934). *Simulium* (*S.*) *fenestratum* (complex) is not included because detailed information for its larva is lacking, although it runs up to couplet 27.

11 The Fauna of Black Flies in East Indonesia (Sulawesi, Maluku, and Irian Jaya)

Overview

From Sulawesi, Maluku, and Irian Jaya, eastern parts of Indonesia, *S.* (*Morops*) *wamenae* described from Irian Jaya by Smart and Clifford (1965) was the only species, until Takaoka and Roberts (1988) reported from Sulawesi nine species consisting of six new species: *S.* (*Gomphostilbia*) *disneyi* (at that time, under the subgenus *Morops*), *S.* (*G.*) *rosemaryae*, *S.* (*G.*) *sulawesiense*, *S.* (*G.*) *torautense*, *S.* (*Simulium*) *dumogaense*, and *S.* (*S.*) *tumpaense*; one new recorded species, *S.* (*Nevermannia*) *aureohirtum*; and two unnamed species: *S.* (*N.*) sp. A and *S.* (*G.*) sp. B.

Takaoka (2003) recorded 87 species of black flies from Sulawesi, Maluku, and Irian Jaya. All species were assigned in the genus *Simulium* s. l., and further placed in five subgenera: two species in the subgenus *Wallacellum*, four species in the subgenus *Nevermannia*, 30 species in the subgenus *Gomphostilbia*, 29 species in the subgenus *Morops*, and 22 species in the subgenus *Simulium* s. str. Among 71 new species, *S.* (*N.*) *minahasaense* and *S.* (*G.*) *manadoense* are the species formerly recorded as *S.* (*N.*) sp. A and *S.* (*G.*) sp. B. Three species-groups, *S. banauense*, *S. sherwoodi*, and *S. trirugosum* species-groups, were transferred from the subgenus *Morops* to the subgenus *Gomphostilbia*.

Takaoka et al. (2019d) added one more new species, *S.* (*G.*) *marosense*, from Sulawesi. Takaoka *et al.* (2020g) described the female, male, and mature larva of *S.* (*S.*) *contractum* Takaoka from Sulawesi.

Takaoka (2003) noted that for black flies, the Weber's line running between Sulawesi and Maluku is the biogeographical boundary separating the Oriental Region from the Australasian Region. The fauna of black flies of Sulawesi represented by 47 species is characterized by a high rate of endemic species (44/47 or 93.6%) and high intra-insular speciation as seen in four species-groups (11 species in the *S. variegatum* species-group, 11 species in the *S. batoense* species-group, nine species in the *S. dumogaense* species-group (at that time, as the *S. melanopus* species-group), and five species in the *S. ceylonicum* species-group). On the other hand, the black fly fauna of Maluku and Irian Jaya is represented by 43 species, of which 12 species are from Maluku, 28 species from Irian Jaya, and three species, *S.* (*G.*) *hemicyclium*, *S.* (*M.*) *papuense*, and *S.* (*Wallacellum*) *alfurense*, from both areas. It is characterized by the dominance of the subgenus *Morops* (29/43 or 67.4%) and poverty of the subgenus *Simulium* s. str. (only one species (2.3%) from Seram Island in Maluku), as well as the relatively high endemism (10/15 or 66.7% in Maluku and 22/31 or 71.0% in Irian Jaya). The species common to both Sulawesi and Maluku are *S.* (*N.*) *aureohirtum* and *S.* (*G.*) *kamimurai*, and no species is common to both Sulawesi and Irian Jaya.

List of 88 Species of Black Flies in East Indonesia (Sulawesi, Maluku, and Irian Jaya)

Subgenus *Gomphostilbia*
S. asakoae species-group
S. *gyorkosae*
S. banauense species-group
S. disneyi
S. batoense species-group
S. abadii, S. brevitruncatum, S. hadiae, S. kolakaense, S. manadoense, S. marosense, S. miyagii, S. singgihi, S. sulawesiense, S. syafruddini, S. torautense, and *S. tumidum*
S. ceylonicum species-group
*S. ambonense**, *S. gimpuense, S. kamimurai***, *S. mogii, S. napuense*, and *S. rosemaryae*
S. heldsbachense species-group
*S. bifurcatum**, *S. heldsbachense**, *S. inermum**, and *S. kalimerahense**
S. hemicyclium species-group
*S. asperum**, *S. hemicyclium**, and *S. kuingingiense**
S. palauense species-group
*S. curvum**
S. varicorne species-group
S. tomae
Gomphostilbia species unplaced to species-group
S. mamasaense and *S. serratum*
Subgenus *Morops*
S. clathrinum species-group
*S. biakense**, *S. hagai**, *S. halmaheraense**, *S. malukuense**, *S. manokwariense**, *S. manuselaense**, *S. ransikiense**, *S. sorongense**, *S. tafulaense**, and *S. watetoense**
S. farciminis species-group
*S. arfakense**, *S. botulus**, *S. farciminis**, *S. jayapuraense**, and *S. pamahaense**
S. melatum species-group
*S. iraiense**, *S. nudipes**, and *S. wamenae**
S. oculatum species-group
*S. anggiense**, *S. baliemense**, *S. demutaense**, *S. irianense**, *S. motoyukii**, *S. norforense**, *S. prafiense**, *S. sutebense**, *S. takae**, and *S. wakrisense**
S. papuense species-group
*S. papuense**
Subgenus *Nevermannia*
S. feuerborni species-group
S. chairuddini and *S. minahasaense*
S. ruficorne species-group
*S. aureohirtum*** and *S. ornatipes**

Subgenus *Simulium* s. str.
S. dumogaense species-group
S. contractum, *S. dumogaense*, *S. fuscitarse*, *S. landonoense*, *S. mediodentatum*, *S. nudifrons*, *S. seramense**, *S. tomohonense*, *S. tumpaense*, and *S. watanabei*
S. variegatum species-group
S. elongatum, *S. flavum*, *S. globosum*, *S. laxum*, *S. linduense*, *S. lorense*, *S. luridum*, *S. malinoense*, *S. molliculum*, *S. palopoense*, and *S. quinquefiliatum*
Simulium s. str. species unplaced to species-group
S. parvulum
Subgenus *Wallacellum*
*S. alfurense** and *S. celebesense*

*Species recorded from Maluku and Irian Jaya (Australasian Region). Species with no asterisk recorded from Sulawesi (Oriental Region).

** Species recorded from both the Oriental and the Australasian Regions.

Keys to Identify 88 Species in East Indonesia (Sulawesi, Maluku, and Irian Jaya)

The keys provided by Takaoka (2003) are modified and updated.

Keys to Identify Five Subgenera of the genus Simulium s. l. *in East Indonesia*

Females

1. Katepisternum haired 2
   Katepisternum bare 4
2. Claw with small or medium-sized subbasal tooth .... *Morops*
   Claw with large basal tooth (Fig. 92K) 3
3. Hind tibia with low ridge on anterior inner surface (Fig. 92G). Calcipala elongate, reaching apex of second tarsomere (Fig. 92I, J) .... *Wallacellum*
   Hind tibia without ridge on anterior inner surface. Calcipala medium-long, not reaching apex of second tarsomere .... *Gomphostilbia*
4. Frons covered with many fine hairs .... *Nevermannia*
   Frons with several hairs along each lateral margin .... *Simulium* s. str.

Males

1. Katepisternum haired 2

Katepisternum bare 4
2. Paramere without hooks .... *Morops*
Paramere with hooks (Fig. 93H) 3
3. Hind tibia with low ridge on anterior inner surface (similar to Fig. 92G). Calcipala elongate, reaching apex of second tarsomere (Fig. 93A, B) .... *Wallacellum*
Hind tibia without ridge on anterior inner surface. Calcipala medium-long, not reaching apex of second tarsomere .... *Gomphostilbia*
4. Coxite longer than style (Fig. 45I) .... *Nevermannia*
Coxite shorter than style (Fig. 55L) .... *Simulium* s. str.

Pupae

1. Abdominal segment 9 with grapnel-like hooklets on each side (Fig. 3D) .... *Gomphostilbia* (except *S. asperum* and *S. kamimurai*)
Abdominal segment 9 without grapnel-like hooklets on each side 2
2. Abdominal segments 5–9 without spine-combs dorsally (Fig. 94E) .... *Wallacellum*
At least abdominal segment 8 with spine-combs dorsally 3
3. Frons with three pairs of trichomes 4
Frons with two pairs of trichomes 5
4. Gill with six or eight slender filaments .... *Gomphostilbia* (*S. asperum* of *S. hemicyclium* species-group and *S. kamimurai* of *S. ceylonicum* species-group)
Gill with four (Fig. 98A) or ca. 45 slender filaments (Fig. 98I), or of one inflated structure with four slender filaments (Fig. 98B–E).. .... *Morops* (except six species of *S. oculatum* species-group)
5. Gill filaments with constriction(s) (Fig. 98F–H).. ....*Morops* (six species of the *S. oculatum* species-group)
Gill filaments without constriction 6
6. Cocoon shoe- or boot-shaped (Fig. 56L, M) ....*Simulium* s. str. (all 10 species of *S. dumogaense* species-group)
Cocoon wall-pocket-shaped 7
7. Most or some of frontal and thoracic trichomes branched ....*Simulium* s. str. (10 species of *S. variegatum* species-group)
All frontal and thoracic trichomes unbranched 8
8. Gill composed of one large globe, two small globes, and two thumb-like short filaments (Fig. 88I, J) ....*Simulium* s. str. (*S. palopoense* of *S. variegatum* species-group)
Gill with four or six slender or somewhat inflated filaments .... *Nevermannia*

Larvae

1. Hypostoma with lateral margins smooth .... *Gomphostilbia*
Hypostoma with lateral margins serrated (Fig. 89V) 2

2. Pectination of uppermost primary ray of labral fan (Fig. 95V) different from those of other primary rays (Fig. 95W) .... *Wallacellum*
   Pectination of uppermost primary ray of labral fan same as those of other primary rays 3
3. Ventral papillae absent ....
   ....*Simulium* s. str. (species of *S. dumogaense* species-group)
   Ventral papillae present 4
4. Postgenal cleft over 1.7 times as long as postgenal bridge ....*Morops*
   Postgenal cleft 0.3–1.5 times as long as postgenal bridge 5
5. Postgenal cleft 0.3–0.4 times as long as postgenal bridge (similar to Fig. 44O). Ventral papillae large ..... *Nevermannia* (species of *S. feuerborni* species-group)
   Postgenal cleft 0.5–1.5 times as long as postgenal bridge (similar to Fig. 46M). Ventral papillae small 6
6. Mandibular serration composed of two large teeth or one large tooth and one medium-sized tooth (Fig. 46I, J)..
   ....*Nevermannia* (species of *S. ruficorne* species-group)
   Mandibular serration composed of one large tooth and one small tooth (Fig. 89U)................*Simulium* s. str. (species of *S. variegatum* species-group)

Keys to Identify 31 Species of the Subgenus Gomphostilbia

Females*

1. Antenna with eight flagellomeres ....*S. tomae*
   Antenna with nine flagellomeres 2
2. Pleural membrane haired ....*S. disneyi*
   Pleural membrane bare 3
3. Frons widely bare medially, with stout hairs in one or two vertical rows submedially or laterally on each side and a few hairs above antennal base 4
   Frons moderately or densely covered with fine hairs, leaving no bare area or narrow mediolongitudinal line bare 9
4. Arms of genital fork of moderate width .... *S. curvum*
   Arms of genital fork widened basally...5
5. Cibarium with M-shaped low projection posteriorly (similar to Fig. 39F) 6
   Cibarium without any projection posteriorly 7
6. Frons:head ratio 1.0:6.2 .... *S. heldsbachense*
   Frons:head ratio 1.0:7.3 .... *S. kalimerahense*
7. Sensory vesicle long, 0.52 times as long as third palpal segment ....*S. asperum*
   Sensory vesicle short, 0.24–0.27 times as long as third palpal segment 8

8. Sensory vesicle with large opening. First flagellomere of antenna 1.5 times as long as second .... *S. inermum*
   Sensory vesicle with medium-sized opening. First flagellomere of antenna 1.9 times as long as second .... *S. bifurcatum*
9. Mandible serrated only on inner margin .... *S. sulawesiense*
   Mandible serrated on both inner and outer margins 10
10. Frons:head ratio 1.0:4.1–5.5 11
    Frons:head ratio 1.0:5.6–6.9 20
11. Hind tibia whitish on basal one-third or more, and darkened on rest 12
    Hind tibia mostly darkened, leaving no or small pale portion basally 17
12. Hind tibia whitish on basal one-third to two-fifths. Labrum 0.67–0.69 times as long as clypeus. Sensory vesicle long, 0.48–0.63 times as long as third palpal segment 13
    Hind tibia whitish on basal half or more. Labrum 0.87–1.05 times as long as clypeus. Sensory vesicle short to medium-long, 0.25–0.34 times as long as third palpal segment 14
13. Sensory vesicle 0.48 times as long as third palpal segment .... *S. ambonense*
    Sensory vesicle 0.56–0.63 times as long as third palpal segment .... *S. kamimurai*
14. Hind tibia whitish on basal half .... *S. rosemaryae*
    Hind tibia whitish on basal two-thirds or three-fifths 15
15. Claw with large basal tooth 0.42 times length of claw .... *S. mogii*
    Claw with large basal tooth 0.50–0.52 times length of claw 16
16. Mandible with ca. 30 inner and 13 outer teeth .... *S. gimpuense*
    Mandible with ca. 20 inner and four or five outer teeth .... *S. gyorkosae*
17. Hind tibia entirely darkened. Frons:head ratio 1.0:4.1. Labrum 0.83 times as long as clypeus .... *S. hadiae*
    Hind tibia mostly darkened with basal small portion pale. Frons:head ratio 1.0: 4.9–5.5. Labrum 0.58–0.71 times as long as clypeus 18
18. Labrum 0.71 times as long as clypeus. Sensory vesicle 0.32 times as long as third palpal segment .... *S. brevitruncatum*
    Labrum 0.58–0.61 times as long as clypeus. Sensory vesicle 0.39–0.45 times as long as third palpal segment 19
19. Frons:head ratio 1.0:5.5. Sensory vesicle 0.39 times as long as third palpal segment .... *S. kolakaense*
    Frons:head ratio 1.0:4.9. Sensory vesicle 0.45 times as long as third palpal segment .... *S. singgihi*
20. Scutum with dark short hairs .... *S. syafruddini*
    Scutum with whitish yellow short hairs 21
21. Labrum 0.81 times as long as clypeus .... *S. tumidum*
    Labrum 0.58–0.75 times as long as clypeus 22

22. Labrum 0.75 times as long as clypeus .... *S. manadoense*
    Labrum 0.58–0.62 times as long as clypeus 23
23. Frons:head ratio 1.0:5.8 .... *S. abadii*
    Frons:head ratio 1.0:6.1–6.9 .... *S. marosense* and *S. torautense*

Males**

1. Hind basitarsus enlarged, wedge-shaped 2
   Hind basitarsus slender, parallel-sided 6
2. Hind tibia darkened except basal one-third white or whitish yellow 3
   Hind tibia darkened except basal two-fifths white or whitish yellow 4
3. Fore basitarsus 5.0 times as long as its greatest width .... *S. ambonense*
   Fore basitarsus 6.3 times as long as its greatest width ....*S. napuense*
4. Hind basitarsus dark brown except basal one-third dark yellow to light brown .... *S. kamimurai* and *S. rosemaryae*
   Hind basitarsus dark brown except basal two-fifths whitish 5
5. Upper-eye (large) facets in 11 or 12 vertical columns .... *S. mogii*
   Upper-eye (large) facets in nine or 10 vertical columns .... *S. gimpuense*
6. Hind tibia darkened with base white 7
   Hind tibia darkened with basal one-fifth or more white or yellow 16
7. Scutum with dark shot hairs 8
   Scutum with white or whitish yellow short hairs 12
8. First flagellomere of antenna twice as long as second .... *S. kuingingiense*
   First flagellomere of antenna 1.56–1.64 times as long as second 9
9. Fore basitarsus 5.3 times as long as its greatest width .... *S. curvum*
   Fore basitarsus 6.8 times as long as its greatest width 10
10. Upper-eye (large) facets in 12 or 13 vertical columns .... *S. hadiae*
    Upper-eye (large) facets in 16 or 17 vertical columns 11
11. Mid basitarsus medium brown except basal one-third pale .... *S. syafruddini*
    Mid basitarsus medium brown except base somewhat pale .... *S. brevitruncatum*
12. Upper-eye (large) facets in 15 or 16 vertical columns 13
    Upper-eye (large) facets in 11–14 vertical columns 14
13. Fore basitarsus 6.7 times as long as its greatest width. First flagellomere of antenna 1.5 times as long as second .... *S. torautense*
    Fore basitarsus 7.5 times as long as its greatest width. First flagellomere of antenna 1.8 times as long as second .... *S. miyagii*
14. Upper-eye (large) facets in 11 vertical columns .... *S. tumidum*
    Upper-eye (large) facets in 12–14 vertical columns 15
15. Fore basitarsus 6.3 times as long as its greatest width .... *S. manadoense*
    Fore basitarsus 6.7–6.8 times as long as its greatest width .... *S. abadii* and *S. singihi*
16. Scutum with dark shot hairs 17
    Scutum with white to yellow short hairs 19

17. Fore basitarsus 5.2 times as long as its greatest width .... *S. hemicyclium*
Fore basitarsus 5.7–6.8 times as long as its greatest width 18
18. Scutum with non-pruinose median vitta ....*S. kalimerahense*
Scutum without non-pruinose median vitta .... *S. heldsbachense*
19. Fore basitarsus 5.0 times as long as its greatest width ....*S. inernum*
Fore basitarsus 6.7–6.8 times as long as its greatest width 20
20. First flagellomere of antenna 1.35 times as long as second ....*S. kolakaense*
First flagellomere of antenna 1.6–1.7 times as long as second 21
21. Ventral plate not produced ventrally .... *S. marosense*
Ventral plate much produced ventrally ....*S. asperum*

Pupae***

1. Gill composed of inflated Y-shaped structure, with six slender filaments basally and two slender filaments apically (Fig. 22D) .... *S. tumidum*
Gill filamentous, not inflated 2
2. Gill with four or six filaments 3
Gill with eight filaments 7
3. Gill with four filaments (Fig. 27D) 4
Gill with six filaments (Fig. 27B) 5
4. Common basal stalk of gill shorter than secondary stalk of ventral paired filaments .... *S. gimpuense*
Common basal stalk of gill longer than secondary stalk of ventral paired filaments .... *S. mogii*
5. Terminal hooks cone-like, bent anteromedially .... *S. curvum*
Terminal hooks wide, plate-like, each with outer margin crenulated 6
6. Grapnel-like hooklets present...*S. rosemaryae*
Grapnel-like hooklets absent... *S. kamimurai*
7. Gill with eight filaments arising close together from long common basal stalk 8
Gill with eight filaments arising from short to medium-long common stalk 9
8. Cocoon with anterodorsal bulge or projection.................................. *S. miyagii*
Cocoon without anterodorsal bulge or projection............................ *S. serratum*
9. Eight gill filaments subequal in length and thickness to one another 10
Six filaments of dorsal and middle groups of gill apparently shorter and thinner than one or two filaments of ventral pair 18
10. Eight gill filaments shorter than pupal body, arising from short common basal stalk and lying close together..*S. brevitruncatum*
Eight gill filaments as long as or longer than pupal body, arising from short to medium-long common basal stalk, and divergent 11
11. Eight gill filaments arranged in four pairs 12
Eight gill filaments arranged in two triplets (dorsal and middle) and one pair (ventral) 13

12. Four pairs of gill filaments arising near base .... *S. kuingingiense*
    Three of four pairs of gill filaments sharing long stalk ....*S. tomae*
13. Dorsal and middle triplets of gill sharing no or extremely short stalk. Common basal stalk of gill extremely short 14
    Dorsal and middle triplets of gill sharing medium-long stalk. Common basal stalk of gill medium-ong 16
14. Dorsal and middle triplets of gill each with long primary stalk ....*S. asperum*
    Dorsal and middle triplets of gill each with short primary stalk 15
15. All gill filaments with distinct annular ridges and furrows on their surface. Cocoon with short, broad anterodorsal projection .... *S. ambonense*
    All gill filaments smooth, without annular ridges and furrows on their surface. Cocoon without anterodorsal projection .... *S. hemicyclium*
16. Terminal hooks triangular, plate-like when viewed posteriorly .... *S. gyorkosae*
    Terminal hooks wide, plate-like when viewed posteriorly 17
17. Grapnel-like hooklets present. Cocoon with short anterodorsal projection .... *S. napuense*
    Grapnel-like hooklets absent. Cocoon without anterodorsal projection .... *S. mamasaense*
18. One of two filaments of ventral pair of gill much longer than seven other filaments .... *S. kalimerahense*
    Two filaments of ventral pair of gill much longer than six other filaments 19
19. Two filaments of ventral pair of gill 1.2–1.5 times as long as six other filaments. Terminal hooks wide, plate-like, each with outer margin crenulated 20
    Two filaments of ventral pair of gill 2.0–4.3 times as long as six other filaments. Terminal hooks triangular or wide but low, each with outer margin not crenulated 22
20. Two filaments of ventral pair of gill nearly of same thickness throughout their length and bluntly ended apically ....*S. bifurcatum*
    Two filaments of ventral pair of gill gradually decreasing in thickness to apex 21
21. Cocoon moderately and compactly woven, with thick anterior margin .... *S. heldsbachense*
    Cocoon roughly woven, with anterior margin not well defined, and with many small open spaces in webs anteriorly .... *S. inermum*
22. Two filaments of ventral pair of gill twice as long as six other filaments 23
    Two filaments of ventral pair of gill 2.4–4.3 times as long as six other filaments 26
23. Dorsal and middle triplets of gill filaments arising separately from dorsal surface of common basal stalk at some distant from base. Cocoon with short or medium-long anterodorsal projection ....*S. torautense*
    Dorsal and middle triplets of gill filaments arising close together near the base. Cocoon with anterodorsal projection 24

24. Two triplets of gill filaments sessile or with extremely short stalk ..... *S. abadii*
    Two triplets of gill filaments with short stalk 25
25. Terminal hooks triangular, when viewed posteriorly .... *S. hadiae*
    Terminal hooks broad, plate-like, when viewed posteriorly ....*S. manadoense*
26. Two filaments of ventral pair of gill subequal in length, and 3.7–4.3 times as long as six other filaments .... *S. singgihi*
    Two filaments of ventral pair of gill somewhat different in length, and longer filament of ventral pair 2.4–3.6 times as long as six other filaments of dorsal and middle triplets 27
27. Outer filament of ventral pair of gill somewhat longer than inner filament .... *S. syafruddini*
    Inner filament of ventral pair of gill somewhat longer than outer filament 28
28. Dorsum of abdominal segments 1 and 2 almost entirely dark gray to light brown .... *S. marosense*
    Dorsum of abdominal segments 1 and 2 unpigmented ....*S. kolakaense*

Larvae****

1. Accessory sclerites present 2
   Accessory sclerites absent 6
2. Accessory sclerites elongate, encircling body, connected to each other anteromedially .... *S. hemicyclium*
   Accessory sclerites elongate, not connected to each other anteromedially 3
3. Body 3.5–4.0 mm long. Rectal organ with five or six secondary lobules per lobe .... *S. heldsbachense*
   Body 5.7–6.5 mm long. Rectal organ with 8–11 secondary lobules per lobe 4
4. Postgenal cleft 1.1 times as long as postgenal bridge .... *S. kalimerahense*
   Postgenal cleft 1.4 or 2.2 times as long as postgenal bridge 5
5. Postgenal cleft 1.4 times as long as postgenal bridge .... *S. inermum*
   Postgenal cleft 2.2 times as long as postgenal bridge .... *S. bifurcatum*
6. Pharate pupal gill composed of two inflated structures with eight slender filaments .... *S. tumidum*
   Pharate pupal gill composed of four, six, or eight slender filaments 7
7. Pharate pupal gill with four or six slender filaments 8
   Pharate pupal gill with eight slender filaments 10
8. Pharate pupal gill with four slender filaments .... *S. mogii*
   Pharate pupal gill with six slender filaments 9
9. Rectal organ with no or one to three secondary lobules per lobe ....*S. rosemaryae*
   Rectal organ with 6–11 secondary lobules per lobe .... *S. kamimurai*

10. Body entirely yellowish or light to dark brown, without reddish-brown markings 11
    Body with reddish-brown markings 13
11. Postgenal cleft medium-long, rounded apically .... *S. asperum*
    Postgenal cleft long, with its apex close to posterior margin of hypostoma 12
12. Pharate pupal gill composed of two long stout filaments and six short slender filaments .... *S. singgihi*
    Pharate pupal gill composed of eight slender filaments of subequal length and thickness .... *S. ambonense*
13. Secondary lobules of rectal organ nodule-like .... *S. curvum*
    Secondary lobules of rectal organ finger-like or thumb-like 14
14. Pharate pupal gill with long common basal stalk 15
    Pharate pupal gill with short or medium-long common basal stalk 16
15. Postgenal cleft medium-long, 2.7 times as long as postgenal bridge .... *S. serratum*
    Postgenal cleft long, 4.3 times as long as postgenal bridge .... *S. miyagii*
16. Pharate pupal gill composed of two long, stout filaments and six short, slender filaments 17
    Pharate pupal gill composed of eight slender filaments of subequal length and thickness 21
17. Cephalic apotome widely darkened on posterior half except narrow area along each lateral margin .... *S. torautense*
    Cephalic apotome entirely whitish yellow 18
18. Postgenal cleft nearly bullet-shaped, only slightly constricted basally, and 3.3–4.6 times as long as postgenal bridge. A few posterior abdominal segments moderately covered with unbranched and bifid dark minute setae dorsally .... *S. syafruddini*
    Postgenal cleft arrow-head-shaped, somewhat constricted basally and 2.0–3.0 times as long as postgenal bridge. A few posterior abdominal segments bare or moderately covered with unbranched minute setae dorsally 19
19. Postgenal cleft twice as long as postgenal bridge. Labral fan with ca. 30 primary rays. Posterior circlet with 84 rows of hooklets .... *S. gyorkosae*
    Postgenal cleft 2.4–3.1 times as long as postgenal bridge. Labral fan with 34–44 primary rays. Posterior circlet with 60–75 rows of hooklets 20
20. Labral fan with 34–36 primary rays .... *S. marosense*
    Labral fan with ca. 44 primary rays .... *S. kolakaense*
21. Postgenal cleft medium-long, 2.4 times as long as postgenal bridge. Third antennal article longer than first one .... *S. kuingingiense*

Postgenal cleft long, 3.5 times as long as postgenal bridge. Third antennal article shorter than first one .... *S. brevitruncatum*

**Simulium hemicyclium*, *S. kuingingiense*, *S. mamasaense*, *S. miyagii*, *S. napuense*, and *S. serratum* are not included because their females are unknown.

***Simulium disneyi*, *S. bifurcatum*, *S. mamasaense*, *S. serratum*, *S. sulawesiense*, and *S. tomae* are not included because their males are unknown.

****Simulium disneyi* and *S. sulawesiense* are not included because their pupae are unknown.

**** *Simulium abadii*, *S. disneyi*, *S. gimpuense*, *S. hadiae*, *S. mamasaense*, *S. manadoense*, *S. napuense*, *S. sulawesiense*, and *S. tomae* are not included because their mature larvae are unknown.

Keys to Identify 29 Species of the Subgenus Morops

Females*

1. Fore coxa whitish yellow 2
   Fore coxa light to medium brown 8
2. Fore basitarsus with thick dorsal hair crest (similar to Fig. 54B) 3
   Fore basitarsus with moderate dorsal hair crest 6
3. Inner margin of ovipositor valve with small projection directed anterodorsally near posterior tip ....*S. jayapuraense*
   Inner margin of ovipositor valve without such projection directed anterodorsally near posterior tip 4
4. Labrum 1.1 times as long as clypeus. Sensory vesicle 0.35 times as long as third palpal segment .... *S. pamahaense*
   Labrum 0.79–0.86 times as long as clypeus. Sensory vesicle 0.41–0.45 times as long as third palpal segment 5
5. Height of frons relative to narrowest width 1.69 ....*S. botulus*
   Height of frons relative to narrowest width 1.96 .... *S. farciminis*
6. Fore basitarsus much dilated, 4.4 times as long as its greatest width ....*S. irianense*
   Fore basitarsus moderately dilated, 6.3–6.5 times as long as its greatest width 7
7. Genital fork with long projection on each arm .... *S. wamenae*
   Genital fork with medium-long projection on each arm ....*S. irayense*
8. Hind basitarsus lacking stout spines on outer surface ....*S. nudipes*
   Hind basitarsus with stout spines on outer surface 9

9. Antenna entirely darkened 10
   Antenna darkened except at least scape and pedicel pale 13
10. Hind basitarsus whitish yellow with basal tip and apical two-fifths medium brown ....*S. arfakense*
   Hind basitarsus light brown with base and apex dark brown 11
11. Claw with subbasal tooth 0.26 times as long as claw. Sensory vesicle 0.25 times as long as third palpal segment ....*S. takae*
   Claw with subbasal tooth 0.35 or 0.36 times as long as claw. Sensory vesicle 0.31–0.33 times as long as third palpal segment 12
12. Labrum 0.90 times as long as clypeus ....*S. motoyukii*
   Labrum 0.75 times as long as clypeus ....*S. baliemense*
13. Hind tibia yellowish on basal half or more and medium brown on rest 14
   Hind tibia entirely light to dark brown or light to dark brown with base yellow 17
14. Fore basitarsus 5.5–5.7 times as long as its greatest width 15
   Fore basitarsus 6.5–6.8 times as long as its greatest width 16
15. Labrum 0.83 times as long as clypeus. Sensory vesicle 0.30 times as long as third palpal segment ....*S. prafiense*
   Labrum 0.90 times as long as clypeus. Sensory vesicle 0.26 times as long as third palpal segment .... *S. anggiense*
16. Sensory vesicle 0.30–0.35 times as long as third palpal segment ....*S. norforense*
   Sensory vesicle 0.24 times as long as third palpal segment .... *S. sutebense*
17. Hind tibia without white hairs on outer and posterior surface .....*S. demutaense*
   Hind tibia covered with white hairs on outer and posterior surface 18
18. Cibarium unarmed 19
   Cibarium with low and broad projection folded forward from posterior margin 20
19. Sensory vesicle 0.36 times as long as third palpal segment .... *S. papuense*
   Sensory vesicle 0.23 times as long as third palpal segment ....*S. malukuense*
20. Frons:head ratio 1.0:5.9 ....*S. manuselaense*
   Frons:head ratio 1.0:4.4–5.0 21
21. Fore tarsus with thick dorsal hair crest (similar to Fig. 54B) 22
   Fore tarsus with moderate dorsal hair crest 23
22. Claw with subbasal tooth 0.3 times as long as claw .... *S. hagai*
   Claw with subbasal tooth 0.4 times as long as claw .... *S. tafulaense*
23. Scutum with white short hairs .... *S. ransikiense*
   Scutum with dark short hairs centrally but whitish short hairs peripherally 24
24. Sensory vesicle 0.22 times as long as third palpal segment .... *S. biakense*

Sensory vesicle 0.26–0.30 times as long as third palpal segment 25
25. Labrum 0.68 times as long as clypeus ....*S. watetoense*
Labrum 0.55 or 0.60 times as long as clypeus 26
26. Labrum 0.55 times as long as clypeus ....*S. halmaheraense*
Labrum 0.60 times as long as clypeus 27
27. Frons:head ratio 1.0:4.6 ....*S. sorongense*
Frons:head ratio 1.0:5.0 ....*S. manokwariense*

Males**

1. Hind basitarsus enlarged 2
Hind basitarsus slender, nearly parallel-sided 7
2. Fore coxa medium brown 3
Fore coxa whitish yellow or yellow 4
3. Fore basitarsus 6.1 times as long as its greatest width ....*S. arfakense*
Fore basitarsus 8.3 times as long as its greatest width ....*S. motoyukii*
4. Fore tarsus with moderate dorsal hair crest. Fore basitarsus 8.2 times as long as its greatest width ....*S. iraiense*
Fore tarsus with thick dorsal hair crest. Fore basitarsus 4.1–5.7 times as long as its greatest width 5
5. Hind tibia dark brown with basal two-fifths yellowish white ....*S. jayapuraense*
Hind tibia medium to dark brown with base dark yellow 6
6. Fore basitarsus 5.7 times as long as its greatest width ....*S. botulus*
Fore basitarsus 4.1 times as long as its greatest width .... *S. farciminis*
7. Fore coxa yellow or dark yellow 8
Fore coxa light to medium brown 9
8. Fore coxa yellow. Fore basitarsus 9.8 times as long as its greatest width .... *S. wamenae*
Fore coxa dark yellow. Fore basitarsus 4.8 times as long as its greatest width . ...*S. irianense*
9. Fore basitarsus 8.6 times as long as its greatest width ....*S. takae*
Fore basitarsus 5.5–7.8 times as long as its greatest width 10
10. Antenna entirely medium brown or medium brown with scape and pedicel light brown 11
Antenna medium brown with at least scape and pedicel yellow or pale yellow 12
11. Scutum with dark short hairs ....*S. halmaheraense*
Scutum with whitish yellow short hairs ....*S. wakrisense*
12. Scutum with dark short hairs 13
Scutum with whitish yellow short hairs 14
13. First flagellomere of antenna 2.2 times as long as second .... *S. sorongense*
First flagellomere of antenna 1.7 times as long as second ...... *S. manuselaense*

14. First flagellomere of antenna 1.3–1.4 times as long as second 15
First flagellomere of antenna 1.6–1.9 times as long as second 16
15. Abdomen with whitish yellow hairs as well as dark ones dorsally ....
.... *S. papuense*
Abdomen with dark hairs dorsally .... *S. nudipes*
16. Abdomen with whitish yellow hairs as well as dark ones dorsally 17
Abdomen with dark hairs dorsally..19
17. Basal scale with fringe of yellow long hairs .... *S. anggiense*
Basal scale with fringe of dark long hairs 18
18. Body ca. 2.4 mm long. Wing ca. 2.1 mm long .... *S. norforense*
Body ca. 3.0 mm long. Wing ca. 2.4 mm long .... *S. sutebense*
19. Upper-eye (large) facets in 12 horizontal rows .... *S. biakense*
Upper-eye (large) facets in 13–15 horizontal rows 20
20. Ventral plate not produced ventrally .... *S. prafiense*
Ventral plate produced ventrally 21
21. Ventral plate narrowed posteriorly from base when viewed ventrally ....
.... *S. hagai*
Ventral plate narrowed posteriorly from middle when viewed ventrally 22
22. Style gradually tapered to apex when viewed ventroposteriorly ....
.... *S. malukuense*
Style nearly parallel-sided on apical half when viewed ventroposteriorly 23
23. Fore basitarsus 5.9 times as long as its greatest width .... *S. manokwariense*
Fore basitarsus 6.3–6.5 times as long as its greatest width 24
24. Ventral plate slightly produced ventrally when viewed laterally ....
.... *S. watetoense*
Ventral plate much produced ventrally when viewed laterally..........................
.... *S. ransikiense*

Pupae

1. Cocoon small, wall-pocket-shaped, covering abdomen and posterior portion of thorax 2
Cocoon otherwise, covering abdomen and most of thorax 6
2. Antennal sheath of usual shape, without any projection. Gill formed of brownish black, short, club-like structure and four slender filaments (Fig. 98B) ....
.... *S. arfakense*
Antennal sheath with eight nodule-like processes or spine-like projections. Gill formed of pale to brown, medium-long or long, horn-like structure and four slender filaments 3
3. Antennal sheath with eight spine-like projections (similar to Fig. 95D). Horn-like structure of gill appearing segmented and smooth on its surface (Fig. 98C)
.... *S. jayapuraense*

Antennal sheath with eight nodule-like processes. Horn-like structure of gill with numerous conical projections in annulations (Fig. 98D, E) 4
4. Frontal trichomes all unbranched, slightly longer than facial ones. Horn-like structure of gill ca. mm long and ca. 2.5 times as thick as interspiracular trunk (Fig. 98D) .... *S. pamahaense*
Frontal trichomes mostly branched, twice to three times as long as facial ones. Horn-like structure of gill 1.5–2.2 mm long and over three times as thick as interspiracular trunk 5
5. Horn-like structure of gill 1.9–2.2 mm long and ca. 4.6 times as thick as interspiracular trunk (Fig. 98E) ....*S. botulus*
Horn-like structure of gill 1.5–1.8 mm long and ca. 3.5 times as thick as interspiracular trunk .... *S. forciminis*
6. Pit-like organ present (Fig. 98A) 7
Pit-like organ absent 16
7. Cocoon wall-pocket-shaped 8
Cocoon shoe-shaped 9
8. Pit-like organ small, *plo* ratio 0.17*** .... *S. hagai*
Pit-like organ medium-sized, *plo* ratio 0.38 .... *S. tafulaense*
9. Frons densely covered with tubercles .... *S. halmaheraense*
Frons sparsely or moderately covered with tubercles 10
10. Stalk of dorsal pair of gill filaments always longer than that of ventral ones ... . *S. manuselaense*
Stalk of dorsal pair of gill filaments as long as that of ventral ones 11
11. Pit-like organ small, *plo* ratio 0.15 .... *S. biakense*
Pit-like organ medium-sized, *plo* ratio 0.34–0.54 12
12. Common basal stalk of gill medium-long, longer than interspiracular trunk 13
Common basal stalk of gill short, shorter than interspiracular trunk 14
13. Gill filaments 2.4–3.5 mm long. *Plo* ratio 0.37 ....*S. watetoense*
Gill filaments 1.6–2.0 mm long. *Plo* ratio 0.43–0.54 .... *S. sorongense*
14. Thorax almost bare along anterior margin ....*S. malukuense*
Thorax sparsely covered with tubercles along anterior margin 15
15. Pit-like organ with numerous and long minute spines along inner margin of dorsal eave-like wall .... *S. ransikiense*
Pit-like organ without minute spines or with few short minute spines along inner margin of dorsaleave-like wall ....*S. manokwariense*
16. Gill filaments with constriction basally or subbasally (Fig. 98F–H) 17
Gill filaments without constriction basally 26
17. Cocoon shoe-shaped 18
Cocoon wall-pocket-shaped 19
18. Gill filaments arising from base, with pale large bulb-like basal fenestra ....*S. irianense*
Gill filaments arising from medium-long common basal stalk, without such large basal fenestra ....*S. prafiense*

19. Cocoon with long anterodorsal projection .... *S. sutebense*
    Cocoon without anterodorsal projection 20
20. Cocoon with marked longitudinal dorsal crest medially 21
    Cocoon without such dorsal crest 22
21. Gill filaments arising from long common basal stalk .... *S. anggiense*
    Gill filaments arising from short common basal stalk ....*S. baliemense*
22. Gill filaments arising directly from base ....*S. motoyukii*
    Gill filaments arising from common basal stalk 23
23. Gill filaments arising from very long common basal stalk (Fig. 98H) ....
    .... *S. norforense*
    Gill filaments arising from short or medium-long stalk 24
24. Four gill filaments each with marked constriction subbasally (Fig. 98F) ....
    ....*S. demutaense*
    Four gill filaments each with marked constriction basally 25
25. Thick-walled basal portion of each gill filament swollen and bulbous (Fig. 98G)
    ....*S. takae*
    Thick-walled basal portion of each gill filament not swollen ....*S. wakrisense*
26. Gill with 30–45 slender filaments (Fig. 98I) .... *S. papuense*
    Gill with four slender filaments 27
27. Cocoon wall-pocket-shaped. Gill filaments with medium-long common basal stalk ....*S. nudipes*
    Cocoon shoe-shaped. Gill filaments sessile or with short common basal stalk ..
   28
28. Frons bare. Frontal trichomes each with six or seven branches. Cuticle of gill filaments with longitudinal ridges .... *S. wamenae*
    Frons covered with tubercles of various sizes. Frontal trichomes each with 10–25 branches. Cuticle of gill filaments with annular ridges ....*S. irayense*

Mature larvae****

1. Abdominal segments 1–5 each with dorsolateral pair of protuberances. Postgenal cleft long, reaching posterior margin of hypostoma .... *S. papuense*
   Abdominal segments 1–5 without dorsolateral pair of protuberances. Postgenal cleft not reaching posterior margin of hypostoma 2
2. Second antennal article longer than third 3
   Second antennal article as long as or shorter than third 16
3. Pharate pupal gill with four slender filaments 4
   Pharate pupal gill with inflated, horn-like or club-like structure and four slender filaments 13
4. Rectal organ with three to six secondary lobules per lobe 5
   Rectal organ with 9–18 secondary lobules per lobe 11
5. Rectal scales absent 6
   Rectal scales present 7

6. Pharate pupal gill with short common basal stalk .... *S. biakense*
Pharate pupal gill with medium-long common basal stalk ....
.... *S. manokwariense* and *S. watetoense*
7. Body 5.8–6.2 mm long .... *S. tafulaense*
Body 3.9–5.6 mm long 8
8. Both sides of postgenal cleft somewhat darkened up to postgenal bridge ....
.... *S. malukuense*
Both sides of postgenal cleft not darkened up to postgenal bridge 9
9. Pharate pupal gill with medium-long common basal stalk .... *S. sorongense*
Pharate pupal gill with short common basal stalk 10
10. Lateral surface of head capsule with widely darkened area dorsal to eye-spot region extended to anterior margin .... *S. hagai*
Lateral surface of head capsule with partially darkened area dorsal to eye-spot region .... *S. halmaheraense*
11. Rectal organ with 16–18 secondary lobules per lobe ....*S. iraiense*
Rectal organ with 9–14 secondary lobules per lobe 12
12. Body 4.4–5.0 mm long. Postgenal cleft ca. 4.3 times as long as postgenal bridge.
.... *S. ransikiense*
Body 5.5–5.8 mm long. Postgenal cleft 2.3–3.2 times as long as postgenal bridge .... *S. manuselaense*
13. Rectal organ with seven to nine secondary lobules per lobe .... *S. jayapuraense*
Rectal organ with 12–20 secondary lobules per lobe 14
14. Pharate pupal gill with club-like inflated structure and four slender filaments...
.... *S. arfakense*
Pharate pupal gill with horn-like inflated structure and four slender filaments ..
....15
15. Rectal organ with 12 or 13 secondary lobules per lobe .... *S. pamahaense*
Rectal organ with 17–20 secondary lobules per lobe ....
.... *S. botulus* and *S. farciminis*
16. Rectal organ with 16–22 secondary lobules per lobe 17
Rectal organ with 4–10 secondary lobules per lobe 18
17. Pharate pupal gill with enlarged common basal stalk ....*S. takae*
Pharate pupal gill with four slender filaments arising from base ....
.... *S. motoyukii*
18. Mandible without supernumerary serration. Major tooth of mandibular serration at acute angle against mandible on apical side ....*S. nudipes*
Mandible with supernumerary serration. Major tooth of mandibular serration at right or obtuse angle against mandible on apical side 19
19. Rectal scales absent. Second antennal article as long as third .....*S. demutaense*
Rectal scales present. Second antennal article shorter than third 20
20. Pharate pupal gill with long common basal stalk bent medially ..*S. norforense*
Pharate pupal gill with medium-long common basal stalk not bent medially ....
....21

21. Four to ten sensilla posterior to anal sclerite .... *S. prafiense*
    Two sensilla posterior to anal sclerite 22
22. Accessory sclerites present, each consisting of one to three ventrolateral streaks. Posterior circlet with 96–100 rows of hooklets .... *S. anggiense*
    Accessory sclerites absent. Posterior circlet with ca. 86 rows of hooklets .... *S. sutebense*

**Simulium wakrisense* is not included because its female is unknown.

***Simulium baliemense*, *S. demutaense*, *S. pamahaense*, and *S. tafulaense* are not included because their males are unknown.

****Plo* ratio is calculated by dividing a diameter of the pit-like organ by a distance from its innermost margin to the mediolongitudinal suture of the thorax (Takaoka and Suzuki, 1995).

*****Simulium baliemense*, *S. irianense*, *S. wakrisense*, and *S. wamenae* are not included because their mature larvae are unknown.

Keys to Identify Four Species of the Subgenus Nevermannia

Females

1. Scutum dark brown, without dark longitudinal vittae. Ovipositor valve tongue-like (Fig. 45C). Paraproct shorter than cercus (Fig. 45E) 2
   Scutum light to medium brown, with faint dark longitudinal vittae. Ovipositor valve triangular (similar to Fig. 43D). Paraproct longer than cercus (similar to Fig. 43H) 3
2. Mandible with 16 or 17 inner and eight or nine outer teeth. Sensory vesicle 0.5 times as long as third palpal segment .... *S. aureohirtum*
   Mandible with 20 inner and nine or 10 outer teeth. Sensory vesicle 0.4 times as long as third palpal segment .... *S. ornatipes*
3. Frons:head ratio 1.0:5.7. Sternite 8 with 14–16 short to long hairs on each side .... *S. chairuddini*
   Frons:head ratio 1.0:4.9–5.4. Sternite 8 with 23 or 24 short to long hairs on each side .... *S. minahasaense*

Males

1. Hind basitarsus enlarged, spindle-shaped (similar to Fig. 43J). Ventral plate lamellate, without median keel (similar to Fig. 43K) .... *S. chairuddini* and *S. minahasaense*
   Hind basitarsus slender, nearly parallel-sided (similar to Fig. 45H). Ventral plate with median keel (similar to Fig. 45I) 2

2. Upper-eye (large) facets in 20 vertical columns and 23 horizontal rows. Fore basitarsus 6.9 times as long as its greatest width .... *S. aureohirtum*
   Upper-eye (large) facets in 22 or 23 vertical columns and 24 or 25 horizontal rows. Fore basitarsus 8.1 times as long as its greatest width .... *S. ornatipes*

Pupae

1. Cocoon wall-pocket-shaped, with broad anterodorsal projection (similar to Fig. 46H) 2
   Cocoon wall-pocket-shaped, without anterodorsal projection 3
2. Gill with six filaments (Fig. 46C) .... *S. aureohirtum*
   Gill with four filaments .... *S. ornatipes*
3. Gill with six long filaments arranged as 4+2 from dorsal to ventral. Ventral pair of gill with short stalk ....*S. chairuddini*
   Gill with six long filaments arranged as 2+2+2 from dorsal to ventral. Ventral pair of gill with long stalk ....*S. minahasaense*

Larvae

1. Mandibular serration composed of two large teeth (Fig. 46I, J). Postgenal cleft 0.8 times as long as postgenal bridge (similar to Fig. 46M) 2
   Mandibular serration composed of one large tooth and one small tooth (similar to Fig. 44M). Postgenal cleft 0.3–0.4 times as long as postgenal bridge (similar to Fig. 44O) 3
2. Pharate pupal gill with four somewhat inflated filaments .... *S. ornatipes*
   Pharate pupal gill with six filaments ....*S. aureohirtum*
3. Rectal organ with 14–17 secondary lobules per lobe ....*S. chairuddini*
   Rectal organ with 6–10 secondary lobules per lobe ....*S. minahasaense*

Keys to Identify 22 Species of the Subgenus **Simulium** *s. str.*

Females*

1. Scutum dark brown to brownish black 2
   Scutum orange yellow 15
2. Abdomen shiny on tergites 6–8 3
   Abdomen shiny on tergites 5–8 6
3. Basal section of radius fully haired 4
   Basal section of radius bare 5
4. Hind tibia medium to dark brown except anterior surface of basal two-fifths and posterior surface of basal two-thirds white .... *S. molliculum*
   Hind tibia medium to dark brown except posterior surface of basal half white . ...*S. laxum*
5. Mid femur yellow with apical half medium to dark brown .... *S. elongatum*

Mid femur yellow with apical two-fifths medium to dark brown .... *S. linduense*

6. Claw simple, without tooth (Fig. 54E) 7
   Claw with basal or subbasal tooth (Fig. 54F, G) 8
7. Hind femur yellow except apical cap darkened .... S. *contractum*
   Hind femur medium to dark brown .... *S. landonoense* and *S. tumpaense*
8. Claw tooth slender, much longer than basal width (Fig. 54F) 9
   Claw tooth stout, slightly shorter than or as long as basal width (Fig. 54G) 12
9. Hind basitarsus with white portion (Fig. 54C) 10
   Hind basitarsus entirely dark brown (Fig. 54D) 11
10. Hind basitarsus white on little less than basal half. Basal section of radius bare on basal one-fourth or two-fifths but haired on apical three-fourths or three-fifths .... *S. nudifrons*
    Hind basitarsus white on basal three-fifths. Basal section of radius entirely bare .... *S. mediodentatum*
11. Ovipositor valve mostly membranous except inner margin, and almost fully covered with long hairs except apical tip.................................... *S. seramense*
    Ovipositor valve mostly sclerotized except anterior portion widely membranous and posterior portion along posterolateral margin narrowly membranous, and covered with long hairs only on basal two-thirds and bare on apical one-third (Fig. 54Q) .... *S. tomohonense*
12. Fore tarsomeres 1–3 with thick dorsal crest of short or medium-long hairs .. 13
    Fore tarsomeres 1–3 with thick dorsal crest of long hairs (Fig. 54B) 14
13. Hind tibia dark brown except basal tip pale yellow. Ovipositor valve attenuated posteromedially (Fig. 54K) ....*S. watanabei*
    Hind tibia dark brown except posterior surface of basal two-thirds white. Ovipositor valve rounded posteriorly .... *S. palopoense*
14. Ovipositor valve nearly triangular, with short projection extended posteriorly from posteromedial corner (Fig. 54H) .... *S. dumogaense*
    Ovipositor valve much produced and attenuated posteriorly, with sharply pointed apical tip .... *S. fuscitarse*
15. Fore basitarsus greatly dilated, 4.3 times as long as its greatest width. Cibarium without any processes (Fig. 86B) ....*S. luridum*
    Fore basitarsus moderately dilated, 5.1–5.7 times as long as its greatest width. Cibarium with numerous processes (Fig. 86A) 16
16. Labrum 0.57 times as long as clypeus ....*S. globosum*
    Labrum over 0.61 times as long as clypeus 17
17. Labrum 0.67 times as long as clypeus ....*S. flavum*
    Labrum 0.61–0.63 times as long as clypeus 18
18. Height of frons 1.7 times as long as its narrowest width. Fore basitarsus 5.1 times as long as its greatest width ....*S. lorense*

Height of frons 1.5 times as long as its narrowest width. Fore basitarsus 5.7 times as long as its greatest width 19

19. Hind basitarsus 6.3 times as long as its greatest width .... *S. malinoense*
    Hind basitarsus 5.8 times as long as its greatest width ....*S. quinquefiliatum*

Males**

1. Scutum dark brown to brownish black 2
   Scutum orange yellow 12
2. Hind basitarsus not enlarged, nearly parallel-sided (Fig. 55I) 3
   Hind basitarsus enlarged, widened to apex (Fig. 55K) 4
3. Hind femur dark brown ....*S. watanabei*
   Hind femur yellow with apical cap light brown .... *S. contractum*
4. Hind basitarsus entirely dark brown (Fig. 55K) .... *S. tomohonense*
   Hind basitarsus bicolored 5
5. Whitish spot on each shoulder of scutum not connected submedially to whitish spot on prescutellar area 6
   Whitish spot on each shoulder of scutum connected submedially to whitish spot on prescutellar area 10
6. Scutum densely covered with dark scale-like short hairs .... *S. palopoense*
   Scutum densely covered with yellow scale-like short hairs 7
7. Hind basitarsus white on basal half or little more 8
   Hind basitarsus white on basal one-fourth or one-third 9
8. Hind femur yellow on little more than basal one-third .... *S. linduense*
   Hind femur yellow on basal half ....*S. elongatum*
9. Mid tibia dark brown except posterior surface of extreme base whitish yellow ....*S. molliculum*
   Mid tibia dark brown except posterior surface of basal two-fifths somewhat pale...*S. laxum*
10. Hind basitarsus white on basal one-fourth .... *S. dumogaense*
    Hind basitarsus white on basal half or little more 11
11. Style with long basal protuberance pointed apically .... *S. nudifrons*
    Style with short basal protuberance having serrate round plate apically (Fig. 55W) .... *S. landonoense* and *S. tumpaense*
12. Abdominal segment 2 entirely light to dark yellow dorsally 13
    Abdominal segment 2 entirely dark brown dorsally 14
13. Hind basitarsus yellow on basal two-thirds (Fig. 86K). Ventral plate with parallel rows of spines lying close together on posterior surface (Fig. 86V) ....*S. luridum*
    Hind basitarsus yellow on basal half. Ventral plate with parallel rows of spines moderately spaced on posterior surface ....*S. lorense*
14. Hind basitarsus yellow on little less than basal half ....*S. flavum*
    Hind basitarsus yellow on little more than basal half 15

15. Fore basitarsus slightly dilated, 7.2 times as long as its greatest width .... *S. globosum*
    Fore basitarsus moderately dilated, 6.0 times as long as its greatest width .... *S. malinoense*

Pupae***

1. Cocoon shoe-shaped or boot-shaped (Fig. 56L, M). Abdomen with dorsal spine-combs only on segment 8 2
   Cocoon otherwise. Abdomen with dorsal spine-combs at least on segments 7 and 8 11
2. Frons with three trichomes on each side (Fig. 56C) .... *S. mediodentatum*
   Frons with two trichomes on each side (Fig. 56A, B) 3
3. Two trichomes on each side of frons widely spaced from each other (Fig. 56B). Dorsum of abdominal segment 2 with four hooks on each side, which are similar in size and shape to those on dorsum of segments 3 and 4 4
   Two trichomes on each side of frons arising close together (Fig. 56A). Dorsum of abdominal segment 2 with five spinous setae on each side, which are smaller in size than those on dorsum of segments 3 and 4 6
4. Cocoon compactly woven, with no open spaces in webs, and with low or medium-high anteroventral collar (Fig. 56M). All trichomes on head and thorax unbranched .... *S. nudifrons*
   Cocoon loosely woven, with small open spaces in webs, and with high anteroventral collar (Fig. 56L). Most trichomes on head and thorax branched 5
5. Gill filaments 1.5–1.8 mm long. Frontal trichomes each branched into 2–4. Tubercles on thorax round .... *S. tomohonense*
   Gill filaments 1.2–1.5 mm long. Frontal trichomes unbranched or bifid. Tubercles on thorax cone-shaped .... *S. serumense*
6. Abdominal segments 6 and 7 each with only inner hook and lacking outer hook on each side 7
   Abdominal segments 6 and 7 each with inner and outer hooks of similar size on each side 9
7. Gill with eight filaments (Fig. 56I) .... *S. landonoense*
   Gill with six filaments 8
8. All gill filaments tapered to apex .... *S. tumpaense*
   All gill filaments widened from base to basal one-third or one-fourth, then tapered to apex (Fig. 56E) .... *S. contractum*
9. Frons bare except anterior portion moderately covered with tubercles .... *S. watanabei*
   Frons densely covered with tubercles 10
10. Frontal trichomes each branched into four or five. Facial trichomes bifid. Hooks on ventral surface of abdominal segment 5 unbranched .... *S. dumogaense*
    Frontal trichomes unbranched of bifid. Facial trichomes unbranched. Hooks on ventral surface of abdominal segment 5 bifid .... *S. fuscitarse*

11. Terminal hooks present 12
    Terminal hooks absent 19
12. Gill with six slender filaments all directed forward 13
    Gill otherwise 14
13. Gill filaments 2.6–3.2 mm long (Fig. 87C). Cocoon roughly woven by distinct threads, with open spaces in webs (Fig. 89G) .... *S. laxum*
    Gill filaments 3.6–4.0 mm long (Fig. 87A). Cocoon thinly and softly woven without open spaces in webs (Fig. 89F) .... *S. molliculum*
14. Gill globose 15
    Gill with five or six inflated filaments each with one or two slender filaments, of which two inflated filaments directed upward 16
15. Gill composed of one large globe, with six filaments (Fig. 88K). Cocoon wall-pocket-shaped without anterodorsal projection .... *S. globosum*
    Gill composed of one large, one medium-sized, and one small globes and two finger-like short filaments (Fig. 88I). Cocoon wall-pocket-shaped with anterodorsal projection .... *S. palopoense*
16. Gill with five inflated filaments, of which four have one slender filament but one has two filaments apically (Fig. 88G) .... *S. quinquefiliatum*
    Gill with six inflated filaments each with one slender filament (Fig. 88C) 17
17. Cocoon with anterodorsal projection .... *S. lorense*
    Cocoon without anterodorsal projection 18
18. Frons bare. Two of four inflated gill filaments directed forward sharing medium-long stalk, and two inflated filaments directed upward subequal in thickness along basal half, then gradually tapered to become slender apical filaments (Fig. 88A) .... *S. flavum*
    Frons with tubercles. Two of four inflated gill filaments directed forward sharing long stalk, and two inflated gill filaments directed upward subequal in thickness throughout their length, then abruptly tapered to become slender apical filaments (Fig. 88E) .... *S. malinoense*
19. Gill with six filaments arising from base, of which five are directed forward and one upward (Fig. 89B). Frons and dorsal surface of thorax densely covered with spinous processes (Fig. 89A) .... *S. luridum*
    Gill with six filaments arising from long inflated common basal stalk. Frons and dorsal surface of thorax densely covered with round tubercles 20
20. Inflated common basal stalk of gill twice as thick as interspiracular trunk (Fig. 87G). Cocoon with long anterodorsal projection extended beyond its anteroventral tips (Fig. 87G) .... *S. elongatum*
    Inflated common basal stalk of gill 3.0 times as thick as interspiracular trunk (Fig. 87E). Cocoon with long anterodorsal projection not extended beyond its anteroventral tips (Fig. 87E) .... *S. linduense*

Mature larvae****

1. Postgenal cleft long, 1.7–3.9 times as long as postgenal bridge 2
   Postgenal cleft short or medium-long, 0.5–1.5 times as long as postgenal bridge 11
2. Postgenal cleft widely rounded (Fig. 56R) 3
   Postgenal cleft triangular or bullet-shaped (Fig. 56P, Q) 5
3. Pharate pupal gill with eight filaments .... *S. landonoense*
   Pharate pupal gill with six filaments 4
4. Six filaments of pharate pupal gill with maximal thickness at base .... *S. tumpaense*
   Six filaments of pharate pupal gill with maximal thickness at some distance from base .... *S. contractum*
5. Sheath of subesophageal ganglion well pigmented (Fig. 56Q) ...... *S. fuscitarse*
   Sheath of subesophageal ganglion unpigmented 6
6. Body 3.8 mm long .... *S. parvulum*
   Body over 5.0 mm long 7
7. Cephalic apotome pale or at most dark yellow with median area just before posterior margin somewhat darkened, and with indistinct or faintly positive posterior mediolongitudinal spots ....*S. watanabei*
   Cephalic apotome darkened along both lateral margins at least posteriorly and along posterior margin, and with distinct dark posterior mediolongitudinal spots 8
8. Posterior mediolongitudinal spot not well demarcated from dark surrounding area .... *S. tomohonense*
   Posterior mediolongitudinal spot well demarcated from whitish surrounding area 9
9. Lateral dark areas on cephalic apotome not reaching posterior margin .... *S. seramense*
   Lateral dark areas on cephalic apotome reaching posterior margin 10
10. Anterior tip of lateral dark areas on cephalic apotome extended somewhat forward, but not beyond anterior tip of anterior mediolongitudinal spot .... *S. dumogaense*
    Anterior tip of lateral dark areas on cephalic apotome extended forward beyond anterior tip of anterior mediolongitudinal spot .... *S. nudifrons*
11. Postgenal cleft triangular (Fig. 56P) ....*S. mediodentatum*
    Postgenal cleft nearly parallel-sided basally 12
12. Postgenal cleft short, 0.54–0.64 times as long as postgenal bridge 13
    Postgenal cleft medium-long, 0.8–1.5 times as long as postgenal bridge 14
13. Common basal stalk of pharate pupal gill much inflated, 0.34 times as wide as long .... *S. linduense*
    Common basal stalk of pharate pupal gill moderately inflated, 0.24 times as wide as long ....*S. elongatum*

14. Pharate pupal gill composed of six thread-like filaments 15
Pharate pupal gill composed of inflated (at least basally) filaments or globular structure 16
15. Six filaments of pharate pupal gill with annular ridges on their surface .... *S. molliculum*
Six filaments of pharate pupal gill with longitudinal or spiral ridges on their surface at least along basal portion ....*S. laxum*
16. Pharate pupal gill with six inflated filaments each having slender apical filament 17
Pharate pupal gill of globe shape 20
17. Pharate pupal gill with large transparent basal fenestra ....*S. luridum*
Pharate pupal gill without such large basal fenestra 18
18. Two outer inflated filaments of pharate pupal gill each gradually tapered medially to become slender apical filaments ....*S. flavum*
At least one of two outer inflated filaments of pharate pupal gill abruptly tapered medially to become slender apical filament 19
19. Two outer inflated filaments of pharate pupal gill each abruptly tapered medially to become slender apical filament .... *S. malinoense*
One of two outer inflated filaments of pharate pupal gill abruptly tapered medially to become slender apical filament .... *S. lorense*
20. Pharate pupal gill composed of three globes and two finger-like short filaments .... *S. palopoense*
Pharate pupal gill composed of one globe and six slender filaments .... *S. globosum*

Simulium parvulum is not included because its female is unknown.

Simulium fuscitarse, *S. mediodentatum*, *S. parvulum*, *S. quinquefiliatum*, and *S. seramense* are not included because their males are unknown.

Simulium parvulum is not included because its pupa is unknown.

Simulium quinquefiliatum is not included because its larva is unknown.

Keys to Identify Two Species of the Subgenus **Wallacellum**

Females

1. No difference between *S. alfurense* and *S. celebesense*

Males

1. Cercus with six hairs ....*S. alfurense*
Cercus with 9–13 hairs ....*S. celebesense*

Pupae

1. Dorsal filament of outer pair of gill 0.4–0.6 times as long as its ventral filament ....*S. alfurense*
   Dorsal filament of outer pair of gill 0.8–1.0 times as long as its ventral filament ....*S. celebesense*

Mature larvae

1. Posterior circlet with 102–106 rows of up to 18 hooklets per row ...*S. alfurense*
   Posterior circlet with 84–90 rows of up to 16 hooklets per row ....*S. celebesense*

12 The Fauna of Black Flies in South China

The fauna of black flies of China is represented by 323 species (including 18 unnamed species), which are placed in six genera of two tribes: *Helodon* Enderlein, *Prosimulium* Roubaud and *Twinnia* Stone and Jamnback in Prosimuliini, and *Metacnephia* Crosskey, *Sulcicnephia* Rubtsov and *Simulium* Latreille s. l. in Simuliini (Chen, 2016; Adler, 2022). South China is zoogeographically regarded as a part of the Oriental Region, whereas the rest areas of China are known to be a part of the Palearctic Region. Among 323 Chinese simuliid species, 88 species belonging to the genus *Simulium* s. l. occur in the Oriental Region, 212 species assigned in all six genera are distributed in the Palearctic Region, and 23 species are common to both regions. The difference in the simuliid fauna between these two regions is remarkable at the genus level.

The species of black flies in the Palearctic Region is out of the scope of this book, thus an overview and keys to identify the species are provided only for the species occurring in southern parts of China (Oriental Region), although the species list and keys to identify six genera and 11 subgenera in the genus *Simulium* s. l., are nationwide, covering the entire areas of China, for reference. The characteristics used in these two keys are those of the species occurring in China and are not necessarily shared with other species of the same genera and subgenera in other countries.

Overview

In Yunnan Province, Xue (1987) reported 15 species including five newly recorded species from China: *Simulium* (*Montisimulium*) *kirgisorum* (at that time, under the genus *Eusimulium*), *S.* (*Nevermannia*) *mie*, *S.* (*N.*) *silvestre*, *S.* (*N.*) *angustitarse* (at that time, these three species under the genus *Eusimulium*), *S.* (*Simulium*) *chiangmaiense* and *S.* (*S.*) *suzukii*. Xue (1991) described a new species, *S.* (*N.*)

sp. [described as *Titanopteryx kirgisorum*, a homonym of *S.* (*M.*) *kirgisorum*]. Xue (1992) described a new species, *S.* (*S.*) *tanae*. Xue (1993a) described a new species, *S.* (*S.*) *nujiangense* and recorded *S.* (*S.*) *chamlongi*. Xue (1993b) described two new species: *S.* (*N.*) *cangshanense* and *S.* (*N.*) *cheni* (at that time, both species under the genus *Eusimulium*). Xue (1993c) described a new species, *S.* (*M.*) sp. (at that time, as *Titanopteryx tanae*, junior homonym of *S.* (*S.*) *tanae*). Chen and Zhang (2003) described a new species, *S.* (*Gomphostilbia*) *bannaense* and recorded *S.* (*S.*) *thailandicum*. Chen (2003) described a new species, *S.* (*S.*) *bahense*. Chen and Zhang (2004) described a new species, *S.* (*G.*) *yunnanense*.

In Guizhou Province, Chen and Zhang (1997a) described a new species, *S.* (*S.*) *caohaiense*. Chen and Zhang (1997b) described two new species, *S.* (*N.*) *leigongshanense* (at that time, under the subgenus *Eusimulium*) and *S.* (*E.*) *weiniengense*. Chen and Zhang (1998) described two new species: *S.* (*Wilhelmia*) *xingyiense* and *S.* (*S.*) *rotifilis*. Chen and Wen (1999) described a new species, *S.* (*S.*) *ephippioidum*. Zhang et al. (1999) recorded 14 species: *S.* (*G.*) *metatarsale*, *S.* (*N.*) *mie*, *S.* (*S.*) *ornatum*, *S.* (*S.*) *hirtipannus*, *S.* (*S.*) *oitanum* (at that time, as *S.* (*S.*) *aokii*), *S.* (*S.*) *quinquestriatum*, *S.* (*S.*) *rufibasis*, *S.* (*E.*) sp. D, *S.* (*G.*) sp. 'A', *S.* (*N.*) sp. C, *S.* (*S.*) sp. E, *S.* (*S.*) sp. F, *S.* (*S.*) sp. G, and *S.* (*S.*) sp. H. Chen (2000) described two new species: *S.* (*S.*) *tenuatum* and *S.* (*S.*) *ramulosum*. Chen et al. (2000) described two new species: *S.* (*G.*) *fanjingshanense* and *S.* (*G.*) *mengi*. Zhang and Chen (2000) described a new species, *S.* (*S.*) *wulingense*. Wen and Chen (2000) described a new species, *S.* (*G.*) *miaolingense*. Zhang and Chen (2001) described a new species, *S.* (*S.*) *uncum*. Chen and Chen (2001) described a new species, *S.* (*S.*) *qianense*. Chen (2001b) described two new species: *S.* (*N.*) *qingshuiense* and *S.* (*N.*) *qiaolaoense*. Chen and Zhang (2002) described a new species, *S.* (*S.*) *prominentum*. Chen et al. (2003a) listed a total of 30 species of black flies from Guizhou Province, including one new species, *S.* (*G.*) *guizhouense*, and two new records: *S.* (*M.*) *nemorivagum* and *S.* (*S.*) *nitidithorax*. Zeng et al. (2006) described the female of *S.* (*S.*) *rotifilis*. Chen et al. (2012) described three new species: *S.* (*M.*) *separatum*, *S.* (*S.*) *neoacontum*, and *S.* (*S.*) *zunyiense*. Chen (2016), in his monograph of Chinese black flies, described a new species, *S.* (*Boophthora*) *guiyangense*.

In Hainan Province, Long et al. (1994) described two new species: *S.* (*G.*) *hainanense* and *S.* (*G.*) *jianfengense*. Chen and Chen (2000) described a new species, *S.* (*G.*) *brivetruncum*. Yang and Chen (2001) described a new species, *S.* (*S.*) *ledongense*. Chen (2001a) described a new species, *S.* (*G.*) *antlerum* (at that time, under the subgenus *Morops*). Chen and Zhang (2001) described three new species: *S.* (*G.*) *curvastylum*, *S.* (*G.*) *nigrofemoratum*, and *S.* (*G.*) *tumum*. Chen et al. (2003b) described two new species: *S.* (*N.*) *qiongzhouense* and *S.* (*S.*) *tianchi*. Chen (2003) described two new species: *S.* (*S.*) *menglaense* and *S.* (*S.*) *wuzhishanense*. Zhang et al. (2003) described two new species: *S.* (*S.*) *waterfallum* and *S.* (*S.*) *parawaterfallum*. Zhan and Chen (2003) described a new species, *S.* (*G.*) *longitruncum* [junior synonym of *S.* (*G.*) *longitruncum* Takaoka and Davies 1995, and no substitute name is available (Adler, 2022)]. Guo et al. (2011) described a new species, *S.* (*G.*) *epauletum*. Cai et al. (2016) described a new species, *S.* (*S.*) *sanyaense*. Chen (2016) included *S.* (*G.*) *pattoni* in his monograph on Chinese black flies, based on a reference, of which the author, title, and journal were not shown, copied the illustrations of the female and male genitalia, hind legs, claw, pupal gill, cocoon, and larval head capsule, and gave a brief description of the female, male, pupa and larva. In Fig. 12–40 by Chen (2016), an illustration of the claw with a basal tooth little less than the half of the claw was shown. It was not correct because the female of *S. pattoni* from India and Sri Lanka is characterized by the claw with a small claw tooth (Davies and Györkös, 1987a).

In Guanxi Province, Chen et al. (2007b) described a new species, *S.* (*S.*) *damingense*. Chen et al. (2007c) described two new species: *S.* (*G.*) *yuanbaoshanense* and *S.* (*G.*) *longshengense*. Sun (2009) described two new species: *S. guangxiense* and *S. penis*.

In Jiangxi Province, Kang et al. (2007) described a new species, *S.* (*G.*) *jiulianshanense*. Chen et al. (2007a) described a new species, *S.* (*S.*) *lushanense*.

In Hunan Province, Chen et al. (2004) described three new species: *S.* (*N.*) *chenzhouense*, *S.* (*N.*) *zhangjiajiense*, and *S.* (*N.*) *wangxianense*. Zhang and Chen (2004) described two new species: *S. hunanense and S. jinbianense*. Sun (2009) described three new species: *S. heteropara, S. huguangense*, and *S. xiangxiense*.

In Fujian Province, Zhang and Wang (1991b) described a new species, *S.* (*S.*) *fuzhouense*. Zhang and Wang (1991a) reported 12 species including one new species and one new subspecies: *S.* (*S.*) *multifurcatum*, and *S.* (*S.*) *sakishimaense fujianense*, and two new records: *S.* (*S.*) *nitidithorax* and *S.* (*S.*) *hirtipannus*. Zhang et al. (2009) reported 15 species from Wuyi Mountain, Fujian Province including a new species, *S.* (*G.*) *wuyishanense* and a new record, *S.* (*G.*) *montiblense*.

In Hong Kong, Takaoka et al. (1995) described two new species: *S.* (*G.*) *dudgeoni* and *S.* (*S.*) *taipokauense*, and newly recorded six species: *S.* (*G.*) *asakoae*, *S.* (*N.*) *aureohirtum*, *S.* (*S.*) *sakishimaense*, *S.* (*S.*) *nodosum*, *S.* (*S.*) *quinquestriatum*, and *S.* (*S.*) *suzukii*. An et al. (2007) described the pupa and larva of *S.* (*G.*) *dudgeoni*. The pupal gill of *S.* (*G.*) *dudgeoni* illustrated by An et al. (2007) is similar in the arrangement to *S.* (*G.*) *hainanense* from Hainan.

The fauna of black flies in South China is represented by 111 species including 11 unnamed species, of which eight species (*S.* sp. 'A'–H) reported from Guizhou Province by Zhang et al. (1999) are dubious for their species status for lack of information on their morphological characteristics, although they are placed to four subgenera. *Simulium* (*G.*) *montiblense* and *S.* (*G.*) *pattoni* originally described from the Philippines and India, respectively, should be reconfirmed for their distribution in China.

List of 323 Species of black flies in China

All 323 species of black flies including 18 unnamed species recorded from China are cited from the world list of black flies by Adler (2022) with slight modification.

Tribe Prosimuliini
Genus *Helodon*
Subgenus *Helodon*
*H. alpestris**, *H. lochmocola**, *H. neomulticaulis**, and *H. sinensis**
Genus *Prosimulium*
P. hirtipes species-group
*P. hirtipes**, *P. irritans**, *P. liaoningense**, and *P. nevexalatamus**
Genus *Twinnia*
*T. changbaiensis**
Tribe Simuliini
Genus *Metacnephia*
*M. bilineata**, *M. brevicollare**, *M. edwardsiana**, *M. kirjanovae**, *M. lui**, and *M. polyfilis**
Genus *Simulium* s. l.
Subgenus *Boophthora*
*S. erythrocephalum**, *S. guiyangense*, and *S. quattuorfile**
Subgenus *Boreosimulium*
S. annulus species-group
*S. konoi**
Subgenus *Byssodon*
S. *meridionale* species-group
*S. maculatum**
Subgenus *Eusimulium*
*S. angustipes**, *S. arenicola**, *S. armeniacum**, *S. aureum**, *S. beiwanense**, *S. kuandianense**, *S. lerbiferum**, *S. liupanshanense**, *S. nigrostriatum**, *S. reginae**, *S. satsumense**, *S. weiningense*, *S.* sp. [*S. taishanense* Xue and An, 2001 (junior homonym of *S. taishanense* Xue and An, 2000, but no available substitute name]*, *S.* sp. [*S. tenuistylum* Yan, Fan and Chen, 2008 (junior homonym of *S. tenuistylum* Datta, 1973, but no available substitute name)]*, and *S.* sp. D (Zhang et al., 1999)
Subgenus *Gomphostilbia*
S. asakoae species-group
S. asakoae, *S. epauletum*, *S. fanjingshanense*, *S. inthanonense*, *S. jianfengense*, *S. kangi**, *S. laoshanstum**, *S. mengi*, *S. miaolingense*, and *S. yunnanense*
S. batoense species-group
S. bannaense, *S. brivetruncum*, *S. chongqingense**, *S. hainanense*, *S. jiulianshanense*, and *S. pattoni*
S. ceylonicum species-group
S. hunanense, *S. nigrofemoralum*, and *S. wuyishanense*
S. darjeelingense species-group
S. curvastylum, *S. guizhouense***, and *S. zayuense**
S. epistum species-group
S. montiblense
S. gombakense species-group
S. antlerum and *S. dudgeoni*

S. varicorne species-group
*S. shogakii** and *S. synanceium**
Gomphostilbia species unplaced to species-group
*S. altayense**, *S. guangxiense*, *S. heteropara*, *S. huguangense*, *S. jinbianense*, *S. longshengense*, *S. meadowi**, *S. metatarsale*, *S. penis*, *S. pingxiangense*, *S. tashikulganense**, *S. tumum*, *S. xiangxiense*, *S. xizangense**, *S. yuanbaoshanense*, *S.* sp. [*S. longitruncum* Zhang and Chen, 2003 (junior homonym of *S. longitruncum* Takaoka and Davies, 1995 but no available substitute name)], and *S.* sp. "A" (Zhang et al., 1999)
Subgenus *Hellichiella*
S. congareenarum species-group
*S. meigeni**
Subgenus *Montisimulium*
*S. alizadei**, *S. chuanbeiense**, *S. concavustylum**, *S. dasguptai**, *S. euryplatamus**, *S. ghoomense***, *S. hailuogouense**, *S. heishuiense**, *S. jiajinshanense**, *S. jisigouense**, *S. kirgisorum***, *S. linghziense**, *S. luliangense**, *S. moxiense**, *S. nemorivagum**, *S. polyprominulum**, *S. schizolomum**, *S. schizostylum**, S. *separatum*, *S. taishanense**, *S. tibetense**, *S. weisiense*, *S. wulaofengense**, *S. xinzhouense**, *S. yuntaiense**, *S.* sp. [*Titanopteryx tanae* Xue 1993, (secondary homonym of *S.* (*S.*) *tanae*, Xue, 1992, but no available substitute name], and *S.* sp. B (Zhang et al., 1999).
Subgenus *Nevermannia*
S. feuerborni species-group
S. chitoense, *S. leigongshanense*, *S. mie***, *S. praelargum***, *S. qiongzhouense*, *S. shandongense**, *S. wangxianense*, and *S.* sp. [*Titanopteryx kirgisorum* Xue, 1991 (junior homonym of *S.* (*Montisimulium*) *kirgisorum* Rubtsov, 1956, but no available substitute name].
S. ruficorne species-group
S. angustitarse, *S. aureohirtum***, *S. lundstromi**, *S. ningxiaum**, and *S. subgriseum**
S. vernum species-group
*S. angustatum**, *S. bicorne**, *S. cangshanense*, *S. cheni*, *S. chenzhouense*, *S. dissimilum**, *S. fluviatile**, *S. gejgelense**, *S. gracile**, *S. henanense**, *S. jilinense**, *S. latifile**, S. *longipile**, *S. longtanstum**, *S. ludingense**, *S. medianum**, *S. novigracile**, *S. purii**, *S. qiaolaoense*, *S. qinlingense**, *S. qingshuiense*, *S. raohense**, *S. silvestre***, *S. subcostatum**, *S. uchidai**, *S. vernum**, *S. wulindongense**, *S. xinbinense**, *S. xueae**, *S. zhangjiajiense*, and *S.* sp. [*S. serratum* Chen and Zhang, 2016 (junior homonym of *S.* (*G.*) *serratum* Takaoka, 2003, but no available substitute name)]*.
Nevermannia species unplaced to species-group
S. sp. "C" (Zhang et al., 1999)
Subgenus *Schoenbaueria*
*S. flavoantennatum**, *S. polyhookum**, *S. nigrum**, and *S.* sp. [*S. tibiale* Tan and Chow, 1976 (secondary homonym of *S. tibiale* Macquart, 1834, but no available substitute name)]*.

Subgenus *Simulium* s. str.
S. argentipes species-group
*S. bidentatum*** and *S. tanae*
S. bezzii species-group
*S. alajense**, *S. atruncum**, *S. coarctatum**, *S. desertorum**, *S. helanshanense**, *S. kozlovi**, *S. longgengen**, *S. rhomboideum**, *S. tachengense**, *S. wutaishanense**, and *S. xiaodaoense**
S. christophersi species-group
*S. acontum**, *S. christophersi**, *S. howletti**, and *S. lichuanense**
S. chungi species-group
S. taipokauense
S. crocinum species-group
*S. ephippioidum***, *S. fuzhouense*, *S. hongpingense**, *S. ledongense*, *S. parawaterfallum*, *S. qini**, *S. ramulosum*, *S. shangchuanense*, *S. tenuatum*, *S. tumidilfilum**, *S. ufengense**, *S. waterfallum*, and *S. wulingense*
S. griseifrons species-group
*S. gravelyi**, *S. japonicum***, *S. kawamurae**, *S. mediaxisus**, and *S. spoonatum*
S. grossifilum species-group
*S. hengshanense***, *S. huangshanense**, and *S. lushanense*
S. indicum species-group
*S. indicum**
S. malyschevi species-group
*S. cholodkovskii**, *S. damingense*, *S. decimatum**, *S. hirtipannus***, *S. jacuticum***, *S. liaoningense**, *S. malyschevi**, *S. nacojapi**, *S. nanyangense**, *S. rufipes**, *S. spiculum**, *S. subvariegatum**, and *S. yui**
S. multistriatum species-group
*S. barraudi**, *S. bifengxiaense**, *S. dentatum**, *S. digitatum**, *S. rotifilis*, *S. sakishimaense***, *S. uncum*, *S. xiaolongtanense**, and *S.* sp. [*S. chongqingense* Zu and Wang, 1995 (junior homonym of *S. chongqingense* Zhu and Wang, 1995, but no available substitute name)]*.
S. nobile species-group
S. nodosum
S. noelleri species-group
*S. nikkoense**, *S. noelleri**, and *S. palustre**
S. ornatum species-group
*S. emeinesis**, *S. ferganicum**, *S. iwatense**, and *S. ornatum*
S. pavlovskii species-group
S. jieyangense, *S. omorii**, and *S. pavlovskii**
S. reptans species-group
*S. gansuense**, *S. remotum**, *S. reptans**, *S. tanetchowi**, and *S. tarnogradskii**
S. slossonae species-group
*S. transiens**
S. striatum species-group
S. bahense, *S. chiangmaiense*, *S. menglaense*, *S. nakhonense*, *S. nujiangense*, *S. quinquestriatum***, *S. saccatum**, *S. thailandicum*, and *S. wuzhishanense*

S. tuberosum species-group
*S. neorufibasis**, *S. nigrifacies**, *S. nitidithorax***, *S. prominentum*, *S. rufibasis***, *S. rugosum**, *S. sanyaense*, *S. splendidum**, *S. suzukii***, *S. tianchi, S. tumulosum**, *S. vulgare**, and *S.* sp. [*S. truncatum* Gao, Zhang, An, Zhang, Zhang, Dong and Zhao, 2008 (junior homonym of *S. truncatum* Lundstrom, 1911, but no available substitute name)]*.
S. variegatum species-group
S. caohaiense, *S. chamlongi***, *S. cognatum**, S. *foliatum**, *S. hanbini**, *S. himalayense**, *S. immortalis**, *S. jingfui**, *S. liubaense**, *S. oitanum***, *S. pelius**, *S. qianense***, *S. shanxiense**, and *S. taiwanicum***
S. venustum species-group
*S. aemulum**, *S. arakawae**, *S. curvitarse**, *S. enhense**, *S. longipalpe**, *S. morsitans**, and *S. promorsitans**
S. yadongense species-group
*S. lonchatum**, *S. pallidofemur**, *S, yadongense**, and *S. zunyiense*
Simulium s. str. species unplaced to any species-group
S. multifurcatum, *S. neoacontum*, *S. peliastrias**, *S. pulanotum**, *S. rubroflavifemur**, *S. shennongjiaense***, *S. triangustum**, *S. truncrosum**, *S. xinbinen**, *S.* sp. E, *S.* sp. F, *S.* sp. G, and *S.* sp. H (all by Zhang et al., 1999)
Subgenus *Trichodagmia*
S. albellum species-group
*S. biseriatum**
Subgenus *Wilhelmia*
S. equinum species-group
*S. angustifurca**, *S. dunhuangense**, *S. equinum**, *S. eurybrachium**, *S. germuense**, *S. lama**, *S. lineatum**, *S. pekingense**, *S. pinnatum**, *S. pseudequinum**, *S. qinghaiense**, *S. qingxilingense**, *S. takahasii***, S. *tongbaishanense**, *S. turgaicum**, *S. veltistshevi**, *S. wushiense**, *S. xingyiense, S. zhangyense**, and *S.* sp. [*S. tachengense* Mahe, An and Yan, 1997 (junior homonym of *S. tachengense* An and Mahe, 1994, but no available substitute name)]*
Simulium s. l. species unplaced to subgenus
*S. cherraense**, *S. dentastylum**, and *S. kariyai**
Genus *Sulcicnephia*
*Su. brevineckoi**, *Su. favipes**, *Su. jeholensis**, *Su. jingpengensis**, *Su. ovtshinnikovi**, *Su. undecimata**, and *Su. vigintistriatum**

*These 212 species are recorded from Palearctic areas of China, whereas 88 species with no asterisk occur in South China (Oriental Region).

**These 23 species are common to both the Palearctic and Oriental areas of China.

Keys to Identify Six Genera of the Family Simuliidae in China

Females

1. Costal vein without spines. Katepisternum shorter than its depth 2
   Costal vein with spines. Katerpisternum longer than its depth 4
2. Ovipositor valve elongate, tapered posteriorly ....*Prosimulium*
   Ovipositor short, broadly rounded, or truncated posteriorly 3
3. Antenna with seven flagellomeres .... *Twinnia*
   Antenna with nine flagellomeres ....*Helodon*
4. Pedisulcus distinct ....*Simulium* s. l.
   Pedisulcus shallow, wrinkled, or absent 5
5. Pleural membrane haired ....*Metacnephia*
   Pleural membrane bare 6
6. Pedisulcus shallow, wrinkled .... *Sulcicnephia*
   Pedisulcus absent .... *Simulium* s. l. (subgenus *Hellichiella*)

Males

1. Costal vein without spines. Katepisternum shorter than its depth 2
   Costal vein with spines. Katerpisternum longer than its depth 4
2. Antenna with seven flagellomeres .... *Twinnia*
   Antenna with nine flagellomeres 3
3. Ventral plate in lateral view flattened, with lip absent, short, or slender ....*Helodon*
   Ventral plate in lateral view not markedly flattened, typically with prominent tip ....*Prosimulium*
4. Calcipala distinct. Pedisulcus distinct ....*Simulium* s. l.
   Calcipala minute or absent. Pedisulcus shallow, wrinkled 5
5. Pleural membrane haired ....*Metacnephia*
   Pleural membrane bare 6
6. Pedisulcus shallow, wrinkled .... *Sulcicnephia*
   Pedisulcus absent .... *Simulium* s. l. (subgenus *Hellichiella*)

Pupae

1. Cocoon rudimentary or shapeless, sack-like, covering various portions of abdomen and thorax 2
   Cocoon shaped like wall-pocket, shoe or boot, with definitely formed, rigid wall (sometimes coarsely woven), typically covering abdomen and thorax 4
2. Abdominal tergites devoid of spine-combs .... *Twinnia*
   Abdominal tergites 6–9 each with transverse row of spine-combs 3
3. Gill with 16 filaments ....*Prosimulium*
   Gill with 33–130 filaments ....*Helodon*
4. Gill with 40–120 filaments ....*Metacnephia*

Gill with 3–24 filaments 5
5. Gill with 10–20 filaments. Cocoon boot-shaped .... *Sulcicnephia*
Gill with 3–16 or 24 filaments. Cocoon wall-pocket-shaped or shoe-shaped with perforations anteriorly when gill with 10–16 filaments .... *Simulium* s. l.

Mature larvae

1. Median tooth of anterior teeth of hypostoma with small tooth on each side 2
Median tooth of anterior teeth of hypostoma without small tooth on each side 4
2. Anal sclerite Y-shaped. Labral fan absent .... *Twinnia*
Anal sclerite X-shaped. Labral fan present 3
3. Head spots on cephalic apotome distinctively positive. Median tooth of anterior teeth of hypostoma as long as or shorter than corner teeth .... *Prosimulium*
Head spots on cephalic apotome obscure or faintly positive. Median tooth of anterior teeth of hypostoma longer than corner teeth .... *Helodon*
4. Anterior margin of hypostoma produced medially, with seven or nine teeth....... .... *Sulcicnephia*
Anterior margin of hypostoma nearly straight or undulate, with nine teeth 5
5. Pharate pupal gill with 40–120 filaments .... *Metacnephia*
Pharate pupal gill with 3–14 and 24 filaments .... *Simulium* s. l.

Keys to Identify 12 Subgenera of the Genus **Simulium** *s. l. in China*

Females

1. Pedisulcus undeveloped .... *Hellichiella*
Pedisulcus developed 2
2. Katepisternum haired .... *Gomphostilbia*
Katepisternum bare 3
3. Pleural membrane haired 4
Pleural membrane bare 6
4. Claw with small subbasal tooth .... *Simulium* s. str. (*S. ornatum* species-group)
Claw without tooth 5
5. Ovipositor valve with slender curled acuminate process posteromedially (Fig. 96D)... *Wilhelmia*
Ovipositor valve without slender curled process posteromedially .... *Trichodagmia*
6. Postnotum with hairs .... *Eusimulium*
Postnotum without hairs 7

7. Claw with large basal tooth 8
   Claw without tooth or with small or medium-sized tooth 13
8. Basal section of radial vein bare ....
   .... *Byssodon* and *Simulium* s. str. (*S. slossonae* species-group)
   Basal section of radial vein haired 9
9. Paraproct much shorter than cercus. Spermatheca oblong.................................
   ....*Nevermannia* (*S. ruficorne* species-group)
   Paraproct as long as or longer than cercus. Spermatheca globular or ovoid ..10
10. Genital fork without projection directed anteriorly from each lateral plate........
    .... *Schoenbaueria* (*S. nigrum*)
    Genital fork with projection directed anteriorly from each lateral plate 11
11. Genital fork with small projection directed anteriorly from each lateral plate ...
    .... *Nevermannia* (*S. vernum* species-group)
    Genital fork with distinct long projection directed anteriorly from each lateral plate 12
12. Scutum with longitudinal vittae .... *Nevermannia* (*S. feuerborni* species-group)
    Scutum without longitudinal vittae .... *Boreosimulium*
13. Basal section of radial vein bare ....
    .... *Simulium* s. str. (except several species) and *Boophthora*
    Basal section of radial vein haired 14
14. Claw without tooth.......... *Schoenbaueria* and *Simulium* s. str. (several species)
    Claw with small or medium-sized basal tooth 15
15. Claw with medium-sized tooth (Fig. 42C) .... *Montisimulium*
    Claw with small tooth .... *Simulium* s. str. (*S. indicum* species-group)

Males

1. Pedisulcus undeveloped ....*Hellichiella*
   Pedisulcus developed 2
2. Katepisternum haired .... *Gomphostilbia*
   Katepisternum bare 3
3. Style with many spines subapically (Fig. 7I) .... *Boophthora*
   Style with or without spine apically or subapically 4
4. Style longer than coxite 5
   Style shorter than coxite 6
5. Pleural membrane haired ....
   ....*Simulium* s. str. (*S. ornatum* species-group) and *Trichodagmia*
   Pleural membrane bare ....*Simulium* s. str.
6. Pleural membrane haired .... *Wilhelmia*
   Pleural membrane bare 7
7. Basal section of radial vein bare .... *Byssodon*
   Basal section of radial vein haired 8
8. Ventral plate with median keel (Figs. 10I and 45I) 9
   Ventral plate without median keel 10

9. Ventral plate narrow, with divergent long arms (Fig. 10I) .... *Eusimulium*
Ventral plate broad, with short arms nearly parallel-sided (Fig. 45I)..................
....*Nevermannia* (*S. ruficorne* species-group)
10. Paramere with one long hook (Fig. 47Q)...
.... *Nevermannia* (*S. vernum* species-group)
Paramere with two or more hooks (Figs. 42O and 43P) 11
11. Hind basitarsus slender, parallel-sided .... *Shoenbaueria*
Hind basitarsus slightly to greatly dilated, spindle-shaped.............................12
12. Hind basitarsus slightly dilated, 0.8 times as wide as hind tibia. Style in ventrolateral view gradually tapered from base to apex .... *Boreosimulium*
Hind basitarsus greatly dilated, nearly as wide as hind tibia. Style in ventrolateral view broad, nearly parallel-sided from base to at least middle....................
.........................*Montisimulium* and *Nevermannia* (*S. feuerborni* species-group)

Pupae

1. Gill filaments with numerous minute black spots (Fig. 42T) ..... *Montisimulium*
Gill filaments without minute black spots 2
2. Abdominal segments with spine-combs dorsally (at least on segment 8) (Fig. 3C)
....3
Abdominal segments 5–9 without spine-combs dorsally (similar to Fig. 95E) .
....12
3. Last abdominal segment with grapnel-like hooklets on each side......................
.... *Gomphostilbia*
Last abdominal segment without grapnel-like hooklets on each side 4
4. Last abdominal segment without terminal hooks..
....*Boophthora* and *Simulium* s. str.
(e.g., *S. malyschevi* species-group and *S. nobile* species-group)
Last abdominal segment with terminal hooks 5
5. Gill with four filaments ....*Boreosimulium*, *Eusimulium*, and *Nevermannia*
(*S. ruficorne* species-group and *S. vernum* species-group),
Schoenbaueria (*S. polyhookum*) and
Simulium s. str. (*S. slossonae* species-group)
Gill with six, eight, nine, 10, 14, or 24 filaments 6
6. Gill with six filaments 7
Gill with eight, nine, 10, 14, or 24 filaments 9
7. Gill filaments as long as or longer than pupal body ....
.... *Nevermannia* (*S. feuerborni* species-group)
Gill filaments much shorter than pupal body (e.g., Fig. 60D–K) 8
8. Cocoon wall-pocket-shaped, without anterodorsal projection (Fig. 60M), or shoe-shaped, with or without anterolateral open space(s) on each side (Fig. 60N–Q), or boot-shaped, roughly woven, entirely covering pupal body and gill filaments (Fig. 60S) ......*Simulium* s. str. (*S. griseifrons* species-group,
S. reptans species-group, *S. tuberosum* species-group,

S. variegatum species-group, and *S. venustum* species-group) and *Trichodagmia*
Cocoon wall-pocket-shaped, with short anterodorsal projection .... *Nevermannia* (*S. aureohirtum* of *S. ruficorne* species-group)
9. Gill with eight filaments .... *Schoenbaueria* (*S. nigrum*) and *Simulium* s. str. (*S. argentipes* species-group, *S. bezzi* species-group, *S. multistriatum* species-group, *S. noelleri* species-group, *S. ornatum* species-group, *S. reptans* species-group, and *S. striatum* species-group)
Gill with nine, 10, 14, or 24 filaments 10
10. Gill with nine or 10 filaments ....*Simulium* s. str. (*S. malyschevi* species-group and *S. striatum* species-group)
Gill with 14 or 24 filaments 11
11. Gill with 14 filaments .... *Hellichiella*
Gill with 24 filaments .... *Byssodon*
12. Gill with six filaments .... *Simulium* s. str. (*S. indicum* species-group)
Gill with eight filaments .... *Wilhelmia*

Mature larvae*

1. Ventral papillae large or medium-sized 2
Ventral papillae small or absent 11
2. Main tooth of mandibular serration at acute angle to mandible apically (Fig. 32T) .... *Gomphostilbia*
Main tooth of mandibular serration at right or obtuse angle to mandible apically (Fig. 42V) 3
3. Postgenal cleft absent or vestigial (Fig. 42X, Y) 4
Postgenal cleft short, medium-long, or long (Figs. 40K and 44O) 5
4. Pharate pupal gill with eight, 10, 12, or 14 filaments .... *Montisimulium*
Pharate pupal gill with four filaments .... *Nevermannia* (*S. angustitarse* of *S. ruficorne* species-group)
5. Postgenal cleft long, reaching posterior margin of hypostoma.......................... .... *Simulium* s. str. (*S. transiens* of *S. slossonae* species-group)
Postgenal cleft short or medium-long 6
6. Postgenal cleft longer than postgenal bridge 7
Postgenal cleft as long as or shorter than postgenal bridge 8
7. Ventral papillae widely separated from each other .... *Boophthora*
Ventral papillae moderately separated from each other............... *Schoenbaueria*
8. Pharate pupal gill with four filaments .... *Eusimulium* and *Nevermannia* (*S. vernum* species-group and *S. ruficorne* species-group)
Pharate pupal gill with six, 10, 11, 12, or 14 filaments 9
9. Pharate pupal gill with 10, 11, 12, or 14 filaments .... *Montisimulium*
Pharate pupal gill with six filaments 10
10. Mandibular serration composed of two large teeth (similar to Fig. 46I) .... *Nevermannia* (*S. aureohirtum* of *S. ruficorne* species-group)

Mandibular serration composed of one large tooth and one small tooth (Fig. 44M) .... *Nevermannia* (*S. feuerborni* species-group)

11. Hypostoma with two or three rows of hypostomal bristles on each side .... *Trichodagmia*
    Hypostoma with one row of hypostomal bristles on each side 12
12. Mandible with intermediate teeth nearly as long as each corner tooth. Pharate pupal gill with eight short tubular filaments .... *Wilhelmia*
    Mandible with intermediate teeth shorter than each corner tooth. Pharate pupal gill with three inflated tubular filaments, four, six, eight, nine, 10, 16, or 24 slender filaments 13
13. Pharate pupal gill with 24 filaments .... *Byssodon*
    Pharate pupal gill with three inflated tubular filaments, four, six, eight, nine, 10, or 16 slender filaments 14
14. Pharate pupal gill with four filaments .... *Boreosimulium*
    Pharate pupal gill with three inflated tubular filaments, six, eight, nine, 10, or 16 slender filaments .... *Simulium* s. str.

*The subgenus *Hellichiella* is not included because its larva is unknown in China.

*Keys to Identify 102 Species of Black Flies in South China**

Females**

1. Pleural membrane haired 2
   Pleural membrane bare 4
2. Claw with small subbasal tooth .... *S.* (*S.*) *ornatum*
   Claw simple, without tooth 3
3. Lateral plates of genital fork widened apically (Fig. 96E) ......*S.* (*W.*) *takahasii*
   Lateral plates of genital fork narrowed apically ....*S.* (*W.*) *xingyiense*
4. Postnotum haired ....*S.* (*E.*) *weiningense*
   Postnotum bare 5
5. Katepisternum haired 6
   Katepisternum bare 21
6. Claw with small subbasal tooth 7
   Claw with large basal tooth 8
7. Head six times as wide as greatest width of frons .... *S.* (*G.*) *pattoni*
   Head four times as wide as greatest width of frons ....*S.* (*G.*) *bannanense*
8. Sensory vesicle 0.6–0.7 times as long as third palpal segment 9
   Sensory vesicle 0.25–0.40 times as long as third palpal segment 12
9. Mandible lacking outer teeth 10
   Mandible with outer teeth 11

10. Genital fork with projection directed forward on each arm .... *S.* (*G.*) *curvastylum*
    Genital fork without projection directed forward on each arm ....*S.* (*G.*) *brevitruncum*
11. Head 3.3 times as wide as greatest width of frons .... *S.* (*G.*) sp.
    Head 4.7–5.1 times as wide as greatest width of frons .....*S.* (*G.*) *inthanonense*
12. Mandible with outer teeth 13
    Mandible lacking outer teeth 20
13. Sensory vesicle 0.4 times as long as third papal segment ....*S.* (*G.*) *dudgeoni*
    Sensory vesicle 0.25–0.32 times as long as third palpal segment 14
14. Hind tibia almost darkened on posterior surface .... *S.* (*G.*) *jinbianense*
    Hind tibia whitish yellow on basal one-third to four-fifths 15
15. Genital fork with projection directed forward from each arm ....*S.* (*G.*) *miaolingense*
    Genital fork without projection directed forward from each arm 16
16. Hind tibia whitish yellow on basal one-third or two-fifths 17
    Hind tibia whitish yellow on basal three-fourths or four-fifths 19
17. Mid tibia whitish yellow on basal three-fifths ....*S.* (*G.*) *fanjingshanense*
    Mid tibia whitish yellow on basal one-third 18
18. Head 3.8 times as wide as greatest width of frons ....*S.* (*G.*) *hunanense*
    Head 5.4 times as wide as greatest width of frons ....*S.* (*G.*) *wuyishanense*
19. Head 5.9 times as wide as greatest width of frons ..... *S.* (*G.*) *yuanbaoshanense*
    Head 4.8 times as wide as greatest width of frons ....*S.* (*G.*) *asakoae* and *S.* (*G.*) *epauletum*
20. Sensory vesicle 0.4 times as long as third palpal segment .... *S.* (*G.*) *nigrofemoralum* and *S.* (*G.*) *pingxiangense*
    Sensory vesicle 0.25–0.33 times as long as third palpal segment...................... .... *S.* (*G.*) *huguangense*, *S.* (*G.*) *heteropara*, *S.* (*G.*) *jianfengense*, *S.* (*G.*) *guangxiense*, *S.* (*G.*) *mengi*, and *S.* (*G.*) *yunnanense*
21. Claw with tooth 22
    Claw simple, without tooth 29
22. Claw with large tooth 23
    Claw with small or medium-sized tooth 25
23. Sensory vesicle 0.4 times as long as third palpal segment ....*S.* (*N.*) *angustitarse* and *S.* (*N.*) *aureohirtum*
    Sensory vesicle 0.5–0.6 times as long as third palpal segment 24
24. Mandible with teeth on both inner and outer margins...................................... ....*S.* (*N.*) *cangshanense*, *S.* (*N.*) *cheni*, *S.* (*N.*) *chenzhouense*, *S.* (*N.*) *henanense*, *S.* (*N.*) *qingshuiense*, and *S.* (*N.*) *silvestre*
    Mandible with teeth on inner margin but no teeth on outer margin................... ....*S.* (*N.*) *chitoense*, *S.* (*N.*) *leigongshanense*, *S.* (*N.*) *mie*, *S.* (*N.*) *qiongzhouense*, *S.* (*N.*) *wangxianense*, and *S.* (*N.*) sp.

25. Basal section of radius haired ....
.....*S.* (*M.*) *nemorivagum*, *S.* (*M.*) *separatum*, *S.* (*M.*) *weisiense*, and *S.* (*M.*) sp.
Basal section of radius bare 26
26. Ovipositor valve flap-like, bluntly truncate posteriorly, forming oblique roof-like ridge parallel to its posterior margin (similar to Fig. 65D) ....
....*S.* (*S.*) *hirtipannus* and *S.* (*S.*) *jacuticum*
Ovipositor valve with inner margin concave 27
27. Ovipositor valve pointed medially (similar to Fig. 84C)..................................
......................................*S.* (*S.*) *caohaiense*, *S.* (*S.*) *chamlongi*, *S.* (*S.*) *oitanum*, *S.* (*S.*) *qianense*, *S.* (*S.*) *taiwanicum*, and *S.* (*S.*) *zunyiense*
Ovipositor valve rounded posteromedially (similar to Fig. 49C) 28
28. Ovipositor valve with inner margin slightly concave ....*S.* (*S.*) *jianfengense*
Ovipositor valve with inner margin deeply concave (similar to Fig. 49C) ....
.... *S.* (*S.*) *bidentatum* and *S.* (*S.*) *tanae*
29. Scutum with longitudinal vittae 30
Scutum without longitudinal vittae 35
30. Ovipositor valve with ventrally produced lobe along inner margin (similar to Fig. 76D, E) 31
Ovipositor valve without such lobe along inner margin 32
31. Basal section of radius bare ....*S.* (*S.*) *menglaense*
Basal section of radius haired ....*S.* (*S.*) *bahense*, *S.* (*S.*) *chiangmaiense*, *S.* (*S.*) *nakhonense*, *S.* (*S.*) *nujiangense*, *S.* (*S.*) *quinquestriatum*, and *S.* (*S.*) *wuzhishanense*
32. Ovipositor valve with long transparent projection directed dorsomedially (similar to Fig. 61B) .... *S.* (*S.*) *hengshanense*
Ovipositor valve without such projection 33
33. Ovipositor valve with inner margin transparent (similar to Fig. 70E)...............
....*S.* (*S.*) *digitatum*, *S.* (*S.*) *rotifilis*, *S.* (*S.*) *sakishimaense*, and *S.* (*S.*) *uncum*
Ovipositor valve triangular, with inner margin somewhat darkened 34
34. Hind femur yellow on basal one-fourth ....*S.* (*S.*) *japonicum*
Hind femur yellow on basal three-fourths .... *S.* (*S.*) *spoonatum*
35. Paraproct with several stout dark hairs apically (Fig. 74L) .... *S.* (*S.*) *nodosum*
Paraproct without such stout dark hairs apically 36
36. Ovipositor valve flap-like, bluntly truncate posteriorly, forming oblique roof-like ridge parallel to its posterior margin ....*S.* (*S.*) *damingense*
Ovipositor valve otherwise 37
37. Hind femur yellow at least on basal three-fourths 38
Hind femur darkened except base yellow 41
38. Basal section of radius haired ....*S.* (*S.*) *ephippioidum*, *S.* (*S.*) *fuzhouense*, *S.* (*S.*) *ledongense*, *S.* (*S.*) *shangchuanense*, and *S.* (*S.*) *tenuatum*
Basal section of radius bare 39
39. Cibarium with dorsomedian projection ....*S.* (*S.*) *shennongjiaense*
Cibarium without dorsomedian projection 40

40. Mandible without outer teeth .... *S.* (*S.*) *ramulosum*
    Mandible with outer teeth .... *S.* (*S.*) *parawaterfallum*, *S.* (*S.*) *ufengense*, *S.* (*S.*) *waterfallum*, and *S.* (*S.*) *wulingense*
41. Abdominal segment 7 with a pair of clustered stout hairs ventrally (similar to Fig. 82C) .... *S.* (*S.*) *rufibasis*
    Abdominal segment 7 without a pair of clustered stout hairs ventrally 42
42. Sensory vesicle long, 0.5–0.6 times as long as third palpal segment .... *S.* (*S.*) *sanyaense* and *S.* (*S.*) *suzukii*
    Sensory vesicle medium-long, 0.3 times as long as third palpal segment 43
43. Cibarium with many tiny processes .... *S.* (*S.*) *tianchi*
    Cibarium without tiny processes .... *S.* (*S.*) *prominentum*

Males***

1. Style longer than coxite 2
   Style shorter than coxite 36
2. Style with many spines subapically (similar to Fig. 7I) ...... *S.* (*B.*) *guiyangense*
   Style with or without one spine apically or subapically 3
3. Pleural membrane haired .... *S.* (*S.*) *ornatum*
   Pleural membrane bare 4
4. Ventral plate not serrated 5
   Ventral plate serrated 26
5. Style without basal protuberance 6
   Style with basal protuberance 7
6. Ventral plate with ventrally produced process .... *S.* (*S.*) *neoacontum*
   Ventral plate without ventrally produced process .... *S.* (*S.*) *damiangense*
7. Style with short protuberance. Ventral plate saddle-shaped (similar to Fig. 76P) 8
   Style with medium-long or long protuberance. Ventral plate otherwise 12
8. Upper-eye (large) facets in 11–14 horizontal rows 9
   Upper-eye (large) facets in 16 horizontal rows..11
9. Scutum with brassy short hairs .... *S.* (*S.*) *bahense*
   Scutum with yellow short hairs 10
10. Hind basitarsus six times as long as its greatest width .... *S.* (*S.*) *menglaense*
    Hind basitarsus four times as long as its greatest width .... *S.* (*S.*) *wuzhishanense*
11. Scutum with yellow short hairs. Hind basitarsus 3.6 times as long as its greatest width .... *S.* (*S.*) *quinquestriatum*
    Scutum with brassy short hairs. Hind basitarsus 5.4 times as long as its greatest width .... *S.* (*S.*) *chiangmaiense* and *S.* (*S.*) *nakhonense*

12. Style with basal protuberance having no spines on its anterior surface 13
Style with basal protuberance having spines on its anterior surface 14
13. Ventral plate with its body longer than width when viewed ventrally ....
....*S.* (*S.*) *digitatum* and *S.* (*S.*) *uncum*
Ventral plate with its body shorter than width when viewed ventrally ....
....*S.* (*S.*) *parawaterfallum*
14. Ventral plate with its body longer than width when viewed ventrally 15
Ventral plate with its body as long as or shorter than greatest width when viewed ventrally 17
15. Upper-eye (large) facets in 17 horizontal rows ....*S.* (*S.*) *japonicum*
Upper-eye (large) facets in 13 or 14 horizontal rows 16
16. Ventral plate with its body gradually narrowed from base to posterior apex when viewed ventrally ....*S.* (*S.*) *tenuatum*
Ventral plate with its body nearly parallel-sided from base to middle, then narrowed posteriorly when viewed ventrally .... *S.* (*S.*) *shennongjiaense*
17. Ventral plate with its body rounded posteriorly when viewed ventrally 18
Ventral plate with its body quadrate or rectangular when viewed ventrally ....
....21
18. Upper-eye (large) facets in 19 horizontal rows ....*S.* (*S.*) *taipokauense*
Upper-eye (large) faces in 10–16 horizontal rows 19
19. Upper-eye (large) facets in 16 horizontal rows ....*S.* (*S.*) *ephippioidum*
Upper-eye (large) facets in 10 or 11 horizontal rows 20
20. Upper-eye (large) facets in 12 vertical columns ....*S.* (*S.*) *lushanense*
Upper-eye (large) facets in 15 vertical columns ....*S.* (*S.*) *ledangense*
21. Ventral plate with arms much shorter than its body ....*S.* (*S.*) *hengchuanense*
Ventral plate with arms as long as or longer than its body 22
22. Ventral plate rectangular 23
Ventral plate quadrate 24
23. First antennal flagellomere 1.6 times as long as second one ....
....*S.* (*S.*) *fuzhouense*
First antennal flagellomere 1.2 times as long as second one ....
.... *S.* (*S.*) *shangchuanense*
24. Ventral plate with arms somewhat longer than its body length ....
.... *S.* (*S.*) *waterfallum*
Ventral plate with arms as long as its body length 25
25. Hind tibia yellow on basal one-third ....*S.* (*S.*) *ufengense*
Hind tibia yellow on basal four-fifths .... *S.* (*S.*) *spoonatum*
26. Style with basal protuberance 27
Style without basal protuberance 30
27. Style with long basal protuberance with pointed apex .....*S.* (*S.*) *sakishimaense*
Style with short rounded basal protuberance 28
28. Ventral plate with its body much longer than wide when viewed ventrally .....
....*S.* (*S.*) *zunyiense*

Ventral plate with its body as long as or little shorter than wide when viewed ventrally 29
29. Mid and hind femora yellow on basal three-fourths .... *S.* (*S.*) *prominentum*
Mid and hind femora almost black .... S. (*S.*) *tianchi*
30. Scutum white pruinose except inverted-V shaped non-pruinose median area (similar to Fig. 74S, T) .... *S.* (*S.*) *nodosum*
Scutum white pruinose on shoulders, along lateral margins and on prescutellar area 31
31. Hind basitarsus nearly parallel-sided. Style with apical half narrower than basal half when viewed ventrolaterally. Ventral plate with two vertical rows of teeth from base to near ventral apex when viewed posteriorly 32
Hind basitarsus spindle-shaped. Style wide, nearly parallel-sided from base to apex when viewed ventrolaterally. Ventral plate with two vertical rows of teeth from base to middle when viewed posteriorly 33
32. Ventral plate with body nearly parallel-sided when viewed ventrally .... *S.* (*S.*) *hirtipannus*
Ventral plate with body narrowed from base to posterior end .... *S.* (*S.*) *jacuticum* and *S.* (*S.*) *jieyangense*
33. Upper-eye (large) facets in 12 horizontal rows .... *S.* (*S.*) *qianense*
Upper-eye (large) facets in 16–22 horizontal rows 34
34. Upper-eye (large) facets in 16 horizontal rows .... *S.* (*S.*) *oitanum*
Upper-eye (large) facets in 18–22 horizontal rows 35
35. Upper-eye (large) facets in 22 horizontal rows .... *S.* (*S.*) *taiwanicum*
Upper-eye (large) facets in 18 or 19 horizontal rows .... *S.* (*S.*) *caohaiense* and *S.* (*S.*) *chamlongi*
36. Katepisternum bare 37
Katepisternum haired 50
37. Pleural membrane haired 38
Pleural membrane bare 39
38. Paramere with eight hooks .... *S.* (*W.*) *xingyiense*
Paramere with 22–24 hooks (Fig. 96O) .... *S.* (*W.*) *takahasii*
39. Hind basitarsus slender. Ventral plate with median keel 40
Hind basitarsus enlarged. Ventral plate without median keel 42
40. Ventral plate with its body narrow and arms long and divergent .... *S.* (*E.*) *weiningense*
Ventral plate with its body broad, and arms short and parallel-sided 41
41. Ventral plate with its body narrowed posteriorly .... *S.* (*N.*) *angustitarse*
Ventral plate with its body slightly widened from base to posterior end .... *S.* (*N.*) *aureohitum*
42. Paramere with one long hook 43
Paramere with three or more hooks 45
43. Ventral plate with its body nearly parallel-sided when viewed ventrally .... *S.* (*N.*) *chenzhouense*
Ventral plate with its body narrowed posteriorly 44

44. Ventral plate with its posterior margin nearly straight .... *S.* (*N.*) *qingshuiense*
Ventral plate with its posterior margin rounded ....
....................................*S.* (*N.*) *cangshanense*, *S.* (*N.*) *cheni*, *S.* (*N.*) *qiaolaoense*, *S.* (*N.*) *silvestre*, and *S.* (*N.*) *zhangjiajiense*
45. Paramere with three to five distinct hooks 46
Paramere with six or seven hooks 49
46. Paramere with three distinct hooks ....*S.* (*M.*) *kirgisorum*
Paramere with four or five distinct hooks 47
47. Paramere with four distinct hooks ....*S.* (*M.*) sp.
Paramere with five distinct hooks 48
48. Ventral plate with posterior margin nearly straight when viewed ventrally ......
.... *S.* (*N.*) sp.
Ventral plate with posterior margin markedly concave medially when viewed ventrally .... *S.* (*M.*) *weisiense*
49. Upper-eye (large) facets in 13 or 14 horizontal rows ....
.... *S.* (*M.*) *nemorivagum*, *S.* (*N.*) *mie*, and *S.* (*N.*) *wangxianense*
Upper-eye (large) facets in 16 horizontal rows ....
..............*S.* (*N.*) *chitoense*, *S.* (*N.*) *leigongshanense*, and *S.* (*N.*) *qiongzhouense*
50. Pleural membrane haired .... *S.* (*G.*) *antlerum*
Pleural membrane bare 51
51. Hind basitarsus enlarged, 3.0–4.8 times as long as its greatest width 52
Hind basitarsus slender, over 5.5 times as long as its greatest width 64
52. Style constricted at apical one-third when viewed ventrolaterally 53
Style gradually narrowed from base to apex, without constriction at apical one-third when viewed ventrolaterally 54
53. Upper-eye (large) facets in 11 horizontal rows. Hind basitarsus 3.3 times as long as its greatest width .... *S.* (*G.*) *curvastylum*
Upper-eye (large) facets in 13 horizontal rows. Hind basitarsus 4.0 times as long as its greatest width .... *S.* (*G.*) *guizhouense*
54. Upper-eye (large) facets in eight or nine horizontal rows 55
Upper-eye (large) facets in 12–18 horizontal rows 56
55. Ventral plate with lateral margins depressed subbasally when viewed ventrally.
....*S.* (*G.*) *miaolingense*
Ventral plate with lateral margins straight when viewed ventrally ....
.... *S.* (*G.*) *mengi*
56. Upper-eye (large) facets in 12–14 horizontal rows 57
Upper-eye (large) facets in 15–18 horizontal rows 61
57. Hind basitarsus 3.0–3.4 times as long as its greatest width ....
.... *S.* (*G.*) *asakoae*, *S.* (*G.*) *epauletum*, *S.* (*G.*) *metatarsale*, and *S.* (*G.*) *pingxiangense*
Hind basitarsus 4.0–4.5 times as long as its greatest width 58
58. Ventral plate with lateral margins depressed subbasally when viewed ventrally
.... *S.* (*G.*) *yunnanense*
Ventral plate with lateral margins nearly straight when viewed ventrally 59

59. Ventral plate with lateral margins nearly parallel-sided when viewed ventrally ....*S.* (*G.*) *hunanense*
Ventral plate with lateral margins somewhat convergent posteriorly when viewed ventrally 60
60. Hind basitarsus 4.5 times as long as its greatest width ....*S.* (*G.*) *nigrofemoralum*
Hind basitarsus 4.0 times as long as its greatest width .....*S.* (*G.*) *wuyishanense*
61. Upper-eye (large) facets in 17 or 18 horizontal rows .... *S.* (*G.*) *inthanonense*
Upper-eye (large) facets in 15 or 16 horizontal rows 62
62. Hind basitarsus 3.0 times as long as its greatest width ....*S.* (*G.*) *jianfengense*
Hind basitarsus 4.8 times as long as its greatest width 63
63. Ventral plate with lateral margins nearly parallel-sided when viewed ventrally ....*S.* (*G.*) *fanjingshanense*
Ventral plate with lateral margins divergent posteriorly when viewed ventrally .... *S.* (*G.*) *dudgeoni*
64. Style constricted at apical one-third when viewed ventrolaterally (similar to Fig. 20L) ....*S.* (*G.*) *bannanense*
Style gradually narrowed from base to apex, without constriction at apical one-third 65
65. Ventral plate with posterior margin rounded posteriorly ....*S.* (*G.*) *tumum*
Ventral plate with posterior margin nearly straight or depressed medially 66
66. Ventral plate with basal bulge directed outward on each side when viewed ventrally .... *S.* (*G.*) *yuanbaoshanense*
Ventral plate without basal bulge directed outward on each side when viewed ventrally 67
67. Ventral plate with lateral margins markedly rounded medially when viewed ventrally ....*S* (*G.*) *brivetruncum* and *S.* (*G.*) *heteropara*
Ventral plate with lateral margins not rounded medially when viewed ventrally 68
68. Ventral plate with lateral margins gradually narrowed from base to posterior end ....*S.* (*G.*) *hainanense*, *S.* (*G.*) *longshengense*, and *S.* (*G.*) *pattoni*
Ventral plate with lateral margins slightly narrowed from base to middle, then slightly widened posteriorly, or slightly or markedly narrowed medially, then nearly parallel-sided 69
69. Ventral plate with lateral margins slightly narrowed from base to middle, then slightly widened posteriorly ....*S.* (*G.*) *guangxiense*
Ventral plate with lateral margins slightly or markedly narrowed medially, then nearly parallel-sided .... *S.* (*G.*) *huguangense* and *S.* (*G.*) *xiangxiense*

Pupae****

1. Gill filaments with dark small particles 2
   Gill filaments without dark small particles 4
2. Gill with 14 filaments .... *S.* (*M.*) *weisiense*
   Gill with 12 filaments 3
3. Cocoon with anterodorsal projection ....*S.* (*M.*) sp.
   Cocoon without anterodorsal projection..
   ....*S.* (*M.*) *kirgisorum, S.* (*M.*) *nemorivagum*, and *S.* (*M.*) *separatum*
4. Gill with three short inflated tubular filaments (Fig. 75F) .... *S.* (*S.*) *nodosum*
   Gill with four or more filaments 5
5. Gill with four filaments 6
   Gill with five or more filaments 12
6. Gill with no or extremely short common basal stalk. Ventral pair of gill filaments sessile 7
   Gill with short common basal stalk. Ventral pair of gill filaments short-stalked 8
7. Cocoon simple, without anterodorsal projection ....*S.* (*E.*) *weiningense*
   Cocoon with short anterodorsal projection .... *S.* (*N.*) *angustitarse*
8. Cocoon with long anterodorsal projection 9
   Cocoon without anterodorsal projection or with anterodorsal bulge 10
9. Stalk of dorsal pair of gill filaments shorter than that of ventral pair ....
   .... *S.* (*N.*) *cheni*
   Stalks of dorsal and ventral pairs of gill filaments subequal in length to each other .... *S.* (*N.*) *cangshanense* and *S.* (*N.*) *silvestre*
10. Four gill filaments subequal in thickness to one another ....
    .... *S.* (*N.*) *chenzhouense*, *S.* (*N.*) *henanense*, and *S.* (*N.*) *zhangjiajiense*
    One or two gill filaments of dorsal pair thicker than those of ventral pair 11
11. Abdominal segment 9 with spine-combs dorsally .... *S.* (*N.*) *qingshuiense*
    Abdominal segment 9 without spine-combs dorsally ....*S.* (*N.*) *qiaoluoense*
12. Gill with five filaments .... *S.* (*B.*) *guiyangense*
    Gill with six or more filaments 13
13. Gill with six filaments 14
    Gill with eight or more filaments 46
14. Cocoon wall-pocket-shaped, extended ventrolaterally, with or without anterodorsal projection 15
    Cocoon wall-pocket-shaped, not extended ventrolaterally, without anterodorsal projection 20
15. Cocoon with anterodorsal projection 16
    Cocoon without anterodorsal projection 18
16. Six gill filaments basally divergent widely. Two filaments of ventral pair of gill with short stalk (Fig. 46C) .... *S.* (*N.*) *aureohirtum*
    Six gill filaments basally directed forward close together. Two filaments of ventral pair of gill with medium-long or long stalk 17

17. Two filaments of ventral pair of gill with medium-long stalk .... *S.* (*N.*) sp.
    Two filaments of ventral pair of gill with long stalk ....*S.* (*N.*) *leigongshanense*
18. Six gill filaments arranged as 2+1+(1+2) from dorsal to ventral .... *S.* (*N.*) *wangxianense*
    Six gill filaments arranged as 2+2+2 from dorsal to ventral 19
19. Stalk of ventral pair of gill filaments about three times as long as common basal stalk .... *S.* (*N.*) *qingzhouense*
    Stalk of ventral pair of gill filaments twice or less times as long as common basal stalk ....*S.* (*N.*) *chitoense* and *S.* (*N.*) *mie*
20. Cocoon simple, with no perforations 21
    Cocoon with one anterolateral open space on each side, or roughly woven anteriorly, yielding several small open spaces along anterior margin 35
21. Abdominal segment 9 without terminal hooks 22
    Abdominal segment with terminal hooks 29
22. Six gill filaments nearly as long as pupal body .... *S.* (*B.*) *guiyangense*
    Six gill filaments shorter than pupal body 23
23. Six gill filaments subequal in thickness to one another 24
    One or two of six gill filaments thicker than other filaments 25
24. Thorax with pit-like organ at the base of gill ....*S.* (*S.*) *tianchi*
    Thorax without pit-like organ at the base of gill ....*S.* (*S.*) *parawaterfallum*
25. One filament of dorsal pair of gill thicker than other filaments 26
    Two filaments of dorsal pair of gill subequal in thickness to each other and thicker than four other filaments 27
26. Middle pair of gill filaments with short stalk ....*S.* (*S.*) *rufibasis*
    Middle pair of gill filaments sessile ....*S.* (*S.*) *prominentum*
27. Frons bare. Abdominal segment 8 with spine-combs dorsally ....*S.* (*S.*) *ufengense*
    Frons with tubercles. Abdominal segments 6–8 or 7 and 8 each with spine-combs dorsally 28
28. Dorsal surface of thorax bare. Abdominal segments 6–8 each with spine-combs dorsally ....*S.* (*S.*) *wulingense*
    Dorsal surface of thorax with tubercles. Abdominal segments 7 and 8 each with spine-combs dorsally .... *S.* (*S.*) *fuzhouense*
29. Dorsal surface of thorax bare ....*S.* (*S.*) *chamlongi*
    Dorsal surface of thorax with tubercles 30
30. Six gill filaments directed forward close together. Gill without common basal stalk ....*S.* (*S.*) *nitidithorax*
    Six gill filaments moderately divergent basally in vertical plane. Gill with short common basal stalk 31
31. Abdominal segment 5 with spine-combs dorsally .... *S.* (*S.*) *qianense*
    Abdominal segment 5 without spine-combs dorsally 32

32. Abdominal segment 6 with spine-combs dorsally .... *S.* (*S.*) *taiwanicum*
    Abdominal segment 6 without spine-combs dorsally 33
33. Middle pair of gill filaments sessile ....*S.* (*S.*) *caohaiense*
    Middle pair of gill filaments with short stalk 34
34. Stalk of dorsal pair of gill filaments basally divergent at obtuse angle from that of ventral pair, when viewed laterally ....*S.* (*S.*) *oitanum*
    Stalk of dorsal pair of gill filaments basally divergent at right angle from that of ventral pair, when viewed laterally ....*S.* (*S.*) *suzukii*
35. Gill with six short filaments arranged as 2+3+1 from dorsal to ventral ....*S.* (*S.*) *lushanense*
    Gill with six filaments arranged as 2+2+2 from dorsal to ventral 36
36. Abdominal segment 9 with terminal hooks 37
    Abdominal segment 9 without terminal hooks 42
37. Frontal trichomes branched ....*S.* (*S.*) *japonicum*
    Frontal trichomes unbranched 38
38. Frons bare .... *S.* (*S.*) *neoacontum*
    Frons with tubercles 39
39. Six gill filaments decreasing in thickness from dorsal to ventral ....*S.* (*S.*) *ledongense*
    Six gill filaments subequal in thickness to one another 40
40. Cocoon with small open spaces along anterior margin .... *S.* (*S.*) *hengshanense*
    Cocoon with one open space on each anterolateral surface 41
41. Middle pair of gill filaments sessile ....*S.* (*S.*) *ephippioidum*
    Middle pair of gill filaments with short stalk ....*S.* (*S.*) *tenuatum*
42. Frons and dorsal surface of thorax bare 43
    Frons and dorsal surface of thorax with tubercles 44
43. Cocoon with small open spaces along anterior margin ....*S.* (*S.*) *acontum*
    Cocoon with one open space on each anterolateral surface ....*S.* (*S.*) *damingense*
44. Frontal trichomes unbranched .... *S.* (*S.*) *hirtipannus*
    Frontal trichomes branched 45
45. Middle pair of gill filaments with short stalk. Abdominal segment 7 without spine-combs dorsally .... *S.* (*S.*) *multifurcatum*
    Middle pair of gill filaments sessile. Abdominal segment 7 with spine-combs dorsally ....*S.* (*S.*) *ramulosum* and *S.* (*S.*) *shennongjiaense*
46. Gill with eight filaments 47
    Gill with nine or more filaments 59
47. Abdominal segments 5–9 without spine-combs dorsally 48
    At least abdominal segments 7 and 8 each with spine-combs dorsally 49
48. Two filaments of ventral pair of gill subequal in thickness to each other (Fig. 97A) ....*S.* (*W.*) *takahasii*
    One of two filaments of ventral pair of gill twice as thick as counter filament ... .*S.* (*W.*) *xingyiense*

49. Abdominal segment 9 without grapnel-like hooklets 50
Abdominal segment 9 with grapnel-like hooklets 53
50. Cocoon wall-pocket-shaped, without open space(s) anterolaterally ....
.... *S.* (*S.*) *ornatum*
Cocoon wall-pocket-shaped, with open space on each anterolateral surface, or shoe-shaped, with several open spaces on each anterolateral surface 51
51. Cocoon wall-pocket-shaped, with open space on each anterolateral surface (similar to Fig. 72J) ....
.... *S.* (*S.*) *digitatum*, *S.* (*S.*) *sakishimaense*, and *S.* (*S.*) *uncum*
Cocoon shoe-shaped, with several open spaces on each anterolateral surface (similar to Fig. 77G) 52
52. Two ventral filaments of gill much thinner than six other filaments ....
.... *S.* (*S.*) *bahense* and *S.* (*S.*) *chiangmaiense*
All eight gill filaments subequal in thickness to one another (similar to Fig. 49Q) .... *S.* (*S.*) *bidentatum* and *S.* (*S.*) *tanae*
53. Cocoon without anterodorsal projection 54
Cocoon with anterodorsal projection 58
54. Gill filaments short, less than half as long as pupal body.................................
.... *S.* (*G.*) *curvastylum* and *S.* (*G.*) *guizhouense*
Gill filaments long (at least one or two filaments of ventral pair), as long as or longer than pupal body 55
55. Gill filaments arranged as (3+3)+2 from dorsal to ventral (dorsal and middle triplets sharing stalk) .... *S.* (*G.*) *asakoae*, *S.* (*G.*) *brevitruncum*, *S.* (*G.*) *epauletum*, *S.* (*G.*) *guangxiense*, *S.* (*G.*) *heteropara*, *S.* (*G.*) *huguangense*, *S.* (*G.*) *hunanense*, *S.* (*G.*) *jianfengense*, *S.* (*G.*) *mengi*, *S.* (*G.*) *metatarsale*, *S.* (*G.*) *miaolingense*, *S.* (*G.*) *penis*, and *S.* (*G.*) *xiangxiense*
Gill filaments arranged as 3+3+2 or 2+4+2, or 1+2+3+2 from dorsal to ventral (dorsal and middle groups not sharing stalk) 56
56. Gill filaments arranged as 2+4+2 from dorsal to ventral .... *S.* (*G.*) sp.
Gill filaments arranged as 3+3+2 or 1+2+3+2 from dorsal to ventral 57
57. Gill filaments arranged as 1+2+3+2 from dorsal to ventral .... *S.* (*G.*) *pattoni*
Gill filaments arranged as 3+3+2 from dorsal to ventral....................................
.... *S.* (*G.*) *hainanense*, *S.* (*G.*) *guangxiense*, and *S.* (*G.*) *longshengense*
58. Dorsum of abdominal segments 1–5 darkened .... *S.* (*G.*) *inthanonense*
Dorsum of abdominal segments1–5 unpigmented or light yellowish ....
.... *S.* (*G.*) *fanjingshanense*, *S.* (*G.*) *jinbianense*, *S.* (*G.*) *nigrofemoralum*, *S.* (*G.*) *tumum*, *S.* (*G.*) *wuyishanense*, *S.* (*G.*) *yuanbaoshanense*, and *S.* (*G.*) *yunnanense*
59. Gill with nine filaments.. *S.* (*S.*) *nujiangense*
Gill with 10 filaments 60
60. Abdominal segment with grapnel-like hooklets. Cocoon wall-pocket-shaped ..
.... 61

Abdominal segment without grapnel-like hooklets. Cocoon shoe-shaped, with several open spaces anteriorly 62
61. Gill composed of large inflated structure and 10 filaments .... *S.* (*G.*) *antlerum*
Gill composed of 10 filaments arranged as 3+5+2 from dorsal to ventral ....*S.* (*G.*) *bannanense* and *S.* (*G.*) *jiulianshanense*
62. Gill filaments arranged as 2+3+3+2 from dorsal to ventral 63
Gill filaments arranged in other ways 64
63. All gill filaments subequal in thickness to one another ....*S.* (*S.*) *quinquestriatum* and *S.* (*S.*) *wuzhishanense*
Two filaments of ventral pair of gill much thinner than two filaments of dorsal pair ....*S.* (*S.*) *menglaense*
64. Head integument with round tubercles. ....*S.* (*S.*) *nakhonense*
Head integument with angulated tubercles.........................*S.* (*S.*) *thailandicum*

Mature larvae*****

1. Main tooth of mandibular serration at acute angle to mandible apically 2
Main tooth of mandibular serration at right or obtuse angle to mandible apically 22
2. Postgenal cleft long, reaching posterior margin of hypostoma 3
Postgenal cleft medium-long to long, not reaching posterior margin of hypostoma 4
3. Labral fan with 26–28 primary rays. Rectal organ with five or six secondary lobules per lobe ....*S.* (*G.*) *bannanense*
Labran fan with 38–40 primary rays. Rectal organ with 10 secondary lobules per lobe ....*S.* (*G.*) *jiulianshanense*
4. Postgenal cleft medium-long, less than 1.5 times length of postgenal bridge 5
Postgenal cleft medium long to long, over 1.8 times length of postgenal bridge 8
5. Rectal organ with 14–16 secondary lobules per lobe .....*S.* (*G.*) *fanjingshanese*
Rectal organ with 7–10 secondary lobules per lobe 6
6. Posterior circlet with 80 rows of hooklets .... *S.* (*G.*) *yuanbaoshanense*
Posterior circlet with 88 rows of hooklets 7
7. Labral fan with 33 primary rays .... *S.* (*G.*) *jinbianense*
Labran fan with 44 primary rays ....*S.* (*G.*) *inthanonense*
8. Postgenal cleft rounded apically 9
Postgenal cleft pointed apically 13
9. Posterior circlet with 92 rows of hooklets .... *S.* (*G.*) *epauletum*
Posterior circlet with 70 or 80 rows of hooklets 10
10. Posterior circlet with 70 rows of hooklets 11
Posterior circlet with 80 rows of hooklets 12
11. Rectal organ with four or five secondary lobules per lobe. Postgenal cleft twice as long as postgenal bridge .... *S.* (*G.*) *brivetruncum*

Rectal organ with 11–13 secondary lobules per lobe. Postgenal cleft 3.0 times as long as postgenal bridge .... S. (G.) *mengi*

12. Postgenal cleft 3.5–4.0 times as long as postgenal bridge .... S. (G.) *nigrofemoralum*
    Postgenal cleft 2.0–2.5 times as long as postgenal bridge ..... S. (G.) *hunanense*
13. Posterior circlet with 50 rows of hooklets .... S. (G.) *huguangense*
    Posterior circlet with 60–87 rows of hooklets 14
14. Posterior circlet with 60 rows of hooklets .... S. (G.) *pattoni*
    Posterior circlet with 70–87 rows of hooklets 15
15. Postgenal cleft over 6.0 times as long as postgenal bridge 16
    Postgenal cleft 1.5–4.0 times as long as postgenal bridge 17
16. Labral fan with 44 primary rays. Posterior circlet with 82 rows of hooklets .... S. (G.) *curvastylum*
    Labral fan with 30 primary rays. Posterior circlet with 72 rows of hooklets .... S. (G.) *tumum*
17. Posterior circlet with 70–75 rows of hooklets 18
    Posterior circlet with 80–87 rows of hooklets 19
18. Rectal organ with three or four secondary lobules per lobe S. (G.) *jianfengense*
    Rectal organ with six to nine secondary lobules per lobe ....S. (G.) *guangxiense*, S. (G.) *hainanense*, and S. (G.) *heteropara*
19. Labral fan with 26–28 primary rays 20
    Labral fan with 36–44 primary rays 21
20. Rectal organ with 8–12 secondary lobules per lobe .... S. (G.) *yunnanense*
    Rectal organ with five or six secondary lobules per lobe .... S. (G.) *wuyishanense*
21. Posterior circlet with 87 rows of hooklets .... S. (G.) *xiangxiense*
    Posterior circlet with 80–84 rows of hooklets .... S. (G.) *asakoae*, S. (G.) *metatarsale*, S. (G.) *miaolingense*, and S. (G.) *penis*
22. Postgenal cleft absent or vestigial (similar to Fig. 42X) 23
    Postgenal cleft small to large 27
23. Pharate pupal gill with four filaments .... S. (N.) *angustitarse*
    Pharate pupal gill with 12 or 14 filaments 24
24. Pharate pupal gill with 14 filaments. Posterior circlet with 50 rows of hooklets .... S. (M.) *weisiense*
    Pharate pupal gill with 12 filaments. Posterior circlet with 94–108 rows of hooklets 25
25. Posterior circlet with 82 rows of hooklets .... S. (M.) sp.
    Posterior circlet with 94–108 rows of hooklets 26
26. Labral fan with 40 primary rays. Posterior circlet with 94 rows of hooklets .... S. (M.) *separatum*
    Labral fan with 20 primary rays. Posterior circlet with 104–108 rows of hooklets .... S. (M.) *kirgisorum*

27. Ventral papillae medium-sized or large 28
Ventral papillae small or absent 38
28. Mandibular serration composed of two large teeth (similar to Fig. 46I) 29
Mandibular serration composed of one large tooth and one small tooth 30
29. Hypostoma with anterior margin moderately depressed between median tooth and each corner tooth. Posterior circlet with ca. 70 rows of hooklets .... *S.* (*N.*) *aureohirtum*
Hypostoma with anterior margin nearly straight or shallowly depressed between median tooth and each corner tooth. Posterior circlet with ca. 80 rows of hooklets .... *S.* (*E.*) *weiningense*
30. Postgenal cleft 2.0–2.5 times as long as postgenal bridge .... *S.* (*B.*) *guiyangense*
Postgenal cleft 0.3–1.5 times as long as postgenal bridge 31
31. Pharate pupal gill with six filaments 32
Pharate pupal gill with four filaments 33
32. Labral fan with 38–40 primary rays .... *S.* (*N.*) *qiongzhouense*
Labral fan with 26–34 primary rays .... *S.* (*N.*) *chitoense*, *S.* (*N.*) *leigongshanense*, *S.* (*N.*) *mie*, *S.* (*N.*) *wangxianense*, and *S.* (*N.*) sp.
33. Postgenal cleft 1.5 times as long as wide .... *S.* (*N.*) *silvestre*
Postgenal cleft nearly as long as wide 34
34. Rectal organ simple, without secondary lobules .... *S.* (*N.*) *henanense*
Rectal organ compound, with secondary lobules 35
35. Posterior circlet with 94 rows of hooklets .... *S.* (*N.*) *zhangjiajiense*
Posterior circlet with 64–83 rows of hooklets 36
36. Rectal organ with 11–13 secondary lobules per lobe .... *S.* (*N.*) *chenzhouense*
Rectal organ with five to eight secondary lobules per lobe 37
37. Labral fan with 28–32 primary rays .... *S.* (*N.*) *qiaolaoense* and *S.* (*N.*) *qingshuiense*
Labral fan with 38–40 primary rays ....*S.* (*N.*) *cangshanense* and *S.* (*N.*) *cheni*
38. Hypostoma with intermediate teeth nearly as long as each corner tooth........... .... *S.* (*W.*) *takahasii* and *S.* (*W.*) *xingyiense*
Hypostoma with intermediate teeth shorter than each corner tooth 39
39. Postgenal cleft reaching posterior margin of hypostoma 40
Postgenal cleft not reaching posterior margin of hypostoma 41
40. Abdominal segments 2–6 each with a pair of conical dorsolateral protuberances. Pharate pupal gill with three inflated tubular filaments .... *S.* (*S.*) *nodosum*
Abdominal segments 2–6 each without a pair of conical dorsolateral protuberances. Pharate pupal gill with six slender filaments ....*S.* (*S.*) *damingense*
41. Rectal organ simple, without secondary lobules .... *S.* (*S.*) *ornatum*
Rectal organ compound, with secondary lobules 42

42. Postgenal cleft long and rounded, over 5.0 times as long as postgenal bridge 43
    Postgenal cleft medium-long, less than 4.5 times as long as postgenal bridge 49
43. Pharate pupal gill with six filaments .... *S.* (*S.*) *chamlongi*, *S.* (*S.*) *jacuticum*, and *S.* (*S.*) *hirtipannus*
    Pharate pupal gill with eight, nine, or 10 filaments 44
44. Pharate pupal gill with eight filaments 45
    Pharate pupal gill with nine or 10 filaments 46
45. Abdominal segments without dorsolateral protuberances ....*S.* (*S.*) *sakishimaense*
    Abdominal segments 1–8 each with a pair of cone-like dorsolateral protuberances ....*S.* (*S.*) *bidentatum*
46. Pharate pupal gill with nine filaments .... *S.* (*S.*) *nujianense*
    Pharate pupal gill with 10 filaments 47
47. Abdominal segments without dorsolateral protuberances ....*S.* (*S.*) *wuzhishanense*
    Abdominal segments 1–8 each with a pair of cone-like dorsolateral protuberances 48
48. Posterior circlet with 120 rows of hooklets ....*S.* (*S.*) *quinquestriatum*
    Posterior circlet with 88 rows of hooklets ....*S.* (*S.*) *nakhonense*
49. Pharate pupal gill with eight filaments ....*S.* (*S.*) *rotifilis*
    Pharate pupal gill with six filaments 50
50. Labral fan with 52–58 primary rays 51
    Labral fan with 24–47 primary rays 53
51. Rectal organ with five secondary lobules per lobe ....*S.* (*S.*) *ledongense*
    Rectal organ with 10–16 secondary lobules per lobe 52
52. Body length 8.0–9.0 mm ....*S.* (*S.*) *taiwanicum*
    Body length 5.0 mm .... *S.* (*S.*) *multifurcatum*
53. Posterior circlet with 40 rows of hooklets ....*S.* (*S.*) *uncum*
    Posterior circlet with 66–115 rows of hooklets 54
54. Postgenal cleft 1.5–2.0 times as long as postgenal bridge 55
    Postgenal cleft 2.5–4.5 times as long as postgenal bridge 58
55. Postgenal cleft pointed apically .... *S.* (*S.*) *zunyiense*
    Postgenal cleft rounded apically 56
56. Labral fan with 24–28 primary rays ....*S.* (*S.*) *caohaiense*
    Labral fan with 32–38 primary rays 57
57. Posterior circlet with 76 rows of hooklets .... *S.* (*S.*) *prominentum*
    Posterior circlet with 84 rows of hooklets ....*S.* (*S.*) *ramulosum*
58. Posterior circlet with 110–115 rows of hooklets 59
    Posterior circlet with 68–100 rows of hooklets 60
59. Rectal organ with six to nine secondary lobules per lobe ......*S.* (*S.*) *japonicum*

Rectal organ with 15–26 secondary lobules per lobe ....
.... *S.* (*S.*) *fuzhouense* and *S.* (*S.*) *shennongjiaense*

60. Posterior circlet with 88–100 rows of hooklets 61
Posterior circlet with 68–82 rows of hooklets 63

61. Postgenal cleft widest in middle .... *S.* (*S.*) *hengshaense*
Postgenal cleft widest basally 62

62. Postgenal cleft 2.5 times as long as postgenal bridge. Posterior circlet with 100 rows of hooklets ....*S.* (*S.*) *ephippioidum*
Postgenal cleft 4.0 times as long as postgenal bridge. Posterior circlet with 88 rows of hooklets ....*S.* (*S.*) *rufibasis*

63. Postgenal cleft 2.2–2.5 times as long as postgenal bridge 64
Postgenal cleft 3.0–4.5 times as long as postgenal bridge 66

64. Postgenal cleft sharply pointed apically ....*S.* (*S.*) *parawaterfallum*
Postgenal cleft rounded apically 65

65. Rectal organ with five or six secondary lobules per lobe .... *S.* (*S.*) *waterfallum*
Rectal organ with 10–13 secondary lobules per lobe ....*S.* (*S.*) *tenuatum*

66. Postgenal cleft widest basally ....*S.* (*S.*) *wulingense*
Postgenal cleft widest in middle 67

67. Body length 3.9–5.5 mm ....
.... *S.* (*S.*) *nitidithorax*, *S.* (*S.*) *suzukii*, and *S.* (*S.*) *tianchi*
Body length 6.3–7.5 mm 68

68. Rectal organ with one to three secondary lobules per lobe ....*S.* (*S.*) *oitanum*
Rectal organ with 13–17 secondary lobules per lobe ....*S.* (*S.*) *qianense*

**Simulium* (*G.*) *montiblense* recorded from Fujian by Zhang et al. (2009) is not included because it is possible that *S. guizhouense* described from Guizhou Province (Chen et al., 2003) was misidentified as *S. montiblense* because the pupal gills of both species are almost indistinguishable from each other. In addition, eight unnamed species, *S.* (*E.*) sp. D, *S.* (*M.*) sp. B, *S.* (*G.*) sp. A, *S.* (*N.*) sp. C, *S.* (*S.*) sp. E, *S.* (*S.*) sp. F, *S.* (*S.*) sp. G, and *S.* (*S.*) sp. H, are not included because no information for their characteristics is available.

***Simulium* (*G.*) *antlerum*, *S.* (*G.*) *jiulianshanense*, *S.* (*G.*) *guizhouense*, *S.* (*G.*) *metatarsale*, *S.* (*G.*) *tumum*, *S.* (*M.*) *kirgisorum*. *S.* (*M.*) *yuntaiense*, *S.* (*N.*) *qiaolaoense*, *S.* (*N.*) *zhangjiajiense*, *S.* (*S.*) *lushanense*, *S.* (*S.*) *neoacontum*, *S.* (*S.*) *thailandicum*, *S.* (*S.*) *taipokauense*, and *S.* (*S.*) *multifurcatum* are not included because their females are unknown. *Simulium* (*G.*) *xiangxiense* is not included for lack of color of legs. *Simulium* (*S.*) *nitidithorax* is not included due to the lack of detailed description of the female.

****Simulium* (*G.*) *jiulianshanense*, *S.* (*G.*) *jinbianense*, *S.* (*G.*) sp., *S.* (*G.*) *penis*, *S.* (*M.*) *separatum*, *S. S.* (*M.*) *yuntaiense*, (*N.*) *henanense*, *S.* (*S.*) *ramulosum*, *S.* (*S.*) *rotifilis*, *S.* (*S.*) *nujiangense*, *S.* (*S.*) *sanyaense*, and *S.* (*S.*) *wulingense* are not included because their males are unknown. *Simulium* (*S.*) *nitidithorax*, *S.* (*S.*)

suzukii, and *S.* (*S.*) *rufibasis* are not included because detailed information for their male characteristics in Chinese specimens is not available.

****Simulium (G.) dudgeoni, S. (G.) pingxianense*, and *S.* (*S.*) *taipokauense* are not included because their pupae are unknown.

*****Simulium (G.) antlerum, S. (G.) guizhouense, S. (G.)* sp., *S. (G.) pingxiangense, S. (S.) bahense, S. (S.) chiangmaiense, S. (S.) jieyangense, S. (S.) lushanense, S. (S.) menglaense, S. (S.) neoacontum, S. (S.) shangchuanense, S. (S.) spoonatum, S. (S.) taipokauense, S. (S.) thailandicum*, and *S.* (*S.*) *ufengense* are not included because their larvae are unknown. *Simulium* (*S.*) *tanae* is not included because most of its morphological characteristics were not described by Xue (1992), though it can run up to couplet 43 in the key, and *S.* (*M.*) *nemorivagum* is also not included for lack of information on the number of rows of the posterior circlet.

13 The Fauna of Black Flies in the Nansei Islands, Japan

Overview

The Nansei Islands, an arc between Kyushu and Taiwan, is biogeographically composed of two areas separated by the Tokara Strait, from which northern islands belong to the Palearctic Region and southern islands belong to the Oriental Region.

Shiraki (1935) described a new species, *Simulium* (*Simulium*) *oshimanum*, from Amamioshima Island. Ogata (1956) recorded four species: *S.* (*Nevermannia*) *aureohirtum*, *S.* (*Gomphostilbia*) *batoense* (at that time, under the subgenus *Eusimulium*), *S.* (*S.*) *japonicum* and *S.* (*S.*) *tuberosum*, and suggested that *S.* (*S.*) *oshimanum* is the same species as *S.* (*S.*) *japonicum*. Ogata (1966) described a new species, *S.* (*S.*) *ryukyuense* [the species formerly regarded as *S.* (*S.*) *tuberosum* by Ogata (1951)] and recorded *S.* (*S.*) *fenestratum* from Ishigaki Island and Iriomote Island. Takaoka (1972, 1973) described three new species: *S.* (*Wallacellum*) *yonakuniense* (at that time, under the subgenus *Morops*), *S.* (*N.*) *morisonoi* (at that time, under the subgenus *Eusimulium*), and *S.* (*G.*) *tokarense*. Takaoka (1976a) described two new species: *S.* (*Eusimulium*) *satsumense* and *S.* (*N.*) *koshikiense* (at that time, as a subspecies of *S.* (*N.*) *subcostatum*) and recorded *S.* (*N.*) *mie* and *S.* (*N.*) *uchidai*. Takaoka (1976b) described a new species, *S.* (*G.*) *okinawense*, and recorded *S.* (*S.*) *bidentatum* (at that time, under the subgenus *Gnus*) and *S.* (*S.*) *oitanum* [at that time, as *S.* (*Odagmia*) *aokii*]. Takaoka (1977a) described a new species, *S.* (*S.*) *sakishimaense* [the species previously regarded as *S.* (*S.*) *fenestratum* by Ogata (1966)], recorded *S.* (*S.*) *arakawae*, *S.* (*S.*) *rufibasis*, *S.* (*S.*) *quinquestriatum*, and synonymized *S.* (*S.*) *ryukyuense* with *S.* (*S.*) *suzukii*. Takaoka (1991) described a new species, *S.* (*G.*) *yaeyamaense* (the species previously misidentified as *S.* (*G.*) *batoense* from Java). Takaoka et al. (2022a) described a new species, *S.* (*S.*) *tanegashimaense* [the species formerly regarded as *S.* (*S.*) *quinquestriatum* by Takaoka (1977a)]. Adler et al. (2021) resurrected *S.* (*S.*) *ryukyuense* from the junior synonym of *S.* (*S.*) *suzukii*. Adler et al. (2019) described a new species, *S.* (*S.*) *yamatoense* [the species formerly regarded as *S.* (*S.*) *rufibasis* in main islands of Japan]. It may suggest that the species regarded as *S.* (*S.*) *rufibasis* in Yakushima Island, the Nansei Islands by Takaoka (1977a) also differs from *S.* (*S.*) *rufibasis* from India (type locality). So, the species in Yakusihima Island is here treated as *S.* (*S.*) sp. nr. *yamatoense* until further studies on the species status of this species are completed.

The black fly fauna of the Islands is thus represented by 18 species of the genus *Simulium* s. l., of which seven species: *S.* (*E.*) *satsumense*, *S.* (*N.*) *koshikiense*, *S.* (*N.*) *uchidai*, *S.* (*S.*) *oitanum*, *S.* (*S.*) *arakawae*, *S.* (*S.*) *tanegashimaense*, and *S.* (*S.*) sp. nr. *yamatoense*, are distributed only in islands north of the Tokara Strait (Palearctic Region), whereas seven species: *S.* (*G.*) *okinawense*, *S.* (*G.*) *tokarense, S.* (*G.*) *yaeyamaense*, *S.* (*N.*) *morisonoi*, *S.* (*S.*) *sakishimaense*, *S.* (*S.*) *ryukyuense*, and *S.* (*W.*) *yonakuniense*, are only in islands south of the Tokara Strait (Oriental Region), and four species: *S.* (*N.*) *aureohirtum*, *S.* (*N.*) *mie*, *S.* (*S.*) *bidentatum*, and *S.* (*S.*) *japonicum*, are distributed in both sides of the strait (Palearctic and Oriental Regions).

List of 18 Species of Black Flies in the Nansei Islands, Japan

Subgenus *Eusimulium*
S. satsumense
Subgenus *Gomphostilbia*
S. asakoae species-group
S. okinawense and *S. tokarense*
S. batoense species-group
S. yaeyamaense
Subgenus *Nevermannia*
S. feuerborni species-group
S. mie and *S. morisonoi*
S. ruficorne species-group
S. aureohirtum
S. vernum species-group
S. koshikiense and *S. uchidai*
Subgenus *Simulium* s. str.
S. argentipes species-group
S. bidentatum
S. griseifrons species-group
S. japonicum
S. multistriatum species-group
S. sakishimaense
S. *striatum* species-group
S. tanegashimaense
S. tuberosum species-group

S. ryukyuense and *S.* sp. nr. *yamatoense*
S. variegatum species-group
S. oitanum
S. venustum species-group
S. arakawae
Subgenus Wallacellum
S. yonakuniense

*Keys to Identify 18 Species of Black Flies in the Nansei Islands, Japan**

Females

1. Katepisternum haired 2
   Katepisternum bare 6
2. Pleural membrane haired ....*S.* (*W.*) *yonakuniense*
   Pleural membrane bare 3
3. Postnotum with fine hairs ....*S.* (*E.*) *satsumense*
   Postnotum bare 4
4. Hind tibia dark brown except base yellow ....*S.* (*G.*) *yaeyamaense*
   Hind tibia yellowish white except apical half dark brown 5
5. Base of radius with tuft of yellow hairs. Genital fork with triangular lobe directed posteromedially on each arm .... *S.* (*G.*) *okinawense*
   Base of radius with tuft of yellow hairs mixed with dark hairs. Genital fork without triangular lobe directed posteromedially on each arm .... *S.* (*G.*) *tokarense*
6. Claw with large basal tooth 7
   Claw with or without small subbasal tooth 11
7. Antenna yellow except flagellomere 1 darkened. Hind tibia with distinct subbasal marking (similar to Fig. 45B). Spermatheca oblong (Fig. 45F) ....*S.* (*N.*) *aureohirtum*
   Antenna entirely darkened or darkened except scape and pedicel yellowish brown. Spermatheca globular (similar to Fig. 43I) 8
8. Scutum without dark longitudinal vittae. Mandible with teeth on both sides 9
   Scutum with three dark longitudinal vittae. Mandible with teeth on inner margin but lacking teeth on outer margin 10
9. Antenna entirely darkened ....*S.* (*N.*) *koshikiense*
   Antenna darkened except scape and pedicel yellowish brown ....*S.* (*N.*) *uchidai*
10. Genital fork with short and slender projection directed anteriorly on each arm ....*S.* (*N.*) *morisonoi*

Genital fork with long and stout projection directed anteriorly on each arm .... *S.* (*N.*) *mie*

11. Claw with small subbasal tooth (similar to Fig. 49B) 12
    Claw without tooth 13
12. Mid femur and tibia entirely yellow. Ovipositor valve rounded posteromedially (similar to Fig. 49C) ....*S.* (*S.*) *bidentatum*
    Mid femur and tibia yellow except their apical caps darkened. Ovipositor valve pointed posteromedially (similar to Fig. 84C) ....*S.* (*S.*) *oitanum*
13. Scutum with longitudinal vittae 14
    Scutum without longitudinal vittae 16
14. Basal portion of radius bare ....*S.* (*S.*) *japonicum*
    Basal portion of radius haired 15
15. Hind tibia yellow except apical one-third darkened ....*S.* (*S.*) *sakishimaense*
    Hind tibia blackish except base yellow .... *S.* (*S.*) *tanegashimaense*
16. Sternite 7 with a pair of stout hairs (similar to Fig. 82C) ....*S.* (*S.*) sp. nr. *yamatoense*
    Sternite 7 without such pair of stout hairs 17
17. Hind tibia yellow except apical half darkened ....*S.* (*S.*) *arakawae*
    Hind tibia darkened except base yellow ....*S.* (*S.*) *ryukyuense*

Males

1. Katepisternum haired 2
   Katepisternum bare 5
2. Pleural membrane haired ....*S.* (*W.*) *yonakuniense*
   Pleural membrane bare 3
3. Hind basitarsus slender, nearly parallel-sided (similar to Fig. 21T) ....*S.* (*G.*) *yaeyamaense*
   Hind basitarsus enlarged, spindle-shaped (similar to Fig. 13O) 4
4. Upper eye (large) facets in 15 vertical columns and 16 horizontal rows on each side .... *S.* (*G.*) *okinawense*
   Upper eye (large) facets in 12 or 13 vertical columns and 14 horizontal rows on each side... *S.* (*G.*) *tokarense*
5. Basal section of radius haired. Coxite longer than style (similar to Fig. 43K) 6
   Basal section of radius bare. Coxite shorter than style (similar to Fig. 49I) 10
6. Hind basitarsus slender, nearly parallel-sided (similar to Fig. 45H) 7
   Hind basitarsus enlarged, spindle-shaped (similar to Fig. 43J) 8
7. Antenna entirely darkened. Hind tibia without subbasal marking .... *S.* (*E.*) *satsumense*

Antenna yellow except flagellomere 1 darkened. Hind tibia with dark subbasal marking .... *S.* (*N.*) *aureohirtum*
8. Paramere with 6–9 hooks .... *S.* (*N.*) *mie* and *S.* (*N.*) *morisonoi*
Paramere with one hook (similar to Fig. 47Q) 9
9. Hind basitarsus ca. 3.5 times as long as its greatest width ... *S.* (*N.*) *koshikiense*
Hind basitarsus ca. 5.0 times as long as its greatest width .... *S.* (*N.*) *uchidai*
10. Scutum with yellow fine short hairs 11
Scutum with brassy to brown short hairs 14
11. Ventral plate with posterior surface smooth 12
Ventral plate with dentate posterior surface 13
12. Style without basal protuberance. Ventral plate quadrate when viewed ventrally .... *S.* (*S.*) *japonicum*
Style with basal protuberance having serrated anterior margin (similar to Fig. 76R). Ventral plate saddle-shaped, rounded posteriorly when viewed ventrally (similar to Fig. 76P) .... *S.* (*S.*) *tanegashimaense*
13. Ventral plate nearly quadrate when viewed ventrally. Style with long basal protuberance having pointed apex (similar to Fig. 71C) .... *S.* (*S.*) *sakishimaense*
Ventral plate Y-shaped when viewed ventrally (similar to Fig. 84I). Style without basal protuberance .... *S.* (*S.*) *oitanum*
14. Ventral plate nearly quadrate when viewed ventrally (similar to Fig. 80L) 15
Ventral plate Y-shaped when viewed ventrally (similar to Fig. 49I) 16
15. Style with short round basal protuberance having many short distinct spines .... *S.* (*S.*) *ryukyuense*
Style without basal protuberance, though having several short spines basally .... *S.* (*S.*) sp. nr. *yamatoense*
16. Style gradually narrowed toward apex when viewed ventrolaterally .... *S.* (*S.*) *bidentatum*
Style nearly parallel-sided toward apex when viewed ventrolaterally .... *S.* (*S.*) *arakawae*

Pupae

1. Gill with four filaments 2
Gill with six, eight, or 10 filaments 5
2. Antennal sheath with eight ridges each having spine (similar to Fig. 95D). Abdomen without spine-combs dorsally (Fig. 95E) .... *S.* (*W.*) *yonakuniense*
Antennal sheath without ridges. Abdomen with spine-combs dorsally 3
3. Two filaments of ventral pair of gill sessile and thinner and shorter than two filaments of dorsal pair (similar to Fig. 10P) .... *S.* (*E.*) *satsumense*
Two filaments of ventral pair of gill with stalk and subequal in length and thickness to two filaments of dorsal pair 4

4. Cocoon wall-pocket-shaped, with short anterodorsal projection .... *S.* (*N.*) *uchidai*
Cocoon wall-pocket-shaped, without anterodorsal projection .... *S.* (*N.*) *koshikiense*
5. Gill with six filaments 6
Gill with eight or 10 filaments 12
6. Cocoon wall-pocket-shaped, with short anterodorsal projection (similar to Fig. 46H) .... *S.* (*N.*) *aureohirtum*
Cocoon wall-pocket-shaped or shoe-shaped, without anterodorsal projection 7
7. Cocoon wall-pocket-shaped or shoe-shaped, with open space(s) anterolaterally on each side .... *S.* (*S.*) *japonicum*
Cocoon wall-pocket-shaped, without open space anterolaterally 8
8. Gill filaments longer than pupal body, directed forward close together, and arising from medium-long common basal stalk .... *S.* (*N.*) *mie* and *S.* (*N.*) *morisonoi*
Gill filaments shorter than pupal body, divergent vertically, and arising from short common basal stalk 9
9. Stalk of dorsal pair and that of ventral pair of gill divergent at angle of 60–70 degrees when viewed laterally. Terminal hooks absent .... *S.* (*S.*) sp. nr. *yamatoense*
Stalk of dorsal pair and that of ventral pair of gill divergent at angle of 90–120 degrees when viewed laterally. Terminal hooks present 10
10. Abdominal segments 6–9 each with spine-combs dorsally ..... *S.* (*S.*) *arakawae*
Abdominal segments 7–9 each with spine-combs dorsally 11
11. Stalk of dorsal pair and that of ventral pair of gill divergent at angle of 120 degrees when viewed laterally. Abdominal segment 1 sparsely covered with tubercles .... *S.* (*S.*) *oitanum*
Stalk of dorsal pair and that of ventral pair of gill divergent at angle of 90 degrees when viewed laterally. Abdominal segment 1 without tubercles .... *S.* (*S.*) *ryukyuense*
12. Gill with 10 filaments .... *S.* (*S.*) *tanegashimaense*
Gill with eight filaments 13
13. Gill with eight filaments arranged as 2+2+2+2 from dorsal to ventral 14
Gill with eight filaments arranged as 3+3+2 from dorsal to ventral 15
14. Eight gill filaments subequal in length to one another (similar to Fig. 49Q). Cocoon shoe-shaped, with high neck and several open spaces anteriorly (similar to Fig. 49R, S) .... *S.* (*S.*) *bidentatum*
Eight gill filaments decreasing in length from dorsal to ventral. Cocoon wall-pocket-shaped or shoe-shaped, with low neck and open space(s) anterolaterally .... *S.* (*S.*) *sakishimaense*
15. Two filaments of ventral pair of gill about twice as long as six other filaments of dorsal and middle triplets .... *S.* (*G.*) *yaeyamaense*
Two filaments of ventral pair of gill slightly longer than six other filaments of dorsal and middle triplets 16

16. All eight filaments of gill directed forward close together, arising from short common basal stalk (Fig. 14D) .... *S.* (*G.*) *tokarense*
    Eight filaments of gill basally divergent vertically, arising from medium-long common basal stalk .... *S.* (*G.*) *okinawense*

Mature larvae

1. Ventral papillae large 2
   Ventral papillae absent or small 11
2. Hypostoma with lateral margins smooth (similar to Fig. 32U) 3
   Hypostoma with lateral margins serrated 5
3. Postgenal cleft long, with its apex close to posterior margin of hypostoma .... *S.* (*G.*) *yaeyamaense*
   Postgenal cleft medium-long, 1.1–2.6 times as long as postgenal bridge 4
4. Rectal organ simple, without secondary lobules .... *S.* (*G.*) *tokarense*
   Rectal organ with four to six finger-like or thumb-like secondary lobules per lobe .... *S.* (*G.*) *okinawense*
5. Mandibular serration composed of two teeth of same size (Fig. 10R) 6
   Mandibular serration composed of one large tooth and one small tooth 7
6. Postgenal cleft medium-long, slightly longer than postgenal bridge .... *S.* (*E.*) *satsumense*
   Postgenal cleft medium-long, 0.8 times as long as postgenal bridge .... *S.* (*N.*) *aureohirtum*
7. Postgenal cleft long, with its apex close to posterior margin of hypostoma .... *S.* (*W.*) *yonakuniense*
   Postgenal cleft short to medium-long, less than twice as long as postgenal bridge 8
8. Mandible with supernumerary serration. Abdomen without reddish brown markings 9
   Mandible without supernumerary serration. Abdomen with reddish brown markings 10
9. Head spot markedly positive. Postgenal cleft twice as long as postgenal bridge .... *S.* (*N.*) *uchidai*
   Head spot faintly positive. Postgenal cleft as long as postgenal bridge .... *S.* (*N.*) *koshikiense*
10. Labral fan with 22–26 primary rays. Rectal organ compound, with three to five thumb-like secondary lobules per lobe .... *S.* (*N.*) *morisonoi*
    Labral fan with 30–34 primary rays. Rectal organ compound, with 8–13 finger-like secondary lobules per lobe .... *S.* (*N.*) *mie*
11. Postgenal cleft rounded on both sides, with its greatest width medially 12
    Postgenal cleft pointed apically, with lateral margins nearly parallel-sided from base to middle 17

12. Posterior abdominal segments covered with dark chisel-like spinous setae (similar to Fig. 77M) .... *S.* (*S.*) *tanegashimaense*
    Posterior abdominal segments without dark chisel-like spinous setae 13
13. Head spots markedly negative ....*S.* (*S.*) *arakawae*
    Head spots faintly positive or obscured 14
14. Rectal organ simple or compound, with one to three secondary lobules per lobe ....*S.* (*S.*) *oitanum*
    Rectal organ compound, with 5–11 secondary lobules per lobe 15
15. Posterior circlet with ca. 72 rows of hooklets ....*S.* (*S.*) *ryukyuense*
    Posterior circlet with ca. 86 or ca. 96 rows of hooklets 16
16. Posterior circlet with ca. 96 rows of hooklets ....*S.* (*S.*) *sakishimaense*
    Posterior circlet with ca. 86 rows of hooklets ....*S.* (*S.*) *bidentatum*
17. Hypostoma with six to nine bristles on each side ....*S.* (*S.*) *japonicum*
    Hypostoma with four or five bristles on each side .... *S.* (*S.*) sp. nr. *yamatoense*

*Keys provided by Takaoka (1977b) are slightly modified.

14 The Fauna of Black Flies in Taiwan

Overview

Shiraki (1935) described nine new species from Taiwan: *Simulium* (*Eusimulium*) *taipei*, *S.* (*Nevermannia*) *falcoe* (at that time, under the subgenus *Eusimulium*), *S.* (*N.*) *geniculare* (at that time, under the subgenus *Eusimulium*), *S.* (*Simulium*) *ambiguum*, *S.* (*S.*) *karenkoense* (at that time, under the genus *Odagmia*), *S.* (*S.*) *katoi*, *S.* (*S.*) *minutum*, *S.* (*S.*) *arisanum*, and *S.* (*S.*) *quinquestriatum* (at that time, as *Stilboplax 5-striatum*). Kono and Takahasi (1940) gave a substitute name, *S.* (*S.*) *shirakii*, for *S.* (*S.*) *minutum*, the name preoccupied.

Takaoka (1979) described eight new species: *S.* (*N.*) *taulingense* (at that time, under the subgenus *Eusimulium*), *S.* (*N.*) *yushangense* (at that time, under the subgenus *Eusimulium*), *S.* (*Montisimulium*) *chowi* (at that time, under the subgenus *Eusimulium*), *S.* (*N.*) *chitoense* (at that time, under the subgenus *Eusimulium*), *S.* (*Gomphostilbia*) *tuenense*, *S.* (*S.*) *taiwanicum*, *S.* (*S.*) *puliense*, and *S.* (*S.*) *ufengense*; and recorded five known species: *S.* (*G.*) *metatarsale*, *S.* (*N.*) *aureohirtum* (at that time, under the subgenus *Eusimulium*), *S.* (*S.*) *rufibasis*, *S.* (*S.*) *suzukii*, and *S.* (*S.*) *sakishimaense*. Chung (1986) recorded *S.* (*Wallacellum*) *yonakuniense* from Lanyu Island.

Takaoka and Huang (2006) described a new species, *S.* (*S.*) *chungi*. Huang and Takaoka (2008a) described a new species, *S.* (*G.*) *syuhaiense*. Huang and Takaoka (2008b) described the female of *S.* (*S.*) *puliense*. Huang and Takaoka (2008c) described a new species, *S.* (*S.*) *pingtungense*. Huang and Takaoka (2009a) described a new species, *S.* (*S.*) *serenum*. Huang and Takaoka (2009b) described the female and larva of *S.* (*S.*) *ufengense*. Huang et al. (2011) described a new species, *S.* (*G.*) *taitungense,* and described the male of *S.* (*G.*) *tuenense*. Huang and Adler (2011) chromosomally showed that *S.* (*S.*) *suzukii* in Taiwan is cytoform 'L' of *S. tani* (complex).

Low et al. (2015) synonymized *S.* (*S.*) *shirakii* with *S.* (*S.*) *nodosum*. Takaoka and Huang (2017) described a new species, *S.* (*G.*) *huangi*. Takaoka and Huang (2018) described a new species, *S.* (*S.*) *taichungense*. Takaoka et al. (2018b) described a new species, *S.* (*S.*) *haiduanense*. Takaoka et al. (2021a) described two new species, *S.* (*G.*) *kaohsiungense* and *S.* (*G.*) *shoufengense,* and recorded *S.* (*G.*) *asakoae* (the species previously regarded as *S.* (*G.*) *metatarsale*). Takaoka et al. (2021b) described two new species: *S.* (*S.*) *hehuanense* (the species formerly regarded as *S.* (*S.*) *rufibasis*) and *S.* (*S.*) *xiulinense*. Takaoka et al. (2021c) described a new species, *S.* (*S.*) *jianshiense* [the species formerly recognized as cytoformn 'L' of *S. tani* (complex)]. Takaoka et al. (2022a) described a new species, *S.* (*S.*) *yanpingense*, and redescribed *S.* (*S.*) *quinquestriatum.*

List of 35 Species of Black Flies in Taiwan

Subgenus *Eusimulium*
S. taipei
Subgenus *Gomphostilbia*
S. asakoae species-group
S. asakoae, *S. kaohsiungense*, *S. shoufengense, S. taitungense*, and *S. tuenense*
S. batoense species-group
S. syuhaiense
S. varicorne species-group
S. huangi
Subgenus *Montisimulium*
S. chowi
Subgenus *Nevermannia*
S. feuerborni species-group
S. chitoense
S. ruficorne species-group
S. aureohirtum, *S. falcoe*, and *S. geniculare*
S. vernum species-group
S. taulingense and *S. yushangense*
Subgenus *Simulium* s. str.
S. argentipes species-group
S. haiduanense
S. chungi species-group
S. chungi
S. crocinum species-group
S. serenum, *S. taichungense*, and *S. ufengense*

S. multistriatum species-group
S. sakishimaense
S. nobile species-group
S. nodosum
S. striatum species-group
S. pingtungense, *S. quinquestriatum*, and *S. yanpingense*
S. tuberosum species-group
S. arisanum, *S. hehuanense*, *S. jianshiense*, *S. puliense*, and *S. xiulinense*
S. variegatum species-group
S. karenkoense and *S. taiwanicum*
Simulium s. str. species unplaced to species-group
S. ambiguum and *S. katoi*
Subgenus *Wallacellum*
S. yonakuniense

Keys to Identify 35 Species of Black Flies in Taiwan

Females*

1. Postnotum with hairs .... *S. (E.) taipei*
   Postnotum without hairs 2
2. Katepisternum with hairs 3
   Katepisternum without hairs 8
3. Pleural membrane with hairs .... *S. (W.) yonakuniense*
   Pleural membrane without hairs 4
4. Antenna with eight flagellomeres .... *S. (G.) huangi*
   Antenna with nine flagellomeres 5
5. Hind tibia dark brown except base yellowish white .... *S. (G.) syuhaiense*
   Hind tibia darkened except basal half or two-thirds yellow 6
6. Sensory vesicle long, 0.53–0.61 times length of third palpal segment .... *S. (G.) kaohsiungense*
   Sensory vesicle short to medium-long, 0.25–0.40 times length of third palpal segment 7
7. Hind tibia yellowish white except apical half darkened .... *S. (G.) taitungense*
   Hind tibia yellowish white except apical one-third or little more darkened........ *S. (G.) asakoae* and *S. (G.) shoufengense*
8. Frons densely covered with short hairs 9
   Frons bare except several hairs along each lateral margin 14
9. Claw with medium-sized tooth (similar to Fig. 42C) .... *S. (M.) chowi*
   Claw with large tooth (similar to Fig. 43C) 10

10. Calcipala 0.7 times as wide as distal part of basitarsus, extended beyond basal half of second tarsomere .... *S.* (*N.*) *chitoense*
    Calcipala half as wide as distal part of basitarsus, extended up to basal half of second tarsomere 11
11. Hind tibia and basitarsus nearly brownish ....*S.* (*N.*) *taulingense*
    Hind tibia and basitarsus bicolored, yellowish on basal half and brownish on rest 12
12. Antenna entirely yellow. Hind tibia yellow on basal half, without dark subbasal band. Frons twice as long as its greatest width. Back of head with numerous brown hairs .... S. (*N.*) *geniculare*
    Antenna yellow with first flagellomere darkened. Frons as long as or 1.5 times as long as its greatest width. Back of head with numerous yellowish hairs 13
13. Frons as long as its greatest width. Antenna yellow with first flagellomere brown .... *S.* (*N.*) *aureohirtum*
    Frons 1.5 times as long as its greatest width. Antenna yellow with two or three basal flagellomeres brown .... *S.* (*N.*) *falcoe*
14. Basal section of radial vein haired 15
    Basal section of radial vein bare 17
15. Hind tibia yellow except apical one-third brownish black ....*S.* (*S.*) *chungi*
    Hind tibia brownish black except base yellow 16
16. Greatest width of frons against narrowest width 1.40–1.45 .... *S.* (*S.*) *pingtungense*
    Greatest width of frons against narrowest width 1.22–1.31 .... *S.* (*S.*) *quinquestriatum* and *S.* (*S.*) *yanpingense*
17. Claw with small subbasal tooth (similar to Fig. 49B) 18
    Claw without subbasal tooth 20
18. Ovipositor valve rounded apically (Fig. 49C) .... *S.* (*S.*) *haiduanense*
    Ovipositor valve narrowed apically (similar to Fig. 84C) 19
19. Scutum with blackish brown short hairs. Hind femur black ....*S.* (*S.*) *karenkoense*
    Scutum with yellow short hairs. Hind femur yellow except apical tip darkened ....*S.* (*S.*) *taiwanicum*
20. Scutum with longitudinal vittae 21
    Scutum without longitudinal vittae 23
21. Hind tibia black except extreme base yellow. Hind basitarsus entirely yellow ....*S.* (*S.*) *ambiguum*
    Hind tibia yellow except apical one-third black. Hind basitarsus yellow on basal half or two-thirds 22
22. Hind tibia as long as hind basitarsus. Hind femur black except basal one-third yellow ....*S.* (*S.*) *katoi*
    Hind tibia longer than hind basitarsus. Hind femur black except basal one-fifth yellow ....*S.* (*S.*) *sakishimaense*

23. Abdominal segment 7 with a pair of clustered hairs ventrally (similar to Fig. 82C) 24
Abdominal segment 7 without a pair of clustered hairs ventrally 25
24. Hind basitarsus yellow except apical two-fifths dark brown .... *S.* (*S.*) *hehuanense*
Hind basitarsus yellow except apical half dark brown .... *S.* (*S.*) *arisanum*
25. Hind tibia entirely black or black with base yellow 26
Hind tibia yellow at least on basal half 28
26. Hind basitarsus entirely yellow .... *S.* (*S.*) *nodosum*
Hind basitarsus yellow except apical two-fifths black 27
27. Sensory vesicle long, half as long as third palpal segment .... *S.* (*S.*) *jianshiense*
Sensory vesicle medium-long, 0.3 times as long as third palpal segment .... *S.* (*S.*) *puliense*
28. Mid tibia entirely yellow .... *S.* (*S.*) *taichungense*
Mid tibia yellow except apical cap light to medium brown 29
29. Hind tibia yellow except apical cap dark brown .... *S.* (*S.*) *serenum*
Hind tibia yellow except apical half dark brown .... *S.* (*S.*) *ufengense*

Males**

1. Katepisternum with hairs 2
Katepisternum without hairs 8
2. Pleural membrane with hairs .... *S.* (*W.*) *yonakuniense*
Pleural membrane without hairs 3
3. Antenna with eight flagellomeres .... *S.* (*G.*) *huangi*
Antenna with nine flagellomeres 4
4. Hind tibia dark brown except base yellowish white. Hind basitarsus parallel-sided .... *S.* (*G.*) *syuhaiense*
Hind tibia darkened except basal half or little less yellowish white. Hind basitarsus enlarged, wedge-shaped or spindle-shaped (similar to Fig. 13O, P) 5
5. Upper-eye (large) facets with 10 or 11 vertical columns and 12–14 horizontal rows .... *S.* (*G.*) *asakoae* and *S.* (*G.*) *shoufengense*
Upper-eye (large) facets with 15 or 16 vertical columns and 15 or 16 horizontal rows 6
6. Hind basitarsus wider than hind femur .... *S.* (*G.*) *kaohsiungense*
Hind basitarsus narrower than hind femur 7
7. Hind tibia with dark subbasal band .... *S.* (*G.*) *tuenense*
Hind tibia without dark subbasal band .... *S.* (*G.*) *taitungense*
8. Basal portion of radius haired 9
Basal portion of radius bare 12
9. Hind basitarsus nearly parallel-sided, much narrower than hind tibia. Mid and hind tibiae yellow on basal half, each with dark subbasal band .... *S.* (*N.*) *aureohirtum*

Hind basitarsus swollen, spindle-shaped, as wide as or wider than hind tibia. Mid and hind tibia darkened 10

10. Paramere with seven hooks. Median sclerite simple club-shaped .... *S.* (*N.*) *chitoense*
Paramere with one hook. Median sclerite with bifid apex 11

11. First flagellomere of antenna 1.6 times length of second ..... *S.* (*N.*) *yushangense*
First flagellomere of antenna twice length of second ....*S.* (*N.*) *taulingense*

12. Scutum white-pruinose except inverted-V-shaped non-pruinose black area. Abdomen with a pair of white-pruinose dorsolateral spots on segments 2 and 4–7 ....*S.* (*S.*) *nodosum*
Scutum white-pruinose on shoulders, along lateral margins and on prescutellar area. Abdomen with a pair of white-pruinose dorsolateral spots on segments 2 and 5–7, or on segments 2, 6, and 7 13

13. Hind femur yellow with apical cap dark brown or brownish black 14
Hind femur dark brown to brownish black or dark brown to brownish black with base yellow 18

14. Fore femur light to medium brown .... *S.* (*S.*) *haiduanense*
Fore femur entirely yellow or yellow with apical cap dark brown 15

15. Fore femur entirely yellow ....*S.* (*S.*) *ufengense*
Fore femur yellow with apical cap dark brown 16

16. Mid tibia brown with apical cap black ....*S.* (*S.*) *taiwanicum*
Mid tibia yellowish white to yellow with apical tip dark yellow 17

17. Fore basitarsus 5.1 times as long as its greatest width ....*S.* (*S.*) *taichungense*
Fore basitarsus 6.0–6.5 times as long as its greatest width ....*S.* (*S.*) *serenum*

18. Fore femur entirely yellow ....*S.* (*S.*) *ambiguum*
Fore femur darkened 19

19. Abdomen with white-pruinose dorsolateral spots on segments 2, 6, and 7 ... 20
Abdomen with white-pruinose dorsolateral spots on segments 2 and 5–7 21

20. Scutum with brown short hairs .... *S.* (*S.*) *puliense*
Scutum with yellow short hairs ....*S.* (*S.*) *hehuanense*

21. Hind basitarsus wider than hind femur ....*S.* (*S.*) *xiulinense*
Hind basitarsus narrower than hind femur 22

22. Scutum with yellow short hairs 23
Scutum with brassy or brown short hairs 24

23. Mid tibia dark brown except basal tip yellowish ....*S.* (*S.*) *quinquestriatum*
Mid tibia yellow on basal half and brownish black on rest ....*S.* (*S.*) *sakishimaense*

24. Upper-eye (large) facets in 17 or 18 vertical columns ....*S.* (*S.*) *yanpingense*
Upper-eye (large) facets in 19 or 20 vertical columns 25

25. Mid tibia medium to dark brown with base yellow .... *S.* (*S.*) *pingtungense*
Mid tibia dark brown with base light brown ....*S.* (*S.*) *jianshiense*

Pupae***

1. Gill with 12 slender filaments, all having numerous tiny black spots (Fig. 42T) .... *S.* (*M.*) *chowi*
   Gill with two or three inflated or four to 10 slender or inflated filaments having no tiny black spots 2
2. Dorsum of abdomen without spine-combs (Fig. 95E) .... *S.* (*W.*) *yonakuniense*
   Dorsum of abdomen with spine-combs 3
3. Frons with three pairs of trichomes. Last abdominal segment with three grapnel-like hooklets on each side 4
   Frons with two pairs of trichomes. Last abdominal segment without grapnel-like hooklets on each side 9
4. Terminal hooks cone-shaped, each with outer margin smooth .... *S.* (*G.*) *syuhaiense*
   Terminal hooks plate-like, each with outer margin crenulated 5
5. Terminal hooks each with outer margin 5.0–5.2 times length of inner margin .... *S.* (*G.*) *huangi*
   Terminal hooks each with outer margin less than 3.0 times length of inner margin 6
6. Dorsomedian portion of thorax sparsely covered with tubercles .... *S.* (*G.*) *taitungense*
   Dorsomedian portion of thorax moderately covered with tubercles 7
7. Dorsum of abdominal segments 1 and 2 light brown .... *S.* (*G.*) *asakoae* and *S.* (*G.*) *tuenense*
   Dorsum of abdominal segments 1 and 2 light yellow 8
8. Two filaments of ventral pair of gill 4.0 mm long .... *S.* (*G.*) *kaohsiungense*
   Two filaments of ventral pair of gill 2.6–3.1 mm long .... *S.* (*G.*) *shoufengense*
9. Terminal hooks well-developed, cone-like (similar to Fig. 44H) 10
   Terminal hooks absent or weakly developed, cone-like 13
10. Gill of inflated dumbbell shape (Fig. 48G) .... *S.* (*N.*) *taulingense*
    Gill of slender filamentous form 11
11. Gill with four filaments .... *S.* (*N.*) *yushangense*
    Gill with six filaments 12
12. Gill filaments longer than pupal body. Ventral pair of filaments with relatively long stalk .... *S.* (*N.*) *chitoense*
    Gill filaments shorter than pupal body. Ventral pair of filaments sessile or short-stalked (Fig. 46C) .... *S.* (*N.*) *aureohirtum*
13. Gill with three inflated tubular filaments (Fig. 75F) .... *S.* (*S.*) *nodosum*
    Gill with six, eight, or 10 filaments 14
14. Gill with 10 filaments (Fig. 77B) 15
    Gill with six or eight filaments 17
15. All gill filaments subequal in thickness (similar to Fig. 77C) .... *S.* (*S.*) *pingtungense*

All gill filaments different in thickness 16
16. Stalk of dorsal pair of gill directed posteriorly (Fig. 77B) ....
....S. (S.) *quinquestriatum*
Stalk of dorsal pair of gill not directed posteriorly ....S. (S.) *yanpingense*
17. Gill with eight filaments (Figs. 49Q and 72A) 18
Gill with six filaments 19
18. Cocoon corbicular-shaped (Fig. 49R, S) .... S. (S.) *haiduanense*
Cocoon with anterolateral open space on each side (similar to Fig. 72J) ....
....S. (S.) *sakishimaense*
19. Terminal hooks present 20
Terminal hooks absent 21
20. Gill filaments subequal in length to one another ....S. (S.) *taiwanicum*
Gill filaments decreasing in length from dorsal to ventral .....S. (S.) *jianshiense*
21. Cocoon with deep anterodorsal incision and large anterolateral open space on each side (Fig. 53L) .... S. (S.) *serenum*
Cocoon without anterodorsal incision and large anterolateral open space on each side 22
22. Cocoon with several small open spaces anterolaterally ......S. (S.) *taichungense*
Cocoon without open spaces anterolaterally 23
23. Head and thoracic integument bare 24
Head and thoracic integument with tubercles 25
24. Inner filaments of dorsal and middle pairs of gill directed upward (Fig. 53C) .....
....S. (S.) *ufengense*
All six gill filaments directed forward ....S. (S.) *arisanum*
25. Longest dorsal filament of gill ca. 3.0 mm ....S. (S.) *xiulinense*
Longest dorsal filament of gill 2.3–2.5 mm ....S. (S.) *hehuanense*

Larvae****

1. Lateral margins of hypostoma smooth (Fig. 32U). Main tooth of mandible at acute angle to mandible apically (Fig. 32T) 2
Lateral margins of hypostoma serrated (Fig. 49U). Main tooth of mandible at right or obtuse angle to mandible apically (Fig. 49T) 6
2. Postgenal cleft long, 7.8–14.3 times length of postgenal bridge ....
.... S. (G.) *syuhaiense*
Postgenal cleft medium-long, less than 5.0 times length of postgenal bridge ....
....3
3. Ventral surface of head capsule with dark X-shaped marking. S. (G.) *taitungense*
Ventral surface of head capsule without such marking 4
4. Abdominal segments 6–8 covered with unbranched setae ....S. (G.) *asakoae*
Abdominal segments 6–8 covered with branched setae 5
5. Postgenal cleft twice as long as postgenal bridge ....S. (G.) *tuenense*
Postgenal cleft 1.3 times as long as postgenal bridge ....S. (G.) *huangi*
6. Postgenal cleft vestigial (similar to Fig. 42X) ....S. (M.) *chowi*

Postgenal cleft small to large 7

7. Pectination of uppermost primary ray of labral fan (Fig. 95V) different from those of other primary rays (Fig. 95W) ....*S.* (*W.*) *yonakuniense*

Pectination of uppermost primary ray of labral fan same as those of other primary rays 8

8. Ventral papillae large 9

Ventral papillae absent or small 12

9. Abdomen with reddish-brown markings dorsally on posterior segments (similar to Fig. 44K) ....*S.* (*N.*) *chitoense*

Abdomen without colored markings dorsally on posterior segments 10

10. Mandible without supernumerary serration ....*S.* (*N.*) *aureohirtum*

Mandible with supernumerary serration (similar to Fig. 48M) 11

11. Accessory sclerites distinct, brownish (Fig. 48P). Pharate pupal gill with four slender filaments ....*S.* (*N.*) *yushangense*

Accessory sclerites weakly developed, indistinct and uncolored. Pharate pupal gill of inflated dumbbell shape ....*S.* (*N.*) *taulingense*

12. Postgenal cleft long reaching posterior margin of hypostoma (Fig. 75N) ....*S.* (*S.*) *nodosum*

Postgenal cleft medium-long or long, but not reaching posterior margin of hypostoma 13

13. Abdominal segments 1–8 each with a pair of dorsolateral protuberances (similar to Fig. 77L) ....*S.* (*S.*) *quinquestriatum*

Abdominal segments 1–8 without such dorsolateral protuberances 14

14. Postgenal cleft medium-long, 1.5–2.5 times as long as postgenal bridge 15

Postgenal cleft long, over 4.0 times as long as postgenal bridge 17

15. Abdomen pale yellow. Postgenal cleft 1.5 times as long as postgenal bridge ....*S.* (*S.*) *taiwanicum*

Abdomen grayish brown. Postgenal cleft 2.1–2.5 times as long as postgenal bridge (Fig. 51Q) 16

16. Sheath of subesophageal ganglion well pigmented (similar to Fig. 53O) ....*S.* (*S.*) *ufengense*

Sheath of subesophageal ganglion not pigmented (Fig. 51Q) ......*S.* (*S.*) *chungi*

17. Postgenal cleft narrow, nearly triangular (similar to Fig. 81P) .... *S.* (*S.*) *puliense*

Postgenal clef rounded (similar to Fig. 77K) 18

18. Pharate pupal gill with 10 filaments .... *S.* (*S.*) *pingtungense*

Pharate pupal gill with eight filaments 19

19. Thoracic and abdominal segments moderately or densely covered with minute colorless setae. Pharate pupal gill with eight filaments of almost same thickness ....*S.* (*S.*) *haiduanense*

Thoracic and abdominal segments almost bare. Pharate pupal gill with eight filaments decreasing in thickness from dorsal to ventral .... *S.* (*S.*) *sakishimaense*

* *Simulium* (*G.*) *tuenense, S.* (*N.*) *yushangense*, and *S.* (*S.*) *xiulinense* are not included because their females are unknown.

** *Simulium* (*E.*) *taipei, S.* (*N.*) *falcoe, S.* (*N.*) *geniculare, S.* (*S.*) *karenkosense*, and *S.* (*S.*) *katoi* are not included because their males are unknown.

*** *Simulium* (*E.*) *taipei, S.* (*N.*) *falcoe, S.* (*N.*) *geniculare S.* (*S.*) *ambiguum, S.* (*S.*) *karenkoense*, and *S.* (*S.*) *katoi* are not included because their pupae are unknown.

**** *Simulium* (*E.*) *taipei, S.* (*N.*) *falcoe, S.* (*G.*) *kaohsiungense, S.* (*G.*) *shoufengense, S.* (*N.*) *geniculare, S.* (*S.*) *ambiguum, S.* (*S.*) *arisanum, S.* (*S.*) *hehuanense, S.* (*S.*) *jianshiense, S.* (*S.*) *karenkoense, S.* (*S.*) *katoi, S.* (*S.*) *serenum, S.* (*S.*) *taichungense, S.* (*S.*) *xiulinense*, and *S.* (*S.*) *yanpingense* are not included because their larvae are unknown.

15 The Fauna of Black Flies in the Philippines

Overview

In the Philippines, *Simulium* (*Simulium*) *melanopus* and *S.* (*Eusimulium*) *aureum* were the only known species of black flies (Edwards, 1929) until Delfinado (1962) described two new species, *S.* (*S.*) *baltazarae* and *S.* (*Gomphostilbia*) *baisasae*, and gave a new name, *S.* (*Nevermannia*) *philippinense*, for the species formerly recognized as *S.* (*E.*) *aureum* by Edwards (1929). Delfinado (1969) described nine new species, recorded one unnamed species, *S.* (*S.*) sp. A, and synonymized *S.* (*S.*) *baltazarae* with *S.* (*S.*) *nobile*. Delfinado (1971) described five new species, recorded one unnamed species, *S.* (*G.*) sp. A, and resurrected *S.* (*S.*) *baltazarae* from a synonym of *S.* (*S.*) *nobile*.

Takaoka (1983) published a monograph of black flies in the Philippines, in which 39 new species and 18 named species were treated; among new species, *S.* (*G.*) *montiblense* and *S.* (*S.*) *delfinadoae* were the species formerly recognized as *S.* (*G.*) sp. A and *S.* (*S.*) sp. A, respectively, by Delfinado (1969, 1971); *S.* (*N.*) *philippinense* was synonymized with *S.* (*N.*) *aureohirtum*; and all these species were classified in four subgenera of the genus *Simulium* s. l.: four species in the subgenus *Nevermannia* (at that time, as the subgenus *Eusimulium*), 21 species in the subgenus *Gomphostilbia* (at that time, eight of 21 species under the subgenus *Morops*), 24 species in the subgenus *Simulium* s. str. and eight species in the new subgenus *Wallacellum*.

Takaoka (2000b) described a new species, *S.* (*G.*) *lilianae* (at that time, under the subgenus *Morops*), from Luzon. Takaoka (2005) described a new species of the subgenus *Gomphostilbia* from Luzon. Takaoka (2006a,b) described five new species of the subgenus *Simulium* s. str. from Luzon. Takaoka (2006c) revised *S.* (*S.*) *forcipatum* and described three new related species from Luzon. Takaoka (2006d) described three new species of the subgenus *Wallacellum* from Luzon. Takaoka (2007a) described a new species of the subgenus *Simulium* s. str. from Luzon. Takaoka and Tenedero (2007a) described two new species from Mindoro. Takaoka and Tenedero (2007b) described a new species of the subgenus *Simulium* s. str. from Negros. Takaoka and Tenedero (2007c, 2008a, b) described four new species of the subgenus *Gomphostilbia* from Palawan. Takaoka and Tenedero (2008c) described two new species of the subgenus *Gomphostilbia* from Samar. Takaoka (2009a) described three new species of the subgenus *Wallacellum*. Takaoka (2009b) described three new species of the *S. banauense* species-group in the subgenus *Gomphostilbia*. Takaoka et al. (2009) described the male of *S.* (*G.*) *binuanense* from Palawan.

The fauna of black flies in the Philippines is represented by 88 species of the genus *Simulium* s. l., of which 33 species are placed in the subgenus *Gomphostilbia*, four species are in the subgenus *Nevermannia*, 37 species are in the subgenus *Simulium* s. str. and 14 species are in the subgenus *Wallacellum*.

It is characterized by the high endemism at the species level (87/88 species), and the presence of six endemic or semi-endemic lineages: *S. alienigenum* species-group, *S. ambigens* species-group, *S. baisasae* species-group, *S. banauense* species-group, *S. melanopus* species-group, and subgenus *Wallacellum*.

List of 88 Species of Black Flies in the Philippines

Subgenus *Gomphostilbia*
S. alienigenum species-group
S. alienigenum
S. ambigens species-group
S. ambigens
S. baisasae species-group
S. baisasae, *S. bicolense*, and *S. visayaense*
S. banauense species-group
S. atuganense, *S. banauense*, *S. bansonae*, *S. cagayanense*, *S. ifugaoense*, *S. katangladense*, *S. lilianae*, *S. liliwense*, *S. manbucalense*, *S. quezonense*, *S. riverai*, and *S. salazarae*
S. batoense species-group
S. babuyanense, *S. binuanense*, *S. davaoense*, *S. mangasepi*, and *S. samarense*
S. ceylonicum species-group
S. apoense, *S. bataksense*, and *S. mindoroense*
S. epistum species-group
S. epistum, *S. lagunaense*, *S. luzonicum*, *S. miblosi*, *S. mindanaoense*, *S. montiblense*, *S. rarum*, and *S. repentinum*
Subgenus *Nevermannia*
S. feuerborni species-group
S. bryopodium and *S. perlucidulum*

S. ruficorne species-group
S. aureohirtum
S. vernum species-group
S. aberrans
Subgenus *Simulium* s. str.
S. argentipes species-group
S. arboreum and *S. canlaonense*
S. melanopus species-group
S. abatanense, *S. acostai*, *S. aquilum*, *S. atrum*, *S. banahauense*, *S. concavum*, *S. conspicuum*, *S. discrepans*, *S. forcipatum*, *S. gatchaliani*, *S. halconense*, *S. iwahigense*, *S. malaibaense*, *S. mayoyaoense*, *S. melanopus*, *S. neodiscrepans*, *S. palawanense*, *S. pandum*, *S. retusum*, *S. salebrosum*, *S. simulacrum*, *S. subatrum*, *S. sumigorense*, *S. suplidoi*, *S. taalense*, and *S. tomentosum*
S. nobile species-group
S. baltazarae, *S. benquetense*, *S. cotabatoense*, *S. delfinadoae*, *S. latistylum*, and *S. leytense*
S. tuberosum species-group
S. mainitense, *S. quasifrenum*, and *S. tukurongense*
Subgenus *Wallacellum*
S. amplum, *S. cabrerai*, *S. carinatum*, *S. claveriaense*, *S. makilingense*, *S. marilogense*, *S. molawinense*, *S. ogonukii*, *S. recurvum*, *S. resimum*, *S. spinosibranchium*, *S. suyoense*, *S. tenederoi*, and *S. tuyense*

*Keys to Identify 88 Species of Black Flies in the Philippines**

Females**

1. Katepisternum haired 2
   Katepisternum bare 37
2. Calcipala elongate, reaching apex of second tarsomere(Fig. 92I, J) 3
   Calcipala medium-long, not reaching apex of second tarsomere 14
3. Calcipala elongate, reaching apex of second tarsomere (Fig. 92I) 4
   Calcipala greatly elongate, reaching at least middle of third tarsomere (Fig. 92J) 7
4. Genital fork without posteromedian projection on each arm (Fig. 92S) .... *S.* (*W.*) *suyoense*
   Genital fork with posteromedian projection on each arm (Fig. 92R, T) 5
5. Height of frons 1.28–1.32 times as long as its narrowest width .... *S.* (*W.*) *tuyense*
   Height of frons 1.42–1.50 times as long as its narrowest width 6
6. Paraproct 0.55 times as long as its greatest width .... *S.* (*W.*) *ogonukii*
   Paraproct 0.67 times as long as its greatest width .... *S.* (*W.*) *marilogense*

7. Sensory vesicle elongate, 0.56 times as long as third palpal segment (similar to Fig. 92B) .... *S.* (*W.*) *cabrerai*
   Sensory vesicle short, 0.22–0.28 times as long as third palpal segment (Fig. 92C) 8
8. Apical portion of each arm of genital fork directed laterally (Fig. 92R, T) 9
   Apical portion of each arm of genital fork directed posteriorly (Fig. 92U) 11
9. Fore basitarsus 7.1 times as long as its greatest width .... *S.* (*W.*) *tenederoi*
   Fore basitarsus 7.8 or 7.9 times as long as its greatest width 10
10. Frons:head ratio 1.0:3.8 .... *S.* (*W.*) *claveriaense*
    Frons:head ratio 1.0:3.3 .... *S.* (*W.*) *makilingense*
11. Fore basitarsus 7.6 times as long as its greatest width .... *S.* (*W.*) *carinatum*
    Fore basitarsus 8.1–9.7 times as long as its greatest width 12
12. Fore basitarsus 9.7 times as long as its greatest width........ *S.* (*W.*) *spinosibranchium*
    Fore basitarsus 8.1–8.4 times as long as its greatest width 13
13. Frons:head ratio 1.0:3.9 or 4.0 .... *S.* (*W.*) *amplum* and *S.* (*W.*) *recurvum*
    Frons:head ratio 1.0:4.4 .... *S.* (*W.*) *resimum*
14. Pleural membrane and/or upper mesopleuron region haired 15
    Pleural membrane and upper mesopleuron region bare 23
15. Claw with small tooth (Fig. 11C) .... *S.* (*G.*) *alienigenum*
    Claw with large tooth (Fig. 17E) 16
16. Tergites of abdominal segment 2 and 5–8 shiny 17
    Tergites of abdominal segment 2 and 6–8 shiny 18
17. Pleural membrane with 20–30 hairs .... *S.* (*G.*) *cagayanense*
    Pleural membrane with three to five hairs .... *S.* (*G.*) *bansonae*
18. Dorsolateral surface of abdominal segment 5 much lighter than those of abdominal segments 3 and 4 19
    Dorsolateral surface of abdominal segment 5 as dark as those of abdominal segments 3 and 4 20
19. Fore basitarsus 6.1 times as long as its greatest width .... *S.* (*G.*) *atuganense*
    Fore basitarsus 7.0 times as long as its greatest width .... *S.* (*G.*) *quezonense*
20. Hind tibia yellow on basal one-third 21
    Hind tibia yellow on basal half 22
21. Tuft hair of base of radius mostly yellow .... *S.* (*G.*) *salazarae*
    Tuft hair of base of radius mostly brown .... *S.* (*G.*) *liliwense*
22. Pleural membrane with ca. 30 hairs. Fore basitarsus 6.0 times as long as its greatest width .... *S.* (*G.*) *riverai*
    Pleural membrane with three to five hairs. Fore basitarsus 6.6–7.0 times as long as its greatest width 23
23. Frons:head ratio 1.0:6.2. Outer margin of mandible serrated ..... *S.* (*G.*) *lilianae*
    Frons:head ratio 1.0:4.6–4.9. Outer margin of mandible not serrated........ *S.* (*G.*) *banauense* and *S.* (*G.*) *ifugaoense*

23. Claw without tooth 24
Claw with small or large tooth 26
24. Tuft hair of base of radius whitish yellow. Outer margin of mandible serrated
.... *S. (G.) baisasae*
Tuft hair of base of radius brown. Outer margin of mandible not serrated 25
25. Sternite 8 with ca 10 long hairs on each side ....*S. (G.) visayaense*
Sternite 8 with ca. 15 long hairs on each side ....*S. (G.) bicolense*
26. Claw with small tooth .... *S. (G.) ambigens*
Claw with large tooth 27
27. Hind tibia medium to dark brown except base yellow 28
Hind tibia medium to dark brown except basal half or basal three-fifths whitish yellow, with or without subbasal dark band 29
28. Sensory vesicle long, 0.56 times as long as third palpal segment ....
....*S. (G.) binuanense*
Sensory vesicle medium-long, 0.36 times as long as third palpal segment ....
....*S. (G.) mangasepi*
29. Base of radius with dark brown hairs. Hind tibia whitish on basal half 30
Base of radius with yellow hairs, or yellow hairs mixed with dark hairs. Hind tibia yellowish on basal half or little less, with dark subbasal marking, or yellowish white on basal three-fifths if dark subbasal marking absent 32
30. Sensory vesicle medium-long, 0.43 times as long as third palpal segment........
.... *S. (G.) bataksense*
Sensory vesicle long, 0.62–0.67 times as long as third palpal segment 31
31. Mandible with ca. 22 inner teeth. Claw with tooth 0.44 times length of claw .....
....*S. (G.) mindoroense*
Mandible with ca. 28 inner teeth. Claw with tooth 0.52 times length of claw
....*S. (G.) apoense*
32. Hind tibia yellowish white on basal three-fifths. Mandible lacking outer teeth (Fig. 32A) ....*S. (G.) miblosi*
Hind tibia white on basal half or little less, with dark subbasal marking. Mandible with outer teeth 33
33. Height of frons twice as long as its greatest width .... *S. (G.) epistum*
Height of frons less than 1.6 times as long as its greatest width 34
34. Outer margin of mandible with ca. 10 weak teeth 35
Outer margin of mandible with ca. 6 weak teeth at some distance from apex ...
....36
35. Body length ca. 2.6 mm. Height of frons 1.2 times as long as its greatest width.
....*S. (G.) lagunaense*
Body length ca. 2.0 mm. Height of frons 1.6 times as long as its greatest width .... *S. (G.) luzonicum*
36. Sensory vesicle with large opening, as wide as sensory vesicle ....
....*S. (G.) mindanaoense*

Sensory vesicle with medium-sized opening, much narrower than sensory vesicle .... *S.* (*G.*) *montiblense*

37. Claw with large basal tooth 38
Claw simple or with small subbasal tooth 40
38. Hind basitarsus yellow on basal half and brownish black on rest. Spermatheca ellipsoidal .... *S.* (*N.*) *aureohirtum*
Hind basitarsus brownish black. Spermatheca globular or ovoid 39
39. Calcipala reaching apical one-third of hind tarsomere 2. Genital fork with projection directed anterodorsally on each arm .... *S.* (*N.*) *perlucidulum*
Calcipala reaching middle of hind tarsomere 2. Genital fork without projection directed anterodorsally on each arm .... *S.* (*N.*) *aberrans*
40. Mid tarsus yellow except tarsomeres 4 and 5 brownish. Subcosta bare 41
Mid tarsus brownish black with basal half or more of basitarsus yellowish. Subcosta with hairs 46
41. Antenna yellow on scape and pedicel and brown to brownish black on flagellomeres (darkened to apex) 42
Antenna yellow or dark yellow except two or three apical flagellomeres blackish 43
42. Head 3.7 times as wide as greatest width of frons. Mandible with ca. 28 inner teeth .... *S.* (*S.*) *benquetense*
Head 4.3 times as wide as greatest width of frons. Mandible with 19 or 20 inner teeth .... *S.* (*S.*) *latistylum*
43. Inner margins of posteriorly produced lobes of sternite 8 deeply concave medially, and each with unpigmented portion on apical one-third but without projection pointed posteriorly .... *S.* (*S.*) *delfinadoae*
Inner margins of posteriorly produced lobes of sternite 8 nearly parallel-sided or slightly curved and each with long unpigmented projection sharply pointed posteriorly (well recognized when viewed laterally) (similar to Fig. 74N) 44
44. Paraproct not produced posteriorly beyond base of cercus .... *S.* (*S.*) *leytense*
Paraproct produced posteriorly beyond base of cercus 45
45. Greatest width of frons nearly as wide as its height .... *S.* (*S.*) *baltazarae*
Greatest width of frons 0.9 times as wide as its height .... *S.* (*S.*) *cotabatoense*
46. Claw simple (Fig. 67C) 47
Claw with small subbasal tooth (Fig. 67E) or minute subbasal swelling (Fig. 67D) 51
47. Dorsum of abdominal segments 2 and 6–8 shiny. Ovipositor valve tongue-like, with few short hairs (Fig. 80C) 48
Dorsum of abdominal segments 2 and 5–8 shiny. Ovipositor valve triangular, with numerous stout hairs (Fig. 67F) 50
48. Sensory vesicle 0.29–0.35 times as long as third palpal segment .... *S.* (*S.*) *quasifrenum*
Sensory vesicle 0.38–0.47 times as long as third papal segment 49
49. Labrum 0.7 times as long as clypeus .... *S.* (*S.*) *mainitense*

Labrum 0.5–0.6 times as long as clypeus .... *S.* (*S.*) *tukurongense*

50. Mid and hind femora brownish black .... *S.* (*S.*) *discrepans*
Mid and hind femora yellow except apical cap light to medium brown .... *S.* (*S.*) *neodiscrepans*

51. Claw with minute subbasal swelling (Fig. 67D) .... *S.* (*S.*) *malaibaense*
Claw with small subbasal tooth (Fig. 67E) 52

52. Mid tibia whitish yellow except apical one-fourth brownish. Paraproct without well-developed ventrointernal plate 53
Mid tibia brownish. Paraproct with well-developed ventrointernal plate 54

53. Cibarium with few processes. Fore basitarsus 4.5 times as long as its greatest width .... *S.* (*S.*) *canlaonense*
Cibarium with several processes. Fore basitarsus 5.5 times as long as its greatest width .... *S.* (*S.*) *arboreum*

54. Fore coxa brown 55
Fore coxa yellow 59

55. Ovipositor valves pointed posteromedially (Fig. 67M) 56
Ovipositor valves rounded posteromedially and strongly sclerotized along inner margin (Fig. 67O) 57

56. Hind basitarsus yellowish white on basal two-thirds. Sensory vesicle short, 0.2 times as long as third palpal segment .... *S.* (*S.*) *banahauense*
Hind basitarsus yellow on basal half. Sensory vesicle short, 0.3 times as long as third palpal segment .... *S.* (*S.*) *subatrum*

57. Basal portion of radial vein with few hairs on apical one-third or one-fourth .. .. *S.* (*S.*) *aquilum*
Basal portion of radial vein bare 58

58. Genital fork with small projection directed anterodorsally on each arm .... *S.* (*S.*) *conspicuum*
Genital fork without projection directed anterodorsally on each arm .... *S.* (*S.*) *atrum*, *S.* (*S.*) *forcipatum*, and *S.* (*S.*) *sumigorense*

59. Scutum with yellow short hairs mixed with brown short hairs .... *S.* (*S.*) *taalense*
Scutum with brown short hairs 60

60. Ovipositor valves short, triangular, though rounded posteromedially .... *S.* (*S.*) *salebrosum*
Ovipositor valves elongate, tapered to apex (Fig. 67K) 61

61. Ovipositor valves acutely bent ventrally (Fig. 67N) 62
Ovipositor valves directed posteriorly or somewhat bent ventrally (Fig. 67L) 64

62. Paraproct widened ventrally (when viewed laterally) ....*S.* (*S.*) *palawanense*
Paraproct narrowed ventrally (when viewed laterally) 63

63. Fore basitarsus 3.9 times as long as its greatest width ....*S.* (*S.*) *pandum*
Fore basitarsus 4.9 times as long as its greatest width ....*S.* (*S.*) *abatanense*

64. Fore basitarsus 5.0 times as long as its greatest width 65
Fore basitarsus 4.0–4.6 times as long as its greatest width 66

65. Sensory vesicle medium-long, 0.36 times as long as third palpal segment ....S. (*S.*) *iwahigense*
Sensory vesicle short, 0.25 times as long as third palpal segment ....S. (*S.*) *suplidoi*
66. Sensory vesicle 0.21–0.23 times as long as third palpal segment ....S. (*S.*) *halconense*
Sensory vesicle 0.28–0.35 times as long as third palpal segment 67
67. Head 3.6–3.8 times as wide as greatest width of frons ...... S. (*S.*) *mayoyaoense*
Head 4.2–4.3 times as wide as greatest width of frons 68
68. Greatest with of frons 1.45 times as wide as its narrowest width ....S. (*S.*) *acostai*
Greatest with of frons 1.30 times as wide as its narrowest width 69
69. Cercus 4.0 times as wide as its length .... S. (*S.*) *melanopus*
Cercus 7.5 times as wide as its length ....S. (*S.*) *concavum*

Males***

1. Style shorter than coxite 2
Style longer than coxite 47
2. Katepisternum haired 3
Katepisternum bare 44
3. Calcipala elongate, reaching apex of second tarsomere (Fig. 93A, B) 4
Calcipala medium-long, not reaching apex of second tarsomere 16
4. Low ridge on anteroinner surface of hind tibia extended from base to little less than middle... S. (*W.*) *tuyense*
Low ridge on anteroinner surface of hind tibia extended from base to middle 5
5. Calcipala elongate, reaching apex of second tarsomere (similar to Fig. 92I) 6
Calcipala greatly elongate, reaching at least middle of third tarsomere (Fig. 93A, B) 7
6. Upper-eye (large) facets in 14 or 15 vertical columns and 15 or 16 horizontal rows ....S. (*W.*) *suyoense*
Upper-eye (large) facets in 13 vertical columns and 13 or 14 horizontal rows. ....S. (*W.*) *marilogense* and S. (*W.*) *ogonukii*
7. Hind basitarsus spindle-shaped (Fig. 93B) .... S. (*W.*) *amplum*
Hind basitarsus parallel-sided (Fig. 93A) 8
8. Upper-eye (large) facets in 12 or 13 vertical columns 9
Upper-eye (large) facets in 14–16 vertical columns 10
9. Ventral plate nearly parallel-sided when viewed ventrally .... S. (*W.*) *makilingense*
Ventral plate somewhat narrowed posteriorly when viewed ventrally ....S. (*W.*) *carinatum*
10. Fore basitarsus 7.8–8.1 times as long as its greatest width 11

Fore basitarsus 8.7–9.7 times as long as its greatest width 13

11. Hind basitarsus 5.5 times as long as its greatest width .... *S.* (*W.*) *tenederoi*
Hind basitarsus 6.3 times as long as its greatest width 12

12. Upper-eye (large) facets in 14 vertical columns and 13 or 14 horizontal rows .... *S.* (*W.*) *claveriaense*
Upper-eye (large) facets in 15 vertical columns and 15 horizontal rows ....*S.* (*W.*) *molawinense*

13. Upper-eye (large) facets in 15 or 16 vertical columns and 16 horizontal rows 14
Upper-eye (large) facets in 14 vertical columns and 14 horizontal rows 15

14. Posterior margin of ventral plate produced medially ....*S.* (*W.*) *cabrerai*
Posterior margin of ventral plate depressed medially ....*S.* (*W.*) *resimum*

15. Fore basitarsus 9.7 times as long as its greatest width ....*S.* (*W.*) *spinosibranchium*
Fore basitarsus 8.7 times as long as its greatest width .... *S.* (*W.*) *recurvum*

16. Pleural membrane and/or upper mesopleuron region haired 17
Pleural membrane and upper mesopleuron region bare 28

17. Upper mesopleuron region haired. Hind tibia yellow except apical one-third darkened .... *S.* (*G.*) *alienigenum*
Upper mesopleuron region bare. Hind tibia brownish at least on apical half 18

18. Antennal flagellum brownish except basal half of first flagellomere yellow .... *S.* (*G.*) *bansonae*
Antennal flagellum yellowish except two to five apical flagellomeres darkened 19

19. Abdominal segment 5 without dark brown pigment dorsolaterally. Hind basitarsus nearly spindle-shaped, at least somewhat tapered apically (Fig. 17N) 20
Abdominal segment 5 with dark brown pigment dorsolaterally. Hind basitarsus parallel-sided (Fig. 17M) 21

20. Upper-eye (large) facets in 11 vertical columns and 12 horizontal rows ....*S.* (*G.*) *atuganense*
Upper-eye (large) facets in 12 or 13 vertical columns and 13 horizontal rows. ....*S.* (*G.*) *quezonense*

21. Fore tibia light to medium brown except extreme base yellow 22
Fore tibia light to medium brown except basal one-fourth to two-fifths yellow 24

22. Tuft hair mostly yellow .... *S.* (*G.*) *salazarae*
Tuft hair mostly brown 23

23. Pleural membrane with three to five hairs ....*S.* (*G.*) *liliwense*
Pleural membrane with 28–30 hairs .... *S.* (*G.*) *cagayanense*

24. Pleural membrane with 16–28 hairs. Upper-eye (large) facets in 11 or 12 vertical columns and 11 or 12 horizontal rows 25
Pleural membrane with three to six hairs. Upper-eye (large) facets in 13–15 vertical columns and 13–15 horizontal rows 26

25. Pleural membrane with 16 hairs. Upper-eye (large) facets in 12 vertical columns and 12 horizontal rows .... *S.* (*G.*) *manbucalense*
    Pleural membrane with 28 hairs. Upper-eye (large) facets in 11 vertical columns and 11 horizontal rows .... *S.* (*G.*) *riverai*
26. Upper-eye (large) facets in 13 vertical columns and 13 horizontal rows. Fore basitarsus 9.4 times as long as its greatest width .... *S.* (*G.*) *ifugaoense*
    Upper-eye (large) facets in 14 or 15 vertical columns and 14 or 15 horizontal rows. Fore basitarsus 7.5–8.4 times as long as its greatest width 27
27. Fore basitarsus 7.5 times as long as its greatest width ....*S.* (*G.*) *katangladense*
    Fore basitarsus 8.4 times as long as its greatest width ....*S.* (*G.*) *banauense*
28. Ventral plate with narrow median keel (Fig. 12G, H) .... *S.* (*G.*) *ambigens*
    Ventral plate without medial keel 29
29. Hind basitarsus enlarged (Fig. 26P) 30
    Hind basitarsus slender 31
30. Scutum entirely pruinose, with three faint dark vittae. Hind basitarsus 1.2–1.3 times as wide as hind femur ....*S.* (*G.*) *apoense*
    Scutum pruinose on shoulders, along lateral margins and on prescutellar area, without dark vittae. Hind basitarsus nearly as wide as hind femur ....
    .... *S.* (*G.*) *mindoroense*
31. Hind tibia medium to dark brown except base yellow or light brown. Antenna dark brown except scape, pedicel, and basal half of first flagellomere yellow .....
   32
    Hind tibia yellow on basal one-third to two-thirds. Antenna yellow except three or four or five apical flagellomeres darkened 37
32. Fore coxae brown. Ventral plate markedly produced ventrally (Fig. 16D,F) ....
    .... *S.* (*G.*) *baisasae*, *S.* (*G.*) *visayaense*, and *S.* (*G.*) *bicolense*
    Fore coxae whitish yellow to dark yellow. Ventral plate nearly flat, not produced ventrally 33
33. Scutum with brassy fine hairs 34
    Scutum with yellow fine hairs 35
34. Scutum whitish gray pruinose except median vitta and submedian round spots not shiny. Upper eye (large) facets in 14 vertical columns and 14 horizontal rows
    ....*S.* (*G.*) *babuyanense*
    Scutum whitish gray pruinose on shoulders, along lateral margins and on prescutellar area. Upper-eye (large) facets in 12 or 13 vertical columns and 12 or 13 horizontal rows ....*S.* (*G.*) *binuanense*
35. Fore basitarsus 5.5 times as long as its greatest width .... *S.* (*G.*) *mangasepi*
    Fore basitarsus 6.6–6.9 times as long as its greatest width 36
36. Opening of sensory vesicle large, as wide as sensory vesicle ....
    ....*S.* (*G.*) *samarense*
    Opening of sensory vesicle medium-sized, narrower than sensory vesicle ....
    .... *S.* (*G.*) *davaoense*
37. Tuft hair of base of radius dark brown ....*S.* (*G.*) *miblosi*
    Tuft hair of base of radius yellow 38
38. Hind tibia whitish yellow on basal three-fifths ....*S.* (*G.*) *repentinum*

Hind tibia whitish yellow on basal one-third to basal half 39

39. Upper-eye (large) facets in 12 or 13 vertical columns and 12 horizontal rows . ...*S.* (*G.*) *rarum*
Upper-eye (large) facets in 10 or 11 vertical columns and 10 or 11 horizontal rows 40

40. Ventral plate nearly parallel-sided when viewed ventrally ....*S.* (*G.*) *mindanaoense*
Ventral plate abruptly narrowed posteriorly when viewed ventrally 41

41. Style broad and nearly parallel-sided from base to apical one-third, then tapered apically when viewed ventrolaterally (similar to Fig. 30R) .... *S.* (*G.*) *epistum*
Style gradually narrowed from base to apex (similar to Fig. 30J) when viewed ventrolaterally...42

42. Upper-eye (large) facets in 10 vertical columns and 10 horizontal rows ....*S.* (*G.*) *montiblense*
Upper-eye (large) facets in 11 vertical columns and 11 horizontal rows 43

43. Wing length 1.4 mm .... *S.* (*G.*) *luzonicum*
Wing length 1.7 mm ....*S.* (*G.*) *lagunaense*

44. Hind basitarsus nearly parallel-sided and much narrower than hind tibia (similar to Fig. 45H) ....*S.* (*N.*) *aureohirtum*
Hind basitarsus swollen, spindle-shaped, and as wide as, or wider than hind tibia (similar to Fig. 43J) 45

45. Paramere with one hook (similar to Fig. 47Q) ....*S.* (*N.*) *aberrans*
Paramere with several hooks (similar to Fig. 43P) 46

46. Hind femur dark brown ....*S.* (*N.*) *bryopodium*
Hind femur yellow with apical tip dark brown ....*S.* (*N.*) *perlucidulum*

47. Scutum broadly silvery-pruinose with transverse, inverted-V-shaped, black band (Fig. 74S, T). Mid tarsus yellow except tarsomeres 4 and 5 brownish 48
Scutum otherwise. Mid tarsus entirely brownish or brownish except basal half or more of basitarsus yellow 52

48. Antenna brownish black with scape, pedicel and base of first flagellomere yellow. Style broad throughout its length when viewed ventrolaterally (similar to Fig. 74V) ....*S.* (*S.*) *latistylum*
Antenna yellow or dark yellow with two apical flagellomeres black (similar to Fig. 74A). Style broad from base to middle, then tapered to apex when viewed ventrolaterally (similar to Fig. 74C') 49

49. Style without apical spine ....*S.* (*S.*) *delfinadoae*
Style with apical spine 50

50. Upper-eye (large) facets in 17 horizontal rows ....*S.* (*S.*) *benquetense*
Upper-eye (large) facets in 13 horizontal rows 51

51. Silvery spots on abdominal segments 5 and 6 separated dorsomedially. Wing length 2.1 mm ....*S.* (*S.*) *leytense*
Silvery spots on abdominal segments 5 and 6 narrowly connected dorsally along anterior margin. Wing length 1.8 mm ....*S.* (*S.*) *baltazarae* and *S.* (*S.*) *cotabatoense*

52. Style with several spines at base (Fig. 80M) .... *S.* (*S.*) *quasifrenum*, *S.* (*S.*) *mainitense*, and *S.* (*S.*) *tukurongense*
    Style without spines at base 53
53. Fore coxa brown 54
    Fore coxa whitish yellow to yellow 58
54. Upper-eye (large) facets in 17 vertical columns and 17 horizontal rows. Aedeagal membrane bare (Fig. 68Y) ....*S.* (*S.*) *banahauense*
    Upper-eye (large) facets in 19–24 vertical columns and 19–26 horizontal rows. Aedeagal membrane with setae 55
55. Hind basitarsus black except subbasal portion somewhat paler. Hind basitarsus 5.5 times as long as its greatest width.......................................*S.* (*S.*) *subatrum*
    Hind basitarsus brownish black except basal two-fifths yellowish white. Hind basitarsus 3.6–4.1 times as long as its greatest width.....................................56
56. Hind basitarsus 3.6 times as long as its greatest width .... *S.* (*S.*) *forcipatum*
    Hind basitartsus 4.1 times as long as its greatest width 57
57. Sensory vesicle with large opening .... *S.* (*S.*) *sumigorense*
    Sensory vesicle with moderate-sized opening ....*S.* (*S.*) *conspicuum*
58. Hind basitarsus whitish on basal one-third 59
    Hind basitarsus whitish or yellowish on basal half to two-thirds 60
59. Scutum white-pruinose with median vitta and two submedian spots non-pruinose (similar to Fig. 68C) .... *S.* (*S.*) *concavum*
    Scutum white-pruinose on shoulders, narrowly along lateral margins and on prescutellar area, with rest median portion non-pruinose (similar to Fig. 68A) . ... *S.* (*S.*) *suplidoi*
60. Hind basitarsus whitish or yellowish on basal two-thirds or three-fifths 61
    Hind basitarsus whitish or yellowish on basal half or little more 64
61. Upper-eye (large) facets in 19 vertical columns and 18 horizontal rows ....*S.* (*S.*) *simulacrum*
    Upper-eye (large) facets in 14–16 vertical columns and 15–17 horizontal rows 62
62. Upper-eye (large) facets in 14 or 15 vertical columns and 15 horizontal rows. Paramere with fine setae on outer surface of base ....*S.* (*S.*) *neodiscrepans*
    Upper-eye (large) facets in 16 vertical columns and 16 or 17 horizontal rows. Paramere without fine setae on outer surface of base 63
63. Hind basitarsus 4.5 times as long as its greatest width .... *S.* (*S.*) *acostai*
    Hind basitarsus 5.0 times as long as its greatest with .... *S.* (*S.*) *iwahigense*
64. Scutum white-pruinose with median vitta and two submedian spots non-pruinose (similar to Fig. 68C) .... *S.* (*S.*) *tomentosum*
    Scutum white-pruinose on shoulders, narrowly along lateral margins and on prescutellar area, with rest median portion non-pruinose (similar to Fig. 68A) 65
65. Hind basitarsus 5.0 times as long as its greatest width .... *S.* (*S.*) *discrepans*
    Hind basitarsus 3.3–4.3 times as long as its greatest width 66
66. Upper-eye (large) facets in 12–14 vertical columns 67

Upper-eye (large) facets in 17–20 vertical columns 69

67. Hind basitarsus 4.3 times as long as its greatest width .... *S.* (*S.*) *salebrosum*
Hind basitarsus 3.3–3.5 times as long as its greatest width 68

68. Upper-eye (large) facets in 12 vertical columns .... *S.* (*S.*) *gatchaliani*
Upper-eye (large) facets in 14 vertical columns ....*S.* (*S.*) *melanopus*

69. Fore basitarsus 6.0 times as long as its greatest width .... S. (*S.*) *taalense*
Fore basitarsus 4.7–5.1 times as long as its greatest width 70

70. Ventral plate triangular, gradually narrowed to apex when viewed posteriorly. ....*S.* (*S.*) *malaibaense*
Ventral plate constricted medially or subbasally when viewed posteriorly 71

71. Two vertical rows of teeth of ventral plate parallel-sided when viewed posteriorly ....*S.* (*S.*) *pandum*
Two vertical rows of teeth of ventral plate convergent ventrally when viewed posteriorly .... *S.* (*S.*) *mayoyaoense*

Pupae****

1. Last abdominal segment with grapnel-like hooklets on each lateral surface. Gill with eight slender filaments arranged as 3+3+2 or (3+3)+2 from dorsal to ventral if grapnel-like hooklets are absent, though rarely 2
Last abdominal segment without grapnel-like hooklets on each lateral surface. Gill otherwise 34

2. Terminal hooks cone-like or triangular plate-like, each with outer margin smooth (similar to Fig. 11R) 3
Terminal hooks wide, plate-like, each with outer margin crenulated 13

3. Gill with four long thread-like filaments (similar to Fig. 11Q) ....*S.* (*G.*) *alienigenum*
Gill with six, eight, nine, or 10 filaments 4

4. Gill with six thread-like filaments 5
Gill with eight, nine, or 10 filaments 6

5. Stalk of middle pair of filaments of gill at acute angle to that of ventral pair of filaments when viewed laterally (Fig. 12J) .... *S.* (*G.*) *ambigens*
Stalk of middle pair of filaments of gill at right angle to that of ventral pair of filaments when viewed laterally (Fig. 16H) ....*S.* (*G.*) *baisasae*

6. Gill with eight, nine or 10 filaments arranged as 2+2+2+2 or 2+2+3+2 or 2+2+3+3 (two paired filaments of triplets short) from dorsal to ventral (Fig. 16I, J) ....*S.* (*G.*) *visayaense*
Gill with eight filaments arranged as 3+3+2, or 10 filaments arranged as 2+2+3+3 (two paired filaments of triplets long) from dorsal to ventral 7

7. Gill with 10 filaments arranged as 2+2+3+3 from dorsal to ventral (Fig. 16K) .... *S.* (*G.*) *bicolense*
Gill with eight filaments arranged as 3+3+2 from dorsal to ventral 8

8. Stalk of ventral pair of gill filaments long 9
Stalk of ventral pair of gill filaments short or medium-long 10

9. Two filaments of ventral pair of gill subequal in length and thickness. Dorsal and middle triplets of filaments of gill close together usually lying horizontally at base. Dorsum of abdominal segment 9 with distinct spine-combs .... *S.* (*G.*) *mangasepi*
Inner filament of ventral pair of gill somewhat longer and thicker than outer filament (Fig. 19R). Dorsal and middle triplets of filaments of gill usually lying vertically at base. Dorsum of abdominal segment 9 with weakly developed spine-combs .... *S.* (*G.*) *binuanense*
10. Gill with eight short slender filaments of same length and thickness, and half as long as pupal body (Fig. 35C). Outer surface of filaments of gill lacking ridges .... *S.* (*G.*) *montiblense*
Gill with eight long filaments, nearly as long as or little longer than pupal body. Outer surface of filaments of gill with ridges 11
11. Stalk of ventral pair of filaments of gill short .... *S.* (*G.*) *apoense*
Stalk of ventral pair of filaments of gill medium-long 12
12. Dorsal and middle triplets of filaments of gill sharing short stalk. Outer filament of ventral pair of gill 1.0–1.1 times as long as inner filament .... *S.* (*G.*) *samarense*
Middle triplet of filaments of gill arising between dorsal triplet and ventral pair of filaments. Outer filament of ventral pair of gill 1.3 times as long as inner filament .... *S.* (*G.*) *davaoense*
13. Gill with six slender or inflated filaments or with two short inflated structures having six slender filaments apically, or with three or four short somewhat inflated filaments having six slender filaments apically 14
Gill with eight slender filaments 25
14. Gill with six slender filaments 15
Gill with six inflated filaments or with two or three inflated structures having six slender filaments 20
15. Frons without tubercles 16
Frons with tubercles 17
16. Six filaments of gill arranged as 1+1+2+2 from dorsal to ventral (two dorsal filaments arising independently) (Fig. 18A) .... *S.* (*G.*) *bansonae*
Six filaments of gill arranged as (2+2)+2 from dorsal to ventral (two dorsal filaments with stalk) .... *S.* (*G.*) *atuganense*
17. Stalk of dorsal and middle paired filaments of gill lying side by side .... *S.* (*G.*) *salazarae*
Stalk of dorsal and middle paired filaments of gill lying vertically 18
18. Two filaments of ventral pair of gill as long as or slightly longer than four other filaments of dorsal and middle pairs (Fig. 18B) .... *S.* (*G.*) *liliwense*
Two filaments of ventral pair of gill 1.3–2.0 times as long as four other filaments of dorsal and middle pairs 19

19. Outer filament of ventral pair of gill nearly as thick as inner filament. Cocoon without any bulge anterodorsally .... *S.* (*G.*) *quezonense*
Outer filament of ventral pair of gill slightly thicker than inner filament. Cocoon with short bulge anterodorsally .... *S.* (*G.*) *katangladense*
20. Gill with six short inflated filaments 21
Gill with two inflated structures having six slender filaments, or with three or four inflated filaments having six slender filaments 22
21. Six inflated filaments of gill widened apically, with large round apices (Fig. 18H) .... *S.* (*G.*) *banauense*
Six inflated filaments of gill slightly narrowed apically, without large round apices (Fig. 18G) .... *S.* (*G.*) *lilianae*
22. Gill with two inflated structures having six slender filaments 23
Gill with three or four short inflated filaments having six slender filaments 24
23. Two inflated structures of gill of equal size, with remarkable pattern on their surface (Fig. 18E) .... *S.* (*G.*) *ifugaoense*
Two inflated structures of gill of different size, without any pattern on their surface (Fig. 18F) .... *S.* (*G.*) *manbucalense*
24. Gill with three short inflated filaments, each with two slender filaments apically (Fig. 18C) .... *S.* (*G.*) *cagayanense*
Gill with four short inflated filaments, each with one or two slender filaments apically (Fig. 18D) .... *S.* (*G.*) *riverai*
25. Last abdominal segment with a few setae (in place of grapnel-like hooklets) on each lateral surface 26
Last abdominal segment with grapnel-like hooklets on each lateral surface 27
26. Stalk of ventral pair of gill filaments much shorter than common basal stalk. Cocoon shoe-shaped, with low anteroventral collar .... *S.* (*G.*) *miblosi*
Stalk of ventral pair of gill filaments much longer than common basal stalk (Fig. 32Q). Cocoon wall-pocket-shaped, without anteroventral collar .... *S.* (*G.*) *rarum*
27. Two filaments of ventral pair of gill subequal in length and thickness 28
Outer filament of ventral pair of gill much longer and thicker than inner filament 30
28. Outer margin of terminal hook 3.6–3.9 times as long as inner margin (Fig. 27F) .... *S.* (*G.*) *mindoroense*
Outer margin of terminal hook 2.5–2.7 times as long as inner margin 29
29. Longest filament of gill 5.3 mm .... *S.* (*G.*) *babuyanense*
Longest filament of gill 2.7 mm .... *S.* (*G.*) *bataksense*
30. Eight filaments of gill arranged as 1+1+1+1+1+1+2 from dorsal to ventral (six dorsal filaments arising independently) (Fig. 31C) .... *S.* (*G.*) *repentinum*
Eight filaments of gill arranged as 3+3+2 from dorsal to ventral 31
31. Outer filament of ventral pair of gill ca. 1.5 times as long as seven other filaments .... *S.* (*G.*) *lagunaense*
Outer filament of ventral pair of gill twice as long as six or seven other filaments 32

32. Inner filament of ventral pair of gill much shorter than outer filament but much longer than six other filaments of dorsal and middle triplets ....*S.* (*G.*) *mindanaoense*
Inner filament of ventral pair of gill subequal in length to six other filaments of dorsal and middle triplets 33
33. Outer filament of ventral pair of gill twice as thick as seven other filaments ... . *S.* (*G.*) *epistum*
Outer filament of ventral pair of gill ca. 1.4 times as thick as seven other filaments ....*S.* (*G.*) *luzonicum*
34. Abdominal segments 6 and 7 each with inner hook and lacking outer hook ventrally on each side 35
Abdominal segments 6 and 7 each with inner hook and outer hook ventrally on each side 48
35. Cocoon shoe-shaped (Fig. 95L–O) 36
Cocoon wall-pocket-shaped (Fig. 95H–K) 39
36. Antennal sheath with spinous projections (Fig. 95D) .... *S.* (*W.*) *ogonukii*
Antennal sheath without spinous projections (Fig. 95C) 37
37. Gill with four filaments in two pairs lying vertically .... *S.* (*W.*) *tenederoi*
Gill with four filaments arising independently, of which two middle filaments lying horizontally 38
38. All gill filaments subequal in length ....*S.* (*W.*) *marilogense*
All gill filaments subequal in length except dorsal filament 0.75 times as long as three other filaments ....*S.* (*W.*) *molawinense*
39. Gill with four slender filaments directed forwards. Cuticle of filaments with annular ridges and furrows 40
Gill with four slender or inflated filaments, of which dorsal filament directed upward, then inward and/or further backward. Cuticle of filaments without annular ridges and furrows 43
40. Gill with four filaments arranged as (2+1)+1 from dorsal to ventral (Fig. 94A) .... *S.* (*W.*) *suyoense*
Gill with four filaments arranged as 2+2 from dorsal to ventral 41
41. Frons with two pairs of unbranched trichomes (similar to Fig. 95A). Antennal sheath with spinous projections (similar to Fig. 95D). Cocoon not produced anterodorsally .... *S.* (*W.*) *tuyense*
Frons with three pairs of branched trichomes (Fig. 95B). Antennal sheath without spinous projections (similar to Fig. 95C). Cocoon much produced anterodorsally, like eaves, often beyond anterior tips of ventrolateral margins (Fig. 95H, I) 42
42. All gill filaments much shorter than pupal body (Fig. 94C) .... *S.* (*W.*) *claveriaense*
All gill filaments slightly longer than pupal body ....*S.* (*W.*) *makilingense*
43. Gill with four slender filaments, of which dorsal filament of dorsal pair directed upward, then inward .... *S.* (*W.*) *amplum*
Gill with four slender or inflated filaments, of which dorsal filament of dorsal pair directed upward, then inward and further backward 44

44. Cuticle of gill filaments with protuberances of various shapes and sizes throughout entire surface 45
Cuticle of gill filaments smooth, without any protuberances 46
45. Antennal sheaths with eight groups of one to three spinous projections. Gill filaments each with large protuberances having one or more black points apically, in two subparallel rows on outer surface (Fig. 94D) .... *S.* (*W.*) *spinosibranchium*
Antennal sheaths without spinous projections. Gill filaments each with numerous tiny nodule-like protuberances throughout its surface (Fig. 94F) .... *S.* (*W.*) *carinatum*
46. All gill filaments much inflated, of which inner middle filament longest, and two middle filaments and one ventral filament with small nipple-like apical cap (Fig. 94G). Cocoon with short, broad anterodorsal projection .... *S.* (*W.*) *cabrerai*
All gill filaments slender or somewhat inflated, without any cap apically, of which outer middle filament longest of all. Cocoon without anterodorsal projection 47
47. All gill filaments somewhat inflated, of which inner middle filament much shorter than half of outer middle filament (Fig. 94H) ....*S.* (*W.*) *resimum*
All gill filaments slender, of which inner middle filament much longer than half of outer middle filament (Fig. 94E) .... *S.* (*W.*) *recurvum*
48. Cocoon wall-pocket-shaped, extended ventrolaterally 49
Cocoon wall-pocket-shaped, not extended ventrolaterally, or shoe-shaped 51
49. Gill with four thread-like filaments ....*S.* (*N.*) *aberrans*
Gill with six thread-like or somewhat inflated filaments 50
50. Gill filaments directed forward close together. Abdominal segments 6–8 each with spine-combs dorsally. Cocoon simple, wall-pocket-shaped, without anterodorsal projection ....*S.* (*N.*) *perlucidulum*
Gill filaments diverged basally. Abdominal segments 7 and 8 each with spine-combs dorsally. Cocoon with short anterodorsal projection .... *S.* (*N.*) *aureohirtum*
51. Gill composed of six inflated tubular filaments (Fig. 75D) ..*S.* (*S.*) *delfinadoae*
Gill composed of 6–38 thread-like filaments 52
52. Gill with 38 thread-like filaments (Fig. 49P) ....*S.* (*S.*) *arboreum*
Gill with 6–12 thread-like filaments 53
53. Gill with 12 filaments (Fig. 75A) ....*S.* (*S.*) *cotabatoense*
Gill with 6–10 filaments 54
54. Gill with 10 filaments (Fig. 75B) ....*S.* (*S.*) *baltazarae*
Gill with six or eight filaments 55
55. Gill with eight filaments (Fig. 69I) .... *S.* (*S.*) *sumigorense*
Gill with six filaments 56
56. Cocoon wall-pocket-shaped 57
Cocoon shoe-shaped 62
57. Terminal hooks present 58

Terminal hooks absent 60

58. Most of dorsal surface of thorax with sharply-pointed cone-like tubercles (similar to Fig. 81B) ....*S.* (*S.*) *quasifrenum*

Most of dorsal surface of thorax with round tubercles (similar to Fig. 81A) ...59

59. Gill filaments 2.0–2.2 mm long ....*S.* (*S.*) *mainitense*

Gill filaments 1.3–1.6 mm long .... *S.* (*S.*) *tukurongense*

60. Frons bare .... *S.* (*S.*) *forcipatum*

Frons with tubercles 61

61. Cocoon with anterodorsal margin much longer than ventrolateral margin when viewed laterally (Fig. 69T) ....*S.* (*S.*) *conspicuum*

Cocoon with anterodorsal margin somewhat shorter than ventrolateral margin when viewed laterally ....*S.* (*S.*) *aquilum*

62. Abdominal segments 5–9 without spine-combs dorsally ....*S.* (*S.*) *discrepans* and *S.* (*S.*) *neodiscrepans*

At least abdominal segment 8 with spine-combs dorsally 63

63. Cuticle of filaments of gill with conspicuous, rather large tubercles on ridges as well as minute ones on inter-ridges 64

Cuticle of filaments of gill with small tubercles of subequal size 65

64. Integument of thorax irregularly raised, forming numerous large ridges dorsally and laterally .... *S.* (*S.*) *salebrosum*

Integument of thorax normal, without such raised ridges ......*S.* (*S.*) *melanopus*

65. Frons moderately or densely covered with tubercles 66

Frons bare or very sparsely covered with tubercles 76

66. Frons densely covered with tubercles on lower portion and along lateral portions but sparsely covered with tubercles on median portion 67

Frons uniformly covered with tubercles 69

67. Frons with three pairs of trichomes ....*S.* (*S.*) *suplidoi*

Frons with two pairs of trichomes 68

68. Angle between uppermost and lowermost filaments of gill 70 degrees when their basal portions viewed laterally (Fig. 69E) .... *S.* (*S.*) *concavum*

Angle between uppermost and lowermost filaments of gill 110–120 degrees when their basal portions viewed laterally ....*S.* (*S.*) *gatchaliani*

69. Cuticle of gill filaments with inconspicuous transverse ridges (not sharply edged) 70

Cuticle of gill filaments with conspicuous transverse ridges (sharply edged) 71

70. Abdominal segment 1 with micro-tubercles dorsally. Abdominal segment 9 without spine-combs dorsally .... *S.* (*S.*) *abatanense*

Abdominal segment 1 without micro-tubercles dorsally. Abdominal segment 9 with spine-combs dorsally ....*S.* (*S.*) *subatrum*

71. Abdominal segment 1 without micro-tubercles dorsally ......*S.* (*S.*) *simulacrum*

Abdominal segment 1 with micro-tubercles dorsally 72

72. Angle between uppermost and lowermost filaments of gill more than 110 degrees when their basal portions viewed laterally 73
    Angle between uppermost and lowermost filaments of gill less than 100 degrees when their basal portions viewed laterally 75
73. Angle between dorsalmost and ventralmost filaments of gill 180 degrees when their basal portions viewed laterally .... *S.* (*S.*) *malaibaense*
    Angle between dorsalmost and ventralmost filaments of gill 110–130 degrees when their basal portions viewed laterally 74
74. Two dorsomedial trichomes absent or very short, if any .... *S.* (*S.*) *taalense*
    Two dorsomedial trichomes medium-long .... *S.* (*S.*) *mayoyaoense*
75. Angle between dorsalmost and ventralmost filaments of gill 90 degrees when their basal portions viewed laterally...... *S.* (*S.*) *acostai* and *S.* (*S.*) *banahauense*
    Angle between dorsalmost and ventralmost filaments of gill 60–70 degrees when their basal portions viewed laterally .... *S.* (*S.*) *halconense* and *S.* (*S.*) *pandum*
76. Abdominal segment 1 with tubercles dorsally ....*S.* (*S.*) *iwahigense*
    Abdominal segment 1 without tubercles dorsally 77
77. Two filaments of dorsal pair of gill short, about half as long as those of ventral pair ....*S.* (*S.*) *latistylum*
    Two filaments of dorsal pair of gill slightly shorter than those of ventral pair 78
78. Angle between dorsalmost and ventralmost filaments of gill 120 degrees when their basal portions viewed laterally ....*S.* (*S.*) *leytense*
    Angle between dorsalmost and ventralmost filaments of gill 90 degrees when their basal portions viewed laterally .... *S.* (*S.*) *benquetense*

Mature larvae****

1. Lateral margins of hypostoma smooth. Main tooth of mandibular serration at acute angle to mandible apically 2
   Lateral margins of hypostoma serrated. Main tooth of mandibular serration at right or obtuse angle to mandible apically 20
2. Rectal organ simple, without secondary lobules 3
   Rectal organ compound, with secondary lobules 6
3. Postgenal cleft ca. 1.7 times as long as postgenal bridge .... *S.* (*G.*) *ambigens*
   Postgenal cleft 2.2–2.5 times as long as postgenal bridge 4
4. Pharate pupal gill with six filaments ....*S.* (*G.*) *baisasae*
   Pharate pupal gill with eight, nine, or 10 filaments 5
5. Pharate pupal gill with eight, nine, or 10 filaments (two paired filaments of triplets short in case of 10 filaments) ....*S.* (*G.*) *visayaense*
   Pharate pupal gill with 10 filaments (two paired filaments of triplets long) ....*S.* (*G.*) *bicolense*
6. Pharate pupal gill with four long filaments .... *S.* (*G.*) *alienigenum*
   Pharate pupal gill with six or eight filaments 7

7. Postgenal cleft narrow, 0.2–0.25 times as wide as head capsule. Pharate pupal gill with six slender or inflated filaments 8
   Postgenal cleft moderate, 0.28–0.31 times as wide as head capsule. Pharate pupal gill with eight filaments arranged as 3+3+2 from dorsal to ventral 13
8. Pharate pupal gill with six slender filaments 9
   Pharate pupal gill with two inflated structures with six slender filaments, or six short inflated filaments, or three or four inflated filaments having six slender filaments 10
9. Postgenal cleft 3.6 times as long as postgenal bridge .... *S.* (*G.*) *liliwense*
   Postgenal cleft 2.2 times as long as postgenal bridge .... *S.* (*G.*) *salazarae*
10. Pharate pupal gill with two inflated structures with six slender filaments .... *S.* (*G.*) *manbucalense*
   Pharate pupal gill with six short inflated filaments, or three or four inflated filaments having six slender filaments 11
11. Pharate pupal gill with six short inflated filaments with large round apices .... *S.* (*G.*) *banauense*
   Pharate pupal gill with three or four inflated filaments having six slender filaments apically 12
12. Pharate pupal gill with three short inflated filaments, each with two slender filaments apically .... *S.* (*G.*) *cagayanense*
   Pharate pupal gill with four short inflated filaments each with one or two slender filaments apically .... *S.* (*G.*) *riverai*
13. Postgenal cleft deep, reaching or with its apex close to posterior margin of hypostoma 14
   Postgenal cleft ca. 6.0 times as long as postgenal bridge 16
14. Postgenal cleft slightly constricted basally .... *S.* (*G.*) *lagunaense*
   Postgenal cleft moderately constricted basally 15
15. Labral fan with 45 primary rays .... *S.* (*G.*) *apoense*
   Labral fan with 35–38 primary rays .... *S.* (*G.*) *mindoroense*
16. Postgenal cleft rounded apically .... *S.* (*G.*) *mangasepi*
   Postgenal cleft pointed apically 17
17. Pharate pupal gill with eight short slender filaments of same length .... *S.* (*G.*) *montiblense*
   Pharate pupal gill with eight filaments, of which one filament of ventral pair much longer than seven other filaments 18
18. Rectal organ with three or four thumb-like secondary lobules per lobe .... *S.* (*G.*) *luzonicum*
   Rectal organ with four to eight finger-like secondary lobules per lobe 19
19. Body length 5.0–5.5 mm. Cuticle of eight filaments of pharate pupal gill with annular ridges .... *S.* (*G.*) *mindanaoense*
   Body length 3.5–3.8 mm. Cuticle of one filament of ventral pair of pharate pupal gill with annular ridges, but those of seven other filaments lacking annular ridges .... *S.* (*G.*) *epistum*
20. Pectination of uppermost primary ray of labral fan (Fig. 95V) different from those of other primary rays (Fig. 95W) 21

Pectination of uppermost primary ray of labral fan similar to those of other primary rays 30

21. Accessory sclerites absent 22
Accessory sclerites present (Fig. 95Z) 25

22. Posterior circlet with 84 rows of 14 hooklets per row .... S. (*W.*) *tuyense*
Posterior circlet with 100–113 rows of 18 or 19 hooklets per row 23

23. Head spots indistinct or very faintly positive .... S. (*W.*) *suyoense*
Head spots distinctly positive 24

24. Posterior circlet with 113 rows of 19 hooklets per row. Labral fan with 36 main rays .... S. (*W.*) *marilogense*
Posterior circlet with 100 rows of 18 hooklets per row. Labral fan with 41 main rays .... S. (*W.*) *ogonukii*

25. Posterior circlet with 152 rows of 21 hooklets per row. Rectal organ simple, without secondary lobules .... (*W.*) *tenederoi*
Posterior circlet with 110–134 rows of 18–20 hooklets per row. Rectal organ compound, with one to seven secondary lobules per lobe 26

26. Body length 8.0 mm. Hypostoma with 10–12 bristles on each side .... S. (*W.*) *cabrerai*
Body length 5.8–7.2 mm. Hypostoma with seven to nine bristles on each side 27

27. Posterolateral spots on cephalic apotome merged with dark background .... S. (*W.*) *resimum*
Posterolateral spots on cephalic apotome well marked, contrasting to clear background 28

28. Pharate pupal gill with four inflated filaments .... S. (*W.*) *carinatum*
Pharate pupal gill with four slender filaments 29

29. All filaments of pharate pupal gill with annular ridges and furrows. Rectal organ simple or with one or two secondary lobules per lobe ...... S. (*W.*) *claveriaense*
All filaments of pharate pupal gill without annular ridges and furrows. Rectal organ with four to seven secondary lobules per lobe .... S. (*W.*) *recurvum*

30. Ventral papillae large 31
Ventral papillae absent or small 33

31. Postgenal cleft medium-long, slightly shorter than postgenal bridge .... S. (*N.*) *aureohirtum*
Postgenal cleft short, about half as long as postgenal bridge 32

32. Abdomen with reddish-brown markings. Accessary sclerites absent .... S. (*N.*) *perlucidulum*
Abdomen without such colored markings. Accessory sclerites present .... S. (*N.*) *aberrans*

33. Postgenal cleft deep, reaching posterior margin of hypostoma 34
Postgenal cleft medium-long, not reaching posterior margin of hypostoma 39

34. Pharate pupal gill with inflated filaments .... S. (*S.*) *delfinadoae*
Pharate pupal gill with slender, thread-like filaments 35

35. Pharate pupal gill with 10 or 12 filaments 36
Pharate pupal gill with six filaments 37

36. Pharate pupal gill with 10 filaments .... *S.* (*S.*) *baltazarae*
Pharate pupal gill with 12 filaments .... *S.* (*S.*) *cotabatoense*
37. Posterior circlet with ca. 70 rows of hooklets .... *S.* (*S.*) *latistylum*
Posterior circlet with ca. 80 rows of hooklets 38
38. Labral fan with 40 primary rays. Rectal organ with one to three secondary lobules per lobe .... *S.* (*S.*) *leytense*
Labral fan with 32 primary rays. Rectal organ with five to eight secondary lobules per lobe .... *S.* (*S.*) *benquetense*
39. Body length 3.5–5.0 mm 40
Body length 5.5–10.3 mm 43
40. Abdomen with reddish-brown transverse band on each segment. Postgenal cleft pointed apically .... *S.* (*S.*) *quasifrenum*
Abdomen without such reddish-brown band on each segment. Postgenal cleft rounded apically (Fig. 69W) 41
41. Posterior circlet with ca. 88 rows of up to 14 hooklets .... *S.* (*S.*) *discrepans*
Posterior circlet with 96–104 rows of up to 16 or 17 hooklets 42
42. Labral fan with ca. 42 primary rays. Rectal organ with six to eight secondary lobules per lobe .... *S.* (*S.*) *iwahigense*
Labral fan with 49–54 primary rays. Rectal organ with zero to four secondary lobules per lobe .... *S.* (*S.*) *neodiscrepans*
43. Hypostoma with 11–14 bristles on each side. Posterior circlet with 216–230 rows of up to 30 hooklets 44
Hypostoma with five to nine bristles on each side. Posterior circlet with 88–160 rows of up to 14–20 hooklets 45
44. Posterior circlet with 216–220 rows of hooklets .... *S.* (*S.*) *forcipatum*
Posterior circlet with 230 rows of hooklets .... *S.* (*S.*) *conspicuum*
45. Posterior circlet with 160 rows of hooklets .... *S.* (*S.*) *subatrum*
Posterior circlet with 88–134 rows of hoolkets 46
46. Postgenal cleft bullet-shaped, rounded apically and nearly parallel-sided or slightly convergent basally 47
Postgenal cleft rounded apically and laterally 48
47. Rectal organ with five or six secondary lobules per lobe. Posterior circlet with 112 rows of up to 16 hooks .... *S.* (*S.*) *taalense*
Rectal organ with 11–14 secondary lobules per lobe. Posterior circlet 124–134 rows of 17–20 hooklets .... *S.* (*S.*) *malaibaense*
48. Pharate pupal gill with ordinary small tubercles on filaments. Rectal organ with 8–12 secondary lobules per lobe .... *S.* (*S.*) *acostai* and *S.* (*S.*) *suplidoi*
Pharate pupal gill with relatively large tubercles on filaments. Rectal organ with five to eight secondary lobules per lobe 49
49. Cephalic apotome pale except portion near posterior margin darkened ....
.... *S.* (*S.*) *salebrosum*
Cephalic apotome darkened on posterior half .... *S.* (*S.*) *melanopus*

*Keys to identify 57 species of black flies in the Philippines provided by Takaoka (1983) are updated.

**_Simulium_ _(G.) davaoense, S. (G.) manbucalense, S. (G.) katangladense, S. (G.) rarum, S. (G.) repentinum, S. (G.) samarense, S. (N.) bryopodium, S. (S.) canlaonense, S. (S.) simulacrum, S. (S.) tomentosum,_ and _S. (W.) molawinense_ are not included because their females are unknown. _Simulium (S.) retusum_ is not included because its original description was superficial.

***_Simulium (G.) lilianae, S. (N.) bryopodium, S. (S.) abatanense, S. (S.) aquilum, S. (S.) arboreum, S. (S.) atrum, S. (S.) canlaonense, S. (S.) halconense, S. (S.) palawanense_, and _S. (S.) retusum_ are not included because their males are unknown.

****_Simulium (N.) bryopodium, S. (S.) atrum, S. (S.) canlaonense, S. (S.) palawanense, S. (S.) retusum_, and _S. (S.) tomentosum_ are not included because their pupae are unknown.

***** _Simulium (G.) atuganense, S. (G.) bansonae, S. (G.) bataksense, S. (G.) binuanense, S. (G.) davaoense, S. (G.) ifugaoense, S. (G.) katangladense, S. (G.) lilianae, S. (G.) miblosi, S. (G.) quezonense, S. (G.) rarum, S. (G.) repentinum, S. (G.) samarense, S. (N.) bryopodium, S. (S.) abatanense, S. (S.) aquilum, S. (S.) arboreum, S. (S.) atrum, S. (S.) banahauense, S. (S.) canlaonense, S. (S.) concavun, S. (S.) halconense, S. (S.) gatchaliani, S. (S.) mainitense, S. (S.) mayoyaoense, S. (S.) palawanense, S. (S.) pandum, S. (S.) retusum, S. (S.) simulacrum, S. (S.) sumigorense, S. (S.) tomentosum, S. (S.) tukurongense, S. (W.) amplum, S. (W.) makilingense, S. (W.) molawinense_, and _S. (W.) spinosibranchium_ are not included because their larvae are unknown.

Figures

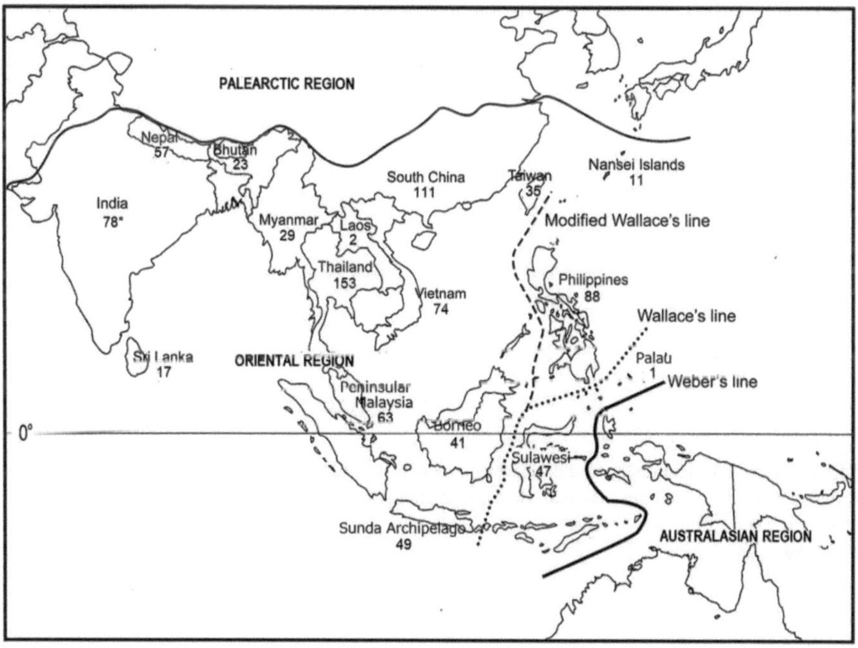

Fig. 1 Map of the Oriental Region showing its northern and eastern boundaries and the numbers of black fly species recorded from each country or area within it. * No. of Indian black fly species does not include 13 species endemic to Kashmir and Himachal Pradesh, which are located in the Palearctic Region

H. Takaoka, *The Black Flies of Subtropical and Tropical Asia*,
https://doi.org/10.1007/978-981-99-5219-9

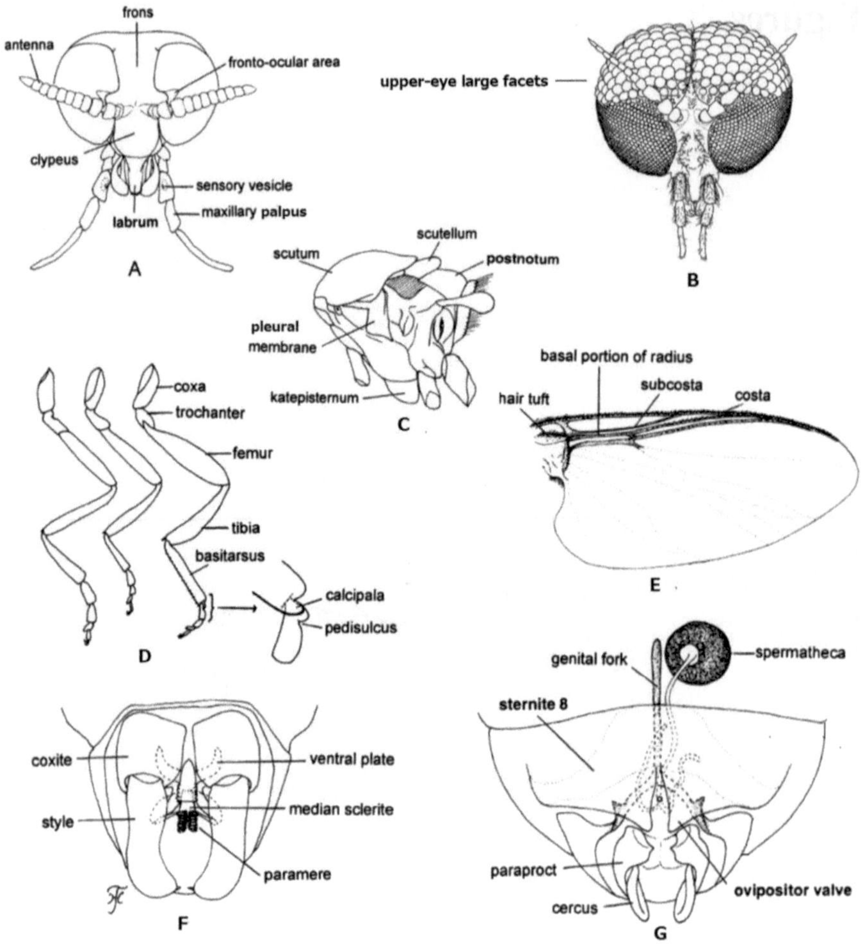

Fig. 2 Morphological features and their terms of adults of black flies. A & B, Heads (front view; A, female; B, male). C, Thorax (lateral view). D, Female legs (fore, mid, and hind legs from left). E, Wing. F, Male genitalia (ventral view). G, Female terminalia (ventral view). [Modified from Takaoka (2003)]

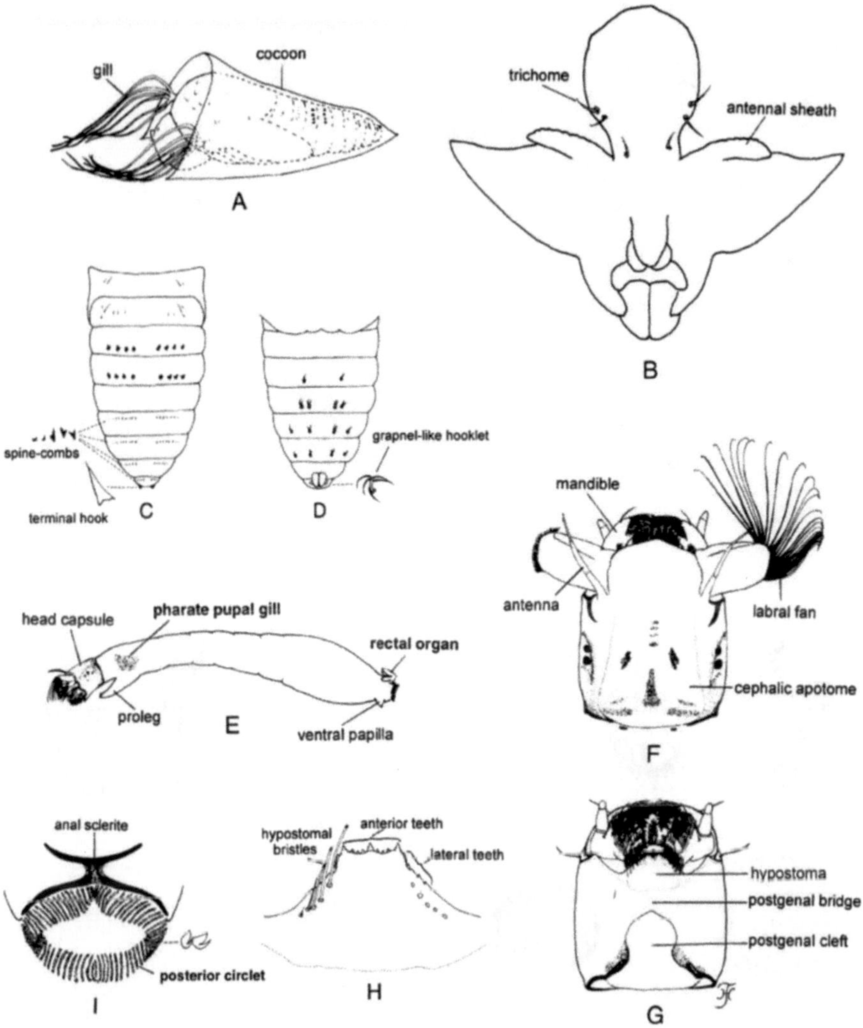

Fig. 3 Morphological features and their terms of the pupa and larva of black flies. A, Pupa in a cocoon. B, Head integument of pupa spread and flattened (front view). C & D, Pupal abdomens (C, dorsal view; D, ventral view). E, Mature larva (lateral view). F & G, Head capsules of larva (F, dorsal view; G, ventral view). H, Hypostoma. I, Posterior tip of the larval abdomen (posterodorsal view) [Reproduced from Takaoka (2003)]

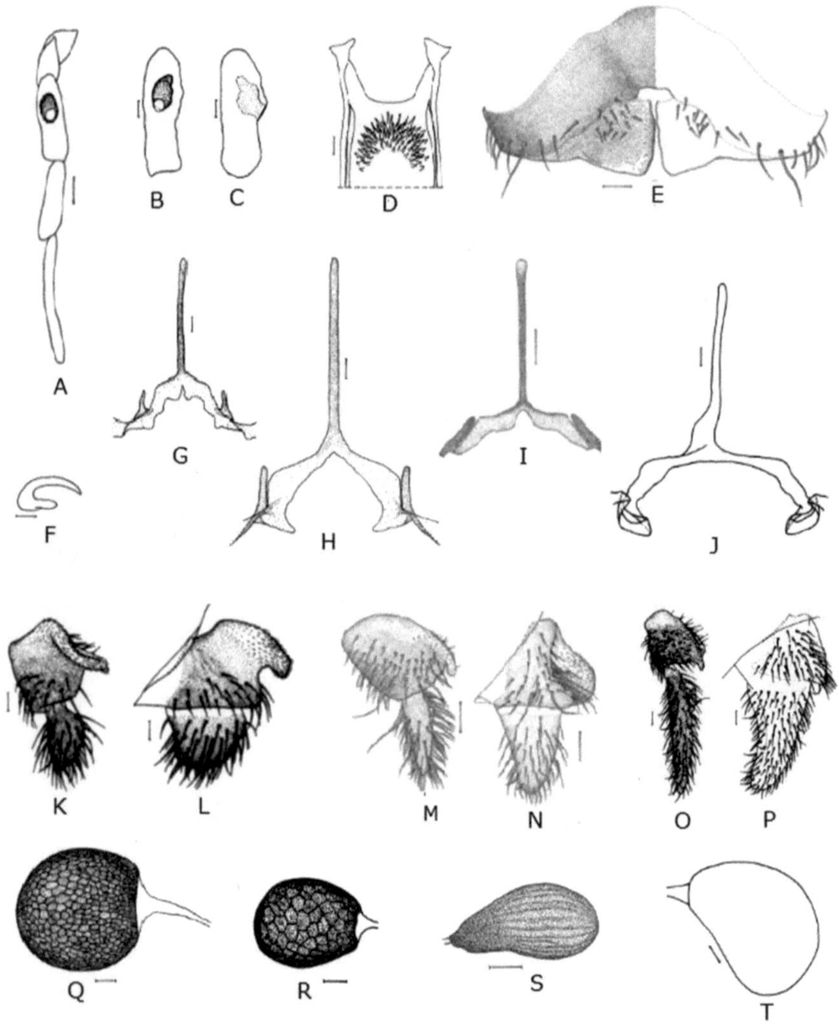

Fig. 4 Female of the subgenus *Asiosimulium.* A, Maxillary palpus (front view). B & C, Third palpal segments with sensory vesicles (B, front view; C, lateral view). D, Cibarium showing a cluster of minute spinous processes. E, Sternite 8 and ovipositor valves (ventral view). F, Claw. G–J, Genital forks of various types. K–P, Paraprocts and cerci of various types (K, M & O, ventral view; L, N & P, lateral view). Q–T, Spermathecae of various types. A, *S. wanchaii*; B, C, J–L & T, *S. suchitrae*; E, I, M, N & S, *S. furvum*; D, F, G, O, P & R, *S. oblongum*; H & Q, *S. saeungae.* Scale bars: 0.05 mm for A; 0.04 mm for E, I, M, N & S; 0.02 mm for B–D, G, H, J–L, O, P & T; 0.01 mm for F

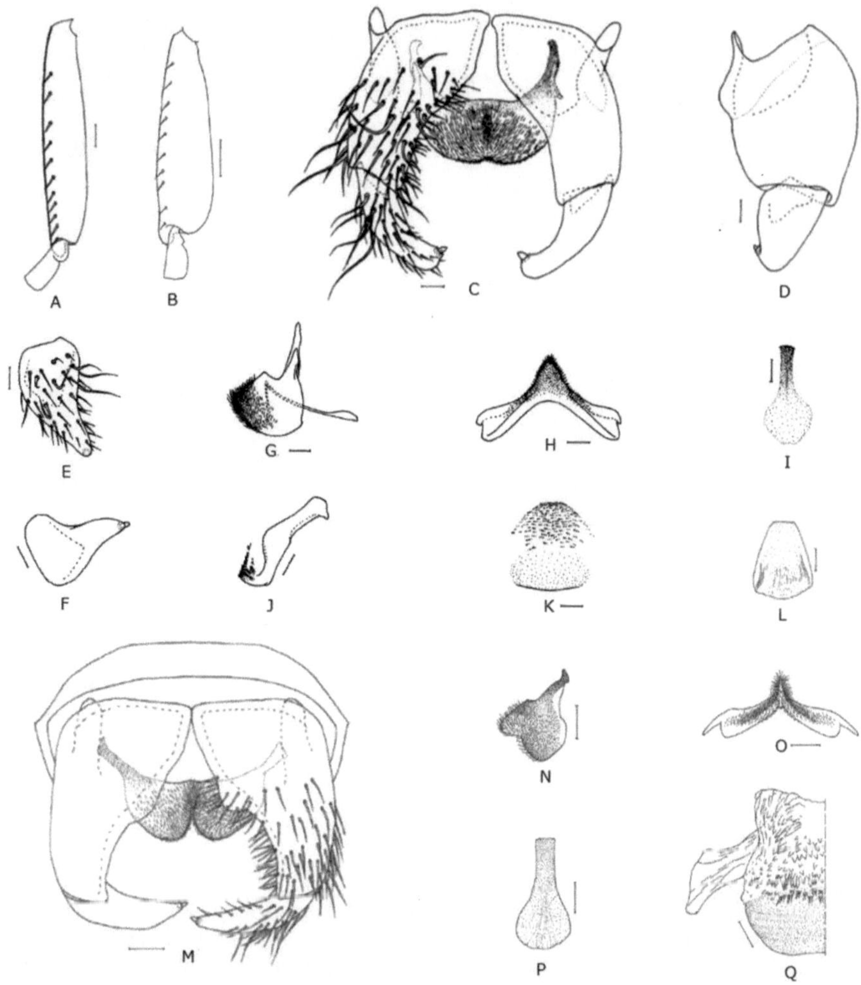

Fig. 5 Male of the subgenus *Asiosimulium*. A & B, Hind basitarsi and second tarsomeres (left side; outer view). C & M, Coxites, styles and ventral plates (and ninth sternal plate in M) (ventral view). D, Coxite and style (right side; lateral view). E & F, Styles (right side; E, ventrolateral view; F, caudal view). G, Ventral plate and median sclerite (lateral view). H & O, Ventral plates (caudal view). I, L & P, Median sclerites (caudal view). J, Paramere (left side; caudal view). K, Aedeagal membrane and dorsal plate (caudal view). N, Ventral plate (lateral view). Q, Paramere, aedeagal membrane, and dorsal plate (right side; caudal view). A & C–K, *S. oblongum*; B & M–Q, *S. furvum*; L, *S. wanchaii*. Scale bars: 0.1 mm for B; 0.05 mm for A; 0.04 mm for M–Q; 0.02 mm for C–L

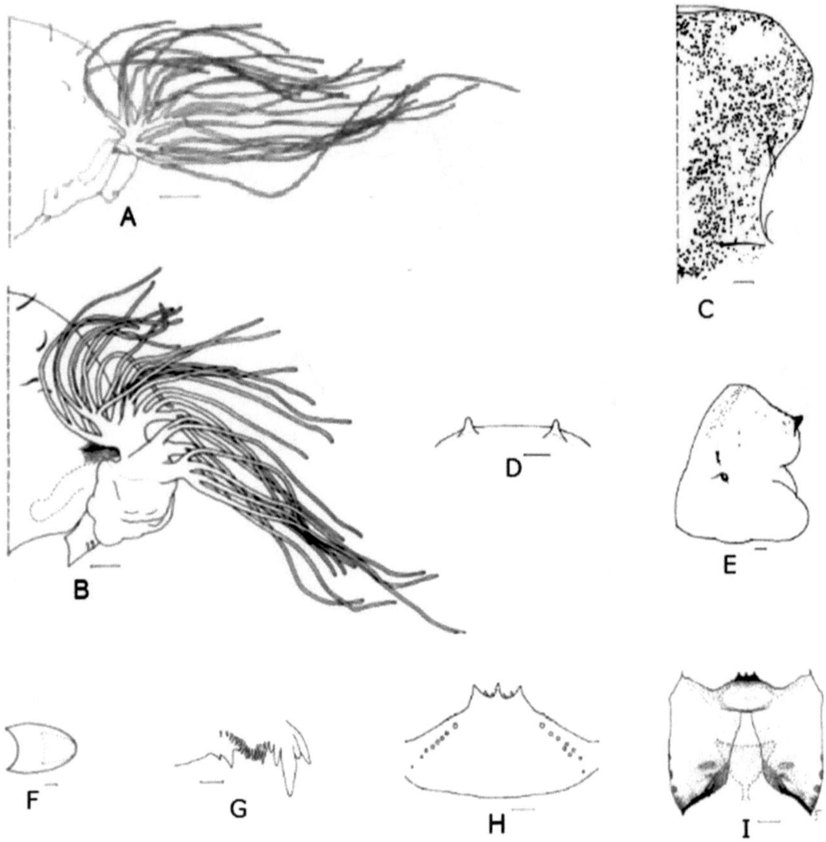

Fig. 6 Pupa and larva of the subgenus *Asiosimulium.* A–F, Pupa; G–I, Larva. A & B, Gill filaments (and inflated basal fenestra in B) (right side; lateral view). C, Frons (left half, anterodorsal view). D, Terminal hooks (caudal view). E, Last abdominal segment showing unbranched and grapnel-like hooklets (lateral view). F, Cocoon (lateral view). G, Apical portion of mandible (lateral view). H, Hypostoma (ventral view). I, Head capsule showing postgenal cleft and dark-pigmented sheath of subesophageal ganglion (ventral view). A, *S. wanchaii*; B, D & F, *S. suchitrae*; C & E, *S. oblongum*; G, H & I, *S. furvum*. Scale bars: 0.5 mm for F; 0.1 mm for A, B & I; 0.05 mm for C; 0.04 mm for H; 0.02 mm for D, E & G

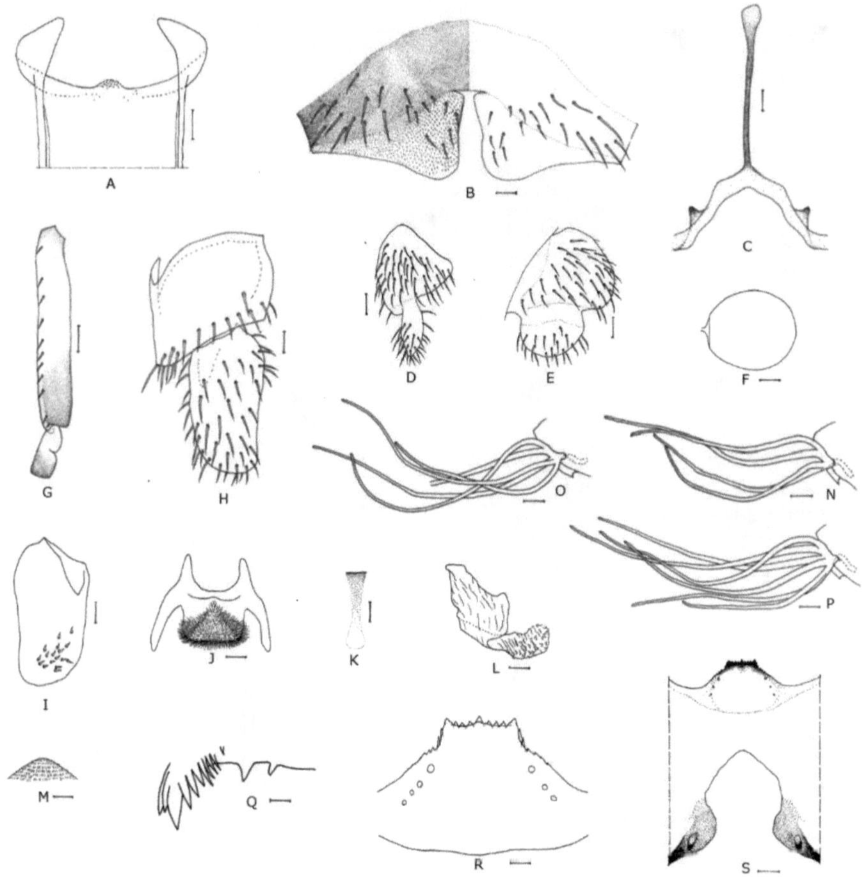

Fig. 7 Female, male, pupa and larva of the subgenus *Boophthora*. A–F, Female; G–M, Male; N–P, Pupa; Q–S, Larva. A, Cibarium. B, Sternite 8 and ovipositor valves (ventral view). C, Genital fork (ventral view). D & E, Paraprocts and cerci (right side; D, ventral view; E, lateral view). F, Spermatheca. G, Hind basitarsus and second tarsomere (left side; outer view). H, Coxite and style (right side; ventrolateral view). I, Style (right side; dorsal view). J, Ventral plate (ventral view). K, Median sclerite (caudal view). L, Paramere (right side; posteroventral view). M, Aedeagal membrane (caudal view). N–P, Gills (left side; outer view). Q, Apical portion of mandible. R, Hypostoma. S, Middle portion of head capsule (ventral view). D, E, G–K & N–Q, *S. yonagoense*: A–C, F, L, M, R & S, *S. makunbei*. Scale bars: 0.2 mm for N–P; 0.05 mm for G & S; 0.02 mm for A–F, H–M & R; 0.01 mm for Q

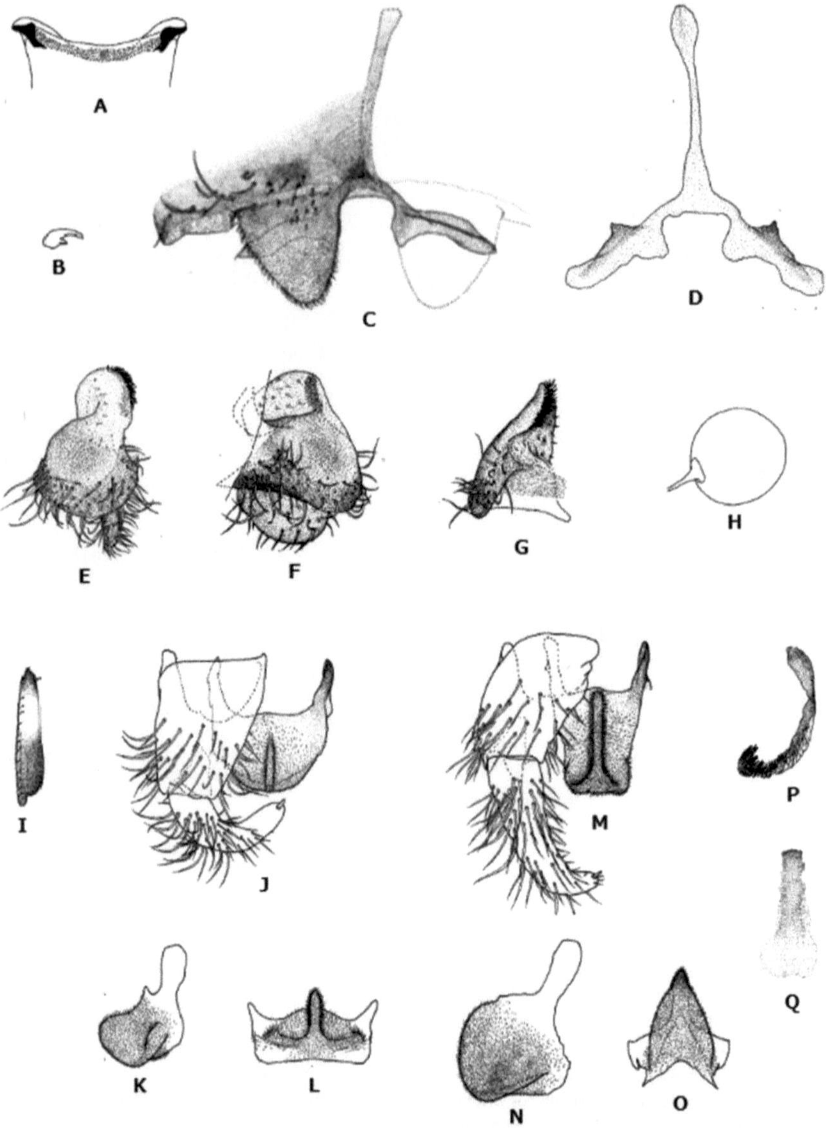

Fig. 8 Female and male of the subgenus *Daviesellum.* A–H, Female; I–Q, Male. A, Cibarium. B, Claw. C, Sternite 8, ovipositor valves and genital fork (ventral view). D, Genital fork. E & F, Paraprocts and cerci (right side; E, ventral view; F, lateral view). G, Paraproct (ventromedial view). H, Spermatheca. I, Hind basitarsus (left side; outer view). J & M, Coxites, styles (only right side) and ventral plates (ventral view). K, L, N & O, Ventral plates (K & N, lateral view; L & O, caudal view). P, paramere (right side; dorsal view). Q, Median sclerite (caudal view). A–C, E–G & I–L, *S. pahangense*; D, H & M–Q, *S. courtneyi*

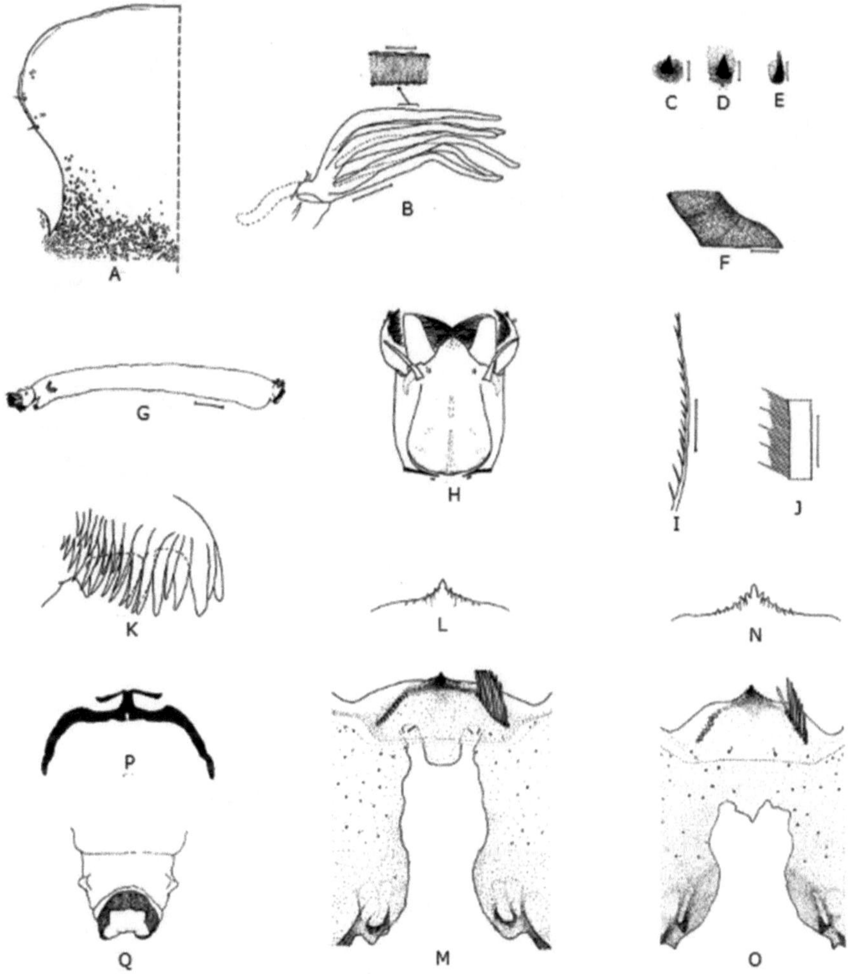

Fig. 9 Pupa and larva of the subgenus *Daviesellum.* A–F, Pupa; G–Q, Larva. A, Frons (right half; dorsal view). B, Gill filaments (right side; lateral view). C–E, Spinous hooks (C, on dorsum of second segment; D, on dorsum of third segment; E, on ventral surface of fifth segment). F, Boot-shaped cocoon (lateral view). G, Whole body (lateral view). H, Head capsule (dorsal view). I, Uppermost primary ray of labral fan. J, One of the other primary rays of labral fan showing different pectinations. K, Apical portion of mandible. L & N, Anterior row of teeth of hypostoma. M & O, Hypostoma and postgenal clefts. P, Anal sclerite. Q, Posterior part of abdomen showing ventral papillae and posterior circlet (ventral view). A–F, H–M, & Q, *S. pahangense*; G & N–P, *S. courtneyi.* Scale bars: 1.0 mm for F & G.; 0.2 mm for B; 0.05 mm for enlargement of B; 0.03 mm for C–E, I & J

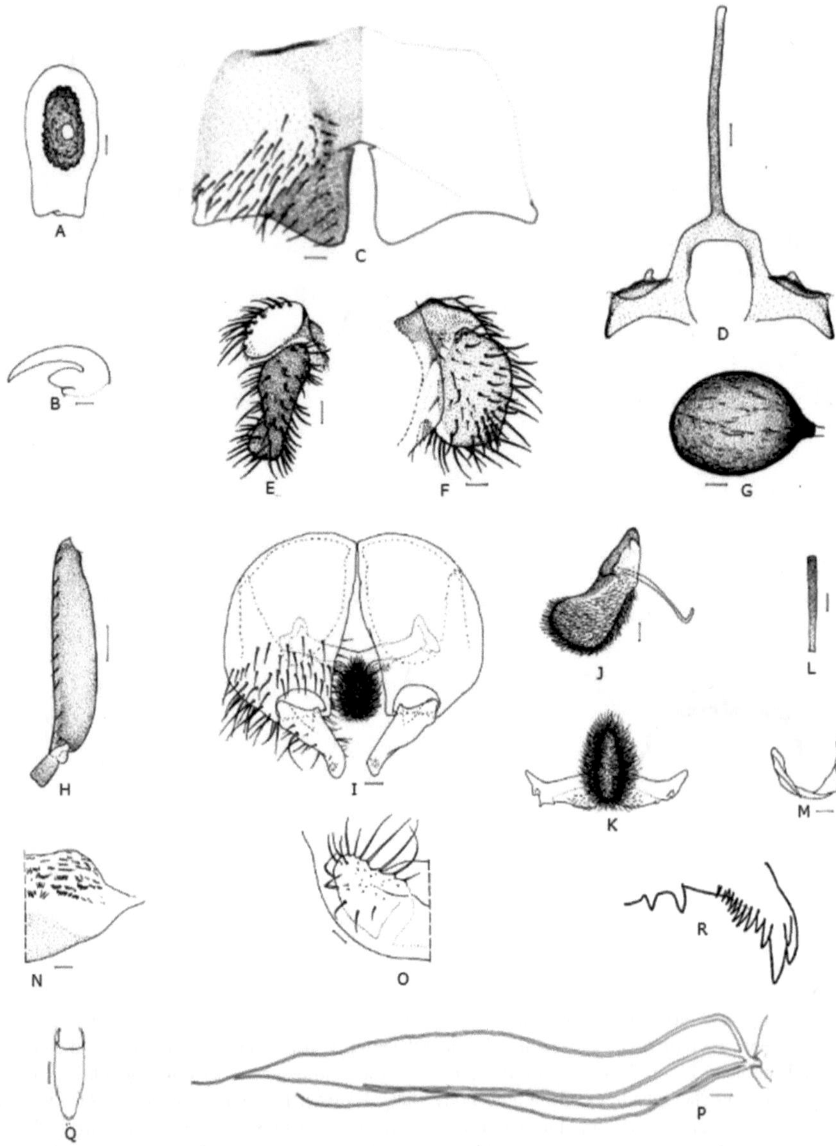

Fig. 10 Female, male, pupa, and larva of the subgenus *Eusimulium.* A–G, Female; H–O, Male, P & Q, Pupa; R. Larva. A, Third palpal segment with long sensory vesicle (front view). B, Claw. C, Sternite 8 and ovipositor valves (ventral view). D, Genital fork (ventral view). E & F, Paraprocts and cerci (right side; E, ventral view; F, lateral view). G, Spermatheca with sclerotized neck. H, Hind basitarsus, and second tarsomere (left side; outer view). I, Coxites, styles and ventral plate (ventral view). J & K, Ventral plates (J, lateral view; K, caudal view). L, Median sclerite (caudal view). M, Paramere (right side; dorsal view). N, Aedeagal membrane and dorsal plate (caudal view). O, Cercus (right side; caudal view). P, Gill filaments (left side; outer view). Q, Cocoon (dorsal view). R, Apical portion of mandible. A–Q, *S. jomsomense*; R, *S. satsumense*

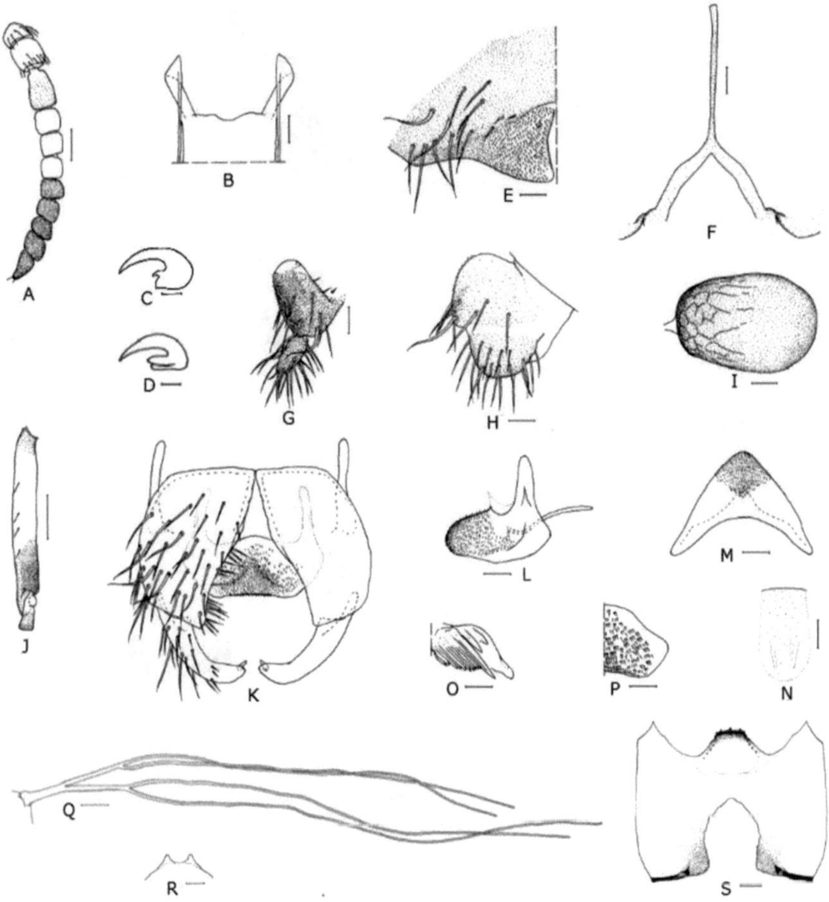

Fig. 11 Female, male, pupa, and larva of the *S. alienigenum* species-group of the subgenus *Gomphostilbia*. A–I, Female; J–P, Male, Q & R, Pupa; S, Larva. A, Antenna (left side; dorsal view). B, Cibarium. C, Claw with small tooth. D, Claw with large tooth. E, Sternite 8 and ovipositor valve (right half; ventral view). F, Genital fork (ventral view). G & H, Paraprocts and cerci (G, right side, ventral view; H, left side, lateral view). I, Spermatheca. J, Hind basitarsus and second tarsomere (left side; outer view). K, Coxites, styles and ventral plate (ventral view). L, Ventral plate and median sclerite (lateral view). M, Ventral plate (caudal view). N, Median sclerite (caudal view). O, Paramere (caudal view). P, Aedeagal membrane (caudal view). Q, Gill filaments (left side; dorsal view). R, Terminal hooks (caudal view). S, Head capsule showing postgenal cleft (ventral view). A, B & D–S, *S. kalimantanense*; C, *S. alienigenum*. Scale bars: 0.2 mm for Q; 0.1 mm for J; 0.05 mm for A & S; 0.02 mm for B, E–I, K–P & R; 0.01 mm for C & D

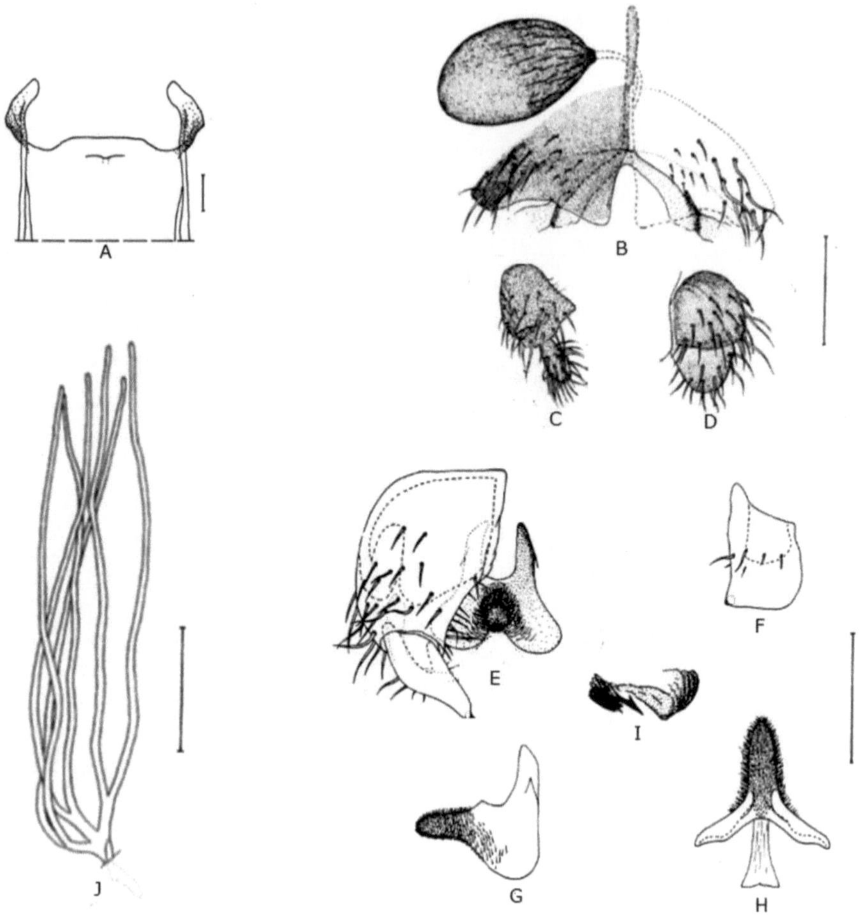

Fig. 12 Female, male, and pupa of the *S. ambigens* species-group of the subgenus *Gomphostilbia*. A–D, Female; E–I, Male; J, Pupa. A, Cibarium. B, Sternite 8, ovipositor valves, genital fork, and spermatheca (ventral view). C & D, Paraprocts and cerci (right side; C, ventral view; D, lateral view). E, Right coxite, right style, and ventral plate (ventral view). F, Style (right side; ventrolateral view). G, Ventral plate (lateral view). H, Ventral plate and median sclerite (caudal view). I, Paramere (left side; caudal view). J, Gill (right side; lateral view). A–J, *S. ambigens*. Scale bars: 0.5 mm for B–I & J; 0.02 mm for A

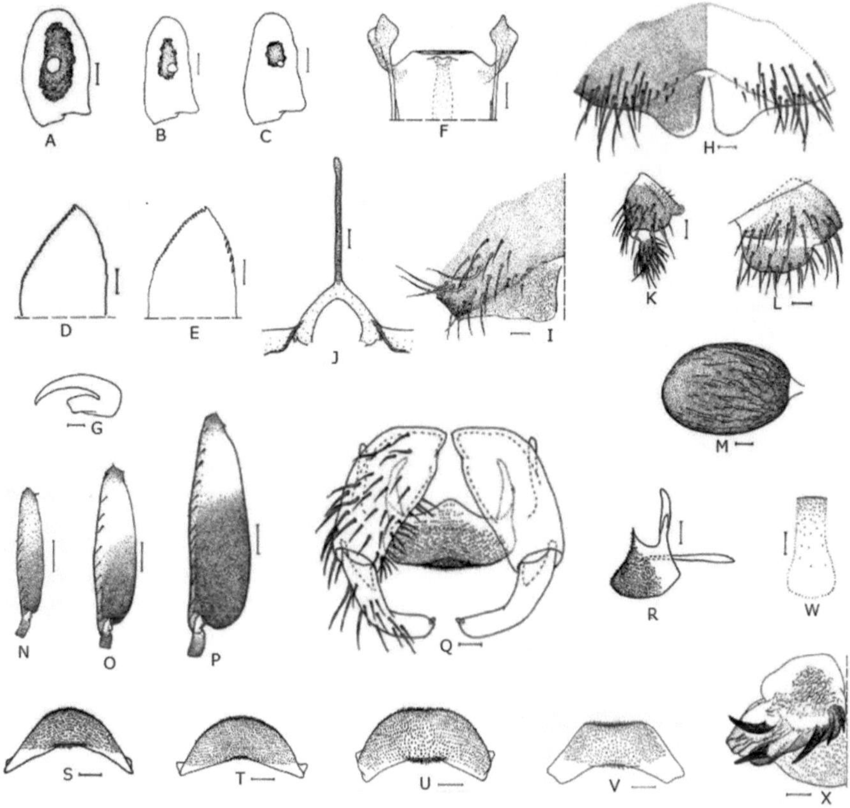

Fig. 13 Female and male of the *S. asakoae* species-group of the subgenus *Gomphostilbia*. A–M, Female; N–X, Male. A–C, Third palpal segments with sensory vesicles of different sizes (right side; front view). D & E, Mandibles (D, without distinct outer teeth; E, with five distinct outer teeth). F, Cibarium. G, Claw. H & I, Sternite 8 and ovipositor valves (ventral view; I, right half only). J, Genital fork (ventral view). K & L, Paraprocts and cerci (right side; K, ventral view; L, lateral view). M, Spermatheca. N–P, Hind basitarsi and second tarsomeres (left side; outer view). Q, Coxites, styles and ventral plate (ventral view). R, Ventral plate and median sclerite (lateral view). S–V, Ventral plates of different shapes (caudal view). W, Median sclerite (caudal view). X, Paramere and aedeagal membrane (right half; caudal view). A, *S. namdanense*; B, E, F, I & U, *S. asakoae*; C, *S. chaudinhense*; D, *S.maehongsonense*; G, H, J–M, P–S, W & X, *S. thungchangense*; N, *S. lurauense*; O, *S. huaimorense*; T, *S. phulocense*; V, *S. sofiani*. Scale bars: 0.1 mm for N–P; 0.02 mm for A–C, F, H–M & Q–X; 0.01 mm for D, E & G

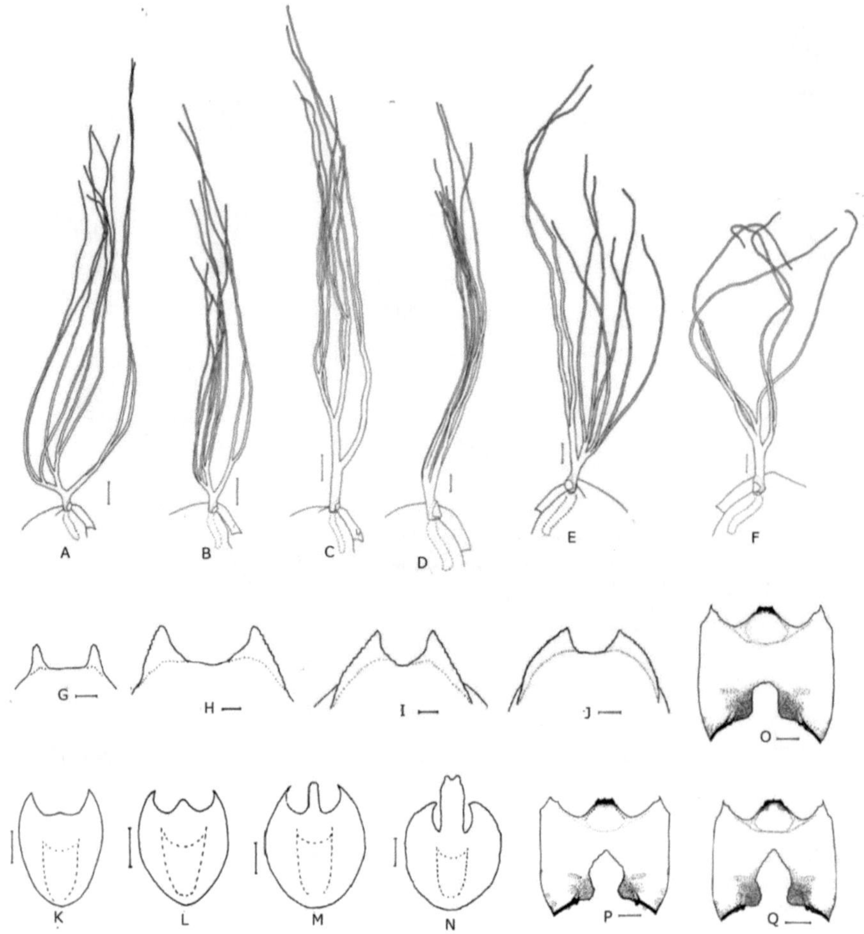

Fig. 14 Pupa and larva of the *S. asakoae* species-group of the subgenus *Gomphostilbia*. A–N, Pupa; O–Q, Larva. A–F, Gills (lateral view; A–D, right side; E & F, left side). G–J, Terminal hooks (caudal view). K–N, Cocoons (dorsal view). O–Q, Head capsules (ventral view). A & O, *S. inthanonense*; B, K & P, *S. asakoae*; C, *S. fuscidorsum*; D, *S. tokarense*; E, *S. sutheppuiense*; F, *S. hoiseni*; G & M, *S. hmongense*; H & N, *S. chiangdaoense*; I & Q, *S. teerachanense*; J, *S. pitasawatae*; L, *S. phulocense*. Scale bars: 1.0 mm for K–N; 0.1 mm for A–F & O–Q; 0.01 mm for G–J

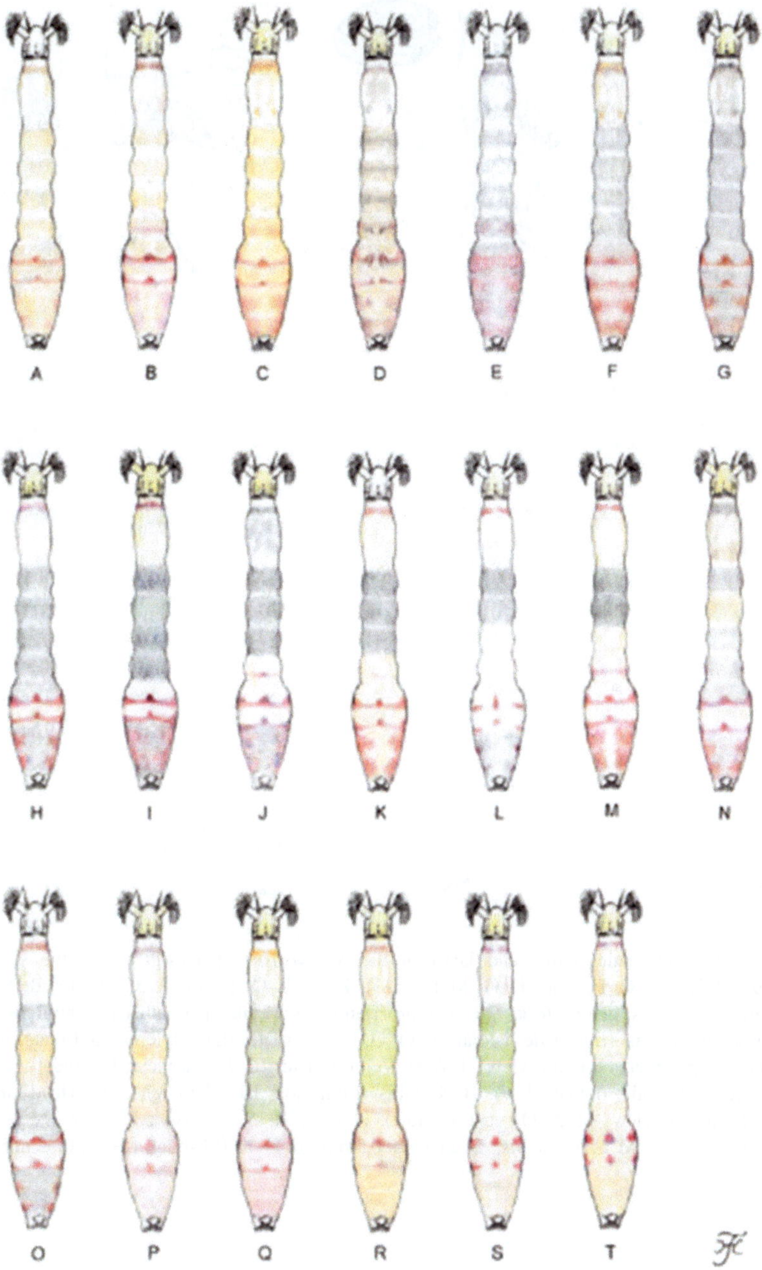

Fig. 15 Larval body color patterns (dorsal view) of the *S. asakoae* species-group of the subgenus *Gomphostilbia*. A, *S. teerachanense*; B, *S. pitasawatae*; C, *S. sutheppuiense*; D, *S. asakoae*; E, *S. klonglanense*; F, *S. namdanense*; G, *S. chaowaense*; H, *S. lociense*; I, *S. huaimorense*; J, *S. nanoiense*; K, *S. junkumae*; L, *S. myanmarense*; M, *S. maewongense*; N, *S. phapeungense*; O, *S. muangpanense*; P, *S. maehongsonense*; Q, *S. thungchangense*; R, *S. chiangdaoense*; S, *S. inthanonense* (morphoform 1) ; T, *S. inthanonense* (morphoform 2). Reproduced from Takaoka et al. (2020h)

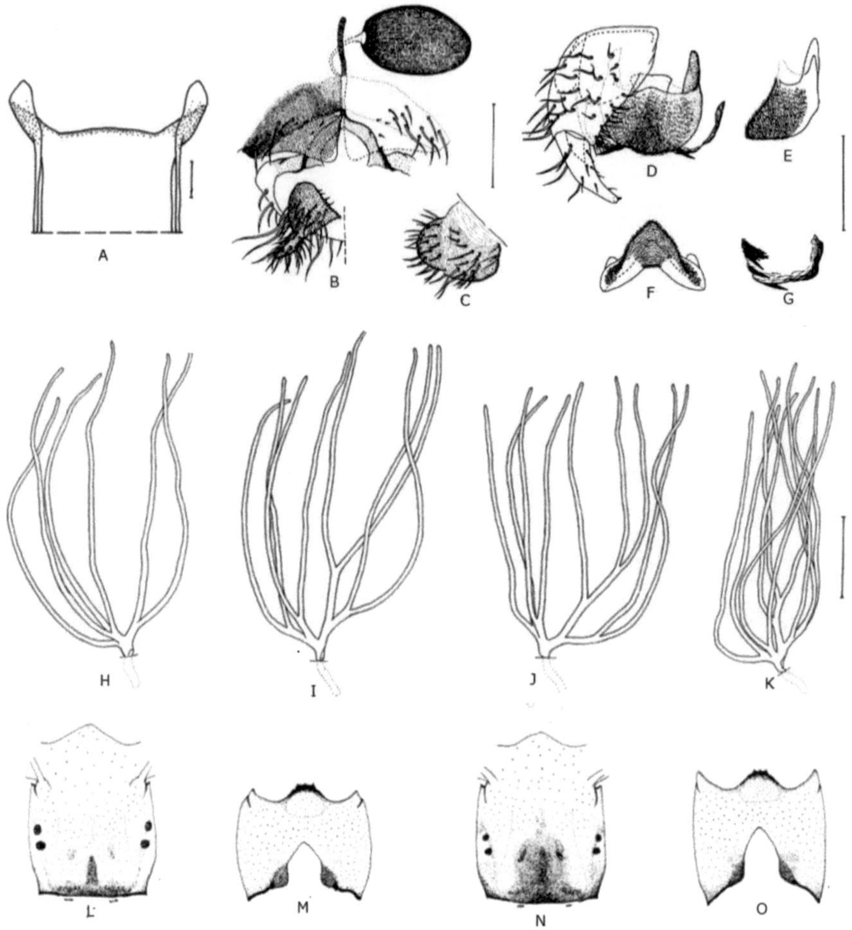

Fig. 16 Female, male, pupa, and larva of the *S. baisasae* species-group of the subgenus *Gomphostilbia*. A–C, Female; D–G, Male; H–K, Pupa; L–O, Larva. A, Cibarium. B, Sternite 8, ovipositor valves, genital fork, right paraproct and cercus, and spermatheca (ventral view). C, Paraproct and cercus (right side; lateral view). D, Right coxite, right style, ventral plate, median sclerite and parameres (ventral view). E & F, Ventral plates (E, lateral view; F, caudal view). G, Paramere (right side; dorsal view). H–K, Gills (right side; lateral view). L–O, Head capsules (L & N, dorsal view; M & O, ventral view). A–C & H, *S. baisasae*; D–G, K, N & O, *S. bicolense*; I, J, L & M, *S. visayaense*. Scale bars: 0.5 mm for H–K; 0.1 mm for B–G; 0.02 mm for A

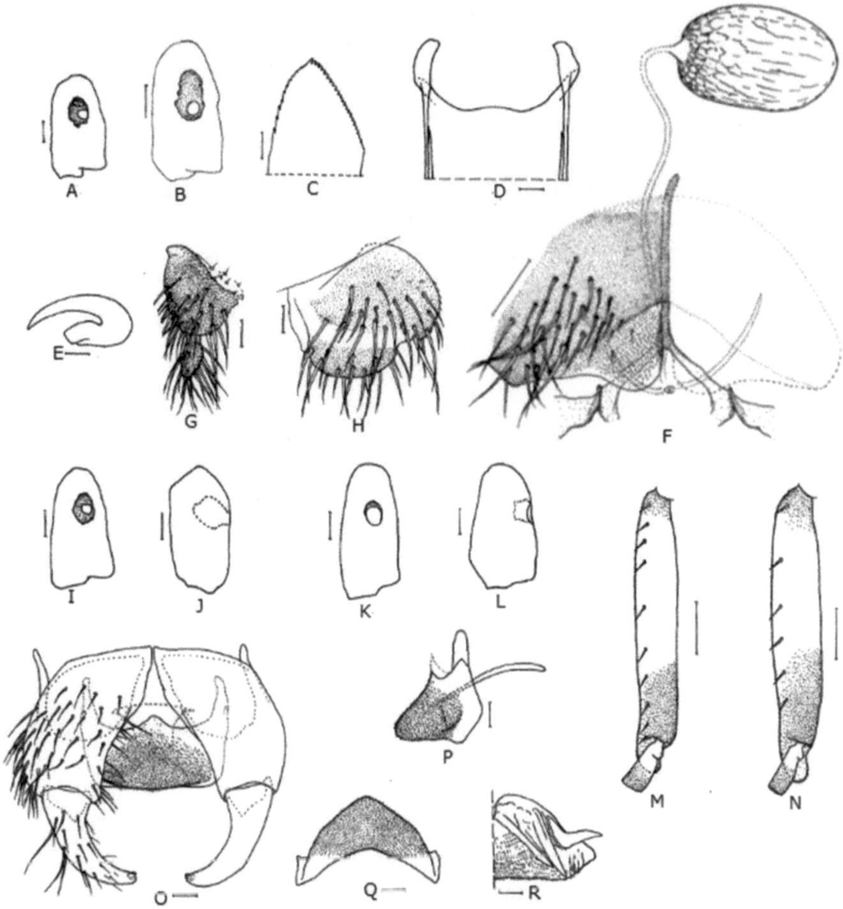

Fig. 17 Female and male of the *S. banauense* species-group of the subgenus *Gomphostilbia*. A–H, Female; I–R, Male. A & B, Third palpal segments with sensory vesicles of different lengths (front view). C, Mandible with teeth on both inner and outer margins. D, Cibarium (front view). E, Claw. F, Sternite 8, ovipositor valves, genital fork and spermatheca (ventral view). G & H, Paraprocts and cerci (right side; G, ventral view; H, lateral view). I–L, Third palpal segments with sensory vesicles having their opening of different sizes (right side; I & K, front view; J & L, lateral view). M & N, Hind basitarsi of different shapes (left side; outer view; M, nearly parallel-sided; N, somewhat tapered apically). O, Coxites, styles and ventral plate (ventral view), P, Ventral plate and median sclerite (lateral view). Q, Ventral plate (caudal view). R, Paramere and aedeagal membrane (left half; caudal view). A & O–R, *S. atuganense*; B, C & F–H, *S. lilianae*; D & E, *S. cagayanense*; I, J & M, *S. liliwense*; K, L & N, *S. katangladense*. Scale bars: 0.1 mm for M & N; 0.05 mm for F; 0.02 mm for A, B, D, G, H, I–L, & O–R; 0.01 mm for C & E

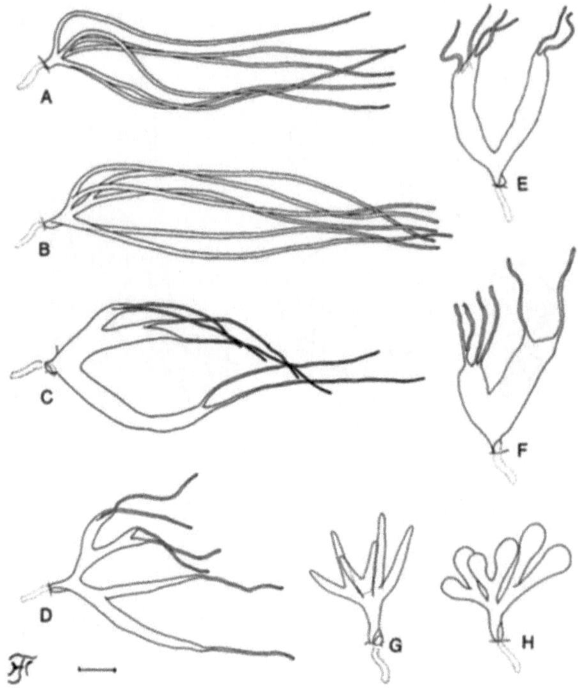

Fig. 18 Pupal gills of the *S. banauense* species-group of the subgenus *Gomphostilbia.* A, *S. bansonae*; B, *S. liliwense*; C, *S. cagayanense*; D, *S. riverai*; E, *S. ifugaoense*; F, *S. manbucalense*; G, *S. lilianae*; H, *S. banauense*. Scale bar: 0.3 mm

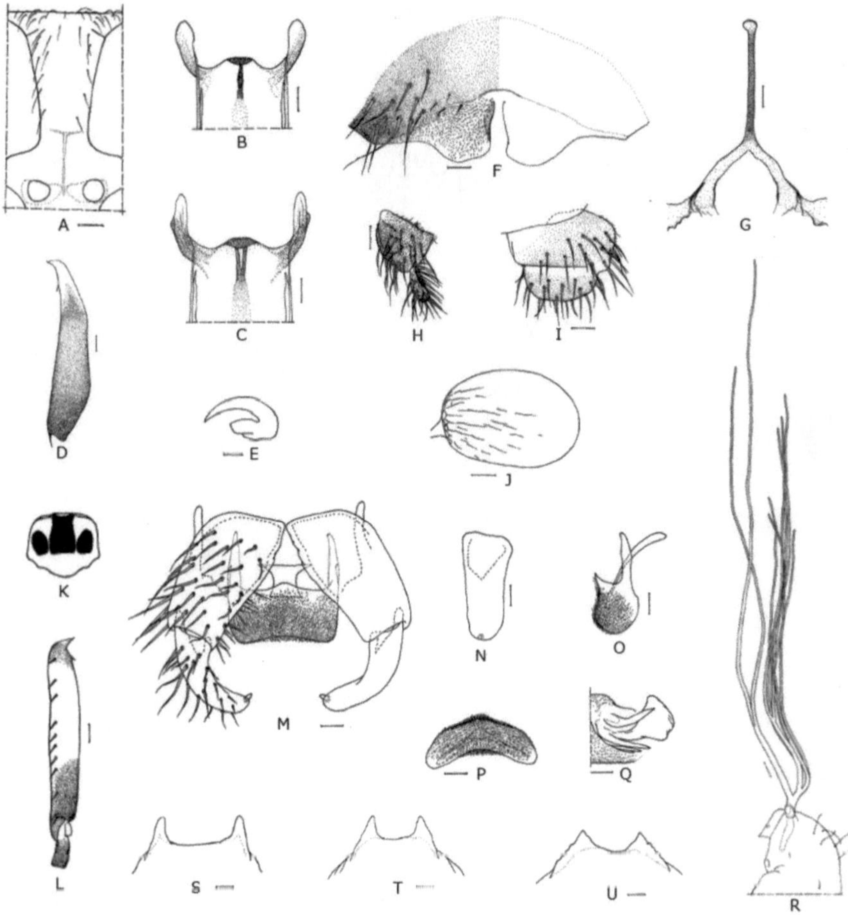

Fig. 19 Female, male, and pupa of the *S. binuanense* subgroup in the *S. batoense* species-group of the subgenus *Gomphostilbia*. A–J, Female; K–Q, Male; R–U, Pupa. A, Frons (front view). B & C, Cibaria. D, Hind tibia (left side; outer view). E, Claw. F, Sternite 8 and ovipositor valves (ventral view). G, Genital fork (ventral view). H & I, Paraprocts and cerci (right side; H, ventral view; I, lateral view). J, Spermatheca. K, Scutum (dorsal view). L, Hind basitarsus and second tarsomere (left side; outer view). M, Coxites, styles, ventral plate, and median sclerite (ventral view). N, Style (right side; ventrolateral view). O, Ventral plate and median sclerite (lateral view). P, Ventral plate (caudal view). Q, Paramere and aedeagal membrane (left half; caudal view). R, Gill (left side; outer view). S–U, Terminal hooks (caudal view). A, B, D–J, R & T, *S. binuanense*; C, *S. tekamense*; K–Q & U, *S. babuyanense*; S, *S. jerantutense*. Scale bars: 0.1 mm for D, L & R; 0.02 mm for A–C, F–J & M–Q; 0.01 mm for E & S–U

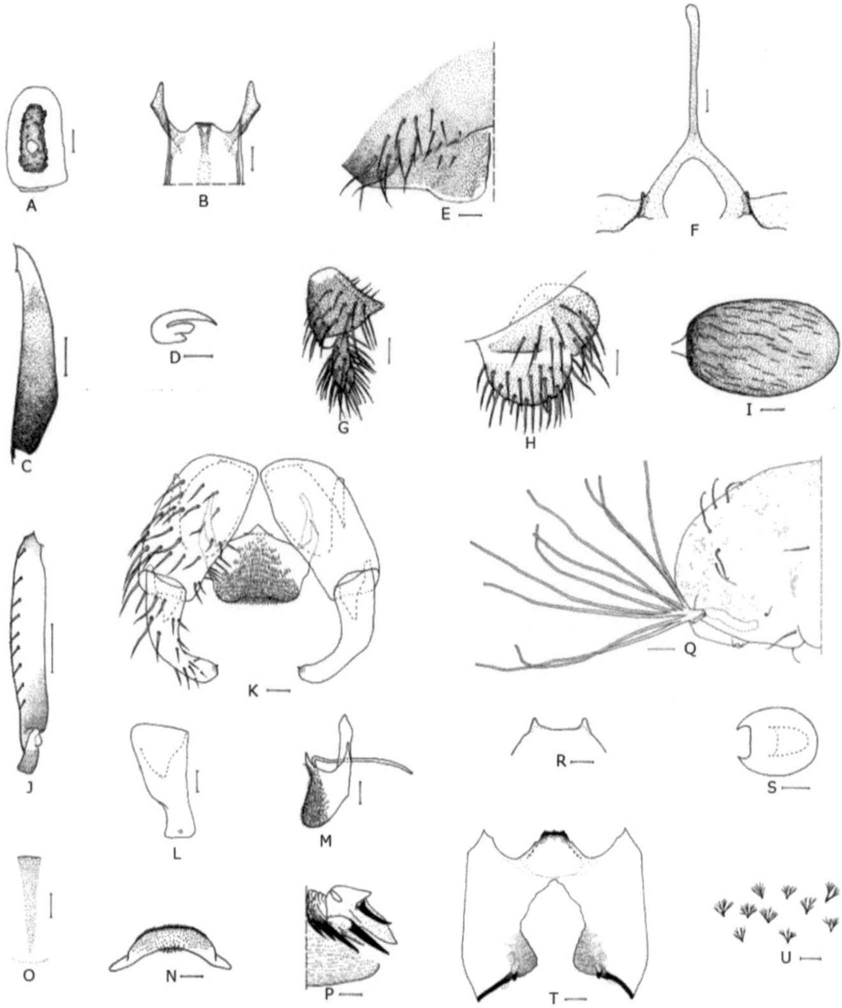

Fig. 20 Female, male, pupa, and larva of the *S. decuplum* subgroup of the *S. batoense* species-group of the subgenus *Gomphostilbia.* A–I, Female; J–P, Male; Q–S, Pupa; T & U, Larva. A, Third palpal segment with long sensory vesicle (right side; front view). B, Cibarium. C, Hind tibia (left side; outer view). D, Claw. E, Sternite 8 and ovipositor valve (right half; ventral view). F, Genital fork (ventral view). G & H, Paraprocts and cerci (right side; G, ventral view; H, lateral view). I, Spermatheca. J, Hind basitarsus and second tarsomere (left side; outer view). K, Coxites, styles and ventral plate (ventral view). L, Style (right side; ventrolateral view). M, Ventral plate and median sclerite (lateral view). N, Ventral plate (caudal view). O, Median sclerite (caudal view). P, Paramere and aedeagal membrane (left half; caudal view). Q, Anterior half of thorax with 10 gill filaments (left side; outer view). R, Terminal hooks (caudal view). S, Cocoon (ventral view). T, Head capsule (ventral view). U, Dark, branched setae on dorsal surface of abdominal segment 8. A–I, *S. chiangraiense*; J–U, *S. lamdongense*. Scale bars: 1.0 mm for S; 0.1 mm for C, J & Q; 0.05 mm for T; 0.02 mm for A, B, D–I, K–P, R & U

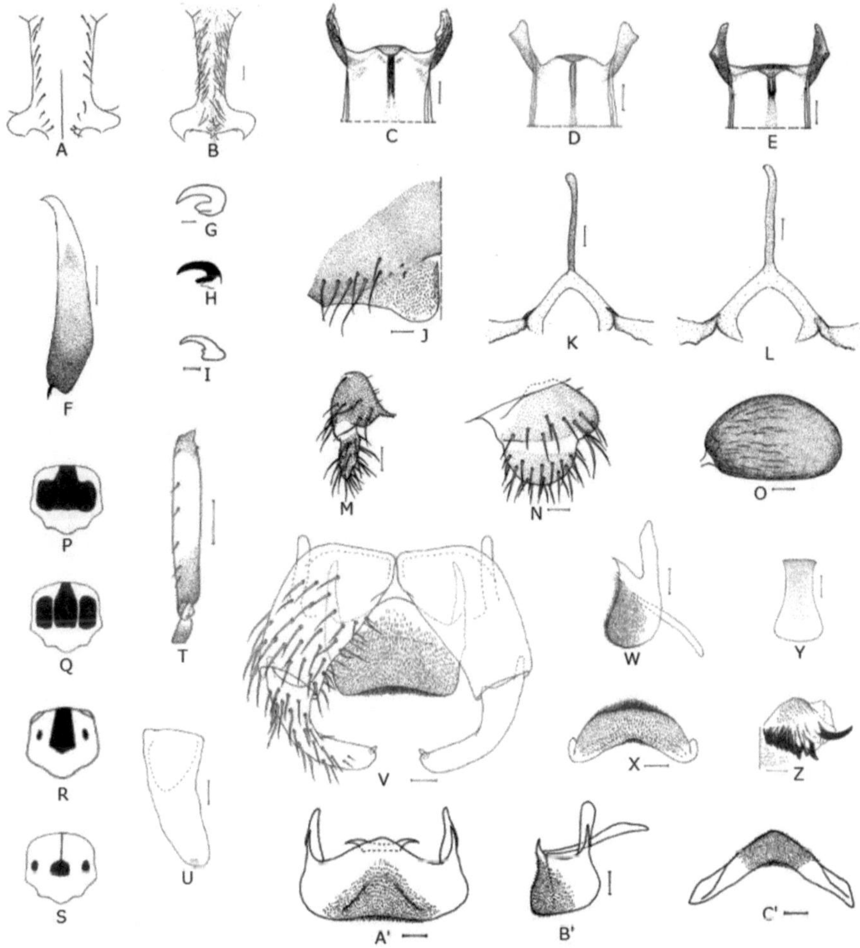

Fig. 21 Female and male of the *S. duolongum* subgroup of the *S. batoense* species-group of the subgenus *Gomphostilbia.* A–O, Female; P–C', Male. A & B, Frons (front view). C–E, Cibaria (front view). F, Hind tibia (left side; outer view). G–I, Claws. J, Sternite 8 and ovipositor valve (right half; ventral view). K & L, Genital forks (ventral view). M & N, Paraprocts and cerci (right side; M, ventral view; N, lateral view). O, Spermatheca. P–S, Scuta (dorsal view). T, Hind basitarsus and second tarsomere (left side; outer view). U, Coxites, styles and ventral plate (ventral view). V, Style (right side; ventrolateral view). W, Ventral plate and median sclerite (lateral view). X, Ventral plate (caudal view). Y, Median sclerite (caudal view). Z, Paramere and aedeagal membrane (left half; caudal view). A' & B', Ventral plates and median sclerites (A', ventral view; B', lateral view). C', Ventral plate (caudal view). A & H, *S. tahanense*; B, C & G, *S. torautense*; D, P, Q & S–Z, *S. johorense*; E, F, I–K & M–O, *S. dachaisense*; L, *S. lemborense*; R, *S. hadiae*; A'–C', *S. samarense*. Scale bars: 0.1 mm for F & T; 0.03 mm for B; 0.02 mm for C–E, J–O, U–Z & A'–C'; 0.01 mm for G & I

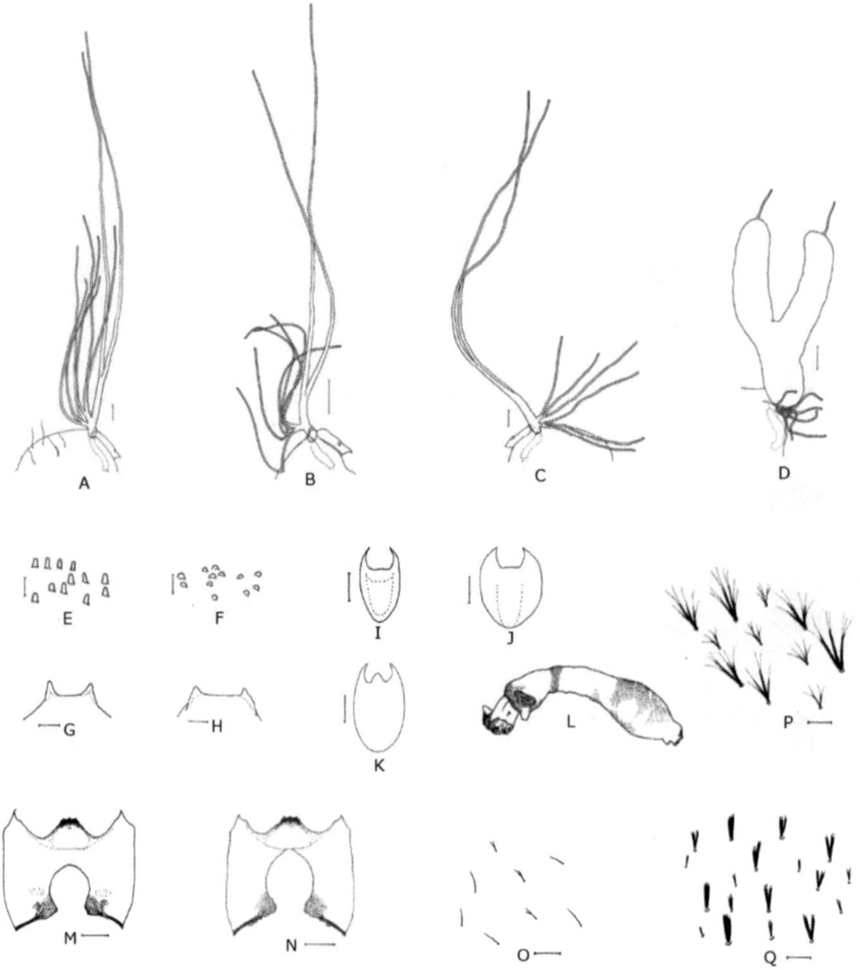

Fig. 22 Pupa and larva of the *S. duolongum* subgroup of the *S. batoense* species-group of the subgenus *Gomphostilbia*. A–K, Pupa; L–Q, Larva. A–D, Gills (A, B & C, lateral view; D, dorsal view; A, B & D, right side; C, left side). E & F, Tubercles (E, on frons; F, on thorax). G & H, Terminal hooks (caudal view). I–K, Cocoons (dorsal view). L, Whole body (lateral view). M & N, Head capsules (ventral view). O–Q, Dark setae of different shapes and sizes on abdominal segment 8. A, M & Q, *S. yvonneae*; B & J, *S. syafruddini*; C & O, *S. lemborense*; D & H, *S. tumidum*; E, F & L, *S. tahanense*; G & I, *S. dachaiense*; K, *S. torautense*; N, *S. johorense*; P, *S. batoense*. Scale bars: 1.0 mm for I–K; 0.2 mm for B; 0.1 mm for A, C, D, M & N; 0.02 mm for G & H; 0.01 mm for E, F & O–Q

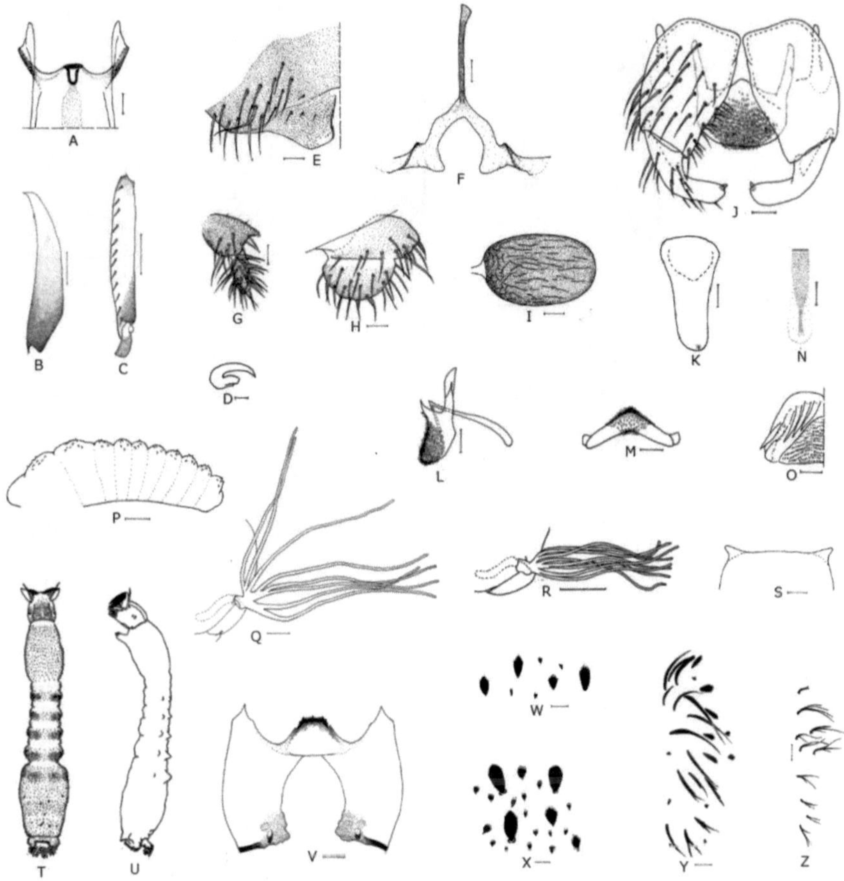

Fig. 23 Female, male, pupa, and larva of the *S. parahiyangum* subgroup of the subgenus *Gomphostilbia*. A, B & D–I, Female; C & J–O, Male; P–S, Pupa; T–Z, Larva. A, Cibarium. B, Hind tibia (left side; outer view). C, Hind basitarsus and second tarsomere (left side; outer view). D, Claw. E, Sternite 8 and ovipositor valve (right half; ventral view). F, Genital fork (ventral view). G & H, Paraprocts and cerci (right side; G, ventral view; H, lateral view). I, Spermatheca. J, Coxites, styles and ventral plate (ventral view). K, Style (right side; ventrolateral view). L, Ventral plate and median sclerite (lateral view). M, Ventral plate (caudal view). N, Median sclerite (caudal view). O, Paramere and aedeagal membrane (right half; caudal view). P, Antennal sheath (lateral view). Q & R, Gills (right side; lateral view). S, Terminal hooks (caudal view). T & U, Whole bodies (T, dorsal view; U, lateral view). V, Head capsule (ventral view). W & X, Dark stout and small setae on dorsal surface of abdominal segment 8. Y & Z, Dark setae of various shapes and sizes on anterolateral surface of abdominal segment 5. A–Q, S, V, W & Y, *S. longlanhense*; R, T & U, *S. parahiyangum*; X & Z, *S. ngaoense*. Scale bars: 0.05 mm for B & C; 0.2 mm for R; 0.1 mm for Q & V; 0.02 mm for A, D–P, S & W–Z

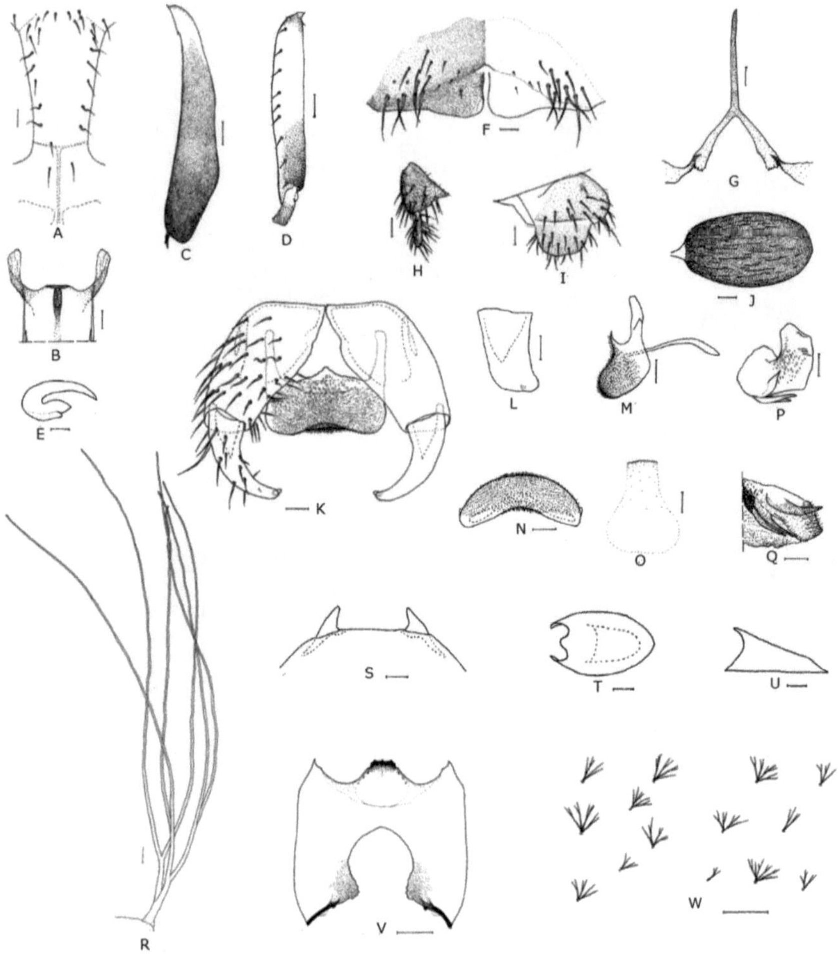

Fig. 24 Female, male, pupa, and larva of the *S. sextuplum* subgroup of the *S. batoense* species-group of the subgenus *Gomphostilbia*. A–C & E–J, Female; D & K–Q, Male; R–U, Pupa; V & W, Larva. A, Frons (frontal view). B, Cibarium. C, Hind tibia (left side; outer view). D, Hind basitarsus and second tarsomere (left side; outer view). E, Claw. F, Sternite 8 and ovipositor valves (ventral view). G, Genital fork (ventral view). H & I, Paraprocts and cerci (right side; H, ventral view; I, lateral view). J, Spermatheca. K, Coxites, styles and ventral plate (ventral view). L, Style (right side; ventrolateral view). M, Ventral plate and median sclerite (lateral view). N, Ventral plate (caudal view). O, Median sclerite (caudal view). P, Paramere (right side; lateral view). Q, Paramere and aedeagal membrane (right half; caudal view). R, Gill (right side; dorsal view). S, Terminal hooks (caudal view). T & U, Cocoons (T, dorsal view; U, lateral view). V, Head capsule (ventral view). W, Dark multibranched setae on dorsal surface of abdominal segment 8. A, C–N & Q–W, *S. paukatense*; B, O & P, *S. fukudae*. Scale bars: 0.05 mm for T & U; 0.1 mm for R & V; 0.05 mm for C & D; 0.02 mm for A, B & F–Q; 0.01 mm for E, S & W

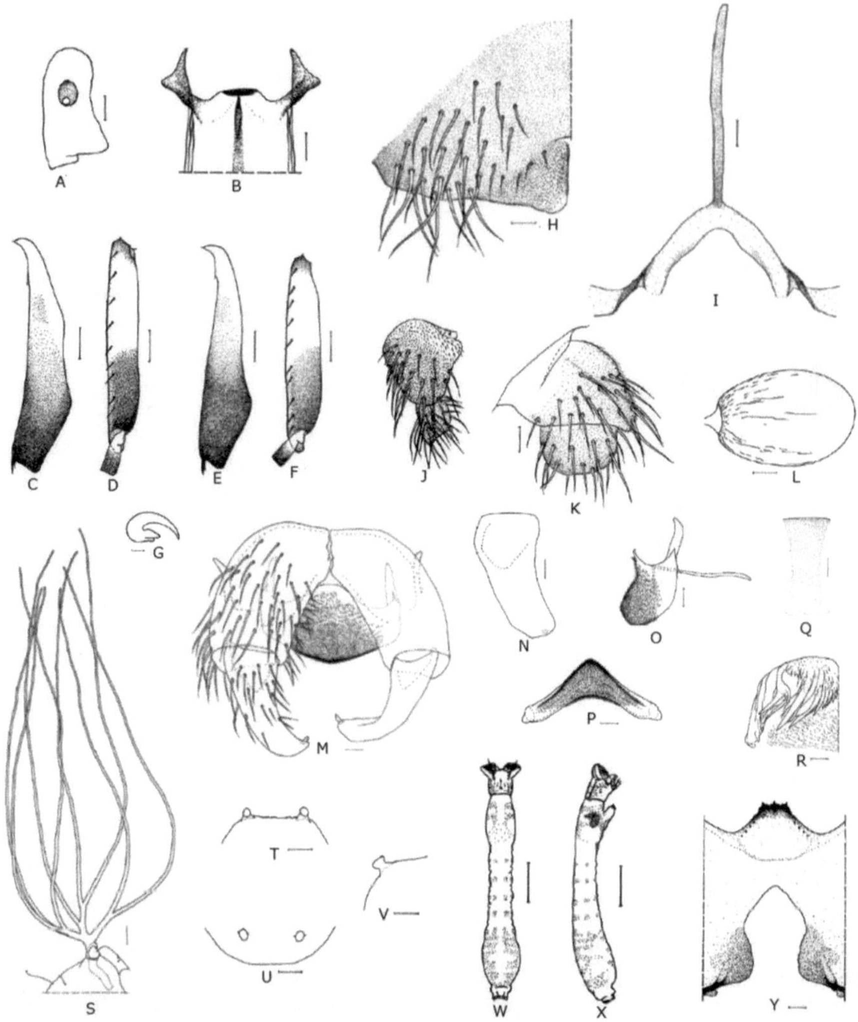

Fig. 25 Female, male, pupa, and larva of the S. *terengganuense* subgroup of the *S. batoense* species-group of the subgenus *Gomphostilbia.* A–D & G–L, Female; E, F & M–R, Male; S–V, Pupa; W–Y, Larva. A, Third palpal segment with small sensory vesicle (right side; frontal view). B, Cibarium. C & E, Hind tibiae (left side; outer view). D & F, Hind basitarsi and second tarsomeres (left side; outer view). G, Claw. H, Sternite 8 and ovipositor valve (right half; ventral view). I, Genital fork (ventral view). J & K, Paraprocts and cerci (right side; J, ventral view; K, lateral view). L, Spermatheca. M, Coxites, styles and ventral plate (ventral view). N, Style (right side; ventrolateral view). O, Ventral plate and median sclerite (lateral view). P, Ventral plate (caudal view). Q, Median sclerite (caudal view). R, Paramere and aedeagal membrane (right half; caudal view). S, Gill (right side; lateral view). T–V, Terminal hooks (T, caudal view; U, dorsal view; V, lateral view). W & X, Whole bodies (W, dorsal view; X, lateral view). Y, Head capsule (ventral view). A–Y, *S. terengganuense*. Scale bars: 1.0 mm for W & X; 0.1 mm for C–F & S; 0.05 mm for Y; 0.02 mm for A, B, H–R, T–V; 0.01 mm for G

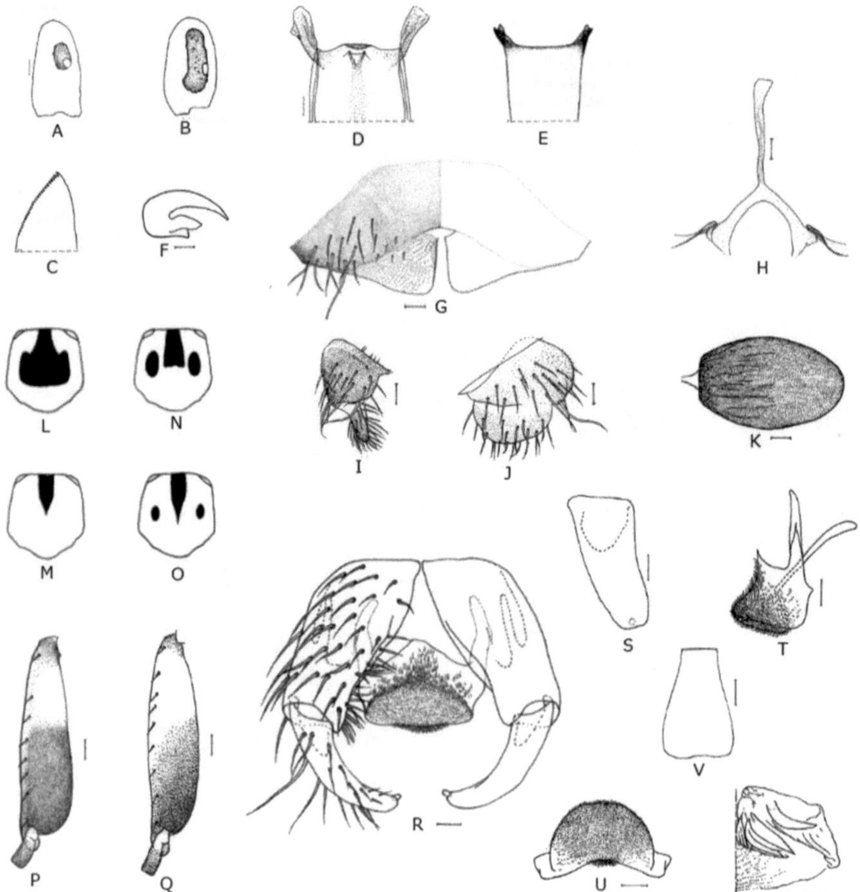

Fig. 26 Female and male of the *S. ceylonicum* species-group of the subgenus *Gomphostilbia.* A–K, Female; L–W, Male. A & B, Third palpal segments with sensory vesicles (right side; frontal view). C, Mandible. D & E, Cibaria. F, Claw. G, Sternite 8 and ovipositor valves (ventral view). H, Genital fork (ventral view). I & J, Paraprocts and cerci (right side; I, ventral view; J, lateral view). K, Spermatheca. L–O, Scuta with different patterns of white pruinose and non-pruinose areas (dorsal view). P & Q, Hind basitarsi and second tarsomeres (left side; outer view; P, wedge-shaped; Q, spindle-shaped). R, Coxites, styles and ventral plate (ventral view). S, Style (right side; ventrolateral view). T, Ventral plate and median sclerite (lateral view). U, Ventral plate (caudal view). V, Median sclerite (caudal view). W, Paramere and aedeagal membrane (left half; caudal view). A, D, N & O, *S. rosemaryae*; B, C & E, *S. atratoides*; F–K, *S. pangsidaense*; L & M, *S. kamimurai*; P & R–W, *S. mindoroense*; Q, *S. capillatum*. Scale bars: 0.05 mm for P & Q; 0.02 mm for A, D, G–K & R–W; 0.01 mm for F

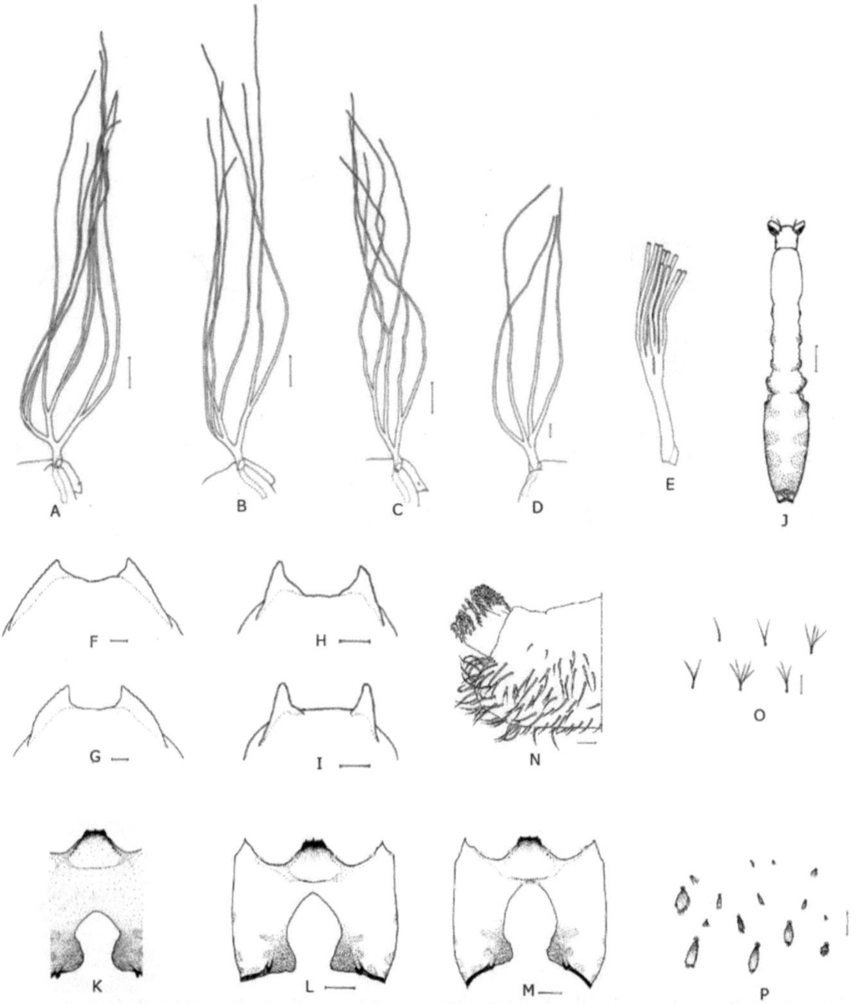

Fig. 27 Pupa and larva of the *S. ceylonicum* species-group of the subgenus *Gomphostilbia*. A–I, Pupa; J–P, Larva. A–E, Gills (A–C & E, right side; lateral view; D, left side; dorsal view; E, only basal portion). F–I, Terminal hooks (caudal view). J, Whole body is markedly constricted between abdominal segments 4 and 5 (dorsal view). K–M, Head capsules showing postgenal cleft of different lengths (ventral view). N, Thoracic proleg covered with hairs (lateral view). O & P, Dark setae of different shapes on abdominal segment 6. A & F, *S. mindoroense*; B & O, *S. rosemaryae*; C, *S. rangatense*; D, *S. gimpuense*; E & K, *S. atratoides*; H, I & M, *S. sheilae*; G, *S. pangsidaense*; L, *S. trangense*; J, N & P, *S. capillatum*. Scale bars: 0.5 mm for J; 0.2 mm for A–C, 0.1 mm for D, L & M; 0.02 mm for H, I & N; 0.01 mm for F, G, O & P

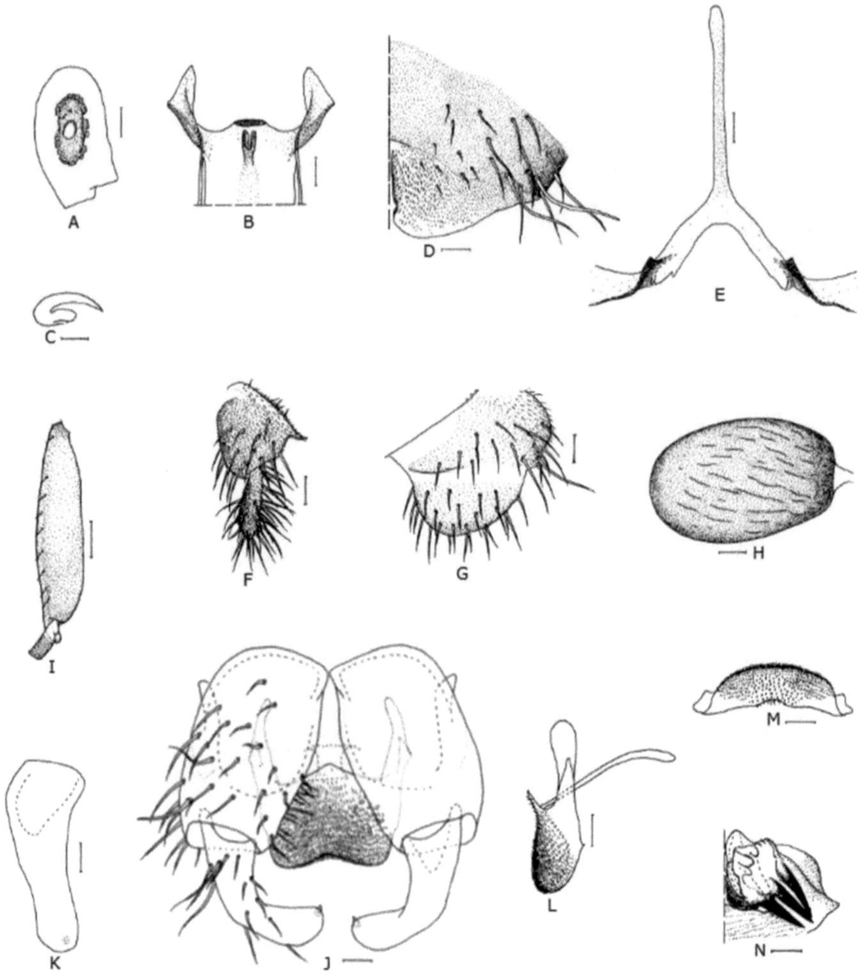

Fig. 28 Female and male of the *S. darjeelingense* species-group of the subgenus *Gomphostilbia*. A–H, Female; I–N, Male. A. Third palpal segment with long sensory vesicle (front view). B, Cibarium. C, Claw. D, Sternite 8 and ovipositor valve (left half; ventral view). E, Genital fork (ventral view). F & G, Paraprocts and cerci (right side; F, ventral view; G, lateral view). H, Spermatheca. I, Hind basitarsus and second tarsomere (left side; outer view). J, Coxites, styles and ventral plate (ventral view). K, Style (right side; ventrolateral view). L, Ventral plate and median sclerite (lateral view). M, Ventral plate (caudal view). N, Paramere and aedeagal membrane (left half; caudal view). A–N, *S. eshimai*. Scale bars: 0.1 mm for I; 0.02 mm for A–H & J–N

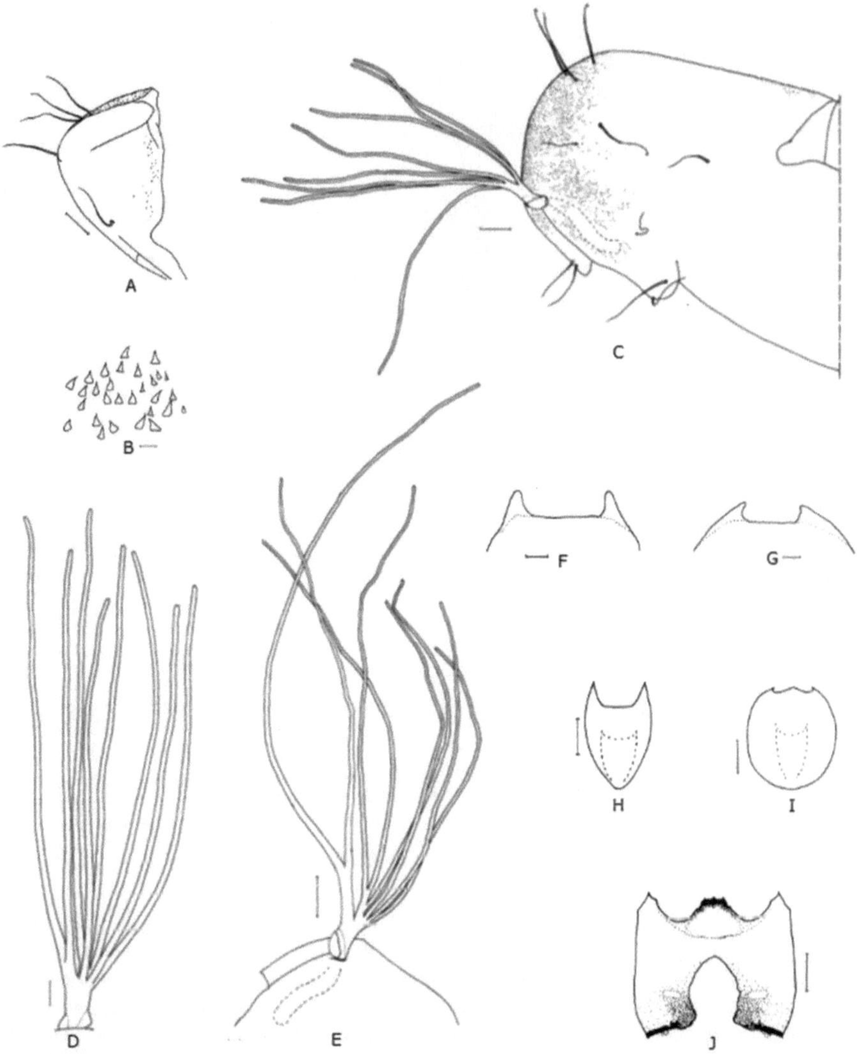

Fig. 29 Pupa and larva of the *S. darjeelingense* species-group of the subgenus *Gomphostilbia*. A–I, Pupa: J, Larva. A, Head showing additional facial trichome as well as three frontal trichomes and one facial trichome on each side (lateral view). B, Tubercles with pointed apices on frons. C, Thorax showing short gill filaments, relatively long trichomes, and tubercles (left side; lateral view). D & E, Gill filaments (D, right side; dorsal view; E, left side; lateral view). F & G, Terminal hooks (F, caudal view; G, anterodorsal view). H & I, Cocoons of different shapes (dorsal view). J, Head capsule showing long postgenal cleft (ventral view). A, C, I & J, *S. eshimai*; B, D & G, *S. chayamaritae*; E, F & H, *S. shresthae*. Scale bars: 1.0 mm for H & I; 0.1 mm for A, C, E & J; 0.04 mm for D; 0.01 mm for B, F & G

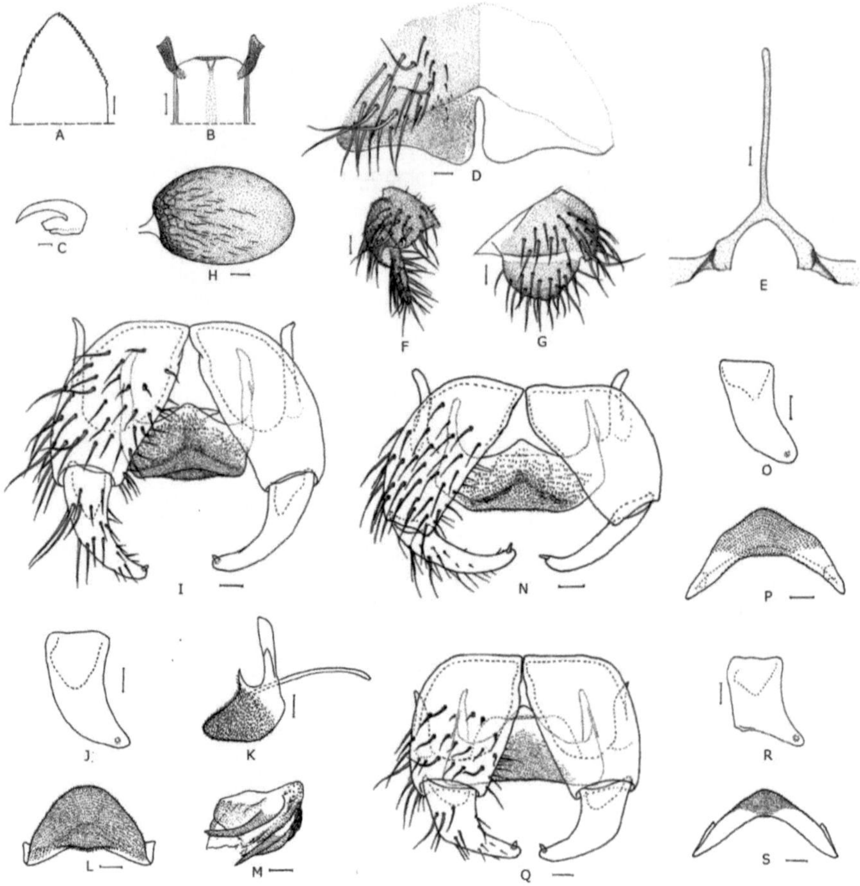

Fig. 30 Female and male of the *S. zonatum* subgroup of the *S. epistum* species-group of the subgenus *Gomphostilbia*. A–H, Female; I–S, Male. A, Mandible. B, Cibarium. C, Claw. D, Sternite 8 and ovipositor valves (ventral view). E, Genital fork (ventral view). F & G, Paraprocts and cerci (right side; F, ventral view; G, lateral view). H, Spermatheca. I, N & Q, Coxites, styles and ventral plates of different shapes (ventral view). J, O & R, Styles of different shapes (right side; ventrolateral view). K, Ventral plate and median sclerite (lateral view). L, P & S, Ventral plates of different shapes (caudal view). M, Paramere (right side; caudal view). A & C–M, *S. kelabitense*; B, *S. sarawakense*; N–P, *S. repentium*; Q–S, *S. auratum*. Scale bars: 0.02 mm for B & D–S; 0.01 mm for A & C

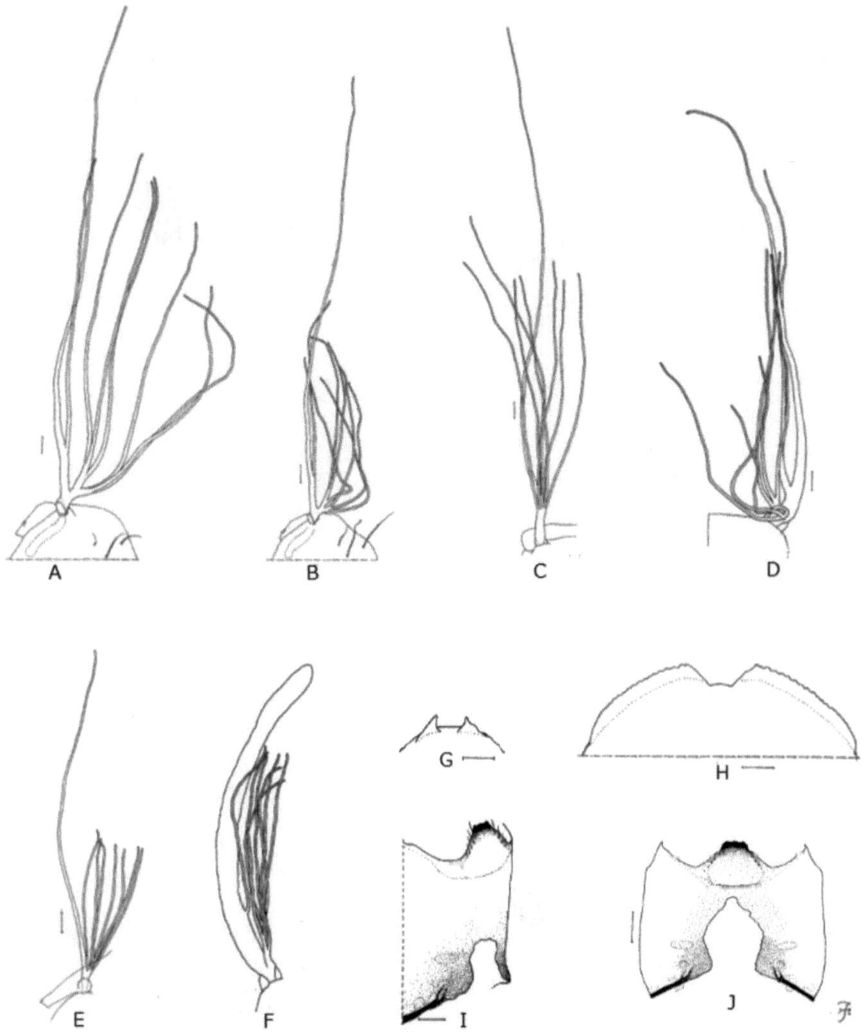

Fig. 31 Pupa and larva of the *S. zonatum* subgroup of the *S. epistum* species-group of the subgenus *Gomphostilbia*. A–H, Pupa; I & J, Larva. A–F, Gills (A & B, left side and lateral view; C, D & F, dorsal view; C & F, left side; D, right side; E, right side; inner view). G & H, Terminal hooks (caudal view; G, simple, triangular; H, wide, plate-like, with outer margins crenulated). I & J, Head capsules (ventral view; I, drawn from a slide-mounted larval specimen). A & J, *S. kelabitense*; B, *S. auratum*; C, *S. repentium*; D, *S. floresense*; G, *S. adleri*; H, *S. datfaense*; E & I, *S. zonatum*; F, *S. padangense*. Scale bars: 0.1 mm for A–E, I & J; 0.02 mm for G & H

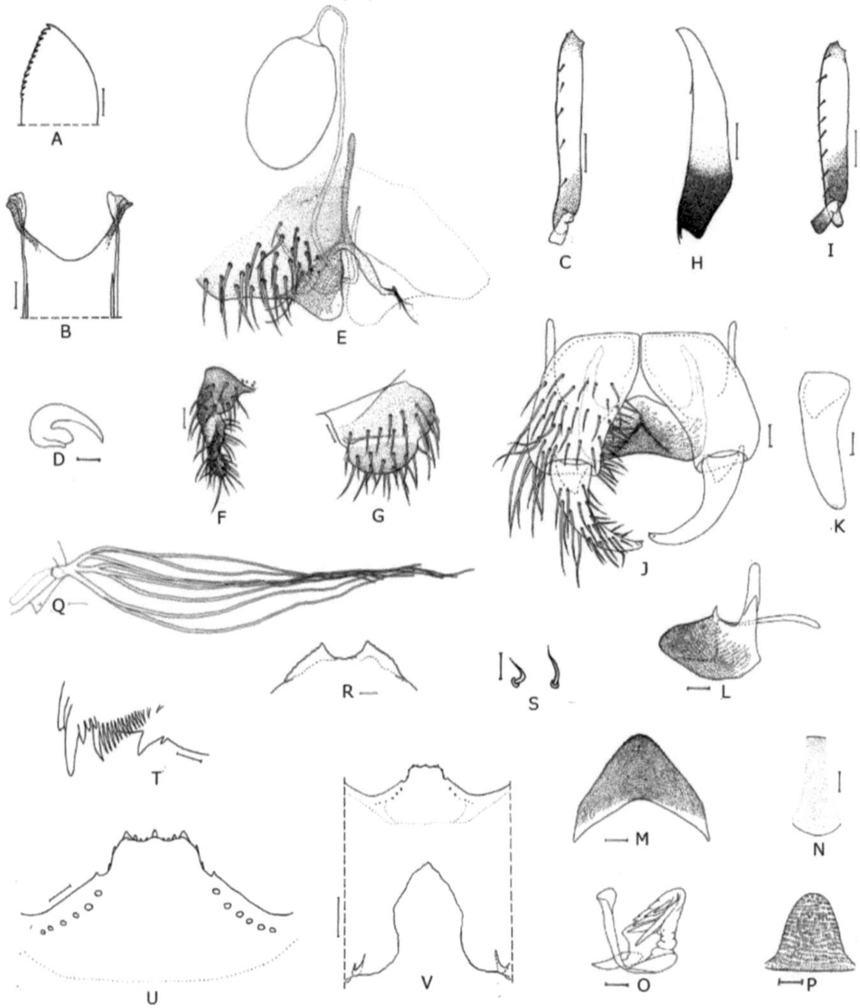

Fig. 32 Female, male, pupa, and larva of the *S. miblosi* subgroup of the *S. epistum* species-group of the subgenus *Gomphostilbia*. A–G, Female; H–P, Male; Q–S, Pupa; T–V, Larva. A, Mandible. B, Cibarium. C, Hind basitarsus and second tarsomere (left side; outer view). D, Claw. E, Sternite 8, ovipositor valves, genital fork, and spermatheca (ventral view). F & G, Paraprocts and cerci (right side; F, ventral view; G, lateral view). H, Hind tibia (left side; outer view). I, Hind basitarsus and second tarsomere (left side; outer view). J, Coxites, styles and ventral plate (ventral view). K, Style (right side; ventrolateral view). L, Ventral plate and median sclerite (lateral view). M, Ventral plate (caudal view). N, Median sclerite (caudal view). O, Paramere (left side; dorsal view). P, Aedeagal membrane (caudal view). Q, Gill (right side; lateral view). R, Terminal hooks (caudal view). S, Setae on each lateral surface of abdominal segment 9. T, Apical portion of mandible. U, Hypostoma. V, Medial portion of head capsule (ventral view). A–G & T–V, *S. lehi*; H–P & S, *S. miblosi*; Q & R, *S. rarum*. Scale bars: 0.1 mm for C, H, I, Q & V; 0.02 mm for B, E–G, J–P & U; 0.01 mm for A, D & R–T

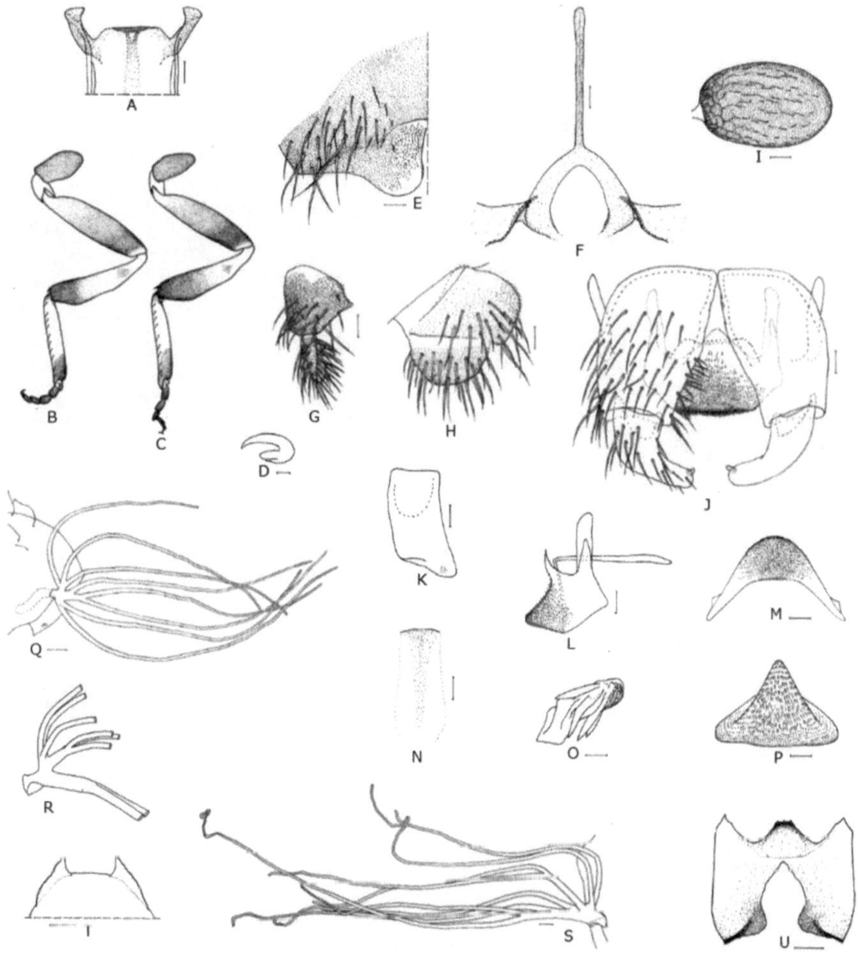

Fig. 33 Female, male, pupa, and larva of the *S. atratum* subgroup of the *S. epistum* species-group of the subgenus *Gomphostilbia.* A, B & D–I, Female; C & J–P, Male; Q–T, Pupa; U, Larva. A, Cibarium. B & C, Hind legs (left side; outer view; B, female; C, male). D, Claw. E, Sternite 8 and ovipositor valves (ventral view). F, Genital fork (ventral view). G & H, Paraprocts and cerci (right side; G, ventral view; H, lateral view). I, Spermatheca. J, Coxites, styles and ventral plate (ventral view). K, Style (right side; ventrolateral view). L, Ventral plate and median sclerite (lateral view). M, Ventral plate (caudal view). N, Median sclerite (caudal view). O, Paramere (right side; caudal view). P, Aedeagal membrane (caudal view). Q–S, Gills (lateral view; Q & R, right side; S, left side; R, basal portion only). T, Terminal hooks (caudal view). U, Head capsule (ventral view). A, D–Q, T & U, *S. isanense*; B, C & R, *S. atratum*; S, *S. otsukai*. Scale bars: 0.1 mm for Q & U; 0.05 mm for S; 0.02 mm for A, E–P & T; 0.01 mm for D

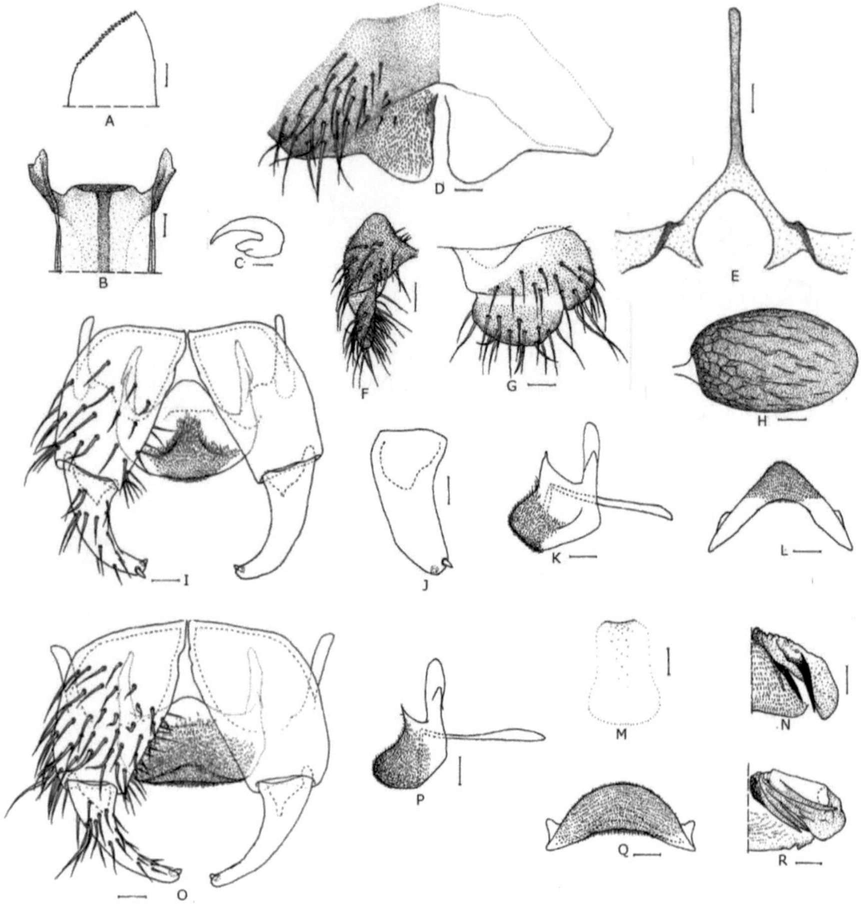

Fig. 34 Female and male of the *S. montiblense* subgroup of the *S. epistum* species-group of the subgenus *Gomphostilbia*. A–H, Female; I–R, Male. A, Mandible. B, Cibarium. C, Claw. D, Sternite 8 and ovipositor valves (ventral view). E, Genital fork (ventral view). F & G, Paraprocts and cerci (right side; F, ventral view; G, lateral view). H, Spermatheca. I & O, Coxites, styles, and ventral plates (ventral view). J, Style (right side; ventrolateral view). K & P, Ventral plates and median sclerites (lateral view). L & Q, Ventral plates (caudal view). M, Median sclerite (caudal view). N & R, Parameres and aedeagal membranes (left half; caudal view). A, B & O–R, *S. barioense*; C–N, *S. fulgidum*. Scale bars: 0.02 mm for B & D–R; 0.01 mm for A & C

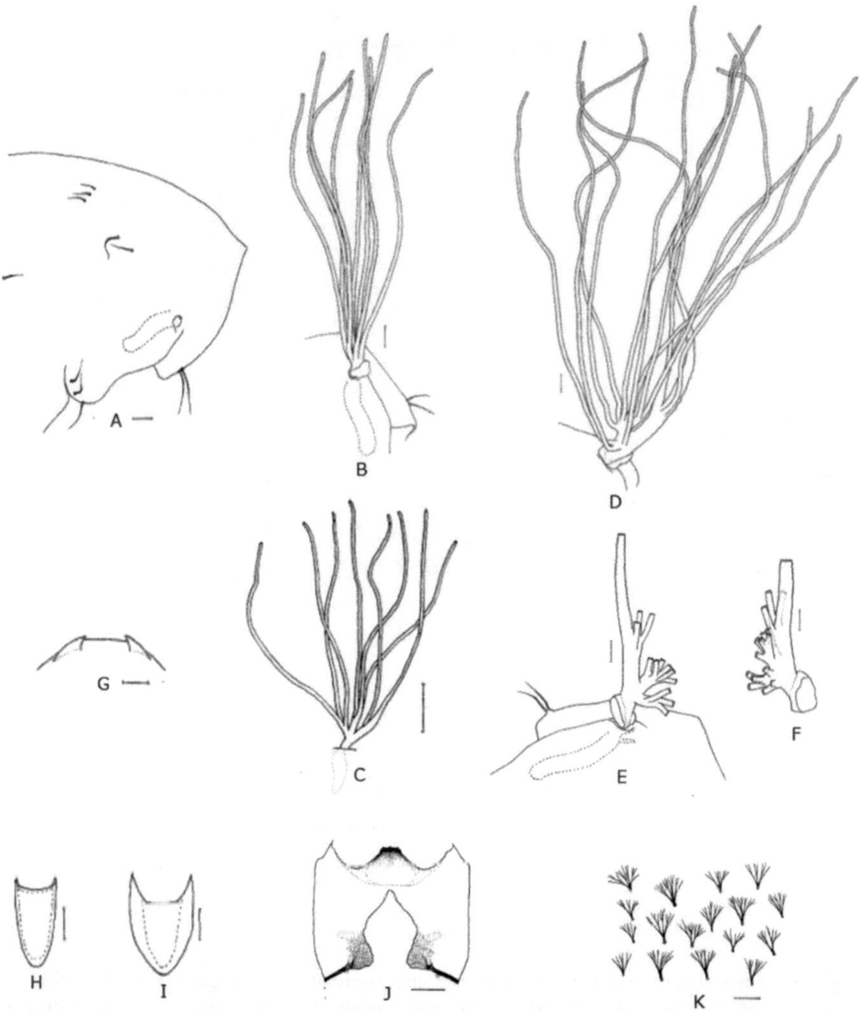

Fig. 35 Pupa and larva of the *S. montiblense* subgroup of the *S. epistum* species-group of the subgenus *Gomphostilbia.* A–I, Pupa; J & K, Larva. A, Anterior half of thorax showing location of trichomes (right side; lateral view). B–D, Gills (B & C, right side; lateral view; D, male; right side; dorsal view). E & F, Basal portions of gills (lateral view; E, male; left side; F, female, right side). G, Terminal hooks (caudal view). H & I, Cocoons (dorsal view). J, Head capsule (ventral view). K, Dark spinous setae on dorsal surface of abdominal segment 8. A, B, H & I, *S. fulgidum*; C, *S. montiblense*; D–G, J & K, *S. barioense*. Scale bars: 0.5 mm for H & I; 0.2 mm for C; 0.1 mm for J; 0.05 mm for A & B; 0.04 mm for D–F; 0.02 mm for G & K

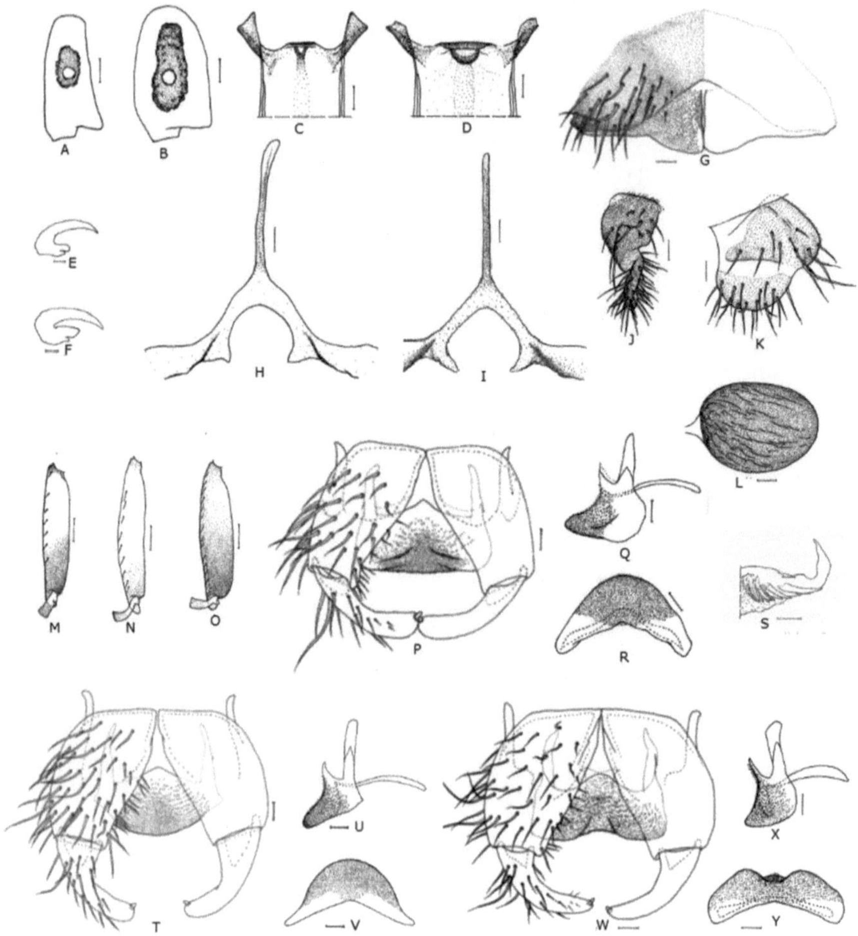

Fig. 36 Female and male of the *S. gombakense* species-group of the subgenus *Gomphostilbia.* A–L, Female; M–Y, Male. A & B, Third palpal segments (right side; front view; A, with medium-long sensory vesicle; B, with long sensory vesicle). C & D, Cibaria (front view). E & F, Claws. G, Sternite 8 and ovipositor valves (ventral view). H & I, Genital forks (ventral view). J & K, Paraprocts and cerci (right side; J, ventral view; K, lateral view). L, Spermatheca. M–O, Hind basitarsi and second tarsomeres (left side; outer view). P, T & W, Coxites, styles and ventral plates (ventral view). Q, U & X, Ventral plates and median sclerites (lateral view). R, V & Y, Ventral plates (caudal view). S, Paramere and aedeagal membrane (caudal view). A, D & H, *S. maleewongae*; B, C, G & I–L, *S. paiense*; S, *S. langkawiense*; E & O, *S. sachini*; F, N & T–V, *S. williei*; M & P–R, *S. gombakense*; W, X & Y, *S. nuwakotense*. Scale bars: 0.1 mm for M–O; 0.02 mm for A–D, G–L & P–Y; 0.01 mm for E & F

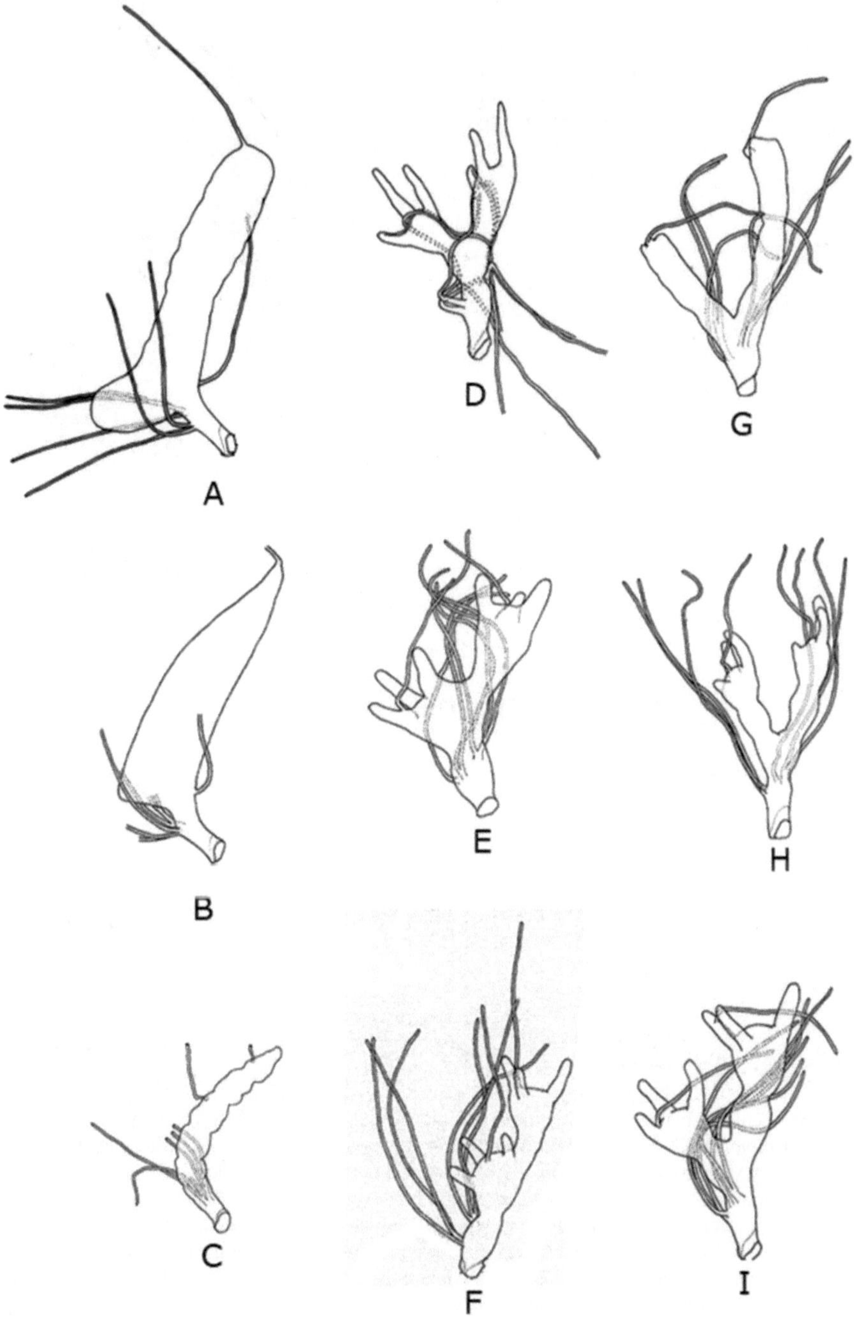

Fig. 37 Pupal gills of the *S. gombakense* species-group of the subgenus *Gomphostilbia*. A, *S. maleewongae*; B, *S. aziruni*; C, *S. williei*; D, *S. thuathienense*; E, *S. gombakense*; F, *S. langkawiense*; G, *S. prayongi*; H, *S. nuwakotense*; I, *S. sachini*

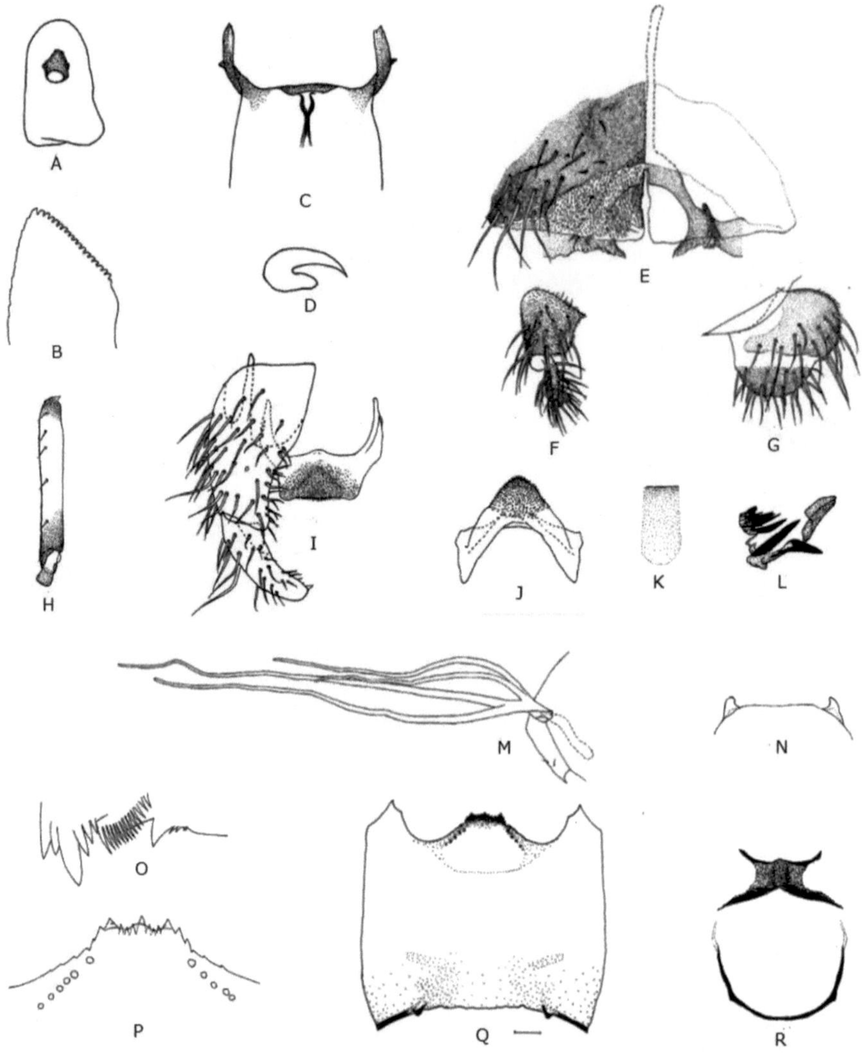

Fig. 38 Female, male, pupa and larva of the *S. palauense* species-group of the subgenus *Gomphostilbia*. A–G, Female; H–L, Male; M & N, Pupa; O–R, Larva. A, Third palpal segment with sensory vesicle (right side; front view). B, Apical portion of mandible. C, Cibarium. D, Tarsal claw. E, Sternite 8, ovipositor valves and genital fork (ventral view). F & G, Paraprocts and cerci (right side; F, ventral view; G, lateral view). H, Hind basitarsus and second tarsomere (left side; outer view). I, Coxite, style and ventral plate (ventral view). J, Ventral plate (caudal view). K, Median sclerite (caudal view). L, Paramere (right side; dorsal view). M, Gill filaments (left side; lateral view). N, Terminal hooks (caudal view). O, Apical portion of mandible. P, Hypostoma. Q, Head capsule (ventral view). R, Anal sclerite and accessory sclerite (caudal view). A–R, *S. palauense*. Scale bar: 0.1 mm for Q

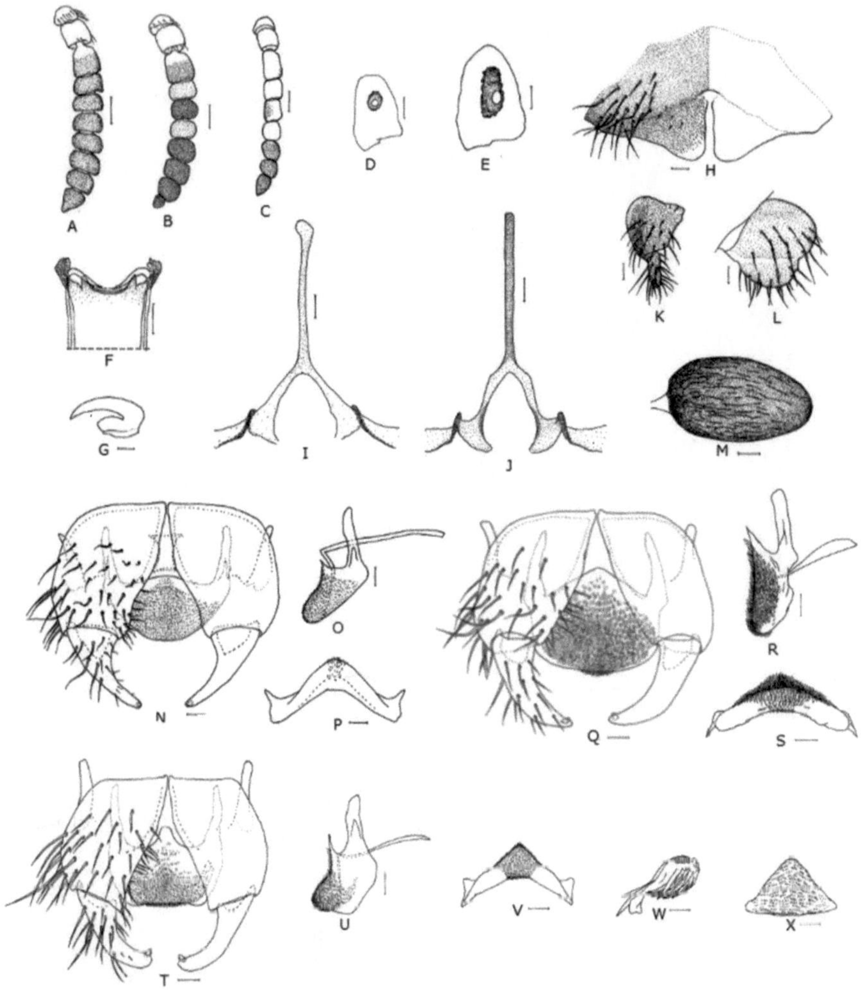

Fig. 39 Female and male of the *S. varicorne* species-group of the subgenus *Gomphostilbia.* A–M, Female; N–X, Male. A–C, Antennae (A & B, with eight flagellomeres showing different color patterns; C, with seven flagellomeres). D & E, Third palpal segments with sensory vesicles of different sizes (right side; front view). F, Cibarium showing two projections on posterodorsal margin (front view). G, Claw. H, Sternite 8 and ovipositor valves (ventral view). I & J, Genital forks (ventral view). K & L, Paraprocts and cerci (right side; K, ventral view; L, lateral view). M, Spermatheca. N, Q & T, Coxites, styles and ventral plates (ventral view; N, *S. novemarticulatum* subgroup; Q, *S. chumpornense* subgroup; T, *S. burtoni* subgroup). O, R & U, Ventral plates and median sclerites (lateral view). P, S & V, Ventral plates (caudal view). W, Paramere (right side; caudal view). X, Aedeagal membrane (caudal view). A, E. G, I, M & T–X, *S. huangi*; B, J & Q–S, *S. kuvangkadilokae*; D, *S. varicorne*; F, *S. chumpornense*; C, G, H, K, L & N–P, *S. charlesi*. Scale bars: 0.05 mm for A & C; 0.04 mm for B; 0.02 mm for D–F & H–X; 0.01 mm for G

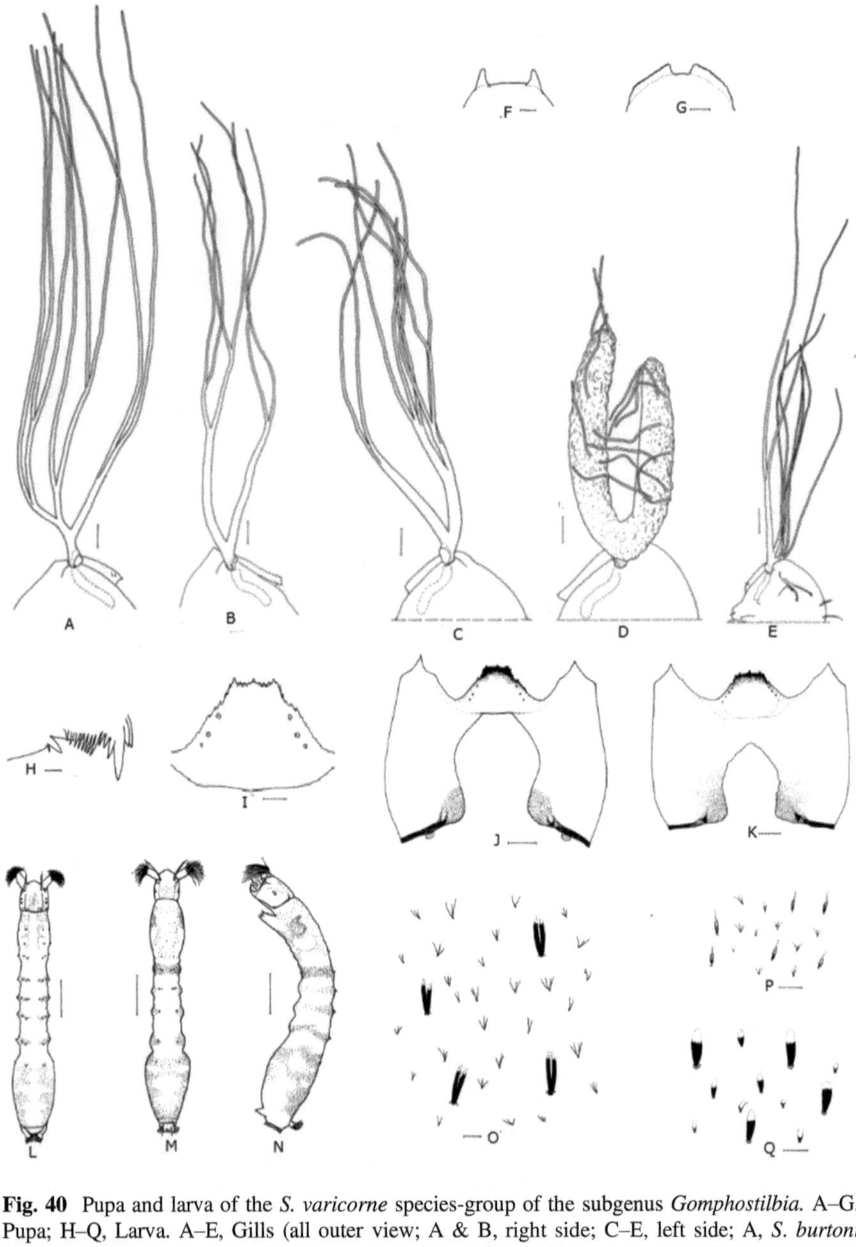

Fig. 40 Pupa and larva of the *S. varicorne* species-group of the subgenus *Gomphostilbia.* A–G, Pupa; H–Q, Larva. A–E, Gills (all outer view; A & B, right side; C–E, left side; A, *S. burtoni* subgroup; B–D, *S. chumpornense* subgroup; E, *S. novemarticulatum* subgroup). F & G, Terminal hooks (caudal view; F, simple, cone-shaped; G, wide, with outer margins crenulated). H, Apical portion of mandible. I, Hypostoma with serrated lateral margins. J & K, Head capsules showing different lengths of postgenal clefts (ventral view). L–N, Whole bodies (L, dorsal view; with multiple pairs of protuberances dorsally and dorsolaterally on each of abdominal segments 1–4; M & N, dorsal and lateral views; with a pair of protuberances dorsally on each of abdominal segments 1–5). O–Q, Setae of various types on dorsum of abdominal segment 8 (O & Q, *S. novemarticulatum* subgroup; P, *S. chumpornense* subgroup). A, G & K, *S. huangi*; B & I, *S. piroonae*; C, J & P, *S. varicorne*; D & L, *S. kuvangkadilokae*; E & Q, *S. charlesi*; F, H & M–O, *S. novemarticulatum*. Scale bars: 0.5 mm for L–N; 0.1 mm for A–E; 0.05 mm for J & K; 0.03 mm for F; 0.02 mm for G, I, P & Q; 0.01 mm for H & O

Fig. 41 *Gomphostilbia* species unplaced to species-group. A–C, Female; D–G, Male; H–M; Pupa; N, Larva. A & B, Cibaria. C, Middle part of abdominal segment 7 (ventral view). D, Coxites, styles and ventral plate (ventral view). E, Style (ventrolateral view). F, Ventral plate and median sclerite (lateral view). G, Ventral plate (caudal view). H–J, Gills (right side; outer view). K, Tubercles on frons. L & M, Terminal hooks. N, Middle part of head capsule (ventral view). A & D–H, *S. azhari*; B, C, J–L & N, *S. guniki*; I & M, *S. dhangi*. Scale bars: 0.2 mm for N; 0.1 mm for H–J; 0.03 mm for B & C; 0.02 mm for A, D–G, I & M; 0.01 mm for K & L

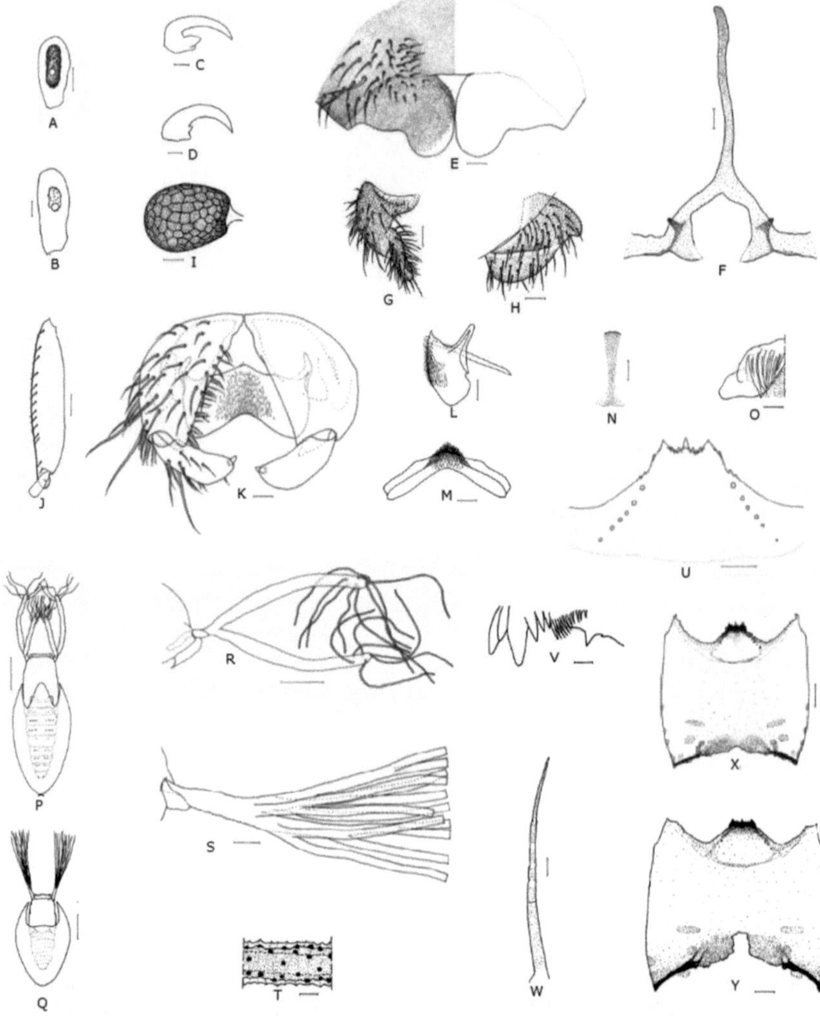

Fig. 42 Female, male, pupa and larva of the subgenus *Montisimulium.* A–I, Female; J–O, Male; P–T, Pupa; U–Y, Larva. A & B, Third palpal segments with sensory vesicle (front view). C & D, Claws (C, with medium-sized basal tooth; D, with small basal tooth). E, Sternite 8 and ovipositor valves (ventral view). F, Genital fork (ventral view). G & H, Paraprocts and cerci (right side; G, ventral view: H, lateral view). I, Spermatheca. J, Hind basitarsus and second tarsomere (left side; outer view). K, Coxites, styles and ventral plate (ventral view). L, Ventral plate and median sclerite (lateral view). M, Ventral plate (caudal view). N, Median sclerite (caudal view). O, Paramere and aedeagal membrane (caudal view). P & Q, Pupae and cocoons (dorsal view). R & S, Gills (R, inflated tubes with 14 slender filaments; S, with 12 thread-like filaments arising from medium-long common basal stalk). T, Middle part of gill filament showing black pigment spots scattered inside the outer cuticular layer. U, Hypostoma. V, Mandible. W, Antenna showing three unpigmented annulations. X & Y, Head capsules (ventral view; X, postgenal cleft vestigial; Y, postgenal cleft small). A, C, J, P, R, U & X, *S. merga*; B, D & E–I, *S. phahompokense*; K–O, *S. laoleense*; Q, S & W, *S. angkaense*; T, V & Y, *S. dattai*. Scale bars: 1.0 mm for P & Q; 0.5 mm for R; 0.1 mm for J, S, X & Y; 0.05 mm for A & U; 0.04 mm for W; 0.02 mm for B, E–I & K–O; 0.01 mm for C, D, T & V

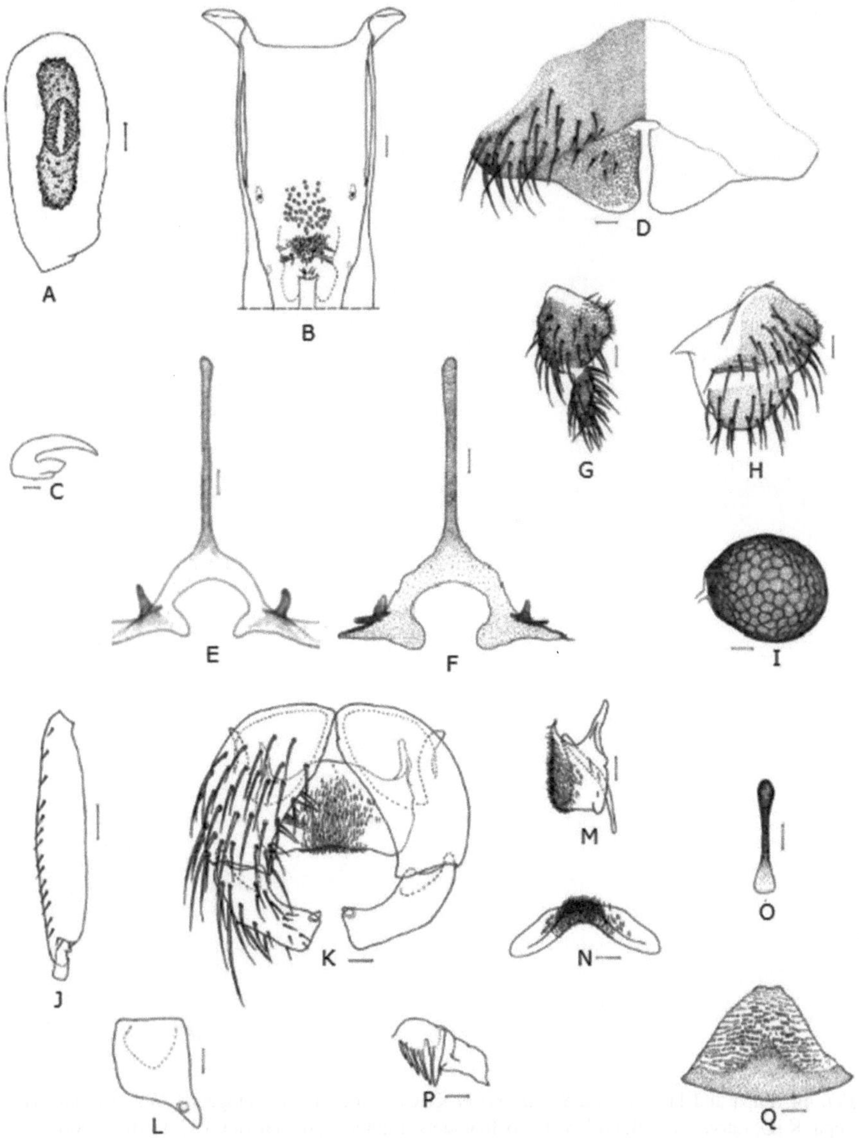

Fig. 43 Female and male of the *S. feuerborni* species-group of the subgenus *Nevermannia*. A–I, Female; J–Q, Male. A, Third palpal segment with long sensory vesicle showing many sensilla inside large opening (right side; front view). B, Cibarium with a cluster of many spinous processes on the surface of its lower part (frontal view). C, Claw. D, Sternite 8 and ovipositor valves (ventral view). E & F, Genital forks (ventral view). G & H, Paraprocts and cerci, (right side; G, ventral view; H, lateral view). I, Spermatheca. J, Hind basitarsus and second tarsomere (left side; outer view). K, Coxites, styles and ventral plate (ventral plate). L, Style (right side; ventrolateral view). M, Ventral plate and median sclerite (lateral view). N, Ventral plate (caudal view). O, Median sclerite (ventral view). P, Paramere (caudal view). Q, Aedeagal membrane and dorsal plate (caudal view). A & E, *S. longbiangense*; B, C, D & F–Q, *S. maeaiense*. Scale bars: 0.1 mm for J; 0.02 mm for A, B, D–I & K–Q; 0.01 mm for C

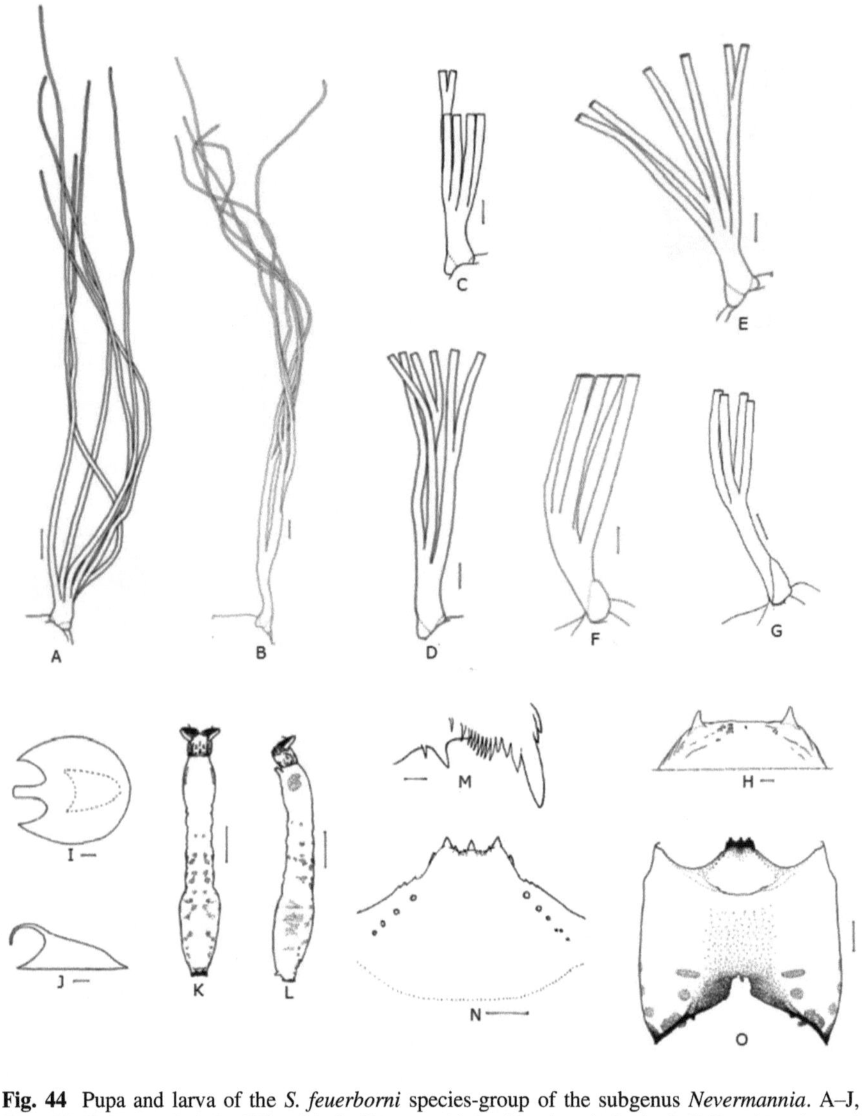

Fig. 44 Pupa and larva of the *S. feuerborni* species-group of the subgenus *Nevermannia*. A–J, Pupa; K–O, Larva. A & B, Gills with six filaments (right side; dorsal view). C–E, Basal portions of six gill filaments (left side; dorsal view). F, Basal portion of gill with five filaments (right side; lateral view). G, Basal portion of gill with four filaments (right side; lateral view). H, Terminal hooks (anterodorsal view). I & J, Cocoons each with anterodorsal projection (I, dorsal view; J, lateral view). K & L, Whole bodies showing dorsal and dorsolateral color markings (K, dorsal view; L, lateral view). M, Apical portion of mandible. N, Hypostoma. O, Head capsule showing small postgenal cleft (ventral view). A, *S. chiangklangense*; B & H, *S. vessabutrae*; C & D, *S. fruticosum*; E, I–L & O, *S. maeaiense*; F, *S. phami*; G, *S. borneoense*; M & N, *S. fangense*. Scale bars: 1.0 mm for K & L; 0.5 mm for I & J; 0.05 mm for N; 0.2 mm for A; 0.1 mm for B, C–G & O; 0.01 mm for H & M

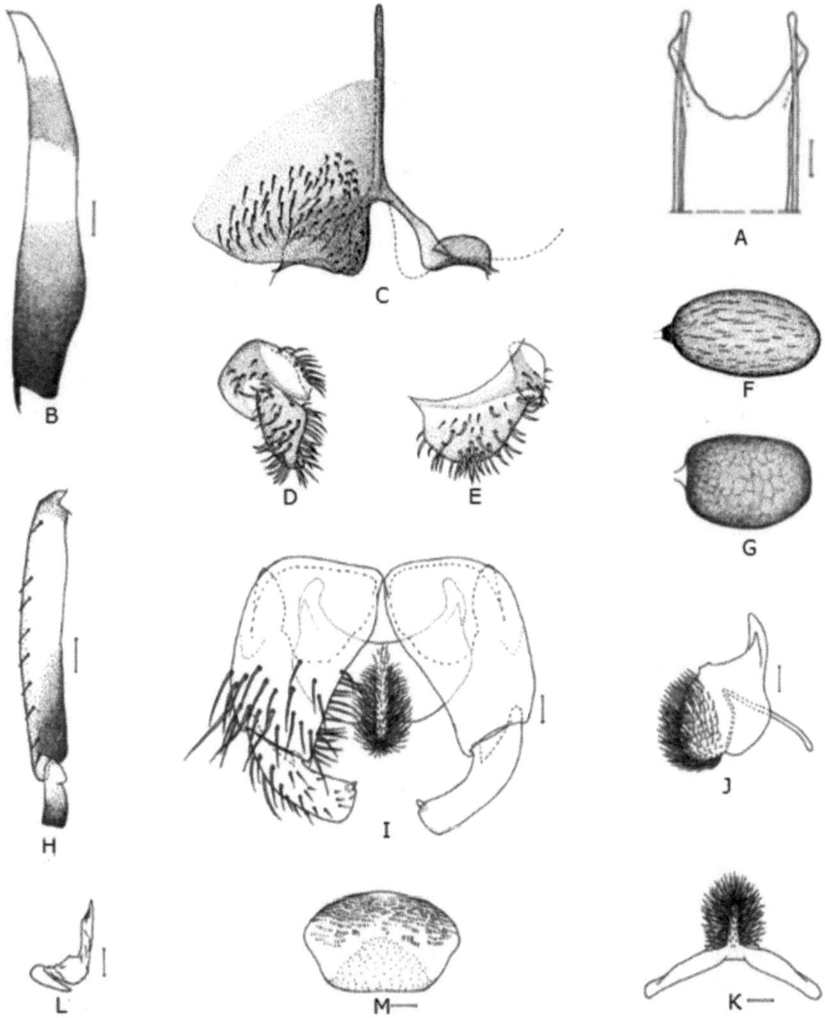

Fig. 45 Female and male of the *S. ruficorne* species-group of the subgenus *Nevermannia*. A–G, Female; H–M, Male. A, Cibarium (front view). B, Hind tibia (left side; outer view). C, Sternite 8, ovipositor valve and genital fork (ventral view). D & E, Paraprocts and cerci, (right side; D, ventral view; E, lateral view). F & G, Spermathecae (F, with pigmented neck; G, without pigmented neck). H, Hind basitarsus and second tarsomere (left side; outer view). I, Coxites, styles and ventral plate (ventral view). J, Ventral plate and median sclerite (lateral view). K, Ventral plate (caudal view). L, Paramere (left side; ventral view). M, Aedeagal membrane and dorsal plate (caudal view). A, B & G–M, *S. wayani*; C–F, *S. aureohirtum*. Scale bars: 0.05 mm for B & H; 0.02 mm for A & I–M

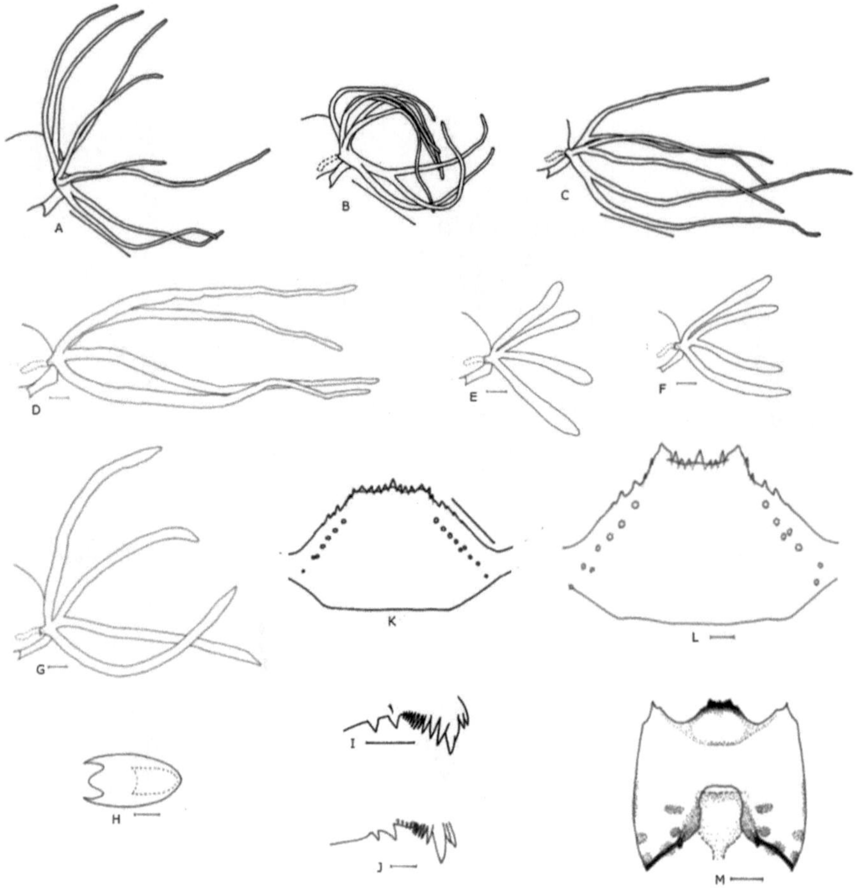

Fig. 46 Pupa and larva of the *S. ruficorne* species-group of the subgenus *Nevermannia*. A–H, Pupa; I–M, Larva. A & B, Gill with eight filaments (right side; lateral view). C, Gill with six filaments (right side; lateral view). D–G, Gill with four filaments (right side; lateral view). H, Cocoon (dorsal view). I & J, Apical portions of mandibles. K & L, Hypostoma. M, Head capsule showing medium-long postgenal cleft (ventral view). A, B, K & I, *S. glatthaari*; C, *S. aureohirtum*; D–H, J, L & M, *S. wayani*. Scale bars: 1.0 mm for A–C & H; 0.2 mm for D–G; 0.1 mm for K & M; 0.05 mm for I; 0.02 mm for J & L

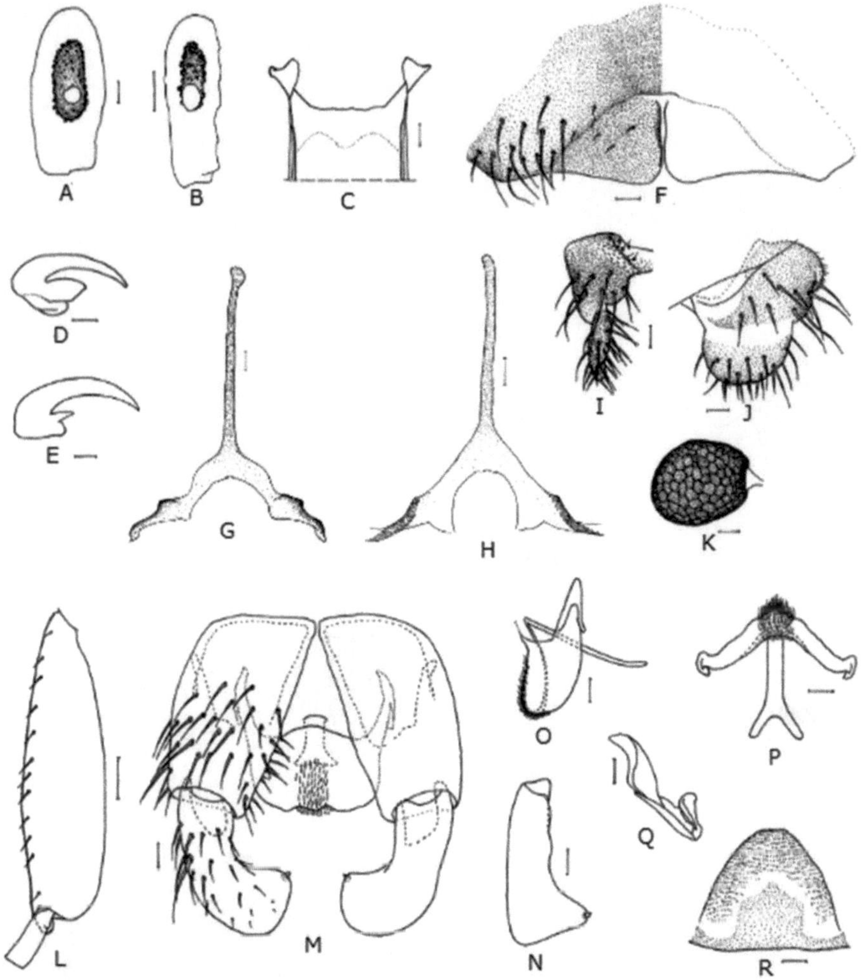

Fig. 47 Female and male of the *S. vernum* species-group of the subgenus *Nevermannia*. A–K, Female; L–R, Male. A & B, Third palpal segments with sensory vesicles (right side; front view). C, Cibarium (front view). D & E, Claws. F, Sternite 8 and ovipositor valves (ventral view). G & H, Genital forks (ventral view). I & J, Paraprocts and cerci (right side; I, ventral view; J, lateral view). K, Spermatheca. L, Hind basitarsus and second tarsomere (left side; outer view). M, Coxites, styles, ventral plate and median sclerite (ventral view). N, Style (right side; ventromedial view). O & P, Ventral plates and median sclerites (O, lateral view; P, caudal view). Q, Paramere (right side; ventral view). R, Aedeagal membrane and dorsal plate (caudal view). A, C, D, F & I–R, *S. thrimshinglaense*; B, E & G, *S. letense*; H, *S. tayense*. Scale bars: 0.1 mm for L; 0.04 mm for B; 0.02 mm for A, C, F–K & M–R; 0.01 mm for D & E

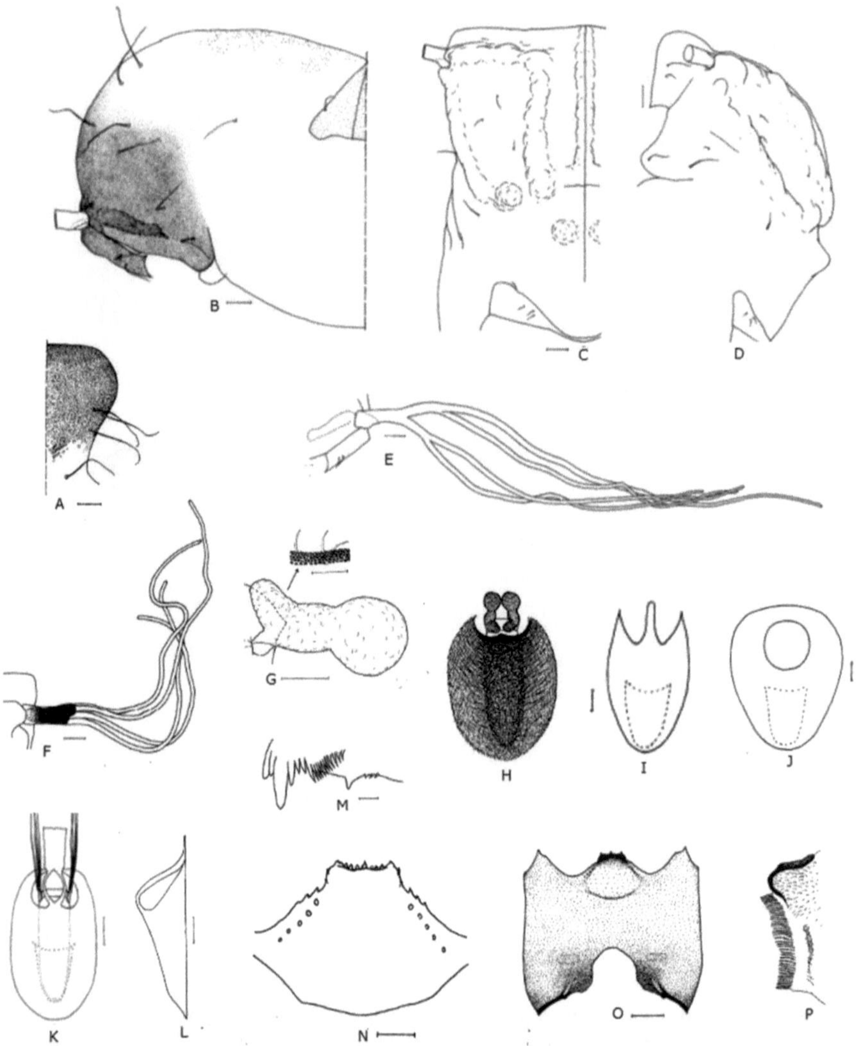

Fig. 48 Pupa and larva of the *S. vernum* species-group of the subgenus *Nevermannia*. A–L, Pupa; M–P, Larva. A, Frons showing dark-pigmented area (left half; dorsal view). B, Thorax showing dark-pigmented area (left half; lateral view). C & D, Thorax showing ridges (left side; C, dorsal view; D, lateral view). E, Gill with four filaments (right side; lateral view). F, Gill with four filaments arising from black-pigmented common basal stalk (right side; dorsal view).G, Gill is composed of inflated, dumbbell-like structure with its surface covered with tubercles and tiny setae (right side; lateral view). H, Pupa and cocoon (dorsal view). I, Cocoon with anterodorsal projection (dorsal view). J, Cocoon rounded, with anteroventral collar (dorsal view). K, Pupa and cocoon with anterodorsal projection having wide open space basally (dorsal view). L, Cocoon (same as cocoon of K; lateral view). M, Apical portion of mandible showing a supernumerary serration. N, Hypostoma. O, Head capsule showing medium-long postgenal cleft (ventral view). P, Posterior tip of abdomen showing anal sclerite, posterior circlet of hooklets and dark-pigmented accessory sclerite (lateral view). A & B, *S. laichauense*; C, D, F & J, *S. letense*; G & H, *S. taulingense*; E, I, M & O, *S. thrimshinglaense*; K & L, *S. lekhaniense*; N, *S. tayense*; P, *S. yushangense*. Scale bars: 1.0 mm for K & L; 0.5 mm for G, I & J, 0.1 mm for A–F & O; 0.04 mm for N; 0.01 mm for M & enlargement of G

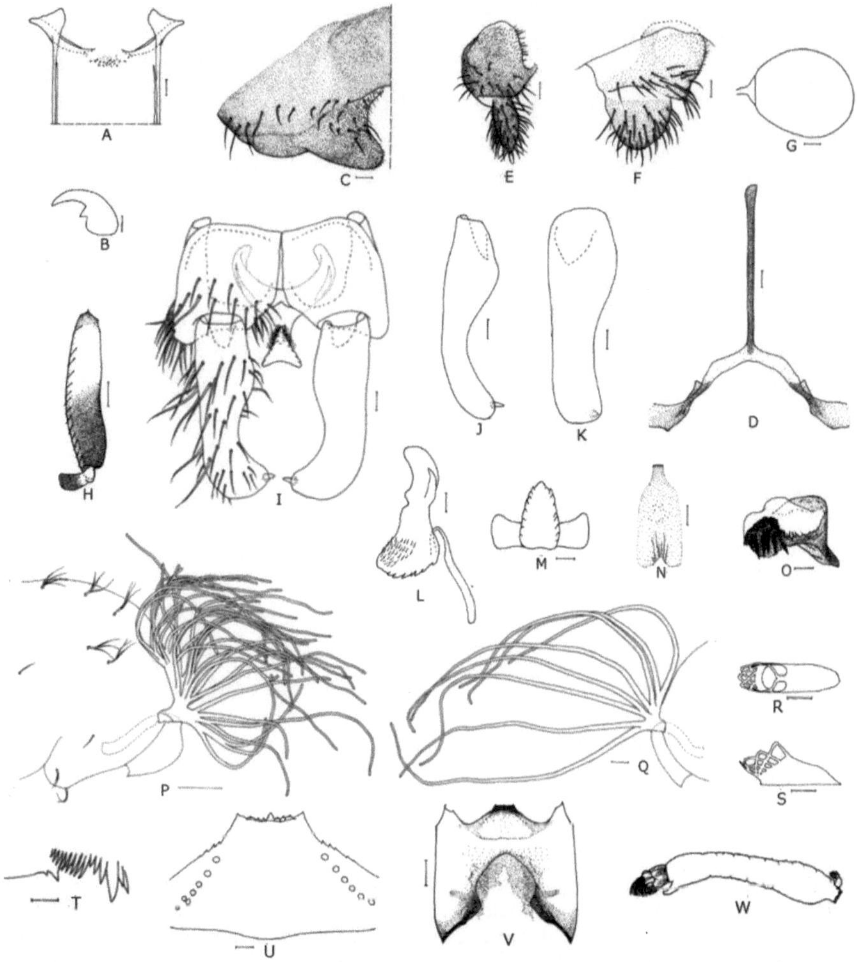

Fig. 49 Female, male, pupa, and larva of the *S. argentipes* species-group of the subgenus *Simulium* s. str. A–G, Female; H–O, Male; P–S, Pupa; T–W, Larva. A, Cibarium. B, Claw. C, Sternite 8 and ovipositor valve (right half; ventral view). D, Genital fork. E & F, Paraprocts and cerci (right side; E, ventral view; F, lateral view). G, Spermatheca. H, Hind basitarsus and second tarsomere (left side; outer view). I, Coxites, styles and ventral plate (ventral view). J & K, Styles (right side; J, medial view; K, ventrolateral view). L, Ventral plate and median sclerite (lateral view). M, Ventral plate (caudal view). N, Median sclerite (ventrocaudal view). O, Paramere (left side; caudal view). P, Anterior one-third of thorax with gill and trichomes (right side; outer view). Q, Gill with eight filaments (left side; lateral view). R & S, Cocoons (R, dorsal view; S, lateral view). T, Apical portion of mandible. U, Hypostoma. V, Head capsule showing mitre-shaped postgenal cleft (ventral view). W, Whole body (lateral view). A, C–O, Q & T–V, *S. haiduanense*; B & P, *S. arboreum*; W, *S. minangkabaum*; R & S, *S. sansahoense*. Scale bars: 1 mm for R & S; 0.2 mm for P; 0.1 mm for H, Q & V; 0.02 mm for A, C–G, I–O, T & U; 0.01 mm for B

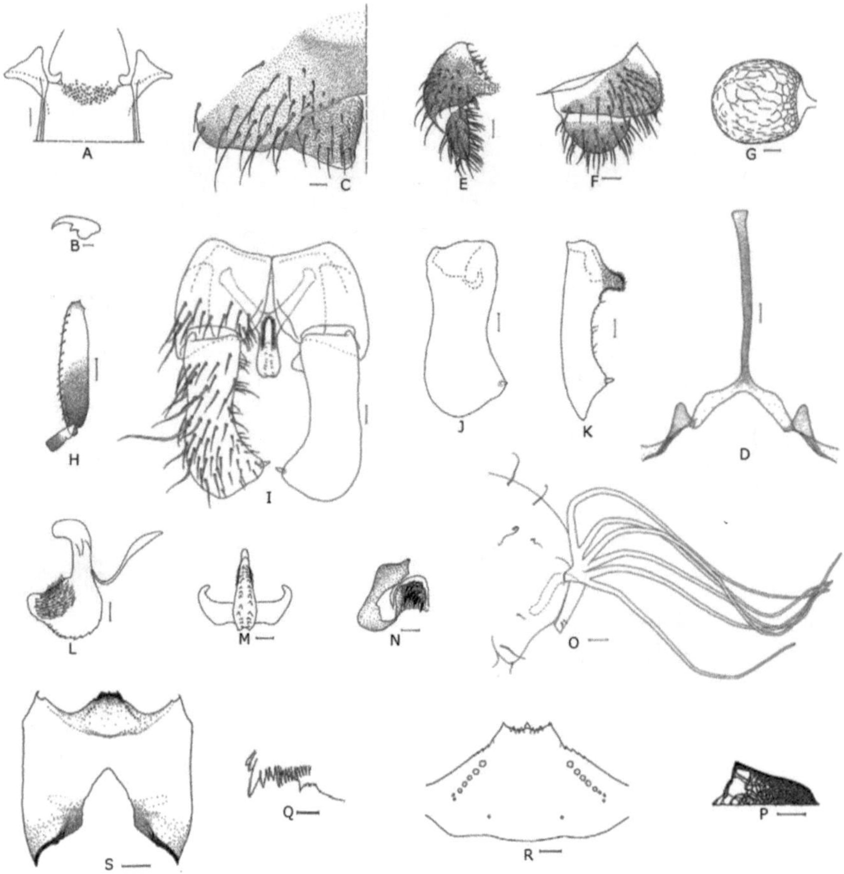

Fig. 50 Female, male, pupa, and larva of the *S. christophersi* species-group of the subgenus *Simulium* s. str. A–G, Female; H–N, Male; O & P, Pupa; Q–S, Larva. A, Cibarium. B, Claw. C, Sternite 8 and ovipositor valve (right half; ventral view). D, Genital fork. E & F, Paraprocts and cerci (right side; E, ventral view; F, lateral view). G, Spermatheca. H, Hind basitarsus and second tarsomere (left side; outer view). I, Coxites, styles, and ventral plate (ventral view). J & K, Styles (right side; J, ventrolateral view; K, medial view). L, Ventral plate and median sclerite (lateral view). M, Ventral plate (caudal view). N, Paramere (right side; ventrocaudal view). O, Anterior one-third of thorax with gill and trichomes (right side; outer view). P, Cocoon (lateral view). Q, Apical portion of mandible. R, Hypostoma. S, Head capsule showing triangular postgenal cleft (ventral view). A–S, *S. atipornae*. Scale bars: 1 mm for P; 0.1 mm for H, J, O & S; 0.03 mm for R; 0.02 mm for A, C–G, I–N & Q; 0.01 mm for B

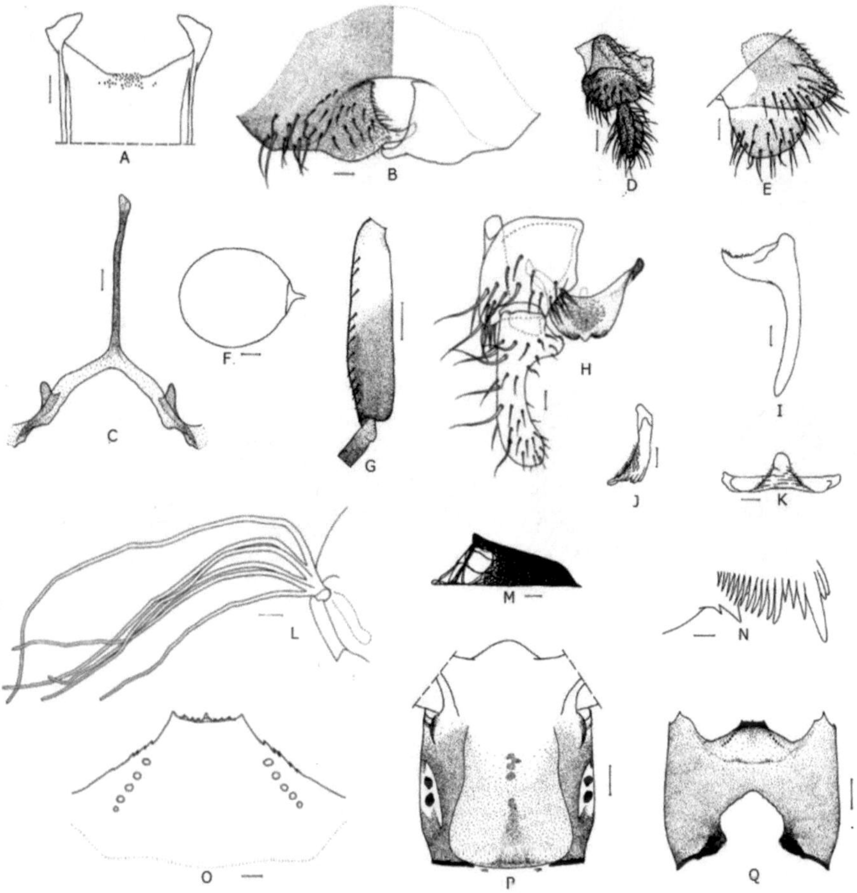

Fig. 51 Female, male, pupa and larva of the *S. chungi* species-group of the subgenus *Simulium* s. str. A–F, Female; G–K, Male; L & M, Pupa; N–Q, Larva. A, Cibarium. B, Sternite 8 and ovipositor valves (ventral view). C, Genital fork. D & E, Paraprocts and cerci (right side; D, ventral view; E, lateral view). F, Spermatheca. G, Hind basitarsus and second tarsomere (left side; outer view). H, Coxite, style (left side) and ventral plate (ventral view). I, Style (right side; lateral view). J & K, Ventral plates (J, lateral view; K, caudal view). L, Gill (left side; outer view). M, Cocoon (lateral view). N, Apical portion of mandible. O, Hypostoma. P & Q, Head capsules (P, dorsal view; Q, ventral view). A–Q, *S. chungi*. Scale bars: 0.5 mm for M; 0.1 mm for G, L, P & Q; 0.02 mm for A–F, H–K & O; 0.01 mm for N

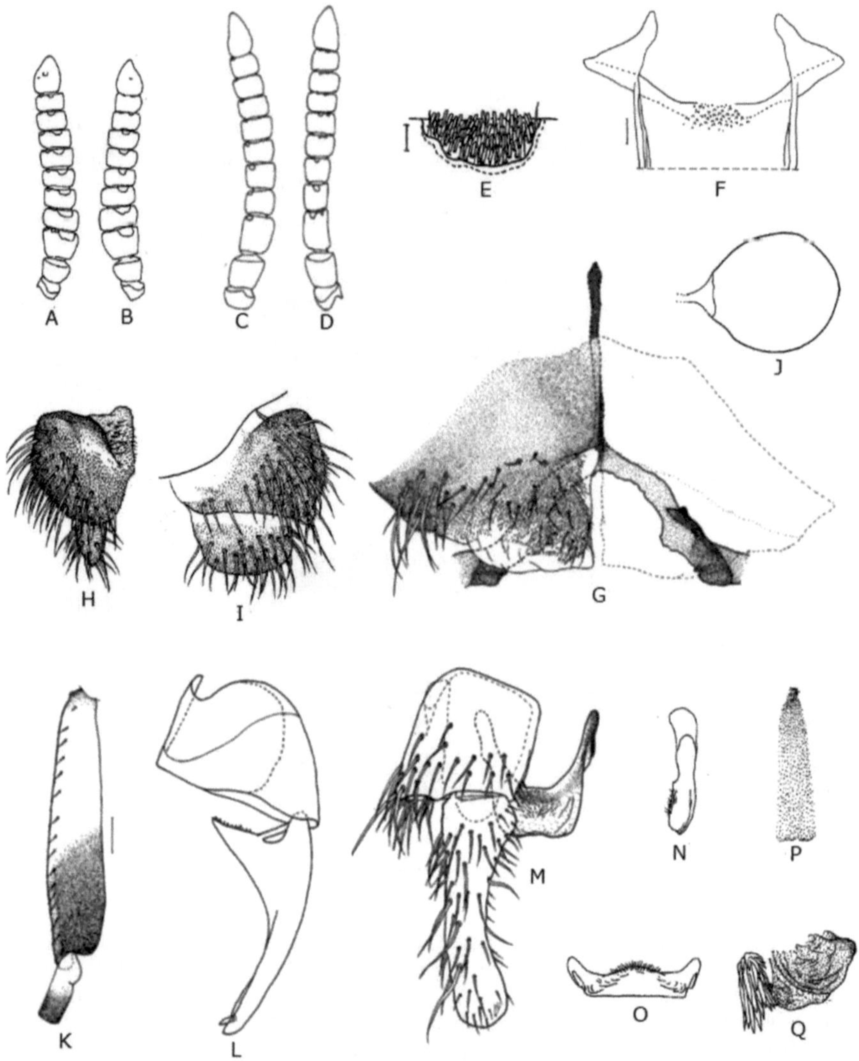

Fig. 52 Female and male of the *S. crocinum* species-group of the subgenus *Simulium* s. str. A, B & E–J, Female; C, D & K–Q, Male. A–D, Antennae showing a well-developed pit on each flagellomere (A & D, outside view; B & C, inside view). E, Pit on second female flagellomere with numerous sensilla. F, Cibarium. G, Sternite 8, ovipositor valves, and genital fork (ventral view). H & I, Paraprocts and cerci (right side; H, ventral view; I, lateral view). J, Spermatheca. K, Hind basitarsus and second tarsomere (left side; outer view). L, Coxite and style (right side; lateral view). M, Coxite, style and ventral plate (ventral view). N & O, Ventral plates (N, lateral view; O, caudal view). P, Median sclerite (caudal view). Q, Paramere (left side; caudal view). A–E, G–J & L–Q, *S. yongi*; F, *S. crocinum*; K, *S. rudnicki*. Scale bars: 0.1 mm for K; 0.02 mm for F; 0.01 mm for E

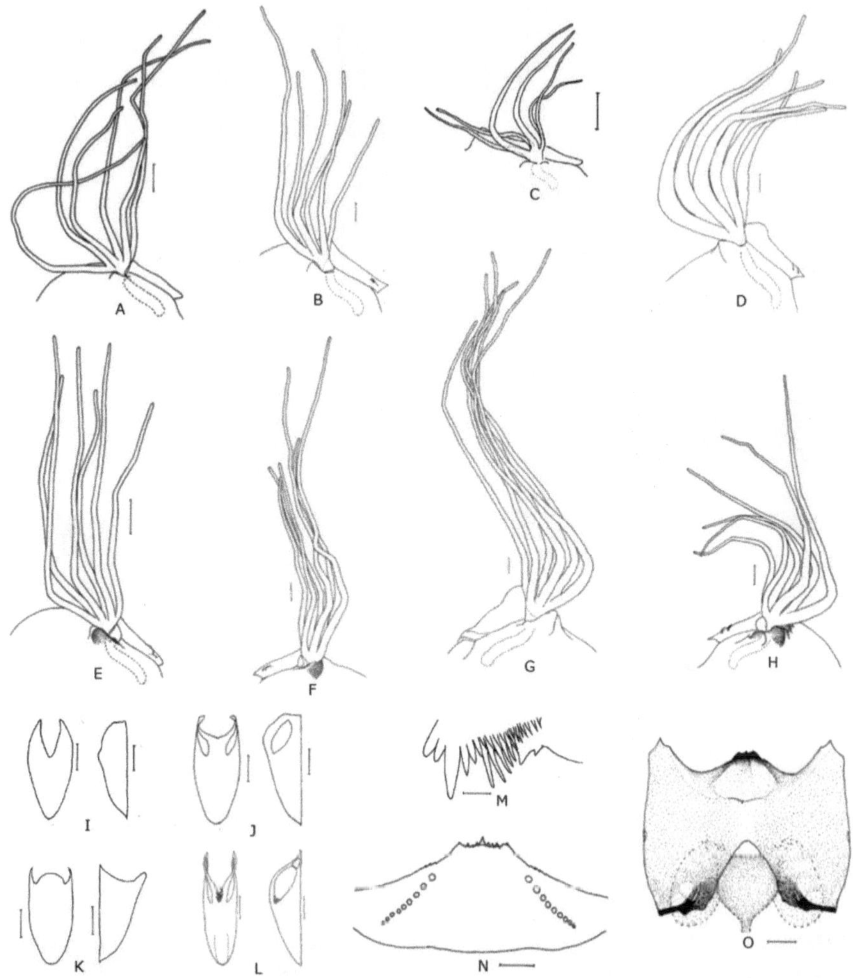

Fig. 53 Pupa and larva of the *S. crocinum* species-group of the subgenus *Simulium* s. str. A–L, Pupa; M–O, Larva. A–H, Gills (lateral view; A–E, right side; F–H, left side). I–L, Cocoons (left side, dorsal view; right side, lateral view). M, Apical portion of mandible. N, Hypostoma. O, Head capsule showing triangular postgenal cleft and dark-pigmented subesophageal ganglion (ventral view). A & M, *S. rudnicki*; C, *S. ufengense*; B, K & N, *S. laocaiense*; E, *S. mediocololatum*; F, *S. yukawai*; G & L, *S. serenum*; H, *S. crocinum*; I, *S. yongi*; D, J & O, *S. turgidum*. Scale bars: 1.0 mm for I–L; 0.2 mm for A, C & E; 0.1 mm for B, D, F–H & O; 0.05 mm for N; 0.02 mm for M

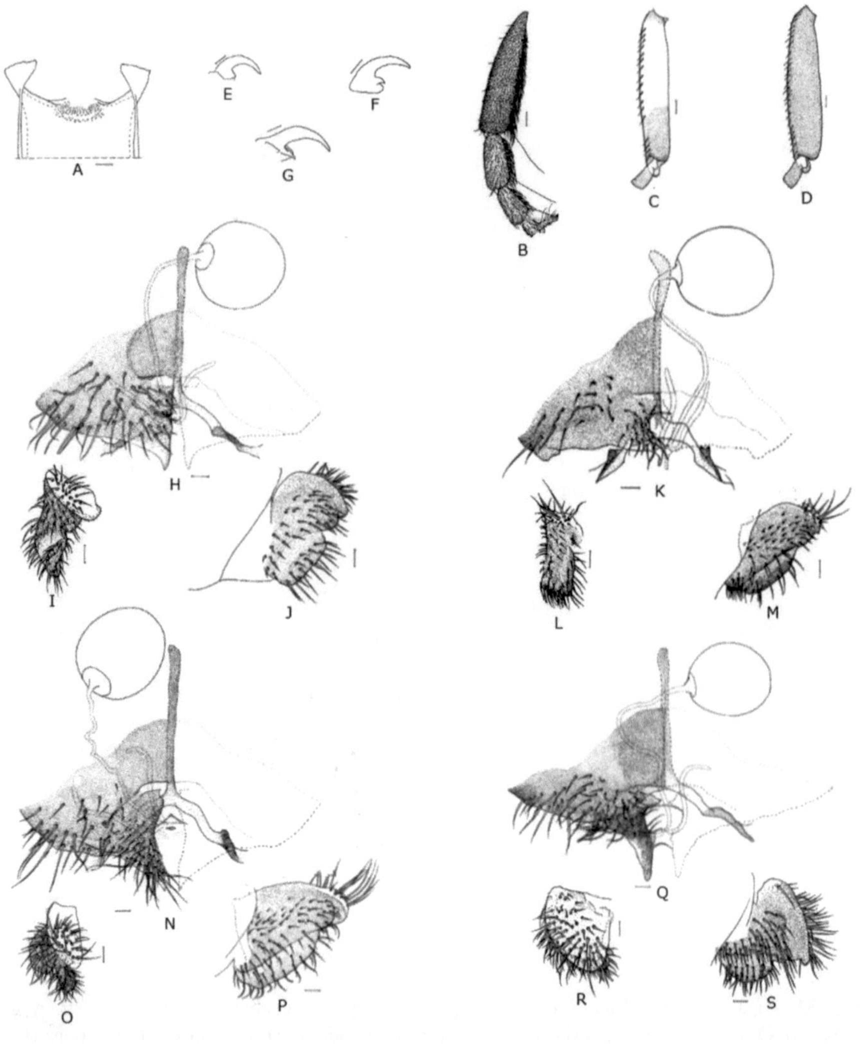

Fig. 54 Female of the *S. dumogaense* species-group of the subgenus *Simulium* s. str. A, Cibarium. B, Fore tarsus with thick hair crest. C & D, Hind basitarsi and second tarsomeres of different color patterns (left side; outer view). E–G, Claws of different shapes. H, K, N & Q, Sternite 8, ovipositor valves, genital forks, and spermathecae (ventral view). I, L, O & R, Paraprocts and cerci (right side; ventral view). J, M, P & S, Paraprocts and cerci (right side; lateral view). B, F & H–J, *S. dumogaense*; C & N–P, *S. mediodentatum*; A, D, G & Q–S, *S. tomohonense*; E, *S. tumpaense*; K–M, *S. watanabei*. Scale bars: 0.05 mm for B–D; 0.02 mm for A & H–S; 0.01 mm for E–G

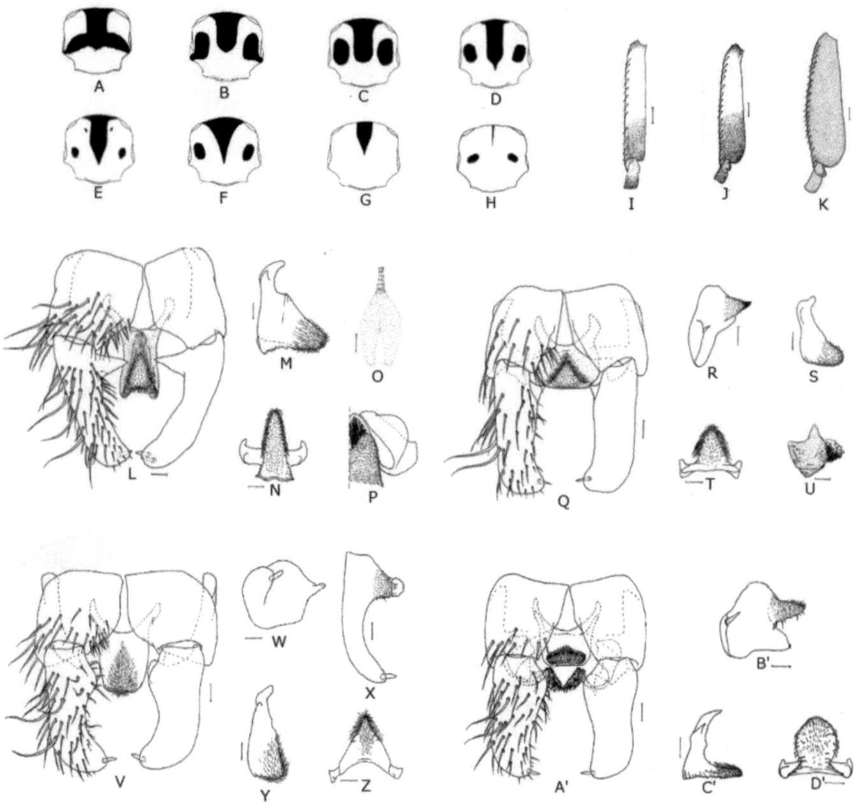

Fig. 55 Male of the *S. dumogaense* species-group of the subgenus *Simulium* s. str. A–H, Scuta with different color patterns (dorsal view). I–K, IIind basitarsi and second tarsomeres of different shapes (left side; outer view). L, Q, V & A', coxites, styles and ventral plates (ventral view). M, S, Y & C', Ventral plates (lateral view). N, T, Z & D', Ventral plates (caudal view). O, Median sclerite (caudal view). P, Paramere and aedeagal membrane (left half; caudal view). U, Paramere showing microsetae on its basal portion (left side; lateral view). R, W & B', Styles with basal protuberance (caudal view). X, Style (right side; medial view). A, H & I, *S. contractum*; B, J & V–Z, *S. tumpaense*; C–G & Q–U, *S. dumogaense*; K & L–P, *S. tomohonense*; A'–D', *S. watanabei*. Scale bars: 0.05 mm for I–K; 0.02 mm for L–D'

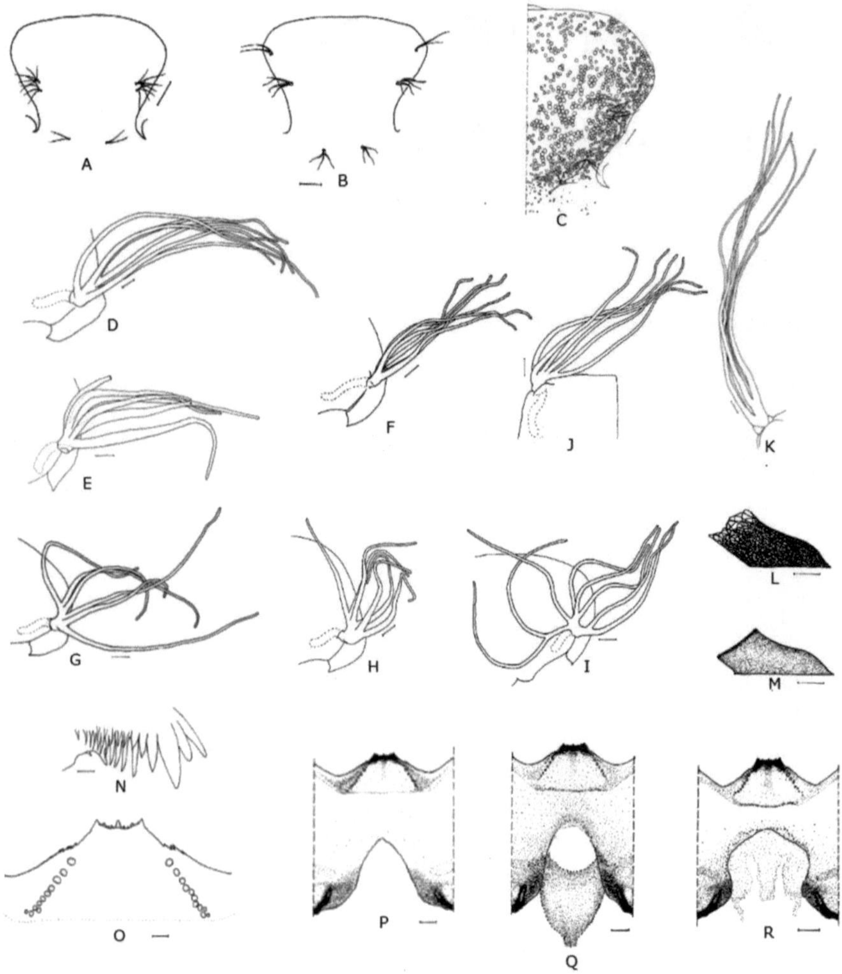

Fig. 56 Pupa and larva of the *S. dumogaense* species-group of the subgenus *Simulium* s. str. A–M, Pupa; N–R, Larva. A–C, Frons and upper parts of faces showing the difference in the location and number of frontal trichomes. D–K, Gills (D–I, right side; lateral view; J & K, left side; dorsal view). L & M, Cocoons (lateral view). N, Apical part of mandible. O, Hypostoma. P–R, Medial portion of the ventral surface of head capsules showing postgenal clefts of different shapes (ventral view). A & F, *S. dumogaense*; J & Q, *S. fuscitarse*; B, D & L, *S tomohonense*; C, K, P, N & O, S. *mediodentatum*; E, *S. contractum*; G & R, *S. tumpaense*; H & M, *S. nudifrons*; I, *S. landonoense*; Scale bars: 1.0 mm for L & M; 0.1 mm for A, B & D–K; 0.05 mm for C & P–R; 0.02 mm for O; 0.01 mm for N

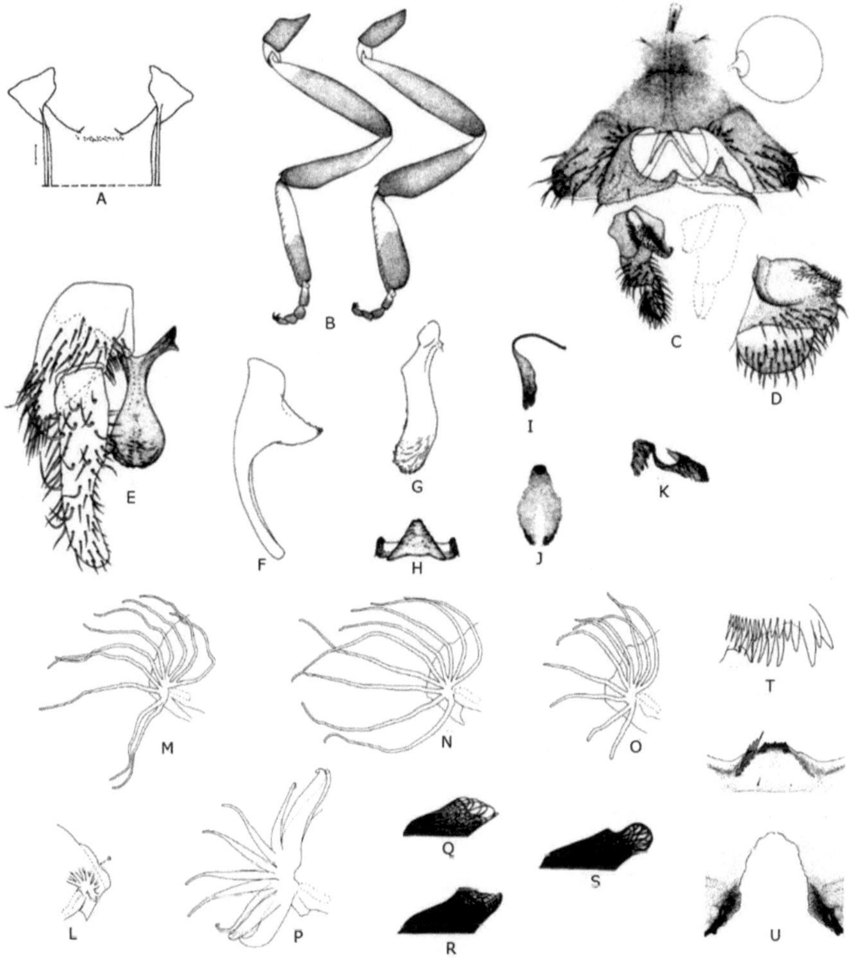

Fig. 57 Female, male, pupa, and larva of the *S. eximium* species-group of the subgenus *Simulium* s. str. A–D, Female; E–K, Male; L–S, Pupa; T & U, Larva. A, Cibarium. B, Left hind legs (outer view; left side, female; right side, male). C, Terminalia (sternite 8, ovipositor valves, paraprocts, cerci, genital fork, and spermatheca) (ventral view). D, Paraproct and cercus (right side; lateral view). E, Genitalia (right coxite, right style, and ventral plate) (ventral view). F, Style (right side; medial view). G & H, Ventral plates (G, lateral view; H, caudal view). I & J, Median sclerites (I, lateral view; J, caudal view). K, Paramere (left side; caudal view). L, Anterior portion of thorax showing rugose swelling (a) on anterior surface (right side; lateral view). M–P, Gills (left side; outer view). Q–S, Cocoons (lateral view). T, Anterior portion of mandible. U, Medial portion of head capsule (ventral view). A, *S. ranauense*; B–D, L–N, Q, T & U, *S. eximium*; E–K, O & R, *S. upikae*; P & S, *S. thienemanni*. Scale bar: 0.02 mm for A

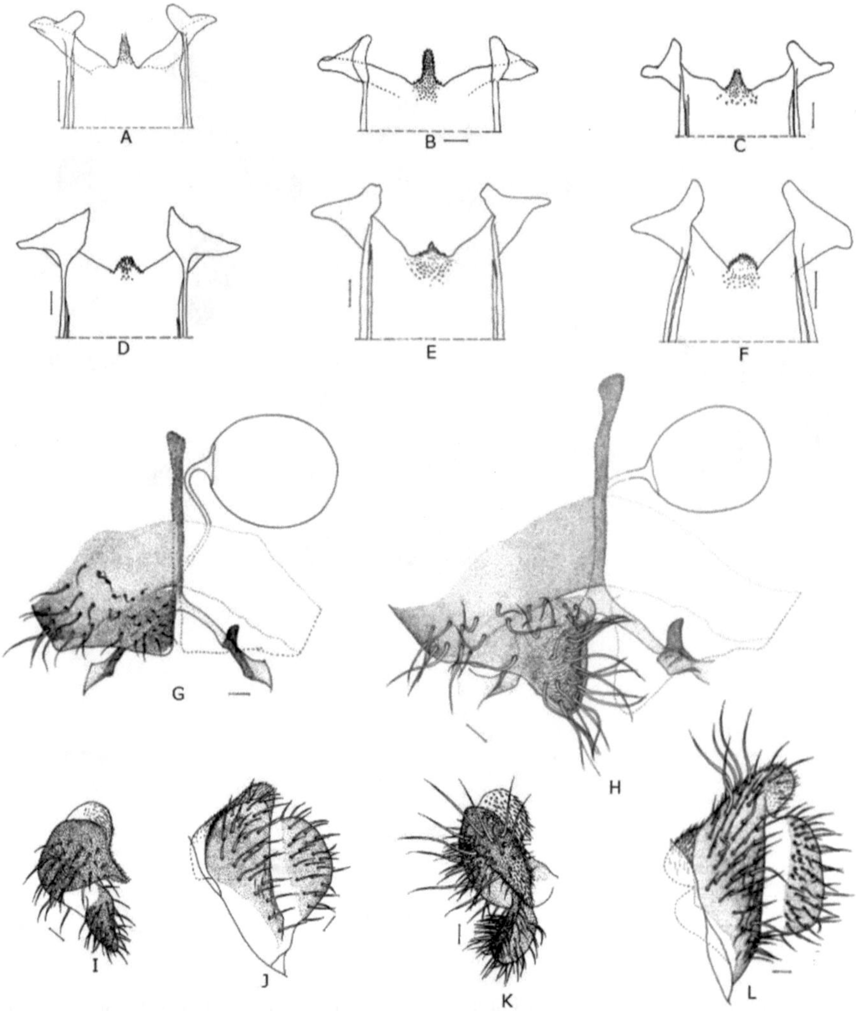

Fig. 58 Female of the *S. griseifrons* species-group of the subgenus *Simulium* s. str. A–F, Cibaria. G & H, Sternite 8, ovipositor valves, genital forks, and spermathecae (ventral view). I–L, Paraprocts and cerci (right side; I & K, ventral view; J & L, lateral view). A, *S. maenoi*; B, *S. pukaengense*; C & G, *S. suchariti*; D, I & J, *S. phukaense*; E, K & L, *S. visuti*; F & H, *S. choochotei*. Scale bars: 0.05 mm for C & G; 0.03 mm for A, F & H; 0.02 mm for B, D, E & I–L

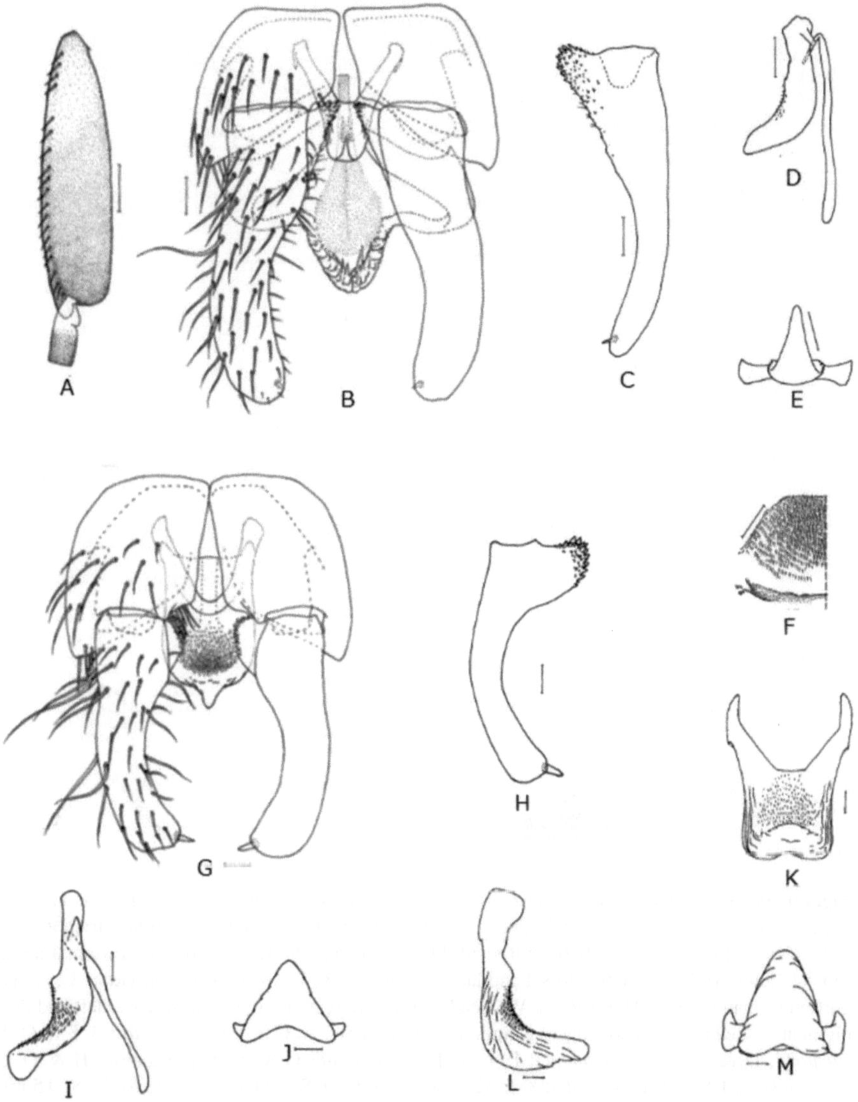

Fig. 59 Male of the *S. griseifrons* species-group of the subgenus *Simulium* s. str. A, Hind basitarsus and second tarsomere (left side; outer view). B, Coxites, styles, ventral plate, median sclerite, and parameres (ventral view). C, Style (left side; medial view). D & I, Ventral plates and median sclerites (lateral view). E, J & M, Ventral plates (caudal view). F, Aedeagal membrane and dorsal plate (caudal view). G, Coxites, styles, and ventral plate (ventral view). H, Style (right side; medial view). K, Ventral plate (ventral view). L, Ventral plate (lateral view). A, *S. digrammicum*; B–F, *S. maenoi*; G–J, *S. suchariti*; K–M, *S. visuti*. Scale bars: 0.1 mm for A; 0.03 mm for B–F; 0.02 mm for G–M.

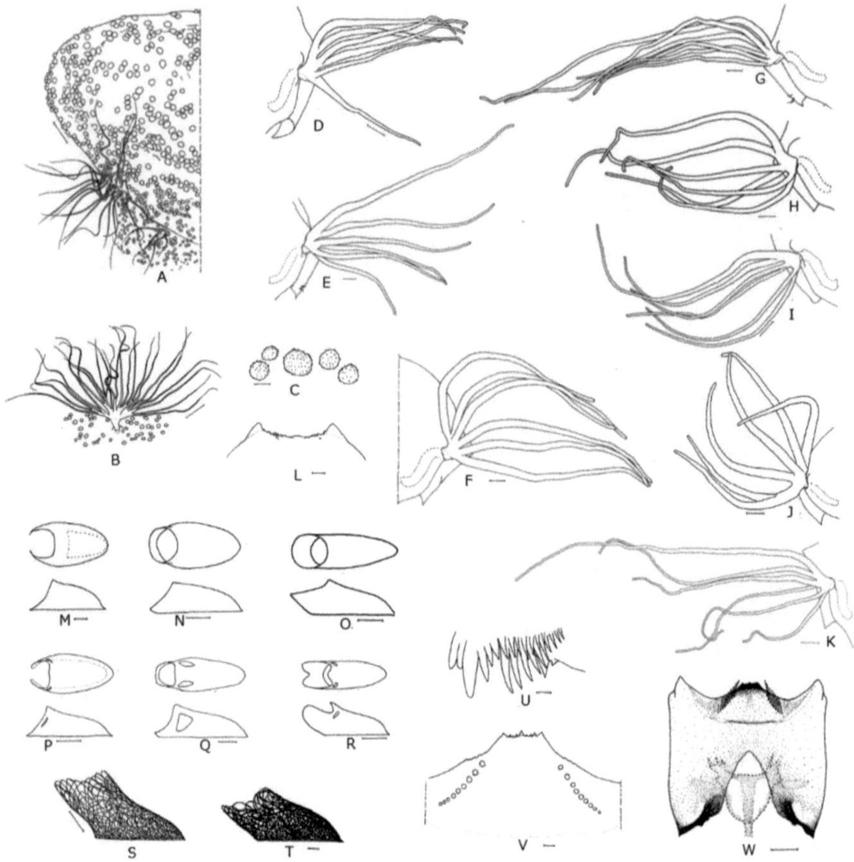

Fig. 60 Pupa and larva of the *S. griseifrons* species-group of the subgenus *Simulium* s. str. A–T, Pupa; U–W, Larva. A, Frons and upper part of the face (dorsal view). B, Frontal trichome. C, Tubercles on frons. D–K, Gills (lateral view; D–F, right side; G–K, left side). L, Terminal hooks (dorsal view). M–R, Cocoons (dorsal and lateral views). S & T, Cocoons (lateral view). U, Apical portion of mandible. V, Hypostoma. W, Head capsule showing triangular postgenal cleft and dark pigmented sheath of subesophageal ganglion (ventral view). A, C, I & P, *S. maenoi*; B, E & Q, *S. suchariti*; D & R, *S. digrammicum*; F, L & T–W, *S. visuti*; G & N, *S. phukaense*; H & O, *S. pukaengense*; J & S, *S. choochotei*; K & M, *S. thongsahuani*. Scale bars: 1.0 mm for N–S; 0.5 mm for M & T; 0.2 mm for J; 0.1 mm for D–I, K & W; 0.05 mm for A & B; 0.02 mm for L & V; 0.01 mm for C & U

Fig. 61 Female, male, pupa, and larva of the *S. grossifilum* species-group of the subgenus *Simulium* s. str. A–F, Female; G–L, Male; M–P, Pupa; Q–T, Larva. A, Cibarium. B, Sternite 8 and ovipositor valves (ventral view). C, Genital fork. D & E, Paraprocts and cerci (right side; D, ventral view; E, lateral view). F, Spermatheca. G, Hind basitarsus and second tarsomere (left side; outer view). H, Coxites, styles, and ventral plate (ventral view). I & J, Styles (right side; I, ventrolateral view; J, medial view). K, Ventral plate and median sclerite (lateral view). L, Ventral plate (caudal view). M & N, Gills (left side; M, dorsal view; N, outer view). O, Spines on gill filaments. P, Cocoon (lateral view). Q, Apical portion of mandible. R, Hypostoma. S & T, Head capsules (S, dorsal view; T, ventral view). A–T, *S. phayaoense*. Scale bars: 1.0 mm for P; 0.1 mm for G, M, N, S & T; 0.02 mm for A–F, H–L & R; 0.01 mm for O & Q

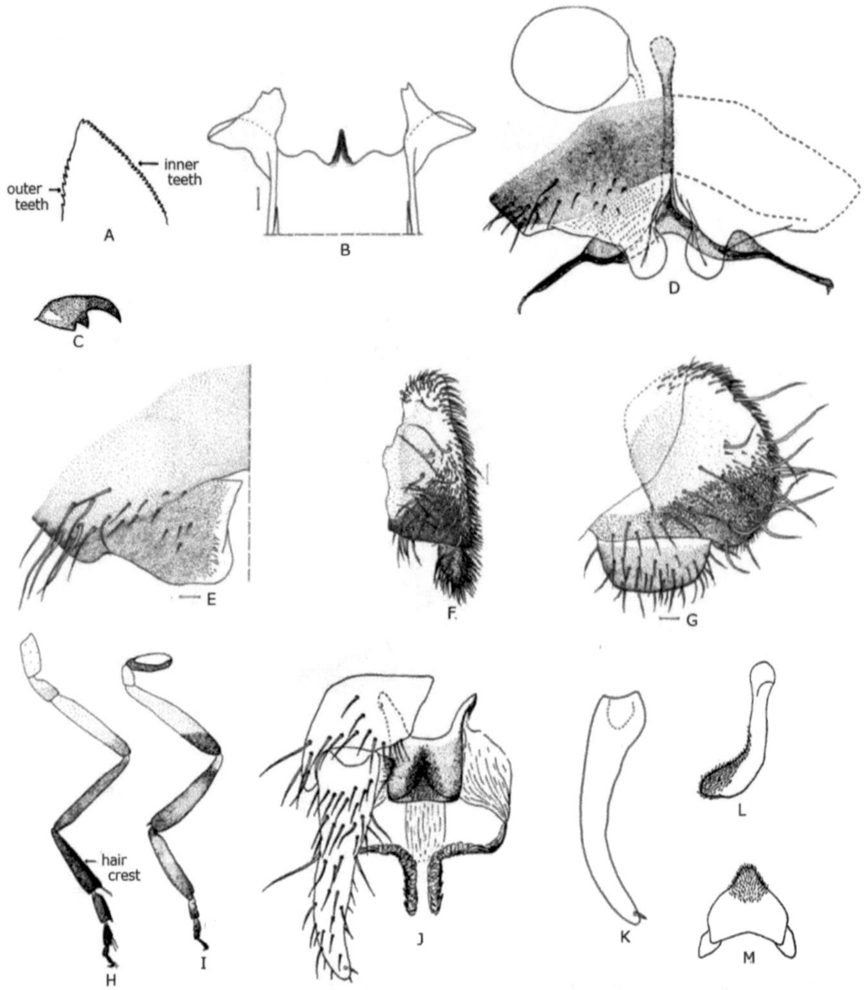

Fig. 62 Female and male of the *S. indicum* species-group of the subgenus *Simulium* s. str. A–G, Female; H–M, Male. A, Apical portion of mandible. B, Cibarium (ventral view). C, Claw. D, Sternite 8, ovipositor valves, genital fork, and spermatheca (ventral view). E, Sternite 8 and ovipositor valve (right half; ventral view). F & G, Paraprocts and cerci (right side; F, ventral view; G, lateral view). H, Foreleg (left side; outer view). I, Hind leg (right side; inner view). J, Coxite, style, ventral plate, parameres, and median sclerite (ventral view). K, Style (right side; medial view). L & M, Ventral plates (L, lateral view; M, caudal view). A, C, D & H–M, *S. nigrogilvum*; B & E–G, *S. umphangense*. Scale bar: 0.02 mm for B & E–G

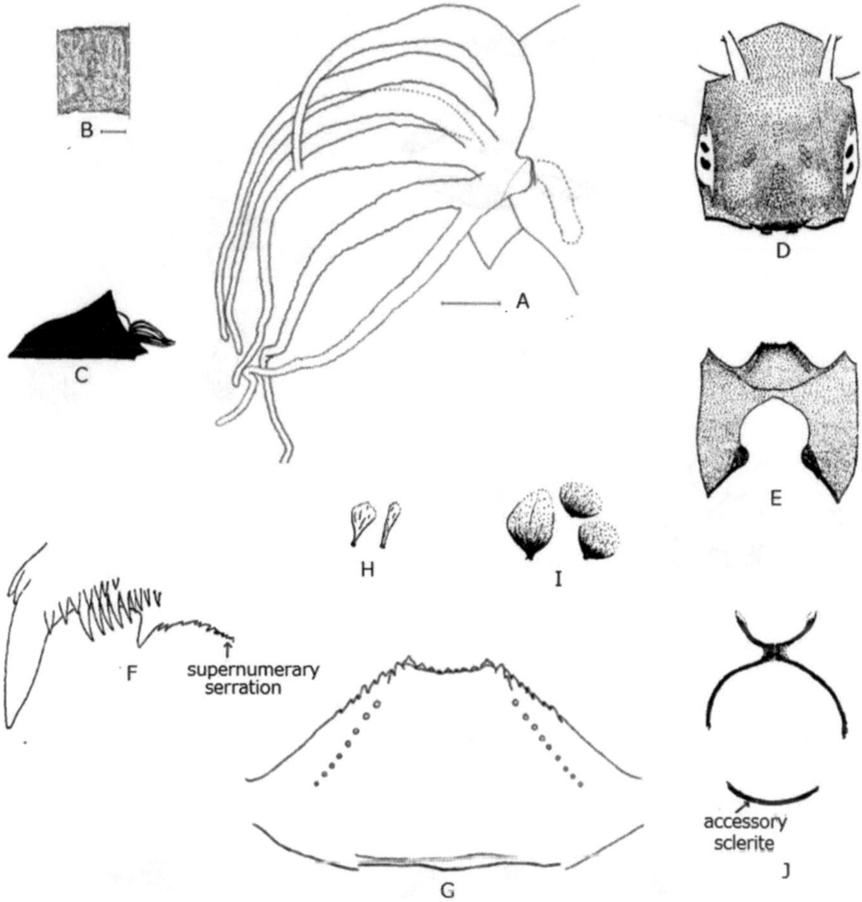

Fig. 63 Pupa and larva of the *S. indicum* species-group of the subgenus *Simulium* s. str. A–C, Pupa; D–J, Larva. A, Gill (left side; outer view). B, Middle part of gill filament (lateral view). C, Pupa and cocoon (lateral view). D & E, Head capsules (D, dorsal view; E, ventral view). F, Apical portion of mandible. G, Hypostoma. H, Unpigmented seta on thorax and abdomen. I, Dark pigmented setae on abdomen. J, Anal sclerite and accessory sclerite (caudal view). A–J, *S. nigrogilvum*. Scale bars: 0.1 mm for A; 0.02 mm for B

Fig. 64 Female, male, pupa, and larva of the *S. iridescens* species-group of the subgenus *Simulium* s. str. A, B & D–F, Female; C & G–L, Male; M–O, Pupa; P–R, Larva. A, Cibarium. B & C, Left hind legs (outer view; B, female; C, male). D, Claw. E, Abdominal segment 7, sternite 8, ovipositor valves, genital fork, paraprocts, cerci, and spermatheca (ventral view). F, Paraprocts and cerci (right side; lateral view). G, Right coxite, right style, ventral plate, and parameres (ventral view). H, Style (right side; outer view). I–K, Ventral plates (I, anteroventral view; J, lateral view; K, caudal view). L, Median sclerite (caudal view). M & N, Gills (left side; outer view). O, Cocoon (dorsal view). P, Apical portion of mandible. Q, Cephalic apotome. R, Median portion of head capsule (ventral view). A–L & N–R, *S. iridescens*; M. *S. javaense*. Scale bar: 0.1 mm for M

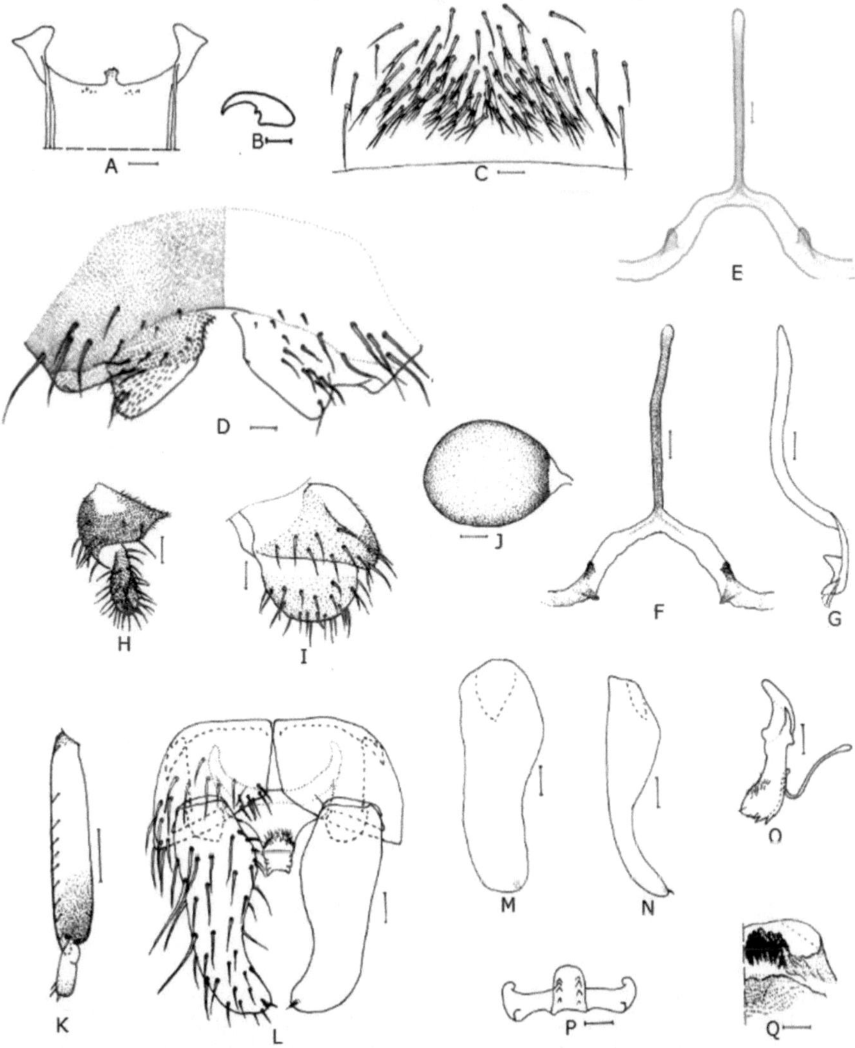

Fig. 65 Female and male of the *S. malyschevi* species-group of the subgenus *Simulium* s. str. A–J, Female; K–Q, Male. A, Cibarium. B, Claw. C, Median portion of ventral surface of abdominal segment 7 showing branched stout hairs (ventral view). D, Sternite 8 and ovipositor valves (ventral view). E, Genital fork with one projection on each arm (ventral view). F & G, Genital forks with two projections on each arm (F, ventral view; G, lateral view). H & I, Paraprocts and cerci (right side; H, ventral view; I, lateral view). J, Spermatheca. K, Hind basitarsus and second tarsomere (left side; outer view). L, Coxites, styles, and ventral plate (ventral plate). M & N, Styles (right side; M, ventrolateral view; N, medial view). O, Ventral plate and median sclerite (lateral view). P, Ventral plate (caudal view). Q, Paramere and aedeagal membrane (left half; caudal view). A–C & F–Q, *S. obliquum*; D, *S. phidimense*; E, *S. lomkaoense*. Scale bars: 0.1 mm for K; 0.02 mm for A, C–J & L–Q; 0.01 mm for B

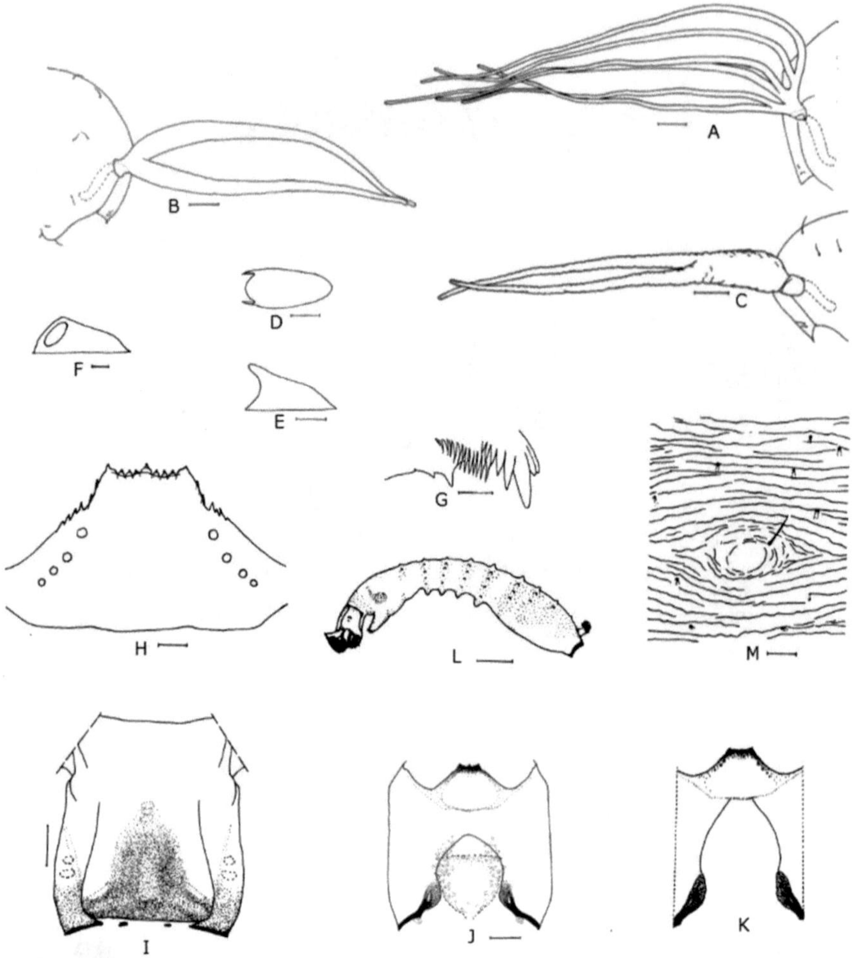

Fig. 66 Pupa and larva of the *S. malyschevi* species-group of the subgenus *Simulium* s. str. A–F, Pupa; G–K, Larva. A–C, Gills (lateral view; A & C, left side; B, right side). D & E, Cocoons simple, wall-pocket-shaped (D, dorsal view; E, lateral view). F, Cocoon with anterolateral open space on each side (lateral view). G, Apical portion of mandible. H, Hypostoma. I–K, Head capsules (I, dorsal view; J & K, ventral view). L, Whole body with multiple pairs of protuberances on the third thoracic segment and first to eighth abdominal segments (lateral view). M, Enlargement of protuberance and surrounding area (dorsal view). A, F, G, H, I, L & M, *S. obliquum*; B, D, E & J, *S. lomkaoense*; C, *S. baimaii*; K, *S. siripoomense*; G, *S. phidimense*. Scale bars: 1.0 mm for D & E; 0.5 mm for F & L; 0.2 mm for C & I; 0.1 mm for A, B & J; 0.02 mm for G, H & M

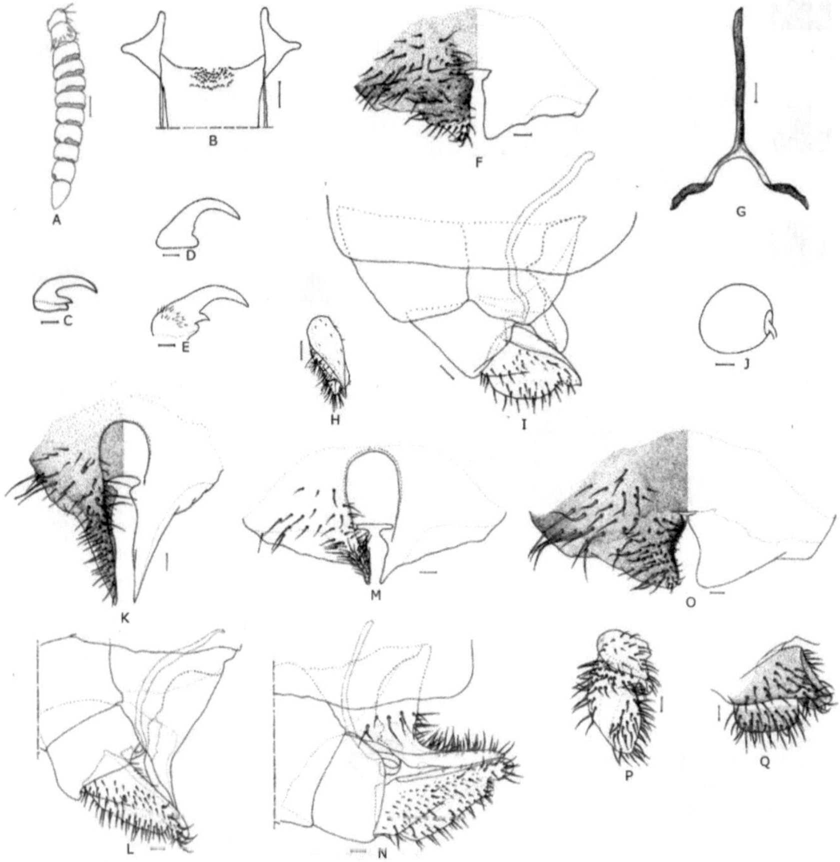

Fig. 67 Female of the *S. melanopus* species-group of the subgenus *Simulium* s. str. A, Antenna showing pits with sensilla on flagellomeres 1–8. B, Cibarium. C–E, Claws. F, K, M & O, Sternite 8 and ovipositor valves (ventral view). G, Genital fork (ventral view). H, P & Q, Paraprocts and cerci (right side; H & P, ventral view; Q, lateral view). I, L & N, Posterior tips of abdomens showing ovipositor valves, genital forks, paraprocts, and cerci (lateral view). A, *S. mirum*; B, & F–J, *S. neodiscrepans*; C, *S. cheedhangi*; D, *S. malaibaense*; E & O–Q, *S. forcipatum*; K & L, *S. concavum*; M & N, *S. pandum*. Scale bars: 0.05 mm for A; 0.02 mm for B & F–Q; 0.01 mm for C–E

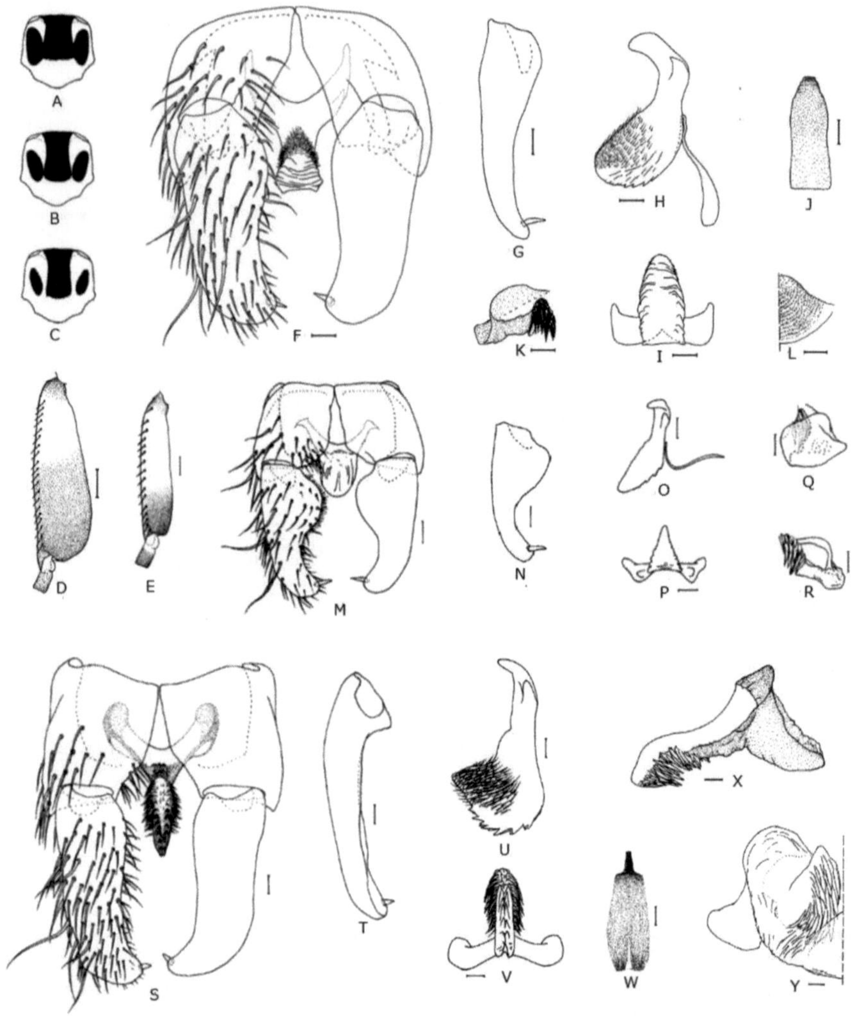

Fig. 68 Male of the *S. melanopus* species-group of the subgenus *Simulium* s. str. A–C, Scuta (dorsal view). D & E, Hind basitarsi and second tarsomeres (left side; outer view). F, M & S, Coxites, styles and ventral plates (ventral view). G, N & T, Styles (G & N, right side; medial view; T, left side; dorsolateral view). H & O, Ventral plate and median sclerites (lateral view). U, Ventral plate (lateral view). I, P & V, Ventral plates (caudal view). J & W, Median sclerites (caudal view). K, R & X, Parameres (K, right side, caudal view; R, left side; caudal view; X, left side; lateral view). L, Aedeagal membrane (caudal view). Q, Paramere showing fine setae on its base (lateral view). Y, Paramere and aedeagal membrane (right side; caudal view). A–C & S–X, *S. forcipatum*; D & F–L, *S. lardizabalae*; E & M–R, *S. neodiscrepans*; Y, *S. banahauense*. Scale bars: 0.1 mm for D; 0.05 mm for E; 0.02 mm for F–Y

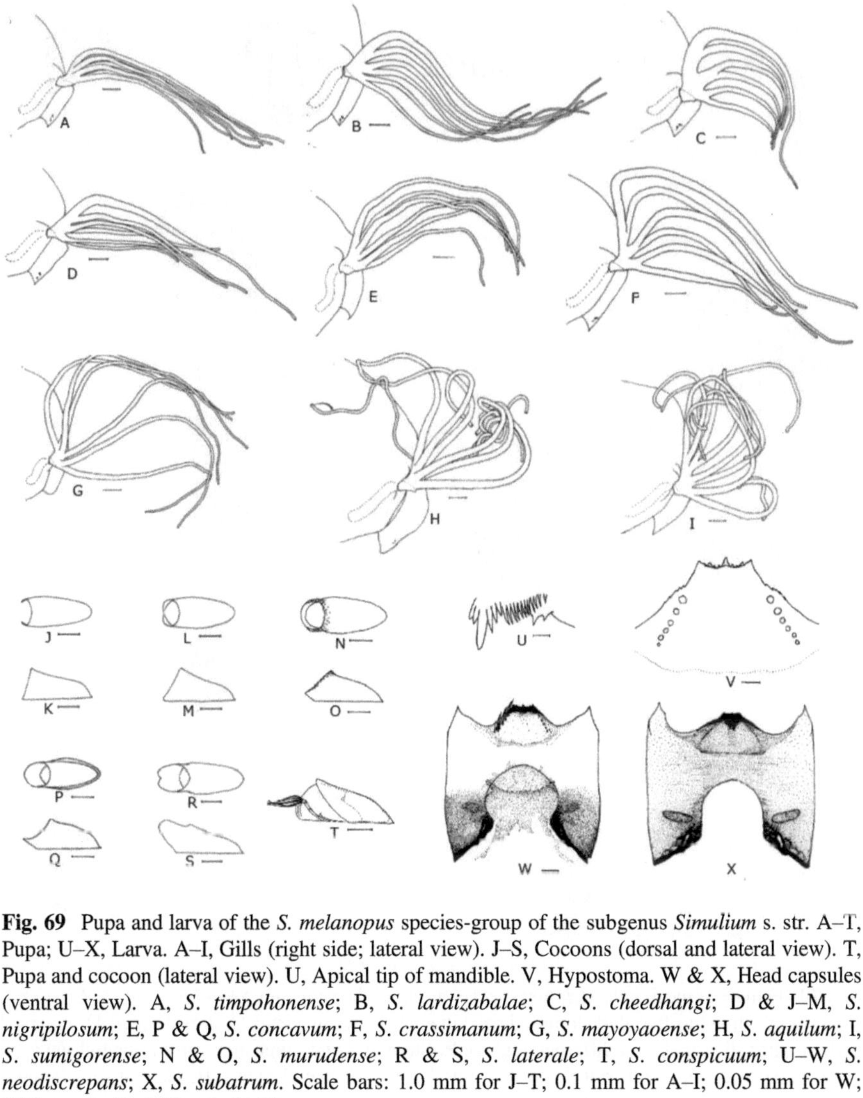

Fig. 69 Pupa and larva of the *S. melanopus* species-group of the subgenus *Simulium* s. str. A–T, Pupa; U–X, Larva. A–I, Gills (right side; lateral view). J–S, Cocoons (dorsal and lateral view). T, Pupa and cocoon (lateral view). U, Apical tip of mandible. V, Hypostoma. W & X, Head capsules (ventral view). A, *S. timpohonense*; B, *S. lardizabalae*; C, *S. cheedhangi*; D & J–M, *S. nigripilosum*; E, P & Q, *S. concavum*; F, *S. crassimanum*; G, *S. mayoyaoense*; H, *S. aquilum*; I, *S. sumigorense*; N & O, *S. murudense*; R & S, *S. laterale*; T, *S. conspicuum*; U–W, *S. neodiscrepans*; X, *S. subatrum*. Scale bars: 1.0 mm for J–T; 0.1 mm for A–I; 0.05 mm for W; 0.02 mm for V; 0.01 mm for U

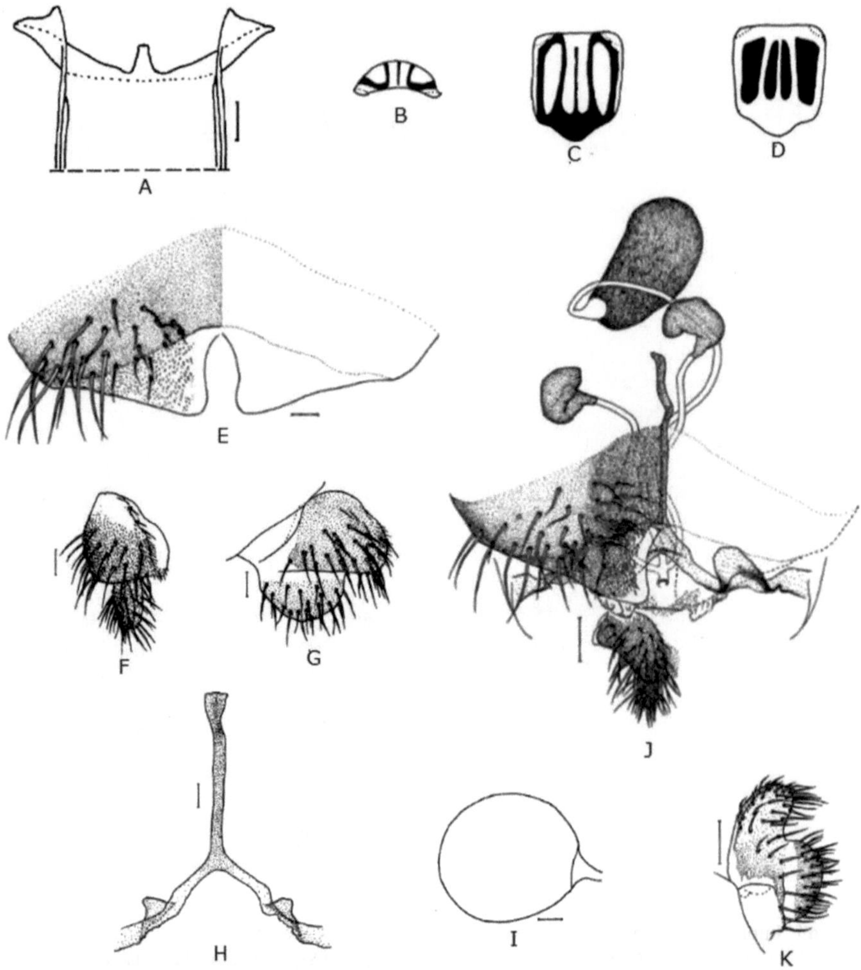

Fig. 70 Female of the *S. multistriatum* species-group of the subgenus *Simulium* s. str. A, Cibarium. B–D, Scuta showing five dark longitudinal vittae (B, anterior view; C, dorsal view; D, reversed pattern when illuminated in different angles). E, Sternite 8 and ovipositor valves (ventral view). F & G, Paraprocts and cerci (right side; F, ventral view; G, lateral view). H, Genital fork (ventral view). I, Spermatheca. J, Sternite 8, ovipositor valves, genital fork, right paraproct and cercus, and three spermathecae (ventral view). K, Paraproct much produced ventrally and cercus (right side; lateral view). B–D, *S. hirtinervis*; A & E–I, *S. lampangense*; J & K, S. *triglobus*. Scale bars: 0.05 mm for J & K; 0.02 mm for A & E–I

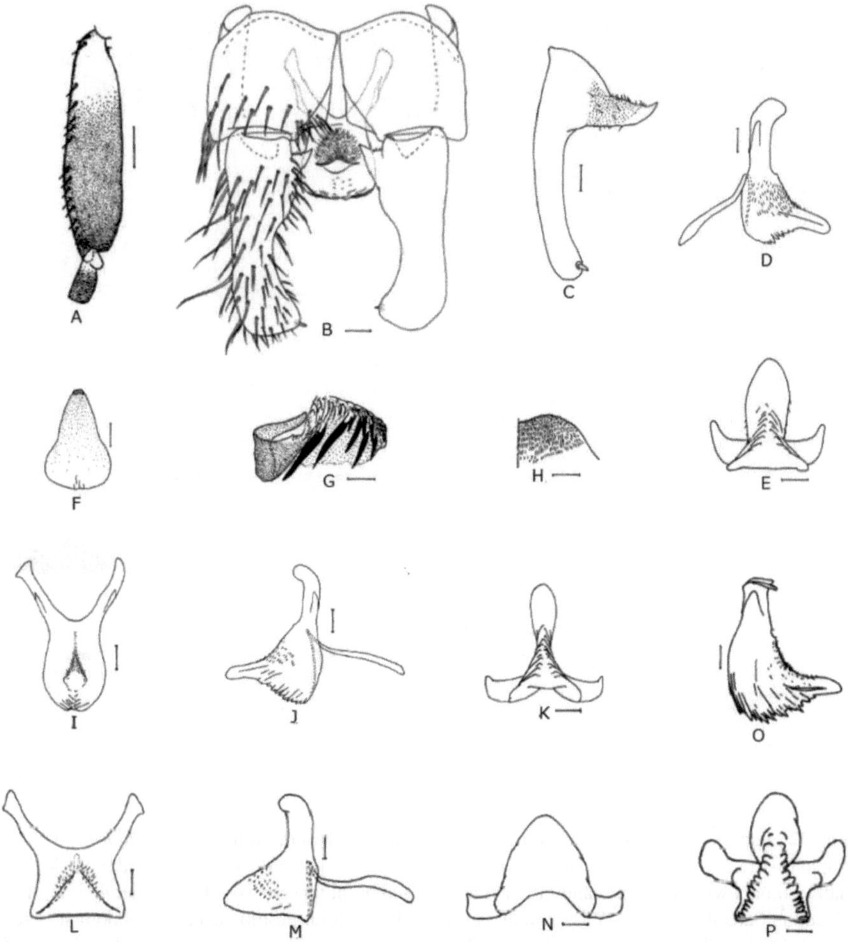

Fig. 71 Male of the *S. multistriatum* species-group of the subgenus *Simulium s. str.* A, Hind basitarsus and second tarsomere (left side; outer view). B, Coxites, styles, and ventral plate (ventral view). C, Style (right side; medial view). D, Ventral plate and median sclerite (lateral view). E, Ventral plate (caudal view). F, Median sclerite (ventrocaudal view). G, Paramere (right side; caudal view). H, Aedeagal membrane (left half; caudal view). I & L, Ventral plates (ventral view). J & M, Ventral plates and median sclerites (lateral view). K, N & P, Ventral plates (caudal view). O, ventral plate (lateral view). A, *S. lampangense*; B–H, *S. laui*; I–K, *S. maskeyae*; L–N, *S. barraudi*; O & P, *S. takense*. Scale bars: 0.1 mm for A; 0.02 mm for B–P

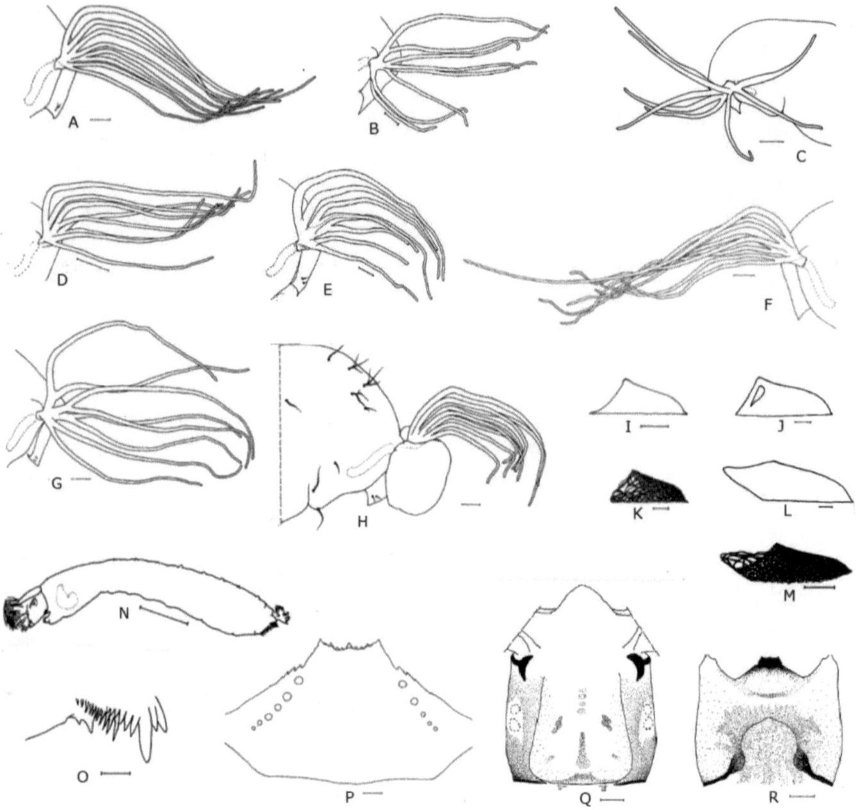

Fig. 72 Pupa and larva of the *S. multistriatum* species-group of the subgenus *Simulium* s. str. A–M, Pupa; N–R, Larva. A–G, Gills (lateral view; A, B, D, E, G, right side; C & F, left side). H, Anterior half of thorax with trichomes and gill having enlarged basal fenestra (right side; lateral view). I–M, Cocoons (lateral view). N, Whole body showing dorsolateral protuberances on abdomen (lateral view). O, Apical portion of mandible. P, Hypostoma. Q & R, Head capsules (Q, dorsal view; R, ventral view). A, *S. demolaense*; B & M, *S. triglobus*; C & L, *S. takense*; D & N, *S. hirtinervis*; E & P, *S. chainarongi*; F & K, *S. thimphuense*; G & J, *S. lampangense*; H, *S. bullatum*; I, *S. chanyae*; O–R, *S. laui*. Scale bars: 1 mm for I, M & N; 0.5 mm for J–L; 0.1 mm for A–H, Q & R; 0.02 mm for O & P

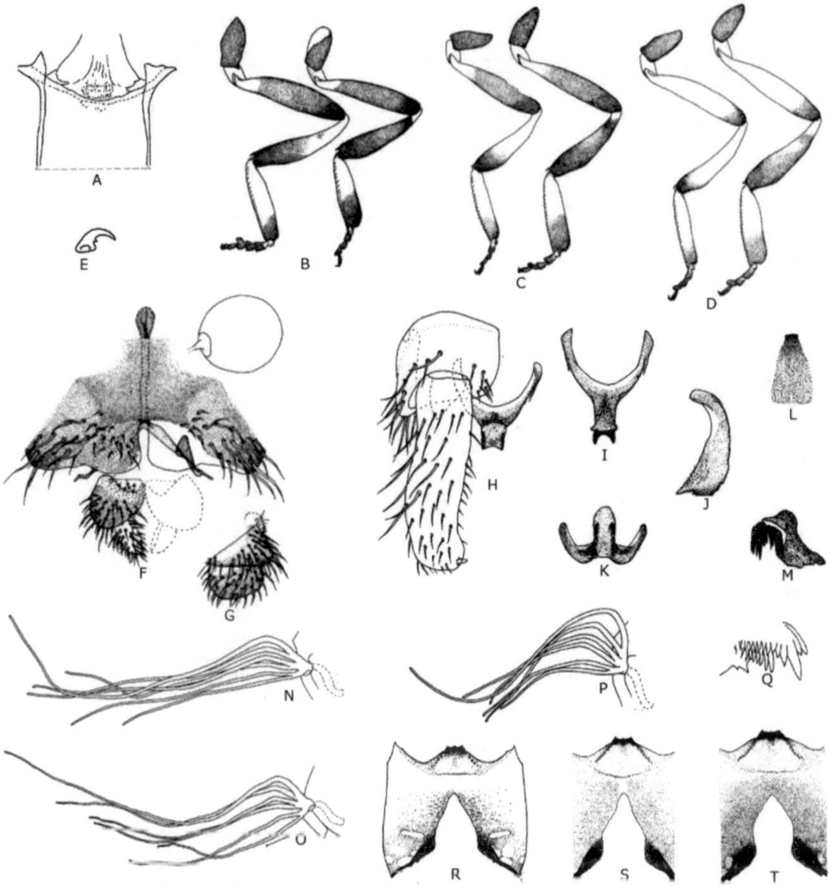

Fig. 73 Female, male, pupa, and larva of the *S. nebulicola* species-group of the subgenus *Simulium* s. str. A, Female cibarium. B–D, Hind legs (all left side and outer view; left side, females; right side, males). E, Female claw. F, Sternite 8, ovipositor valves, genital fork, paraproct, cercus, and spermatheca of female terminalia (ventral view). G, Paraproct and cercus (right side; lateral view). H, Right coxite, right style, and ventral plate of male genitalia (ventral view). I–K, Ventral plates (I, anteroventral view; J, lateral view; K, caudal view). L, Median sclerite (caudal view). M, Paramere (left side; caudal view). N–P, Pupal gills (left side; outer view). Q, Anterior portion of larval mandible. R, Larval head capsule (ventral view). S & T, Median portions of larval head capsules (ventral view). A, B, E, L, M, O, Q & R, *S. sumatraense*; C, P & T, *S. nebulicola*; D, F–K, N & S, *S. celsum*. Scale bar: 0.1 mm for O

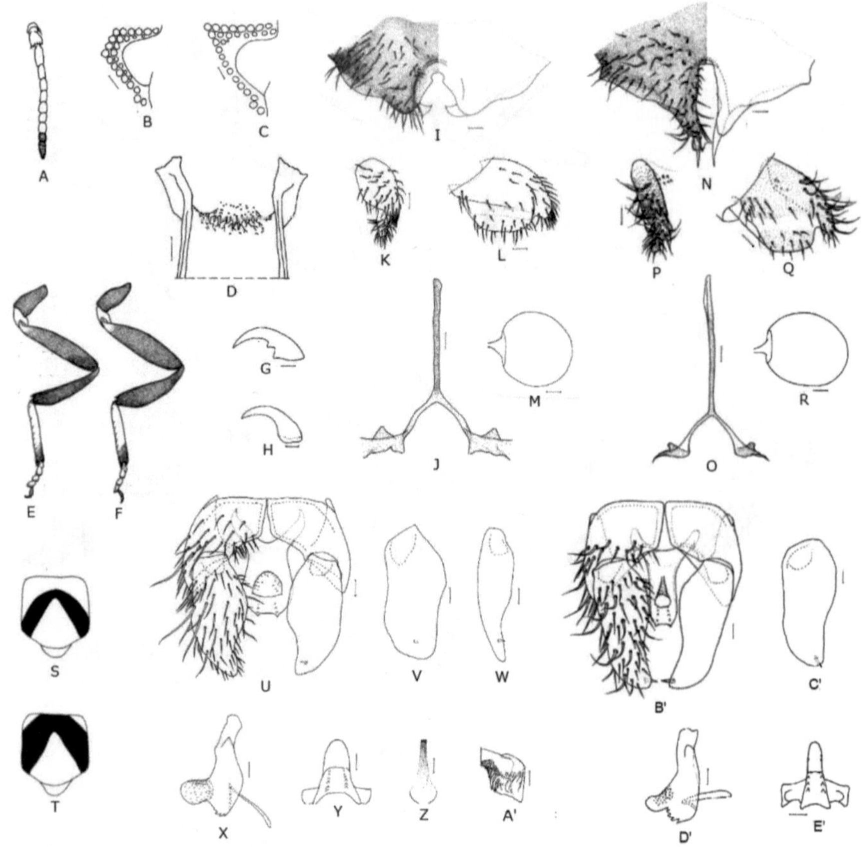

Fig. 74 Female and male of the *S. nobile* species-group of the subgenus *Simulium* s. str. A–E & G–R, Female; F & S–E', Male. A, Antenna. B & C, Fronto-ocular areas (right side; frontal view). D, Cibarium. E & F, Hind legs (left side; outer view). G & H, Claws. I & N, Sternite 8 and ovipositor valves (ventral view). J & O, Genital forks (ventral view). K, L, P & Q, Paraprocts and cerci (right side; K & P, ventral view; L & Q, lateral view). M & R, Spermathecae. S & T, Scuta (dorsal view). U & B', Coxites, styles, and ventral plates (ventral view). V, W & C', Styles (right side; V & C', ventrolateral view; W, medial view). X & D', Ventral plates and median sclerites (lateral view). Y & E', Ventral plates (caudal view). Z, Median sclerite (caudal view). A', Paramere (left side; caudal view). A, E & F, *S. nobile*; B, D, G, N–R & B'–E', *S. timorense*; C, H–M & U–A', *S. nodosum*; S & T, *S. delfinadoae*. Scale bars: 0.03 mm for A; 0.02 mm for B, C, D, I, R & U–E'; 0.01 mm for G & H

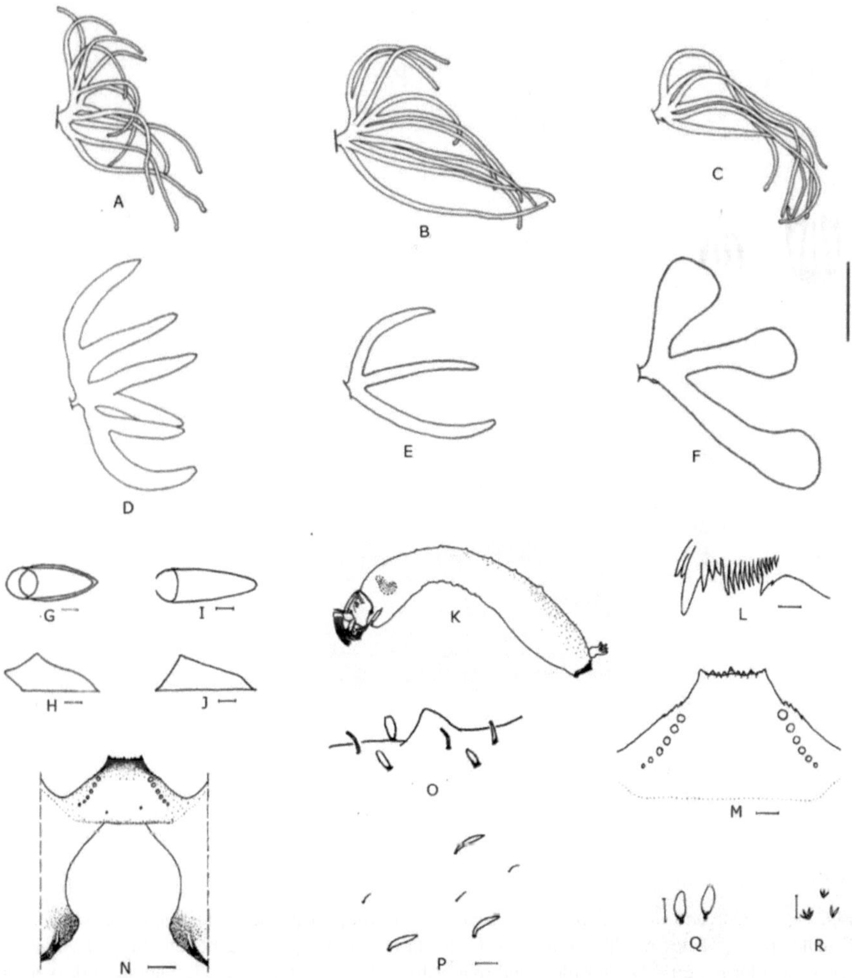

Fig. 75 Pupa and larva of the *S. nobile* species-group of the subgenus *Simulium* s. str. A–J, Pupa; K–R, Larva. A–F, Gills (right side; outer view). G–J, Cocoons (G & I, dorsal view; H & J, lateral view). K, Whole body (lateral view). L, Apical portion of mandible. M, Hypostoma. N, Middle part of head capsule (ventral view). O, Conical protuberance and five spatulate minute setae on dorsum of abdominal segment 5 (lateral view). P, Spatulate setae and microsetae on dorsum of abdominal segment 5. Q, Spatulate setae (anterodorsal view). R, Branched setae on dorsum of abdominal segment 6. A, *S. cotabatoense*; B, *S. baltazarae*; C, *S. benquetense*; D, *S. delfinadoae*; E, *S. asishi*; F, I, J & O, *S. nodosum*; G, H, L–N & P, *S. timorense*; K, Q & R, *S. nobile*. Scale bars: 0.5 mm for A–J; 0.05 m for N; 0.02 mm for M; 0.01 mm for L & P–R

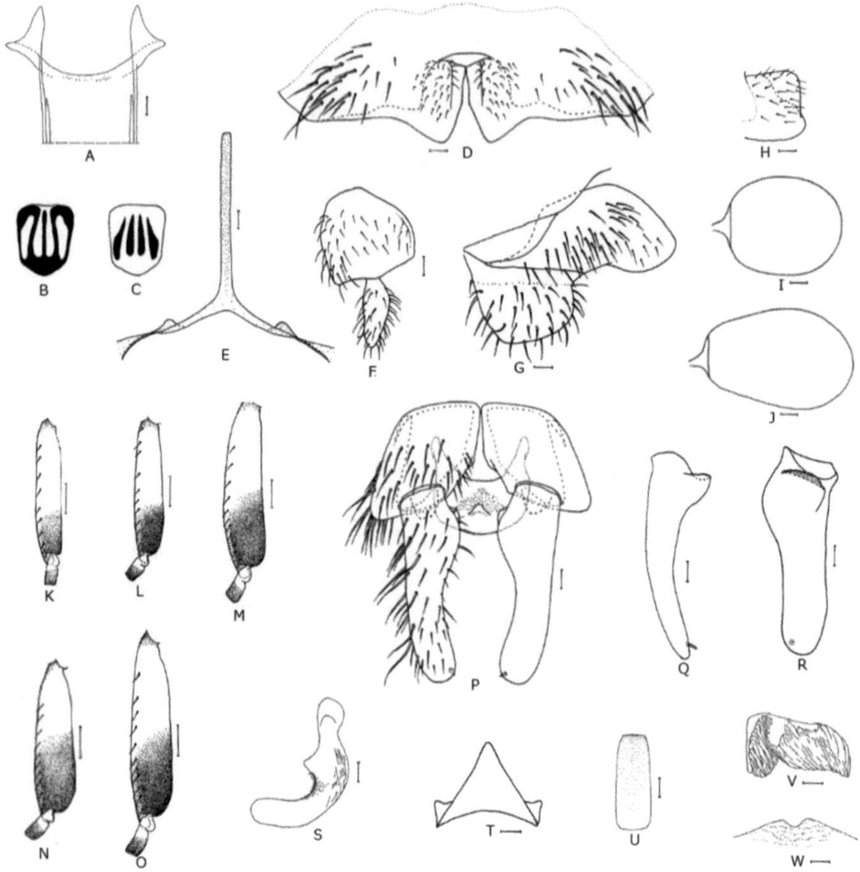

Fig. 76 Female and male of the *S. striatum* species-group of the subgenus *Simulium* s. str. A–J, Female; K–W, Male. A. Cibarium. B & C, Scuta (B, showing five dark longitudinal vittae; C, reversed color pattern). D, Sternite 8 and ovipositor valves (ventral view). E, Genital fork (ventral view). F & G, Paraprocts and cerci (right side; F, ventral view; G, lateral view). H, Ventrally produced lobe of ovipositor valve (right side; lateral view). I & J, Spermathecae. K–O, Hind basitarsi and second tarsomeres (left side; outer view). P, Coxites, styles, and ventral plate (ventral view). Q & R, Styles (right side; Q, medial view; R, dorsal view). S & T, Ventral plates (S, lateral view; T, caudal view). U, Median sclerite (caudal view). V, Paramere (left side; caudal view). W, Aedeagal membrane (caudal view). A, D–J, M & P–W, *S. yanpingense*; B & C, *S. jeffreyi*; K, *S. phraense*; L, *S. nakhonense*; N, *S. perakense*; O, *S. wangkwaiense*. Scale bars: 0.1 mm for K–O; 0.02 mm for A, D–J & P–W

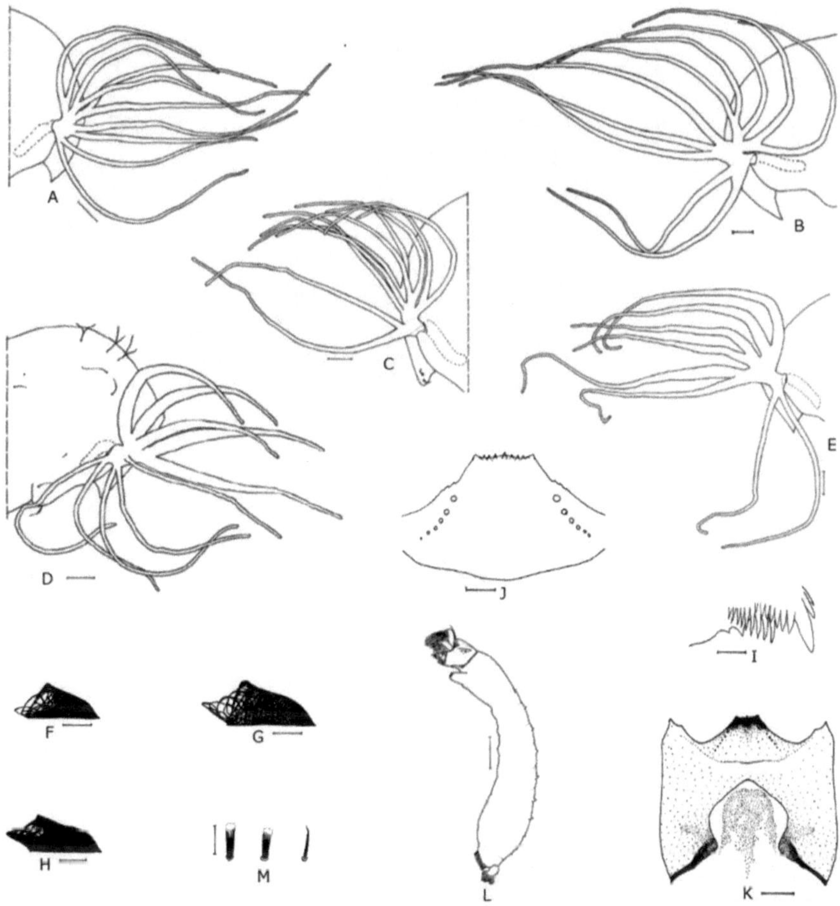

Fig. 77 Pupa and larva of the *S. striatum* species-group of the subgenus *Simulium* s. str. A–H, Pupa; I–M, Larva. A–E, Gills (lateral view; A & D, right side; B, C & E, left side). F–H, Cocoons (lateral view). I, Apical portion of mandible. J, Hypostoma. K, Head capsule (ventral view). L, Whole body showing paired dorsolateral protuberances (lateral view). M, Dark, chisel-shaped spinous setae (left and middle, anterior view; right, lateral view). A, *S. baliense*; C, G & K, *S. perakense*; D, *S. phraense*; E, *S. chiangmaiense*; F, *S. tadtonense*; B, I, J & M, *S. quinquestriatum*; H, *S. maeklongkeense*; L, *S. jeffreyi*. Scale bars: 1.0 mm for F–H & L; 0.1 mm for A–E & K; 0.04 mm for J; 0.02 mm for I; 0.01 mm for M

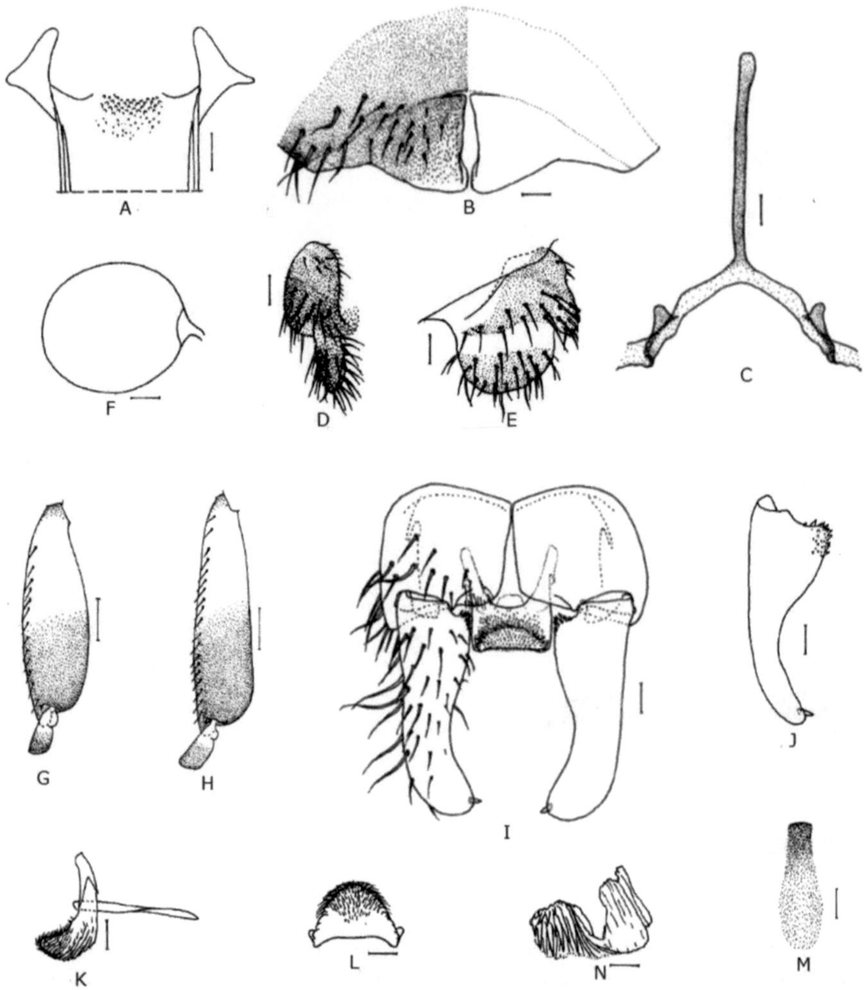

Fig. 78 Female and male of the *S. brevipar* subgroup of the *S. tuberosum* species-group of the subgenus *Simulium* s. str. A–F, Female; G–N, Male. A. Cibarium. B, Sternite 8 and ovipositor valves (ventral view). C, Genital fork (ventral view). D & E, Paraprocts and cerci (right side; D, ventral view; E, lateral view). F, Spermatheca. G & H, Hind basitarsi and second tarsomeres (left side; outer view). I, Coxites, styles, and ventral plate (ventral view). J, Style (right side; medial view). K, Ventral plate and median sclerite (lateral view). L, Ventral plate (caudal view). M, Median sclerite (caudal view). N, Paramere (caudal view). A–F & I–N, *S. yuphae*; G, *S. huense*; H, *S. sapaense*. Scale bars: 0.1 mm for G & H; 0.02 mm for A–F & I–N

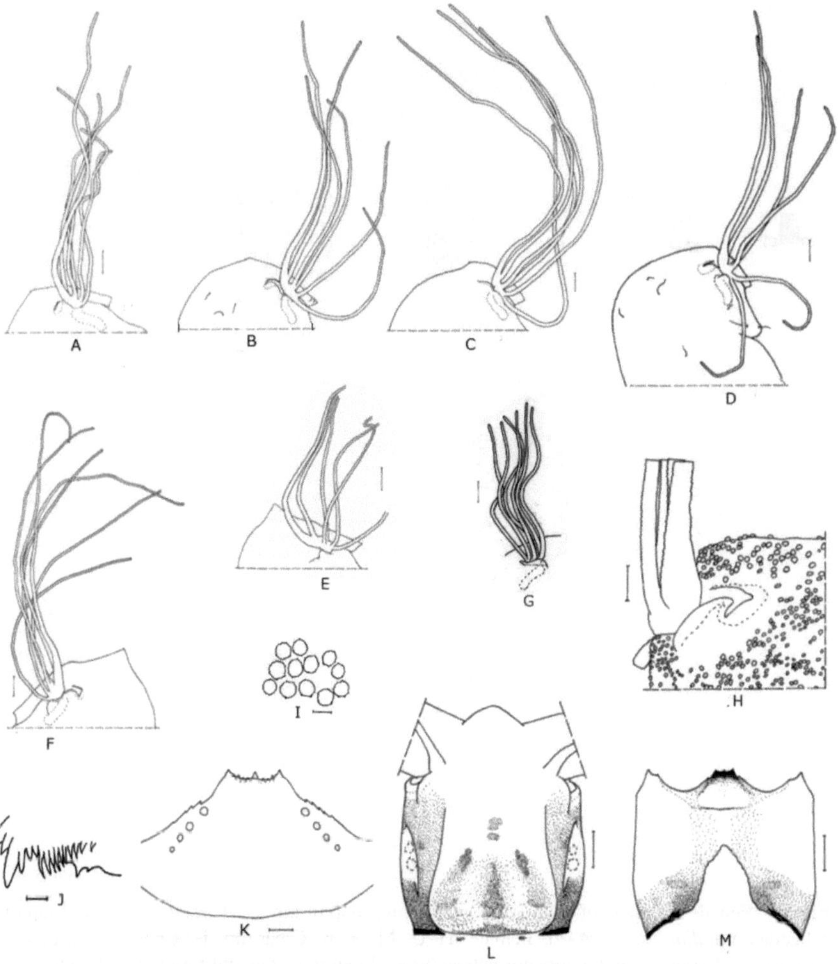

Fig. 79 Pupa and larva of the *S. brevipar* subgroup of the *S. tuberosum* species-group of the subgenus *Simulium* s. str. A–I, Pupa; J–M, Larva. A–G, Gills (lateral view; A–E, right side; F & G, left side). H, Anterolateral portion of thorax showing pit-like organ near base of gill (dorsal view). I, Relatively large tubercles with tiny secondary projections on frons and thorax. J, Apical portion of mandible. K, Hypostoma. L & M, Head capsules (L, dorsal view; M, ventral view). A, *S. lowi*; B, H & J–M, *S. huense*; C, *S. sapaense*; D, *S. yuphae*; E, *S. tiomanense*; F & I, *S. tatopaniense*; G, *S. sigiti*. Scale bars: 0.1 mm for A–G, L & M; 0.05 mm for H; 0.02 mm for K; 0.01 mm for I & J

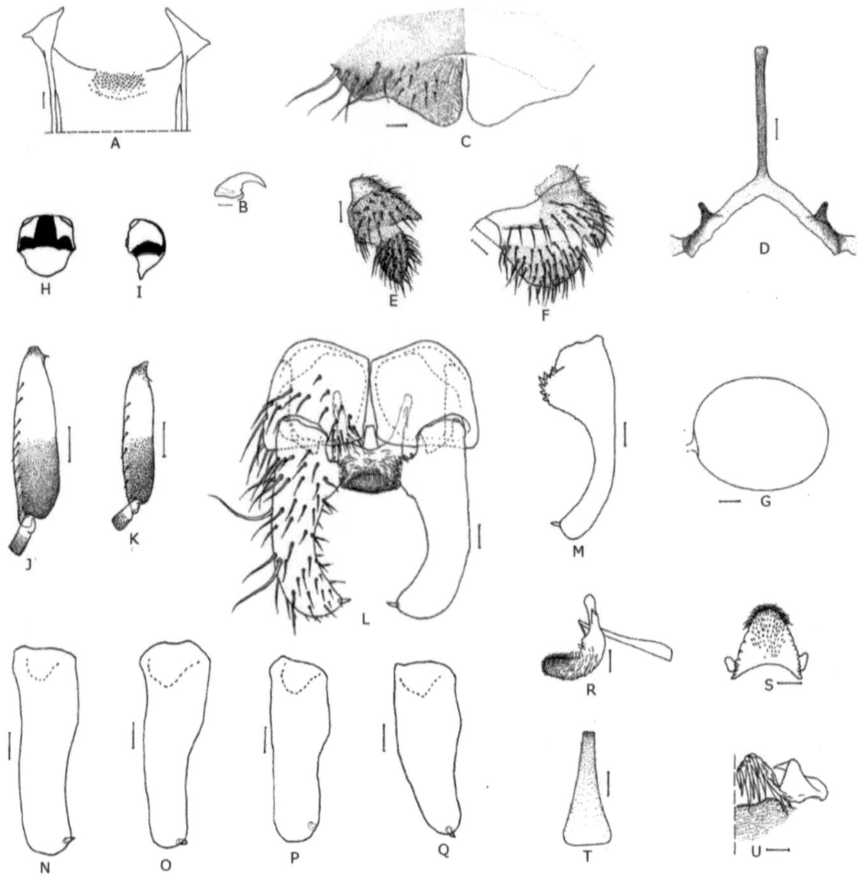

Fig. 80 Female and male of the *S. nitidithorax* subgroup of the *S. tuberosum* species-group of the subgenus *Simulium* s. str. A–G, Female; H–U, Male. A, Cibarium. B, Claw with a tiny basal swelling. C, Sternite 8 and ovipositor valves (ventral view). D, Genital fork (ventral view). E & F, Paraprocts and cerci (right side; E, ventral view; F, lateral view). G, Spermatheca. H & I, Scuta showing a patten of white pruinose areas (H, dorsal view; I, lateral view). J & K, Hind basitarsi and second tarsomeres (left side; outer view). L, Coxites, styles, and ventral plate (ventral view). M, Style (left side; medial view). N–Q, Styles of different shapes (right side; ventrolateral view). R, Ventral plate and median sclerite (lateral view). S, Ventral plate (caudal view). T, Median sclerite (caudal view). U, Paramere and aedeagal membrane (left half; caudal view). C–G, J, L, M & R–U, *S. alberti*; A, H & I, *S. rangjungense*; B, *S. keningauense*; K & Q, *S. lunduense*; N, *S. masilauense*; O, *S. aeneifacies*; P, S. *sabahense*. Scale bars: 0.1 mm for J & K; 0.02 mm for C–G & L–U; 0.01 mm for A & B

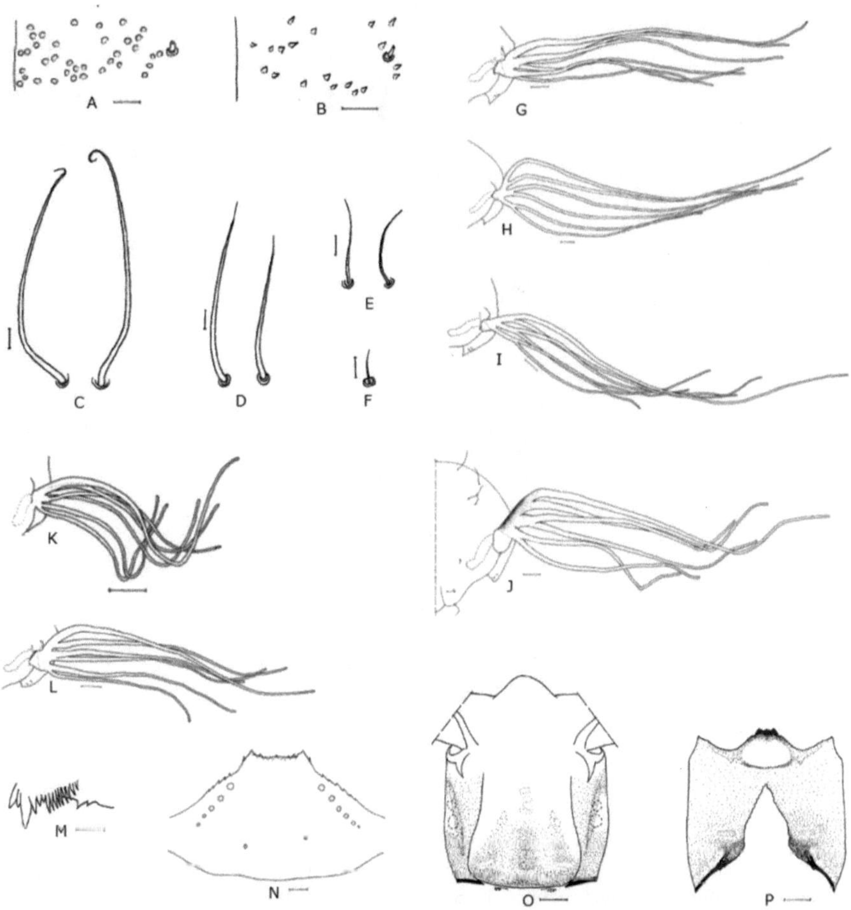

Fig. 81 Pupa and larva of the *S. nitidithorax* subgroup of the *S. tuberosum* species-group of the subgenus *Simulium* s. str. A–L, Pupa; M–P, Larva. A & B, Tubercles on dorsal surface of thorax (A, round; B, conical). C–F, Dorsomedial trichomes of different sizes. G–L, Gills (lateral view; right side). M, Apical portion of mandible. N, Hypostoma. O & P, Head capsules (O, dorsal view; P, ventral view). A, C & I, *S. alberti*; B, D & G, *S. keningauense*; E, *S. aeneifacies*; F, *S. rangjungense*; H, *S. jianshiense*; J & M–P, *S. congi*; K, *S. puliense*; L, *S. sabahense*. Scale bars: 0.2 mm for K; 0.1 mm for G–J, H, L, O & P; 0.02 mm for A, B, M & N; 0.01 mm for C–F

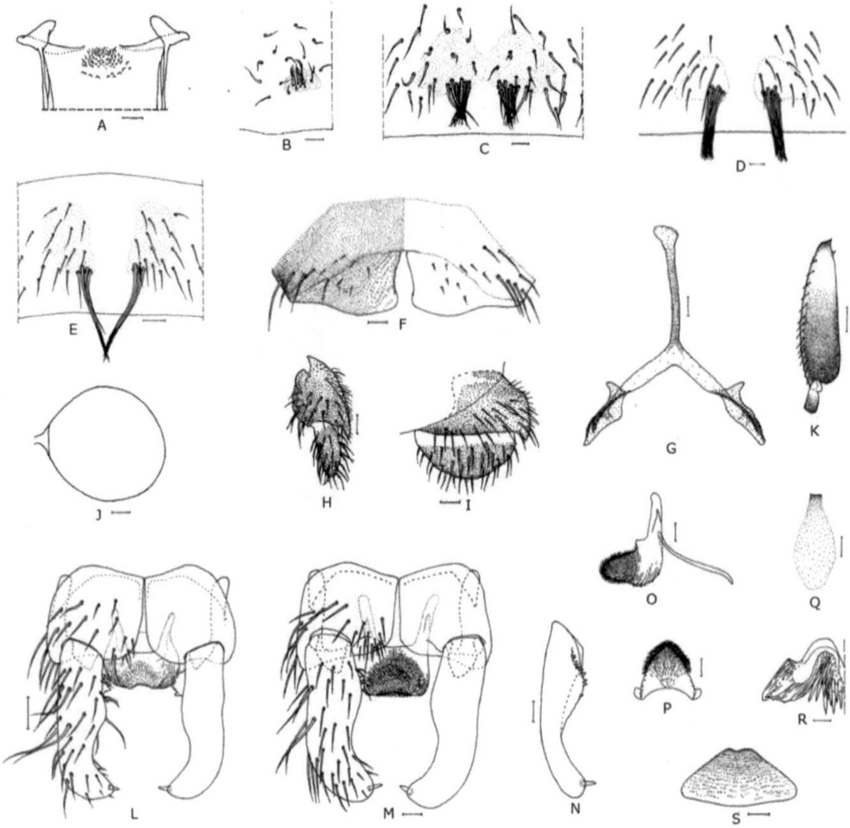

Fig. 82 Female and male of the *S. rufibasis* subgroup of the *S. tuberosum* species-group of the subgenus *Simulium* s. str. A–J, Female; K–S, Male. A, Cibarium. B–E, Paired stout tuft hairs on sternite 7 (ventral view; B, right half only). F, Sternite 8 and ovipositor valves (ventral view). G, Genital fork (ventral view). H & I, Paraprocts and cerci (right side; H, ventral view; I, lateral view). J, Spermatheca. K, Hind basitarsus and second tarsomere (left side; outer view). L & M, Coxites, styles, and ventral plates (ventral view). N, Style (right side; medial view). O, Ventral plate and median sclerite (lateral view). P, Ventral plate (caudal view). Q, Median sclerite (caudal view). R, Paramere (right side; caudal view). S, Aedeagal membrane (caudal view). A & C, *S. manooni*; B, *S. setsukoae*; F–J, *S. dailekhense*; D, *S. nakanoi*; E & L, *S. weji*; K & M–S, *S. tenebrosum*. Scale bars: 0.1 mm for K; 0.03 mm for E & L; 0.02 mm for A–D, F–J & M–S

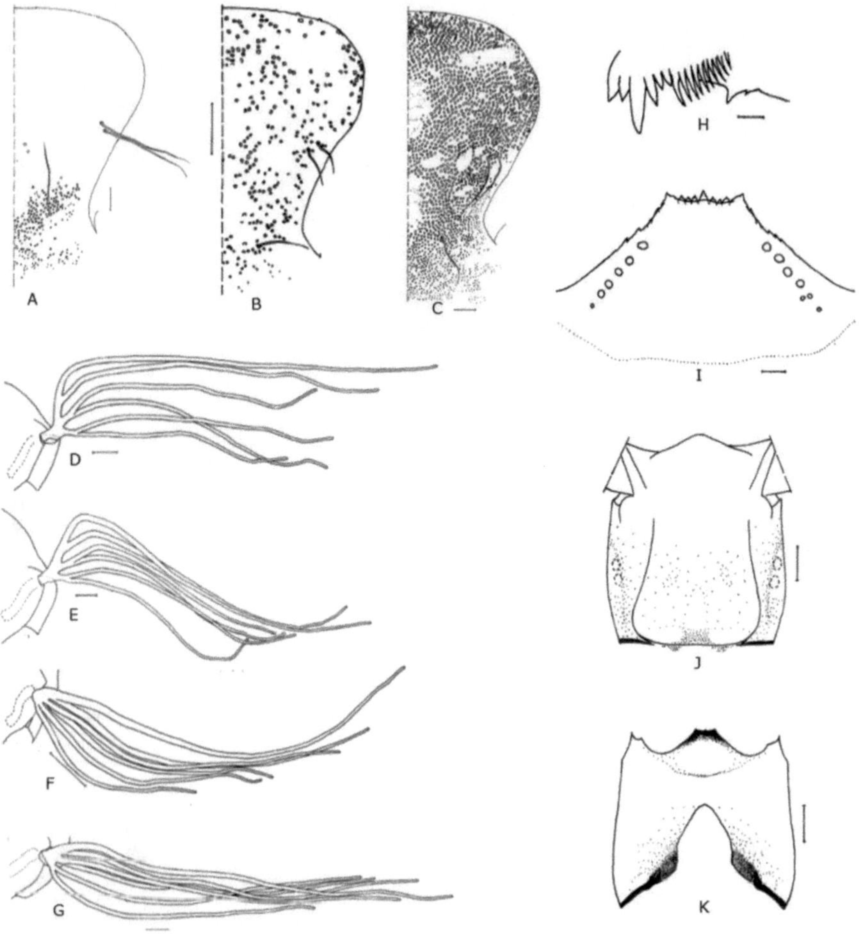

Fig. 83 Pupa and larva of the *S. rufibasis* subgroup of the *S. tuberosum* species-group of the subgenus *Simulium* s. str. A–G, Pupa; H–K, Larva. A–C, Frons and upper part of the face (left half; anterodorsal view; A, frons almost bare; B, frons moderately covered with tubercles; C, frons densely covered with tubercles). D–G, Gills (right side; lateral view). H, Apical portion of mandible. I, Hypostoma. J & K, Head capsules (J, dorsal view; K, ventral view). A, *S. rosliramlii*; B, G, H & I, *S. manooni*; C, *S. giayense*; D, *S. nakanoi*; E, *S. langtangense*; F, *S. weji*; J & K, *S. tenebrosum*. Scale bars: 0.2 mm for F; 0.1 mm for B, D, E, G, J & K; 0.05 mm for A & C; 0.02 mm for I; 0.01 mm for H

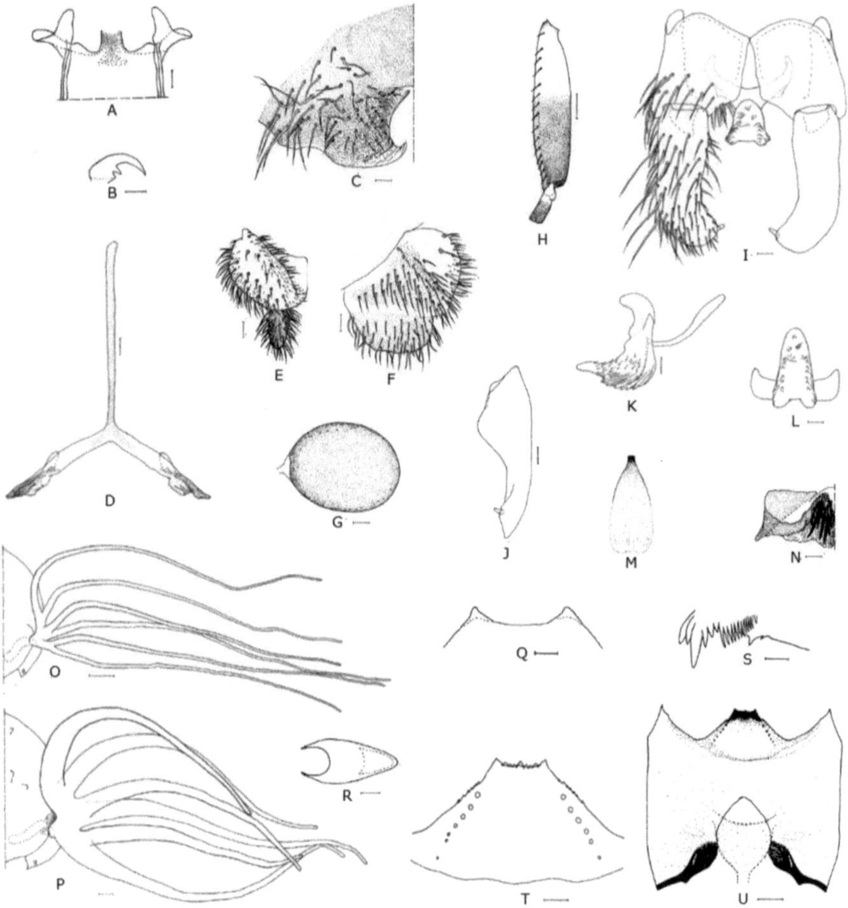

Fig. 84 Female, male, pupa, and larva of the *S. variegatum* species-group of the subgenus *Simulium* s. str. A–G, Female, H–N, Male; O–R, Pupa; S–U, Larva. A, Cibarium. B, Claw. C, Sternite 8 and ovipositor valve (right half; ventral view). D, Genital fork (ventral view). E & F, Paraprocts and cerci (right side; E, ventral view; F, lateral view). G, Spermatheca. H, Hind basitarsus and second tarsomere (left side; outer view). I, Coxites, styles, and ventral plate (ventral view). J, Style (left side; medial view). K, Ventral plate and median sclerite (lateral view). L, Ventral plate (caudal view). M, Median sclerite (caudal view). N, Paramere (right side; caudal view). O & P, Gills (right side; outer view). Q, Terminal hooks (caudal view). R, Cocoon (dorsal view). S, Apical portion of mandible. T, Hypostoma. U, Head capsule (ventral view). A, P & R, *S. srisukai*; B–O, Q & S–U, *S. phuluense*. Scale bars: 1.0 mm for R; 0.1 mm for H, O & U; 0.05 mm for P; 0.04 mm for T; 0.02 mm for A–G, I–N, Q & S

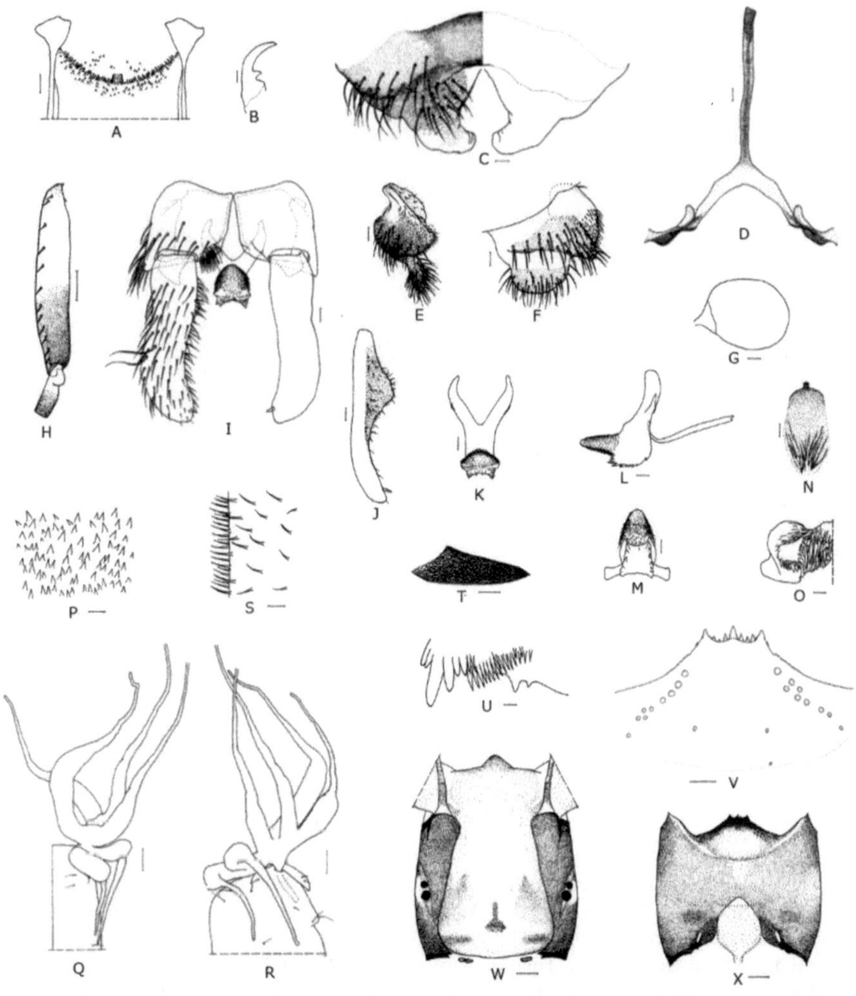

Fig. 85 Female, male, pupa, and larva of the *S. variegatum* species-group of the subgenus *Simulium* s. str. from Nepal. A–G, Female; H–O, Male; P–T, Pupa; U–X, Larva. A, Cibarium. B, Claw; C, Sternite 8 and ovipositor valves (ventral view). D, Genital fork (ventral view). E & F, (right side; E, ventral view; F, lateral view). G, Spermatheca. H, Hind basitarsus and second tarsomere (left side; outer view). I, Coxites, styles, and ventral plate (ventral view). J, Style (right side; medial view). K & M, Ventral plates (K, anteroventral view; M, caudal view). L, Ventral plate and median sclerite (lateral view). N, Median sclerite (caudal view). O, Paramere (right side; caudal view). P, Pointed processes on frons. Q & R, Gills (right side; Q, dorsal view; R, lateral view). S, Fine spinous processes on inflated portions of gill filaments. T, Cocoon (lateral view). U, Apical portion of mandible. V, Hypostoma. W & X, Head capsules (W, dorsal view; X, ventral view). A–X, *S. tulshii*. Scale bars: 1.0 mm for T; 0.1 mm for H, Q, R, W & X; 0.04 mm for V; 0.02 mm for A, C–G, I–O; 0.01 mm for B, P, S & U

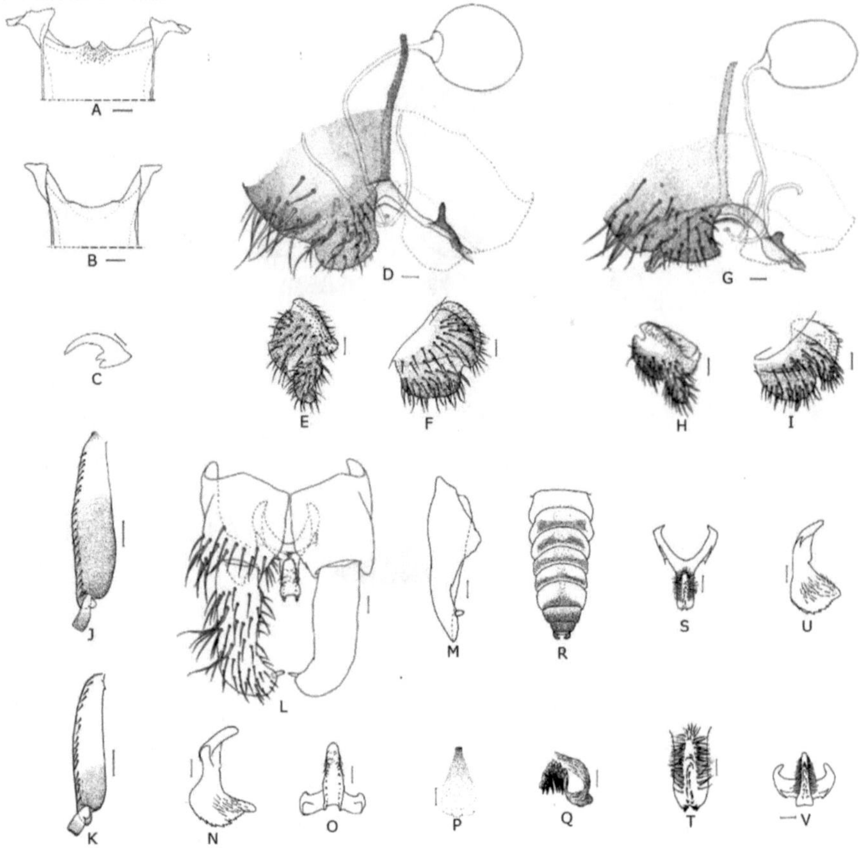

Fig. 86 Female and male of the *S. variegatum* species-group of the subgenus *Simulium* s. str. from Sulawesi. A–I, Female; J–V, Male. A & B, Cibaria. C, Claw. D & G, Sternite 8, ovipositor valves, genital forks, and spermathecae (ventral view). E–I, Paraprocts and cerci (right side; E & H, ventral view; F & I, lateral view). J & K, Hind basitarsi and second tarsomeres (left side; outer view). L, Coxites, styles, and ventral plate (ventral view). M, Style (right side; medial view). N & O, Ventral plates (N, lateral view; O, caudal view). P, Median sclerite (caudal view). Q, Paramere (left side; caudal view). R, Abdomen (dorsal view). S, U & V, Ventral plates (S, ventral view; U, lateral view; V, caudal view). T, Body of ventral plate (ventral view). A, C–F, J & L–Q, *S. molliculum*; B, G–I, K & R–V, *S. luridum*. Scale bars: 0.1 mm for J & K; 0.02 mm for A, B, D–I, L–Q, S, U & V; 0.01 mm for C & T

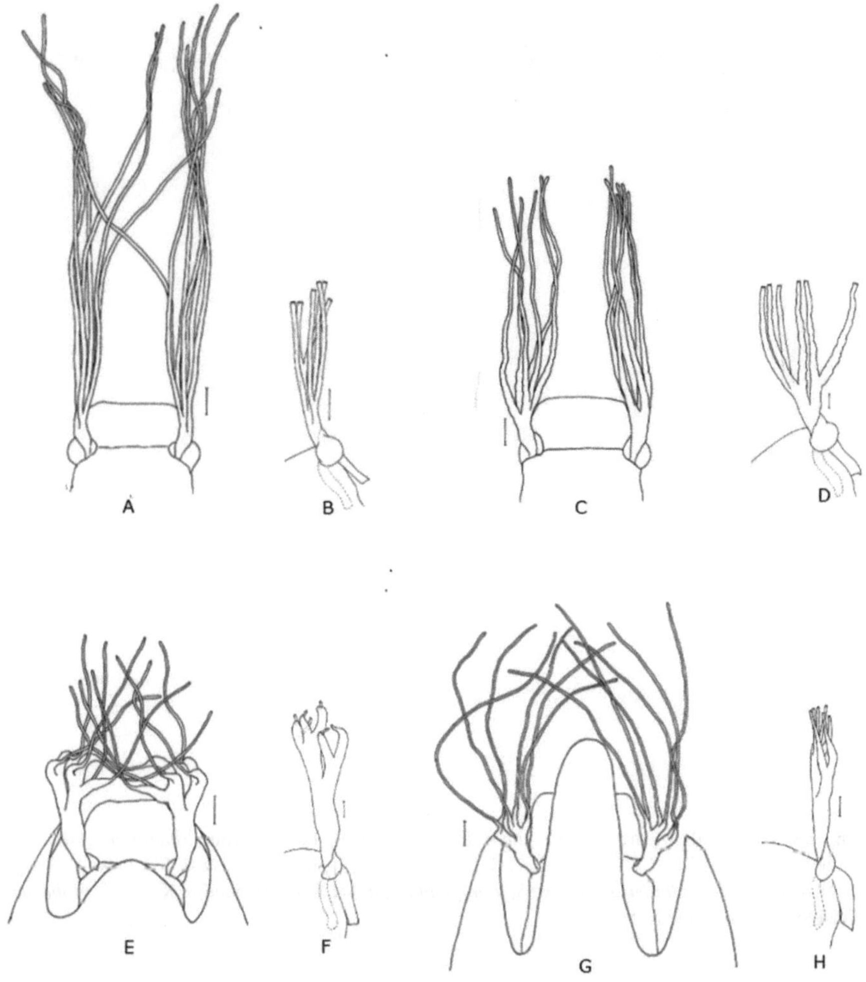

Fig. 87 Pupal gills of the *S. variegatum* species-group of the subgenus *Simulium* s. str. from Sulawesi. A, C, E & G, dorsal view; B, D, F & H, lateral view. A & B, *S. molliculum*; C & D, *S. laxum*; E & F, *S. linduense*; G & H, *S. elongatum*. Scale bars: 0.02 mm for A, B, C, E & G; 0.1 mm for D, F & H

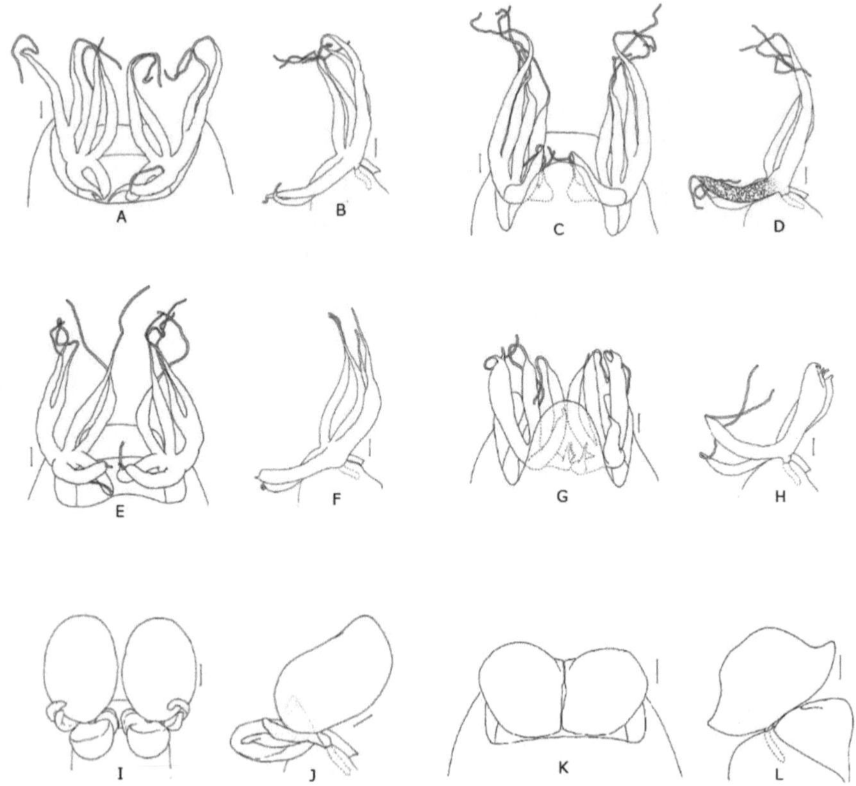

Fig. 88 Pupal gills of the *S. variegatum* species-group of the subgenus *Simulium* s. str. from Sulawesi. A, C, E, G, I & K, dorsal view; B, D, F, H, J & L, lateral view. A & B, *S. flavum*; C & D, *S. lorense*; E & F, *S. malinoense*; G & H, *S. quinquefiliatum*; I & J, *S. palopoense*; K & L, *S. globosum*. Scale bar: 0.2 mm for A–L

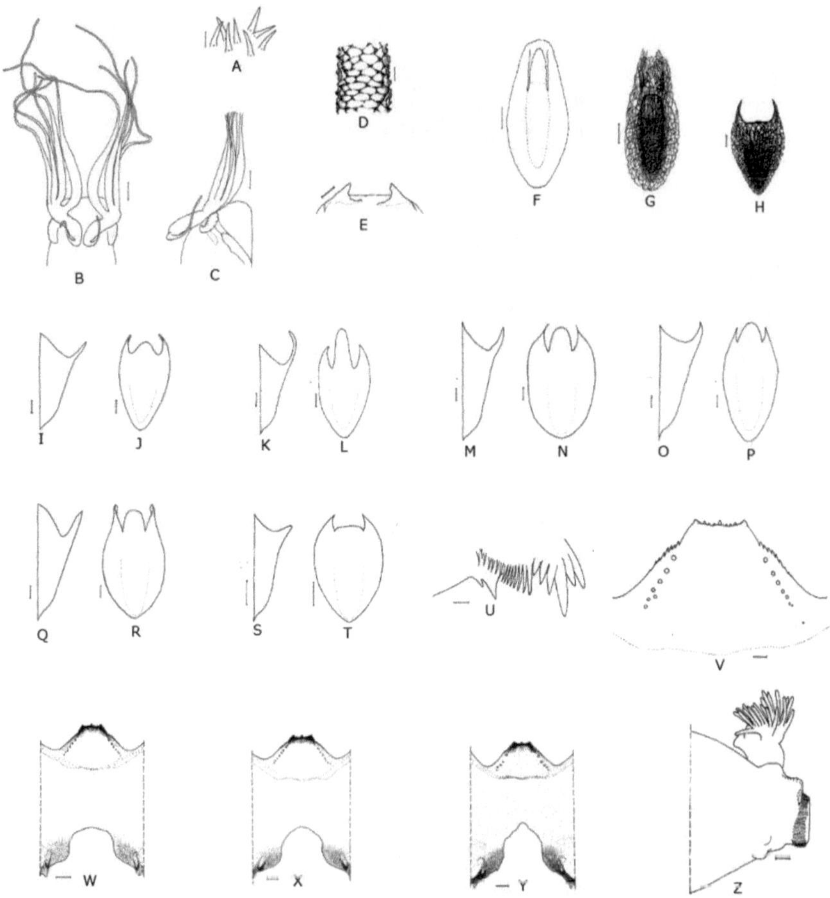

Fig. 89 Pupa and larva of the *S. variegatum* species-group of the subgenus *Simulium* s. str. from Sulawesi. A–T, Pupa; U–Z, Larva. A, Spinous processes on frons. B & C, Gills (B, dorsal view; C, lateral view). D, Enlargement of part of gill filament. E, Terminal hooks (caudal view). F–T, Cocoons of various shapes (F–H, J, L, N, P, R & T, dorsal view; I, K, M, O, Q & S, lateral view). U, Apical portion of mandible. V, Hypostoma. W–Y, Head capsules (ventral view). Z, Posterior tip of the abdomen showing small ventral papilla (lateral view). A–E, *S. luridum*; F, U, V & Z, *S. molliculum*; G, *S. laxum*; H, *S. flavum*; I & J, *S. linduense*; K, L & W, *S. elongatum*; M & N, *S. lorense*; O & P, *S. quinquefiliatum*; Q, R & X, *S. palopoense*; S & T, *S. globosum*; Y, *S. malinoense*. Scale bars: 1.0 mm for F, G, S & T; 0.5 mm for H–R; 0.2 mm for B & C; 0.1 mm for Z; 0.05 mm for W–Y; 0.02 mm for D, E & V; 0.01 mm for A & U

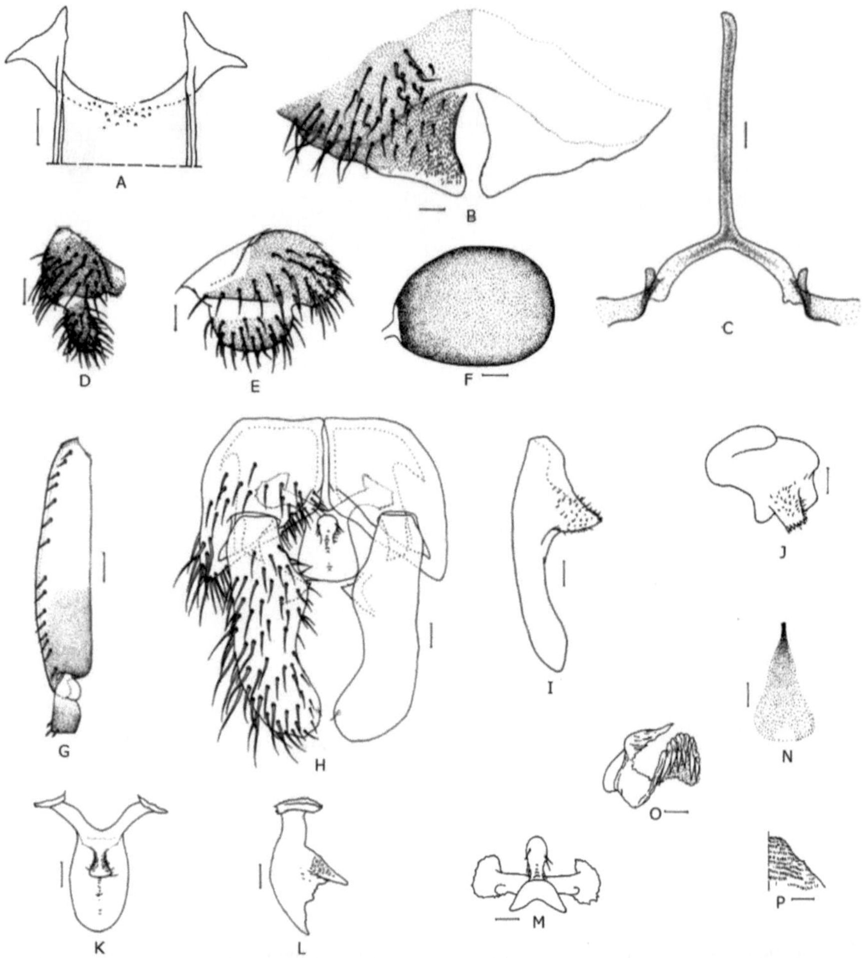

Fig. 90 *Simulium* s. str. species unplaced to species-group (1). A–F, Female; G–P, Male. A, Cibarium. B, Sternite 8 and ovipositor valves (ventral view). C, Genital fork (ventral view). D & E, Paraprocts and cerci (right side; D, ventral view; E, lateral view). F, Spermatheca. G, Hind basitarsus and second tarsomere (left side; outer view). H, Coxites, styles and ventral plate (ventral view). I & J, Styles (right side; I, medial view; J, caudal view). K–M, Ventral plates (K, ventral view; L, lateral view; M, caudal view). N, Median sclerite (caudal view). O, Paramere (right side; caudal view). P, Aedeagal membrane (left half; caudal view). A–P, *S. tamorense*. Scale bars: 0.05 mm for G; 0.02 mm for A–F & H–P

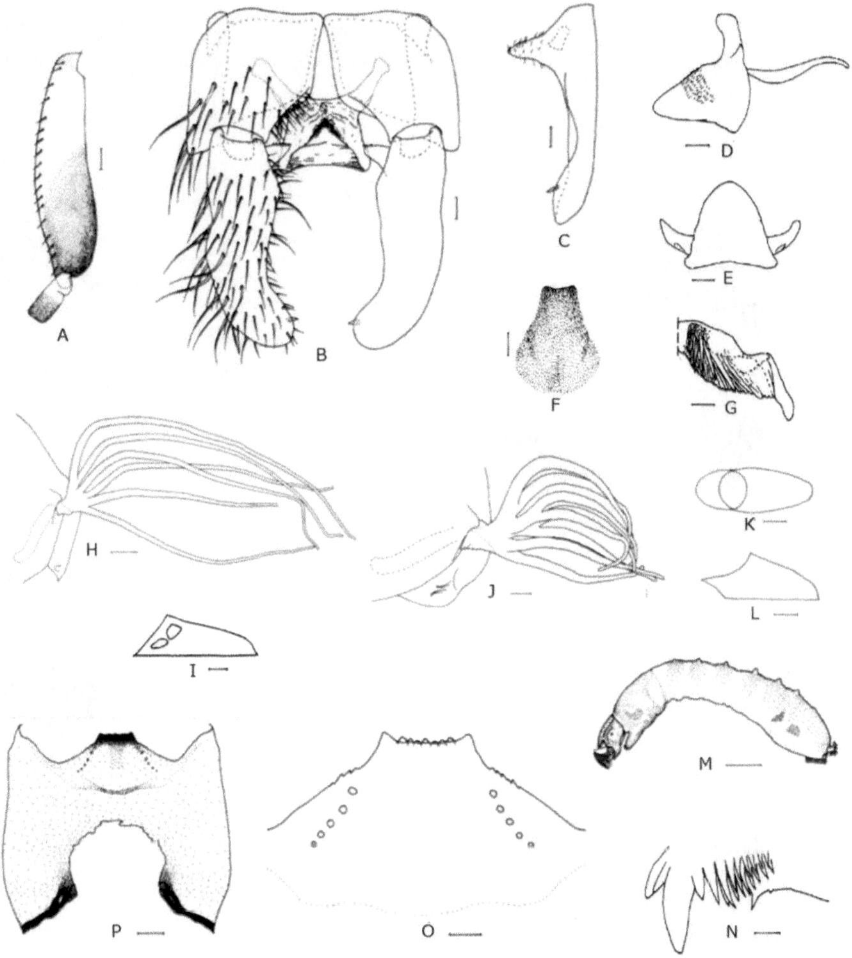

Fig. 91 *Simulium* s. str. species unplaced to species-group (2). A–G, Male; H–L, Pupa; M–P, Larva. A. Hind basitarsus and second tarsomere (left side; outer view). B, Coxites, styles, and ventral plate (ventral view). C, Style (left side; medial view). D, Ventral plate and median sclerite (lateral view). E, Ventral plate (caudal view). F, Medial sclerite (caudal view). G, Paramere (left side; caudal view). H & J, Gills (right side; outer view). I, K & L, Cocoons (I & L, lateral view; K, dorsal view). M, Whole body (lateral view). N, Apical portion of mandible. O, Hypostoma. P, Head capsule (ventral view). A–I, *S. titarense*; J–P, *S. tamorense*. Scale bars: 0.5 mm for I & K–M; 0.1 mm for H; 0.05 mm for A, J & P; 0.02 mm for B–G & O; 0.01 mm for N

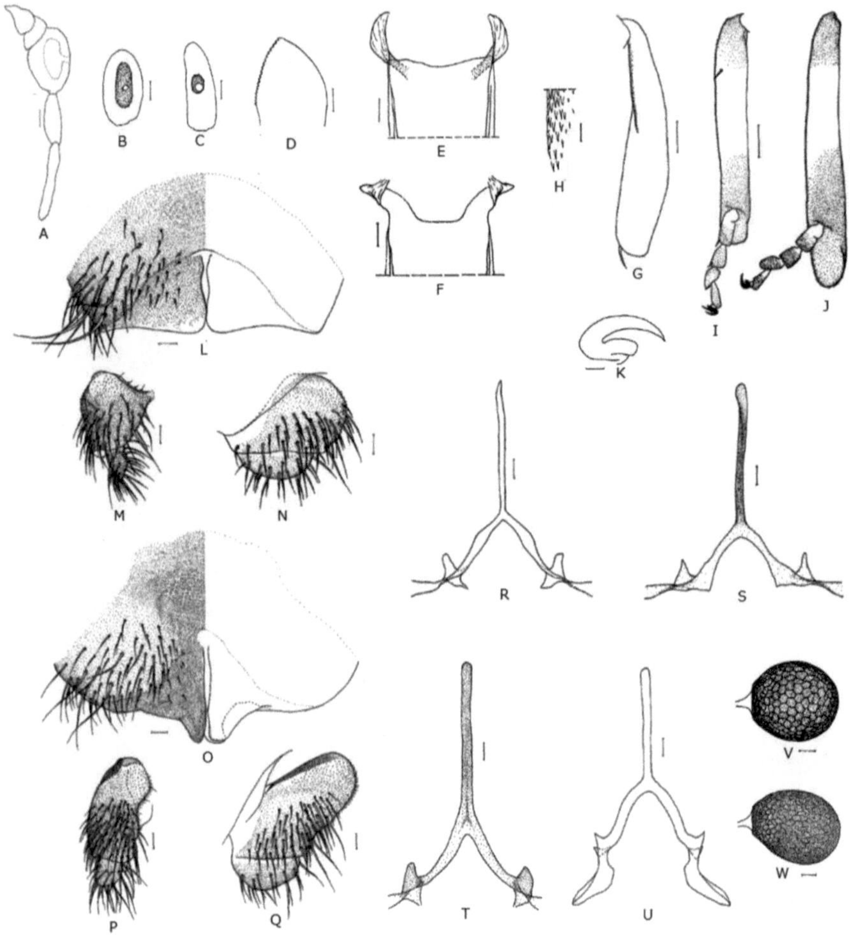

Fig. 92 Female of the subgenus *Wallacellum.* A, Palpus (lateral view). B & C, Third palpal segments with sensory vesicles (right side; front view). D, Mandible. E & F, Cibaria. G, Hind tibia showing narrowly constricted low ridge (right side; inner view). H, Spinous scales on surface of low ridge of hind tibia. I & J, Hind tarsi with calcipala of different lengths (left side; outer view). K, Claw. L & O, Sternite 8 and ovipositor valves (ventral view). M, N, P & Q, Paraprocts and cerci (right side; M & P, ventral view; N & Q, lateral view). R–U, Genital forks (ventral view). V & W, Spermathecae. A, B, G–I, K, L–N & R, *S. marilogense*; D & E, *S. alfurense*; C, O–Q, T & W, *S. tenederoi*; F & S, *S. suyoense*; J & V, *S. claveriaense*; U, *S. amplum*. Scale bars: 0.1 mm for G, I & J; 0.04 mm for A–C; 0.02 mm for E, F & L–W; 0.01 mm for D, H & K

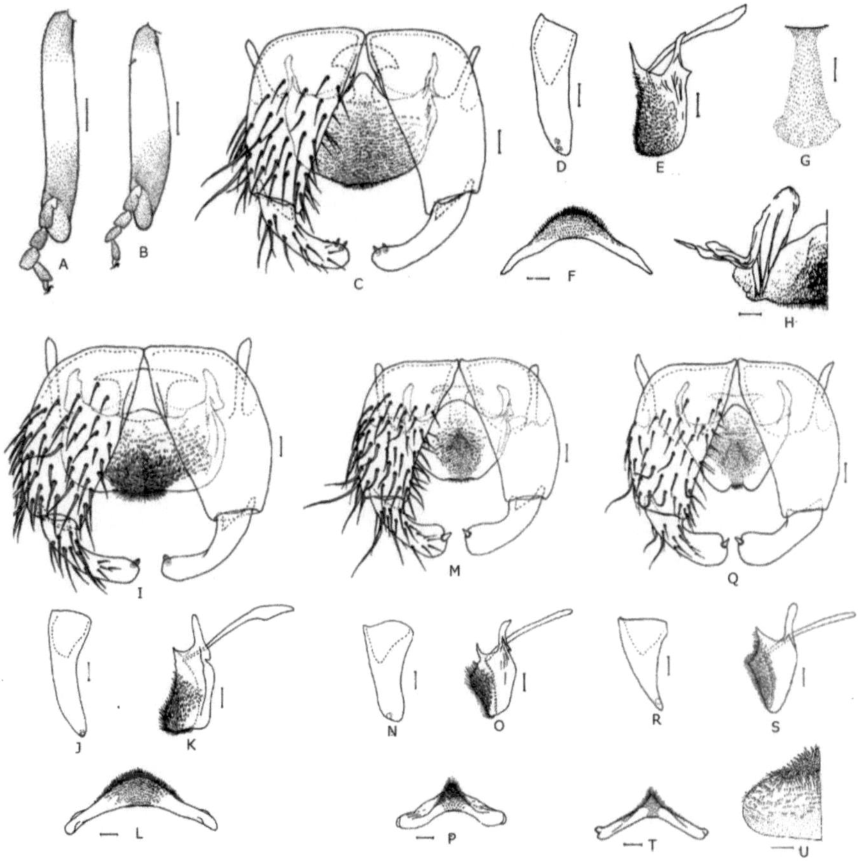

Fig. 93 Male of the subgenus *Wallacellum.* A & B, Hind basitarsi and second tarsomeres (left side; outer view). C, I, M & Q, Coxites, styles, ventral plates, and median sclerites (ventral view). D, J, N & R, Styles (right side; ventrolateral view). E, K, O & S, Ventral plates and median sclerites (lateral view). F, L, P & T, Ventral plates (caudal view). G, Median sclerite (caudal view). H, Paramere and aedeagal membrane (caudal view). U, Aedeagal membrane and dorsal plate (right half; caudal view). A, *S. recurvum*; B & Q–T, *S. amplum*; C–H, *S. suyoense*; I–L, *S. claveriaense*; M–P, *S. molawinense*; U, *S. marilogense*. Scale bars: 0.1 mm for A & B; 0.02 mm for C–U

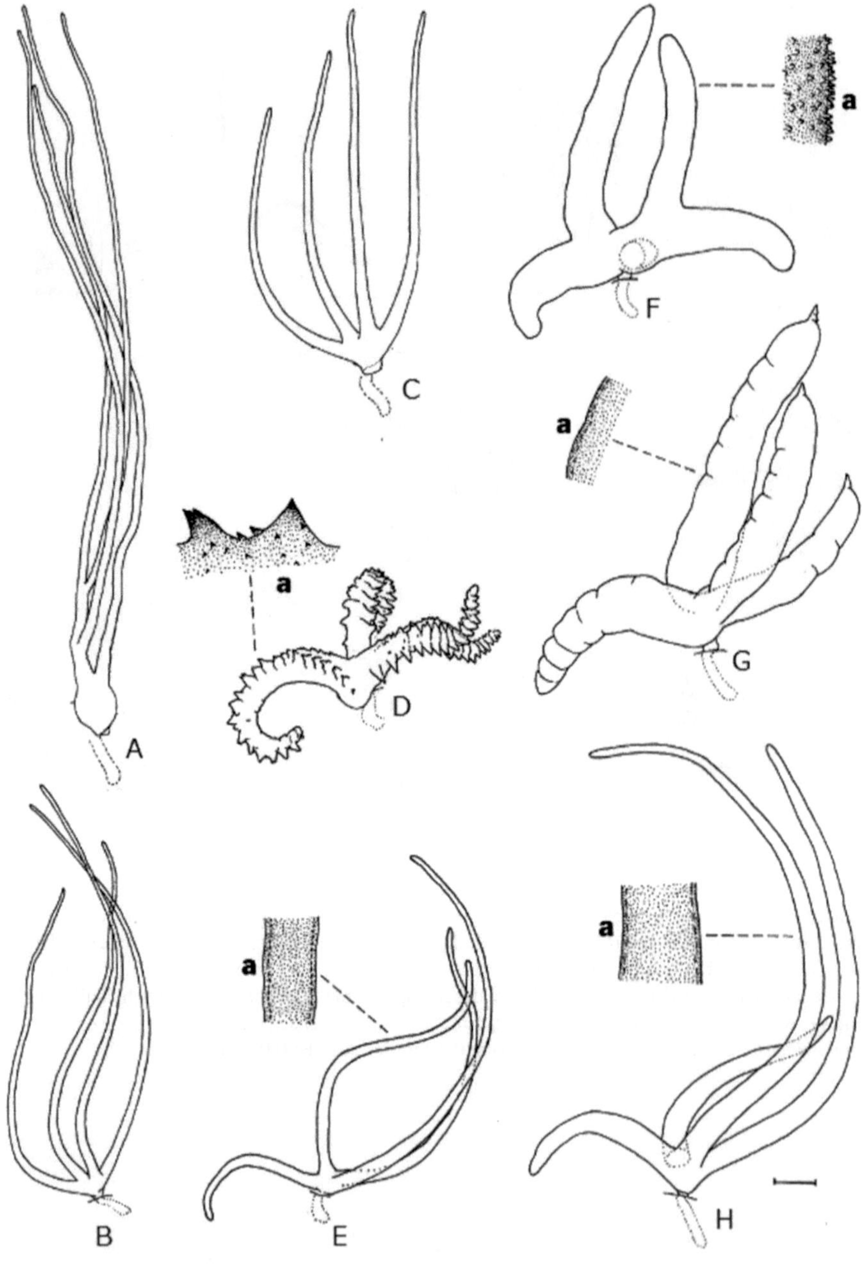

Fig. 94 Pupal gills of the subgenus *Wallacellum*. A, *S. suyoense*; B, *S. tuyense*; C, *S. claveriaense*; D, *S. spinosibranchium*; E, *S. recurvum*; F, *S. carinatum*; G, *S. cabrerai*; H, *S. resimum*. a, Enlargements of parts of gill filaments. Scale bar: 0.2 mm for A–H

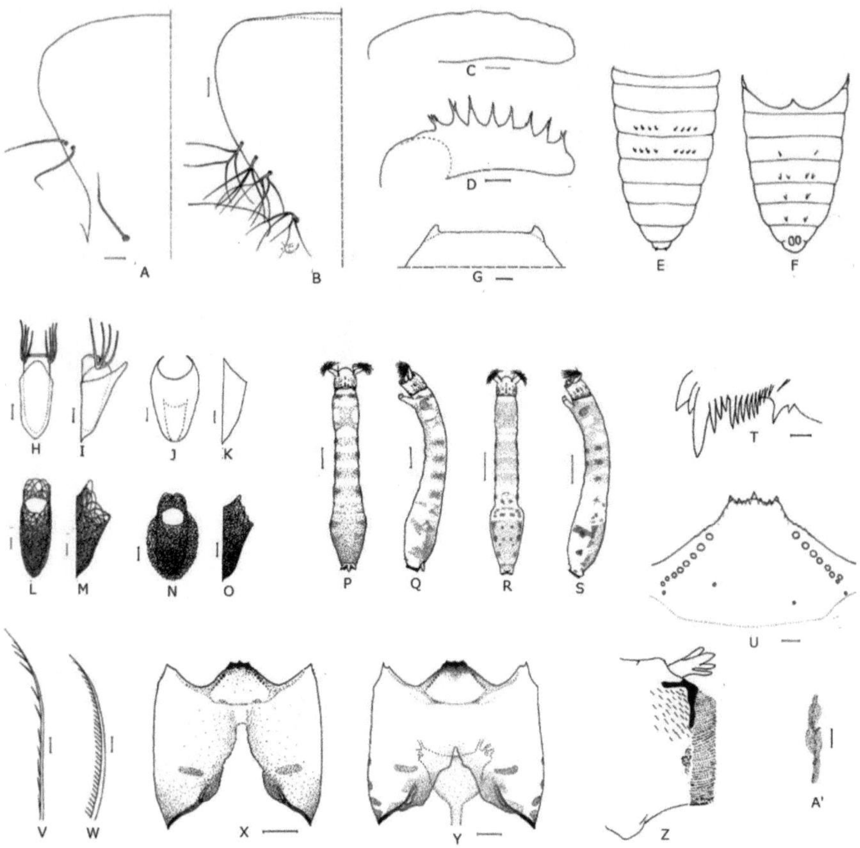

Fig. 95 Pupa and larva of the subgenus *Wallacellum*. A–O, Pupa; P–A', Larva. A & B, Frons and upper part of the face (frontal view). C & D, Antennal sheaths (left side; outer view). E & F, Abdomens (E, dorsal view; F, ventral view). G, Terminal hooks (caudal view).H–O, Cocoons (H, J, L & N, dorsal view; I, K, M & O, lateral view). P–S, Whole bodies (P & R, dorsal view; Q & S, lateral view). T, Apical portion of mandible. U, Hypostoma. V & W, Rays of labral fan (V, uppermost primary ray; apical half; W, one of other primary rays; apical two-fifths). X & Y, Head capsules (ventral view). Z, Posterior tip of abdomen showing rectal organ, ventral papilla, and accessory sclerite (lateral view). A', Accessory sclerite. A, C, G, P & Q, *S. marilogense*; B, H, I, & A', *S. claveriaense*; D & X, *S. alfurense*; J & K, *S. amplum*; L, M, R, S & Y, *S. tenederoi*; N & O, *S. molawinense*; T–W, *S. suyoense*. E, F & Z, *S. carinatum*. Scale bars: 1.0 mm for R & Q; 0.5 mm for H–Q; 0.1 mm for X & Y; 0.05 mm for A–D; 0.02 mm for G, U, V & W; 0.01 mm for T & A'

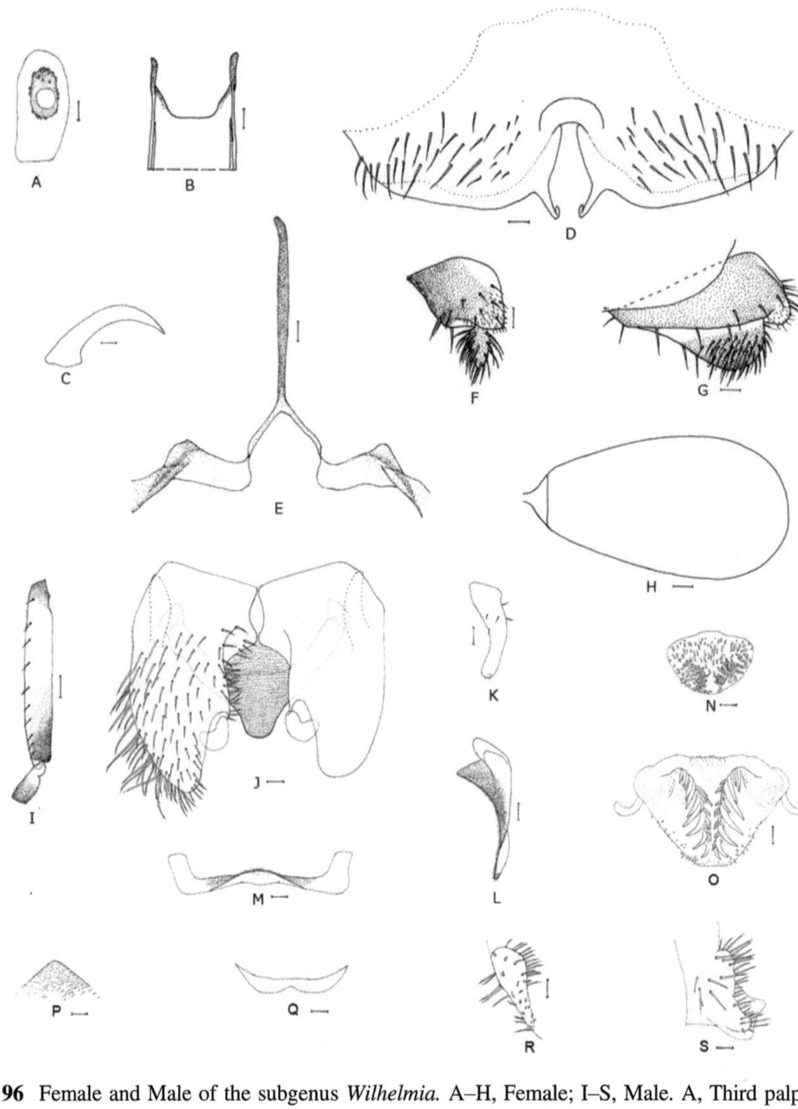

Fig. 96 Female and Male of the subgenus *Wilhelmia*. A–H, Female; I–S, Male. A, Third palpal segment with sensory vesicle (front view). B, Cibarium (front view). C, Claw. D, Sternite 8 and ovipositor valves (ventral view). E, Genital fork. F & G, Paraprocts and cerci (right side; F, ventral view; G, lateral view). H, Spermatheca. I, Hind basitarsus and second tarsomere (left side; outer view). J, Coxites, styles, and ventral plate (ventral view). K, Style (right side; caudal view). L & M, Ventral plates (L, lateral view; M, caudal view). N, Median sclerite (ventral view). O, Paramere (caudal view). P, Aedeagal membrane (caudal view). Q, Dorsal plate (ventral view). R & S, Cerci (right side; R, caudal view; S, lateral view). A–S, *S. takahasii*. Scale bars: 0.05 mm for I; 0.02 mm for A, B, D–H & J–S; 0.01 mm for C

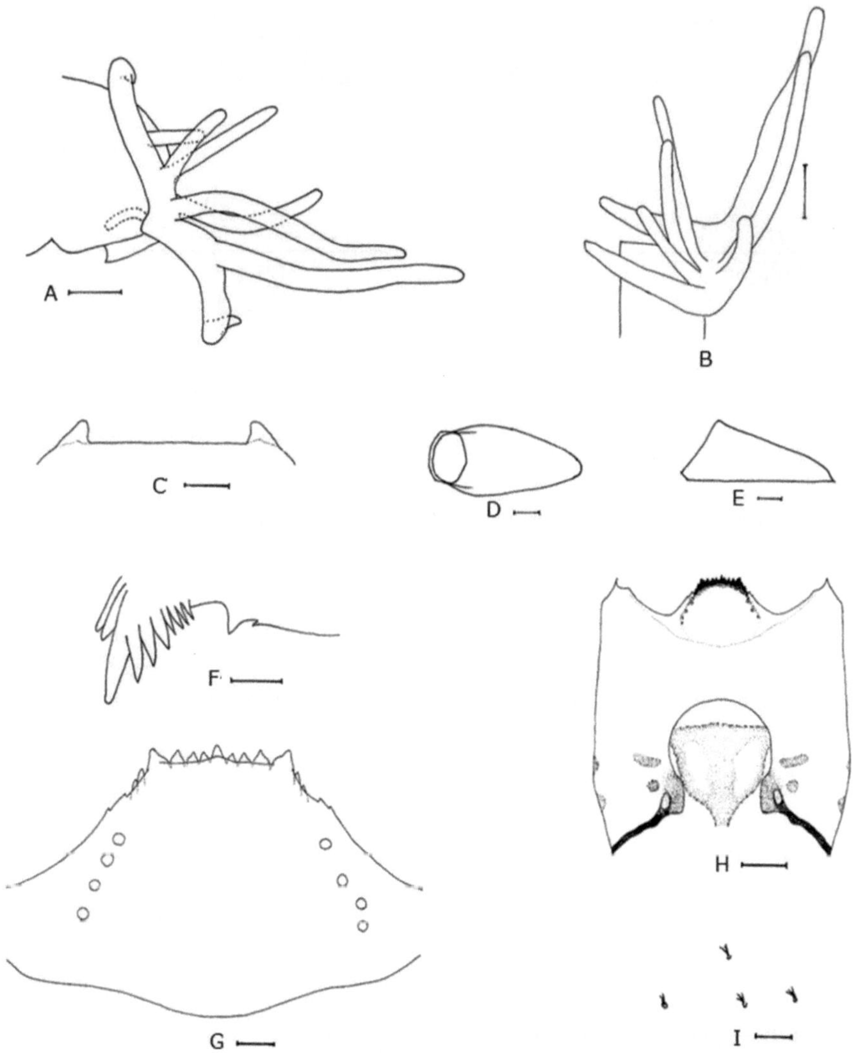

Fig. 97 Pupa and mature Larva of the subgenus *Wilhelmia.* A–E, Pupa; F–I, Larva. A & B, Gills (right side; A, lateral view; B, dorsal view). C, Terminal hooks (caudal view). D & E, Cocoons (D, dorsal view; E, lateral view). F, Apical portion of mandible. G, Hypostoma. H, Head capsule (ventral view). I, Dark tiny setae on dorsum of abdominal segment 8. A–I, *S. takahasii.* Scale bars: 0.5 mm for D & E; 0.2 mm for A & B; 0.1 mm for H; 0.02 mm for C, F & G; 0.01 mm for I

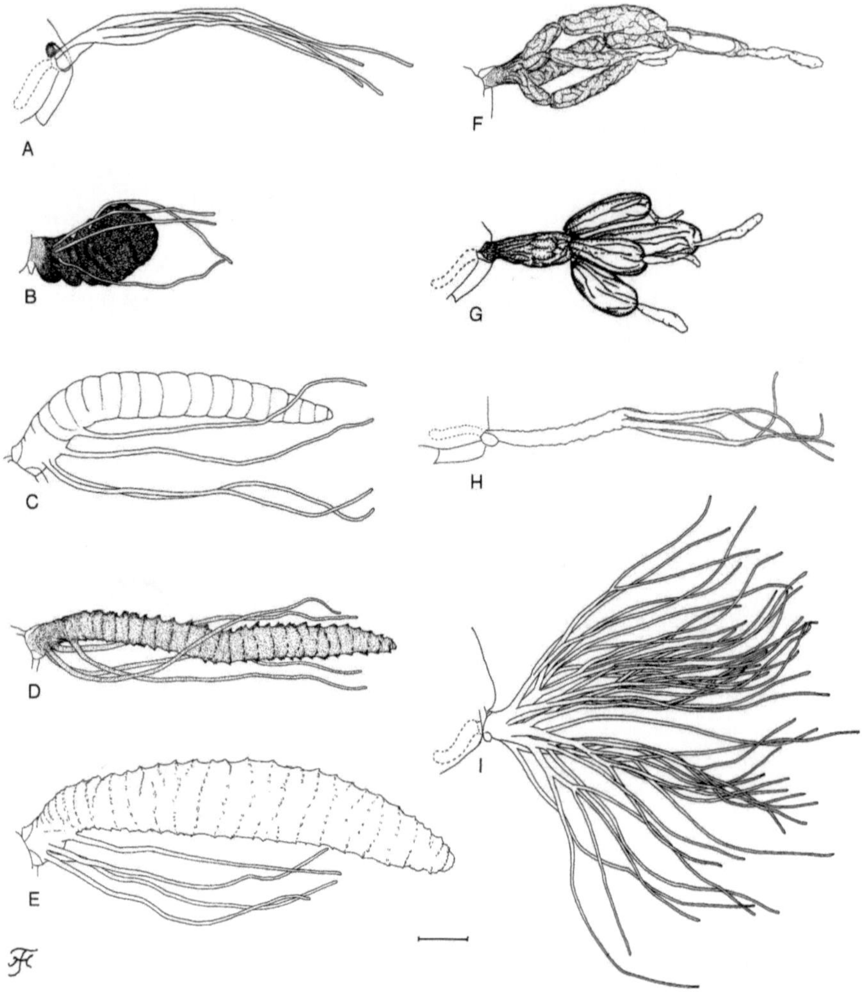

Fig. 98 Pupal gills of the subgenus *Morops*. A, *S. clathrinum* species-group. B–E, *S. farciminis* species-group. F–H, *S. oculatum* species-group. I, *S. papuense* species-group. A, *S. selwynense*; B, *S. arfakense*; C, *S. jayapuraense*; D, *S. pamahaense*; E, *S. botulus*; F, *S. demutaense*; G, *S. takae*; H, *S. norforense*; I, *S. papuense*. Scale bar: 0.2 mm for A–I

Epilog

For over 50 years, it has been a joy for me to investigate black flies in subtropical and tropical Asia (Oriental Region), the last unexplored zoogeographic region, and to accumulate the knowledge on the fauna of the Oriental black flies. It is so fascinating to realize that the fauna of black flies of the Oriental Region is so rich in the number of species and so diverse in lineages and certain morphological features such as female terminalia (Fig. 4G–T) and pupal gills (Figs. 18, 37, 87–89, and 94). How wonderful it is to know a new world of black flies! How nice it is if I could share my joy and excitement with many young researchers and students!

Considering what they need most as a foundation for their future works on black flies, I aimed to revise the current classification, to give a synopsis of each of 719 species (675 named and 44 unnamed) and to construct keys for species identification. However, as I felt while constructing these keys, it was not easy to morphologically identify each of the related species of certain groups, especially in the adult and larval stages. So, these keys are not perfect. The readers should keep this in mind when using the keys.

Many taxonomic problems and biological aspects of Oriental black flies remain to be investigated. Integrated approaches using chromosomes and/or DNA sequences and morphology have been used successfully to resolve the taxonomic problems relevant to intra- and inter-specific genetic variations, generalist species, species-complexes, and phylogenetic relationships within and among species-groups and among subgenera (e.g., Adler et al., 2015; Low et al., 2015a, 2016b; Otsuka et al., 2001, 2007). These combined approaches should prove successful in future investigations of the Oriental simuliid fauna.

Remaining taxonomic problems include (1) revision of the classification scheme, in particular, treatment of different lineages in large species-groups such as the *S. batoense* species-group and the *S. epistum* species-group in the subgenus G*omphostilbia*; (2) clarification of the species status of several long-standing nominal species, of which ambiguity is caused by insufficient description [e.g., *S.* (*E.*) *taipei* from Taiwan], unavailability of type specimens [e.g., *S.* (*S.*) *ambiguum* from Taiwan], unavailability of the adult stage [e.g., *S.* (*G.*) *doisaketense* from

H. Takaoka, *The Black Flies of Subtropical and Tropical Asia*,
https://doi.org/10.1007/978-981-99-5219-9

Thailand] or unavailability of additional specimens from type localities [e.g., *S.* (*G.*) *metatarsale* from India] (Takaoka, 2016); (3) interpretation of morphospecies with two or more genetic lineages [e.g., *S.* (*G.*) *paiense* from Thailand]; and (4) defining the range of intra- and inter-specific variations of certain numerical features [e.g., the number of male upper-eye (large) facets] used for species identification.

Studies on the biological and physiological aspects of black flies have been limited to the following: voltinism and over-wintering for several Indian species (Datta, 1983); autogeny of *S.* (*N.*) *aureohirtum* (Takaoka and Noda, 1979; Takaoka, 1989a); mating, blood-feeding and oviposition in captivity by *S.* (*Wilhelmia*) *takahasii* (Takaoka, 1985); life history from egg to adult of *S.* (*S.*) *bidentatum* (Hadi and Takaoka, 1995); natural infections of mermithids, microsporidans and other fungi [e.g., in larvae of Indonesian species by Takaoka et al., 2000 and Takaoka, 2003]; analysis of salivary gland proteins in relation to anticoagulation activity in *S.* (*S.*) *indicum* (Borah et al., 2014); description of the morphology of salivary glands and analysis of their proteins in five Thai species (Jariyapan et al., 2006); and ultrastructure of sensilla on the antennae and maxillary palpi in *S.* (*S.*) *nigrogilvum* and *S.* (*S.*) *umphangense* (Huang et al., 2022).

Ecological studies of black flies in relation to population dynamics, seasonal abundance, and vertical and horizontal distributions are also few [e.g., in Thailand by Choochote et al., 2005 and Srisuka et al., 2015b, 2017 and in Peninsular Malaysia by Ya'cob et al., 2016]. Further ecological studies of black flies related to global warming, deforestation, urbanization, role in the food-web of the terrestrial ecosystem, and conservation of natural environments also should be considered.

Studies of anthropophilic species of black flies, which are most important for human and animal health and welfare, are also limited to the following: preferences of blood-feeding for humans and other animals [e.g., in *S.* (*S.*) *indicum* in northern India by Lewis, 1974 and in several Thai species by Choochote et al., 2005], and natural infections with filariae and protozoan parasites in adult females of several Thai species (Takaoka et al., 2003; Fukuda et al., 2003; Aupalee et al., 2020b; Huang et al., 2021; Ishii et al., 2008; Jumpato et al., 2018; Thaijarern et al., 2019a; Pramual et al., 2020).

In the last decade, DNA barcoding of individual species of black flies has been performed [e.g., in Thailand by Pramual et al., 2021, in Malaysia by Izwan-Anas et al., 2023, in Indonesia by Hew et al., 2023a, and in Vietnam by Putt et al., 2023]. However, many species of black flies in other countries are still awaiting DNA barcoding. In the near future, identification of female black flies caught in the field and of parasites they carry will be simultaneously possible by DNA barcoding of both the black fly species and their potential parasites.

Readers may understand how much work on black flies in the Oriental Region remains to be done and the importance of accurate species identification as a basis for further biological, ecological, chromosomal, and genetic studies as well as transmission studies of parasites or pathogens that females of black flies transmit.

The fauna of black flies in the Oriental Region is distinguished from that in five other zoogeographic regions by the absence of many genera that are present in these other regions. However, *Simulium,* the only genus recognized in the Oriental

Region, is the largest and most specialized among 31 genera of the world and is widely distributed in all five of the other regions. In this regard, information about black flies in the Oriental Region is, for the most part, common to those of black flies in the five other regions. For example, speciation with morphological diversification of the pupal gills is remarkably similar in all six regions.

It is hoped that this book will prove helpful for future studies on black flies in the Oriental Region and, consequently, in the five other zoogeographic regions.

References

Adler, P. H. (2020). *World Blackflies (Diptera: Simuliidae): A Comprehensive Revision of the Taxonomic and Geographical Inventory [2020]* (p. 142).

Adler, P. H. (2022). *World Blackflies (Diptera: Simuliidae): A Comprehensive Revision of the Taxonomic and Geographical Inventory [2022]* (p. 145). Retrieved from http://entweb.clemson.edu/biomia/pdfs/blackflyinventory.pdf

Adler, P. H., & Crosskey, R. W. (2008). *World Blackflies (Diptera: Simuliidae): A Comprehensive Revision of the Taxonomic and Geographical Inventory [2008]* (p. 105).

Adler, P. H., & Crosskey, R. W. (2009). *World Blackflies (Diptera: Simuliidae): A Comprehensive Revision of the Taxonomic and Geographical Inventory [2009]* (p. 100).

Adler, P. H., & Crosskey, R. W. (2010). *World Blackflies (Diptera: Simuliidae): A Comprehensive Revision of the Taxonomic and Geographical Inventory [2010]* (p. 112).

Adler, P. H., & Crosskey, R. W. (2012). *World Blackflies (Diptera: Simuliidae): A Comprehensive Revision of the Taxonomic and Geographical Inventory [2012]* (p. 119).

Adler, P. H., & Crosskey, R. W. (2014). *World Blackflies (Diptera: Simuliidae): A Comprehensive Revision of the Taxonomic and Geographical Inventory [2014]* (p. 122).

Adler, P. H., & Crosskey, R. W. (2015). *World Blackflies (Diptera: Simuliidae): A Comprehensive Revision of the Taxonomic and Geographical Inventory [2015]* (p. 123).

Adler, P.H. & Crosskey, R.W. (2016). *World Blackflies (Diptera: Simuliidae): A Comprehensive Revision of the Taxonomic and Geographical Inventory [2016]* (p. 126).

Adler, P. H., & Crosskey, R. W. (2017). *World Blackflies (Diptera: Simuliidae): A Comprehensive Revision of the Taxonomic and Geographical Inventory [2017]* (p. 131).

Adler, P. H., Currie, D. C., & Wood, D. M. (2004). *The Black Flies (Simuliidae) of North America* (p. xv + 941). Cornell University Press.

Adler, P. H., Fukuda, M., Takaoka, H., Reeves, W. K., Kim, S. K., & Otsuka, Y. (2019a). Revision of *Simulium rufibasis* (Diptera: Simuliidae) in Japan and Korea: Chromosomes, DNA, and Morphology. *Journal of Medical Entomology, 57*, 388–403.

Adler, P. H., Huang, Y. T., Reeves, W. K., Kim, S. K., Otsuka, Y., & Takaoka, H. (2013). Macrogenomic evidence for the origin of the black fly *Simulium suzukii* (Diptera: Simuliidae) on Okinawa Island, Japan. *PLoS One, 8*, 1–9.

Adler, P. H., Inci, A., Yildirim, A., Duzlu, O., Macreadie, J. W., Kudela, M., Khazeni, A., Bruderova, T., Setiz, G., Takaoka, H., Otsuka, Y., & Bass, J. (2015). Geographic generalist or multiple species?—Evidence from the macrogenome of black flies in the *Simulium* subgenus *Wilhelmia* (Diptera: Simuliidae). *Biological Journal of the Linnean Society, 114*, 163–183.

Adler, P. H., Srisuka, W., & Saeung, A. (2020a). Cryptic species of black flies (Diptera: Simuliidae) at high elevation in the Oriental Region: the *Simulium vernum* species group in Thailand. *Acta Tropica, 205*(1), 1–7. (105393).

H. Takaoka, *The Black Flies of Subtropical and Tropical Asia*,
https://doi.org/10.1007/978-981-99-5219-9

Adler, P. H., Srisuka, W., Low, V. L., Takaoka, H., & Saeung, A. (2019b). High-elevation chromosomal diversity of black flies (Diptera: Simuliidae) in Thailand. *Insect Systematics and Diversity, 3*, 1–10.

Adler, P. H., Takaoka, H., & Ruiz-Arrondo, I. (2020b). Chromosomal relationships of two new species in the *Simulium vernum* species-group (Diptera: Simuliidae) from Nepal. *Proceedings of the Entomological Society of Washington, 122*, 735–749.

Adler, P. H., Takaoka, H., Sofian-Azirun, M., Low, V. L., Ya'cob, Z., Chen, C. D., Lau, K. W., & Pham, X. D. (2016). Vietnam, a hotspot for chromosomal diversity and cryptic species in black flies (Diptera: Simuliidae). *PLoS One, 11*(10), 1–29. e0163881. https://doi.org/10.1371/journal.pone.0163881

Adler, P. H., Takaoka, H., Sofian-Azirun, M., Chen, C. D., & Suana, W. (2019c). Evolutionary and biogeographic history of the black fly *Simulium wayani* (Diptera: Simuliidae) on the island of Timor. *Acta Tropica, 193*, 1–6.

Almet, J., Hadi, U. K., Soviana, S., & Supriyono. (2016). Species and population diversity of *Simulium* (Diptera: Simuliidae) in Kupang, district of East Nusa Tengara. *Proceedings of The First Asian Simuliidae Symposium*, 16.

An, J. Y. (1995). List of Chinese Simuliidae (Diptera: Simuliidae). *Chinese Journal of Vector Biology and Control, 7*, 470–476. (In Chinese).

An, J. Y., & Chen, H. B. (2007). Studies of subgenus *Gomphostilbia* on divide into groups (Diptera: Simuliidae). *Acta Parasitologica et Medica Entomologica Sinica, 11*, 110–113.

An, J. Y., Chen, J. L., & Chen, R. D. (2007). A supply description of the pupae and larvae stages of *Simulium* (*Gomphostilbia*) *dudgeoni* from Hong Kong, China (Diptera: Simuliidae). *Acta Parasitologica Et Medica Entomologica Sinica, 14*, 249–251.

An, J. Y., Hao, B. S., & Mai, Z. Q. (1990). Description of *Simulium* (*Gomphostilbia*) *pingxiangensis* sp. nov. from Guangxi Zhuangzu Aut. Reg. China. *Control of Blood-Sucking Dipteran Insects, 2*, 100–102.

An, J. Y., Hao, B. S., & Yan, G. (1998). Two new blackflies of *Simulium* (*Simulium*) from Guangdong Province, China (Diptera: Simuliidae). *Acta Entomologica Sinica, 41*, 187–193.

An, J. Y., Yan, G., Yang, L. X., & Hao, B. S. (1994). A new species of Simuliidae from Guangdong Province, China (Diptera: Simuliidae). *Sichuan Journal of Zoology, 13*, 4–6.

Anbalagan, S., Prasanna, V. A., Dinakaran, S., & Krishnan, M. (2014). Two new species of *Simulium* (*Gomphostilbia*) (Diptera: Simuliidae) from Peninsular India, with keys to Peninsular Indian members of the genus *Simulium*. *Zootaxa, 3861*, 451–465.

Anbalagan, S., Balachandran, C., Prasanna, V. A., Kannan, M., Dinakaran, S., & Krishnan, M. (2015a). A new species of *Simulium* (*Gomphostilbia*) (Diptera: Simuliidae) from South India, with keys to Indian members of the subgenus *Gomphostilbia*. *Zootaxa, 3974*(4), 555–563.

Anbalagan, S., Prasanna, V. A., Kannan, M., Dinakaran, S., & Krishnan, M. (2015b). *Simulium* (*Gomphostilbia*) (Diptera: Simuliidae) from Southern Western Ghats, India: Two new species and DNA barcoding. *Acta Tropica, 149*, 94–105.

Anbalagan, S., Vijayan, S., Balachandran, C., Dinakaran, S., & Krishnan, M. (2018). Two new black fly species of *Simulium* (*Simulium*) (Diptera: Simuliidae) from South India. *Journal of Asia-Pacific Entomology, 21*, 1406–1415.

Anbalagan, S., Vijayan, S., Balachandran, C., & Dinakaran, S. (2019a). A new species of *Simulium* (*Simulium*) (Diptera: Simuliidae), with keys to *S. striatum* species-group from India. *Journal of Threatened Taxa, 11*(5), 13573–13578.

Anbalagan, S., Vijayan, S., Dinakaran, S., & Krishnan, M. (2019b). A new species of *Simulium* (*Gomphostilbia*) (Diptera: Simuliidae) from South India. *Zootaxa, 4551*(4), 479–486.

Anbalagan, S., Rekha, K., Vijayan, S., Balachandran, C., Dinakaran, S., & Krishnan, M. (2020a). A new black fly species of *Simulium* (*Nevermannia*) (Diptera: Simuliidae) from the Southern Eastern India. *Zootaxa, 4768*(3), 374–382.

Anbalagan, S., Vijayan, S., Balachandran, C., Thiyonila, B., & Surya, A. (2020b). Two new species of *Simulium* (*Gomphostilbia*) (Diptera: Simuliidae) from South India. *Zootaxa, 4742*(1), 57–72.

Audley-Charles, M. G. (1981). Geological history of the region of Wallace's line. Pp. 24–35. In T. C. Whitmore (Ed.), *Wallace's line and plate tectonics* (p. 91). Clarendon Press.

Aupalee, K., Saeung, A., Srisuka, W., Fukuda, M., Junkum, A., Pitasawat, B., & Takaoka, H. (2020a). Three new species of the *Simulium* (*Simulium*) *striatum* species-group (Diptera: Simuliidae) from Thailand, with their phylogenetic relationships. *Acta Tropica, 211*, 1–14.

Aupalee, K., Saeung, A., Srisuka, W., Fukuda, M., Streit, A., & Takaoka, H. (2020b). Seasonal filarial infections and their black fly vectors in Chiang Mai Province, northern Thailand. *Pathogens, 512*, 1–18.

Aupalee, K., Srisuka, W., Saeung, A., & Takaoka, H. (2022). A new species of the *Simulium* (*Simulium*) *striatum* species-group (Diptera: Simuliidae) from Thailand: description and phylogenetic relationship with its Thai members. *Acta Tropica, 230*, 1–9. (106387).

Becher, E. (1885[1884]). A new species of *Simulium* from Assam. *Journal of the Asiatic Society of Bengal, 53*, 199–200.

Bentinck, W. (1955). *The black flies of Japan and Korea* (p. 23). 406 Medical General Laboratory U.S.A. Army.

Bi, G. H., & Chen, H. B. (2004). A new species of the genus *Simulium* Latreille from Hunan Province, China (Diptera, Simuliidae). *Acta Zootaxonomica Sinica, 29*, 569–571.

Borah, S., Naglot, A., Goswami, S., Rahman, I., & Deka, M. (2014). Anticoagulation activity of salivary gland extract of Oriental black fly *Simulium indicum*. *Asian Pacific Journal of Tropical Biomedicine, 4*(Suppl. 1), 5252–5257.

Borah, S., Rahman, I., Goswami, S., Deka, M., & Takaoka, H. (2012). Notes on black flies (Diptera: Simuliidae) from North-East India: New records of five species from Arunachal Pradesh and taxonomic reviews of two species from Assam. *Tropical Biomedicine, 29*, 92–97.

Brunetti, E. (1911). New Oriental Nemocera. *Records of the Indian Museum, 4*, 282–288.

Brunetti, E. (1912). A new species of blood-sucking fly (*Simulium*) from Ceylon. *Spolia Zelan, 8*, 90–91.

Cai, R., Liu, Y., & An, J. (2016). A new species of *Simulium* (*Simulium*) (Diptera: Simuliidae) from Hainan Island, China. *Entomotaxonomica, 38*, 213–216.

Chen, H. B. (2000). Descriptions of two new *Simulium* species from Leigong Mountain, Guizhou Province, China (Diptera: Simuliidae). *Acta Zootaxonomica Sinica, 25*, 100–105.

Chen, H. B. (2001a). Studies on black flies of Hainan Island II. Records of subgenus *Morops* Enderlein in China with description of a new species (Diptera: Simuliidae). *Acta Zoologica Sinica, 26*, 565–568.

Chen, H. B. (2001b). Two new species of *Simulium* (*Nevermannia*) from Guizhou Province, China (Diptera: Simuliidae). *Acta Zootaxonomica Sinica, 44*, 560–566.

Chen, H. B. (2003). Notes on Chinese members of the *Simulium striatum*-group, with descriptions of three new species (Diptera, Simuliidae). *Acta Zootaxonomica Sinica, 28*, 336–342.

Chen, H. B. (2016). *Chinese blackflies (Diptera: Simuliidae)* (p. 673). Guizhou Province Science and Technology Publication. (In Chinese except the descriptions of new species written also in English).

Chen, H. B., & An, J. Y. (2003). *The blackflies of China (Diptera: Simuliidae)* (p. 448). Science Press.

Chen, H. B., & Chen, H. (2000). A new species of *Simulium* (*Gomphostilbia*) from Hainan Island, China. Pp. 129–133. In *Special issue of Chinese Society of Entomology for 21st century*.

Chen, H., & Chen, H. B. (2001). A new blackfly species of *Simulium* (Diptera: Simuliidae) from Guizhou Province, China. *Acta Entomologica Sinica, 8*, 208–212.

Chen, H. B., Kang, Z., & Zhang, C. L. (2007a). A new species of the genus *Simulium* from Guangxi Province (Diptera, Simuliidae). *Acta Zootaxonomica Sinica, 32*, 579–580.

Chen, H. B., & Wen, X. J. (1999). A new species of *Simulium* (*Simulium*) from Fanjing Mountain, Guizhou Province, China (Diptera: Simuliidae). *Acta Zootaxonomica Sinica, 24*, 436–439.

Chen, H. B., Xiu, J. F., & Zhang, C. L. (2012). A survey of blackflies with three new species from Kuankuoshui, Guizhou, China (Diptera, Simuliidae). *Acta Zootaxonomica Sinica, 37*, 382–388.

Chen, H. B., & Zhang, C. L. (1997a). A new species of *Odagmia* Enderlein from China (Diptera: Simuliidae). *Acta Zootaxonomica Sinica, 22*, 194–196. (In Chinese with an English abstract).

Chen, H. B., & Zhang, C. L. (1997b). Two new species of *Eusimulium* from Guizhou Province, China (Diptera: Simuliidae). *Acta Zootaxonomica Sinica, 22*, 301–303.

Chen, H. B., & Zhang, C. L. (1998). Two new blackflies of the genus *Simulium* Latreille from Guizhou Province, China (Diptera: Simuliidae). *Acta Zootaxonomica Sinica, 23*, 57–61.

Chen, H. B., & Zhang, C. L. (2001). Studies on blackflies of Hainan Island, China. I. Description of three new species of *Simulium* (*Gomphostilbia*) (Diptera: Simuliidae). *Acta Zootaxonomica Sinica, 26*, 361–368.

Chen, H.B. & Zhang, C.L. (2002). A new species of *Simulium (Simulium)* from Guizhou Province, China (Diptera: Simuliidae). *Acta Entomologica Sinica., 45* (Supplement), 77–78.

Chen, H. B., & Zhang, C. L. (2003). Review of taxonomy of blackflies from Xishuangbanna, China, with a description of a new species (Diptera: Simuliidae). *Acta Zootaxonomica Sinica, 28*, 542–545.

Chen, H. B., & Zhang, C. L. (2004). A new species of *Simulium* (*Gomphostilbia*) from Yunnan Province, China (Diptera: Simuliidae). *Acta Parasitology et Medica Entomologica Sinica, 11*, 87–90.

Chen, H. B., Zhang, C. L., & Bi, G. H. (2004). Descriptions of three new species of *Simulium* (subg. *Nevermannia*) from Hunan Province, China (Diptera, Simuliidae). *Acta Zootaxonomica Sinica, 29*, 365–371.

Chen, H. B., Zhang, J. Q., & Huang, L. (2005). Two new black fly species of *Simulium* (*Simulium*) from Sichuan Province, China (Diptera: Simuliidae). *Acta Zootaxonomica Sinica, 30*(2), 430–435.

Chen, H. B., Zhang, J. Q., & Wen, X. J. (2000). Two new species of *Simulium* (*Gomphostilbia*) from Fanjing Mountain, China (Diptera, Simuliidae). *Acta Entomologica Sinica, 7*, 21–28.

Chen, H. B., Zhang, C. L., & Yang, M. (2003a). Checklist of Guizhou blackflies with descriptions of a new species (Diptera: Simuliidae). *Guizhou Science, 21*, 46–50.

Chen, H. B., Zhang, C. L., & Yang, M. (2003b). Descriptions of two new species of the subgenus *Nevermannia* Enderlein and *Simulium* Latreille from Hainan Island, China (Diptera, Simuliidae). *Acta Zootaxonomica Sinica, 28*, 745–750.

Chen, H. B., Zhang, J. Q., & Zhang, C. L. (2007b). A new blackfly species of the genus *Simulium* from Guangxi, China (Diptera, Simuliidae). *Acta Zootaxonomica Sinica, 32*, 779–781.

Chen, H. B., Zhang, J. Q., & Zhang, C. L. (2007c). Two new species of the genus *Simulium* (*Gomphostilbia*) from Guangxi, China (Diptera, Simuliidae). *Acta Zootaxonomica Sinica, 32*, 782–786.

Choochote, W., Takaoka, H., Fukuda, M., Otsuka, Y., Aoki, C., & Eshima, N. (2005). Seasonal abundance and daily flying activity of black flies (Diptera: Simuliidae) attracted to human baits in Doi Inthanon National Park, northern Thailand. *Medical Entomology and Zoology, 56*, 335–348.

Chung, C. L. (1986). A new record of *Simulium* (*Wallacellum*) *yonakuniense* from Lanyu Is., Taitung County, Taiwan (Diptera: Simuliidae). *Journal of the Taiwan Museum, 39*, 1–10.

Crosskey, R. W. (1967). The classification of *Simulium* Latreille (Diptera: Simuliidae) from Australia, New Guinea and the Western Pacific. *Journal of Natural History, 1*, 23–51.

Crosskey, R. W. (1969). A re-classification of the Simuliidae (Diptera) in Africa and its islands. *Bulletin of the British Museum (Natural History) Entomology, 14*, 1–194. United States Department of Agriculture, London, U.K.

Crosskey, R. W. (1973). Family Simuliidae. Pp. 423–430. In M. D. Delfinado & D. E. Hardy (Eds.), *A Catalog of the Diptera of the Oriental Region. Vol. I. Suborder Nematocera* (p. 618). University Press of Hawaii.

Crosskey, R. W. (1987). An annotated checklist of the world black flies (Diptera: Simuliidae). Pp. 425–520. In K. C. Kim & R. W. Merritt (Eds.), *Black flies: Ecology, population management, and annotated world list* (p. 528). The Pennsylvania State University Press.

Crosskey, R. W. (1989). Family Simuliidae. Pp. 221–225. In J. Evenhuis (Ed.), *Catalog of the Diptera of the Australasian and Oceanian Regions* (p. 1155). Bishop Museum Special Publication No. 86. Bishop Museum Press.

Crosskey, R. W. (1990). *The Natural History of Blackflies* (p. ix + 711). John Wiley & Sons Inc.

Crosskey, R. W., & Howard, T. (1997). *A New Taxonomic and Geographical Inventory of World Black Flies (Diptera: Simuliidae)* (p. 114). Natural History Museum.

Currie, D. C. (1997). Black flies (Diptera: Simuliidae) of the Yukon, with reference to the black-fly fauna of northwestern North America. Pp. 563–614. In H. V. Danks & J. A. Downs (Eds.), *Insects of the Yukon* (p. 1034). Biological survey of Canada (Terrestrial Arthropods), Ottawa.

Currie, D. C., & Adler, P. H. (2008). Global diversity of black flies (Diptera: Simuliidae) in freshwater. *Hydrobiologia, 595*, 469–475.

Dasgupta, S. K., Sharma, I., Mitra, A., & Dasgupta, B. (1968). On the incidence and blood meal of biting insects (Diptera, Ceratopogonidae, Simuliidae) of Darjeeling, India. *Entomologiceskoe Obozrenie, 48*, 770–773.

Datta, M. (1973). New species of black flies (Diptera: Simuliidae) of the subgenera *Eusimulium* Roubaud and *Gomphostilbia* Enderlein from the Darjeeling area, India. *Oriental Insects, 7*, 363–402.

Datta, M. (1974a). Some black flies (Diptera: Simuliidae) of the subgenus *Simulium* Latreille (s. str.) from the Darjeeling area (India). *Oriental Insects, 8*, 15–27.

Datta, M. (1974b). New species of black flies (Diptera: Simuliidae) from Darjeeling area, India. *Oriental Insects, 8*, 457–468.

Datta, M. (1975a). Simuliidae (Diptera) from Assam foot-hills, India. *Japanese Journal of Sanitary Zoology, 26*, 31–40.

Datta, M. (1975b). A new black fly species (Simuliidae-Diptera) from the Darjeeling area, India. *Proceedings of the Indian Academy of Sciences, 81*, 67–74.

Datta, M. (1983). A review of the Simuliidae (Diptera) from the Oriental Region. *Oriental Insects, 17*, 215–267.

Datta, M. (1985a). The subgenus *Odagmia* Enderlein (Diptera: Simuliidae) with a new species from Himachal Pradesh, India. *Bulletin of the Zoological Survey of India, 7*, 25–32.

Datta, M. (1985b). Notes on blackflies (Diptera: Simuliidae) from Kashmir, India. *Bulletin of the Zoological Survey of India, 7*, 231–236.

Datta, M. (1988). *Simulium* (*S.*) *asishi* spec. nov., a new black fly from the Terai of West Bengal, India (Diptra: Simuliidae). *Opuscula Zoologica Fluminensia, 31*, 1–8.

Datta, M. (1991a). A preliminary report on the black flies (Diptera: Simuliidae) from Bhutan. *Journal of the Bengal Natural History Society. New Series, 10*, 29–37.

Datta, M. (1991b). Occurrence of blackflies in Bankuru and Purulia District in West Bengal, India (Diptera: Simuliidae). *Journal of the Bengal Natural History Society, 10*(1), 11–15.

Datta, M. (1992a). An overview of the Simuliidae (Diptera) of West Bengal, India. *Journal of the Bengal Natural History Society. New Series, 11*, 41–62.

Datta, M. (1992b). Note on the evolution of *Simulium asishi* Datta from *Simulium nodosum* Puri (Diptera: Simuliidae) in West Bengal, India. *Proceedings of the Zoological Society, 45*, 123–127.

Datta, M. (1997). Insecta: Diptera: Simuliidae. In *State Fauna Series: Fauna of West Bengal*, Part 3. Pp. 127–162.

Datta, M., Dey, R. K., & Paul, A. K. (1975). Rare black fly species (Diptera: Simuliidae) of the subgenus *Eusimulium* Roubaud from the Darjeeling area, India. *Proceedings of the Indian Academy of Sciences-Section B, 82B*, 92–99.

Datta, M., & Pal, T. K. (1975). Few rare black fly species (Diptera: Simuliidae) of the subgenus *Simulium* Latreille (s. str.) from the Darjeeling area, India. *Proceedings of the Indian Academy of Sciences, 81B*, 154–161.

Davies, D. M., & Györkös, H. (1987a). The Simuliidae (Diptera) of Sri Lanka. Descriptions of species in the subgenera *Eusimulium* and *Gomphostilbia* of the genus *Simulium*. *Canadian Journal of Zoology, 65*, 1483–1502.

Davies, D. M., & Györkös, H. (1987b). The Simuliidae (Diptera) of Sri Lanka. Descriptions of three new species of *Simulium* (*Simulium*). *Canadian Journal of Zoology, 65*, 2734–2746.

Davies, D. M., & Györkös, H. (1988a). The Simuliidae (Diptera) of Sri Lanka. Description of a new species of *Simulium* (*Byssodon*). *Canadian Journal of Zoology, 66*, 2150–2154.

Davies, D. M., & Györkös, H. (1988b). The Simuliidae (Diptera) of Sri Lanka. Description of a new species of *Simulium* (*Morops*). *Canadian Journal of Zoology, 66*, 605–610.

Davies, D. M., & Györkös, H. (1992). The Simuliidae (Diptera) of Sri Lanka. Description of additional species of *Simulium* (*Simulium*), with a key for Sri Lankan species in the subgenus and a checklist for the country. *Canadian Journal of Zoology, 70*, 1029–1046.

Delfinado, M. D. (1962). The Philippine species of *Simulium* (Diptera, Simuliidae). *Philippine Journal of Science, 89*, 47–62.

Delfinado, M. D. (1969). Notes on Philippine black flies (Diptera: Simuliidae). *Journal of Medical Entomology, 6*, 199–207.

Delfinado, M. D. (1971). Some Simuliidae and Curtonotidae from the Philippines and the Bismarck Islands (Insecta, Diptera). *Steenstrupia, 1*, 131–139.

De Meijere, J. C. H. (1907). Studien über Südostasiatische Dipteren I. *Tijdschrift voor Entomologie, 50*, 196–264.

De Meijere, J. C. H. (1913). Studien über Südostasiatische Dipteren II. *Tijdschrift voor Entomologie, 56*, 317–354.

Deng, C. Y., Xue, Q. L., Zhang, Y. Z., & Chen, H. B. (1994). A new species and new records of blackflies from Tibet, China. Pp. 10–16. In *Sixtieth Anniversary of the Funding of the China Zoological Society Memorial Volume dedicated to the Hundredth Anniversary of the Birth of the late Professor Sisan Chen*. China Science Technology Publication. [In Chinese with an English summary].

Deng, C. Y., Xue, H. D., & Chen, H. B. (2005). One new species of the subgenus *Montisimulium* from Yunnan. *Chinese Journal of Vector Biology and Control, 16*, 191–192. (In Chinese with an English abstract).

Dickerson, R. E. (1928). *Distribution of life in the Philippines* (p. 322). Philippine Bureau of Sciences Monograph No. 21.

Edwards, F. W. (1925). Diptera Nematocera from Dutch East Indies. *Treubia, 6*, 154–157.

Edwards, F. W. (1927). Two new *Simulium* from Kashmir. *Bulletin of Entomological Research, 18*, 169–170.

Edwards, F. W. (1928). Diptera Nematocera from the Federated Malay States Museums. *Journal of the Federated Malay States Museums, 14*, 1–139.

Edwards, F. W. (1929). Philippine nematocerous Diptera II. *Notulae Entomologicae, 9*, 1–14.

Edwards, F. W. (1933). Diptera Nematocera from Mount Kinabalu. *Journal of the Federated Malay States Museums, 17*, 223–296.

Edwards, F. W. (1934). The Simuliidae (Diptera) of Java and Sumatra. Deutsche Limnologische Sunda-Expedition. *Archiv für Hydrobiologie, 13*, 92–138.

Enderlein, G. (1921). Das System del kriebelmücken (Simuliidae). *Deutsche Tierärztliche Wochenshrift, 29*, 197–200.

Friederichs, K. (1925). Beobachtungen an Simuliiden in Ost-Java. *Archiv fur Schiffs-und Tropenhygiene, 29*, 119–125.

Fukuda, M., Choochote, W., Bain, O., Aoki, C., & Takaoka, H. (2003). Natural infections with filarial larvae in two species of black flies (Diptera: Simuliidae) in northern Thailand. *Japanese Journal of Tropical Medicine and Hygiene, 31*(2), 99–102.

Fukuda, M., Otsuka, Y., Uni, S., Bain, O., & Takaoka, H. (2010). Molecular identification of infective larvae of three species of *Onchocerca* found in wild-caught females of *Simulium bidentatum* in Japan. *Parasite, 17*(1), 39–45.

Fukuda, M., Takaoka, H., Uni, S., & Bain, O. (2008). Infective larvae of five *Onchocerca* species from experimentally infected *Simulium* species in an area of zoonotic onchocerciasis in Japan. *Parasite, 15*, 111–119.

Guo, X., Zhang, H., & An, J. (2011). A new species of *Simulium* (*Gomphostilbia*) from Hainan Island, China (Diptera: Simuliidae). *Oriental Insects, 45*, 87–92.

Hadi, U. K., & Takaoka, H. (1995). Effects of constant temperatures on oviposition and immature development of *Simulium bidentatum* (Diptera: Simuliidae), a vector of bovine *Onchocerca* (Nematoda: Onchocercidae) in central Kyushu, Japan. *Journal of Medical Entomology, 32*, 801–806.

Hadi, U. K., Takaoka, H., & Aoki, C. (1995). Larval salivary gland chromosomes of the blackfly, *Simulium* (*Gomphostilbia*) *yaeyamaense* (Diptera: Simuliidae) from Ryukyu Islands, Japan. *Japanese Journal of Sanitary Zoology, 46*, 235–239.

Hadi, U. K., Aoki, C., & Takaoka, H. (1996a). Leg color variation of female *Simulium aokii* (Diptera: Simuliidae). *Japanese Journal of Tropical Medicine and Hygiene, 24*, 129–131.

Hadi, U. K., Takaoka, H., Kondo, K., & Hirai, H. (1996b). Larval salivary gland chromosomes of *Simulium* (*Simulium*) *eximium* (Diptera: Simuliidae) from Java and Sumatra, Indonesia, with implication of sibling speciation. *Medical Entomology and Zoology, 47*, 139–144.

Halt, B. G., Lessard, J. P., Borregaard, M. K., Fritz, S. A., Araujo, M. B., Dimitrov, D., Fabre, P. H., Graham, C. M., Graves, G. R., Jensson, K. A., Nogues-Bravo, D., Wang, Z. H., Whittaker, R. J., Fjeldsa, J., & Rahbek, C. (2013). An update of Wallace's zoological regions of the world. *Science, 339*(6115), 74–78. https://doi.org/10.1126/scien-78ce.1228282

Henry, W., Dey, S. K., Varma, R., Thapa, S., & Procunier, W. S. (2010). Polytene chromosomes of an Indian Himalayan black fly *Simulium* (*Nevermannia*) *praelargum* (Diptera: Simuliidae). *Current Zoology, 56*, 436–443.

Henry, W., Thapa, S., Adler, P. H., Subrata, D., & Varma, R. (2011). Cytotaxonomy of *Simulium (Montisimulium) ghoomense* (Diptera: Simuliidae) from the Darjeeling Hills, India. *Zootaxa, 2872*, 49–57.

Hew, Y. X., Ya'cob, Z., Chen, C. D., Lau, K. W., Sofian-Azirun, M., Putt, Q. Y., Izwan-Anas, N., Hadi, U. K., Suana, I. W., Takaoka, H., & Low, V. L. (2023a). DNA barcoding of black flies (Diptera: Simuliidae) in Indonesia. *Parasites & Vectors, 16*(1), 248–262. https://doi.org/10.1186/s13071-023-05875-1

Hew, Y. X., Ya'cob, Z., Chen, C. D., Lau, K. W., Sofian-Azirun, M., Halim, M. R. A., Putt, Q. Y., Tan, T. K., Hadi, U. K., Suana, I. W., Takaoka, H., & Low, V. L. (2023b). Co-occurrence of dual lineages within *Simulium (Gomphostilbia) atratum* De Meijere in the Indonesian Archipelago along Wallace's Line. *Acta Tropica, 250*(7), 107097. https://doi.org/10.1016/j.actatropica.2023.107097

Huang, Y. T., & Adler, P. H. (2011). Chromosomal relationships of *Simulium suzukii*, an Old World member of the *Simulium tuberosum* species group (Diptera: Simuliidae). *Medical Entomology and Zoology, 62*, 23–30.

Huang, Y. T., Phasuk, J., Champaisaeng, J., & Adler, P. H. (2010). A new species of black fly in the subgenus *Simulium* (Diptera: Simuliidae) from Thailand. *Medical Entomology and Zoology, 61*, 49–58.

Huang, F., Srisuka, W., Aupalee, K., Streit, A., Fukuda, M., Pitasawat, B., Junkum, A., Saingamsook, J., Somboon, P., Takaoka, H., & Saeung, A. (2021). Diversity of nematodes infecting the human-biting black fly species, *Simulium nigrogilvum* (Diptera: Simuliidae) in central Thailand. *Acta Tropica, 224*, 1–10. (106140).

Huang, F., Srisuka, W., Aupalee, K., Yasanga, T., Phuackchantuck, R., Pitasawat, B., Junkum, A., Limsopatham, K., Sanit, S., Saingamsook, J., Takaoka, H., & Saeung, A. (2022). Ultrastructure of sensilla on the antennae and maxillary palps of the human-biting black flies, *Simulium nigrogilvum* and *Simulium umphangense*, (Diptera: Simuliidae) in Thailand. *Acta Tropica, 232*, 1–10. (166494).

Huang, Y. T., & Takaoka, H. (2008a). A new species of *Simulium* (*Gomphostilbia*) (Diptera: Simuliidae) from Taiwan. *Medical Entomology and Zoology, 59*, 171–179.

Huang, Y. T., & Takaoka, H. (2008b). Description of the female of *Simulium* (*Simulium*) *puliense* (Diptera: Simuliidae) from Taiwan. *Medical Entomology and Zoology, 59*, 305–308.

Huang, Y. T., & Takaoka, H. (2008c). *Simulium* (*Simulium*) *pingtungense*, a new black fly (Diptera: Simuliidae) from Taiwan. *Medical Entomology and Zoology, 59*, 309–317.

Huang, Y. T., & Takaoka, H. (2009a). *Simulium (Simulium) serenum*, a new species of black fly (Diptera: Simuliidae) from Taiwan. *Medical Entomology and Zoology, 60*, 33–38.

Huang, Y. T., & Takaoka, H. (2009b). Description of the female and mature larva of *Simulium (Simulium) ufengense* (Diptera: Simuliidae) from Taiwan. *Medical Entomology and Zoology, 60*, 91–95.

Huang, Y. T., Takaoka, H., Otsuka, Y., & Aoki, C. (2011). *Simulium (Gomphostilbia) taitungense*, a new species of black fly (Diptera: Simuliidae) from Taiwan, with description of the male of *S. (G.) tuenense* Takaoka. *Tropical Biomedicine, 28*, 577–588.

Ishii, Y., Choochote, W., Bain, O., Fukuda, M., Otsuka, Y., & Takaoka, H. (2008). Seasonal and diurnal biting activities and zoonotic filarial infections of two *Simulium* species (Diptera: Simuliidae) in northern Thailand. *Parasite, 15*, 121–129.

Izwan-Anas, N., Low, V. L., Ya'cob, Z., Lourdes, E. Y., Halim, M. R. A., Sofian-Azirun, M., Takaoka, H., & Adler, P. H. (2023). DNA barcodes and species boundaries of black flies (Diptera: Simuliidae) in Malaysia. *Arthropod Systematics & Phylogeny, 81*, 931–943. https://doi.org/10.3897/asp.81.e104426

Jariyapan, N., Takaoka, H., Choochote, W., Poovorawan, Y., Otsuka, Y., Jitpakdi, A., & Junkum, A. (2006). Morphology and electrophoretic protein profiles of female salivary glands in four Oriental black fly species (Diptera: Simuliidae). *Journal of Vector Ecology, 31*(2), 406–411.

Jitklang, S., Ahantarig, A., Kuvangkadilok, C., Baimai, V., & Adler, P. H. (2012). Parasites of larval black flies (Diptera: Simuliidae) in Thailand. *Songklanakarin Journal of Science and Technology, 34*, 597–599.

Jitklang, S., & Kuvangkadilok, C. (2008). (2007). A new species of *Simulium (Gomphostilbia)* (Diptera: Simuliidae) from southern Thailand, with description of its polytene chromosomes. *Studia Dipterologica, 14*, 1–7.

Jitklang, S., Kuvangkadilok, C., Baimai, V., Takaoka, H., & Adler, P. H. (2008). Cytogenetics and morphotaxonomy of the *Simulium (Gomphostilbia) ceylonicum* species group (Diptera: Simuliidae) in Thailand. *Zootaxa, 1917*, 1–28.

Jomkumsing, P., Tangkawanit, U., Wongpakam, K., & Pramual, P. (2019). Who is biting you? DNA barcodes reveal cryptic diversity in human-biting black flies (Diptera: Simuliidae). *Acta Tropica, 196*, 22–29.

Jumpato, W., Tangkawanit, U., Wongpakam, K., & Pramual, P. (2018). Molecular detection of *Leucocytozoon* (Apicomplexa: Haemosporida) in black flies (Diptera: Simuliidae) from Thailand. *Acta Tropica, 190*, 228–234.

Kani, T. (1944). On pupae of 11 species of the family Simuliidae in Kamogawa River, Kyoto. *Physiology and Ecology Research, Ohtsu Lake laboratory, Department of Zoology, Faculty of Science, Kyoto University, 11*, 1–8. (In Japanese).

Kang, Z., Zhang, C. L., & Chen, H. B. (2007). A new species of *Simulium (Gomphostilbia)* from Jiangxi Province, China (Diptera: Simuliidae). *Acta Parasitologica Et Medica Entomologica Sinica, 14*, 185–187.

Kuvangkadilok, C., Boonkemtong, C., & Phayuhasena, S. (1998). C-banding in polytene chromosomes of six *Simulium* species (Diptera: Simuliidae) from Doi Inthanon National Park, northern Thailand. *Journal of the Science Society Thailand, 24*, 215–230.

Kuvangkadilok, C., Lualon, U., & Baimai, V. (2008). Cytotaxonomy of *Simulium siamense* Takaoka & Suzuki (Diptera: Simuliidae) in Thailand. *Genome, 51*, 972–987.

Kuvangkadilok, C., Phayuhasena, S., & Baimai, V. (1999a). Population cytogenetic studies on *Simulium feuerborni* Edwards (Diptera: Simuliidae) from northern Thailand. *Genome, 42*, 80–86.

Kuvangkadilok, C., Phayuhasena, S., & Boonkemtong, C. (1999b). Larval polytene chromosomes of five species of black flies (Diptera: Simuliidae) from Doi Inthanon National Park, northern Thailand. *Cytologia, 64*, 197–207.

Kuvangkadilok, C., & Takaoka, H. (2000). Taxonomic notes on Simuliidae (Diptera) from Thailand: Description of a new species and new distributional records of nine known species. *Japanese Journal of Tropical Medicine and Hygiene, 28*, 167–175.

Lewis, D. J. (1964). Diptera from Nepal. Simuliidae. *Bulletin of the British Museum (Natural History), 15*, 291–294.

Lewis, D. J. (1973). The Simuliidae (Diptera) of Pakistan. *Bulletin of Entomological Research, 62*, 453–470.

Lewis, D. J. (1974). Man-biting Simuliidae (Diptera) of Northern India. *Israel Journal of Entomology, 9*, 23–53.

Long, Z. M., An, J. Y., Hao, B. S., Yan, G., & Yuan, W. H. (1994). Two new blackflies of *Simulium* (*Gomphostilbia*) from Hainan Island, China (Diptera, Simuliidae). *Chinese Journal of Vector Biology and Control, 5*, 405–409.

Lourdes, E. Y., Ya'cob, Z., Low, V. L., Takaoka, H., & Adler, P. H. (2022). Natural infections and distributions of parasitic mermithids (Nematoda: Mermithidae) infecting larval black flies (Diptera: Simuliidae) in tropical streams of Malaysia. *Acta Tropica, 230*(1), 1–7. (106386).

Low, V. L., Adler, P. H., Sofian-Azirun, M., Srisuka, W., Saeung, A., Huang, Y. T., Hadi, U. K., Pham, X. D., & Takaoka, H. (2015a). Tests of conspecificity for allopatric vectors: *Simulium nodosum* and *Simulium shirakii* (Diptera: Simuliidae) in Asia. *Parasites & Vectors, 8*(295), 1–6. https://doi.org/10.1186/s13071-015-1911-5

Low, V. L., Adler, P. H., Takaoka, H., Ya'cob, Z., Lim, P. E., Tan, T. K., Lim, Y. A., Chen, C. D., Norma-Rashid, Y., & Sofian-Azirun, M. (2014). Mitochondrial markers reveal high genetic diversity but low genetic differentiation in the black fly *Simulium tani* Takaoka & Davies along an elevational gradient in Malaysia. *PLoS One, 9*(6), 1–10. e100512. https://doi.org/10.1371/journal.pone.0100512

Low, V. L., Srisuka, W., Saeung, A., Tan, T. K., Ya'cob, Z., Yeong, Y. S., & Takaoka, H. (2020). DNA barcoding of *Simulium asakoae* Takaoka & Davies (Diptera: Simuliidae) from northern Thailand. *Journal of Medical Entomology, 57*, 1675–1678.

Low, V. L., Takaoka, H., Adler, P. H., Tan, T. K., Weng, C. H., Chen, C. Y., Lim, Y. A. L., Ya'cob, Z., Chen, C. D., Sofian-Azirun, M., & Wang, D. (2018). A novel molecular and chromosomal lineage of the anthropophilic *Simulium* (*Simulium*) *rufibasis* subgroup (Diptera: Simuliidae) in Taiwan. *Parasitology Research, 117*, 3137–3143.

Low, V. L., Takaoka, H., Adler, P. H., Ya'cob, Z., Norma-Rashid, Y., Chen, C. D., & Sofian-Azirun, M. (2015b). A multilocus approach resolves phylogenetic relationships of the *Simulium asakoae* and *Simulium ceylonicum* species-groups (Diptera: Simuliidae) in Malaysia: evidence for distinct evolutionary lineages. *Medical and Veterinary Entomology, 29*, 330–337. https://doi.org/10.1111/mve.12120

Low, V. L., Takaoka, H., Pramual, P., Adler, P. H., Ya'cob, Z., Chen, C. D., Yotopranoto, S., Zaid, A., Hadi, U. K., & Sofian-Azirun, M. (2016a). Three taxa in one: cryptic diversity in the blackfly *Simulium nobile* (Diptera: Simuliidae) in South East Asia. *Journal of Medical Entomology, 53*, 972–976.

Low, V. L., Takaoka, H., Pramuel, P., Adler, P. H., Ya'cob, Z., Huang, Y. T., Pham, X. D., Ramli, R., Chen, C. D., Wannaket, A., & Sofian-Azirun, M. (2016b). Delineating taxonomic boundaries in the largest species complex of black flies (Simuliidae) in the Oriental Region. *Scientific Reports, 6*, 1–8. (20346).

Lundström, C. (1911). Beiträge zur Kenntnis der Dipteren Finlands. VII. Melusinidae (Simuliidae). *Acta Societatis pro Fauna et Flora Fennica, 34* (part 12), 1–23.

Maskey, M. (1987[1985]). A new species of black fly (Diptera: Simuliidae) from Nepal. *Journal of Natural History Museum (Nepal), 9*, 15–22.

Maskey, M. (1989). New species of black flies (Diptera: Simuliidae) from Nepal. *Journal of Entomology Nepal, 1*, 37–45.

Maskey, M. (1998). New species of subgenus *Odagmia* (Diptera: Simuliidae) from Nepal. *Journal of Natural History Museum (Nepal), 17*, 7–22.

Matsumura, S. (1931). Family Simuliidae. 6000 Ill. Ins. Jap. Empire. pp. 406–407. Toe Shoin, Tokyo (In Japanese).

Meigen, J. W. (1804). *Klassifikazion und Beschreibung der europäischen zweiflügeligen Insekten (Diptera Linn.). Erster Band* (p. 314). K. Reichard, Braunschweig [=Brunswick].

Meigen, J. W. (1818). *Systematische Beschreibung der Bekannten Europäischen Zweiflügeligen Insekten 1* (p. 332). Forstmann.

Nie, W. Z., Chen, A. J., Li, J. C., An, J. Y., & Li, D. X. (2006). A new species of *Simulium* (*Gomphostilbia*) from Sumatra, Indonesia (Diptera: Simuliidae). *Acta Parasitology et Medica Entomologica Sinica, 13*(3), 158–160.

Ogata, K. (1956). Notes on Simuliidae of the Ryukyu Islands (Diptera). *Japanese Journal of Medical Science & Biology, 9*, 59–69.

Ogata, K. (1966). Additional notes on Simuliidae of the Ryukyu Islands (Diptera). *Kontyu, 34*, 123–130.

Ogata, K., & Sasa, M. (1954). Taxonomic notes on Simuliidae or black flies of Japan, with special references on the subgenera *Eusimulium* Roubaud and *Nevermannia* Enderlein (Diptera). *The Japanese Journal of Experimental Medicine, 24*, 325–333.

Ogata, K., Sasa, M., & Suzuki, T. (1956). *The Japanese black flies and their control* (p. 126). DDT Kyokai.

Orii, T., Uemoto, K., & Onishi, O. (1969). Key to larval stage of black flies (Simuliidae) of Japan. *Sanitary Injurious Insects, 13*, 1–13. (In Japanese).

Otsuka, Y., Aoki, C., Choochote, W., De las Llagas, L., & Takaoka, H. (2007). Phylogenetic analysis of three subgenera: *Asiosimulium, Daviesellum* and *Wallacellum*, of the genus *Simulium* s. l. endemic in the Oriental Region. *Medical Entomology and Zoology, 58*, 329–333.

Otsuka, Y., Aoki, C., Saito, K., Hadi, U. K., Suzuki, H., & Takaoka, H. (2001). Phylogenetic analyses of a blackfly subgenus *Simulium* (*Nevermannia*) based on mithochondrial 16S ribosomal RNA gene sequences. *Japanese Journal of Tropical Medicine and Hygiene, 29*, 261–266.

Otsuka, Y., & Takaoka, H. (2015). Biogeographical distribution and phylogenetic analysis of *Simulium* (*Wallacellum*) (Diptera: Simuliidae) based on the mitochondrial sequences. *South Pacific Studies, 35*, 45–56.

Otsuka, Y., Takaoka, H., Aoki, C., & Choochote, W. (2003). Phylogenetic analysis of the subgenus *Himalayum* within the genus *Simulium* s. l. (Diptera: Simuliidae) using mitochondrial 16S rRNA gene sequences. *Medical Entomology and Zoology, 54*, 113–120.

Pangjanda, S., & Pramual, P. (2017). Tests of conspecificity for closely related black fly (Diptera: Simuliidae) species of the *Simulium striatum* group in Thailand. *Zootaxa, 4231*(3), 421–430.

Pham, X. D. (1998). New records of six black fly species (Diptera: Simuliidae) from Vietnam. *Medical Entomology and Zoology, 49*, 121–123.

Pham, X. D. (1999). Additional records of three black fly species (Diptera: Simuliidae) from Vietnam. *Medical Entomology and Zoology, 50*, 335–336.

Phasuk, J., Chanpaisaeng, J., Adler, P. H., & Courtney, G. W. (2005). Chromosomal and morphological taxonomy of larvae of *Simulium* (*Gomphostilbia*) (Diptera: Simuliidae) in Thailand. *Zootaxa, 1052*, 49–60.

Phayuhasena, S., Colgan, D. J., Kuvangkadilok, C., Pramual, P., & Baimai, V. (2010). Phylogenetic relationships among the black fly species (Diptera: Simuliidae) of Thailand based on multiple gene sequences. *Genetica, 138*, 633–648.

Pramual, P. (2014). Association of black fly (Diptera: Simuliidae) life stages using DNA barcode. *Journal of Asia-Pacific Entomology, 17*(3), 549–554.

Pramual, P., Jomkumsing, P., Tangkawanit, U., Wongpakam, K., Bunchom, N., & Takaoka, H. (2022a). Description of the male of *Simulium triglobus* Takaoka & Kuvangkadilok (Diptera: Simuliidae) from Thailand. *Tropical Biomedicine, 39*, 60–65.

Pramual, P., Jomkumsing, P., Thongyan, T., Wongpakam, K., & Takaoka, H. (2022b). A new species of black fly, formerly cytoform C of the *Simulium angulistylum* complex (Diptera: Simuliidae), from the high mountain in northeastern Thailand. *Tropical Biomedicine, 39*, 281–290.

Pramual, P., & Kuvangkadilok, C. (2012). Integrated cytogenetic, ecological, and DNA barcode study reveals cryptic diversity in *Simulium* (*Gomphostilbia*) *angulistylum* (Diptera: Simuliidae). *Genome, 55*, 1–12.

Pramual, P., & Tangkawanit, U. (2008). A new species of *Simulium* (*Gomphostilbia*) (Diptera: Simuliidae) from Northeastern Thailand. *Medical Entomology and Zoology, 59*, 297–303.

Pramual, P., Thaijarern, J., Sofian-Azirun, M., Ya'cob, Z., Hadi, U. K., & Takaoka, H. (2015). Cytogenetic and molecular evidence of additional cryptic diversity in high elevation black fly *Simulium feuerborni* (Diptera: Simuliidae) populations in Southeast Asia. *Journal of Medical Entomology, 52*(5), 829–836.

Pramual, P., Thaijarern, J., & Wongpakam, K. (2016). DNA barcoding of human-biting black flies (Diptera: Simuliidae) in Thailand. *Acta Tropica, 164*, 33–40.

Pramual, P., Tangkawanit, U., Kunprom, C., Vaisusk, K., Chatan, W., Wongpakam, K., & Thongboonma, S. (2020). Seasonal population dynamics and a role as natural vector of *Leucocytozoon* of black fly, *Simulium chumpornense* Takaoka & Kuvangkadilok. *Acta Tropica, 211*(6), 1–8. (105617).

Pramual, P., Jomkumsing, P., Wongpakam, K., & Wongwian, P. (2021). DNA barcoding of tropical black flies (Diptera: Simuliidae) in Thailand: one decade of progress. *Acta Tropica, 224*, 1–11. (106116).

Pramual, P., & Wongpakam, K. (2011). Cytogenetics of *Simulium siamense* Takaoka and Suzuki, 1984 (Diptera: Simuliidae) in northern Thailand. *Aquatic Insects, 33*, 171–184.

Pramual, P., & Wongpakam, K. (2013). Population genetics of the high elevation black fly *Simulium (Nevermannia) feuerborni* Edwards in Thailand. *Entomological Science, 16*, 298–308.

Pramual, P., Wongpakam, K., & Adler, P. H. (2011). Cryptic biodiversity and phylogenetic relationships revealed by DNA barcoding of Oriental black flies in the subgenus *Gomphostilbia* (Diptera: Simuliidae). *Genome, 54*, 1–9.

Puri, I. M. (1932a). Studies on Indian Simuliidae. Part I. *Simulium himalayense* sp. n.; *Simulium gurneyae* Senior-White; and *Simulium nilgiricum* sp. n. *The Indian Journal of Medical Research, 19*, 883–898.

Puri, I. M. (1932b). Studies on Indian Simuliidae. Part II. Descriptions of males, females, and pupae of *Simulium rufibasis* Brunetti, its variety *fasciatum* nov. var. and of three new species from the Himalayas. *The Indian Journal of Medical Research, 19*, 899–913.

Puri, I. M. (1932c). Studies on Indian Simuliidae. Part III. Descriptions of males, females, and pupae of *Simulium griseifrons* Brunetti (1911), and of four new species with striped thorax. *The Indian Journal of Medical Research, 19*, 1125–1143.

Puri, I. M. (1932d). Studies on Indian Simuliidae. Part IV. Descriptions of two new species from North-East India *Simulium howletti* sp. n. and *Simulium hirtipannus* sp. n., with a note on *S. ornatum* Meigen. *The Indian Journal of Medical Research, 20*, 505–514.

Puri, I. M. (1932e). Studies on Indian Simuliidae. Part V. Species and varieties of the *striatum* series. *The Indian Journal of Medical Research, 20*, 515–532.

Puri, I. M. (1933a). Studies on Indian Simuliidae. Part VI. Descriptions of males, females, and pupae of two new species from Palni Hills and of male and pupa of *S. tenuitarsus* sp. n. from Bengal Terai. *The Indian Journal of Medical Research, 20*, 803–812.

Puri, I. M. (1933b). Studies on Indian Simuliidae. Part VII. Descriptions of larva, pupa and female of *Simulium nodosum* sp. nov. with an appendix dealing with *S. novolineatum* nov. nom. (= *S. lineatum* Puri). *The Indian Journal of Medical Research, 20*, 813–817.

Puri, I. M. (1933c). Studies on Indian Simuliidae. Part VIII. Descriptions of larvae, pupae males and females of *S. aureohirtum* Brunetti and *S. aureum* Fries. *The Indian Journal of Medical Research, 21*, 1–10.

Puri, I. M. (1933d). Studies on Indian Simuliidae. Part IX. *S. equinum* var. *mediterraneum* Puri and *S. paraequinum* sp. n. *The Indian Journal of Medical Research, 21*, 11–16.

Putt, Q.Y., Ya'cob, Z., Adler, P.H., Chen, C.D., Hew, Y.X., Izwan-Anas, N., Lau, K.W., Sofian-Azirun, M., Pham, X. D., Takaoka, H., & Low, V. L. (2023). From bites to barcodes: Uncovering the hidden diversity of black flies (Diptera: Simuliidae) in Vietnam. *Parasites & Vectors, 16*(1), 266–287. https://doi.org/10.1186/s13071-023-05892-0

Rubtsov, I. A. (1940). Blackflies. *Fauna USSR, Diptera, 6*(6), 145–148.

Rubtsov, I. A. (1956). Blackflies (fam. Simuliidae). Fauna of the USSR. New Series 64, Insects, Diptera. Volume 6. Part 6. 859 pp. In *Academia Nauka SSSR*. [In Russian, and translated to English and republished in 1989 by Amerind Publishing Co., New Delhi, 1,042 pp.].

Rubtsov, I. A. (1959–1964). Simuliidae (Melusinidae). In E. Lindner (Ed.), *Die Fliegen der Palaearktischen Region* (Vol. 14, pp. 1–689).

Rubtsov, I. A. (1974). On the evolution, phylogeny and classification of the family of blackflies (Simuliidae, Diptera) (p. 83). *Trudy zoologicheskogo Instituta, 53*, 230–281. [In Russian; English translation: Evolution, phylogeny and classification of the family Simuliidae (Diptera). British Library Lending Division, Boston Spa, UK.].

Saeung, A., Srisuka, W., Aupalee, K., Fukuda, M., Otsuka, Y., Taai, K., Maleewong, W., & Takaoka, H. (2020). Natural infections with larvae of *Onchocerca* species type I in the human-biting black fly, *Simulium nigrogilvum* (Diptera: Simuliidae), in western Thailand. *Acta Tropica, 204*, 1–7. (105344).

Saeung, A., Srisuka, W., Maleewong, W., Low, V. L., & Takaoka, H. (2017). Descriptions of the female and larva of *Simulium* (*Gomphostilbia*) *udomi* from Thailand, and their transfer to the *Simulium asakoae* species-group. *Acta Tropica, 172*, 14–19. https://doi.org/10.1016/j.actatropica.2017.04.014

Sato, H., Saito, K., & Takaoka, H. (2005). Revised description of *Simulium* (*Nevermannia*) *sasai*, a rare and poorly known species in Japan (Diptera: Simuliidae). *Medical Entomology and Zoology, 56*, 299–307.

Senior-White, R. (1922). Notes on Indian Diptera. *Memoirs of Department of Agriculture in India (Entomological Series), 7*, 126–131.

Shiraki, T. (1935). Simuliidae of the Japanese Empire. *Memoirs of the Faculty of Agriculture, Taihoku Imperial University, 16*, 1–90.

Shrestha, S., & Takaoka, H. (2009). New records of ten species of black flies (Diptera: Simuliidae) from Nepal. *Medical Entomology and Zoology, 60*, 253–258.

Simuliid Research Group of the Japan Society of Sanitary Zoology. (1974). List of Japanese black flies. *Japanese Journal of Sanitary Zoology, 25*, 191–193. (In Japanese).

Smart, J., & Clifford, E. A. (1965). Simuliidae (Diptera) of the territory of Papua and New Guinea. *Pacific Insects, 7*, 505–619.

Smart, J., & Clifford, E. A. (1969). Simuliidae (Diptera) of Sabah (British North Borneo). *Zoological Journal of the Linnean Society, 48*, 9–47.

Srisuka, W., Aupalee, K., Fukuda, M., Saeung, A., & Takaoka, H. (2022a). Morphological description and phylogenetic position of a new cryptic species in the *Simulium asakoae* species-group, *Simulium* (*Gomphostilbia*) *mokroense* from Western Thailand. *Acta Tropica, 228*, 1–10. (106313).

Srisuka, W., Aupalee, K., Low, V. L., Ya'cob, Z., Fukuda, M., Saeung, A., & Takaoka, H. (2021). A new species of *Simulium* (*Gomphostilbia*) (Diptera: Simuliidae) from Northern Thailand, with its genetic relationship in the *S. asakoae* species-group. *Acta Tropica, 218*, 1–8. (105889).

Srisuka, W., Aupalee, K., Otsuka, Y., Fukuda, M., Takaoka, H., & Saeung, A. (2022b). A new species of *Simulium* (*Gomphostilbia*) (Diptera: Simuliidae) from Thailand, with a key to identify females of 14 species of the *Simulium varicorne* species-group. *ZooKeys, 1083*, 1–12.

Srisuka, W., Takaoka, H., Aupalee, K., & Saeung, A. (2023a). Discovery of a new black fly species in the *Simulium* (*Simulium*) *striatum* species-group (Diptera: Simuliidae) from northeastern Thailand and phylogenetic relationships within its species-group. *Acta Tropica, 238*, 1–10. (106790).

Srisuka, W., Takaoka, H., Aupalee, K., & Saeung, A. (2023b). *Simulium* (*Gomphostilbia*) *wijiti*, a new species of blackfly (Diptera: Simuliidae) from northern Thailand and its phylogenetic relationships with related species in the *Simulium ceylonicum* species-group. *Acta Tropica, 244*, 1–10. (106947).

Srisuka, W., Takaoka, H., Fukuda, M., Otsuka, Y., & Saeung, A. (2019). A new species of the *Simulium* (*Gomphostilbia*) (Diptera: Simuliidae) from Thailand, with its phylogenetic relationships with related species in the *Simulium asakoae* species-group. *Acta Tropica, 197*, 1–6. (105043).

Srisuka, W., Takaoka, H., Otsuka, Y., Fukuda, M., Thongsahuan, S., Choochote, W., & Saeung, A. (2015b). Seasonal biodiversity of black flies (Diptera: Simuliidae) and evaluation of ecological factors influencing species distribution at Doi Pha Hom Pok National Park, Thailand. *Acta Tropica, 149*, 212–219.

Srisuka, W., Takaoka, H., Otsuka, Y., Fukuda, M., Thongsahuan, S., Taai, K., Choochote, W., & Saeung, A. (2017). Biodiversity, seasonal abundance and distribution of black flies (Diptera: Simuliidae) in six ecologically and geographically different regions in Thailand. *Parasites & Vectors, 10*(574), 1–12. https://doi.org/10.1186/s13071-017-2492-y

Srisuka, W., Takaoka, H., & Saeung, A. (2015a). Description of the male, pupa and larva of *Simulium (Asiosimulium) wanchaii* (Diptera: Simuliidae) from Thailand, with keys to identify four species of the subgenus *Asiosimulium*. *Tropical Biomedicine, 32*, 504–510.

Stauder, A., & Kiel, F. (2004). New observations on phoretic associations of Simuliidae (Diptera) with Heptageniidae (Ephemeroptera) in India. *Acta Zoologica Universitatis Comenianae, 46*, 15–21.

Stone, A. (1964). Diptera: Simuliidae. *Insects of Micronesia, 12*, 629–635.

Summers, S. L. M. (1911). Notes from the Entomological Department of the London School of Tropical Medicine. No. II. Description of a new species of *Simulium* from the Siamese hills. *Annals and Magazine of Natural History, 8*, 586–588.

Sun, Y. X. (2009). Description of five new species of *Simulium* (*Gomphostilbia*) Enderlein subgenus from China (Diptera: Simuliidae). *Chinese Journal of Vector Biology and Control, 20*, 545–549.

Takahasi, H. (1941). Description of a new species of Simuliidae from Nippon, with other notes. *Insecta Matsumurana, 15*, 86–88.

Takaoka, H. (1972). A new species of Simuliidae from Yonakuni Island, Ryukyu Islands, Japan (Diptera: Simuliidae). *Journal of Medical Entomology, 9*, 521–523.

Takaoka, H. (1973). Descriptions of 2 new species of blackflies, *Simulium* (*Gomphostilbia*) *tokarense* and *S.* (*Eusimulium*) *morisonoi* (Diptera: Simuliidae), from the Tokara Islands, Japan. *Japanese Journal of Sanitary Zoology, 23*, 201–207.

Takaoka, H. (1976a). Studies on blackflies of the Nansei Islands, Japan (Simuliidae; Diptera). I. On six species of the subgenus *Eusimulium* Roubaud, with the descriptions of *Simulium* (*E.*) *satsumense* sp. nov. and *S.* (*E.*) *subcostatum koshikiense* ssp. nov. *Japanese Journal of Sanitary Zoology, 27*, 163–180.

Takaoka, H. (1976b). Studies on blackflies of the Nansei Islands, Japan (Simuliidae; Diptera). II. On six species of the subgenera *Gomphostilbia* Enderlein, *Morops* Enderlein, *Odagmia* Enderlein and *Gnus* Rubzov, with the description of *Simulium* (*Gomphostilbia*) *okinawense* sp. nov. *Japanese Journal of Sanitary Zoology, 27*, 385–398.

Takaoka, H. (1977a). Studies on blackflies of the Nansei Islands, Japan (Simuliidae; Diptera). III. On six species of the subgenus *Simulium* Latreille. *Japanese Journal of Sanitary Zoology, 28*(2), 193–217.

Takaoka, H. (1977b). Studies on blackflies of the Nansei Islands, Japan (Simuliidae; Diptera). IV. Keys to females, males, pupae and larvae of all the 18 species of the genus *Simulium* of the Nansei Islands. *Japanese Journal of Sanitary Zoology, 28*, 219–224.

Takaoka, H. (1979). The black flies of Taiwan (Diptera: Simuliidae). *Pacific Insects, 20*, 365–403.

Takaoka, H. (1982). Mermithid, microsporidan and fungal parasitism of larval blackflies from Oita in Japan. *Japanese Journal of Sanitary Zoology, 33*, 149–154.

Takaoka, H. (1983). *The Blackflies (Diptera: Simuliidae) of the Philippines* (p. xi + 199). Japan Society for The Promotion of Science.

Takaoka, H. (1985). Observations on the mating, blood feeding and oviposition of *Simulium takahasii* (Rubtsov) (Simuliidae, Diptera) in the laboratory. *Japanese Journal of Sanitary Zoology, 36*, 211–217.

Takaoka, H. (1989a). Further observations on the autogeny of *Simulium aureohirtum* Brunetti (Diptera: Simuliidae) in the Ryukyu Islands. *Japanese Journal of Sanitary Zoology, 40*, 21–28.

Takaoka, H. (1989b). Notes on blackflies (Diptera: Simuliidae) from Myanmar (formerly Burma). *Japanese Journal of Tropical Medicine and Hygiene, 17*, 243–257.

Takaoka, H. (1991). A revisionary note on the blackfly, hitherto called "*Simulium* (*Gomphostilbia*) *batoense* Edwards, 1934" from the Ryukyu Islands, Japan (Diptera: Simuliidae). *Japanese Journal of Tropical Medicine and Hygiene, 19*, 265–273.

Takaoka, H. (1994). Natural vectors of three bovine *Onchocerca* species (Nematoda: Onchocercidae) and seasonal transmission by three blackfly species (Diptera: Simuliidae) in central Kyushu, Japan. *Journal of Medical Entomology, 31*, 404–416.

Takaoka, H. (1995). The Simuliidae (Diptera) from Bougainville Island, Papua New Guinea. *Japanese Journal of Tropical Medicine and Hygiene, 23*, 239–252.

Takaoka, H. (1996a). The geographical distribution of the genus *Simulium* Latreille in the Oriental and Australasian Regions. *Japanese Journal of Tropical Medicine and Hygiene, 24*, 113–124.

Takaoka, H. (1996b). Description of a new species of *Simulium* (*Simulium*) from Sabah, Malaysia (Diptera: Simuliidae). *Japanese Journal of Tropical Medicine and Hygiene, 24*, 157–161.

Takaoka, H. (2000a). Taxonomic notes on *Simulium gombakense* (Diptera: Simuliidae) from Peninsular Malaysia: descriptions of male and pupa, and subgeneric transfer from *Morops* to *Gomphostilbia*. *Japanese Journal of Tropical Medicine and Hygiene, 28*, 111–114.

Takaoka, H. (2000b). A new black-fly species of *Simulium* (*Morops*) from Luzon Island, Philippines (Diptera: Simuliidae). *Japanese Journal of Tropical Medicine and Hygiene, 28*, 361–364.

Takaoka, H. (2001a). Two new and three newly recorded species of black flies (Diptera: Simuliidae) from Sabah, Malaysia. *Japanese Journal of Tropical Medicine and Hygiene, 29*, 221–230.

Takaoka, H. (2001b). Description of two new species of black flies (Diptera: Simuliidae) from Sarawak, Malaysia. *Japanese Journal of Tropical Medicine and Hygiene, 29*, 243–252.

Takaoka, H. (2001c). *Simulium* (*Simulium*) *weji* sp. nov. (Diptera: Simuliidae) from Thailand. *Japanese Journal of Tropical Medicine and Hygiene, 29*, 349–354.

Takaoka, H. (2003). *The Black Flies (Diptera: Simuliidae) of Sulawesi, Maluku and Irian Jaya* (p. xxii + 581). Kyushu University Press.

Takaoka, H. (2005). A new species of *Simulium* (*Gomphostilbia*) from Luzon Island, Philippines (Diptera: Simuliidae). *Medical Entomology and Zoology, 56*, 211–218.

Takaoka, H. (2006a). A new species of *Simulium* (*Simulium*) (Diptera: Simuliidae) from Luzon Island, Philippines. *Medical Entomology and Zoology, 57*, 99–103.

Takaoka, H. (2006b). Four new species of *Simulium* (*Simulium*) from Luzon Island, Philippines (Diptera: Simuliidae). *Medical Entomology and Zoology, 57*, 287–307.

Takaoka, H. (2006c). Revised description of *Simulium* (*Simulium*) *forcipatum* Delfinado, and description of three new related species from Luzon Island, Philippines (Diptera: Simuliidae). *Medical Entomology and Zoology, 57*, 309–326.

Takaoka, H. (2006d). Three new species of *Simulium* (*Wallacellum*) from Luzon Island, Philippines (Diptera: Simuliidae). *Medical Entomology and Zoology, 57*, 327–346.

Takaoka, H. (2007a). A new species of *Simulium* (*Simulium*) from Luzon Island, the Philippines (Diptera: Simuliidae). *Medical Entomology and Zoology, 58*, 11–17.

Takaoka, H. (2007b). Two new species of *Simulium* (*Simulium*) from Sabah, Malaysia (Diptera: Simuliidae). *Medical Entomology and Zoology, 58*, 291–301.

Takaoka, H. (2008a). *Simulium* (*Gomphostilbia*) *hoiseni* sp. nov. (Diptera: Simuliidae): a new species from Peninsular Malaysia. *Medical Entomology and Zoology, 59*, 9–14.

Takaoka, H. (2008b). Taxonomic revision of *tuberosum* species-group of *Simulium* (*Simulium*) in Sabah and Sarawak, Malaysia (Diptera: Simuliidae). *Medical Entomology and Zoology, 59*, 55–80.

Takaoka, H. (2008c). Four new species of *Simulium* (*Gomphostilbia*) (Diptera: Simuliidae) from Sarawak, Malaysia. *Medical Entomology and Zoology, 59*, 181–211.

Takaoka, H. (2009a). Three new species of *Simulium* (*Wallacellum*) (Diptera: Simuliidae) from the Philippines. *Medical Entomology and Zoology, 60*, 39–63.

Takaoka, H. (2009b). Three new species of the *Simulium* (*Gomphostilbia*) *banauense* species-group (Diptera: Simuliidae) in the Philippines. *Medical Entomology and Zoology, 60*, 65–79.
Takaoka, H. (2009c). Three new species of *Simulium* (*Gomphostilbia*) (Diptera: Simuliidae) from Sabah and Sarawak, Malaysia. *Medical Entomology and Zoology, 60*, 97–112.
Takaoka, H. (2012). Morphotaxonomic revision of *Simulium* (*Gomphostilbia*) (Diptera: Simuliidae) in the Oriental Region. *Zootaxa, 3577*, 1–42.
Takaoka, H. (2015a). Zoonotic onchocerciasis: its causative *Onchocerca* species and black fly vector species. *Medical Entomology and Zoology, 66*, 23–30. (In Japanese with an English abstract).
Takaoka, H. (2015b). The Simuliidae (Diptera) in the Oriental Region: Classification, morphological characters, geographical distributions, and transmission of filariae. *Medical Entomology and Zoology, 66*, 143–165. (In Japanese with an English abstract).
Takaoka, H. (2016). *Black flies and Parasitic Diseases* (p. 54). Institute of Biological Sciences. Faculty of Science, University of Malaya.
Takaoka, H. (2017a). Morphotaxonomic revision of species-groups of *Simulium* (*Simulium*) (Diptera: Simuliidae) in the Oriental Region. *Zootaxa, 4353*(3), 425–446.
Takaoka, H. (2017b). Speciation, faunal affinities and geographical dispersal of black flies (Diptera: Simuliidae) in the Oriental Region. *Acta Tropica, 166*, 234–240.
Takaoka, H., & Adler, P. H. (1997). A new subgenus, *Simulium* (*Daviesellum*), and a new species, *S.* (*D.*) *courtneyi*, (Diptera: Simuliidae) from Thailand and Peninsular Malaysia. *Japanese Journal of Tropical Medicine and Hygiene, 25*, 17–27.
Takaoka, H., Aoki, C., & Hayakawa, H. (1992). Natural infections of blackflies with larvae of zoonotic *Onchocerca* spp. in northeast Japan. *Japanese Journal of Tropical Medicine and Hygiene, 20*, 1–9.
Takaoka, H., & Baba, M. (1987). Experimental infection of *Simulium takahasii* with a canine filaria *Dirofilaria immitis*. *Japanese Journal of Sanitary Zoology, 38*, 1–6.
Takaoka, H., Baba, M., & Aoki, Y. (1986). Experimental infection of the blackfly *Simulium takahasii* with *Brugia pahangi*. *Japanese Journal of Parasitology, 35*, 237–241.
Takaoka, H., & Bain, O. (1990). Infections of blackflies (Diptera: Simuliidae) with three types of zoonotic *Onchocerca* larvae in Oita, Japan. *Japanese Journal of Tropical Medicine and Hygiene, 18*, 1–10.
Takaoka, H., & Choochote, W. (2002). Taxonomic notes on the *griseifrons* species-group in *Simulium* (*Simulium*) (Diptera: Simuliidae) from Thailand: Descriptions of two new species and description of the male, pupa and larva of *S.* (*S.*) *digrammicum* Edwards. *Tropical Medicine and Health, 30*, 115–132.
Takaoka, H., & Choochote, W. (2004a). A list of and keys to black flies (Diptera: Simuliidae) in Thailand. *Tropical Medicine and Health, 32*, 189–197.
Takaoka, H., & Choochote, W. (2004b). Two new species of *Simulium* (*Simulium*) (Diptera: Simuliidae) from Thailand. *Tropical Medicine and Health, 32*, 31–36.
Takaoka, H., & Choochote, W. (2004c). Taxonomic notes on the *griseifrons* species-group of *Simulium* (*Simulium*) (Diptera: Simuliidae) in northern Thailand. *Tropical Medicine and Health, 32*, 311–327.
Takaoka, H., & Choochote, W. (2005a). Two new species of *Simulium* (*Montisimulium*) (Diptera: Simuliidae) from Northern Thailand. *Medical Entomology and Zoology, 56*, 21–31.
Takaoka, H., & Choochote, W. (2005b). A new subgenus and a new species of *Simulium* s. l. (Diptera: Simuliidae) from Thailand. *Medical Entomology and Zoology, 56*, 33–41.
Takaoka, H., & Choochote, W. (2005c). A new species of *Simulium* (*Simulium*) (Diptera: Simuliidae) from Thailand. *Medical Entomology and Zoology, 56*, 43–47.
Takaoka, H., & Choochote, W. (2005d). Two new species of *Simulium* (*Simulium*) (Diptera: Simuliidae) from northern Thailand. *Medical Entomology and Zoology, 56*, 99–110.
Takaoka, H., & Choochote, W. (2005e). Two new species of *Simulium* Latreille (Diptera: Simuliidae) from northwestern Thailand. *Medical Entomology and Zoology, 56*, 123–133.

Takaoka, H., & Choochote, W. (2005f). Two new species of the *griseifrons* species-group of *Simulium* (*Simulium*) (Diptera: Simuliidae) in northern Thailand. *Medical Entomology and Zoology, 56*, 219–235.

Takaoka, H., & Choochote, W. (2005g). A new species of *Simulium* (*Simulium*) from northern Thailand (Diptera: Simuliidae). *Tropical Medicine and Health, 33*, 95–101.

Takaoka, H., & Choochote, W. (2005h). Two new species of *Simulium* (*Nevermannia*) (Diptera: Simuliidae) from northern Thailand. *Tropical Medicine and Health, 33*, 133–141.

Takaoka, H., & Choochote, W. (2005i). Two new species of black flies (Diptera: Simuliidae) from northern Thailand. *Medical Entomology and Zoology, 56*, 319–334.

Takaoka, H., & Choochote, W. (2005j). Discovery of two more new species of *Simulium* (*Montisimulium*) (Diptera: Simuliidae) in Doi Inthanon National Park, Chiang Mai, Thailand. *Tropical Medicine and Health, 33*, 209–215.

Takaoka, H., & Choochote, W. (2006a). A new species of the subgenus *Simulium* (*Asiosimulium*) (Diptera: Simuliidae) from Thailand. *Medical Entomology and Zoology, 57*, 45–48.

Takaoka, H., & Choochote, W. (2006b). Descriptions of adults of *Simulium* (*Simulium*) *baimaii* from Thailand (Diptera: Simuliidae) and its assignment to the *malyschevi* species-group. *Medical Entomology and Zoology, 57*, 49–53.

Takaoka, H., & Choochote, W. (2006c). A new species of *Simulium* (*Nevermannia*) from northern Thailand (Diptera: Simuliidae). *Medical Entomology and Zoology, 57*, 83–92.

Takaoka, H., & Choochote, W. (2006d). A new species of the *griseifrons* species-group of *Simulium* (*Simulium*) (Diptera: Simuliidae) in northern Thailand. *Medical Entomology and Zoology, 57*, 115–124.

Takaoka, H., & Choochote, W. (2006e). A new species of *Simulium* (*Gomphostilbia*) (Diptera: Simuliidae) from northern Thailand. *Medical Entomology and Zoology, 57*, 229–233.

Takaoka, H., & Choochote, W. (2007). A new species of the *multistriatum* species-group of *Simulium* (*Simulium*) (Diptera: Simuliidae) from Northern Thailand. *Tropical Medicine and Health, 35*, 19–22.

Takaoka, H., & Choochote, W. (2009). Description of the female and male of *Simulium* (*Montisimulium*) *angkaense* (Diptera: Simuliidae) from Thailand. *Medical Entomology and Zoology, 60*, 113–118.

Takaoka, H., Choochote, W., Aoki, C., Fukuda, M., & Bain, O. (2003). Black flies (Diptera: Simuliidae) attracted to humans and water buffalos and natural infections with filarial larvae, probably *Onchocerca* sp., in northern Thailand. *Parasite, 10*, 3–8.

Takaoka, H., & Craig, D. A. (1999). Taxonomic and ecological notes on *Simulium* (*Gomphostilbia*) *palauense* (Diptera: Simuliidae) from Palau, Micronesia, with redescriptions of adults and descriptions of the pupa and mature larva. *Japanese Journal of Tropical Medicine and Hygiene, 27*, 195–201.

Takaoka, H., & Davies, D. M. (1995a). *The Black Flies (Diptera: Simuliidae) of West Malaysia* (p. viii + 175). Kyushu University Press.

Takaoka, H., & Davies, D. M. (1995b). A new black fly species of *Simulium* (*Nevermannia*) from Sumatra, Indonesia. *Japanese Journal of Tropical Medicine and Hygiene, 23*, 127–131.

Takaoka, H., & Davies, D. M. (1996). *The Black Flies (Diptera: Simuliidae) of Java, Indonesia* (p. viii + 81). Bishop Museum Bulletin in Entomology No. 6. Bishop Museum Press.

Takaoka, H., & Davies, D. M. (1997). *Simulium* (*Simulium*) *yongi* sp. nov. (Diptera: Simuliidae) from Peninsular Malaysia. *Japanese Journal of Tropical Medicine and Hygiene, 25*, 11–16.

Takaoka, H., Davies, D. M., & Dudgeon, D. (1995). Black flies (Diptera: Simuliidae) from Hong Kong: taxonomic notes with descriptions of two new species. *Japanese Journal of Tropical Medicine and Hygiene, 23*, 189–196.

Takaoka, H., Fukuda, M., Otsuka, Y., Aoki, C., Uni, S., & Bain, O. (2012a). Blackfly vectors of zoonotic onchocerciasis in Japan. *Medical and Veterinary Entomology, 26*, 372–378.

Takaoka, H., Fukuda, M., Otsuka, Y., Low, V. L., & Ya'cob, Z. (2021e). Redescription of *Simulium* (*Gomphostilbia*) *omutaense* Ogata & Sasa from Japan, with its phylogenetic relationship with other members of the *S. batoense* species-group. *Acta Tropica, 225*, 1–9. (106207).

Takaoka, H., & Hadi, U. K. (1991). Two new blackfly species of *Simulium* (*Simulium*) from Java, Indonesia (Diptera: Simuliidae). *Japanese Journal of Tropical Medicine and Hygiene, 19*, 357–370.

Takaoka, H., Hadi, U. K., & Sigit, S. H. (2006a). The black flies (Diptera: Simuliidae) of Flores and Timor, Indonesia. *Medical Entomology and Zoology, 57*, 1–26.

Takaoka, H., Hadi, U. K., & Sigit, S. H. (2006b). A new species of *Simulium* (*Simulium*) from Sumatra, Indonesia (Diptera: Simuliidae). *Medical Entomology and Zoology, 57*, 27–34.

Takaoka, H., Hayakawa, H., & Ya'cob, Z. (2019a). New record of *Simulium* (*Simulium*) *thongsahuani* (Diptera: Simuliidae) from Peninsular Malaysia. *Medical Entomology and Zoology, 70*, 33–34.

Takaoka, H., & Huang, Y. T. (2006). A new species of *Simulium* (*Simulium*) (Diptera: Simuliidae) from Taiwan. *Medical Entomology and Zoology, 57*, 219–227.

Takaoka, H., & Huang, Y. T. (2017). A new species of *Simulium* (*Gomphostilbia*) (Diptera: Simuliidae) from Taiwan. *Zootaxa, 4312*(3), 438–448.

Takaoka, H., & Huang, Y. T. (2018). A new black fly species of *Simulium* (*Simulium*) (Diptera: Simuliidae) from Taiwan. *Zootaxa, 4422*(1), 141–150.

Takaoka, H., & Kuvangkadilok, C. (1999). Four new species of black flies (Diptera: Simuliidae) from Thailand. *Japanese Journal of Tropical Medicine and Hygiene, 27*, 497–509.

Takaoka, H., Low, V. L., Huang, Y. T., Fukuda, M., & Ya'cob, Z. (2021a). Two new black fly species of the *Simulium* (*Gomphostilbia*) *asakoae* species-group (Diptera: Simuliidae) from Taiwan. *Acta Tropica, 222*, 1–9.

Takaoka, H., Low, V. L., Huang, Y. T., Fukuda, M., & Ya'cob, Z. (2021b). Two new black fly species of the *Simulium* (*Simulium*) *tuberosum* species-group (Diptera: Simuliidae) from Taiwan. *Tropical Biomedicine, 38*, 403–412.

Takaoka, H., Low, V. L., Sofian-Azirun, M., Otsuka, Y., Ya'cob, Z., Chen, C. D., Lau, K. W., & Lardizabal, M. L. (2016a). Dimorphic male scutal patterns and upper-eye facets of *Simulium mirum* sp. nov. (Diptera: Simuliidae) from Malaysia. *Parasites &Vectors, 9: 136*, 1–12.

Takaoka, H., Low, V. L., Srisuka, W., Ya'cob, Z., & Saeung, A. (2018a). Morphological and molecular evidence for multiple taxa in *Simulium parahiyangum* (Diptera: Simuliidae), a geographic generalist black fly in South-East Asia. *Journal of Medical Entomology, 55*, 1453–1463.

Takaoka, H., Low, V. L., Tan, T. K., Huang, Y. T., Fukuda, M., & Ya'cob, Z. (2018b). A new species of the *Simulium* (*Simulium*) *argentipes* species-group (Diptera: Simuliidae) from Taiwan, and its phylogenetic relationships with four related species. *Journal of Medical Entomology, 55*, 884–892.

Takaoka, H., Low, V. L., Tan, T. K., Sofian-Azirun, M., Chen, C. D., Lau, K. W., & Pham, X. D. (2019b). A new species of the *Simulium* (*Nevermannia*) (Diptera: Simuliidae) from Vietnam. *Acta Tropica, 190*, 320–328.

Takaoka, H., Low, V. L., Tan, T. K., Ya'cob, Z., Sofian-Azirun, M., Chen, C. D., Lau, K. W., & Pham, X. D. (2019c). A new species of the *Simulium* (*Gomphostilbia*) *duolongum* subgroup (Diptera: Simuliidae) from Vietnam. *Journal of Medical Entomology, 56*, 432–440.

Takaoka, H., & Mulla, M. S. (2000). Taxonomic notes on *Simulium siripoomense* from Thailand (Diptera: Simuliidae): descriptions of female and pupa, and change of its subgeneric status from *Byssodon* to *Simulium* s. str. *Japanese Journal of Tropical Medicine and Hygiene, 28*, 25–28.

Takaoka, H., Namgay, R., & Somboon, P. (2021c). A new species and four newly recorded species of black flies (Diptera: Simuliidae) from Bhutan. *Tropical Biomedicine, 38*, 81–85.

Takaoka, H., & Noda, S. (1979). Autogeny of the black fly *Simulium* (*Eusimulium*) *aureohirtum* (Diptera: Simuliidae). *Journal of Medical Entomology, 15*, 183–184.

Takaoka, H., & Okazawa, T. (1988). List of black fly species (Diptera: Simuliidae) in Japan. *Japanese Journal of Sanitary Zoology*, 97–103. (In Japanese).

Takaoka, H., Otsuka, Y., Choochote, W., Aoki, C., Hayakawa, H., & Thongsahuan, S. (2010a). Descriptions of the male, pupa and larva of *Simulium* (*Gomphostilbia*) *novemarticulatum*

(Diptera: Simuliidae) from Peninsular Malaysia and Thailand. *Medical Entomology and Zoology, 61*, 59–67.

Takaoka, H., Otsuka, Y., Choochote, W., & Thongsahuan, S. (2009a). Two new and one newly recorded species of *Simulium* (*Gomphostilbia*) (Diptera: Simuliidae) from southern Thailand. *Medical Entomology and Zoology, 60*, 259–268.

Takaoka, H., Otsuka, Y., Choochote, W., & Thongsahuan, S. (2010b). A new species of *Simulium* (*Simulium*) (Diptera: Simuliidae) from southern Thailand. *Medical Entomology and Zoology, 61*, 17–29.

Takaoka, H., Otsuka, Y., Fukuda, M., Low, V. L., & Ya'cob, Z. (2023a). Morphological and genetic analysis of *Simulium* (*Gomphostilbia*) *okinawense* Takaoka and *S.* (*G.*) *tokarense* Takaoka (Diptera: Simuliidae) from the Nansei Islands, Japan: redescription and transfer from the *S. ceylonicum* species-group to the *S. asakoae* species-group. *Tropical Biomedicine, 40*(1), 1–13.

Takaoka, H., Otsuka, Y., Fukuda, M., Low, V. L., & Ya'cob, Z. (2023b). Morphological redescription of *Simulium takahasii* (Rubtsov), the first species of the subgenus *Wilhelmia* Enderlein (Diptera: Simuliidae) recognized in East Asia. *Tropical Biomedicine, 40*(2), 1–7.

Takaoka, H., Otsuka, Y., Huang, Y. T., Low, V. L., Fukuda, M., & Ya'cob, Z. (2022a). Two new species of the *Simulium* (*Simulium*) *striatum* species-group (Diptera: Simuliidae) from Taiwan and Japan, with redescription of *S.* (*S.*) *quinquestriatum* (Shiraki). *Acta Tropica, 227*(2022), 1–14.

Takaoka, H., Otsuka, Y., Low, V. L., Adler, P. H., Huang, Y. T., Fukuda, M., & Ya'cob, Z. (2021d). Morphological and phylogenetic revision of cytoform 'L' of *Simulium* (*Simulium*) *tani* complex (Diptera: Simuliidae) from Taiwan, and cytoform 'D' of *S.* (*S.*) *suzukii* complex from the Nansei Islands, Japan. *Acta Tropica, 224*(2021), 106124.

Takaoka, H., & Roberts, D. M. (1988). Notes on blackflies (Diptera: Simuliidae) from Sulawesi, Indonesia. *Japanese Journal of Tropical Medicine and Hygiene, 16*, 191–219.

Takaoka, H., & Saito, K. (1996). A new species and new records of black flies (Diptera: Simuliidae) from Thailand. *Japanese Journal of Tropical Medicine and Hygiene, 24*, 163–169.

Takaoka, H., & Saito, K. (2000). Description of a new species of *Simulium* (*Nevermannia*) from Japan (Diptera: Simuliidae). *Japanese Journal of Tropical Medicine and Hygiene, 28*, 19–24.

Takaoka, H., & Saito, K. (2007). List of species of Japanese black flies (Diptera: Simuliidae). *Yugai Seibutsu, 4*, 44–60. (In Japanese with an English abstract).

Takaoka, H., Sato, H., & Higashihara, J. (2018c). Five new records of black flies (Diptera: Simuliidae) from Myanmar. *Tropical Biomedicine, 35*, 804–814.

Takaoka, H., & Shrestha, S. (2010). New species of black flies (Diptera: Simuliidae) from Nepal. *Zootaxa, 2731*, 1–62.

Takaoka, H., Shrestha, S., & Dangi, N. (2020a). Three new species and two new records of the *Simulium* (*Simulium*) *striatum* species-group (Diptera: Simuliidae) from Nepal. *Medical Entomology and Zoology, 71*, 39–50.

Takaoka, H., Shrestha, S., & Dangi, N. (2020b). A new species and a new record of the *Simulium* (*Simulium*) *striatum* species-group (Diptera: Simuliidae) from Nepal. *Medical Entomology and Zoology, 71*, 91–100.

Takaoka, H., Shrestha, S., & Dangi, N. (2020c). Three new black fly species of the genus *Simulium* (Diptera: Simuliidae) from Nepal. *Medical Entomology and Zoology, 71*, 201–218.

Takaoka, H., Shrestha, S., & Dangi, N. (2020d). Three new species and one newly recorded species of the *Simulium* (*Simulium*) *multistriatum* species-group (Diptera: Simuliidae) from Nepal. *Medical Entomology and Zoology, 71*, 289–306.

Takaoka, H., Shrestha, S., & Dangi, N. (2020e). Four new species of the *Simulium* (*Simulium*) *tuberosum* species-group (Diptera: Simuliidae) from Nepal. *Medical Entomology and Zoology, 71*, 307–328.

Takaoka, H., Shrestha, S., Dangi, N., & Adler, P. H. (2020f). Six new species of black flies (Diptera: Simuliidae) from Nepal. *Medical Entomology and Zoology, 71*, 177–200.

Takaoka, H., & Sigit, S. H. (1992). A new blackfly species of *Simulium* (*Gomphostilbia*) from Java, Indonesia (Diptera: Simuliidae). *Japanese Journal of Tropical Medicine and Hygiene, 20*, 135–142.

Takaoka, H., & Sigit, S. H. (1997). Three new black fly species of *Simulium* (Diptera: Simuliidae) from Sumatra, Indonesia. *Japanese Journal of Tropical Medicine and Hygiene, 25*, 69–80.

Takaoka, H., Sofian-Azirun, M., & Belabut, D. M. (2012b). Descriptions of two new species of *Simulium* (*Simulium*) (Diptera: Simuliidae) from Tioman Island, Peninsular Malaysia. *Raffles Bulletin of Zoology, 60*, 399–409.

Takaoka, H., Sofian-Azirun, M., Chen, C. D., Halim, M. R. A., Lau, K. W., Low, V. L., & Suana, I. W. (2020g). Descriptions of the female, male and mature larva of *Simulium contractum* Takaoka (Diptera: Simuliidae) from Sulawesi, Indonesia. *Tropical Biomedicine, 37*, 683–690.

Takaoka, H., Sofian-Azirun, M., Chen, C. D., Lau, K. W., Halim, M. R. A., Low, V. L., & Suana, I. W. (2018d). Three new species of black flies (Diptera: Simuliidae) from the Lesser Sunda Archipelago, Indonesia. *Tropical Biomedicine, 35*, 951–974.

Takaoka, H., Sofian-Azirun, M., Chen, C. D., Lau, K. W., Halim, M. R. A., Low, V. L., Ya'cob, Z., Ashikin, N., & Suana, I. W. (2017a). Two new species of *Simulium* (*Gomphostilbia*) (Diptera: Simuliidae) from Flores, Indonesia. *Journal of Medical Entomology, 54*, 576–586.

Takaoka, H., Sofian-Azirun, M., Chen, C. D., Lau, K. W., Halim, M. R. A., Low, V. L., Ya'cob, Z., Ashikin, N., & Suana, I. W. (2017b). A new species of *Simulium* (*Gomphostilbia*) (Diptera: Simuliidae) from Flores, Indonesia. *Zootaxa, 4236*(1), 149–156.

Takaoka, H., Sofian-Azirun, M., Chen, C. D., Lau, K. W., Halim, M. R. A., & Suana, I. W. (2019d). A new black fly species of *Simulium* (*Gomphostilbia*) (Diptera: Simuliidae) from Sulawesi, Indonesia. *Zootaxa, 4651*(2), 392–400.

Takaoka, H., Sofian-Azirun, M., & Hashim, R. (2011a). *Simulium* (*Gomphostilbia*) *sofiani*, a new species of black fly (Diptera: Simuliidae) from Peninsular Malaysia. *Tropical Biomedicine, 28*, 389–399.

Takaoka, H., Sofian-Azirun, M., & Hashim, R. (2013a). Description of a new species of *Simulium* (*Gomphostilbia*) (Diptera: Simuliidae) from Cameron's Highlands, Peninsular Malaysia, with keys to 18 species of the *Simulium asakoae* species-group. *Journal of Medical Entomology, 50*, 1179–1189.

Takaoka, H., Sofian-Azirun, M., Hashim, R., & Ya'cob, Z. (2011b). Two new species of *Simulium* (*Gomphostilbia*) (Diptera: Simuliidae) from Peninsular Malaysia, with keys to Peninsular Malaysian members of the *Simulium ceylonicum* species group. *ZooKeys, 118*, 53–74.

Takaoka, H., Sofian-Azirun, M., Hashim, R., Ya'cob, Z., & Chen, C. D. (2012c). Two new species of *Simulium* (*Gomphostilbia*) (Diptera: Simuliidae) from Peninsular Malaysia. *Journal of Medical Entomology, 49*, 803–812.

Takaoka, H., Sofian-Azirun, M., & Ya'cob, Z. (2012d). A new species of *Simulium* (*Simulium*) (Diptera: Simuliidae) from Langkawi Island, Peninsular Malaysia. *Journal of Medical Entomology, 49*, 1198–1205.

Takaoka, H., Sofian-Azirun, M., & Ya'cob, Z. (2013b). A new species of *Simulium* (*Gomphostilbia*) (Diptera: Simuliidae) from Langkawi Island, Peninsular Malaysia. *Journal of Medical Entomology, 50*, 701–708.

Takaoka, H., Sofian-Azirun, M., & Ya'cob, Z. (2014a). Two new species of *Simulium* (*Gomphostilbia*) (Diptera: Simuliidae) from Peninsular Malaysia, with keys to 10 Peninsular Malaysian species of the *Simulium batoense* species-group. *Journal of Medical Entomology, 51*, 10–26.

Takaoka, H., Sofian-Azirun, M., & Ya'cob, Z. (2014b). Two new species of the *Simulium batoense* species-group of *Simulium* (*Gomphostilbia*) (Diptera: Simuliidae) from Peninsular Malaysia. *Zootaxa, 3774*, 473–480.

Takaoka, H., Sofian-Azirun, M., & Ya'cob, Z. (2015a). Two new species of *Simulium* (*Simulium*) (Diptera: Simuliidae) from Mount Murud, Sarawak, Malaysia. *Journal of Medical Entomology, 52*, 38–49.

Takaoka, H., Sofian-Azirun, M., Ya'cob, Z., Chen, C. D., Lau, K. W., Fernandez, K., & Lardizabal, L. M. (2015b). Revision of the *Simulium melanopus* species-group of *Simulium* (*Simulium*) (Diptera: Simuliidae) in Sabah, Malaysia. *Zootaxa, 3985*(1), 1–30.

Takaoka, H., Sofian-Azirun, M., Ya'cob, Z., Chen, C. D., Lau, K. W., Low, V. L., Pham, X. D., & Adler, P. H. (2017c). The black flies (Diptera: Simuliidae) of Vietnam. *Zootaxa, 4261*(1), 1–165.

Takaoka, H., Sofian-Azirun, M., Ya'cob, Z., Chen, C. D., Lau, K. W., & Pham, H. T. (2014c). Female black flies of *Simulium* (Diptera: Simuliidae) collected on humans in Tam Dao National Park, Vietnam: Description of a new species and notes on four species newly recorded from Vietnam. *Tropical Biomedicine, 31*(4), 742–748.

Takaoka, H., Sofian-Azirun, M., Ya'cob, Z., Chen, C. D., Lau, K. W., & Pham, H. T. (2014d). New species and records of black flies (Diptera: Simuliidae) from Vinh Phuc Province, Vietnam. *Zootaxa, 3838*(3), 347–366.

Takaoka, H., Sofian-Azirun, M., Ya'cob, Z., Chen, C. D., Lau, K. W., & Pham, X. D. (2014e). Three new species of *Simulium* (*Nevermannia*) (Diptera: Simuliidae) from Vietnam. *Zootaxa, 3866*(4), 555–571.

Takaoka, H., Sofian-Azirun, M., Ya'cob, Z., Chen, C. D., Low, V. L., & Harmonis. (2017d). A new species and seven new records of black flies (Diptera: Simuliidae) from Kalimantan, Indonesia. *Tropical Biomedicine, 34*(2), 437–448.

Takaoka, H., Sofian-Azirun, M., Ya'cob, Z., Chen, C. D., Lau, K. W., & Pham, X. D. (2015c). The black flies (Diptera: Simuliidae) from Thua Thien Hue and Lam Dong Provinces, Vietnam. *Zootaxa, 3961*(1), 1–96.

Takaoka, H., Sofian-Azirun, M., Ya'cob, Z., Chen, C. D., Lau, K. W., Low, V. L., & Suana, I. W. (2017e). The black flies (Diptera: Simuliidae) of the Lesser Sunda Archipelago, Indonesia. *Acta Tropica, 169*, 170–186.

Takaoka, H., Sofian-Azirun, M., Ya'cob, Z., Chen, C. D., Low, V. L., & Harmonis. (2016b). A new species of *Simulium* (*Gomphostilbia*) (Diptera: Simuliidae) from Kalimantan, Indonesia, with keys to identify 19 Bornean species of the subgenus *Gomphostilbia*. *Journal of Medical Entomology, 53*, 798–806.

Takaoka, H., Sofian-Azirun, M., Ya'cob, Z., Chen, C. D., Low, V. L., & Zaid, A. (2016c). *Simulium* (*Gomphostilbia*) *merapiense* sp. nov. (Diptera: Simuliidae) from Java, Indonesia. *Journal of Medical Entomology, 53*, 76–82.

Takaoka, H., Sofian-Azirun, M., Ya'cob, Z., & Hashim, R. (2014f). Two new species of *Simulium* (*Gomphostilbia*) (Diptera: Simuliidae) from Cameron's Highlands, Peninsular Malaysia, with keys to 21 species of the *Simulium asakoae* species-group. *Zootaxa, 3765*(1), 54–68.

Takaoka, H., Sofian-Azirun, M., Ya'cob, Z., Hashim, R., & Otsuka, Y. (2014g). A new species of *Simulium* (*Gomphostilbia*) (Diptera: Simuliidae) from Malaysia, with keys to 32 species of the *Simulium ceylonicum* species-group. *Journal of Medical Entomology, 51*, 517–528.

Takaoka, H., & Somboon, P. (2008). Eleven new species and one new record of black flies (Diptera: Simuliidae) from Bhutan. *Medical Entomology and Zoology, 59*(3), 213–262.

Takaoka, H., & Srisuka, W. (2009). *Simulium* (*Gomphostilbia*) *chiangdaoense* sp. nov. (Diptera: Simuliidae) from northern Thailand. *Medical Entomology and Zoology, 60*(4), 269–276.

Takaoka, H., & Srisuka, W. (2010a). Description of the female of *Simulium* (*Gomphostilbia*) *kuvangkadilokae* (Diptera: Simuliidae) from Thailand. *Medical Entomology and Zoology, 61*, 39–47.

Takaoka, H., & Srisuka, W. (2010b). *Simulium* (*Gomphostilbia*) *chayamaritae*, a new species of black fly (Diptera: Simuliidae) from northern Thailand. *Medical Entomology and Zoology, 61*, 265–271.

Takaoka, H., & Srisuka, W. (2010c). *Simulium* (*Nevermannia*) *vessabutrae,* a new black fly species (Diptera: Simuliidae) from Thailand. *Medical Entomology and Zoology, 61*, 97–104.

Takaoka, H., & Srisuka, W. (2010d). Description of a new species of *Simulium* (*Montisimulium*) (Diptera: Simuliidae) from Thailand. *Medical Entomology and Zoology, 61*, 261–264.

Takaoka, H., & Srisuka, W. (2010e). *Simulium* (*Nevermannia*) *wichaii*, a new species of black fly (Diptera: Simuliidae) from northern Thailand. *Medical Entomology and Zoology, 61*, 273–279.

Takaoka, H., & Srisuka, W. (2011). A new species of *Simulium* (*Nevermannia*) (Diptera, Simuliidae) from Thailand, with keys to members of the *Simulium feuerborni* species-group in Thailand. *ZooKeys, 89*, 57–70.

Takaoka, H., Srisuka, W., Fukuda, M., & Saeung, A. (2020h). Twenty-one new species of the *Simulium* (*Gomphostilbia*) *asakoae* species-group (Diptera: Simuliidae) in Thailand, with their genetic relationships. *ZooKeys, 950*, 51–152.

Takaoka, H., Srisuka, W., Low, V. L., & Saeung, A. (2017f). A new species of the *Simulium* (*Gomphostilbia*) *epistum* species-group (Diptera: Simuliidae) from Thailand. *Acta Tropica, 176*, 373–379.

Takaoka, H., Srisuka, W., Low, V. L., & Saeung, A. (2018e). A new species of the *Simulium* (*Simulium*) *striatum* species-group (Diptera: Simuliidae) from Thailand, and its differentiation from two related species based on a fast-evolving nuclear gene. *Journal of Medical Entomology, 55*, 561–568.

Takaoka, H., Srisuka, W., Low, V. L., & Saeung, A. (2018f). Five new species of the *Simulium decuplum* subgroup of the *Simulium* (*Gomphostilbia*) *batoense* species-group (Diptera: Simuliidae) from Thailand. *Acta Tropica, 182*, 271–284.

Takaoka, H., Srisuka, W., Low, V. L., & Saeung, A. (2018g). A new species and a new record of the *Simulium* (*Gomphostilbia*) *gombakense* species-group (Diptera: Simuliidae) from Thailand. *Acta Tropica, 185*, 156–166.

Takaoka, H., Srisuka, W., Low, V. L., & Saeung, A. (2018h). Morphological analysis and phylogenetic relationships of *Simulium rufibasis* Brunetti (Diptera: Simuliidae) in Thailand. *Journal of Medical Entomology, 56*, 408–415.

Takaoka, H., Srisuka, W., Low, V. L., & Seung, A. (2019e). A new species of the *Simulium* (*Simulium*) *multistriatum* species-group (Diptera: Simuliidae) from Thailand, and its phylogenetic relationships with 10 related species. *Journal of Medical Entomology, 56*, 86–94.

Takaoka, H., Srisuka, W., Maleewong, W., Low, V. L., & Saeung, A. (2017g). Two new species of the *Simulium asakoae* species-group of *Simulium* (*Gomphostilbia*) (Diptera: Simuliidae) from Myanmar and their phylogenetic relationships with related species in the *S. asakoae* species-group. *Acta Tropica, 176*, 39–50.

Takaoka, H., Srisuka, W., Maleewong, W., & Saeung, A. (2017h). A new species of *Simulium* (*Asiosimulium*) (Diptera: Simuliidae) from Myanmar. *Medical Entomology and Zoology, 68*, 1–4.

Takaoka, H., Srisuka, W., Maleewong, W., & Saeung, A. (2018j). New record of 13 species of black flies (Diptera: Simuliidae) from Myanmar. *Acta Tropica, 178*, 281–289.

Takaoka, H., Srisuka, W., Otsuka, Y., & Choochote, W. (2014h). A new species and species-group of *Simulium* (*Simulium*) (Diptera: Simuliidae) from Thailand. *Journal of Medical Entomology, 51*, 725–732.

Takaoka, H., Srisuka, W., & Saeung, A. (2010c). Description of the female of *Simulium* (*Gomphostilbia*) *gombakense* (Diptera: Simuliidae) from Thailand. *Medical Entomology and Zoology, 61*, 111–114.

Takaoka, H., Srisuka, W., & Saeung, A. (2017i). *Simulium maleewongae*, a new species of *Simulium* (*Gomphostilbia*) (Diptera: Simuliidae) from Thailand. *Journal of Medical Entomology, 54*, 91–99.

Takaoka, H., Srisuka, W., & Saeung, A. (2017j). A new human-biting black fly species of *Simulium* (*Simulium*) (Diptera: Simuliidae) from Thailand. *Journal of Medical Entomology, 54*, 945–948.

Takaoka, H., Srisuka, W., & Saeung, A. (2017k). Description of the male of *Simulium* (*Simulium*) *suchariti* from Thailand. *Medical Entomology and Zoology, 68*, 27–30.

Takaoka, H., Srisuka, W., & Saeung, A. (2017l). Two new species of the *Simulium* (*Simulium*) *variegatum* species-group of black flies (Diptera: Simuliidae) from Thailand. *Journal of Medical Entomology, 54*, 1213–1223.

Takaoka, H., Srisuka, W., & Saeung, A. (2017m). A new black fly species of *Simulium* (*Gomphostilbia*) from Thailand. *Journal of Medical Entomology, 54*, 1552–1559.

Takaoka, H., Srisuka, W., & Saeung, A. (2018k). A new species of *Simulium* (*Asiosimulium*) (Diptera: Simuliidae) from Thailand. *Journal of Medical Entomology, 55*, 569–574.

Takaoka, H., Srisuka, W., & Saeung, A. (2018l). Description of the female of *Simulium chayamaritae* (Diptera: Simuliidae) from Thailand. *Tropical Biomedicine, 35*, 975–980.

Takaoka, H., Srisuka, W., & Saeung, A. (2019f). A new species of the *Simulium* (*Simulium*) *variegatum* species-group (Diptera: Simuliidae) from Thailand. *Acta Tropica, 191*, 155–161.

Takaoka, H., Srisuka, W., & Saeung, A. (2019g). Checklist and keys for the black flies (Diptera: Simuliidae) of Thailand. *Medical Entomology and Zoology, 70*, 53–77.

Takaoka, H., Srisuka, W., Saeung, A., Maleewong, W., & Low, V. L. (2017n). A new species of *Simulium* (*Gomphostilbia*) from Laos. *Journal of Medical Entomology, 54*, 1543–1551.

Takaoka, H., Srisuka, W., Saeung, A., Otsuka, Y., Aoki, C., & Choochote, W. (2010d). Taxonomic notes on *Simulium* (*Montisimulium*) *laoleense* and a close related new species (Diptera: Simuliidae) from Thailand. *Medical Entomology and Zoology, 61*, 363–374.

Takaoka, H., Srisuka, W., Saeung, A., Otsuka, Y., & Choochote, W. (2012e). *Simulium* (*Nevermannia*) *chomthongense*, a new species of black fly (Diptera: Simuliidae) from Chiang Mai, Thailand. *Tropical Biomedicine, 29*, 381–390.

Takaoka, H., Srisuka, W., Saeung, A., & Choochote, W. (2013c). *Simulium* (*Asiosimulium*) *furvum*, a new species of black fly (Diptera: Simuliidae) from Thailand. *Journal of Medical Entomology, 50*, 493–500.

Takaoka, H., Srisuka, W., Saeung, A., Otsuka, Y., & Choochote, W. (2013d). *Simulium* (*Nevermannia*) *khunklangense,* a new species of black fly (Diptera: Simuliidae) from Chiang Mai, Thailand. *Zootaxa, 3694*(3), 280–288.

Takaoka, H., Srisuka, W., Saeung, A., & Choochote, W. (2014i). A new species of *Simulium* (*Gomphostilbia*) (Diptera: Simuliidae) from Thailand, with keys to 11 species of the *Simulium varicorne* species-group. *Journal of Medical Entomology, 51*, 314–322.

Takaoka, H., Srisuka, W., Saeung, A., & Otsuka, Y. (2014j). *Simulium* (*Simulium*) *lomkaoense*, a new species of black fly (Diptera: Simuliidae) from Thailand. *Journal of Medical Entomology, 51*, 1109–1115.

Takaoka, H., Srisuka, W., Ya'cob, Z., Low, V. L., & Saeung, A. (2021f). A new species of the *Simulium* (*Gomphostilbia*) *ceylonicum* species-group (Diptera: Simuliidae) from Thailand. *Acta Tropica, 217*, 1–7.

Takaoka, H., & Suzuki, H. (1984). The blackflies (Diptera: Simuliidae) from Thailand. *Japanese Journal of Sanitary Zoology, 35*(1), 7–45.

Takaoka, H., & Suzuki, H. (1995). The Simuliidae (Diptera) from the Solomon Islands, South Pacific. *Japanese Journal of Tropical Medicine and Hygiene, 23*, 253–272.

Takaoka, H., & Tenedero, V. F. (2007a). Two new and four newly recorded species of *Simulium* (Diptera: Simuliidae) from Mindoro Island, the Philippines. *Medical Entomology and Zoology, 58*, 29–43.

Takaoka, H., & Tenedero, V. F. (2007b). A new species of *Simulium* (*Simulium*) from Negros Island, the Philippines (Diptera: Simuliidae). *Medical Entomology and Zoology, 58*, 89–97.

Takaoka, H., & Tenedero, V. F. (2007c). Two new and one newly recorded species of *Simulium* (Diptera: Simuliidae) from Palawan Island, the Philippines. *Medical Entomology and Zoology, 58*, 207–228.

Takaoka, H., & Tenedero, V. F. (2008a). A new species of *Simulium* (*Gomphostilbia*) (Diptera: Simuliidae) from Palawan Island, the Philippines. *Medical Entomology and Zoology, 59*, 1–7.

Takaoka, H., & Tenedero, V. F. (2008b). *Simulium* (*Gomphostilbia*) *rarum* sp. nov., a new species of black fly (Diptera: Simuliidae) from Palawan Island, the Philippines. *Medical Entomology and Zoology, 59*, 319–324.

Takaoka, H., & Tenedero, V. F. (2008c). New species and record of black flies (Diptera: Simuliidae) from Samar, the Philippines. *Medical Entomology and Zoology, 59*, 325–336.

Takaoka, H., & Tenedero, V. F. (2019). Two new species of the *Simulium* (*Simulium*) *tuberosum* species-group (Diptera: Simuliidae) from Palawan, the Philippines. *Zootaxa, 4568*(2), 383–393.

Takaoka, H., & Tenedero, V. F. (2022a). Description of the males of *Simulium* (*Gomphostilbia*) *miblosi* Takaoka and *S.* (*G.*) *apoense* Takaoka (Diptera: Simuliidae) from Mindanao, the Philippines. *Tropical Biomedicine, 39*, 332–337.

Takaoka, H., & Tenedero, V. F. (2022b). Description of the female of *Simulium* (*Wallacellum*) *tuyense* Takaoka (Diptera: Simuliidae) from the Philippines. *Tropical Biomedicine, 39*, 451–454.

Takaoka, H., Tenedero, V. F., & Bertuso, A. G. (2009b). Description of the male of *Simulium* (*Gomphostilbia*) *binuanense* (Diptera: Simuliidae) from Palawan Island, Philippines. *Medical Entomology and Zoology, 60*, 213–216.

Takaoka, H., Thapa, S., Chhetri, S., & Henry, W. (2022b). Descriptions of adults of *Simulium* (*Gomphostilbia*) *sachini* Takaoka & Henry and *S.* (*G.*) *williei* Takaoka & Thapa in the *S.* (*G.*) *gombakense* species-group (Diptera: Simuliidae) from India. *Tropical Biomedicine, 39*, 78–85.

Takaoka, H., Thapa, S., & Henry, W. (2009c). Description of the male of *Simulium* (*Montisimulium*) *dasguptai* (Diptera: Simuliidae) from Darjeeling, India. *Medical Entomology and Zoology, 60*, 217–220.

Takaoka, H., Thapa, S., & Henry, W. (2010e). Description of two new species of *Simulium* (*Gomphostilbia*) (Diptera: Simuliidae) from Darjeeling, India. *Medical Entomology and Zoology, 61*, 105–110.

Takaoka, H., Thapa, S., & Henry, W. (2011c). A new species of black fly (Diptera: Simuliidae) with an entirely yellow thorax from Darjeeling, India. *Zootaxa, 2824*, 62–68.

Takaoka, H., Ya'cob, Z., & Sofian-Azirun, M. (2018m). Classification, annotated list and keys for the black flies (Diptera: Simuliidae) in Peninsular Malaysia. *Zootaxa, 4498*, 1–65.

Takaoka, H., Ya'cob, Z., Sofian-Azirun, M., Chen, C. D., Low, V. L., Lau, K. W., & Pham, X. D. (2019h). A new species of the *Simulium* (*Simulium*) *crocinum* species-group (Diptera: Simuliidae) from Vietnam. *Tropical Biomedicine, 36*, 245–251.

Takaoka, H., Yunus, M., Hadi, U. K., Sigit, S. H., & Miyagi, I. (2000). Preliminary report of faunistic surveys on black flies (Diptera: Simuliidae) in Sumatra, Indonesia. *Japanese Journal of Tropical Medicine and Hygiene, 28*, 157–166.

Tangkawanit, U., Kuvangkadilok, C., Baimai, V., & Adler, P. H. (2009a). Morphotaxonomy of the *Simulium* (*Simulium*) *tuberosum* species group (Diptera: Simuliidae) in Thailand. *Zootaxa, 155*(2048), 31–46.

Tangkawanit, U., Kuvangkadilok, C., Baimai, V., & Adler, P. H. (2009b). Cytosystematics of the *Simulium tani* group (Diptera: Simuliidae) in Thailand. *Zoological Journal of the Linnean Society, 155*, 289–315.

Tangkawanit, U., Wongpakam, K., & Pramual, P. (2018). A new species of black fly (Diptera: Simuliidae) species of the subgenus *Asiosimulium* Takaoka & Choochote from Thailand. *Zootaxa, 4388*(1), 111–122.

Thaijarern, J., Adler, P. H., & Pramual, P. (2018). Limited differentiation among black flies in the *Simulium multistriatum* species group (Diptera: Simuliidae) in Thailand: cryptic species, homosequential species and homosequential cryptic species. *Zoological Journal of the Linnean Society, 184*, 1024–1054.

Thaijarern, J., Pramual, P., & Adler, P. H. (2017). Life-stage association of black flies, using a fast-evolving nuclear gene sequence, and description of the larva of *Simulium lampangense* Takaoka & Choochote (Diptera: Simuliidae) from Thailand. *Zootaxa, 4299*(2), 263–270.

Thaijarern, J., Sopaladawan, P. N., Wongpakam, K., & Pramual, P. (2014). Phylogeography of the black fly *Simulium aureohirtum* (Diptera: Simuliidae) in Thailand. *Genome, 57*(8), 411–417. https://doi.org/10.1139/gen-2014-0103

Thaijarern, J., Tangkawanit, U., Wongpakam, K., & Pramual, P. (2019a). Molecular detection of *Trypanosoma* (Kinetoplastida: Trypanosomatidae) in black flies (Diptera: Simuliidae) from Thailand. *Acta Tropica, 200*(1), 1–5. (105196).

Thaijarern, J., Wongpakam, K., Kangrang, A., & Pramual, P. (2019b). A new species of black fly (Diptera: Simuliidae) in the *Simulium (Simulium) multistriatum* species-group from Thailand. *Zootaxa, 4586*(3), 461–474.

Thanwisai, A., Kuvangkadilok, C., & Baimai, V. (2006). Molecular phylogeny of black flies (Diptera: Simuliidae) from Thailand, using ITS2 rDNA. *Genetica, 128*, 177–204.

Thapa, S., Adler, P. H., Chhetri, S. Varma, R., & Henry, W. (2017). Chromosomal evidence for a new cryptic species of black fly in the *Simulium praelargum* complex (Diptera: Simuliidae) from West Bengal, India. *Zootaxa 4244*(1):137–144.

Thapa, S., Procunier, W. S., Henry, W., & Chhetri, S. (2014a). Heterochromatin and sibling species of *Simulium praelargum* s.l. (Diptera: Simuliidae). *Genome, 57*, 1–10.

Thapa, S., Procunier, W., Henry, W., & Chhetri, S. (2014b). Heterochromatin and sibling species of *Simulium praelargum* s. l. (Diptera: Simuliidae). *Genome, 57*, 223–232.

Uemoto, K. (1978). Taxonomic problems on Japanese black flies. *Proceedings of Simuliid Research Group of the Japan Society of Sanitary Zoology, 27*, 1–3. (In Japanese).

Vijayan, S., Anbalagan, S., Rekha, K., Dinakaran, S., & Krishnan, M. (2019). A new black fly species of the subgenus *Gomphostilbia* (*Simulium*: Diptera: Simuliidae) from India. *Journal of Asia-Pacific Entomology, 22*, 568–574.

Wallace, A. R. (1876). *The Geographical Distribution of Animals*. Cambridge University Press. (Reproduced by Harper, in 1962: 2 volumes, 503 pp. and 553 pp.).

Wen, X. J., & Chen, H. B. (2000). A new species of *Simulium* (*Gomphostilbia*) from China (Diptera: Simuliidae). *Guizhou Science, 18*, 112–115.

Wen, X. J., Wei, J., & Chen, H. B. (2007). A new species of the subgenus *Simulium* (Diptera: Simuliidae). *Entomotaxonomia, 29*, 290–292.

Xue, H. T. (1987). Records of black flies from Yunnan Province, China (Diptera: Simuliidae). *Acta Zootaxonomica Sinica, 17*, 110–112. (In Chinese with an English abstract).

Xue, H. T. (1991). A new species of black fly (Simuliidae: Diptera) from Yunnan, China. *Chinese Journal of Vector Biology and Control, 2*, 93–94. (In Chinese with an English abstract).

Xue, H. T. (1992). A new species of blackfly *Simulium* from Yunnan Province, China (Diptera: Simuliidae). *Acta Zootaxonomica Sinica, 17*, 93–96. (In Chinese with an English abstract).

Xue, H. T. (1993a). Two new species of blackflies (*Eusimulium*) from Yunnan Province, China (Diptera: Simuliidae). *Acta Zootaxonomica Sinica, 18*, 229–232. (In Chinese with an English abstract).

Xue, H. T. (1993b). On a new species and a new record of *Simulium* from Yunnan Province, China (Diptera: Simuliidae). *Acta Zootaxonomica Sinica, 18*, 234–236. (In Chinese with an English abstract).

Xue, H. T. (1993c). A new species of blackfly from Yunnan Province, China (Diptera: Simuliidae). *Acta Zootaxonomica Sinica, 18*, 466–468. (In Chinese with an English abstract).

Ya'cob, Z., Dawood, M. M., Jainih, L., Takaoka, H., & Sofian-Azirun, M. (2020). Review of Malaysian blackflies (Diptera: Simuliidae): note on new record of *Simulium kalimantanense* from Sabah. *Journal of Tropical Biology & Conservation, 17*, 11–18.

Ya'cob, Z., Low, V. L., Tan, T. K., Noor-Izwan, A., Lourdes, E. Y., Ramli, R., Takaoka, H., & Adler, P. H. (2021). Sexually anomalous individuals of the black fly *Simulium trangense* (Diptera: Simuliidae) infected with mermithid parasites (Nematoda: Mermithidae). *Parasitology Research*, 1–7.

Ya'cob, Z., Takaoka, H., Low, V. L., & Sofian-Azirun, M. (2017a). First description of a new cryptic species, *Simulium vanluni* from Peninsular Malaysia: An integrated morpho-taxonomical and genetic approach for naming cryptic species in the Family Simuliidae. *Acta Tropica, 165*, 31–39.

Ya'cob, Z., Takaoka, H., Low, V. L., & Sofian-Azirun, M. (2017b). Uncovering the mask of the *Simulium feuerborni* complex (Diptera: Simuliidae): description of *Simulium pairoti* sp. nov. from Malaysia based on integrative taxonomic evidence. *Acta Tropica, 169*, 133–141.

Ya'cob, Z., Takaoka, H., Low, V. L., Tan, T. K., & Sofian-Azirun, M. (2019). Description of the female of *Simulium* (*Gomphostilbia*) *aziruni* (Diptera: Simuliidae) and its genetic relationships with members of the *Simulium gombakense* species-group. *Acta Tropica, 193*, 66–70.

Ya'cob, Z., Takaoka, H., Pramual, P., Low, V. L., & Sofian-Azirun, M. (2016). Distribution patterns of blackfly (Diptera; Simuliidae) assemblages along an altitudinal gradient in Peninsular Malaysia. *Parasites & Vectors, 9*, 219. 1–219.16.

Ya'cob, Z., Takaoka, H., & Sofian-Azirun, M. (2014). *Simulium ledangense*, a new species of the *Simulium feuerborni* species-group of the subgenus *Nevermannia* (Diptera: Simuliidae) from Mount Ledang, Peninsular Malaysia. *Zootaxa, 3881*(3), 228–236.

Ya'cob, Z., Takaoka, H., & Sofian-Azirun, M. (2015a). *Simulium hiroyukii*, a new species of *Simulium* (*Gomphostilbia*) (Diptera: Simuliidae) from Sarawak, Malaysia. *Zootaxa, 3911*(3), 424–432.

Ya'cob, Z., Takaoka, H., & Sofian-Azirun, M. (2015b). *Simulium bakalalanense,* a new species of *Simulium* (*Gomphostilbia*) and three new records (Diptera: Simuliidae) from Sarawak, Malaysia. *Tropical Biomedicine, 32*, 783–790.

Ya'cob, Z., Takaoka, H., & Sofian-Azirun, M. (2018). A new species of *Simulium* (*Simulium*) (Diptera: Simuliidae) from Peninsular Malaysia. *Acta Tropica, 182*, 1–3.

Yang, M., & Chen, H. B. (2001). A new species of *Simulium* (*Simulium*) from Jianfeng Mountain, Hainan Province, China (Diptera, Simuliidae). *Acta Zootaxonomica Sinica, 26*, 90–92.

Yang, M., Luo, H. B., & Chen, H. B. (2005). Description of a new species of the genus *Simulium* Latreille from Shennongjia, Hubei, China. *Acta Zootaxonomica Sinica, 30*, 839–841.

Zeng, Y. C., Kang, Z., & Chen, H. B. (2006). A supplementary description of the female of *Simulium* (*Simulium*) *rotifils* from China (Diptera: Simuliidae). *Acta Parasitology et Medica Entomologica Sinica, 13*, 161–162. (In Chinese with an English abstract).

Zhang, C. L., & Chen, H. B. (2000). A new species of *Simulium* (*Simulium*) blackfly from Fan Jing Mountain of Guizhou (Diptera, Simuliidae). *Journal of Guiyang Medical College, 25*, 221–223.

Zhang, C. L., & Chen, H. B. (2001). A new blackfly species of the genus *Simulium* Latreille from Guizhou Province, China (Diptera: Simuliidae). *Acta Zootaxonomica Sinica, 26*(2), 216–218. (In Chinese with an English abstract).

Zhang, C. L., & Chen, H. B. (2003). Taxonomy of blackflies in Hainan, China, with a new species (Diptera: Simuliidae). *Chinese Journal of Vector Biology and Control, 14*, 354–355. (In Chinese with an English abstract).

Zhang, C. L., & Chen, H. B. (2004). Two new species of subg. *Gomphostilbia* of *Simulium* Latreille from Hunan Province, China (Diptera: Simuliidae). *Acta Zootaxonomica Sinica, 29*, 372–376.

Zhang, C. L., Yang, M., & Chen, H. B. (2003). Two new species of the genus *Simulium* breeding adjacent on a waterfall from Wuzhishan of Hainan Island, China (Diptera, Simuliidae). *Acta Zootaxonomica Sinica, 28*, 741–744.

Zhang, J. Q., Gao, B., Cai, H. Z., Huang, E. J., & Chen, H. B. (2009). Notes on the blackflies of the Wui mountain, China with a new species description (Diptera: Simuliidae). *Acta Parasitology et Medica Entomologica Sinica, 16*, 101–110.

Zhang, T., & Wang, D. Q. (1991a). A preliminary survey of blackflies (Diptera: Simuliidae) from Fujian Province. *Acta Entomologica Sinica, 34*, 483–491.

Zhang, T., & Wang, D. Q. (1991b). A new species of *Simulium* from Fujian, China (Diptera: Simuliidae). *Acta Zootaxonomica Sinica, 16*, 109–113.

Zhang, T., Wen, X. J., & Chen, H. B. (1999). A survey of taxonomy of blackflies from Guizhou, China (Diptera: Simuliidae). *Guizhou Science, 17*, 231–235. In Chinese with an English abstract).

Index[1]

[1]The following index contains all specific, subgroup, species-group, subgeneric and generic names of the Simuliidae quoted in this book. Principal references are shown in bold-faced type.

H. Takaoka, *The Black Flies of Subtropical and Tropical Asia*,
https://doi.org/10.1007/978-981-99-5219-9

C

D

E

N

Q

T

U

Y

Z

MIX
Papier aus verantwortungsvollen Quellen
Paper from responsible sources
FSC® C105338

If you have any concerns about our products,
you can contact us on
ProductSafety@springernature.com

In case Publisher is established outside the EU,
the EU authorized representative is:
Springer Nature Customer Service Center GmbH
Europaplatz 3, 69115 Heidelberg, Germany

Printed by Libri Plureos GmbH
in Hamburg, Germany